System Identification, Environmental Modelling, and Control System Design

Liuping Wang · Hugues Garnier
Editors

System Identification, Environmental Modelling, and Control System Design

Springer

Editors
Liuping Wang
Computer Engineering
RMIT University
Swanston Street 10
3000 Melbourne, Victoria
Australia
liuping.wang@rmit.edu.au

Hugues Garnier
Centre de Recherche en
Automatique de Nancy
CRAN UMR7039 CNRS-UHP-INPL
Université Nancy I
PO Box 239
54506 Vandoeuvre-les-Nancy CX
France
hugues.garnier@cran.uhp-nancy.fr

ISBN 978-0-85729-973-4 e-ISBN 978-0-85729-974-1
DOI 10.1007/978-0-85729-974-1
Springer London Dordrecht Heidelberg New York

British Library Cataloguing in Publication Data
A catalogue record for this book is available from the British Library

Library of Congress Control Number: 2011938794

Cover design: VTeX UAB, Lithuania

Printed on acid-free paper

Springer is part of Springer Science+Business Media (www.springer.com)

Preface

Dedication: Peter Young (1939–), Engineer, Academic and Polymath

This book is dedicated to Professor Peter Young's 70th birthday. The majority of the authors in this book are Peter Young's friends, collaborators, former colleagues, and former students.

Professor Peter Young is a major pioneer in the development of recursive estimation and its use in adaptive forecasting, data assimilation and adaptive control system design. He has over 40 years experience in academic and industrial research, with more than 250 publications in the open literature including several books. He has made important research contributions to the areas of time series analysis, environmental modelling and computer-aided control system design. He is the leading expert on the identification and estimation of data-based transfer function models and has successfully promoted their use in forecasting and control system design.

It is through a strong contextual focus on applications in diverse fields that Peter has made such innovative contributions to generic methods, algorithms and their associated software. His environmental applications experience includes: water quality modelling and control, as well as rainfall-flow modelling and its use in adaptive forecasting, flood warning and data assimilation. In the earth sciences he has contributed to: weather radar calibration; climate modelling and data analysis. Other contributions include climate control in agricultural buildings; and the modelling and control of inter-urban traffic systems.

Peter is an archetypal example of the ever-inquiring charismatic researcher, always passionately challenging themselves and colleagues with new questions and ideas. His resultant collaborations have been with co-workers in many research institutions across the globe. This book dedicated to Peter is a testament to, and celebration of, the depth and breadth of his influences.

Theory of System Identification

Peter Young became interested in data-based modelling of dynamic systems when he was a student apprentice in the UK aircraft industry (English Electric Aviation, now BAe Systems). In 1961, he realised the limitations of least squares linear regression analysis when there was noise on the regressors and this led him to look at various approaches to this problem, particularly in relation to the estimation of parameters in dynamic systems. Because of its simplicity, he was attracted to the, at that time, little known *Instrumental Variable* (IV) method. The rest is, as they say, history. Peter started serious research on IV methods, first at Loughborough University of Technology in 1963, then in the Engineering Department of Cambridge University in 1965. During this time, he realised the importance of prefiltering data when estimating parameters in *Transfer Function* (TF) models, both to avoid the direct differentiation of noisy data, in the case of continuous-time TF models, and to improve the statistical efficiency of the estimates in both continuous and discrete-time TF models. Finally, after moving to take up a Research Chair at the Australian National University in 1975, he put all of his previous results together [1], and showed that, under the usual statistical assumptions (noise-free input variables and a rational spectral density noise process described by an ARMA process), his iterative or recursive-iterative *Refined Instrumental Variable* (RIV) estimation procedure, involving appropriate adaptively updated prefilters, was asymptotically equivalent to statistically efficient maximum likelihood estimation and prediction error minimization. Moreover, both then and now, it is unique in its ability to estimate parameters in both discrete *and* continuous-time TF models from sampled data. This RIV approach was implemented and thoroughly evaluated later by Peter and Professor Tony Jakeman, who was Peter's research collaborator at the time [2]. And recent papers [3, 4] have extended these results to include an improved IV method of ARMA noise model estimation and a three stage RIV procedure for estimating a TF model in a closed loop control system.

Given this abiding interest in the concept of instrumental variable estimation and its implementation in parameter estimation algorithms of various types, it is appropriate in this Festschrift for Peter that the first two chapters deal with aspects of this topic. The book begins with a tutorial-style chapter on instrumental variables by Professor Torsten Söderström, who has also had a life-long interest in IV estimation. A general derivation of the covariance matrix of the IV parameter estimates is presented and it is shown how this matrix is influenced by a number of user choices regarding the nature of the instrumental variables and the estimation algorithm. The chapter discusses how these user choices can be made in order to ensure that the covariance matrix is as small as possible, in a well-defined sense, and compares optimal instrumental variable algorithms with the alternative prediction error method. In particular, it discusses optimal instrumental variable methods that yield statistically efficient parameter estimates and shows that the iterative Refined RIV algorithm (referred to in the chapter as a 'multistep' algorithm) proposed by Peter in the context of the maximum likelihood estimation of Box-Jenkins transfer function models, possesses these optimal properties.

The original papers on the implementation of the RIV method by Peter Young and Tony Jakeman included results that demonstrated how the Simplified RIV (SRIV) method could be used for estimating the parameters in a hybrid continuous-time transfer function model (i.e. a continuous time system model with an additive, discrete-time noise model) from sampled data. In recent years, Peter has worked closely with Professor Hugues Garnier and Dr. Marion Gilson on the development of the full RIV version of this hybrid algorithm. In Chap. 2, Marion and Hugues, together with their colleague Vincent Laurain, investigate instrumental variables in the context of nonlinear system identification. In particular, they present RIV estimation methods for discrete or hybrid continuous-time Hammerstein models with coloured additive noise described by a discrete-time ARMA process. In order to use a regression form of solution and avoid gradient optimization, the Hammerstein model is reformulated as a linear, augmented multi-input-single-output model. The performance of the proposed methods are demonstrated by relevant Monte Carlo simulation examples.

Identifiability is a very important aspect of model identification and parameter estimation. In a useful, tutorial-style Chap. 3, Professor Eric Walter investigates identifiability with the aim of helping readers decide whether identifiability and the closely connected property of distinguishability are theoretically important and practically relevant for their research or teaching. The chapter discusses methods that can be used to test models for these properties and shows that measures of identifiability can be maximized, provided that there are some degrees of freedom in the procedure for data collection that allow for optimal experimental design. Finally, the paper shows that interval analysis and bounded parameter estimation can provide useful procedures when the model of interest cannot be made identifiable. Consistent with the tutorial nature of the chapter, simple illustrative examples are included and worked out in detail.

Professor Bruce Beck was one of Peter Young's first research students in the Control and Systems Division of the Engineering Department at Cambridge University, UK, and both of their careers were profoundly affected by the joint work they did together at this time. Although their subsequent research has moved in somewhat different directions, they have both pursued an underlying inductive approach to the identification of model structure from real experimental or monitored data, mostly in relation to environmental applications. In Chap. 4, Bruce, together with his colleagues Z. Lin, and Hans Stigter, continue with this theme and address the important issues involved in model structure identification and the growth of knowledge, including novel recent research into the possibility of diverting the software of molecular graphics into serving the purpose of scientific visualization in supporting the procedural steps of model structure identification.

Inspired by the work of Peter Young, who has made a life time of contributions to parameter estimation for real world systems, several chapters follow on the general identification concepts and technology. Professor Graham Goodwin is one of the most important contributors to the theory and practice of automatic control over the past forty years and is a long-time and valued friend of Peter Young. In Chap. 5, Graham combines with Mauricio Cea to consider the problem of joint state and parameter estimation for continuous time systems in the important practical situation where data are collected with non-uniform sampling intervals. This problem is formulated in the context of nonlinear filtering and the chapter shows how a new class of nonlinear filtering algorithm *Minimum Distortion Filtering* (MDF) can be applied to this problem. A simple example is used to illustrate the performance of the algorithm and the results are compared with those obtained using numerically intensive *Particle Filtering*. It is clear that, in this example, the MDF approach has distinct advantages in both computational and estimation terms.

In the nineteen seventies, Victor Solo was a research student at the Australian National University supervised by the noted expert on time series analysis, Professor Ted Hannan, and Peter Young. Since then, Professor Solo has worked at the cutting edge of research on novel aspects of both the theory and practice of time series analysis. In Chap. 6, Victor first notes that adaptive signal processing and adaptive control developed slowly and independently until the 1970s. And he points out that Peter Young was one of the pioneers in this area of study and that Peter's 1984 book *Recursive Estimation and Time Series Analysis* (a heavily revised and expanded version of which has just been published: see http://www.springer.com/engineering/control/book/978-3-642-21980-1) is one of the few books of this era that discusses the use of fixed gain recursive algorithms for *Time Varying Parameter* (TVP) estimation, as well as TVP estimation in an off-line setting, exploiting recursive smoothing. In the former context, Victor's main aim is to discuss the powerful tool of 'averaging analysis' that can be used to evaluate the stability of recursive estimation algorithms. He points out that, although adaptive or learning algorithms have found wide use in control, signal processing and machine learning, the use of averaging analysis is not as well known as it should be. He reviews this approach within the context of the adaptive *Least Mean Square* (LMS) type of algorithm and develops averaging in a heuristic manner, illustrating its use on a number of illustrative examples.

In Chap. 7, Professor Manfred Deistler, another old friend of Peter Young's and one of the most important time series analysts of his day, combines with his colleagues Christoph Flamm, Ulrike Kalliauer, Markus Waser and Andreas Graef, to describe measures for dependence and causality between component processes in multivariate time series in a stationary context. Symmetric measures, such as the partial spectral coherence, as well as directed measures, such as the partial directed coherence and the conditional Granger causality index, are described and discussed. These measures are used for deriving undirected and directed graphs (where the vertices correspond to the one-dimensional component processes), showing the inner structure of a multivariate time series. The authors' interest in these graphs originates from the problem of detecting the focus of an epileptic seizure, based on the analysis of invasive EEG data and an example for such an analysis is given in the last section of the chapter.

Although they are not in the same Department at Lancaster, Professor Granville Tunnicliffe-Wilson and Peter Young have been friends and colleagues for many years. Together, they helped Peter Armitage, from the Civil Service College in London, with courses on forecasting for Civil Servants that were held in London and

Lancaster. It is notable that Granville was a research student of Professor Gwilym Jenkins at Lancaster and contributed much to the writing of the famous 1970 book by Jenkins and George Box on time series analysis, forecasting and control. It is appropriate, therefore, that Chap. 8, by Granville and Peter Armitage, is a tutorial-style chapter on Box-Jenkins methods; methods that are now used across the world, not least because they have been incorporated into standard software, such as the X11-ARIMA seasonal adjustment package developed by the US Bureau of the Census. The exposition is not, however, limited to the Box and Jenkins approach and other methods of model structure identification are suggested. The chapter is based on several time series case studies, ranging from the airline series example presented by Box and Jenkins, to an example of half hourly electricity demand. This chapter also serves as a fitting memorial to Peter Armitage, who died recently but who did so much to further the adoption of advanced forecasting methods in the UK Civil Service.

State Dependent Parameter (SDP) modelling was developed by Peter Young in the 1990s to identify non-linearities in the context of dynamic transfer function models [5]. SDP estimation is based on exploiting the recursive Kalman Filter (KF) and Fixed Interval Smoothing (FIS) algorithms to produce non-parametric estimates (graphs) of the model parameters as a function of other measured variables. This approach, which is very useful for locating the position and form of the nonlinearities prior to their parameterization, has been applied successfully in many application areas, especially to identify the structure of nonlinear Data-Based Mechanistic (DBM) models (see Chap. 16) from observed time series data. In Chap. 9, Drs. Marco Ratto and Andrea Pagano, highlight other applications of SDP modelling, where fruitful co-operation with Peter has led to a series of joint papers in which *State-Dependent Regression* (SDR) analysis has been applied to perform various useful functions in sensitivity analysis, dynamic model reduction and emulation ('meta') modelling, where a linked set of reduced order models is capable of reproducing closely the main static and dynamic features of large computer simulation models (see also Chaps. 10 and 16). The chapter also describes how SDR algorithms can be used to identify and improve the performance of tensor product smoothing spline ANOVA models.

SDP modelling is also considered in Chap. 10, which is contributed by Peter Young's colleagues from Lancaster, Drs. Wlodek Tych, Paul Smith, Arun Chotai and James Taylor, together with a former research student, Dr. Jafar Sadeghi. This describes and develops Jafar and Wlodek's generalization of the SDP approach to include *Multi-State Dependent Parameter* (MSDP) nonlinearities. The recursive estimation of the MSDP model parameters in a multivariable state space occurs along a multi-path trajectory, again employing the KF and FIS algorithms. The novelty of the method lies in redefining the concepts of sequence (predecessor, successor), so allowing for their use in a multi-state dependent context and facilitating the subsequent efficient parameterisation for a fairly wide class of non-linear, stochastic dynamic systems. The approach is illustrated by two worked examples in MATLAB. The format of the estimated SDP model also allows its direct use in new methods of SDP control system design within a *Non-Minimal State Space* (NMSS) control system design framework, as originally suggested by Peter Young (see Chap. 27).

Peter Young has been friends with Professor Liuping Wang for the last decade and, over the last few years, he has worked with her on both a NMSS-based formulation of model predictive control and SDP model estimation using wavelets. In Chap. 11, Liuping and her colleague Nguyen-Vu Truong continue with the SDP theme and apply a new SDP-based approach to the important problem of electrical demand forecasting. Such forecasting is critical to power system operation, since it serves as an input to the management and planning of activities such as power production, transmission and distribution, the dispatch and pricing process, as well as system security analysis. From the system's point of view, this is a *complex nonlinear dynamic system* in which the power demand is a highly nonlinear function of the historical data and various external variables. The chapter describes an application of an SDP model based on a two-dimensional wavelet (2-DWSDP) to the forecasting of daily peak electrical demand in the state of Victoria, Australia. The parsimonious structure of the identified model enhances the model's generalization capability, and it shows the advantages of SDP estimation in providing very descriptive views and interpretations about the interactions and relationships between various components which affect the system's behaviour.

In Chap. 12, Professor David Hendry and his colleague Jennifer Castle consider approaches to the automatic selection of nonlinear models within an econometric context. The strategy is: first, to test for non-linearity in the unrestricted linear formulation; then, if this test is rejected, a general model is specified using polynomials that are simplified to a minimal congruent representation; finally, model selection is by encompassing tests of specific non-linear forms against the selected model. The authors propose solutions to some of the many problems that non-linearity poses: extreme observations leading to non-normal (fat-tailed) distributions; collinearity between non-linear functions; situations when there are more variables than observations in approximating the non-linearity; and excess retention of irrelevant variables. Finally, an empirical application concerned with a 'returns-to-education' demonstrates the feasibility of the non-linear automatic model selection algorithm *Autometrics*.

The theme of model structure selection in nonlinear system identification is continued in Chap. 13 by X. Hong, S. Chen and Professor Chris Harris, this time using radial-basis functions for the modelling of the nonlinear systems. From the angle of the diversified RBF topologies, they consider three different topologies; (i) the RBF network with tunable nodes; (ii) the Box-Cox output transformation based RBF network (Box-Cox RBF); and (iii) the RBF network with boundary value constraints (BVC-RBF). These proposed RBF topologies enhance the modelling capabilities in various ways and it is shown to be advantageous if the linear learning algorithms, e.g. the orthogonal forward selection (OFS) algorithm based leave-one-out (LOO) criteria, are still applicable as part of the proposed algorithms.

Applications of System Identification

Band-pass, Kalman, and adaptive filters are used for the removal of resuscitation artifacts from human ECG signals. Chapter 14 by Professor Ivan Markovsky, Anton

Amann and Sabine Van Huffel is a tutorial-style chapter that clarifies the rationale for applying these methods in this particular biomedical context. The novel aspects of the exposition are the deterministic interpretation and comparative study of the methods using a database of separately recorded human ECG and animal resuscitation artifact signals. The performance criterion used in this analysis is the signal-to-noise ratio (SNR) improvement, defined as the ratio of the SNRs of the filtered signal and the given ECG signal. The empirical results show that for low SNR, a band-pass filter yields the best performance; while for high SNR, an adaptive filter yields the best performance.

Professor Eric Rogers and Peter Young have worked for many years on the Editorial Board of the *International Journal of Control*. Chapter. 15, by Fengmin Le, Chris Freeman, Ivan Markovsky and Eric, reports recent work involving the use of robots in stroke rehabilitation, where model-based algorithms have been developed to control the application of functional electrical stimulation to the upper limb of stroke patients with incomplete paralysis, in order to assist them in reaching tasks. This, in turn, requires the identification of the response of a human muscle to electrical stimulation. The chapter provides an overview of the progress reported in the literature, together with some currently open research questions.

Data-based Mechanistic Modelling and Environmental Systems

The term *Data-Based Mechanistic* (DBM) *modelling* was first used by Peter Young in the early nineteen nineties, but the basic concepts of this approach to modelling dynamic systems have been developed by Peter and various colleagues over many years. For example, they were first applied seriously within a hydrological context by Peter and Bruce Beck in the early 1970s, with application to the modelling of water quality and flow in rivers, and set by Peter within a more general framework shortly thereafter. Since then, they have been applied to many different systems in diverse areas of application from ecology, through engineering to economics. The next several chapters present various applications, mainly in the area of water resources where DBM modelling, as well as other systems modelling and control procedures, are used to good effect.

From a philosophical standpoint, DBM modelling stresses the need to rely, whenever possible, on inductive inference from time series data, without over-reliance on pre-conceived notions about the structure of the model that can often lead to over-large computer simulation models with severe identifiability problems. Indeed, it was a reaction to such large, over-parameterized models that gave birth to DBM modelling. In Chap. 16, Peter Young briefly outlines of the main stages and procedures involved in DBM modelling. Its main aim, however, is to put the DBM approach to modelling in a philosophical context and demonstrate how this is reflected in an illustrative example, where DBM modelling is applied to the investigation of solute transport and dispersion in water bodies. By providing a *Dynamic Emulation Model* (DEM) bridge between large computer simulation models, produced in a hypothetico–deductive manner, and parsimonious DBM models that are normally identifiable from the available data, it emphasises the need to utilise both approaches, in an integrated manner, in order to meet multiple modelling objectives.

Peter Young's research on flood forecasting techniques based on both rainfall-flow (run-off generation) and flow-flow (flow routing) modelling goes back a long way to the early nineteen seventies. However, since the nineteen eighties it has been heavily influenced by collaboration with his friend and colleague Professor Keith Beven, one of the foremost contributors to the theory and practice of hydrology. In this flood forecasting context, the DBM modeling approach normally identifies a non-linear SDP (see above) transformation of the input rainfall signal that is dependent on the current state (river flow or level) of the system. In Chap. 17, Keith, David Leedal, Paul Smith and Peter, discuss four methods of parameterizing and optimizing the input non-linearity function, each of which have associated advantages and disadvantages: a simple power law; a radial basis function network; piecewise cubic Hermite data interpolation; and, finally, the Takagi-Sugeno Fuzzy Inference method, which employs *human-in-the-loop* interaction during the parameter estimation process.

The *Aggregated Dead Zone* (ADZ) model[1] was one of the first DBM to be developed, initially by Peter Young and Tom Beer in the early nineteen eighties and later by Dr. Steve Wallis, Peter and Keith Beven, who extended it to include the concept of a 'dispersive fraction'. In Chap. 18, Dr. Sarka Blazkova, together with Keith Beven and Dr. Paul Smith, use the ADZ model for the analysis of tracer data from larger rivers. The model provides excellent explanation of the observed concentrations, with a dispersive fraction parameter that varies relatively little with flow (discharge), making the model applicable over a wide range of flow variations. It is also shown how the information on transport and dispersion at different flows can be augmented by pollution incident and continuously logged water quality data. The model can then be applied to predict the downstream dispersion of pollutants at any arbitrary flow, taking account of the uncertainty in the SRIV estimation (see above) of the ADZ model parameters.

Peter Young has worked with Drs. Andrea Castelletti and Francesa Pianosi on the DBM modeling of river catchments affected by snow melt. However, Andrea and Francesca, together with Professor Rodolfo Soncini-Sessa are also concerned with the wider topic of water resources management to effectively cope with all the key drivers of global change (climate, demographic, economic, social, policy/law/institutional, and technology changes). Here, it is essential that the traditional sectoral management approach to water resources is transformed into a new paradigm, where water is considered as the principal and cross cutting medium for balancing food, energy security, and environmental sustainability. One major technical challenge, in expanding the scope of water resources management across sectors and to the river basin level, is to develop new methodologies and tools to cope with the increasing complexity of water systems. In Chap. 19, Andrea, Francesa and Rodolfo consider the management and control of a large water system composed of reservoirs, natural catchments feeding the reservoirs, diversion dams, water users

[1] Also called the *Aggregated Mixing Volume* (AMV) model when applied in a more general context.

(e.g. hydropower plants, irrigation districts), and artificial and natural canals that connect all the above components. In particular, they review some of the recent, and in their opinion, more promising alternatives to stochastic dynamic programming in designing sub-optimal control policies.

The theme of river basin management is continued in Chap. 20. Here Professors Rob Evans and Ivan Mareels, together with N. Okello, M. Pham, W. Qiu and S.K. Saleem, point out that river basins are key components of water supply grids. As a result, river basin operators must handle a complex set of objectives, including runoff storage, flood control, supply for consumptive use, hydroelectric power generation, silting management, and maintenance of river basin ecology. At present, operators rely on a combination of simulation and optimization tools to help make operational decisions. However, the complexity associated with this approach makes it unsuitable for real-time (daily or hourly) operation. The consequence is that between longer-term optimized operating points, river basins are largely operated in an open loop manner. This leads to operational inefficiencies, most notably wasted water and poor ecological outcomes. In the chapter, the authors propose a systematic approach for the real-time operation of entire river basin networks, employing simple low order models on which to design optimal model predictive control strategies.

Agriculture is the world-wide biggest consumer of water. However, a large portion of the water is wasted due to inefficient distribution from lakes and reservoirs via rivers to farms. While more efficient water distribution can be achieved with the help of improved control and decision support systems, this requires the identification and estimation relatively simple river models, such as the DBM models used above in river flood forecasting applications. Traditionally, the partial differential Saint Venant equations have been used for modelling flow in rivers but they are not suitable for use in control design and simpler alternative models of the DBM type are required. Such an approach is described in Chap. 21 by Mathias Foo, Su Ki Ooi and Professor Erik Weyer. Based on operational river data and physical considerations, they estimate simple 'time-delay' and 'integrator-delay' models and compare them with the Saint Venant equation model. The efficacy of these simple models is then illustrated in a simulation exercise where they are used to design a control system that is applied successfully to the full Saint Venant equation model.

Professor Howard Wheater, who is very well known for his research and development work in this area, has been one of Peter Young's friends since the nineteen seventies, when both were in different Divisions of the Engineering Department at Cambridge University. Peter has also worked recently with Howard's colleague, Dr. Neil McIntyre, and Howard on the DBM modelling of the non-linear dynamic processes that are active in an experimental catchment. In Chap. 22, Howard, Neil and their colleagues consider the insights developed from Peter's research on DBM modelling in the context of predicting the effects of land use and land management change across multiple scales. This is a particularly challenging problem and they review the strengths and weaknesses of alternative modelling approaches, including: physics-based modelling; conceptual models, conditioned by regionalised indices; and DBM modelling, showing the latter's utility in identifying appropriate model structures that can guide the hydrological application.

It is clear from Chaps. 19, 20, 21 and 22 that hydrological models have an important role to play in supporting water management. Another important problem in catchment hydrology is river catchment classification. This remains a significant challenge for hydrologists, with available schemes not providing a sufficient basis for consistently distinguishing between different types of hydrological behaviour. However, in Chap. 23, Professors Thorsten Wagener and Neil McIntyre show how the DBM approach to time-series modelling is an eminently suitable approach to this problem since it is designed to extract the dominant modes (signatures) of a system response. They develop a classification procedure based on this idea and apply it to 278 catchments distributed across the Eastern USA, with the aim of exploring whether the catchments may be classified according to their dominant mode responses, including identifying both the type of response (the transfer function structure) and the scale of the response (the associated parameter values). They conclude that the approach holds considerable promise in this kind of application but that more research is required to establish better which of DBM model signatures, or combinations of these, are most powerful in the classification role.

Previous chapters show how the ADZ/AMV models can provide a theoretically elegant and practically useful approach to water quality and pollutant transport modelling that can used both in assessing the risk from pollution incidents and the sustainable management of water resources. An ability to predict the concentrations of a pollutant travelling along the river is necessary in assessing the ecological impact of the pollutant and to plan a remedy against possible damage to humans and the environment. The risk from a pollutant at a given location along the river depends on the maximum concentration of any toxic component, the travel times of the pollutant from the release point and the duration over which its concentration exceeds feasible threshold levels. In Chap. 24, Peter Young's previous research colleague at Lancaster, Dr. Renata Romanowicz (now at the Institute of Geophysics in Warsaw), outlines on-going research on DBM models and compares the results with physically-based approaches using worked examples from pollutant transport modelling. In addition to steady-state examples, a transfer function pollutant transport model for transient flows, that can be interpreted directly in ADZ/AMV terms, is presented and used in a tutorial-style case study on the application of a multi-rate transfer function models to the identification of environmental processes.

Dr. Peter Minchin has known Peter Young since the nineteen seventies, when they worked together on the application of RIV estimation (see above) to the analysis and modelling of phloem translocation data. In Chap. 25, Peter Minchin describes the application of such techniques to the problem of modelling Phloem vasculature within higher plants functions at very high hydrostatic pressure (*circa* 10 atmospheres). A detailed time sequence of phloem sap movement through a plant is possible with in vivo measurement of 11C tracer, which is ideal for input-output transfer function modelling within a DBM context. The resulting estimates of transport distribution times, pathway leakage, and partitioning between competing sinks have led to the first mechanistic understanding of phloem partitioning between competing sinks, from which sink priority has been shown to be an emergent property.

Dr. Bellie Sivakumar and Peter Young have maintained contact for some years because of their mutual interest in the use of low order nonlinear models and DBM modelling in environmental systems analysis. In this regard, the last two decades have witnessed a significant momentum in the promising application of *Chaos Theory* (CT) to environmental systems. Nevertheless, there have also been persistent skepticism and criticism of such studies, motivated by the potential limitations in the data-based modelling of chaotic systems. In Chap. 26, Sivakumar offers a balanced perspective of chaos studies in environmental systems: between the philosophy of CT at one extreme, to the down-to-earth pragmatism that is needed in its application at the other. After briefly reviewing the development of CT, some basic identification and estimation methods are described and their reliability for determining system properties are evaluated. A brief review of CT studies in environmental systems as well as the progress and pitfalls is then made. Analysis of four river flow series lend support to the contention that environmental systems are neither deterministic nor stochastic, but a combination of the two; and that CT can offer a middle-ground approach to these extreme deterministic and stochastic views. It is concluded that, in view of the strengths of both CT and DBM concepts (commonalities as well as differences), the coupling of these two data-based modelling approaches seems to be a promising way of formulating a much-needed general framework for environmental modelling.

Control System Design

Peter Young's early research career was concerned with data-based modelling applied in the context of automatic control system design and he has retained an abiding interest in control system design for the past fifty years. His most novel contributions in this area are concerned with the exploitation of *Non-Minimal State Space* descriptions of dynamic systems based on their estimated transfer function models. This began with early research carried out while he was working as a civilian for the U.S. Navy in California and it culminated with a series of research studies beginning at Lancaster in the nineteen eighties and extending to recent research on NMSS-based model predictive control carried out with Professor Liuping Wang at RMIT in Melbourne. In this last section of the book, the first three chapters cover aspects of NMSS and model predictive control system design, while the final one discusses a recently developed MATLAB Toolbox for the analysis of more general state space systems.

The largely tutorial Chap. 27 by Peter Young's long-time colleagues, Drs. James Taylor, Arun Chotai and Wlodek Tych, use case studies based on recent engineering applications, to illustrate the NMSS approach to feedback control system design. The paper starts by reviewing the subject and pointing out that the NMSS representation is a rather natural state space description of a discrete-time transfer function, since its dimension is dictated by the complete structure of the transfer function model. Also, it notes that the resulting *Proportional-Integral-Plus* (PIP) control algorithm can be interpreted as a logical extension of the conventional *Proportional-Integral* (PI) controller, facilitating its straightforward implementation using a standard hardware-software arrangement. Finally, the chapter shows how the basic NMSS approach is readily extended into multivariable, model-predictive and nonlinear control system design contexts and gives pointers to the latest research results in this regard.

Professor Neville Rees and Peter Young have been very close friends for over forty years since they worked together in California between 1968 and 1970. In Chap. 28, Neville and his colleague Chris Lu join with Peter to describe a joint project they have been involved with recently. They briefly introduce the concept of large computer model reduction using dynamic emulation modelling (DEM), as discussed in Chap. 16. SRIV identification and estimation methods (see earlier) available in the *CAPTAIN* Toolbox are exploited to develop a nominal, reduced order DEM for a large Simulink model of a complex, nonlinear, dynamic power plant system, using data obtained from planned experiments performed on this large simulation model. The authors then show how this single, three input, three output, linear emulation model can form the basis for multivariable, NMSS control systems design. The control simulation results cover a wide range of operating conditions and show significant performance improvements in relation to the standard, multichannel PID control system performance. This is despite the fact that the design is based on the single multivariable model and the simulation model has numerous nonlinear elements.

One of the key components in a renewable energy system, such as wind energy generator, is a three-phase regenerative PWM converter, which is both nonlinear and time-varying by nature. In Chap. 29, Dae Yoo, together with Professor Liuping Wang and another of Peter Young's long-time friends, Professor Peter Gawthrop, consider the model predictive control of such a converter. In particular, with the classical synchronous frame transformation, the nonlinear PWM model is linearized to obtain a continuous-time state-space model. Then, based on this linearized model, a continuous-time model predictive control system for the converter is designed and implemented successfully on a laboratory scale test-bed built by the authors. The proposed approach includes a prescribed degree of stability in the algorithm that overcomes the performance limitation caused by the existing right-half-plant zero in the system. This also provides an effective tuning parameter for the desired closed-loop performance.

The final Chap. 30 of this book is written by another former research colleague of Peter Young, Professor Diego Pedregal, and Dr. James Taylor (see previously). It illustrates the utility of, and provides the basic documentation for, *SSpace*, a recently developed Matlab™ toolbox for the analysis of State Space systems. The key strength of the toolbox is its generality and flexibility, both in terms of the particular state space form selected and the manner in which generic models are straightforwardly translated into MATLAB code. With the help of a relatively small number of functions, it is possible to fully exploit the power of state space systems, performing operations such as filtering, smoothing, forecasting, interpolation, signal extraction and likelihood estimation. The chapter provides an overview of *SSpace* and demonstrates its usage with several worked examples.

References

1. Young, P.C.: Some observations on instrumental variable methods of time-series analysis. Int. J. Control **23**, 593–612 (1976)
2. Young, P.C., Jakeman, A.J.: Refined instrumental variable methods of time-series analysis: Parts I, II and III. Int. J. Control **29**, 1–30; **29**, 621–644; **31**, 741–764 (1979–1980)
3. Young, P.C.: The Refined instrumental variable method: unified estimation of discrete and continuous-time transfer function models. J. Eur. Syst. Automat. **42**, 149–179 (2008).
4. Young, P.C.: Gauss, Kalman and advances in recursive parameter estimation. J. Forecast. **30**, 104–146 (2011) (Special issue celebrating 50 years of the Kalman Filter)
5. Young, P.C.: Stochastic, dynamic modelling and signal processing: time variable and state dependent parameter estimation. In: Fitzgerald, W.J., Walden, A., Smith, R., Young, P.C. (eds.) Nonlinear and Nonstationary Signal Processing, pp. 74–114. Cambridge University Press, Cambridge (2000)

Melbourne, Australia — Liuping Wang
Nancy, France — Hugues Garnier

Contents

Contributors

Anton Amann Innsbruck Medical University and Department of Anesthesia and General Intensive Care, Anichstr 35, 6020 Innsbruck, Austria, Anton.Amann@i-med.ac.at

Peter Armitage The Civil Service College, London, England, UK

C. Ballard Imperial College London, London SW7 2AZ, UK

M.B. Beck University of Georgia, Athens, GA, USA, mbbeck@uga.edu

Keith J. Beven Lancaster Environment Centre, Lancaster University, Lancaster, UK, k.beven@lancaster.ac.uk

Sarka D. Blazkova T G Marsaryk Water Resource Institute, Prague, Czech Republic

N. Bulygina Imperial College London, London SW7 2AZ, UK

Andrea Castelletti Politecnico di Milano, Milano, Italy, castelle@elet.polimi.it

Jennifer L. Castle Magdalen College & Institute for New Economic Thinking at the Oxford Martin School, University of Oxford, Oxford, UK, jennifer.castle@magd.ox.ac.uk

Mauricio G. Cea School of Electrical Engineering and Computer Science, University of Newcastle, University Drive NSW 2308, Australia, Mauricio.Cea@uon.edu.au

S. Chen School of Electronics and Computer Science, University of Southampton, Southampton, UK, sqc@ecs.soton.ac.uk

Arun Chotai Lancaster Environment Centre, Lancaster University, Lancaster, UK, a.chotai@lancaster.ac.uk

Manfred Deistler Institute for Mathematical Methods in Economics, Vienna University of Technology, Vienna, Austria, manfred.deistler@tuwien.ac.at

R. Evans National ICT Australia Ltd, Eveleigh, Australia, rob.evans@nicta.com.au

Christoph Flamm Institute for Mathematical Methods in Economics, Vienna University of Technology, Vienna, Austria, christoph.flamm@tuwien.ac.at

Mathias Foo National ICT Australia, Victoria Research Lab, Department of Electrical and Electronic Engineering, The University of Melbourne, Parkville, VIC 3010, Australia, mfoo@ee.unimelb.edu.au

Chris T. Freeman School of Electronics and Computer Science, University of Southampton, Southampton SO17 1BJ, UK

Hugues Garnier CNRS, Nancy-Université, Vandoeuvre-lès-Nancy Cedex, France, hugues.garnier@cran.uhp-nancy.fr

Peter Gawthrop School of Engineering, University of Glasgow, Glasgow, UK, Peter.Gawthrop@glasgow.ac.uk

Marion Gilson CNRS, Nancy-Université, Vandoeuvre-lès-Nancy Cedex, France, marion.gilson@cran.uhp-nancy.fr

Graham C. Goodwin School of Electrical Engineering and Computer Science, University of Newcastle, University Drive NSW 2308, Australia, Graham.Goodwin@newcastle.edu.au

Andreas Graef Institute for Mathematical Methods in Economics, Vienna University of Technology, Vienna, Austria, andreas.graef@tuwien.ac.at

C.J. Harris School of Electronics and Computer Science, University of Southampton, Southampton, UK

David F. Hendry Economics Department & Institute for New Economic Thinking at the Oxford Martin School, University of Oxford, Oxford, UK, david.hendry@nuffield.ox.ac.uk

X. Hong School of Systems Engineering, University of Reading, Reading, UK, x.hong@reading.ac.uk

Sabine Van Huffel ESAT-SCD, K.U. Leuven, Kasteelpark Arenberg 10, 3001 Leuven, Belgium, Sabine.VanHuffel@esat.kuleuven.be

B.M. Jackson Imperial College London, London SW7 2AZ, UK

Ulrike Kalliauer VERBUND Trading AG, Vienna, Austria, ulrike.kalliauer@verbund.com

Vincent Laurain CNRS, Nancy-Université, Vandoeuvre-lès-Nancy Cedex, France, vincent.laurain@cran.uhp-nancy.fr

Fengmin Le School of Electronics and Computer Science, University of Southampton, Southampton SO17 1BJ, UK

Dave T. Leedal Lancaster Environment Centre, Lancaster University, Lancaster, UK, d.t.leedal@lancaster.ac.uk

L. Li National ICT Australia Ltd, Eveleigh, Australia, li.li@nicta.com.au

Z. Lin North Dakota State University, Fargo, ND, USA, zhulu.lin@ndsu.edu

C.X. Lu School of Electrical Engineering and Telecommunications, University of New South Wales, Sydney, Australia, c.lu@unsw.edu.au

I. Mareels The University of Melbourne, Melbourne, Australia, iven.mareels@unimelb.edu.au

Ivan Markovsky School of Electronics and Computer Science, University of Southampton, Southampton SO17 1BJ, UK, im@ecs.soton.ac.uk

Neil McIntyre Department of Civil and Environmental Engineering, Imperial College London, London SW72AZ, UK, n.mcintyre@imperial.ac.uk

Peter E.H. Minchin The New Zealand Institute for Plant and Food Research Limited, Te Puke, 412 No. 1 Rd, RD2, Te Puke 3182, New Zealand

N. Okello National ICT Australia Ltd, Eveleigh, Australia, nickens.okello@nicta.com.au

Su Ki Ooi Department of Electrical and Electronic Engineering, The University of Melbourne, Parkville, VIC 3010, Australia, skoo@ee.unimelb.edu.au

Andrea Pagano JRC, Joint Research Centre, The European Commission, TP 361, 21027 Ispra (VA), Italy, andrea.pagano@jrc.ec.europa.eu

Diego J. Pedregal E.T.S. de Ingenieros Industriales and Institute of Applied Mathematics to Science and Engineering (IMACI), University of Castilla, La Mancha, Ciudad Real, Spain, Diego.Pedregal@uclm.es

M. Pham National ICT Australia Ltd, Eveleigh, Australia, minh.pham@nicta.com.au

Francesca Pianosi Politecnico di Milano, Milano, Italy, pianosi@elet.polimi.it

W. Qiu National ICT Australia Ltd, Eveleigh, Australia, wanzhi.qiu@nicta.com.au

Marco Ratto JRC, Joint Research Centre, The European Commission, TP 361, 21027 Ispra (VA), Italy, marco.ratto@jrc.ec.europa.eu

N.W. Rees School of Electrical Engineering and Telecommunications, University of New South Wales, Sydney, Australia, n.rees@unsw.edu.au

Eric Rogers School of Electronics and Computer Science, University of Southampton, Southampton SO17 1BJ, UK, etar@ecs.soton.ac.uk

Renata J. Romanowicz Institute of Geophysics, Polish Academy of Sciences, ul. Ksiecia Janusza 64, 01-452 Warsaw, Poland, Romanowicz@igf.edu.pl

Jafar Sadeghi Lancaster Environment Centre, Lancaster University, Lancaster, UK

S.K. Saleem National ICT Australia Ltd, Eveleigh, Australia, khusro.saleem@nicta.com.au

Bellie Sivakumar The University of New South Wales, Sydney, NSW 2052, Australia, s.bellie@unsw.edu.au; University of California, Davis, CA 95616, USA, sbellie@ucdavis.edu

Paul J. Smith Lancaster Environment Centre, Lancaster University, Lancaster, UK, p.j.smith@lancaster.ac.uk

Victor Solo School of Electrical Engineering, University of New South Wales, Sydney, Australia, v.solo@unsw.edu.au

Rodolfo Soncini-Sessa Politecnico di Milano, Milano, Italy, soncini@elet.polimi.it

J.D. Stigter Wageningen University, Wageningen, The Netherlands, hans.stigter@wur.nl

Torsten Söderström Division of Systems and Control, Department of Information Technology, Uppsala University, PO Box 337, 75105 Uppsala, Sweden, torsten.soderstrom@it.uu.se

C. James Taylor Engineering Department, Lancaster University, Lancaster, UK, c.taylor@lancaster.ac.uk

Nguyen-Vu Truong Institute of Applied Mechanics and Informatics, Vietnam Academy of Science and Technology, Hanoi, Vietnam

Granville Tunnicliffe Wilson Lancaster University, Lancaster, England, UK, g.tunnicliffe-wilson@lancaster.ac.uk

Włodek Tych Lancaster Environment Centre, Lancaster University, Lancaster, UK, w.tych@lancaster.ac.uk

Thorsten Wagener Department of Civil and Environmental Engineering, The Pennsylvania State University, University Park, PA 16802, USA, thorsten@engr.psu.edu

Eric Walter Laboratoire des Signaux et Systèmes, CNRS–SUPELEC–Univ Paris-Sud, 91192 Gif-sur-Yvette, France, Eric.Walter@lss.supelec.fr

Liuping Wang School of Electrical and Computer Engineering, RMIT University, Melbourne, Australia, liuping.wang@rmit.edu.au

Markus Waser Institute for Mathematical Methods in Economics, Vienna University of Technology, Vienna, Austria, markus.waser@tuwien.ac.at

Erik Weyer Department of Electrical and Electronic Engineering, The University of Melbourne, Parkville, VIC 3010, Australia, ewey@unimelb.edu.au

H.S. Wheater Imperial College London, London SW7 2AZ, UK, h.wheater@imperial.ac.uk

Dae Keun Yoo RMIT University, Victoria 3000, Australia, dkyo@hotmail.com

Peter C. Young Centre for Research on Environmental Systems and Statistics, University of Lancaster, Lancaster, UK, p.young@lancaster.ac.uk; Fenner School of Environment and Society, Australian National University, Canberra, Australia

Part I
Theory of System Identification

Chapter 1
How Accurate Can Instrumental Variable Models Become?

Torsten Söderström

1.1 Introduction

Instrumental variable (IV) methods for system identification has been a popular technique for several decades. It has its roots in statistics, [8], and early applications appeared in econometrics. Some early works in the engineering literature include [6, 7, 20]. Some more recent papers in the field are, for example, [2, 3, 19]. Within (control) engineering, many papers and other publications by Peter Young has been pioneering, see [21–25] to mention just a few out of many more. A more comprehensive list of his many publications in the field appear elsewhere. The IV method will be presented in the chapter along with its user parameters. It has low computational complexity, comparable to the least squares (LS) method, but in contrast to the LS method it has the ability to give consistent parameter estimates for very arbitrary type of disturbances.

The accuracy of the estimated IV model, measured for example in terms of the covariance matrix of the parameter estimates, can be greatly influenced by the choice of the user parameters in the algorithm. In many cases this influence is studied using Monte Carlo simulations, which no doubt is a very useful tool. The aim of this chapter though is to give a theoretical treatment of the accuracy aspects.

Based on discussions with the editors, the presentation will be tutorial in style. Is largely based on work by the author and colleagues, see for example [11, 14–16, 18]. Basic key results are given with derivations and proofs. Linked to the tutorial style, several extensions are presented as exercises for the reader. A reader who prefers

Dedicated to Professor Peter Young on the occasion of his 70th anniversary, with thanks for many years of discussions on the instrumental variable method.

T. Söderström (✉)
Division of Systems and Control, Department of Information Technology, Uppsala University, PO Box 337, 75105 Uppsala, Sweden
e-mail: torsten.soderstrom@it.uu.se

L. Wang, H. Garnier (eds.), *System Identification, Environmental Modelling, and Control System Design*,
DOI 10.1007/978-0-85729-974-1_1, © Springer-Verlag London Limited 2012

the challenges is thereby welcome to test and generalize the ideas from the basic case. For those who like to know the details more directly, proper references are also provided.

A general background on system identification can be found in many textbooks, for example, [4, 12]. These books as well as [11, 24] contain many references to the work by various authors on instrumental variable estimators.

1.2 Instrumental Variable Methods

1.2.1 The Least Squares Method

The least squares (LS) method is applicable to models of the form

$$A(q^{-1})y(t) = B(q^{-1})u(t) + \varepsilon(t), \tag{1.1}$$

with

$$\begin{aligned} A(q^{-1}) &= 1 + a_1 q^{-1} + \cdots + a_{na} q^{-na}, \\ B(q^{-1}) &= b_1 q^{-1} + \cdots + b_{nb} q^{-nb}. \end{aligned} \tag{1.2}$$

Here $y(t)$ denotes the discrete-time output at time t, $u(t)$ is the input, and $\varepsilon(t)$ denotes an equation error, which can describe disturbances or unmodelled dynamics. Further q^{-1} is the backward shift operator q^{-1}, so that $q^{-1}u(t) = u(t-1)$.

The model (1.1) can be equivalently expressed as the *linear regression* model

$$y(t) = \varphi^T(t)\theta + \varepsilon(t), \tag{1.3}$$

where the regressor vector $\varphi(t)$ and the parameter vector θ are given by

$$\varphi^T(t) = \big(-y(t-1) \ \dots \ -y(t-na) \ u(t-1) \ \dots \ u(t-nb)\big), \tag{1.4}$$

$$\theta = (a_1 \ \dots \ a_{na} \ b_1 \ \dots \ b_{nb})^T. \tag{1.5}$$

Assume that data $u(1), y(1), \dots, u(N), y(N)$ are available. The LS estimate $\hat{\theta}$ of the parameter vector θ is defined as the minimizing argument of the sum of squared equation errors $V_N(\theta) = \frac{1}{N}\sum_{t=1}^{N} \varepsilon^2(t)$. By setting the gradient of $V_N(\theta)$ to zero we get the so-called normal equations

$$\left[\frac{1}{N}\sum_{t=1}^{N} \varphi(t)\varphi^T(t)\right]\hat{\theta} = \frac{1}{N}\sum_{t=1}^{N} \varphi(t)y(t). \tag{1.6}$$

A parameter estimate $\hat{\theta}$ is said to be *consistent* if

$$\hat{\theta} \to \theta_o \quad \text{as } N \to \infty \text{ (with prob 1)} \tag{1.7}$$

where θ_o is the 'true' parameter vector, which is assumed to describe the data, and this is a desirable property. When the data set is large enough, the obtained model is then arbitrarily accurate. Let us now examine the LS estimate for consistency. Consider a linear system of an arbitrary order and write it as

$$A_o(q^{-1})y(t) = B_o(q^{-1})u(t) + v(t) \tag{1.8}$$

or, equivalently

$$y(t) = \varphi^T(t)\theta_o + v(t). \tag{1.9}$$

Assume that $v(t)$ is a stationary stochastic process that is independent of the input signal. The estimation error becomes

$$\begin{aligned}\hat{\theta} - \theta_o &= \left[\frac{1}{N}\sum_{t=1}^{N}\varphi(t)\varphi^T(t)\right]^{-1}\left[\frac{1}{N}\sum_{t=1}^{N}\varphi(t)y(t) - \left\{\frac{1}{N}\sum_{t=1}^{N}\varphi(t)\varphi^T(t)\right\}\theta_o\right] \\ &= \left[\frac{1}{N}\sum_{t=1}^{N}\varphi(t)\varphi^T(t)\right]^{-1}\left[\frac{1}{N}\sum_{t=1}^{N}\varphi(t)v(t)\right]. \end{aligned} \tag{1.10}$$

Under weak conditions, the sums in (1.10) tend to the corresponding expected values as the number of data points, N, tends to infinity, [4, 12]. Hence $\hat{\theta}$ is consistent if

$$\mathsf{E}\{\varphi(t)\varphi^T(t)\} \quad \text{is nonsingular,} \tag{1.11}$$

$$\mathsf{E}\{\varphi(t)v(t)\} = 0. \tag{1.12}$$

The condition (1.11) is satisfied in most cases. There are a few exceptions:

- The input has a spectral density that is nonzero at *less* than nb frequencies ($u(t)$ is not persistently exciting of order nb).
- The data are completely noise-free ($v(t) \equiv 0$ and the model order is chosen too high (which implies that $A_o(q^{-1})$ and $B_o(q^{-1})$ have common factors).
- The input $u(t)$ is generated by a linear low order feedback from the output.

The condition (1.12) is much more restrictive than (1.11). In case the disturbance $v(t)$ is white noise it will be independent of *all the past data values* and (1.12) will be satisfied. If $v(t)$ is correlated noise, it will be correlated with the delayed output values present in $\varphi(t)$ and (1.12) will be violated. To get consistent and accurate parameter estimates *we must*, hence, *require that $v(t)$ in (1.8) is an uncorrelated disturbance.*

The LS method is a simple method for system identification that has some attractive properties. The estimate is easy to compute and has good robustness properties. The restrictive consistency properties are the main drawback and can be seen as the main reason for considering more advanced methods, including the IV method.

1.2.2 The Instrumental Variable Method

Instrumental variable methods can be seen as generalizations of the LS estimates. The main idea can be said to modify the estimate so that it is consistent for an arbitrary disturbance. We consider ARX models (1.1), (1.2). Next we modify the normal equations of (1.6) into

$$\left[\frac{1}{N}\sum_{t=1}^{N} z(t)\varphi^T(t)\right]\hat{\theta} = \left[\frac{1}{N}\sum_{t=1}^{N} z(t)y(t)\right], \tag{1.13}$$

where $z(t)$ is a vector of *instrumental variables*. This vector can be chosen in different ways (as exemplified below) subject to certain conditions guaranteeing the consistency of the estimate (1.13). These conditions will be specified later. Evidently the IV estimate defined by (1.13) is a generalization of the LS estimate: For $z(t) = \varphi(t)$, (1.13) reduces to (1.6). The basic IV method can be generalized in different ways.

The *extended* IV estimates of θ_o are obtained by generalizing (1.13) in two directions. Such IV estimation methods allow for an augmented $z(t)$ vector (i.e. one can have $\dim z(t) > \dim \varphi(t)$), as well as a prefiltering of the data. The extended IV estimate is given by

$$\hat{\theta} = \arg\min_{\theta} \left\| \left[\sum_{t=1}^{N} z(t)F(q^{-1})\varphi^T(t)\right]\theta - \left[\sum_{t=1}^{N} z(t)F(q^{-1})y(t)\right] \right\|_Q^2. \tag{1.14}$$

Here $z(t)$ is the IV vector of dimension $nz \geq \dim\theta$, $F(q^{-1})$ is an asymptotically stable (pre-)filter, and $\|x\|_Q^2 = x^T Qx$, where Q is a positive definite weighting matrix. When $F(q^{-1}) \equiv 1$ and $nz = n\theta$, $(Q = I)$, the basic IV estimate (1.13) is obtained. Note that the estimate (1.14) is the weighted least squares solution of an overdetermined linear system of equations. The solution can readily be found to be

$$\hat{\theta} = (R_N^T Q R_N)^{-1} R_N^T Q r_N \tag{1.15}$$

where

$$R_N = \frac{1}{N}\sum_{t=1}^{N} z(t)F(q^{-1})\varphi^T(t), \qquad r_N = \frac{1}{N}\sum_{t=1}^{N} z(t)F(q^{-1})y(t) \tag{1.16}$$

even if this is not the *numerically* best way to implement it.

Another generalization is to model multi-input multi-output (MIMO) systems. We write these as

$$y(t) = \Phi^T(t)\theta + \varepsilon(t) \tag{1.17}$$

where now $y(t)$ is a column vector and $\Phi(t)$ a regressor matrix of the form

$$\Phi^T(t) = \begin{pmatrix} \varphi^T(t) & & 0 \\ & \ddots & \\ 0 & & \varphi^T(t) \end{pmatrix}, \tag{1.18}$$

$$\varphi^T(t) = \big(-y^T(t-1)\ \dots\ -y^T(t-na)\ u^T(t-1)\ \dots\ u^T(t-nb)\big). \tag{1.19}$$

The extended IV estimate in the MIMO case is quite similar to (1.14). It now holds

$$\hat{\theta} = \arg\min_{\theta} \left\| \left[\sum_{t=1}^{N} z(t)F(q^{-1})\Phi^T(t)\right]\theta - \left[\sum_{t=1}^{N} z(t)F(q^{-1})y(t)\right] \right\|_Q^2. \tag{1.20}$$

1.2.3 Consistency Analysis and Conditions

Consider the extended IV estimate (1.14) applied to a system of the form (1.9). Assume that the data are stationary, $z(t)$ uncorrelated with the disturbances $v(s)$ for all t and s. We then have, similarly to (1.10)

$$\begin{aligned} \hat{\theta} - \theta_o &= (R_N^T Q R_N)^{-1}(R_N^T Q r_N - R_N^T Q R_N \theta_o) \\ &= (R_N^T Q R)^{-1} R_N^T Q \frac{1}{N}\sum_{t=1}^{N} z(t)F(q^{-1})\{y(t) - \varphi^T(t)\theta_o\} \\ &\to (R^T Q R)^{-1} R^T Q\, \mathsf{E}\big\{z(t)F(q^{-1})v(t)\big\} = 0, \end{aligned} \tag{1.21}$$

where

$$R \overset{\Delta}{=} \mathsf{E}\big\{z(t)\varphi^T(t)\big\} \tag{1.22}$$

is assumed to have full rank. For any reasonable choice of instrumental vector $z(t)$ this rank condition is satisfied for almost any (but not *all*) systems. We can hence say that the estimate $\hat{\theta}$ is *generically consistent*. We see that the consistency conditions (1.11), (1.12) in the general case become here

$$R \text{ must have full column rank} \tag{1.23}$$

$$\mathsf{E}\big\{z(t)F(q^{-1})v(t)\big\} = 0. \tag{1.24}$$

For any reasonable choice of instrumental vector $z(t)$ the rank condition (1.23) is satisfied for almost any (but not *all*) systems. We can hence say that the estimate $\hat{\theta}$ is *generically consistent*. There are though counter-examples based on certain combinations of input spectrum and system parameter θ that lead to a singular matrix R. Details are given in [11, 12].

1.2.4 User Choices. Examples of Instrumental Vectors

The IV estimator contains some user parameters, in addition to the choice of model order. These user parameters are the following:

1. A first choice is the IV vector $z(t)$. This choice concerns both its dimension, and how the elements are formed from the measured data. As an example of an IV vector, consider the case

$$z(t) = \big(-\eta(t-1) \; \ldots \; -\eta(t-na) \; u(t-1) \; \ldots \; u(t-nb)\big)^T \tag{1.25}$$

where the signal $\eta(t)$ is obtained by filtering the input,

$$C(q^{-1})\eta(t) = D(q^{-1})u(t). \tag{1.26}$$

The coefficients of the polynomials C and D can be chosen in many ways. One special choice is to let C and D be *a priori* estimates of A and B, respectively. Another special case is $C(q^{-1}) \equiv 1$, $D(q^{-1}) \equiv -q^{-nb}$, in which case $z(t)$ consists of just delayed inputs. The choice of IV vector can have a significant impact on the quality of the estimate. Due to the consistency conditions (1.23), (1.24), it can be said that the IV vector $z(t)$ should be well correlated with the regressor $\varphi(t)$, and uncorrelated with the disturbances $v(t)$.
2. A second user choice is the prefilter $F(q^{-1})$.
3. The third user choice applies when the system of IV equations is overdetermined, that is when the IV vector has higher dimension than the regressor vector. In that case the weighting matrix Q of the equations, see (1.15), need to be chosen.

Vectors like $z(t)$ in (1.25) above, and more generally those whose elements are obtained by filtering and delaying the input signal will, *for open loop operation*, be independent of the disturbances and mostly also satisfy the rank condition on R. For *closed loop operations* some modifications can be done to achieve consistent estimates provided as external signal, independent of the disturbances, can be measured.

Example 1.1 Consider a system operating in closed loop, where the governing equations are

$$y(t) = \varphi^T(t)\theta + v(t), \tag{1.27}$$

$$u(t) = -\frac{S(q^{-1})}{R(q^{-1})}y(t) + \frac{T(q^{-1})}{R(q^{-1})}r(t), \tag{1.28}$$

where $r(t)$ is a measurable external signal, and the polynomials R, S and T may not be known. Construct an IV estimator for identifying θ based on the measurements of the signals $y(t)$, $u(t)$ and $r(t)$.

Hint. Details for such a construction are given in [13].

1.3 How Accurate are IV Estimates?

The purpose of this section is to analyze the covariance matrix of the parameter error $\hat{\theta} - \theta_o$. Trivially, the error will depend on the disturbances, and it will be necessary to introduce some assumptions about the statistical properties of the disturbances. It should be emphasized that these assumptions are not needed to *apply* the IV method, but only when we want to analyze and characterize the statistical quality of the estimates.

For this aim, we assume that the disturbance $v(t)$ in (1.9) is a stationary stochastic process, and introduce an innovations description

$$v(t) = H(q^{-1})e(t), \qquad Ee(t)e(s) = \lambda^2 \delta_{t,s}, \qquad H(q^{-1}) = 1 + \sum_{k=1}^{\infty} h_k q^{-k}. \tag{1.29}$$

Note that this is equivalent to a spectral factorization of the disturbance spectrum $\phi_v(\omega)$. We assume that coefficients h_k decay at an exponential rate, that is, there is a constant C, and a number α, $0 \leq \alpha < 1$, such that

$$|h_k| \leq C\alpha^k, \quad \forall k \geq 0. \tag{1.30}$$

1.3.1 The Basic IV Estimator

Consider the basic IV estimator (1.13). The normalized parameter error can be written as,

$$\sqrt{N}\left(\hat{\theta} - \theta_o\right) = \left[\frac{1}{N}\sum_{t=1}^{N} z(t)\varphi^T(t)\right]^{-1} \left[\frac{1}{\sqrt{N}}\sum_{t=1}^{N} z(t)v(t)\right]. \tag{1.31}$$

Under weak assumptions, the underlying signals are ergodic, [4, 12], and the normalized sum converges to its expected value, cf. (1.22)

$$\frac{1}{N}\sum_{t=1}^{N} z(t)\varphi^T(t) \to \mathsf{E}\{z(t)\varphi^T(t)\} \triangleq R, \quad N \to \infty. \tag{1.32}$$

Further, one can show that the remaining factor in (1.31) is asymptotically Gaussian distributed in the sense

$$\frac{1}{\sqrt{N}}\sum_{t=1}^{N} z(t)v(t) \xrightarrow{\text{dist}} \mathcal{N}(0, S), \quad N \to \infty, \tag{1.33}$$

$$S = \lim_{N\to\infty} \mathsf{E}\left[\frac{1}{\sqrt{N}}\sum_{t=1}^{N} z(t)v(t)\right]\left[\frac{1}{\sqrt{N}}\sum_{s=1}^{N} z(s)v(s)\right]^T. \tag{1.34}$$

Expressions for the covariance matrix S in various situations will be given below. It follows from Slutsky's theorem that the normalized parameter error is also asymptotically Gaussian distributed, [12], as

$$\sqrt{N}(\hat{\theta} - \theta_o) \xrightarrow{\text{dist}} \mathscr{N}(0, P_{\text{IV}}), \quad P_{\text{IV}} = R^{-1} S R^{-T}. \tag{1.35}$$

Remark 1.1 Recall the consistency condition (1.23). If R is almost singular (more precisely, R has a large condition number), the estimate is almost not consistent. In (1.35) this shows up in that R^{-1} and hence P_{IV} will have large elements.

To characterize the matrix S in (1.35), we have in particular:

Lemma 1.1 *Assume that*

1. $z(t)$ *and* $v(s)$ *are jointly Gaussian,*
2. *The signals* $z(t)$ *and* $v(t)$ *are (at least) partly independent in the sense*

$$\mathsf{E}\{z(t)v(s)\} = 0 \quad \text{if either } t \le s \text{ or } t \ge s. \tag{1.36}$$

Then

$$S = \lambda^2 \mathsf{E}\big\{[H(q^{-1})z(t)][H(q^{-1})z(t)]^T\big\}. \tag{1.37}$$

Proof See Appendix A.1. □

Remark 1.2 If the system operates in open loop, and the vector of instruments is formed from filtered and delayed input values, for example as in (1.25), then the equality in (1.36) holds for all t and s.

Remark 1.3 In an errors-in-variables setting, [9], the case (1.36) may appear. Assume

$$y(t) = y_0(t) + \tilde{y}(t), \tag{1.38}$$

$$u(t) = u_0(t) + \tilde{u}(t), \tag{1.39}$$

$$v(t) = A(q^{-1})\tilde{y}(t) - B(q^{-1})\tilde{u}(t), \tag{1.40}$$

where $\tilde{y}(t)$ and $\tilde{u}(t)$ are independent white noise sequences, of zero mean and variances λ_y^2 and λ_u^2, respectively. Further, $y(t)$, $u(t)$ denote the measured variables, while $y_0(t) = \varphi_0^T(t)\theta_0$. Set $n = \max(na, nb)$. Then a possible instrumental vector is

$$z(t) = \varphi(t - n) + \varphi(t + n) \tag{1.41}$$

for which the condition (1.36) applies.

Remark 1.4 The condition (1.36) is fairly general. A main point of IV estimators is that they should work for quite general disturbances. It is therefore hard to construct realistic conditions on the instrumental vector and the disturbances that are substantially weaker than (1.36).

Another result is the following, where we drop the assumption on Gaussian signals, and instead require the instrumental vector to depend linearly on the data.

Lemma 1.2 *Assume that $z(t)$ is a linear process in the sense (μ is an arbitrary index)*

$$z_\mu(t) = \sum_{i=0}^{\infty} g_i^{(\mu)} e(t-i), \tag{1.42}$$

$$v(t) = \sum_{i=0}^{\infty} h_i e(t-i) \tag{1.43}$$

and that

$$\mathsf{E}\{z(t)v(s)\} = 0 \quad \textit{if either } t \le s \textit{ or } t \ge s. \tag{1.44}$$

Then

$$S = \lambda^2 \mathsf{E}\{[H(q^{-1})z(t)][H(q^{-1})z(t)]^T\}. \tag{1.45}$$

Proof See Appendix A.2. □

1.3.2 Extensions

We now consider various extensions of the simple and basic case treated above.

Exercise 1.1 Consider the extended IV estimator (1.15). Show that (1.35) generalizes to

$$\sqrt{N}(\hat{\theta} - \theta_o) \xrightarrow{\text{dist}} \mathscr{N}(0, P_{\text{IV}}) \tag{1.46}$$

with the covariance matrix P_{IV} given by

$$P_{\text{IV}} = \lambda^2 (R^T Q R)^{-1} R^T Q S Q R (R^T Q R)^{-1}, \tag{1.47}$$

$$S = \mathsf{E}\{[F(q^{-1})H(q^{-1})z(t)][F(q^{-1})H(q^{-1})z(t)]^T\}. \tag{1.48}$$

Hint. For a derivation, see [11, 12].

Exercise 1.2 Generalize the result of Exercise 1.3.1 to the MIMO case. Show that the result is as in (1.46), (1.47), but that (1.48) has to be modified. In this case, write the innovations form for the disturbances as

$$v(t) = H(q^{-1})e(t), \qquad H(0) = I, \qquad \mathsf{E}\{e(t)e^T(s)\} = \Lambda \delta_{t,s}. \tag{1.49}$$

Further, introduce matrix coefficients $\{K_i\}_{i=0}^{\infty}$ by

$$\sum_{i=0}^{\infty} K_i z^i = F(z)H(z). \tag{1.50}$$

Then show that in this case

$$S = \mathsf{E}\left\{\left[\sum_{i=0}^{\infty} Z(t+i)K_i\right]\Lambda\left[\sum_{j=0}^{\infty} K_j^T Z^T(t+j)\right]\right\}. \tag{1.51}$$

Hint. A derivation of the result appears in [12].

Remark 1.5 There are cases when an IV estimator is based on higher-order statistics, and then the instrumental vector $z(t)$ is no longer a linear function of the measured input-output data. Such examples appear in errors-in-variables problems. Consider again the situation given in (1.38)–(1.40). Assume that the noise-free input $u_0(t)$ is independent of the noise sequences $\tilde{u}(t)$, $\tilde{y}(t)$, and that these noises are white and Gaussian. Assume further that the noise-free input $u_0(t)$ has nonzero third-order moments

$$\mathsf{E}\{u_0^3(t)\} = \mu_u \neq 0. \tag{1.52}$$

Then a possible instrumental vector can be constructed as

$$z(t) = \left(u^2(t-1) \dots u^2(t-na-nb)\right)^T. \tag{1.53}$$

The general analysis leading to (1.46), (1.47) is still valid. To derive an explicit expression for S becomes fairly complicated, as many high order moments of the data are involved. Furthermore, in this case the instruments $z(t)$ are *not* independent of the noise $v(t)$, which also complicates the analysis. A detailed treatment can be found, for example, in [18].

1.4 How to Get Optimal Accuracy

1.4.1 General Results

By user choices, such as $F(q^{-1})$, $z(t)$ and Q, the covariance matrix P_{IV} can be affected. In this section we discuss *how to choose* these variables so that the covariance matrix P_{IV} of the parameter estimates becomes small, or even as small as possible.

First there is a result on choosing the weighting matrix Q optimally for a given instrumental vector $z(t)$ and a fixed prefilter $F(q^{-1})$.

Lemma 1.3 *Consider the covariance matrix $P_{\mathrm{IV}} = P_{\mathrm{IV}}(Q)$ given by (1.47), where the dependence on Q is highlighted. Then it holds*

$$P(Q) \geq P(S^{-1}) \tag{1.54}$$

(in the sense that $P(Q) \geq P(S^{-1})$ is a nonnegative definite matrix).

Proof See Appendix A.3. □

Remark 1.6 The condition $Q = S^{-1}$ is sufficient, but not always necessary to get optimal accuracy. An example in a particular IV situation, where even the choice of no weighting ($Q = I$) gives optimal accuracy, is given in [10].

Next we discuss the choice of the instrumental vector and treat first the length of the vector $z(t)$. Of course, the dimension of $z(t)$ must be at least as large as the dimension of the regressor vector $\varphi(t)$, as otherwise the matrix R will not have full column rank, and the consistency condition on R would not be fulfilled. Does it pay to make the dimension of $z(t)$ large? The answer depends on what weighting is applied.

Lemma 1.4 *Consider the general IV estimator (1.15), with the instrumental vector $z(t)$, and assume that the weighting is taken optimally as $Q = S^{-1}$. Consider also an augmented IV estimator with the instrumental vector taken as*

$$\overline{z}(t) = \begin{pmatrix} z(t) \\ \tilde{z}(t) \end{pmatrix} \tag{1.55}$$

leading to

$$\overline{R} = \begin{pmatrix} R \\ \tilde{R} \end{pmatrix}, \qquad \overline{S} = \begin{pmatrix} S & S_{12} \\ S_{21} & S_{22} \end{pmatrix}. \tag{1.56}$$

Let the weighting be taken as

$$\overline{Q} = \overline{S}^{-1} \tag{1.57}$$

Then it holds that

$$P_{\mathrm{IV}}^{(z)} \geq P_{\mathrm{IV}}^{(\overline{z})}. \tag{1.58}$$

Further, equality in (1.58) applies if and only if the columns of $\overline{R}$ lies in the range space of the left block of $\overline{S}$,

$$\overline{R} \in \mathscr{R}\left(\begin{pmatrix} S \\ S_{21} \end{pmatrix}\right). \tag{1.59}$$

Proof See Appendix A.4. □

The lemma means that it pays to include additional instrumental vector elements *provided that optimal weighting is applied.* Next we provide an exercise that shows that the assumption of optimal weighting is essential.

Exercise 1.3 Consider the ARMA(1,1) process

$$y(t) + ay(t-1) = e(t) + ce(t-1) \tag{1.60}$$

and estimation of the AR parameter a using instrumental variables. Hence, the regressor vector will become a scalar, $\varphi(t) = -y(t-1)$. Consider, and compare, three different IV vectors.

1. $z_1(t) = -y(t-2)$.
2. $z_2(t) = -y(t-3)$.
3. $z_3(t) = (-y(t-2) \; -y(t-3))^T$, with no weighting, that is $Q = I$.

Evaluate the variance $P_{\text{IV}}(a)$ in the three cases. Use $c = 0$ for simplicity, and show that

$$P_{\text{IV}}(\hat{a}_1) < P_{\text{IV}}(\hat{a}_3) < P_{\text{IV}}(\hat{a}_2). \tag{1.61}$$

As the variance of $\hat{a}_3$ is larger than that of $\hat{a}_1$, the accuracy is not improved when augmenting $z_1(t)$ to $z_3(t)$.

Hint. Expressions for the variances are given in Appendix A.5.

For further optimization we have the following result.

Lemma 1.5 *Consider the general extended IV estimator, with optimal weighting. Then the covariance matrix P_{IV} has a lower bound*

$$P_{\text{IV}} \geq \lambda^2 \mathsf{E}\big\{[H^{-1}(q^{-1})\varphi_0(t)][H^{-1}(q^{-1})\varphi_0^T(t)]\big\}^{-1} \triangleq P_{\text{IV}}^{\text{opt}}. \tag{1.62}$$

The lower bound is achieved if

$$F(q^{-1}) = H^{-1}(q^{-1}), \qquad z(t) = \varphi_0(t) \quad (Q \textit{ irrelevant}). \tag{1.63}$$

Above $\varphi_0(t)$ denotes the "noise-free" part of $\varphi(t)$

$$\begin{aligned}\varphi_0(t) &= \mathsf{E}\{[\varphi(t)|u(t-1), u(t-2), \ldots]\} \\ &= \big(-y_0(t-1) \; \ldots \; -y_0(t-na) \; u(t-1) \; \ldots \; u(t-nb)\big)^T,\end{aligned} \tag{1.64}$$

where $y_0(t)$, the noise-free part of the output, is given by

$$A(q^{-1})y_0(t) = B(q^{-1})u(t). \tag{1.65}$$

Proof See Appendix A.6. □

1.4.2 A Multistep Algorithm

Even if the optimal choice of instruments and prefilter in (1.63) cannot be applied in practice (since $H(q^{-1})$ and $\tilde{\varphi}(t)$ are not known, but rather the goal of the estimation), a multistep algorithm can be constructed where the quantities in (1.63) are substituted by estimated values. Such an algorithm turns out to give as accurate result as the truly optimal IV method, which in turn often is comparable to the accuracy that can be achieved with a prediction error method (PEM).

When constructing a multistep algorithm we first need to specify a model structure that also takes the noise correlation into account. Let it be given by

$$y(t) = \varphi^T(t)\theta + H(q^{-1}; \theta, \eta)e(t) \tag{1.66}$$

where η is a vector of some additional parameters. We will give two typical examples of the parameterization of the noise filter $H(q^{-1})$ later, see Example 1.2.

The multistep algorithm is as follows.

1. Use any IV method to get a first estimate of θ. Denote the result $\hat{\theta}_1$.
2. Use the model structure (1.66) and set $\theta = \hat{\theta}_1$. Estimate η using any method producing a consistent estimate. Denote the result $\hat{\eta}_1$.
3. Apply the optimal IV method, as given by (1.63), using $H(q^{-1}) = H(q^{-1}; \hat{\theta}_1, \hat{\eta}_1)$. Denote the result $\hat{\theta}_2$.
4. Possibly repeat Step 2 using $\theta = \hat{\theta}_2$. Denote the result $\hat{\eta}_2$.

The last two steps may be further repeated a number of times.

Remark 1.7 The multistep algorithm is essentially the same as Young's refined IV method, [25]. In that method, the starting point is rather the likelihood equations. Designing an iterative scheme for solving these equations for the so called Box-Jenkins model, cf. (1.75), leads to a repeated use of Steps 3 and 4 in the multistep algorithm.

1.5 Influence of Noise Model Parameterization and Analysis of the Multistep Algorithm

The multistep algorithm is analyzed in [5, 11, 15], and the following result is derived.

Lemma 1.6 *The following results apply.*

(a) *The estimates $\hat{\theta}_1, \hat{\eta}_1, \hat{\theta}_2, \hat{\eta}_2$ are consistent and Gaussian distributed.*
(b) *The asymptotic normalized covariance matrix (for large values of N) of the estimate $\hat{\theta}_2$ is equal to that of the optimal IV estimate of Lemma 1.5*

$$N \operatorname{cov}(\hat{\theta}_2) = P_{\mathrm{IV}}^{\mathrm{opt}}. \tag{1.67}$$

(c) *Assume that noise filter parameterization fulfills*

$$H(q^{-1}; \hat{\theta}, \hat{\eta}) = A(q^{-1})\overline{H}(q^{-1}; \hat{\theta}, \hat{\eta}) \tag{1.68}$$

and that $\overline{H}(q^{-1})$ does not depend on the parameter vector θ. Assume further that a prediction error (or any other statistically efficient) estimator is used in

Step 2. Then, asymptotically as $N \to \infty$, the normalized covariance matrices fulfill

$$N \operatorname{cov}(\hat{\eta}_2) = N \operatorname{cov}(\hat{\eta}_1) \tag{1.69}$$

and

$$N \operatorname{cov}\begin{pmatrix} \hat{\theta}_2 \\ \hat{\eta}_1 \end{pmatrix} = N \operatorname{cov}\begin{pmatrix} \hat{\theta}_{\text{PEM}} \\ \hat{\eta}_{\text{PEM}} \end{pmatrix}. \tag{1.70}$$

(d) *If the noise parameterization is as in (1.68) but $\overline{H}(q^{-1})$ does depend on the parameter vector θ, then the results are instead that*

$$N \operatorname{cov}(\hat{\eta}_2) \leq N \operatorname{cov}(\hat{\eta}_1) \tag{1.71}$$

and

$$N \operatorname{cov}\begin{pmatrix} \hat{\theta}_2 \\ \hat{\eta}_1 \end{pmatrix} \leq N \operatorname{cov}\begin{pmatrix} \hat{\theta}_{\text{PEM}} \\ \hat{\eta}_{\text{PEM}} \end{pmatrix}. \tag{1.72}$$

The inequalities (1.71), (1.72) are to be interpreted in a matrix sense, as in (1.54).

Remark 1.8 The relation (1.69) means that (for large enough values of N) it is enough to take three steps of the algorithm. The accuracy of the parameter estimates will not improve by taking also further steps.

Let us now consider two typical cases of noise filter parameterizations.

Example 1.2 Consider first an ARMAX model

$$A(q^{-1})y(t) = B(q^{-1})u(t) + C(q^{-1})e(t). \tag{1.73}$$

Then η will consist of the polynomial coefficients of $C(q^{-1})$. As in this case

$$H(q^{-1}) = C(q^{-1}) = A(q^{-1})\overline{H}(q^{-1}) \tag{1.74}$$

we find that in this case $\overline{H}(q^{-1})$ does depend on θ.

Next consider the so called 'Box-Jenkins' (BJ) model, made popular in [1],

$$y(t) = \frac{B(q^{-1})}{A(q^{-1})}u(t) + \frac{C(q^{-1})}{D(q^{-1})}e(t). \tag{1.75}$$

In this case the parameter vector η consists of the polynomial coefficients of both $C(q^{-1})$ and $D(q^{-1})$. Further in this case we have instead of (1.74),

$$H(q^{-1}) = \frac{A(q^{-1})C(q^{-1})}{D(q^{-1})} = A(q^{-1})\overline{H}(q^{-1}) \tag{1.76}$$

so in this case the filter $\overline{H}(q^{-1})$ does not depend on θ.

The crucial difference between the two models (1.73) and (1.75) is that the dynamics from the input and from the noise are described with independent parameters in the Box-Jenkins model (1.75), while they are described with common poles in the ARMAX model (1.73). It is worth stressing that any finite-order linear model can be transformed into any of these two model forms.

So, what model parameterization of the noise dynamics should be chosen? Further, will use of a PEM give more accurate estimates than an optimal IV method? Apparently, two possible model candidates are the ARMAX structure, and the BJ structure. There are, in the opinion of this author, no clear cut answer which one to prefer. Rather, we provide a number of comments and aspects on this choice.

- The theory has its limitations. For example, an appropriate model structure is assumed for all results on the covariance matrix of the parameter estimates, such as (1.35). The theory also concerns the asymptotic case, $N \to \infty$. How large N should be for the results to 'apply', can vary from case to case.
- Any finite-order linear system can be transformed to an ARMAX structure as well as to a BJ structure.
- When applying the multistep algorithm to achieve the accuracy given by $P_{\mathrm{IV}}^{\mathrm{opt}}$, it does not matter in what way the noise filter is estimated, as long as a consistent estimate of $H(q^{-1})$ is found, cf. (1.67).
- Different choices of the model parameterization of $H(q^{-1})$ may lead to different computational complexities when deriving a consistent estimate of this filter.
- When applying a PEM, the use of BJ has some advantages from the consistency point of view. To get consistent parameter estimates of θ, the model parameterization of the noise dynamics must cover the 'true filter' when an ARMAX model is chosen, but this is not required for consistency when a BJ model is used.
- If there are disturbances 'early in the process' (refer, for example, to process noise in a state space description), then there are common poles in the dynamics from the input to the output and in the dynamics from the noise innovations $e(t)$ to the output. If there are common poles in the dynamics from $u(t)$ and $e(t)$ and a PEM is applied, then best accuracy is achieved if this fact is exploited, that is, an ARMAX structure is employed.
- If there are no common poles in the dynamics from $u(t)$ and $e(t)$, and a PEM is applied, then an ARMAX model structure will have higher order than a BJ structure, and the estimated model has to be tested for pole-zero cancellations. This can be done in a statistically optimal way, see [17], and then the use of an ARMAX structure leads to the same accuracy as a BJ structure.

Consider for illustration the following simple example.

Exercise 1.4 The dynamics is a first order ARMAX system

$$(1 + aq^{-1})y(t) = bq^{-1}u(t) + (1 + cq^{-1})e(t). \tag{1.77}$$

The input signal is assumed to be white noise, with zero mean and variance σ^2. The white noise source $e(t)$ is assumed to have zero mean and variance λ^2. Apply estimators and evaluate the asymptotic normalized covariance matrices for the following cases.

(a) A prediction error method is applied with an ARMAX model structure. Show that this gives the covariance matrix

$$P_{\text{PEM}}^{\text{ARMAX}} = \frac{\lambda^2}{\sigma^2} \frac{(1-a^2)(1-ac)^2(1-c^2)}{[b^2 + (\lambda^2/\sigma^2)(a-c)^2]} \times \begin{pmatrix} \frac{1}{1-c^2} & -\frac{bc}{(1-ac)(1-c^2)} \\ -\frac{bc}{(1-ac)(1-c^2)} & \frac{b^2(1-a^2c^2)+(\lambda^2/\sigma^2)(c-a)^2(1-c^2)}{(1-a^2)(1-ac)^2(1-c^2)} \end{pmatrix}. \tag{1.78}$$

Hint. See [12] for a derivation.

(b) A PEM is applied with a BJ model structure. (This means that in the estimation we do not make any assumption of the input and noise dynamics to have a joint pole.) Show that the covariance matrix fulfills

$$P_{\text{PEM}}^{\text{BJ}} = \frac{\lambda^2}{\sigma^2} \frac{(1-a^2)(1-ac)^2(1-c^2)}{b^2} \begin{pmatrix} \frac{1}{1-c^2} & -\frac{bc}{(1-ac)(1-c^2)} \\ -\frac{bc}{(1-ac)(1-c^2)} & \frac{b^2(1+ac)}{(1-a^2)(1-ac)(1-c^2)} \end{pmatrix}, \tag{1.79}$$

$$P_{\text{PEM}}^{\text{BJ}} - P_{\text{PEM}}^{\text{ARMAX}} = \left(\frac{\lambda^2}{\sigma^2}\right)^2 \frac{(c-a)^2(1-a^2)(1-ac)^2}{b^2[b^2 + (\lambda^2/\sigma^2)(c-a)^2]} \begin{pmatrix} 1 \\ \frac{-bc}{1-ac} \end{pmatrix} \begin{pmatrix} 1 & \frac{-bc}{1-ac} \end{pmatrix} \geq 0. \tag{1.80}$$

(c) The optimal IV method is applied. Show that the covariance matrix will be

$$P_{\text{IV}} = P_{\text{PEM}}^{\text{BJ}}. \tag{1.81}$$

This exercise hence illustrates a case where using a PEM with an ARMAX model structure gives somewhat better accuracy than the optimal IV method.

Appendix: Proofs and Derivations

A.1 Proof of Lemma 1.1

Using the assumption on joint Gaussian distribution we can apply the general rule for product of Gaussian variables

$$\mathsf{E}\{x_1x_2x_3x_4\} = \mathsf{E}\{x_1x_2\}\mathsf{E}\{x_3x_4\} + \mathsf{E}\{x_1x_3\}\mathsf{E}\{x_2x_4\} + \mathsf{E}\{x_1x_4\}\mathsf{E}\{x_2x_3\}. \tag{1.82}$$

Using the result (1.82) in (1.34) leads to

$$S = \lim_{N\to\infty} \frac{1}{N} \sum_{t=1}^{N} \sum_{s=1}^{N} \{[\mathsf{E}\{z(t)z^T(s)\}][\mathsf{E}\{v(t)v(s)\}]$$

$$+ [\mathsf{E}\{z(t)v(s)\}][\mathsf{E}\{z^T(s)v(t)\}]\}$$

$$= \lim_{N\to\infty} \sum_{\tau=-N}^{N} \left(1 - \frac{\tau}{N}\right)\{R_z(\tau)r_v(\tau) + r_{zv}(\tau)r_{zv}^T(-\tau)\}. \tag{1.83}$$

Recall that the covariance function $r_v(\tau)$ decays exponentially with τ. Therefore we can write

$$\left\| \lim_{N\to\infty} \frac{1}{N} \sum_{\tau=-N}^{N} |\tau| R_z(\tau) r_v(\tau) \right\| \le \lim_{N\to\infty} \frac{1}{N} \sum_{\tau=0}^{N} 2\tau C \alpha^{\tau} = 0 \tag{1.84}$$

for some $|\alpha| < 1$. Using this result, we get

$$S = \sum_{\tau=-\infty}^{\infty} [R_z(\tau)r_v(\tau) + r_{zv}(\tau)r_{zv}^T(-\tau)]. \tag{1.85}$$

Now use the conventions

$$h_0 = 1, \qquad h_i = 0 \quad \text{for } i < 0 \tag{1.86}$$

The assumption (1.36) now implies that

$$r_{zv}(\tau)r_{zv}^T(-\tau) = 0 \quad \forall \tau \tag{1.87}$$

as at least one of the factors is zero. Therefore

$$S = \sum_{\tau=-\infty}^{\infty} R_z(\tau)r_v(\tau) = \sum_{\tau=-\infty}^{\infty} \left[R_z(\tau)\lambda^2 \sum_{i=0}^{\infty} h_i h_{i+\tau} \right]$$

$$= \lambda^2 \sum_{\tau=-\infty}^{\infty} \sum_{i=-\infty}^{\infty} h_i h_{i+\tau} \mathsf{E}\{ z(t+\tau)z^T(t)\}$$

$$= \lambda^2 \sum_{\tau=-\infty}^{\infty} \sum_{i=-\infty}^{\infty} h_i h_{i+\tau} \mathsf{E}\{ z(t-i)z^T(t-i-\tau)\}$$

$$= \lambda^2 \mathsf{E}\left\{ \sum_{i=-\infty}^{\infty} h_i z(t-i) \sum_{\tau=-\infty}^{\infty} h_{i+\tau} z^T(t-i-\tau) \right\}$$

$$= \lambda^2 \mathsf{E}\left\{\left(\sum_{i=-\infty}^{\infty} h_i z(t-i)\right)\left(\sum_{k=-\infty}^{\infty} h_k z^T(t-k)\right)\right\}$$

$$= \lambda^2 \mathsf{E}\left\{\left[H(q^{-1})z(t)\right]\left[H(q^{-1})z(t)\right]^T\right\} \tag{1.88}$$

which is (1.37).

A.2 *Proof of Lemma 1.2*

Using the definitions, one find that an arbitrary element of the matrix S is given by

$$S_{\mu,\nu} = \lim_{N\to\infty} \frac{1}{N} \sum_{t=1}^{N} \sum_{s=1}^{N} \mathsf{E}\left\{\sum_{k=0}^{\infty} g_k(\mu)e(t-k) \sum_{i=0}^{\infty} h_i e(t-i)\right.$$
$$\left. \times \sum_{j=0}^{\infty} g_j(\nu)e(t-j) \sum_{\ell=0}^{\infty} h_\ell e(t-\ell)\right\}. \tag{1.89}$$

To proceed we need to evaluate the expectation of products of the white noise sequence. Set $m_e = \mathsf{E}\{e^4(t)\}$. As $e(t)$ has zero mean, the expected value of a product of four factors of the noise is nonzero if either the time arguments are pairwise equal, or all are equal. This principle gives

$$\mathsf{E}\{e(t-k)e(t-i)e(t-j)e(t-\ell)\}$$
$$= \lambda^4 \delta_{i,k}\delta_{j,\ell} + \lambda^4 \delta_{t-k,s-j}\delta_{t-i,s-\ell}$$
$$+ \lambda^4 \delta_{t-k,s-\ell}\delta_{t-i,s-j} + (m_e - 3\lambda^4)\delta_{i,k}\delta_{j,\ell}\delta_{t-k,s-j}. \tag{1.90}$$

Inserting this into (1.89) leads to

$$S_{\mu,\nu} = \lambda^4 \lim_{N\to\infty} \sum_{\tau=-N}^{N} \left(1 - \frac{|\tau|}{N}\right) \sum_j g_j^{(\nu)} g_{\tau+j}^{(\mu)} \sum_\ell h_{\tau+\ell} h_\ell$$
$$+ \lambda^4 \lim_{N\to\infty} \sum_{\tau=-N}^{N} \left(1 - \frac{|\tau|}{N}\right) \sum_j g_j^{(\nu)} h_{\tau+j} \sum_\ell h_\ell g_{\tau+\ell}^{(\mu)}$$
$$+ (m_e - 3\lambda^4) \lim_{N\to\infty} \sum_{\tau=-N}^{N} \sum_\ell h_{\tau+\ell} g_j \ell^{(\nu)} g_{\tau+\ell}^{(\mu)} h_\ell$$
$$\stackrel{\Delta}{=} T_1 + T_2 + T_3. \tag{1.91}$$

Comparing the calculations in the proof of Lemma 1.3.1, we find that the first term in (1.91) is precisely

$$T_1 = \mathsf{E}\{[H(q^{-1})z(t)][H(q^{-1})z(t)]^T\}. \tag{1.92}$$

Further, the second term turns out to be

$$T_2 = \sum_{\tau=-\infty}^{\infty} r_{zv}(\tau) r_{zv}^T(-\tau) \tag{1.93}$$

which vanishes due to the assumption (1.44).

The last term can be written as

$$\begin{aligned} T_3 &= (m_e - 3\lambda^4) \sum_{\tau} \sum_{\ell} h_\ell h_{\tau+\ell} g_j \ell^{(\nu)} g_{\tau+\ell}^{(\mu)} \\ &= (m_e - 3\lambda^4) \Big[\sum_k h_k g_k^{(\mu)}\Big]\Big[\sum_j h_k g_j^{(\nu)}\Big]. \end{aligned} \tag{1.94}$$

However, we know that

$$\begin{aligned} \mathsf{E}\, z_\mu(t) v(t) &= \mathsf{E}\Big[\sum_k h_k e(t-k)\Big]\Big[\sum_j g_j^{(\mu)} e(t-j)\Big] \\ &= \lambda^2 \sum_k \sum_j h_k g_j^{(\mu)} \delta_{j,k} = \lambda^2 \sum_k h_k g_k^{(\mu)} = 0. \end{aligned} \tag{1.95}$$

This implies that

$$T_3 = 0 \tag{1.96}$$

and the lemma is proven.

A.3 *Proof of Lemma 1.3*

We first write from (1.47)

$$P(S^{-1}) = \lambda^2 (R^T S^{-1} R)^{-1}. \tag{1.97}$$

The inequality (1.54) can equivalently be written as

$$\lambda^2 (R^T Q R)^{-1} R^T Q S Q R (R^T Q R)^{-1} \geq \lambda^2 (R^T S^{-1} R)^{-1} \tag{1.98}$$

which can be rewritten as

$$(R^T Q R)(R^T Q S Q R)^{-1}(R^T Q R) \leq R^T S^{-1} R. \tag{1.99}$$

This in turn follows from the theory of partitioned matrices, cf. Lemma A.3 of [12], as

$$\begin{pmatrix} R^T S R & R^T Q R \\ R^T Q R & R^T S Q Q R \end{pmatrix} = \begin{pmatrix} R^T S^{-1} \\ R^T Q \end{pmatrix} S \begin{pmatrix} S^{-1} R & Q R \end{pmatrix} \geq 0. \tag{1.100}$$

A.4 Proof of Lemma 1.4

Using the theory of partitioned matrices, see for example Lemma A.2 in [12] and (1.56)

$$\begin{aligned}
\left[P_{\mathrm{IV}}^{(\bar{z})}/\lambda^2\right]^{-1} &= \bar{R}^T \bar{S}^{-1} \bar{R} \\
&= \begin{pmatrix} R^T & \tilde{R}^T \end{pmatrix} \begin{pmatrix} S & S_{12} \\ S_{21} & S_{22} \end{pmatrix}^{-1} \begin{pmatrix} R \\ \tilde{R} \end{pmatrix} \\
&= \begin{pmatrix} R^T & \tilde{R}^T \end{pmatrix} \left[\begin{pmatrix} S^{-1} & 0 \\ 0 & 0 \end{pmatrix} + \begin{pmatrix} -S^{-1} S_{12} \\ I \end{pmatrix} \right. \\
&\quad \left. \times \left(S_{22} - S_{21} S^{-1} S_{12}\right)^{-1} \begin{pmatrix} -S_{21} S^{-1} & I \end{pmatrix} \right] \begin{pmatrix} R \\ \tilde{R} \end{pmatrix} \\
&= R^T S^{-1} R \\
&\quad + \left(-R^T S^{-1} S_{12} + \tilde{R}^T\right)\left(S_{22} - S_{21} S^{-1} S_{12}\right)^{-1}\left(-S_{21} S^{-1} R + \tilde{R}\right) \\
&\geq R^T S^{-1} R = \left[P_{\mathrm{IV}}^{(z)}/\lambda^2\right]^{-1}.
\end{aligned} \tag{1.101}$$

Equality in (1.101) applies if and only if

$$-S_{21} S^{-1} R + \tilde{R} = 0. \tag{1.102}$$

The condition (1.59) is equivalent to

$$R = S\alpha, \qquad \tilde{R} = S_{21}\alpha \tag{1.103}$$

for some matrix α. As S is nonsingular, this is in turn equivalent to $\alpha = S^{-1} R$, and

$$\tilde{R} = S_{21} S^{-1} R \tag{1.104}$$

which is (1.102).

A.5 Answer to Exercise 1.3

Use the notation

$$r_k = \mathsf{E}\{y(t+k)y(t)\} = \frac{\lambda^2}{1-a^2}(-a)^{|k|}. \tag{1.105}$$

Then

$$\operatorname{var}(\hat{a}_1) = \frac{1}{r_1^2} \mathsf{E}\{(y(t-2)e(t))^2\} = \frac{r_0\lambda^2}{r_1^2} = \frac{1-a^2}{a^2}, \tag{1.106}$$

$$\operatorname{var}(\hat{a}_2) = \frac{1}{r_2^2} \mathsf{E}\{(y(t-3)e(t))^2\} = \frac{r_0\lambda^2}{r_2^2} = \frac{1-a^2}{a^4} = \frac{1}{a^2}\operatorname{var}(\hat{a}_1), \tag{1.107}$$

$$\operatorname{var}(\hat{a}_3) = \frac{1}{(r_1^2+r_2^2)^2} \begin{pmatrix} r_1 & r_2 \end{pmatrix} \lambda^2 \begin{pmatrix} r_0 & r_1 \\ r_1 & r_0 \end{pmatrix} \begin{pmatrix} r_1 \\ r_0 \end{pmatrix} = \frac{1-a^2}{a^2} \frac{1+3a^2}{(1+a^2)^2}. \tag{1.108}$$

A.6 Proof of Lemma 1.5

Using the definition (1.50) of $K(z)$ introduce the notations

$$\alpha(t) = R^T Q \sum_{i=0}^{\infty} K_i z(t+i), \tag{1.109}$$

$$\beta(t) = H^{-1}(q^{-1})\varphi_0(t). \tag{1.110}$$

Then it holds

$$\begin{aligned} R^T Q R &= R^T Q \mathsf{E}\{z(t)F(q^{-1})\varphi_0^T(t)\} \\ &= R^T Q \mathsf{E}\{z(t)K(q^{-1})H^{-1}(q^{-1})\varphi_0^T(t)\} \\ &= R^T Q \mathsf{E}\Big\{z(t)\sum_i K_i H^{-1}(q^{-1})\varphi_0^T(t-i)\Big\} \\ &= \mathsf{E}\{\alpha(t)\beta^T(t)\}. \end{aligned} \tag{1.111}$$

Using (1.48) leads to

$$\lambda^2 R^T Q S Q R = \mathsf{E}\{\alpha(t)\alpha^T(t)\}. \tag{1.112}$$

The stated inequality (1.62) then reads

$$\begin{aligned} P_{\text{IV}} &= \big(\mathsf{E}\{\alpha(t)\beta^T(t)\}\big)^{-1}\big(\mathsf{E}\{\alpha(t)\alpha^T(t)\}\big)\big(\mathsf{E}\{\beta(t)\alpha^T(t)\}\big)^{-1} \\ &\geq \lambda^2\big(\mathsf{E}\{\beta(t)\beta^T(t)\}\big)^{-1}. \end{aligned} \tag{1.113}$$

Now, (1.113) is equivalent to

$$\big(\mathsf{E}\{\beta(t)\alpha^T(t)\}\big)\big(\mathsf{E}\{\alpha(t)\alpha^T(t)\}\big)^{-1}\big(\mathsf{E}\{\alpha(t)\beta^T(t)\}\big) \leq \big(\mathsf{E}\{\beta(t)\beta^T(t)\}\big), \tag{1.114}$$

which follows from the theory of partitioned matrices, cf. Lemma A.4 in [12], as

$$\mathsf{E}\left\{\begin{pmatrix} \beta(t)\beta^T(t) & \beta(t)\alpha^T(t) \\ \alpha(t)\beta^T(y) & \alpha(t)\alpha^T(t) \end{pmatrix}\right\} = \mathsf{E}\left\{\begin{pmatrix} \beta(t) \\ \alpha(t) \end{pmatrix} \begin{pmatrix} \beta^T(t) & \alpha^T(t) \end{pmatrix}\right\} \geq 0. \tag{1.115}$$

Further, we see that with the specific choice

$$z(t) = H^{-1}(q^{-1})\varphi_0(t), \qquad F(q^{-1}) = H^{-1}(q^{-1}), \qquad Q = I \tag{1.116}$$

it holds that

$$R = \mathsf{E}\{[H^{-1}(q^{-1})\varphi_0(t)][H^{-1}(q^{-1})\varphi_0^T(t)]\} = \mathsf{E}\{\beta(t)\beta^T(t)\} = S \tag{1.117}$$

from which the equality in (1.62) follows.

References

1. Box, G.E.P., Jenkins, G.W.: Time Series Analysis, Forecasting and Control, 2nd edn. Holden-Day, San Francisco (1976)
2. Friedlander, B.: The overdetermined recursive instrumental variable method. IEEE Trans. Autom. Control **AC-29**, 353–356 (1984)
3. Gilson, M., Van den Hof, P.: Instrumental variable methods for closed-loop identification. Automatica **41**(2), 241–249 (2005)
4. Ljung, L.: System Identification—Theory for the User, 2nd edn. Prentice Hall, Upper Saddle River (1999)
5. Ljung, L., Söderström, T.: Theory and Practice of Recursive Identification. MIT Press, Cambridge (1983)
6. Mayne, D.Q.: A method for estimating discrete time transfer functions. In: Advances in Computer Control, Second UKAC Control Convention, Bristol, UK (1967)
7. Peterka, V., Halousková, Q.: Tally estimate of åström model for stochastic systems. In: Proc. 2nd IFAC Symposium on Identification and System Parameter Estimation, Prague, Chechoslovakia (1970)
8. Reiersøl, O.: Confluence analysis by means of lag moments and other methods of confluence analysis. Econometrica **9**, 1–24 (1941)
9. Söderström, T.: Errors-in-variables methods in system identification. Automatica **43**(6), 939–958 (2007). Survey paper
10. Söderström, T., Hong, M.: Identification of dynamic errors-in-variables systems with periodic data. In: Proc. 16th IFAC World Congress, Prague, Czech Republic, July 4–8 (2005)
11. Söderström, T., Stoica, P.: Instrumental Variable Methods for System Identification. Springer, Berlin (1983)
12. Söderström, T., Stoica, P.: System Identification. Prentice Hall International, Hemel Hempstead (1989)
13. Söderström, T., Stoica, P., Trulsson, E.: Instrumental variable methods for closed loop systems. In: 10th IFAC World Congress, Munich, Germany (1987)
14. Stoica, P., Friedlander, B., Söderström, T.: Instrumental variable methods for ARMA models. In: Leondes, C.T. (ed.) Control and Dynamic Systems—Advances in Theory and Applications. System Identification and Adaptive Control, vol. 25, pp. 79–150. Academic Press, New York (1987)
15. Stoica, P., Söderström, T.: Optimal instrumental variable estimation and approximate implementation. IEEE Trans. Autom. Control **AC-28**, 757–772 (1983)
16. Stoica, P., Söderström, T., Friedlander, B.: Optimal instrumental variable estimates of the AR parameters of an ARMA process. IEEE Trans. Autom. Control **AC-30**, 1066–1074 (1985)
17. Söderström, T., Stoica, P., Friedlander, B.: An indirect prediction error method. Automatica **27**, 183–188 (1991)
18. Thil, S., Hong, M., Söderström, T., Gilson, M., Garnier, H.: Statistical analysis of a third-order cumulants based algorithm for discrete errors-in-variables identification. In: IFAC 17th World Congress Seoul, Korea, July 6–11 (2008)

19. Van Huffel, S., Vandewalle, J.: Comparison of total least squares and instrumental variable methods for parameter estimation of transfer function models. Int. J. Control **50**, 1039–1056 (1989)
20. Wong, K.Y., Polak, E.: Identification of linear discrete time systems using the instrumental variable approach. IEEE Trans. Autom. Control **12**, 707–718 (1967)
21. Young, P.C.: An instrumental variable method for real-time identification of a noisy process. Automatica **6**, 271–287 (1970)
22. Young, P.C., Jakeman, A.J.: Refined instrumental variable methods of time series analysis: Part III extensions. Int. J. Control **31**, 741–764 (1980)
23. Young, P.C.: Parameter estimation for continuous-time models—a survey. Automatica **17**, 23–29 (1981)
24. Young, P.C.: Recursive Estimation and Time-Series Analysis. Springer, Berlin (1984)
25. Young, P.C., Jakeman, A.J.: Refined instrumental variable methods of recursive time-series analysis. Part I: Single input, single output systems. Int. J. Control **29**, 1–30 (1979)

Chapter 2
Refined Instrumental Variable Methods for Hammerstein Box-Jenkins Models

Vincent Laurain, Marion Gilson, and Hugues Garnier

2.1 Introduction

Hammerstein block diagram model is widely represented for modelling nonlinear systems [3, 6, 8, 26]. The nonlinear block can be represented as a piecewise linear function [2] or as a sum of basis functions [7, 21].

Among the very recent work on discrete-time (DT) Hammerstein models in the time domain, the most exposed methods are the extended least squares for Hammerstein ARMAX models [6] which were further extended to Hammerstein OE models [7]. E.R Bai exposed a two stage algorithm involving least squares and single value decomposition used in different configurations [3, 4, 19] and was very recently analysed for Hammerstein Box-Jenkins models [28]. Nonetheless, the convergence properties of the algorithm are studied but there was no study driven in case of noise modelling error. Suboptimal Hammerstein model estimation in case of a bounded noise was studied in [5]. A blind maximum likelihood method is derived in [27] but the output signal is considered to be errorless.

In the continuous-time (CT) case, an exhaustive survey by Rao and Unbehauen [25] shows that CT model identification methods applied to Hammerstein models are poorly studied in the literature. In [22], the authors focus on the time-derivative approximation problems while solving the optimization problem using least squares. A non-parametric method can be found in [14] while an approach

V. Laurain (✉) · M. Gilson · H. Garnier
CNRS, Nancy-Université, Vandoeuvre-lès-Nancy Cedex, France
e-mail: vincent.laurain@cran.uhp-nancy.fr

M. Gilson
e-mail: marion.gilson@cran.uhp-nancy.fr

H. Garnier
e-mail: hugues.garnier@cran.uhp-nancy.fr

L. Wang, H. Garnier (eds.), *System Identification, Environmental Modelling, and Control System Design*,
DOI 10.1007/978-0-85729-974-1_2,

dedicated to periodic input signals can be found in [33]. To the best of the authors' knowledge, the parametric estimation problem has not been addressed yet for CT Hammerstein models with colored added noise.

Section 2.2 shows how the refined instrumental variable (RIV) method introduced in [29] can be extended in order to deal with Hammerstein BJ models. Moreover, the development of instrumental variable techniques able to cope with the direct continuous-time model estimation in colored noise conditions are exposed in Sect. 2.3. All presented methods are statistically analyzed through relevant Monte Carlo simulations and the features of the proposed method are studied in the different pre-cited contexts.

2.2 Discrete-Time Hammerstein Model Identification

2.2.1 System Description

Consider the Hammerstein system represented in Fig. 2.1 and assume that both input and output signals, $u(t_k)$ and $y(t_k)$ are uniformly sampled at a constant sampling time $\mathrm{T_s}$ over N samples. The Hammerstein system $\mathscr{S}_{\mathrm{o}}$, is described by the following input-output relationship:

$$\mathscr{S}_{\mathrm{o}} \quad \begin{cases} \bar{u}(t_k) = f(u(t_k)), \\ \chi_{\mathrm{o}}(t_k) = G_{\mathrm{o}}(q)\bar{u}(t_k), \\ y(t_k) = \chi_{\mathrm{o}}(t_k) + v_{\mathrm{o}}(t_k), \end{cases} \tag{2.1}$$

where u and y are the deterministic input and noisy output respectively, χ_{o} is the noise-free output and v_{o} the additive noise with bounded spectral density. $G_{\mathrm{o}}(q)$ is the linear transfer function which can be written as

$$G_{\mathrm{o}}(q) = \frac{B_{\mathrm{o}}(q^{-1})}{A_{\mathrm{o}}(q^{-1})}, \tag{2.2}$$

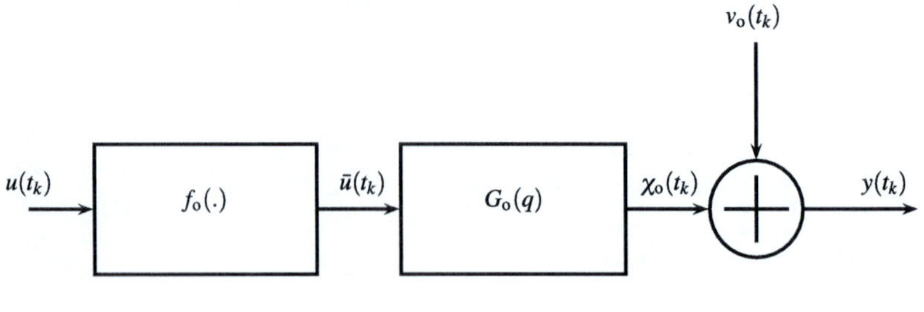

Fig. 2.1 Hammerstein block representation

where $B_o(q^{-1})$ and $A_o(q^{-1})$ are polynomial in q^{-1} of degree n_b and n_a respectively:

$$A_o(q^{-1}) = 1 + \sum_{i=1}^{n_a} a_i^o q^{-i}, \quad \text{and} \quad B_o(q^{-1}) = \sum_{j=0}^{n_b} b_j^o q^{-j}, \tag{2.3}$$

where the coefficients a_i^o and $b_j^o \in \mathbb{R}$. The most general case is considered where the colored noise associated with the sampled output measurement $y(t_k)$ is assumed to have a rational spectral density which might have no relation to the actual process dynamics of $\mathscr{S}_o$. Therefore, v_o is represented by a discrete-time *autoregressive moving average* (ARMA) model:

$$v_o(t_k) = H_o(q)e_o(t_k) = \frac{C_o(q^{-1})}{D_o(q^{-1})} e_o(t_k), \tag{2.4}$$

where $C_o(q^{-1})$ and $D_o(q^{-1})$ are monic polynomials with constant coefficients and with respective degree n_c and n_d. Furthermore, all roots of $z^{n_d} D_o(z^{-1})$ are inside the unit disc. It can be noticed that in case $C_o(q^{-1}) = D_o(q^{-1}) = 1$, (2.4) defines an OE noise model. It can be noticed that the same theory could be straightforwardly used if some pure delay was present on the input but this case is not exposed here for clarity's sake.

2.2.2 *Model Considered*

Next we introduce a discrete-time Hammerstein Box-Jenkins (BJ) type of model structure that we propose for the identification of the data-generating system (2.1) with noise model (2.4). In the chosen model structure, the noise model and the process model are parameterized separately.

2.2.2.1 Linear Part of the Hammerstein Model

The linear process model is denoted by $\mathscr{L}_{\rho_L}$ and is defined in a linear representation form as:

$$\mathscr{L}_{\rho_L} : \big(A(q^{-1}, \rho_L), B(q^{-1}, \rho_L)\big), \tag{2.5}$$

where the polynomials A and B are parameterized as

$$\mathscr{L}_{\rho_L} \quad \left\{ A(q^{-1}, \rho_L) = 1 + \sum_{i=1}^{n_a} a_i q^{-i}, \quad \text{and} \quad B(q^{-1}, \rho_L) = \sum_{j=0}^{n_b} b_j q^{-j}. \right.$$

The associated model parameters ρ_L are stacked columnwise:

$$\rho_L = \begin{bmatrix} a_1 \dots a_{n_a} \; b_0 \dots b_{n_b} \end{bmatrix}^\top \in \mathbb{R}^{n_a+n_b+1}. \tag{2.6}$$

Introduce also $\mathcal{L} = \{\mathscr{L}_{\rho_{\mathrm{L}}} \mid \rho \in \mathbb{R}^{n_{\rho_{\mathrm{L}}}}\}$, as the collection of all process models in the form of (2.5).

2.2.2.2 Nonlinear Part of the Hammerstein Model

The static nonlinearity model is denoted by $\mathscr{F}_{\rho_{\mathrm{NL}}}$ and defined:

$$\mathscr{F}_{\rho_{\mathrm{NL}}} : (f(u, \rho_{\mathrm{NL}})) \tag{2.7}$$

where $f(u, \rho_{\mathrm{NL}})$ is parameterized as a sum of basis functions

$$f(u(t_k), \rho_{\mathrm{NL}}) = \sum_{i=1}^{l} \alpha_i(\rho_{\mathrm{NL}})\gamma_i(u(t_k)). \tag{2.8}$$

In this parametrization, $\{\gamma_i\}_{i=1}^{l}$ are meromorphic functions[1] of $u(t_k)$ which are assumed to be *a priori* known. Furthermore, they have a static dependence on u, and are chosen such that they allow the identifiability of the model (pairwise orthogonal functions on $\mathbb{R}$ for example). The associated model parameters ρ_{NL} are stacked columnwise:

$$\rho_{\mathrm{NL}} = \begin{bmatrix} \alpha_1 \ldots \alpha_l \end{bmatrix}^{\top} \in \mathbb{R}^l, \tag{2.9}$$

Introduce also $\mathcal{F} = \{\mathscr{F}_{\rho_{\mathrm{NL}}} \mid \rho_{\mathrm{NL}} \in \mathbb{R}^l\}$, as the collection of all process models in the form of (2.7).

Remark Note that the Hammerstein model $(\beta f(u, \rho_{\mathrm{NL}}), \frac{G(q, \rho_{\mathrm{L}})}{\beta})$ produces the same input-output data for any β. Therefore, to get a unique parametrization, the gain of $(\beta f(u, \rho_{\mathrm{NL}})$ or $G(q, \rho_{\mathrm{L}})/\beta$ has to be fixed [1, 6]. Hence, the first coefficient of the function $f(.)$ is fixed to 1, i.e. $\alpha_1 = 1$ in (2.9).

2.2.2.3 Noise Model

The noise model denoted by $\mathscr{H}$ is defined as a *linear time invariant* (LTI) transfer function:

$$\mathscr{H}_\eta : (H(q, \eta)), \tag{2.10}$$

where H is a monic rational function given in the form of

$$H(q, \eta) = \frac{C(q^{-1}, \eta)}{D(q^{-1}, \eta)} = \frac{1 + c_1 q^{-1} + \cdots + c_{n_c} q^{-n_c}}{1 + d_1 q^{-1} + \cdots + d_{n_d} q^{-n_d}}. \tag{2.11}$$

[1] $f : \mathbb{R}^n \mapsto \mathbb{R}$ is a real meromorphic function if $f = g/h$ with g, h analytic and $h \neq 0$.

The associated model parameters η are stacked columnwise in the parameter vector,

$$\eta = \begin{bmatrix} c_1 \ldots c_{n_c} \; d_1 \ldots d_{n_d} \end{bmatrix}^\top \in \mathbb{R}^{n_\eta}, \tag{2.12}$$

where $n_\eta = n_c + n_d$. Additionally, denote $\mathcal{H} = \{\mathscr{H}_\eta \mid \eta \in \mathbb{R}^{n_\eta}\}$, the collection of all noise models in the form of (2.10).

2.2.2.4 Whole Hammerstein Model

With respect to a given nonlinear, linear process and noise part $(\mathscr{F}_{\rho_{NL}}, \mathscr{L}_{\rho_L}, \mathscr{H}_\eta)$, the parameters can be collected as

$$\theta_H = [\rho_L^\top \; \rho_{NL}^\top \; \eta^\top], \tag{2.13}$$

and the signal relations of the Hammerstein BJ model, denoted in the sequel as $\mathscr{M}_\theta$, are defined as:

$$\mathscr{M}_{\theta_H} \begin{cases} \bar{u}(t_k) = \sum_{i=1}^{l} \alpha_i(\rho_{NL})\gamma_i(u(t_k)), \\ A(q^{-1}, \rho_L)\chi(t_k) = B(q^{-1}, \rho_L)\bar{u}(t_k), \\ v(t_k) = \dfrac{C(q^{-1}, \eta)}{D(q^{-1}, \eta)} e(t_k), \\ y(t_k) = \chi(t_k) + v(t_k). \end{cases} \tag{2.14}$$

Based on this model structure, the model set, denoted as $\mathcal{M}$, with the linear process $(\mathscr{L}_{\rho_L})$, the nonlinearity $(\mathscr{F}_{\rho_{NL}})$ and noise $(\mathscr{H}_\eta)$ models parameterized independently, takes the form

$$\mathcal{M} = \left\{ (\mathscr{F}_{\rho_{NL}}, \mathscr{L}_{\rho_L}, \mathscr{H}_\eta) \mid \operatorname{col}(\rho_{NL}, \rho_L, \eta) = \theta_H \in \mathbb{R}^{n_{\rho_{NL}} + n_{\rho_L} + n_\eta} \right\}. \tag{2.15}$$

2.2.2.5 Reformulation of the Model

The optimization problem is not convex in general. However, it can be clearly seen from the parametrization (2.8) that the model (2.14) can be rewritten in order to obtain a linear regression structure. By combining the first two equations in (2.14), the model can be rewritten as:

$$\mathscr{M}_{\theta_H} \begin{cases} A(q^{-1}, \rho_L)\chi(t_k) = B(q^{-1}, \rho_L) \sum_{i=1}^{l} \alpha_i(\rho_{NL})\gamma_i(u(t_k)), \\ v(t_k) = \dfrac{C(q^{-1}, \eta)}{D(q^{-1}, \eta)} e(t_k), \\ y(t_k) = \chi(t_k) + v(t_k), \end{cases} \tag{2.16}$$

which can be expanded as (note that for clarity's sake $\gamma_i(u(t_k))$ is denoted $u_i(t_k)$ in the sequel)

$$\mathcal{M}_{\theta_{\mathrm{H}}} \begin{cases} A(q^{-1}, \rho_{\mathrm{L}})\chi(t_k) = \sum_{i=1}^{l} \underbrace{\alpha_i(\rho_{\mathrm{NL}})B(q^{-1}, \rho_{\mathrm{L}})}_{B_i(q^{-1}, \rho_{\mathrm{NL}}, \rho_{\mathrm{L}})} \underbrace{\gamma_i(u(t_k))}_{u_i(t_k)}, \\ v(t_k) = \dfrac{C(q^{-1}, \eta)}{D(q^{-1}, \eta)} e(t_k), \\ y(t_k) = \chi(t_k) + v(t_k). \end{cases} \tag{2.17}$$

Under these modelling settings, the nonlinearity model and the linear process model can be combined into the process model, denoted by $\mathcal{G}_\rho$ and defined in the form:

$$\mathcal{G}_\rho : \big(A(q^{-1}, \rho), B_i(q^{-1}, \rho)\big), \tag{2.18}$$

where the polynomials A and B_i are given by

$$\mathcal{G}_\rho \begin{cases} A(q^{-1}, \rho) = 1 + \sum_{i=1}^{n_\mathrm{a}} a_i q^{-i}, \\ B_i(q^{-1}, \rho) = \alpha_i \sum_{j=0}^{n_\mathrm{b}} b_j q^{-j}, \quad i = 1 \ldots l, \ \alpha_1 = 1. \end{cases}$$

The associated model parameters are stacked columnwise in the parameter vector ρ,

$$\rho = \begin{bmatrix} \mathbf{a} \\ \alpha_1 \mathbf{b} \\ \vdots \\ \alpha_l \mathbf{b} \end{bmatrix} \in \mathbb{R}^{n_\rho}, \quad \mathbf{a} = \begin{bmatrix} a_1 \\ a_2 \\ \vdots \\ a_{n_a} \end{bmatrix} \in \mathbb{R}^{n_a}, \ \mathbf{b} = \begin{bmatrix} b_0 \\ b_1 \\ \vdots \\ b_{n_b} \end{bmatrix} \in \mathbb{R}^{n_b+1}, \tag{2.19}$$

with $n_\rho = n_a + l(n_b + 1)$. Introduce also $\mathcal{G} = \{\mathcal{G}_\rho \mid \rho \in \mathbb{R}^{n_\rho}\}$, as the collection of all process models in the form of (2.18). Finally, with respect to the given process and noise part $(\mathcal{G}_\rho, \mathcal{H}_\eta)$, the parameters can be collected as $\theta = [\,\rho^\top \ \eta^\top\,]$ and the signal relations of the Hammerstein BJ model, denoted in the sequel as $\mathcal{M}_\theta$, are defined as:

$$\mathcal{M}_\theta : y(t_k) = \frac{\sum_{i=1}^{l} B_i(q^{-1}, \rho) u_i(t_k)}{A(q^{-1}, \rho)} + \frac{C(q^{-1}, \eta)}{D(q^{-1}, \eta)} e(t_k), \tag{2.20}$$

with $B_i(q^{-1}, \rho) = \alpha_i B(q^{-1}, \rho)$ and $u_i(t_k) = \gamma_i(u(t_k))$. Based on this model structure, the whole model set including the process $(\mathcal{G}_\rho)$ and noise $(\mathcal{H}_\eta)$ models parameterized independently, is denoted as $\mathcal{M}$ and takes finally the form

$$\mathcal{M} = \big\{(\mathcal{G}_\rho, \mathcal{H}_\eta) \mid \mathrm{col}(\rho, \eta) = \theta \in \mathbb{R}^{n_\rho + n_\eta}\big\}. \tag{2.21}$$

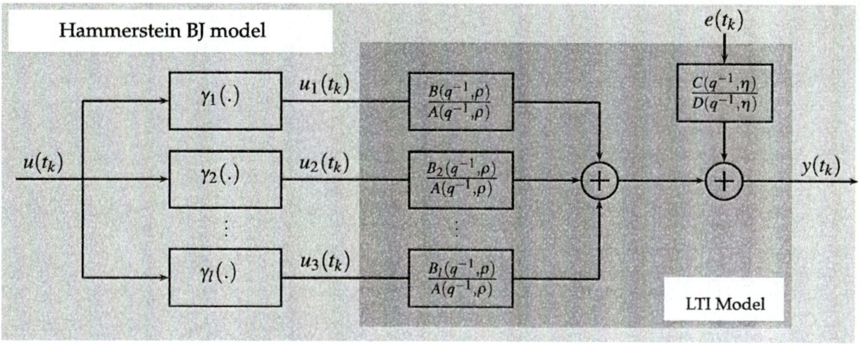

Fig. 2.2 Hammerstein augmented model

The set (2.21) corresponds to the set of candidate models in which we seek the best fitting model using data gathered from $\mathscr{S}_o$ under a given identification criterion (cost function).

Remarks It has to be noticed that this model transforms the Hammerstein structure into an augmented LTI *Multi Input Single Output* model structure such as presented in Fig. 2.2. Consequently, the number of parameters to be estimated is not minimal as $n_\rho = n_a + l(n_b + 1)$ which is in general greater than $n_{\rho_L} + n_{\rho_{NL}} = n_a + l + (n_b + 1)$. Therefore, as the model is not minimal, the optimal estimation of this augmented MISO model does not correspond to the optimal estimates of the true Hammerstein model. Nonetheless, the gain granted using this modelling is the possible linear regression form and therefore, the convexification of the optimization problem. In order to define the identification problem it is firstly necessary to define a minimization criterion. Nonetheless, the augmented model structure given in (2.20) is now an LTI structure, and therefore, the PEM framework from [20] can be directly used here.

2.2.3 *Identification Problem Statement*

Based on the previous considerations, the identification problem addressed can now be stated.

Problem 2.1 Given a discrete-time Hammerstein data generating system $\mathscr{S}_o$ defined as in (2.1) and a data set $\mathcal{D}_N$ collected from $\mathscr{S}_o$. Based on the Hammerstein BJ model structure $\mathscr{M}_\theta$ defined by (2.20), estimate the parameter vector θ using $\mathcal{D}_N$ under the following assumptions:

HA1 $\mathscr{S}_o \in \mathcal{M}$, i.e. there exits a θ_o defining a $\mathscr{G}_{\rho_o} \in \mathcal{G}$ and a $\mathscr{H}_{\eta_o} \in \mathcal{H}$ such that $(\mathscr{G}_{\rho_o}, \mathscr{H}_{\eta_o})$ is equal to $\mathscr{S}_o$.
HA2 $u(t_k)$ is not correlated to $e_o(t_k)$.

HA3 $\mathcal{D}_N$ is informative with respect to $\mathcal{M}$.
HA4 $\mathcal{S}_o$ is BIBO stable, i.e. for any bounded input signal u, the output of $\mathcal{S}_o$ is bounded.

2.2.4 Refined IV for Hammerstein Models

The *Hammerstein RIV* (HRIV) method derives from the RIV algorithm for DT linear systems. This was evolved by converting the maximum likelihood estimation equations to a pseudo-linear form involving optimal prefilters [29, 32]. A similar analysis can be utilised in the present situation since the problem is very similar, in both algebraic and statistical terms. The linear-in-the-parameters model (2.20) then takes the linear regression form [31]:

$$y(t_k) = \varphi^\top(t_k)\rho + \tilde{v}(t_k), \tag{2.22}$$

where ρ is as described in (2.19), $\tilde{v}(t_k) = A(q^{-1}, \rho)v(t_k)$ and

$$\varphi(t_k) = \begin{bmatrix} -\mathbf{y}(t_k) \\ \mathbf{u_1}(t_k) \\ \vdots \\ \mathbf{u_l}(t_k) \end{bmatrix}, \quad \mathbf{y}(t_k) = \begin{bmatrix} y(t_{k-1}) \\ \vdots \\ y(t_{k-n_a}) \end{bmatrix}, \quad \mathbf{u_i}(t_k) = \begin{bmatrix} u_i(t_k) \\ \vdots \\ u_i(t_{k-n_b}) \end{bmatrix}.$$

Using the conventional PEM approach on (2.22) leads to the prediction error $\varepsilon_\theta(t_k)$ given as:

$$\varepsilon_\theta(t_k) = \frac{D(q^{-1}, \eta)}{C(q^{-1}, \eta)} \left\{ y(t_k) - \sum_{i=1}^{l} \frac{B_i(q^{-1}, \rho)}{A(q^{-1}, \rho)} u_i(t_k) \right\}, \tag{2.23}$$

which can be written as

$$\varepsilon_\theta(t_k) = \frac{D(q^{-1}, \eta)}{C(q^{-1}, \eta)A(q^{-1}, \rho)} \left\{ A(q^{-1}, \rho)y(t_k) - \sum_{i=1}^{l} B_i(q^{-1}, \rho)u_i(t_k) \right\}, \tag{2.24}$$

where the prefilter $D(q^{-1}, \eta)/C(q^{-1}, \eta)$ will be recognised as the inverse of the ARMA(n_c, n_d) noise model. However, since the polynomial operators commute in this linear case, (2.24) can be considered in the alternative form:

$$\varepsilon_\theta(t_k) = A(q^{-1}, \rho)y_f(t_k) - \sum_{i=1}^{l} B_i(q^{-1}, \rho)u_{if}(t_k) \tag{2.25}$$

where $y_f(t_k)$ and $u_{if}(t_k)$ represent the outputs of the prefiltering operation using the filter:

$$Q(q, \theta) = \frac{D(q^{-1}, \eta)}{C(q^{-1}, \eta)A(q^{-1}, \rho)}. \tag{2.26}$$

Therefore, from (2.25), the associated linear-in-the-parameters model then takes the form:

$$y_f(t_k) = \varphi_f^\top(t_k)\rho + \tilde{v}_f(t_k), \tag{2.27}$$

where

$$\varphi_f(t_k) = \begin{bmatrix} -\mathbf{y}_f(t_k) \\ \mathbf{u}_{1f}(t_k) \\ \vdots \\ \mathbf{u}_{lf}(t_k) \end{bmatrix}, \quad \mathbf{y}_f(t_k) = \begin{bmatrix} y_f(t_{k-1}) \\ \vdots \\ y_f(t_{k-n_a}) \end{bmatrix}, \quad \mathbf{u}_{if}(t_k) = \begin{bmatrix} u_{if}(t_k) \\ \vdots \\ u_{if}(t_{k-n_b}) \end{bmatrix}, \tag{2.28}$$

and $\tilde{v}_f(t_k) = Q(q,\theta)\tilde{v}(t_k) = e(t_k)$ which is a white noise.

Therefore, according to the conditions for optimal IV estimates [30], the optimal instrument and filter for the augmented LTI MISO model structure (2.20) depicted in Fig. 2.2 are given as:

$$\begin{aligned} \zeta^{\text{opt}}(t_k) = \big[&-\chi_o(t_{k-1}) \ldots -\chi_o(t_{k-n_a})\; u_1(t_k) \ldots u_1(t_{k-n_b}) \\ &\ldots u_l(t_k) \ldots u_l(t_{k-n_b})\big]^\top, \end{aligned} \tag{2.29}$$

and

$$L^{\text{opt}}(q) = Q(q,\theta_o) = \frac{D_o(q^{-1})}{C_o(q^{-1})A_o(q^{-1})}. \tag{2.30}$$

2.2.5 The Hammerstein RIV (HRIV) Algorithm for BJ Models

Of course none of $A(q^{-1},\rho_o)$, $B_i(q^{-1},\rho_o)$, $C(q^{-1},\eta_o)$ or $D(q^{-1},\eta_o)$ is known and only their estimates are available. Therefore, neither the optimal prefilter nor the optimal instrument can be accessed and they can only be estimated. The 'auxiliary model' used to generate the noise-free output as well as the computation of the associated prefilter (2.26), are updated based on the parameter estimates obtained at the previous iteration to overcome this problem.

Algorithm 2.1 (HRIV)

Step 1 Generate an initial estimate of the process model parameter $\hat{\rho}^{(0)}$(*e.g.* using the LS method). Set $C(q^{-1},\hat{\eta}^{(0)}) = D(q^{-1},\hat{\eta}^{(0)}) = 1$. Set $\tau = 0$.

Step 2 Compute an estimate of $\chi(t_k)$ via

$$\hat{\chi}(t_k) = \frac{\sum_{i=1}^{l} B_i(q^{-1},\hat{\rho}^{(\tau)})u_i(t_k)}{A(q^{-1},\hat{\rho}^{(\tau)})},$$

where $\hat{\rho}^{(\tau)}$ is the estimate obtained at the previous iteration. According to assumption HA4 each $\hat{\chi}$ is bounded.

Step 3 Compute the filter as in (2.26):

$$L(q,\hat{\theta}^{(\tau)}) = \frac{D(q^{-1},\hat{\eta}^{(\tau)})}{C(q^{-1},\hat{\eta}^{(\tau)})A(q^{-1},\hat{\rho}^{(\tau)})}$$

and the associated filtered signals $\{u_{i\mathrm{f}} = \gamma_i(u)_{\mathrm{f}}\}_{i=1}^{l}$, y_{f} and $\{\hat{\chi}_{\mathrm{f}}\}_{i=1,l=0}^{n_{\mathrm{a}},n_\alpha}$.

Step 4 Build the filtered regressor $\varphi_{\mathrm{f}}(t_k)$ and the filtered instrument $\hat{\zeta}_{\mathrm{f}}(t_k)$ which equal in the given context:

$$\begin{aligned}\varphi_{\mathrm{f}}(t_k) &= \big[-y_{\mathrm{f}}(t_{k-1}) \dots -y_{\mathrm{f}}(t_{k-n_{\mathrm{a}}})\ u_{1\mathrm{f}}(t_k) \dots u_{1\mathrm{f}}(t_{k-n_b}) \\ &\quad \dots u_{l\mathrm{f}}(t_k) \dots u_{l\mathrm{f}}(t_{k-n_b})\big]^{\mathsf{T}}, \\ \hat{\zeta}_{\mathrm{f}}(t_k) &= \big[-\hat{\chi}_{\mathrm{f}}(t_{k-1}) \dots -\hat{\chi}_{\mathrm{f}}(t_{k-n_{\mathrm{a}}})\ u_{1\mathrm{f}}(t_k) \dots u_{1\mathrm{f}}(t_{k-n_b}) \\ &\quad \dots u_{l\mathrm{f}}(t_k) \dots u_{l\mathrm{f}}(t_{k-n_b})\big]^{\mathsf{T}}. \end{aligned} \tag{2.31}$$

Step 5 The IV optimization problem can be stated in the form

$$\hat{\rho}^{(\tau+1)}(N) = \arg\min_{\rho\in\mathbb{R}^{n_\rho}} \left\| \left[\frac{1}{N}\sum_{k=1}^{N}\hat{\zeta}_{\mathrm{f}}(t_k)\varphi_{\mathrm{f}}^{\mathsf{T}}(t_k)\right]\rho - \left[\frac{1}{N}\sum_{k=1}^{N}\hat{\zeta}_{\mathrm{f}}(t_k)y_{\mathrm{f}}(t_k)\right]\right\|^2, \tag{2.32}$$

where the solution is obtained as

$$\hat{\rho}^{(\tau+1)}(N) = \left[\sum_{k=1}^{N}\hat{\zeta}_{\mathrm{f}}(t_k)\varphi_{\mathrm{f}}^{\mathsf{T}}(t_k)\right]^{-1}\sum_{k=1}^{N}\hat{\zeta}_{\mathrm{f}}(t_k)y_{\mathrm{f}}(t_k).$$

The resulting $\hat{\rho}^{(\tau+1)}(N)$ is the IV estimate of the process model associated parameter vector at iteration $\tau+1$ based on the prefiltered input/output data.

Step 6 An estimate of the noise signal v is obtained as

$$\hat{v}(t_k) = y(t_k) - \hat{\chi}(t_k,\hat{\rho}^{(\tau)}). \tag{2.33}$$

Based on $\hat{v}$, the estimation of the noise model parameter vector $\hat{\eta}^{(\tau+1)}$ follows, using in this case the ARMA estimation algorithm of the MATLAB identification toolbox (an IV approach can also be used for this purpose, see [30]).

Step 7 If $\theta^{(\tau+1)}$ has converged or the maximum number of iterations is reached, then stop, else increase τ by 1 and go to Step 2.

At the end of the iterative process, coefficients $\hat{\alpha}_i$ are not directly accessible. They are however deduced from polynomial $\hat{B}_i(q^{-1})$ as $B_i(q^{-1},\rho) = \alpha_i B(q^{-1},\rho)$. The hypothesis $\alpha_1 = 1$ guarantees that $B_1(q^{-1},\rho) = B(q^{-1},\rho)$ and $\hat{\alpha}_i$ can be computed from:

$$\hat{\alpha}_i = \frac{1}{n_b+1}\sum_{j=0}^{n_b}\frac{\hat{b}_{i,j}}{\hat{b}_{1,j}}, \tag{2.34}$$

where $\hat{b}_{i,j}$ is the jth coefficient of polynomial term $B_i(q^{-1}, \rho)$ for $i = 2 \ldots l$.

Moreover, after the convergence is complete, it is possible to compute the estimated parametric error covariance matrix $\hat{\mathbf{P}}_\rho$ from the expression:

$$\hat{\mathbf{P}}_\rho = \hat{\sigma}_e^2 \left(\sum_{k=1}^{N} \hat{\zeta}_{\mathrm{f}}(t_k) \hat{\zeta}_{\mathrm{f}}^{\mathrm{T}}(t_k) \right)^{-1}, \tag{2.35}$$

where $\hat{\zeta}$ is the IV vector obtained at convergence and $\hat{\sigma}_e^2$ is the estimated residual variance.

Comments By using the described algorithm, if convergence occurs, the `HRIV` estimates might be statistically optimal for the augmented model proposed, but the minimal number of parameters needed for representing the MISO structure and the Hammerstein structure are not equal. Consequently, the HRIV estimates cannot be statistically optimal for the Hammerstein model structure. Nonetheless, even if not optimal, the `HRIV` estimates are unbiased with a low variance as it will be seen in the result Sect. 2.2.7.

2.2.6 *`HSRIV` Algorithm for OE Models*

A simplified version of `HRIV` algorithm named `HSRIV` follows the exact same theory for estimation of Hammerstein output error models. It is mathematically described by, $C(q^{-1}, \eta^j) = C_\mathrm{o}(q^{-1}) = 1$ and $D(q^{-1}, \eta^j) = D_\mathrm{o}(q^{-1}) = 1$. All previous given equations remain true, and it suffices to estimate ρ^j as $\theta^j = \rho^j$. The implementation of `HSRIV` is much simpler than `HRIV` as there is no noise model estimation in the algorithm.

2.2.7 *Performance Evaluation of the Proposed `HRIV` and `HSRIV` Algorithms*

This section presents numerical evaluation of both suggested `HRIV` and `HSRIV` methods. For the presented example, the nonlinear block has a polynomial form, i.e. $\gamma_i(u(t_k)) = u^i(t_k), \forall i$ and the system to identify is given by

$$\mathcal{S}_\mathrm{o} \quad \begin{cases} \bar{u}(t_k) = u(t_k) + 0.5u^2(t_k) + 0.25u^3(t_k), \\ G_\mathrm{o}(q) = \dfrac{0.5q^{-1} + 0.2q^{-2}}{1 + q^{-1} + 0.5q^{-2}}, \\ H_\mathrm{o}(q) = \dfrac{1}{1 - q^{-1} + 0.2q^{-2}}, \end{cases}$$

where $u(t_k)$ follows a uniform distribution with values between -2 and 2.

The models considered for estimation are:

$$\mathscr{M}_{\mathtt{HRIV}} \quad \begin{cases} G(q,\rho) = \dfrac{b_1 q^{-1} + b_2 q^{-2}}{1 + a_1 q^{-1} + a_2 q^{-2}}, \\ H(q,\eta) = \dfrac{1}{1 + d_1 q^{-1} + d_2 q^{-1}}, \\ f(u(t_k)) = u(t_k) + \alpha_1 u^2(t_k) + \alpha_2 u^3(t_k) \end{cases} \tag{2.36}$$

for the HRIV method which fulfills [HA1] and

$$\mathscr{M}_{\mathtt{HSRIV}} \quad \begin{cases} G(q,\rho) = \dfrac{b_1 q^{-1} + b_2 q^{-2}}{1 + a_1 q^{-1} + a_2 q^{-2}}, \\ H(q,\eta) = 1, \\ f(u(t_k)) = u(t_k) + \alpha_1 u^2(t_k) + \alpha_2 u^3(t_k) \end{cases} \tag{2.37}$$

for the HSRIV method which only fulfills $G_o \in \mathscr{G}$ ($H_o \notin \mathscr{H}$).

The result of a Monte Carlo simulation (MCs) analysis is shown in Table 2.1 and the algorithms considered are: HRIV, HSRIV and LSQNONLIN. The LSQNONLIN is a nonlinear optimization algorithm from the MATLAB® optimization toolbox. It assumes the same model as the HRIV method ($S_o \in \mathscr{M}$) and hands out the statistically optimal estimates if the method is properly initialized. In order to place the LSQNONLIN method at its advantage, it is initialized with the true parameter values and therefore this method can be considered as the ground truth.

The MCs results are based on $N_{\text{run}} = 100$ random realization, with the Gaussian white noise input to the ARMA noise model being selected randomly for each realization. In order to compare the statistical performance of the different approaches, the computed mean and standard deviation of the estimated parameters are presented. The noise added at the output is adjusted such that it corresponds to a *Signal-to-Noise-Ratio* (SNR) of 5dB using:

$$\text{SNR} = 10 \log\left(\frac{P_\chi}{P_{v_o}}\right), \tag{2.38}$$

where P_g is the average power of signal g. The number of samples is chosen as $N = 2000$.

As expected, Table 2.1 shows that the proposed algorithms produce unbiased estimates of the Hammerstein model parameters. It can be further noticed that the standard deviation of the estimates remains low even under the unrealistic noise level of 5 dB. Even though, the ratio between the HSRIV and HRIV estimate standard deviation equals to 2. This can be logically explained by the fact that the HSRIV algorithm assumes a wrong noise model and such result remains acceptable in practical applications. Finally it can be depicted that the HRIV provides the statistical optimal estimates for the parameters which are not replicated inside the parameter vector ρ, that is a_1, a_2, d_1 and d_2. Concerning the other coefficients the standard

Table 2.1 Estimation results of the proposed algorithm

Method		b_0	b_1	a_1	a_2	α_1	α_2	d_1	d_2
	True value	0.5	0.2	1	0.5	0.5	0.25	-1	0.2
LSQNONLIN	$mean(\hat{\theta})$	0.4991	0.1983	0.9984	0.4992	0.5011	0.2512	-1.0004	0.2001
	$std(\hat{\theta})$	0.0159	0.0109	0.0114	0.0059	0.0187	0.0194	0.0224	0.0219
HSRIV	$mean(\hat{\theta})$	0.4992	0.1975	0.9944	0.4976	0.4956	0.2657	X	X
	$std(\hat{\theta})$	0.0402	0.0471	0.0186	0.0071	0.1107	0.0999	X	X
HRIV	$mean(\hat{\theta})$	0.5004	0.2009	0.9984	0.4992	0.5006	0.2487	-1.0011	0.2007
	$std(\hat{\theta})$	0.0193	0.0208	0.0114	0.0059	0.0384	0.0397	0.0224	0.0220

deviation is approximately multiplied by 2 but the absolute value remains acceptable considering the level of noise added. It can be concluded that the presented algorithms, even if not optimal in the Hammerstein case, constitute good candidates for practical applications where the noise is unknown, and can be a strong help for initializing optimal methods such as LSQNONLIN.

2.3 Continuous-Time Hammerstein Model Identification

Even if measured data are sampled, the underlying dynamic of a real system is continuous and direct continuous-time model identification methods regained interest in the recent years [15]. The advantage of using direct continuous-time model identification has been pointed out in many different contexts in the LTI framework [9–13, 15, 24]. Nonetheless, a survey by Rao and Unbehauen [25] shows that CT model identification methods applied to Hammerstein models are poorly represented in literature and only a few methods can be found. In [22], the authors focus on the time-derivative approximation problems while solving the optimization problem using least squares. A non-parametric method can be found in [14] while an approach dedicated to periodic input signals can be found in [33]. To the best of the authors' knowledge, the parametric estimation problem has not been addressed yet for CT Hammerstein models which focus with some colored added noise. Consequently, this section presents an RIV algorithm for direct CT model identification for CT Hammerstein models.

2.3.1 System Description

Consider the CT Hammerstein data generating system depicted in Fig. 2.3 corresponding to the following input-output relationship:

$$\mathscr{S}_o \quad \begin{cases} \bar{u}(t) = f_o(u(t)), \\ \chi_o(t) = G_o(d)\bar{u}(t), \\ y(t) = \chi(t) + v_o(t), \end{cases} \tag{2.39}$$

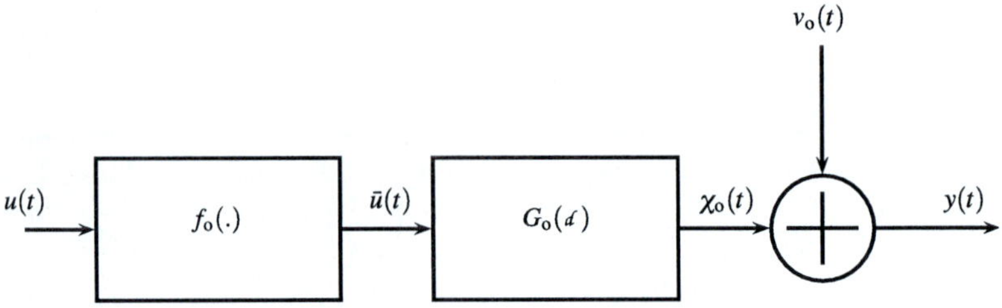

Fig. 2.3 CT Hammerstein block representation

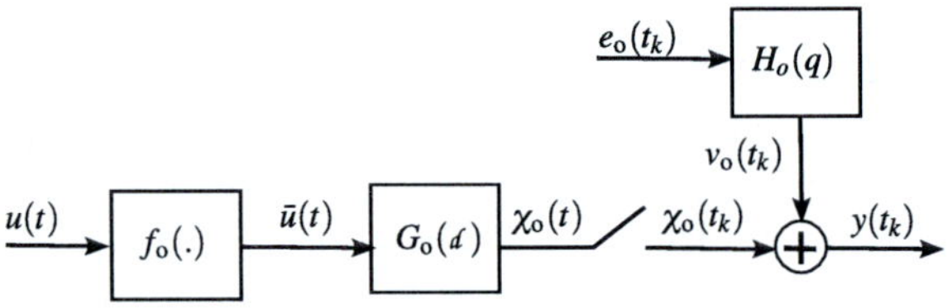

Fig. 2.4 Hybrid Hammerstein block representation

where

$$G_o(d) = \frac{B_o(d)}{A_o(d)} \tag{2.40}$$

and $B_o(d)$ and $A_o(d)$ are polynomials in the differential operator d ($d^i x(t) = \frac{d^i x(t)}{dt^i}$) of respective degree n_b and n_a ($n_a \geq n_b$).

In terms of identification we can assume that sampled measurements of (y, u) are available at a sampling time $k\text{T}_s > 0$. Hence, we will denote the discrete time samples of these signals as $u(t_k) = u(k\text{T}_s)$, where $k \in \mathbb{Z}$. The basic idea to solve the noisy *continuous-time* (CT) modelling problem is to assume that the CT noise process $v_o(t)$ can be written at the sampling instances as a *discrete-time* (DT) white noise process filtered by a DT transfer function [16, 23]. The practically general case is considered where the colored noise associated with the sampled output measurement $y(t_k)$ is assumed to have a rational spectral density which might has no relation to the actual process dynamics. Therefore, v_o is represented by a discrete-time *autoregressive moving average* (ARMA) model:

$$v_o(t_k) = H_o(q)e_o(t_k) = \frac{C_o(q^{-1})}{D_o(q^{-1})} e_o(t_k), \tag{2.41}$$

where $e_o(t_k)$ is a DT zero mean white noise process, q^{-1} is the backward time shift operator, i.e. $q^{-i}u(t_k) = u(t_{k-i})$, and C_o with D_o are monic polynomials with constant coefficients. This avoids the rather difficult mathematical problem of treating sampled CT random process [9] and their equivalent in terms of a filtered piecewise constant CT noise source (see [23]). Therefore, we will consider the Hammerstein system represented in Fig. 2.4 where it is assumed that both input and output signals, $u(t)$ and $y(t)$ are uniformly sampled at a constant sampling time T_s over N samples.

Consequently, in terms of (2.41), the Hammerstein system $\mathscr{S}_o$ (2.39), is described by the following input-output relationship:

$$\mathscr{S}_o \quad \begin{cases} \bar{u}(t) = f_o(u(t)), \\ \chi_o(t) = G_o(d)\bar{u}(t), \\ v_o(t_k) = H_o(q)e(t_k), \\ y(t_k) = \chi_o(t_k) + v_o(t_k). \end{cases} \tag{2.42}$$

This corresponds to a so-called Hammerstein hybrid Box-Jenkins system concept already used in CT identification of LTI systems (see [16, 23, 31]). Furthermore, in terms of (2.4), exactly the same noise assumption is made as in the classical DT Box-Jenkins models [20].

2.3.2 *Model Considered*

2.3.2.1 Process Modelling

Similarly to the discrete-time case, by aiming at the convexification of the optimization problem, the static nonlinearity model is modelled as the linear sum of basis functions:

$$f(u(t), \rho) = \sum_{i=1}^{l} \alpha_i(\rho)\gamma_i(u(t)), \quad \alpha_1 = 1 \tag{2.43}$$

while the CT linear part can be parameterized such that:

$$\chi(t) = G(d, \rho)\bar{u}(t) = \frac{B(d, \rho)}{A(d, \rho)} f(u(t), \rho), \tag{2.44}$$

with

$$A(d, \rho) = d^{n_a} + \sum_{i=1}^{n_a} a_i d^{n_a - i} \quad \text{and} \quad B(d, \rho) = \sum_{j=0}^{n_b} b_j d^{n_b - j}. \tag{2.45}$$

Just as in the DT case, both equations (2.43) and (2.44) can be combined such that:

$$\chi(t) = \frac{B(d, \rho)}{A(d, \rho)} \sum_{i=1}^{l} \alpha_i(\rho)\gamma_i(u(t)) = \frac{1}{A(d, \rho)} \sum_{i=1}^{l} \underbrace{\alpha_i(\rho)B(d, \rho)}_{B_i(d,\rho)}\underbrace{\gamma_i(u(t))}_{u_i(t)}. \tag{2.46}$$

Under these modelling settings, the nonlinearity model and the linear process model can be combined into a process model, denoted by $\mathscr{G}_\rho$ and defined in the form:

$$\mathscr{G}_\rho : (A(d, \rho), B_i(d, \rho)) \tag{2.47}$$

where the polynomials A and B_i are parameterized as

$$\mathcal{G}_\rho \quad \begin{cases} A(d,\rho) = 1 + \sum_{i=1}^{n_\mathrm{a}} a_i d^{n_a - i}, \\ B_i(d,\rho) = \alpha_i \sum_{j=0}^{n_\mathrm{b}} b_j d^{n_b - j}, \quad i = 1 \dots l. \end{cases}$$

The associated model parameters are stacked columnwise in the parameter vector ρ,

$$\rho = \begin{bmatrix} \mathbf{a} \\ \alpha_1 \mathbf{b} \\ \vdots \\ \alpha_l \mathbf{b} \end{bmatrix} \in \mathbb{R}^{n_\rho}, \quad \mathbf{a} = \begin{bmatrix} a_1 \\ a_2 \\ \vdots \\ a_{n_a} \end{bmatrix} \in \mathbb{R}^{n_a}, \ \mathbf{b} = \begin{bmatrix} b_0 \\ b_1 \\ \vdots \\ b_{n_b} \end{bmatrix} \in \mathbb{R}^{n_b+1}, \tag{2.48}$$

with $n_\rho = n_a + l(n_b + 1)$. Introduce also $\mathcal{G} = \{\mathcal{G}_\rho \mid \rho \in \mathbb{R}^{n_\rho}\}$, as the collection of all process models in the form of (2.47).

2.3.2.2 Noise Model

The noise model being expressed in discrete-time, it is denoted by $\mathcal{H}$ and defined as in the DT case (see Sect. 2.2.2.3) Additionally, denote $\mathcal{H} = \{\mathcal{H}_\eta \mid \eta \in \mathbb{R}^{n_\eta}\}$, the collection of all noise models in the form of (2.10).

2.3.2.3 Whole Model

With respect to the given process and noise parts $(\mathcal{G}_\rho, \mathcal{H}_\eta)$, the parameters can be collected as $\theta = [\,\rho^\top \ \eta^\top\,]$ and the signal relations of the CT Hammerstein BJ model, denoted in the sequel as $\mathcal{M}_\theta$, are defined as:

$$\mathcal{M}_\theta \quad \begin{cases} \chi(t) = \dfrac{\sum_{i=1}^{l} B_i(d,\rho) u_i(t)}{A(d,\rho)}, \\ v(t_k) = \dfrac{C(q^{-1},\eta)}{D(q^{-1},\eta)} e(t_k), \\ y(t_k) = \chi(t_k) + v(t_k), \end{cases} \tag{2.49}$$

with $B_i(d,\rho) = \alpha_i(\rho) B(d,\rho)$ and $u_i(t) = \gamma_i(u(t))$. Based on this model structure, the model set, denoted as $\mathcal{M}$, with the linear process $(\mathcal{G}_\rho)$ and noise $(\mathcal{H}_\eta)$ models parameterized independently, takes the form

$$\mathcal{M} = \left\{ (\mathcal{G}_\rho, \mathcal{H}_\eta) \mid \mathrm{col}(\rho, \eta) = \theta \in \mathbb{R}^{n_\rho + n_\eta} \right\}. \tag{2.50}$$

Again, this set corresponds to the set of candidate models in which we seek the best fitting model using data gathered from $\mathcal{S}_o$ under a given identification criterion (cost function). The identification problem can be stated in the exact same way as in the DT case (see Sect. 2.2.3).

Remarks It has to be noticed that, just as in the DT case, this model transforms the Hammerstein structure into an augmented LTI *Multi Input Single Output* model similarly to the DT case. Consequently, the number of parameters to be estimated is not minimal as $n_\rho = n_a + l(n_b + 1)$ which is in general greater than $n_{\rho_L} + n_{\rho_{NL}} = n_a + l + (n_b + 1)$. Therefore, as the model is not minimal, and therefore the optimal estimation of this augmented MISO model does not correspond to the optimal estimates of the true Hammerstein model. Nonetheless, the gain granted using this modelling is the possible linear regression form and therefore, the convexification of the optimization problem.

2.3.3 *Refined IV for CT Hammerstein BJ Models*

Using the LTI model (2.49), $y(t_k)$ can be written in the regression form:

$$y^{(n_a)}(t_k) = \varphi^\top(t_k)\rho + \tilde{v}(t_k), \tag{2.51}$$

where

$$\varphi(t_k) = \left[-y^{(n_\mathrm{a}-1)}(t_k) \dots -y(t_k)\; u_1^{(n_\mathrm{b})}(t_k) \dots u_1(t_k) \dots u_l^{(n_\mathrm{b})}(t_k) \dots u_l(t_k)\right]^\top$$

$$\rho = \left[a_1 \dots a_{n_\mathrm{a}}\; b_0 \dots b_{n_\mathrm{b}} \dots \alpha_l b_0 \dots \alpha_l b_{n_\mathrm{b}}\right]^\top$$

and

$$\tilde{v}(t_k) = A(d, \rho)v(t_k),$$

where $x^{(n)}(t_k)$ denotes the sample of the nth derivative of the signal $x(t)$ sampled at time t_k.

By driving the exact same discussion as in Sect. 2.2.4 it can be shown that the optimal filtered instrument for the augmented LTI MISO model structure (2.49) is given as:

$$\zeta^{\mathrm{opt}}(t_k) = \left[-\chi_\mathrm{o}^{(n_\mathrm{a}-1)}(t_k) \dots -\chi_\mathrm{o}(t_k)\; u_1^{(n_\mathrm{b})}(t_k) \dots u_1(t_k) \dots u_l^{(n_\mathrm{b})}(t_k) \dots u_l(t_k)\right]^\top \tag{2.52}$$

while the optimal filter is given as the filter chain involving the continuous-time filtering operation using the filter (see [31]):

$$L_\mathrm{c}^{\mathrm{opt}} = Q_\mathrm{c}(d, \rho_\mathrm{o}) = \frac{1}{A_\mathrm{o}(d)}, \tag{2.53}$$

and the discrete-time filtering operation using the filter:

$$L_{\mathrm{d}}^{\mathrm{opt}} = Q_{\mathrm{d}}(q, \eta_{\mathrm{o}}) = \frac{D_{\mathrm{o}}(q^{-1})}{C_{\mathrm{o}}(q^{-1})}. \tag{2.54}$$

2.3.4 Hammerstein RIVC (HRIVC) Algorithm for BJ Models

For space and redundancy's sake the HRIVC algorithm is not described here, but the interested reader can find a detailed algorithm in [18]. By using the HRIVC algorithm, if convergence occurs, the HRIVC estimates might be statistically optimal for the augmented model proposed, but the minimal number of parameters needed for representing the MISO structure and the Hammerstein structure are not equal. Consequently, the HRIVC estimates cannot be statistically optimal for the CT Hammerstein model structure. Nonetheless, even if not optimal, the HRIVC estimates are unbiased with a low variance as it will be seen in the result Sect. 2.3.5. A simplified version of HRIVC algorithm named HSRIVC follows the exact same theory for estimation of Hammerstein output error models.

2.3.5 Performance Evaluation of the Proposed HRIVC and HSRIVC Algorithms

This section presents numerical evaluation of both suggested HRIVC and HSRIVC methods. For the presented example, the nonlinear block has a polynomial form, i.e. $\gamma_i(u(t)) = u^i(t), \forall i$ and

$$\bar{u}(t) = u(t) + 0.5u^2(t) + 0.25u^3(t), \tag{2.55}$$

where $u(t)$ follows a uniform distribution with values between -2 and 2. The system is simulated using a zero-order-hold on the input.

The system considered is a hybrid Hammerstein Box-Jenkins model in which the linear dynamic block is first a second-order system described by:

$$G_{\mathrm{o}}(d) = \frac{10d + 30}{d^2 + d + 5}, \tag{2.56}$$

and the noise is given by

$$H_{\mathrm{o}}(q) = \frac{1}{1 - q^{-1} + 0.2q^{-2}}.$$

Table 2.2 Estimation results for different noise models

SNR	Method		b_0	b_1	a_1	a_2	α_1	α_2	d_1	d_2
		True value	10	30	1	5	0.5	0.25	−1	0.2
15 dB	HSRIVC	$mean(\hat{\theta})$	9.9957	29.8760	1.0001	4.9991	0.5026	0.2523	X	X
		$std(\hat{\theta})$	0.3670	1.5660	0.0170	0.0436	0.0201	0.0180	X	X
		RMSE	0.0367	0.0523	0.0169	0.0087	0.0405	0.0723	X	X
	HRIVC	$mean(\hat{\theta})$	9.9906	30.0172	1.0006	5.0020	0.5008	0.2506	−1.0002	0.2005
		$std(\hat{\theta})$	0.2497	0.8954	0.0119	0.0265	0.0118	0.0115	0.0219	0.0223
		RMSE	0.0250	0.0298	0.0119	0.0053	0.0236	0.0460	0.0218	0.1112
5 dB	HSRIVC	$mean(\hat{\theta})$	10.0882	29.6146	1.0010	4.9814	0.5080	0.2604	X	X
		$std(\hat{\theta})$	1.0764	4.4585	0.0517	0.1291	0.0610	0.0542	X	X
		RMSE	0.1079	0.1490	0.0517	0.0261	0.1230	0.2208	X	X
	HRIVC	$mean(\hat{\theta})$	10.049	30.0277	0.9998	4.9980	0.5015	0.2522	−0.9997	0.1994
		$std(\hat{\theta})$	0.7861	2.8278	0.0379	0.0871	0.0369	0.0366	0.0227	0.0219
		RMSE	0.0787	0.0942	0.0378	0.0174	0.0738	0.1466	0.0227	0.1096

The models considered for estimation are:

$$\mathcal{M}_{\mathtt{HRIVC}} \quad \begin{cases} G(d,\rho) = \dfrac{b_0 d + b_1}{d^2 + a_1 d + a_2}, \\ H(q,\eta) = \dfrac{1}{1 + d_1 q^{-1} + d_2 q^{-2}}, \\ f(u(t)) = u(t) + \alpha_1 u^2(t) + \alpha_2 u^3(t) \end{cases} \tag{2.57}$$

for the HRIVC method and

$$\mathcal{M}_{\mathtt{HSRIVC}} \quad \begin{cases} G(d,\rho) = \dfrac{b_0 d + b_1}{d^2 + a_1 d + a_2}, \\ H(q,\eta) = 1, \\ f(u(t)) = u(t) + \alpha_1 u^2(t) + \alpha_2 u^3(t) \end{cases} \tag{2.58}$$

for the HSRIVC method.

The result of a Monte Carlo simulation (MCs) analysis is shown in Table 2.2 for the algorithms considered. The MCs results are based on $N_{\text{run}} = 500$ random realization, with the Gaussian white noise input to the ARMA noise model being selected randomly for each realization. In order to compare the statistical performance of the different approaches, the computed mean, standard deviation and *Root Mean Squared Error* of the estimated parameters are presented. The noise added at the output is adjusted such that it corresponds to a SNR of 15 dB and 5 dB. The number of samples is $N = 2000$.

Table 2.2 shows that according to the theory, the HRIVC and HSRIVC methods provide similar, unbiased estimates of the model parameters. Both methods seem to be robust even at unrealistic noise level of 5 dB as the RMSE remain under 22% for both methods. Results obtained using the HRIVC algorithm, have standard deviations which are always smaller than the ones produced by HSRIVC. Even though, the HSRIVC algorithm based on an Output Error model is a reasonable alternative to the full HRIVC algorithm based on a Box-Jenkins model: in practice the noise model cannot be exactly known and therefore the use of the HRIVC algorithm would simply raise the number of parameters to be estimated. If the noise model is correctly assumed, it is as well correctly estimated as shown in Table 2.2.

2.4 Conclusion

In this chapter, some methods dedicated to Hammerstein CT and DT nonlinear models in open-loop were investigated. Extension to the closed-loop case has been published and can be found in [17]. Through a relevant set of examples, it was possible to show that the HRIV approach is robust to noise conditions and to noise error modelling. The presented methods are suboptimal as they estimate a larger number of parameters than the minimum needed for the system description. Nonetheless, the variance in the estimated parameters is acceptable in practical conditions, and if not satisfactory, the estimates can be used as initialisation values for some optimal method which usually are posed into some nonlinear optimization problems and often rely on some robust initialisation. The refined instrumental variable approach for Hammerstein models remains an interesting estimation method for practical applications where the noise condition are unknown.

References

1. Bai, E.-W.: A blind approach to the Hammerstein–Wiener model identification. Automatica **38**(6), 967–979 (2002)
2. Bai, E.-W.: Identification of linear systems with hard input nonlinearities of known structure. Automatica **38**(5), 853–860 (2002)
3. Bai, E.-W., Chan, K.-S.: Identification of an additive nonlinear system and its applications in generalized Hammerstein models. Automatica **44**(2), 430–436 (2008)
4. Bai, E.W.: An optimal two-stage identification algorithm for Hammerstein–Wiener nonlinear systems. Automatica **34**(3), 333–338 (1998)
5. Cerone, V., Regruto, D.: Parameter bounds for discrete-time Hammerstein models with bounded errors. IEEE Trans. Autom. Control **48**(10), 1855–1860 (2003)
6. Ding, F., Chen, T.: Identification of Hammerstein nonlinear ARMAX systems. Automatica **41**(9), 1479–1489 (2005)
7. Ding, F., Shi, Y., Chen, T.: Auxiliary model-based least-squares identification methods for Hammerstein output-error systems. Syst. Control Lett. **56**(5), 373–380 (2007)
8. Giri, F., Bai, E.-W. (eds.): Block-Oriented Nonlinear System Identification. Springer, London (2010)
9. Gillberg, J., Ljung, L.: Frequency domain identification of continuous-time ARMA models from sampled data. Automatica **45**(6), 1371–1378 (2009)
10. Gillberg, J., Ljung, L.: Frequency domain identification of continuous-time output error models, Part I: Uniformly sampled data and frequency function approximation. Automatica **46**(1), 1–10 (2010)

11. Gillberg, J., Ljung, L.: Frequency domain identification of continuous-time output error models, part II: Non-uniformly sampled data and B-spline output approximation. Automatica **46**(1), 11–18 (2010)
12. Gilson, M., Garnier, H., Young, P.C., Van den Hof, P.: Refined instrumental variable methods for closed-loop system identification methods. In: Proceedings of the 15th IFAC Symposium on System Identification, Saint-Malo, France (2009)
13. Gilson, M., Van den Hof, P.M.J.: Instrumental variable methods for closed-loop system identification. Automatica **41**, 241–249 (2005)
14. Greblicki, W.: Continuous-time Hammerstein system identification. IEEE Trans. Autom. Control **45**(6), 1232–1236 (2000)
15. Garnier, H., Wang, L. (eds.): Identification of Continuous-Time Models from Sampled Data. Springer, London (2008)
16. Johansson, R.: Identification of continuous-time models. IEEE Trans. Signal Process. **42** (1994)
17. Laurain, V., Gilson, M., Garnier, H.: Refined instrumental variable methods for identifying Hammerstein models operating in closed loop. In: Proceedings of the 48th IEEE Conference on Decision and Control, Shanghai, China (2009)
18. Laurain, V., Gilson, M., Garnier, H., Young, P.C.: Refined instrumental variable methods for identification of Hammerstein continuous time Box-Jenkins models. In: Proceedings of the 47th IEEE Conference on Decision and Control, Cancun, Mexico (2008)
19. Liu, Y., Bai, E.W.: Iterative identification of Hammerstein systems. Automatica **43**(2), 346–354 (2007)
20. Ljung, L.: System Identification: Theory for the User, 2nd edn. Prentice-Hall, New York (1999)
21. Palanthandalam-Madapusi, H.J., Edamana, B., Bernstein, D.S., Manchester, W., Ridley, A.J.: NARMAX identification for space weather prediction using polynomial radial basis functions. In: 46th IEEE Conference on Decision and Control, New Orleans, LA, USA (2007)
22. Patra, A., Unbehauen, H.: Identification of a class of nonlinear continuous-time systems using Hartley modulating functions. Int. J. Control **62**(6), 1431–1451 (1995)
23. Pintelon, R., Schoukens, J., Rolain, Y.: Box-Jenkins continuous-time modeling. Automatica **36** (2000)
24. Rao, G.P., Garnier, H.: Identification of continuous-time systems: direct or indirect? Syst. Sci. **30**(3), 25–50 (2004)
25. Rao, G.P., Unbehauen, H.: Identification of continuous-time systems. IEE Proc. Control Theory Appl. **153**(2) (2006)
26. Schoukens, J., Widanage, W.D., Godfrey, K.R., Pintelon, R.: Initial estimates for the dynamics of a Hammerstein system. Automatica **43**(7), 1296–1301 (2007)
27. Vanbeylen, V., Pintelon, R., Schoukens, J.: Blind maximum likelihood identification of Hammerstein systems. Automatica **44**(11), 3139–3146 (2008)
28. Wang, J., Zhang, Q., Ljung, L.: Revisiting Hammerstein system identification through the two-stage algorithm for bilinear parameter estimation. Automatica **45**(9), 2627–2633 (2009)
29. Young, P.C.: Some observations on instrumental variable methods of time-series analysis. Int. J. Control **23**, 593–612 (1976)
30. Young, P.C.: The refined instrumental variable method: unified estimation of discrete and continuous-time transfer function models. J. Eur. Syst. Autom. **42**, 149–179 (2008)
31. Young, P.C., Garnier, H., Gilson, M.: Refined instrumental variable identification of continoustime hybrid Box-Jenkins models. In: Garnier, H., Wang, L. (eds.) Identification of Continuous-Time Models from Sampled Data, pp. 91–131. Springer, London (2008)
32. Young, P.C., Jakeman, A.: Refined instrumental variable methods of recursive time-series analysis—Part III. Extensions. Int. J. Control **31**(4), 741–764 (1980)
33. Zhai, D., Rollins, D.K., Bhandari, N., Wu, H.: Continuous-time Hammerstein and Wiener modeling under second-order static nonlinearity for periodic process signals. Comput. Chem. Eng. **31**, 1–12 (2006)

Chapter 3
Identifiability, and Beyond

Eric Walter

3.1 Colliding with (a Lack of) Identifiability

During the preparation of my PhD dissertation [50], I wrote a quasilinearization routine to estimate the parameters of a class of continuous-time state-space models known as compartmental models [18, 23]. These models, widely used in biology, pharmacokinetics and the modeling of ecosystems, consist of a finite number of homogeneous subsystems, called compartments, which exchange material with each other and the environment. To test this routine, I generated data by simulating the two-compartment model

$$\dot{\mathbf{x}}_{\mathrm{m}}^{*} = \begin{bmatrix} -0.1 & 0.15 \\ 0.1 & -0.2 \end{bmatrix} \mathbf{x}_{\mathrm{m}}^{*} + \begin{bmatrix} 0 \\ 1 \end{bmatrix} u, \quad \mathbf{x}_{\mathrm{m}}^{*}(0_{-}) = \begin{bmatrix} 0 \\ 0 \end{bmatrix}, \tag{3.1}$$

with the observation equation

$$y_{\mathrm{m}}^{*} = \begin{bmatrix} 1 & 0 \end{bmatrix} \mathbf{x}_{\mathrm{m}}^{*}. \tag{3.2}$$

In (3.1) and (3.2) and it what follows, the star indicates a quantity that corresponds to the "true" system assumed to have generated the data. This model has an input in Compartment 2 and the content x_1 of Compartment 1 is directly observed. The sum of the entries of the first column of the matrix in (3.1) is zero, which implies that there is no direct flow from Compartment 1 to the environment, so what is in Compartment 1 can only leave the system after passing through Compartment 2. Data $y(t_i)$, $i = 1, \ldots, n$, were generated taking u as a unit delta impulse at time $t = 0$, which is equivalent to assuming that $\mathbf{x}_{\mathrm{m}}^{*}(0_{+}) = [0\ 1]^{\mathrm{T}}$, with no input. No noise

E. Walter (✉)
Laboratoire des Signaux et Systèmes, CNRS–SUPELEC–Univ Paris-Sud, 91192 Gif-sur-Yvette, France
e-mail: Eric.Walter@lss.supelec.fr

L. Wang, H. Garnier (eds.), *System Identification, Environmental Modelling, and Control System Design*,
DOI 10.1007/978-0-85729-974-1_3,

was added to the output, except for the errors induced by numerical integration. These data were then used to estimate the parameters of the model $\mathscr{M}(\boldsymbol{\theta})$ defined by

$$\dot{\mathbf{x}}_{\mathrm{m}} = \begin{bmatrix} -\theta_1 & \theta_2 \\ \theta_1 & -(\theta_2+\theta_3) \end{bmatrix} \mathbf{x}_{\mathrm{m}} + \begin{bmatrix} 0 \\ 1 \end{bmatrix} u, \quad \mathbf{x}_{\mathrm{m}}(0_-) = \begin{bmatrix} 0 \\ 0 \end{bmatrix}, \tag{3.3}$$

$$y_{\mathrm{m}} = \begin{bmatrix} 1 & 0 \end{bmatrix} \mathbf{x}_{\mathrm{m}}, \tag{3.4}$$

by minimizing the quadratic cost function

$$J(\boldsymbol{\theta}) = \sum_{i=1}^{f} \left(y^*_{\mathrm{m}}(t_i) - y_{\mathrm{m}}(t_i, \boldsymbol{\theta}) \right)^2. \tag{3.5}$$

Depending on the initial value given to $\boldsymbol{\theta}$, the quasilinearization routine would converge either to a model close to (3.1) or to something like

$$\dot{\mathbf{x}}_{\mathrm{m}} = \begin{bmatrix} -0.05 & 0.15 \\ 0.05 & -0.25 \end{bmatrix} \mathbf{x}_{\mathrm{m}} + \begin{bmatrix} 0 \\ 1 \end{bmatrix} u, \quad \mathbf{x}_{\mathrm{m}}(0_-) = \begin{bmatrix} 0 \\ 0 \end{bmatrix}. \tag{3.6}$$

In both cases, the optimal value for the cost was very low, and the fit excellent. After some time spent hunting nonexistent bugs in the software, I finally understood that the problem came from some defect in the model structure, which led to several values of the parameter vector corresponding to the same output behavior, and realized it corresponded to a problem of *lack of identifiability*. This problem was already pointed out, e.g., in [29], which stressed its importance in biology, psychology, sociology and economics. If most recent applications of the concept are to biological models, it seems to have been known much earlier in economy [14, 15], physics and chemistry [3].

This chapter is organized as follows. Identifiability is defined in Sect. 3.2, before presenting methods of test in Sect. 3.3. The case when one hesitates between several models structures is considered in Sect. 3.4. Section 3.5 deals with the design of experiments so as to maximize some measure of identifiability. Section 3.6 very briefly presents some tools that can be used to characterize, in a global manner, the set of all feasible estimates of the parameter vector, thereby making it possible to bypass the identifiability study. Conclusions and perspectives are in Sect. 3.7.

3.2 Defining Identifiability

Assume that some parametric model structure $\mathscr{M}(\boldsymbol{\theta})$ has been chosen, based on prior knowledge on the system to be modeled. Before attempting to estimate $\boldsymbol{\theta}$ from experimental data, one would like to check that hidden defects do not render the exercise hopeless. The notion of identifiability can be used for this purpose, in a somewhat idealized setting where no noise is present and the data are generated by a model $\mathscr{M}(\boldsymbol{\theta}^*)$, with $\boldsymbol{\theta}^*$ the true value for the parameter vector. Write

$$\mathscr{M}(\hat{\boldsymbol{\theta}}) = \mathscr{M}(\boldsymbol{\theta}^*) \tag{3.7}$$

to denote that the model with parameter vector $\hat{\boldsymbol{\theta}}$ has *exactly* the same output as the model with parameter vector $\boldsymbol{\theta}^*$ for any input and time (time may be replaced by a vector of independent variables, when appropriate). Identifiability then depends on the number of solutions of (3.7) for $\hat{\boldsymbol{\theta}}$. If there is only one solution (which can only be $\hat{\boldsymbol{\theta}} = \boldsymbol{\theta}^*$), then $\mathscr{M}$ is uniquely identifiable at $\boldsymbol{\theta}^*$.

Because the conclusion may depend on the value of $\boldsymbol{\theta}^*$, which is of course unknown in any real-life problem, we are led to the following definitions [51].

The model $\mathscr{M}$ is *structurally globally identifiable* (s.g.i.) if, for almost any value of $\boldsymbol{\theta}^*$, (3.7) has only one solution $\hat{\boldsymbol{\theta}} = \boldsymbol{\theta}^*$. It is *structurally locally identifiable* (s.l.i.) if the set of solutions for $\hat{\boldsymbol{\theta}}$ is finite or at least countable, which means that a neighborhood can be defined around $\boldsymbol{\theta}^*$ in which the solution for $\hat{\boldsymbol{\theta}}$ is unique. The identifiability of the ith entry of $\boldsymbol{\theta}$ may be defined similarly: θ_i is s.g.i. if, for almost any value of $\boldsymbol{\theta}^*$, (3.7) implies that $\hat{\theta}_i = \theta_i^*$; it is s.l.i. if the solution for $\hat{\theta}_i$ is unique when search is restricted to some neighborhood of $\boldsymbol{\theta}^*$.

With these definitions, it becomes possible to test a model structure for identifiability even before data collection, and sometimes to modify this structure so as to eliminate ambiguity, if any, for instance by introducing new sensors or actuators. Section 3.3 will illustrate the fact that there may exist a manifold of atypical values for $\boldsymbol{\theta}^*$ on which an s.g.i. or s.l.i. model becomes unidentifiable. The probability of hitting this manifold when picking $\boldsymbol{\theta}$ at random is zero (and models with parameter vectors on this manifold are generally so pathological that one would never consider them anyway).

Before presenting methods to test models for identifiability in the next section, it may be useful to stress that identifiability is not always an important property to be requested of models. If one is only interested in reproducing an observed input-output behavior, then it does not matter if they are several models that do the job in exactly the same manner. (The lack of local identifiability may create numerical difficulties to some optimization algorithms, but these difficulties are easily solved by regularization.) On the other hand, it is important to test models for identifiability if the parameters have a physical interpretation, or if decisions are to be taken on the basis of their numerical values. It is also important when some physically meaningful state variables must be estimated from the input-output data, for instance using a Kalman filter, as the different models with the same input-output behavior will correspond to as many state estimates.

3.3 Testing Identifiability

We restrict ourselves here to continuous-time finite-dimensional deterministic state-space models, and assume throughout the chapter, for the sake of simplicity, that

$\mathbf{x}(0_-) = \mathbf{0}$. The model output vector $\mathbf{y}$ may be *linear in the input* vector $\mathbf{u}$, as in

$$\dot{\mathbf{x}} = \mathbf{A}(\boldsymbol{\theta})\mathbf{x} + \mathbf{B}(\boldsymbol{\theta})\mathbf{u}, \qquad \mathbf{y} = \mathbf{C}(\boldsymbol{\theta})\mathbf{x}, \tag{3.8}$$

or *nonlinear in the input*, as in

$$\dot{\mathbf{x}} = \mathbf{f}(\mathbf{x}, \boldsymbol{\theta}, \mathbf{u}), \qquad \mathbf{y} = \mathbf{h}(\mathbf{x}, \boldsymbol{\theta}, \mathbf{u}). \tag{3.9}$$

In both cases, the model output is in general *nonlinear in the parameters*, and this is the only situation considered here. It corresponds to most knowledge-based models, for which the concept of identifiability is particularly pertinent. (When the model output is linear in the parameters, the notions of local and global identifiability become equivalent, and the test for identifiability boils down to a rank condition on a regression matrix.)

3.3.1 Linear Case

When the model is described by (3.8), the most commonly used routes are the Laplace transform approach of [1], and the similarity transformation approach, which can be traced back to [3], see also [17, 56].

3.3.1.1 Laplace Transform Approach

Take the Laplace transform of (3.8) and eliminate the state variables to get

$$\mathbf{Y}(s, \boldsymbol{\theta}) = \mathbf{H}(s, \boldsymbol{\theta})\mathbf{U}(s), \tag{3.10}$$

where s is the Laplace variable and $\mathbf{H}(s, \boldsymbol{\theta})$ is the transfer matrix, which satisfies

$$\mathbf{H}(s, \boldsymbol{\theta}) = \mathbf{C}(\boldsymbol{\theta}) \left[s\mathbf{I} - \mathbf{A}(\boldsymbol{\theta})\right]^{-1} \mathbf{B}(\boldsymbol{\theta}), \tag{3.11}$$

with $\mathbf{I}$ the identity matrix. Note that directly implementing (3.11) is a fairly inefficient way of computing the transfer matrix from the state-space representation, to be avoided on large-scale problems.

Equation (3.7) will be satisfied for any input if and only if

$$\mathbf{H}(s, \hat{\boldsymbol{\theta}}) - \mathbf{H}(s, \boldsymbol{\theta}^*) = \mathbf{0} \quad \forall s. \tag{3.12}$$

Expressing $\mathbf{H}(s, \boldsymbol{\theta})$ in some canonical form $\mathbf{H}_c(s, \boldsymbol{\theta})$, i.e., a form that can be written in only one way, simplifies computation considerably. A canonical form is for instance obtained by (i) writing each entry of the transfer matrix as a ratio of polynomials ordered in s, (ii) simplifying each numerator and corresponding denominator by their GCD (only the controllable and observable part of the model remains), and (iii) setting the coefficient of the denominator monomial with highest (or lowest) degree in s equal to one. Equation (3.7) will then be satisfied for any input if and only if the *coefficients* of $\mathbf{H}_c(s,\boldsymbol{\theta})$ have the same values for $\boldsymbol{\theta} = \hat{\boldsymbol{\theta}}$ and for $\boldsymbol{\theta} = \boldsymbol{\theta}^*$. One thus gets a set of algebraic equations in $\hat{\boldsymbol{\theta}}$ and $\boldsymbol{\theta}^*$, to be solved for $\hat{\boldsymbol{\theta}}$ for almost any value of $\boldsymbol{\theta}^*$.

Example 3.1 Consider the model structure defined by (3.3) and (3.4). Its transfer function can be written in canonical form as

$$H_c(s,\boldsymbol{\theta}) = \frac{\theta_2}{s^2 + (\theta_1 + \theta_2 + \theta_3)s + \theta_1\theta_3}. \tag{3.13}$$

Equation (3.7) thus translates into three algebraic constraints linking $\hat{\boldsymbol{\theta}}$ and $\boldsymbol{\theta}^*$:

$$\begin{aligned} \hat{\theta}_2 &= \theta_2^*, \\ \hat{\theta}_1 + \hat{\theta}_2 + \hat{\theta}_3 &= \theta_1^* + \theta_2^* + \theta_3^*, \\ \hat{\theta}_1\hat{\theta}_3 &= \theta_1^*\theta_3^*. \end{aligned} \tag{3.14}$$

The first of these constraints establishes that θ_2 is s.g.i., while the last two only add that the sum and product of θ_1 and θ_3 are s.g.i. We thus have two solutions, namely $\hat{\boldsymbol{\theta}} = \boldsymbol{\theta}^*$ and $\hat{\boldsymbol{\theta}} = [\theta_3^*, \theta_2^*, \theta_1^*]^{\mathrm{T}}$, so θ_1 and θ_3 are s.l.i. but not s.g.i. A rational choice between the two possible values of θ_1 and θ_3 would be impossible on the sole basis of the data, even if these data were free of noise. These findings are corroborated by the numerical experiment reported in Sect. 3.1. Note that if $\theta_2^* = 0$, then the output y_{m}^* is identically zero and the output contains no information on θ_1 and θ_3, which become completely unidentifiable. This is an example of an atypical hyper-surface, and illustrates why it is necessary to take such pathological values of the parameters into account when defining structural identifiability.

3.3.1.2 Similarity Transformation Approach

Consider again the model structure defined by (3.8), and assume that the data are generated by the model $\mathscr{M}(\boldsymbol{\theta}^*)$

$$\dot{\mathbf{x}}^* = \mathbf{A}(\boldsymbol{\theta}^*)\mathbf{x}^* + \mathbf{B}(\boldsymbol{\theta}^*)\mathbf{u}, \qquad \mathbf{y} = \mathbf{C}(\boldsymbol{\theta}^*)\mathbf{x}^*. \tag{3.15}$$

If $\mathbf{T}$ is the time-invariant invertible matrix of a state-space similarity transformation such that $\hat{\mathbf{x}} = \mathbf{T}\mathbf{x}^*$, then the model

$$\dot{\hat{\mathbf{x}}} = \mathbf{T}\mathbf{A}(\boldsymbol{\theta}^*)\mathbf{T}^{-1}\hat{\mathbf{x}} + \mathbf{T}\mathbf{B}(\boldsymbol{\theta}^*)\mathbf{u}, \qquad \mathbf{y} = \mathbf{C}(\boldsymbol{\theta}^*)\mathbf{T}^{-1}\hat{\mathbf{x}} \tag{3.16}$$

has the same input-output behavior as $\mathscr{M}(\boldsymbol{\theta}^*)$. It will correspond to a model $\mathscr{M}(\hat{\boldsymbol{\theta}})$ with the same structure as $\mathscr{M}(\boldsymbol{\theta}^*)$ if and only if there exists $\hat{\boldsymbol{\theta}}$ such that

$$\mathbf{A}(\hat{\boldsymbol{\theta}}) = \mathbf{T}\mathbf{A}(\boldsymbol{\theta}^*)\mathbf{T}^{-1}, \qquad \mathbf{B}(\hat{\boldsymbol{\theta}}) = \mathbf{T}\mathbf{B}(\boldsymbol{\theta}^*), \quad \text{and} \quad \mathbf{C}(\hat{\boldsymbol{\theta}}) = \mathbf{C}(\boldsymbol{\theta}^*)\mathbf{T}^{-1}. \tag{3.17}$$

Then $\mathscr{M}(\hat{\boldsymbol{\theta}}) = \mathscr{M}(\boldsymbol{\theta}^*)$. Kalman's algebraic equivalence theorem, on the other hand, states that if two minimal (i.e., controllable and observable) state-space models have the same input-output behavior, then they can be deduced from one another by a state-space similarity transformation. Provided that the model structure considered is observable and controllable, the number of solutions of $\mathscr{M}(\hat{\boldsymbol{\theta}}) = \mathscr{M}(\boldsymbol{\theta}^*)$ for $\hat{\boldsymbol{\theta}}$ is thus equal to the number of solutions of (3.17) for $\hat{\boldsymbol{\theta}}$ *and* $\mathbf{T}$. If the dimension of the state is n, then the number of unknowns in $\mathbf{T}$ is n^2, and one may wonder what is to be gained by augmenting the number of unknowns of the problem by such a potentially large number. It turns out that the computations are sometimes simpler than with the Laplace transform approach, so the two approaches should be viewed as complementary. Partial results obtained by different approaches can also be combined to reach a conclusion.

Example 3.2 Consider a model defined by (3.8), in which each component of the input vector $\mathbf{u}$ acts directly on a single component of the state and each component of the output vector $\mathbf{y}$ consists of a single component of the state. The control and observation matrices thus do not depend on the parameters to be estimated. Possibly after re-indexing the state, input and output variables, they can be written as

$$\mathbf{B} = \begin{bmatrix} \mathbf{I}_q & \mathbf{0}_{q\times(m-q)} \\ \mathbf{0}_{(p-q)\times q} & \mathbf{0}_{(p-q)\times(m-q)} \\ \mathbf{0}_{(m-q)\times q} & \mathbf{I}_{m-q} \\ \mathbf{0}_{(n-m-p+q)\times q} & \mathbf{0}_{(n-m-p+q)\times(m-q)} \end{bmatrix}, \tag{3.18}$$

and

$$\mathbf{C} = \begin{bmatrix} \mathbf{I}_q & \mathbf{0}_{q\times(p-q)} & \mathbf{0}_{q\times(m-q)} & \mathbf{0}_{q\times(n-m-p+q)} \\ \mathbf{0}_{(p-q)\times q} & \mathbf{I}_{p-q} & \mathbf{0}_{(p-q)\times(m-q)} & \mathbf{0}_{(p-q)\times(n-m-p+q)} \end{bmatrix}, \tag{3.19}$$

where $n = \dim \mathbf{x}$, $m = \dim \mathbf{u}$, $p = \dim \mathbf{y}$ and q is the number of state variables that are directly observed and acted upon. (Matrices $\mathbf{B}$ and $\mathbf{C}$ that do not satisfy (3.18) and (3.19) can always be put in this standard form by a series of linear transformations provided that they have full rank [51, 56]; q is then equal to $\operatorname{rank}(\mathbf{CB})$, an s.g.i. quantity.) Assume that there is no constraint on the drift matrix $\mathbf{A}$, all entries of which have to be estimated, so $\dim \boldsymbol{\theta} = n^2$. We want to know under which conditions $\mathbf{A}$ is s.g.i. With no prior constraint on $\mathbf{A}$, the model structure is observable and controllable, so the similarity transformation approach applies. (If it were not the case, we could restrict ourselves to studying the controllable and observable part of the model, which would be similar to considering canonical transfer functions with the Laplace transform approach.) Partition the similarity transformation matrix into 16 blocks with dimensions compatible with those of the blocks of $\mathbf{B}$ and $\mathbf{C}$, to get

$$\mathbf{T} = \begin{bmatrix} \mathbf{T}_{11} & \mathbf{T}_{12} & \mathbf{T}_{13} & \mathbf{T}_{14} \\ \mathbf{T}_{21} & \mathbf{T}_{22} & \mathbf{T}_{23} & \mathbf{T}_{24} \\ \mathbf{T}_{31} & \mathbf{T}_{32} & \mathbf{T}_{33} & \mathbf{T}_{34} \\ \mathbf{T}_{41} & \mathbf{T}_{42} & \mathbf{T}_{43} & \mathbf{T}_{44} \end{bmatrix}. \tag{3.20}$$

Given (3.18) and (3.19), the last two equations in (3.17) imply that

$$\mathbf{T} = \begin{bmatrix} \mathbf{I}_q & \mathbf{0}_{q\times(p-q)} & \mathbf{0}_{q\times(m-q)} & \mathbf{0}_{q\times(n-m-p+q)} \\ \mathbf{0}_{(p-q)\times q} & \mathbf{I}_{p-q} & \mathbf{0}_{(p-q)\times(m-q)} & \mathbf{0}_{(p-q)\times(n-m-p+q)} \\ \mathbf{0}_{(m-q)\times q} & \mathbf{T}_{32} & \mathbf{I}_{m-q} & \mathbf{T}_{34} \\ \mathbf{0}_{(n-m-p+q)\times q} & \mathbf{T}_{42} & \mathbf{0}_{(n-m-p+q)\times(m-q)} & \mathbf{T}_{44} \end{bmatrix}, \tag{3.21}$$

and the first equation in (3.17) does not bring any more information on $\mathbf{T}$, as $\mathbf{A}$ is free. So $\mathbf{A}$ is s.g.i. if and only if the total number of entries in the blocks $\mathbf{T}_{32}$, $\mathbf{T}_{34}$, $\mathbf{T}_{42}$ and $\mathbf{T}_{44}$, which corresponds to the number of degrees of freedom of the model, is equal to zero. Since $\mathbf{T}_{32}$ is $(m-q)\times(p-q)$, $\mathbf{T}_{34}$ is $(m-q)\times(n-m-p+q)$, $\mathbf{T}_{42}$ is $(n-m-p+q)\times(p-q)$ and $\mathbf{T}_{44}$ is $(n-m-p+q)\times(n-m-p+q)$, this number of degrees of freedom is equal to $(n-m)(n-p)$. $\mathbf{A}$ is therefore s.g.i. if and only if $(n-m)(n-p)=0$, i.e., if either $\dim\mathbf{x}=\dim\mathbf{y}$ or $\dim\mathbf{x}=\dim\mathbf{u}$, which shows a nice symmetry between action and observation from an informational point of view.

3.3.2 Nonlinear Case: Local State Isomorphism Approach

Nonlinear models such as (3.9) tend to be more identifiable than linear ones, and introducing a non-linearity into an otherwise unidentifiable linear model very often makes it s.g.i., to the point that some even questioned the interest of further developing methods for testing nonlinear models for identifiability. Indeed, finding a nonlinear model that was relevant and unidentifiable remained a challenge for quite some time. We shall limit ourselves here to *very* briefly presenting one of the most powerful approaches, namely the local state isomorphism approach [48]. It applies to models described by

$$\dot{\mathbf{x}} = \mathbf{f}(\mathbf{x},\boldsymbol{\theta}) + \mathbf{g}(\mathbf{x},\boldsymbol{\theta})u, \qquad \mathbf{y} = \mathbf{h}(\mathbf{x},\boldsymbol{\theta}), \tag{3.22}$$

with $\mathbf{f}$, $\mathbf{g}$ and $\mathbf{h}$ analytic. Under conditions of observability and controllability, $\mathscr{M}(\hat{\boldsymbol{\theta}})$ will have the same input-output behavior as $\mathscr{M}(\boldsymbol{\theta}^*)$ up to some strictly positive time if and only if there exists a local state isomorphism $\hat{\mathbf{x}} = \boldsymbol{\varphi}(\mathbf{x}^*)$ such that, for any $\mathbf{x}^*$ in a neighborhood of the origin, the drift terms correspond

$$\mathbf{f}(\hat{\mathbf{x}},\hat{\boldsymbol{\theta}}) = \frac{\mathrm{d}\boldsymbol{\varphi}}{\mathrm{d}\mathbf{x}}(\mathbf{x}^*)\mathbf{f}(\mathbf{x}^*,\boldsymbol{\theta}^*), \tag{3.23}$$

the control terms correspond

$$\mathbf{g}(\hat{\mathbf{x}},\hat{\boldsymbol{\theta}}) = \frac{\mathrm{d}\boldsymbol{\varphi}}{\mathrm{d}\mathbf{x}}(\mathbf{x}^*)\mathbf{g}(\mathbf{x}^*,\boldsymbol{\theta}^*), \tag{3.24}$$

and the observations correspond

$$\mathbf{h}(\hat{\mathbf{x}},\hat{\boldsymbol{\theta}}) = \mathbf{h}(\boldsymbol{\varphi}(\mathbf{x}^*),\hat{\boldsymbol{\theta}}) = \mathbf{h}(\mathbf{x}^*,\boldsymbol{\theta}^*). \tag{3.25}$$

After checking observability and controllability, one may thus look for the set of all solutions of (3.23), (3.24) and (3.25) for $\hat{\boldsymbol{\theta}}$ and $\boldsymbol{\varphi}$. Recall that these equations must hold true for any $\mathbf{x}^*$ in a neighborhood of the origin, so whenever one equation turns out to be polynomial in the indeterminates forming $\mathbf{x}^*$ it can be split into equations for each of the monomials. If, for almost any $\boldsymbol{\theta}^*$, this solution set reduces to a singleton where $\hat{\boldsymbol{\theta}} = \theta^*$ and $\boldsymbol{\varphi}$ is the identity transformation, then the model is s.g.i.

An important special case is when $\mathbf{f}$ and $\mathbf{g}$ are polynomials in $\mathbf{x}$, parametrized by $\boldsymbol{\theta}$ and the observation equation is linear in the state, i.e., $\mathbf{h}(\mathbf{x}, \boldsymbol{\theta}) = \mathbf{C}(\boldsymbol{\theta})\mathbf{x}$. The state isomorphism can then be directly written under the form of a linear transformation $\boldsymbol{\varphi}(\mathbf{x}^*) = \mathbf{T}\mathbf{x}^*$, a considerable simplification [9].

Example 3.3 Although this method is dedicated to nonlinear models, it can of course be applied to linear models, so consider the model defined by

$$\dot{\mathbf{x}} = \mathbf{A}(\boldsymbol{\theta})\mathbf{x} + \mathbf{b}(\boldsymbol{\theta})u, \qquad \mathbf{y} = \mathbf{C}(\boldsymbol{\theta})\mathbf{x}. \tag{3.26}$$

The conditions of the special case apply, and the local state isomorphism becomes a similarity transformation $\boldsymbol{\varphi}(\mathbf{x}^*) = \mathbf{T}\mathbf{x}^*$, so (3.23), (3.24) and (3.25) translate into

$$\mathbf{A}(\hat{\boldsymbol{\theta}})\mathbf{T}\mathbf{x}^* = \mathbf{T}\mathbf{A}(\boldsymbol{\theta}^*)\mathbf{x}^*, \qquad \mathbf{b}(\hat{\boldsymbol{\theta}}) = \mathbf{T}\mathbf{b}(\boldsymbol{\theta}^*), \quad \text{and} \quad \mathbf{C}(\hat{\boldsymbol{\theta}})\mathbf{T}\mathbf{x}^* = \mathbf{C}(\boldsymbol{\theta}^*)\mathbf{x}^*. \tag{3.27}$$

Since these equations must be valid for any $\mathbf{x}^*$ around the origin and since $\mathbf{T}$ must be invertible to correspond to a diffeomorphism, these equations are equivalent to those that would have been obtained via the similarity transformation approach.

3.3.3 Using Elimination Theory and Computer Algebra

As illustrated by Example 3.1, (3.7) often translates into a set of polynomial equations in $\hat{\boldsymbol{\theta}}$ parametrized by $\boldsymbol{\theta}^*$, which is unfortunately often not as easy to solve as on this toy example. Elimination theory makes it possible to transform this set of equations into one much simpler to solve [45, 57]. In a way reminiscent of Gaussian elimination for solving linear sets of equations, Buchberger's algorithm [8], for instance, provides a method for transforming the initial set of polynomial equations into a triangular set

$$P_1(\hat{\theta}_1, \boldsymbol{\theta}^*) = 0, \qquad P_2(\hat{\theta}_2, \boldsymbol{\theta}^*) = 0, \quad \ldots, \quad P_{\dim \boldsymbol{\theta}}(\hat{\theta}_{\dim \boldsymbol{\theta}}, \boldsymbol{\theta}^*) = 0, \tag{3.28}$$

where the components of $\hat{\boldsymbol{\theta}}$ may have been re-indexed. For the model to be s.g.i., each P_i should have first order in $\hat{\theta}_i$. Although the method is systematic, with conceptually simple steps, it may lead to very complicated algebraic manipulation that are best left to computer algebra systems such as Maple. It does not take a very complex model, however, to exceed the ability of even today's supercomputers, so the automated solution of systems of polynomial equations is still an active research topic.

Differential algebra [28, 46], in which differentiation is added to the usual axioms of algebra, makes it possible to eliminate state variables from (3.9) so as to get differential input-output relations. These relations, which involve only known variables and their derivatives and the parameters to be estimated, can also be used to test nonlinear models for identifiability [2, 4, 32, 41]. The Maple package `diffalg`, based on results presented in [6], turns out to be very useful in this context. It can be shown [30] that any s.g.i. model can be rearranged to a linear regression (since high-order derivatives of noisy measurements may appear in the regressors, this should be used with caution in the context of actual parameter estimation).

3.4 Testing Model Structures for Distinguishability

It was assumed so far that the structure of the model had been chosen. When one hesitates between several competing model structures for the description of the same system, one would also like to check that there is some hope of using experimental data to select the best of them. The notion of structural distinguishability can then be used, under similar idealized conditions as for identifiability [58]. Consider a pair of competing model structures $\hat{\mathscr{M}}(\cdot)$ and $\mathscr{M}^*(\cdot)$, and write

$$\hat{\mathscr{M}}(\hat{\boldsymbol{\theta}}) = \mathscr{M}^*(\boldsymbol{\theta}^*) \tag{3.29}$$

to denote that the model with structure $\hat{\mathscr{M}}(\cdot)$ and parameter vector $\hat{\boldsymbol{\theta}}$ has the same output as the model with structure $\mathscr{M}^*(\cdot)$ and parameter vector $\boldsymbol{\theta}^*$ for any input and time. ($\hat{\boldsymbol{\theta}}$ and $\boldsymbol{\theta}^*$are now completely different parameter vectors, and may even differ in their dimensions.) $\hat{\mathscr{M}}(\cdot)$ is *structurally distinguishable* (s.d.) from $\mathscr{M}^*(\cdot)$ if, for almost all values of $\boldsymbol{\theta}^*$, there is no $\hat{\boldsymbol{\theta}}$ such that (3.29) is satisfied. When $\hat{\mathscr{M}}(\cdot)$ is s.d. from $\mathscr{M}^*(\cdot)$ and $\mathscr{M}^*(\cdot)$ s.d. from $\hat{\mathscr{M}}(\cdot)$, $\hat{\mathscr{M}}(\cdot)$ and $\mathscr{M}^*(\cdot)$ become s.d. Although it is easy to show via counterexamples that identifiability of two model structures is neither necessary nor sufficient for their distinguishability, trivial adaptations of the methods presented above make it possible to test models for structural distinguishability. The main difference is that we now hope that there is no solution to (3.29), instead of hoping for a unique solution to (3.7).

3.5 Maximizing Identifiability

Answers to questions of structural identifiability are of a qualitative nature. Once the structural identifiability of the parameters of interest has been established, one would like to know *how much* they are identifiable, and this clearly depends on the conditions of data collection. An s.g.i. parameter may actually turn out to be estimated with such an imprecision that it is not identifiable at all in practice, if the experimental conditions are badly chosen. To design experiments optimally [12, 19, 60, 61] so as to maximize practical identifiability, we first need an approach to assess parameter uncertainty as a function of the experimental conditions.

3.5.1 Quantifying Identifiability

The approach based on the Fisher information matrix (FIM) is by far the simplest and most used method for assessing the uncertainty in the parameters that results from noise corrupting the data. Under fairly general technical conditions (which include global identifiability of the model parameters), when the number of data points tends to infinity, the maximum-likelihood estimate $\hat{\boldsymbol{\theta}}_{\text{ML}}$ of the parameter vector tends to be normally distributed $\mathcal{N}(\boldsymbol{\theta}^*, \mathbf{F}^{-1}(\boldsymbol{\theta}^*, \Xi))$, with Ξ describing the experimental conditions under which the data were collected, and $\mathbf{F}$ the FIM. The FIM is thus strongly connected with the (asymptotic) dispersion of the parameter estimates around the true value. $\mathbf{F}(\boldsymbol{\theta}, \Xi)$ can be computed as the expectation of the product of the gradient of the log likelihood of the data by its transpose, with the expectation taken over all possible values of the data under the hypothesis that they are generated by a model with parameter vector $\boldsymbol{\theta}$. As long as the FIM is invertible, the parameters are at least locally identifiable. In the important special case where Ξ consists of the instants of time t_i ($i = 1, \ldots, f$) at which a scalar output $y(t_i)$ is collected, if one assumes that

$$y(t_i) = y_{\text{m}}(t_i, \boldsymbol{\theta}^*) + \varepsilon_i, \tag{3.30}$$

with the ε_is independently identically distributed $\mathcal{N}(0, \sigma^2)$, computation boils down to

$$\mathbf{F}(\boldsymbol{\theta}, \Xi) = \frac{1}{\sigma^2} \sum_{i=1}^{f} \left(\frac{\partial y_{\text{m}}(t_i, \boldsymbol{\theta})}{\partial \boldsymbol{\theta}} \right) \left(\frac{\partial y_{\text{m}}(t_i, \boldsymbol{\theta})}{\partial \boldsymbol{\theta}} \right)^{\text{T}}, \tag{3.31}$$

which corresponds to the Gauss-Newton approximation of the Hessian, often used for parameter optimization. Unless y_{m} is linear or affine in $\boldsymbol{\theta}$, a very special case not considered here, the FIM thus depends on $\boldsymbol{\theta}$, so evaluating the (asymptotic) covariance $\mathbf{F}^{-1}(\boldsymbol{\theta}^*, \Xi)$ is impossible in practice. Instead, one usually evaluate $\mathbf{F}^{-1}(\hat{\boldsymbol{\theta}}_{\text{ML}}, \Xi)$. This characterization of the uncertainty on the parameter estimates is more credible if the number of data points is large, the model output is only mildly nonlinear in $\boldsymbol{\theta}$ and the measurement errors are independently distributed with small magnitudes. Because it is much less computer-intensive than alternative Monte-Carlo methods, the FIM approach plays a prominent role in optimal experiment design for parameter estimation, which may be seen as a maximization of practical identifiability, just as optimal experiment design for model discrimination could be seen as a maximization of practical distinguishability.

3.5.2 Optimal Experiment Design

Experiment design is of course only feasible if there are some degrees of freedom on the experiments to be carried out, for instance in terms of location of sensors or actuators, of input shape or of measurement times. We assume here that the design consists of f elementary experiments leading to scalar observations y_i, $i = 1, \dots, f$, and denote the experimental conditions of the ith scalar observation by $\boldsymbol{\xi}^i$. For instance, $\boldsymbol{\xi}^i$ may simply correspond to the ith measurement time. When f such observations are collected, the concatenation of their experimental conditions yields the experimental design

$$\varXi = [\boldsymbol{\xi}^1, \boldsymbol{\xi}^2, \dots, \boldsymbol{\xi}^f]. \tag{3.32}$$

It is important to realize that there are always practical constraints on $\varXi$ (e.g., on the duration of the experiment, the energy or amplitude of the inputs, the minimum time between samples, the total number of samples, working hours...), which define a feasible experimental domain. These constraints *must* be taken into account for the solution to be relevant. The definition of an optimality criterion then makes it possible to cast experiment design as *constrained optimization*, as opposed to parameter estimation, usually carried out via unconstrained optimization. In the context of experiment design for parameter estimation, one almost invariably optimizes some scalar function of the FIM, because the FIM is simple enough to be repeatedly evaluated, as required by the optimization algorithms. Disregarding, for the time being, the problem of the dependency of the FIM in the parameters, the most commonly used criterion is *D-optimality*, where

$$\begin{aligned}\hat{\varXi}_{\mathrm{D}} &= \arg\min_{\varXi} \det \mathbf{F}^{-1}(\boldsymbol{\theta}^*, \varXi) = \arg\max_{\varXi} \det \mathbf{F}(\boldsymbol{\theta}^*, \varXi) \\ &= \arg\max_{\varXi} \ln \det \mathbf{F}(\boldsymbol{\theta}^*, \varXi). \end{aligned} \tag{3.33}$$

$\hat{\varXi}_{\mathrm{D}}$ thus minimizes the volume of the asymptotic confidence ellipsoids for $\boldsymbol{\theta}^*$. It can be shown to be invariant under any non-singular reparametrization that does not depend on the experiment. Thus, for instance, the optimal experiment does not depend on the units in which the parameters are expressed. This seems natural to ask for, and is one of the reasons for the popularity of D-optimality. However, a D-optimal experiment may correspond to very elongated confidence ellipsoids, with large confidence intervals for the parameters. This is why alternative optimality criteria also based on the FIM may be worth considering, at the cost of losing invariance under reparametrization.

Example 3.4 Consider the one-compartment model defined by the state equation

$$\dot{x} = -\frac{Cl}{V}x + u, \quad x(0_-) = 0, \tag{3.34}$$

where the two parameters to be estimated are the clearance Cl and the volume of distribution V. These parameters are classical in pharmacokinetics, and called *micro-parameters*. Assume an impulsive input of drug $u(t) = d_0\delta(t)$ (with the dose d_0 known). This is equivalent to assuming that there is no input and $x(0_+) = d_0$. Assume further that the observation equation is

$$y(t_i) = \frac{1}{V}x(t_i) + \varepsilon_i, \quad i = 1, \ldots, f, \tag{3.35}$$

with the ε_is independently identically distributed $\mathcal{N}(0, \sigma^2)$. We cannot do with less than two measurements ($f = 2$, $\Xi = [t_1, t_2]$) to estimate the two parameters, and wish to position these measurements in a D-optimal manner. Since

$$y(t_i) = \frac{d_0}{V}\exp\left(-\frac{Cl}{V}t_i\right) + \varepsilon_i, \quad i = 1, \ldots, f, \tag{3.36}$$

and D-optimal design is invariant by design-independent reparametrization, we can parametrize the model in terms of its *macro-parameters* as

$$y_{\mathrm{m}}(t_i, \mathbf{p}) = p_1 \exp(-p_2 t_i), \quad \text{with } p_1 = \frac{d_0}{V} \text{ and } p_2 = \frac{Cl}{V}, \tag{3.37}$$

and use (3.31) to compute the FIM as

$$\mathbf{F}(\mathbf{p}, \Xi) = \frac{1}{\sigma^2}\sum_{i=1}^{2}\begin{bmatrix} \exp(-p_2 t_i) \\ -p_1 t_i \exp(-p_2 t_i) \end{bmatrix}\begin{bmatrix} \exp(-p_2 t_i) & -p_1 t_i \exp(-p_2 t_i) \end{bmatrix}. \tag{3.38}$$

The objective function to be maximized with respect to Ξ is then

$$\det \mathbf{F}(\mathbf{p}, \Xi) = \frac{1}{\sigma^4} p_1^2 (t_2 - t_1)^2 \exp\left[-2p_2(t_1 + t_2)\right]. \tag{3.39}$$

Obviously the measurements must take place after the drug is introduced, and we can arbitrarily label the first measurement time as t_1, so we take $t_2 \geq t_1 \geq 0$ as the design constraints. The D-optimal design is then $\hat{\Xi}_{\mathrm{D}} = (0, 1/p_2^*)$. It thus depends on the very parameters to be estimated! This is typical of models whose output is nonlinear in their parameters, and a major difficulty with most knowledge-based models.

Various strategies can be followed to address this difficulty. The first one corresponds to choosing some (hopefully reasonable) nominal value $\boldsymbol{\theta}_0$ for the parameter vector, and designing the experiment to be D-optimal at $\boldsymbol{\theta}_0$. This is called *local design*. It can be carried out using either generic algorithms for constrained optimization or more specific algorithms, such as DETMAX [34].

With *sequential design* [10, 16], one cycles through experimentation, estimation and experiment-design steps. One may start experimenting with a local design (or any reasonable experimental design). Each estimation step improves knowledge about the parameters, and this knowledge is taken advantage of during the next experiment-design step. To ensure convergence, a key point is that each estimation step should make use of all previous observations.

When repetition of experiments as required by sequential design is impossible, so that a single (one-shot) experiment must be designed, a first possible approach is *average optimality* [13, 42], where parameter dependence is removed by averaging on $\boldsymbol{\theta}$. One may, for instance, compute an ELD-optimal design [11]

$$\hat{\Xi}_{\mathrm{ELD}} = \arg\max_{\Xi} E_{\boldsymbol{\theta}}\left\{\ln\det \mathbf{F}(\boldsymbol{\theta}, \Xi)\right\}, \tag{3.40}$$

where the expectation is taken with respect to some prior distribution for $\boldsymbol{\theta}$, assumed to be available. Note that introducing the expectation operator in the three equivalent expressions for D-optimality provided in (3.33) produces three *different* criteria. The one reported above has the advantage over the other two of being justifiable by information-theoretic arguments. Specific algorithms such as stochastic gradient make it possible to find $\hat{\Xi}_{\mathrm{ELD}}$ without having to evaluate any mathematical expectation. See [60] for more details.

If the best experiment in the worst circumstances should be preferred to one that is best on average, then *maximin optimality* can be considered [43], such that

$$\hat{\Xi}_{\mathrm{MMD}} = \arg\max_{\Xi} \min_{\boldsymbol{\theta}} \det \mathbf{F}(\boldsymbol{\theta}, \Xi). \tag{3.41}$$

One must then assume that a set of prior admissible values for $\boldsymbol{\theta}$ is available. Again, specific algorithms must be employed, such as Shimizu and Aiyoshi's relaxation algorithm [47].

Example 3.5 Consider again the one-compartment model of Example 3.4, parametrized in terms of its macro-parameters, and assume that the prior distribution of p_2 is uniform over $[1, 10]$. Then $\hat{\Xi}_{\mathrm{ELD}} = (0, 0.182)$ and $\hat{\Xi}_{\mathrm{MMD}} = (0, 0.1)$.

3.6 Beyond Identifiability

One may want to bypass the study of the structural identifiability of the model being considered for at least two reasons. The first is when one is unable to reach a conclusion, either because the calculations involved are too complicated or because no generic conclusion is possible (for instance, because there are two regions of parameter space where the conclusion differ, none of them corresponding to an atypical manifold). The second, and more fundamental one, is that lack of identifiability of the model structure is just one of the possible motives for ambiguity or an unacceptable uncertainty in the estimated parameters. Even if the model is s.g.i., there may exist radically different parameter vectors that are associated to acceptable behavior of the model, so there is a need for methods to characterize the set $\mathscr{S}$ of *all* values of the parameter estimates that are acceptable (in a sense to be specified) given the data. This is in contrast with the usual methods for nonlinear parameter estimation, which look for a single parameter vector minimizing some cost function by local iterative methods, with the well-known risk of getting trapped at a local minimizer.

Interval analysis, the main tool to be used to provide an approximate but guaranteed approximation of $\mathscr{S}$, will be very briefly presented, before describing applications of guaranteed set characterization in two contexts, namely optimal estimation and bounded-error estimation.

3.6.1 Interval Analysis

To allow guaranteed statements on calculations involving real numbers x, interval analysis (IA) [24, 35, 38] computes on intervals $[x]$, described by machine-representable lower bound $\underline{x}$ and upper bound $\overline{x}$. Thus an interval $[\underline{x}, \overline{x}]$ is represented by a pair of real numbers (just as a complex number). As regards arithmetical operations, it is easy to derive rules for the addition, subtraction or multiplication of intervals. For instance, computing $[c] = [b] + [a]$ simply means computing $\underline{c}$ as the largest machine-representable number that is smaller than $\underline{b} + \underline{a}$, and $\overline{c}$ as the smallest machine-representable number that is larger than $\overline{b} + \overline{a}$. Division requires more care, to deal with the case where the interval at the denominator contains zero. Interval vectors $[\mathbf{x}]$ (also called *boxes*) and interval matrices $[\mathbf{M}]$ can be defined as Cartesian products of scalar intervals, and operations on matrices and vectors such as addition, subtraction or multiplication are trivially extended to operations on interval matrices and interval vectors.

An interval guaranteed to contain the image of an interval by an elementary function such as the exponential or any trigonometric function is easy to compute. For instance, $\exp([x])$ is included in the interval $[\exp(\underline{x}), \exp(\overline{x})]$, which is rounded outward to get a machine-representable interval.

For any function $\mathbf{f}(\cdot)$ defined by combining arithmetical operators and elementary functions, IA makes it possible to build *inclusion functions* $[\mathbf{f}](\cdot)$ such that $\mathbf{f}([\mathbf{x}]) \subset [\mathbf{f}]([\mathbf{x}])$, where $[\mathbf{f}]([\mathbf{x}])$ is a box. It thus becomes possible to make guaranteed statements about the image of a box by a function, even though this image is usually impossible to compute exactly. If, for instance, $\mathbf{0}$ does not belong to the box $[\mathbf{f}]([\mathbf{x}])$, then we know that it does not belong to $\mathbf{f}([\mathbf{x}])$ either. Many types of inclusion functions can be defined (and combined). *Natural inclusion functions*, for example, are obtained by replacing all real variables, operators and elementary functions by their interval counterparts.

The construction of inclusion functions for the solutions of systems of ordinary differential equations for which no closed-form is available is slightly more complicated. It can be achieved through the use of *guaranteed ODE solvers*, such as AWA [31], COSY [5, 22] or VNODE [37]. Since these solvers cannot provide accurate enclosures of the solutions when the equations are uncertain, it may be necessary to bound the solutions of uncertain ODEs by those of deterministic ODEs, on which guaranteed ODE solvers will prove much more accurate. The notion of cooperativity or Müller's theorems [36] are then extremely helpful [26, 53–55].

3.6.2 Optimal Estimation

Hansen's algorithm is representative of the deterministic global optimization algorithms that can be used when parameter estimation translates into optimization. Its presentation here will be more than sketchy, and the reader is invited to consult [20] for more details.

Let $c(\boldsymbol{\theta})$ be the cost to be minimized, assumed to be twice differentiable (it may correspond, for instance, to minus the log likelihood). Let $\mathbf{g}(\boldsymbol{\theta})$ be its gradient and $\mathbf{H}(\boldsymbol{\theta})$ its Hessian. Assume that search must take place within some (possibly very large) box $[\boldsymbol{\theta}_0]$ of parameter space, and that we have inclusion functions $[c](\cdot)$ for the cost, $[\mathbf{g}](\cdot)$ for its gradient and $[\mathbf{H}](\cdot)$ for its Hessian (or at least $[h_{ii}](\cdot)$ for the ith diagonal entry of its Hessian, $i = 1, \ldots, \dim \boldsymbol{\theta}$). The global minimizers are not expected to lie on the boundary of $[\boldsymbol{\theta}_0]$, so this is unconstrained minimization, and any local or global minimizer $\hat{\boldsymbol{\theta}}$ should be such that $\mathbf{g}(\hat{\boldsymbol{\theta}}) = \mathbf{0}$ (stationarity condition) and $h_{ii}(\hat{\boldsymbol{\theta}}) \geq 0, \;\; i = 1, \ldots, \dim \boldsymbol{\theta}$ (convexity condition). The idea is to eliminate (or reduce) sub-boxes of $[\boldsymbol{\theta}_0]$ that cannot contain any global minimizer. Let $[\boldsymbol{\theta}]$ be one such sub-box. It can be eliminated

- if the lower bound of $[c]([\boldsymbol{\theta}])$, which is the best value of the cost that one can hope for on $[\boldsymbol{\theta}]$, is greater (i.e., worse) than the best value obtained so far,
- if $[\mathbf{g}]([\boldsymbol{\theta}])$ does not contain $\mathbf{0}$, which proves that $[\boldsymbol{\theta}]$ contains no stationary point,
- if there is a diagonal entry h_{ii} of the Hessian such that the upper bound of $[h_{ii}]([\boldsymbol{\theta}])$ is strictly negative, which proves that the cost is not locally convex anywhere on $[\boldsymbol{\theta}]$.

It can be reduced by a *contractor*, i.e., an operator that transforms it into a smaller box without losing any minimizer. Contractors are particularly important in the struggle against the curse of dimensionality, because they reduce the size of the search region without bisection. A number of possible contractors are presented in [24]. Hansen's algorithm uses the Newton contractor, an interval counterpart to the Newton method for the solution of the equation $\mathbf{g}(\boldsymbol{\theta}) = \mathbf{0}$. The basic (and beautiful) idea of the Newton contractor is as follows. The mean-value theorem implies that, for any $\boldsymbol{\theta}$ in $[\boldsymbol{\theta}]$, there exists $\mathbf{z}$ also in $[\boldsymbol{\theta}]$ such that

$$\mathbf{g}(\boldsymbol{\theta}) = \mathbf{g}(\mathbf{m}) + \mathbf{H}(\mathbf{z})(\boldsymbol{\theta} - \mathbf{m}), \tag{3.42}$$

where $\mathbf{m}$ is the center of the box $[\boldsymbol{\theta}]$. Now assume that $\hat{\boldsymbol{\theta}} \in [\boldsymbol{\theta}]$ is an unconstrained minimizer. Since $\mathbf{g}(\hat{\boldsymbol{\theta}}) = \mathbf{0}$,

$$\mathbf{g}(\mathbf{m}) + \mathbf{H}(\mathbf{z})(\hat{\boldsymbol{\theta}} - \mathbf{m}) = \mathbf{0} \tag{3.43}$$

and, if the Hessian is invertible,

$$\hat{\boldsymbol{\theta}} = \mathbf{m} - \mathbf{H}^{-1}(\mathbf{z})\mathbf{g}(\mathbf{m}). \tag{3.44}$$

So, assuming that the Hessian is invertible at any $\mathbf{z}$ in $[\boldsymbol{\theta}]$,

$$\hat{\boldsymbol{\theta}} \in \mathbf{m} - \left[\mathbf{H}^{-1}\right]([\boldsymbol{\theta}])\mathbf{g}(\mathbf{m}). \tag{3.45}$$

As $\hat{\boldsymbol{\theta}}$ also belongs to $[\boldsymbol{\theta}]$, we have

$$\hat{\boldsymbol{\theta}} \in [\boldsymbol{\theta}] \cap \left[\mathbf{m} - \left[\mathbf{H}^{-1}\right]([\boldsymbol{\theta}])\mathbf{g}(\mathbf{m})\right], \tag{3.46}$$

which may turn out to be much smaller than $[\boldsymbol{\theta}]$, or even empty. In practice, one avoids attempting to invert an interval matrix, and looks instead for an outer approximation to the set of all solutions for $\hat{\boldsymbol{\theta}}$ in $[\boldsymbol{\theta}]$ of the linear system of equations

$$\mathbf{g}(\mathbf{m}) + \left[\mathbf{H}([\boldsymbol{\theta}])\right](\hat{\boldsymbol{\theta}} - \mathbf{m}) = \mathbf{0}. \tag{3.47}$$

Hansen's algorithm manages a list of sub-boxes of $[\boldsymbol{\theta}_0]$ the union of which is guaranteed to contain all global minimizers within $[\boldsymbol{\theta}_0]$. Initially this list only contains $[\boldsymbol{\theta}_0]$. Whenever a sub-box fails to be eliminated or reduced to the empty set, it is bisected into two sub-boxes, possibly after contraction, unless its width is lower than some prespecified threshold. The algorithm terminates when all the boxes left in the list have a width lower than the threshold.

Results obtained by application of this algorithm (with a different contractor) to the example of Sect. 3.1, using (3.5) as the cost function, can be found in [27]. The algorithm computes a guaranteed outer approximation of the set of all global minimizers that presents a symmetry around the plane $\theta_1 = \theta_3$, consistent with the identifiability analysis carried out in Example 3.1 but obtained without taking this identifiability analysis into account.

3.6.3 Bounded-Error Estimation

In this alternative approach [25, 33, 39, 40, 44, 52], instead of looking for the set of all global minimizers of the cost function, we look for the set of all parameter vectors that are consistent with some prior bounds on the errors that we are prepared to accept. With each vector of experimental data $\mathbf{y}(t_i)$, we assume that is associated a known box $[\underline{\mathbf{e}}_i, \overline{\mathbf{e}}_i]$ of acceptable errors, and look for the set

$$\mathscr{S} = \left\{\boldsymbol{\theta} \mid \underline{\mathbf{e}}_i \le \mathbf{y}(t_i) - \mathbf{y}_{\mathrm{m}}(t_i, \boldsymbol{\theta}) \le \overline{\mathbf{e}}_i,\ i = 1, \ldots, f\right\}. \tag{3.48}$$

Let $\mathbf{y}$, $\underline{\mathbf{e}}$, $\overline{\mathbf{e}}$ and $\mathbf{f}(\boldsymbol{\theta})$ be the vectors obtained by concatenating all $\mathbf{y}(t_i)$, $\underline{\mathbf{e}}_i$, $\overline{\mathbf{e}}_i$ and $\mathbf{y}_{\mathrm{m}}(t_i, \boldsymbol{\theta})(i = 1, \ldots, f)$, respectively, and take $[\mathbf{y}] = [\mathbf{y} - \overline{\mathbf{e}}, \mathbf{y} - \underline{\mathbf{e}}]$. Then $\mathscr{S}$ can be defined as

$$\mathscr{S} = \{\boldsymbol{\theta} \mid \mathbf{f}(\boldsymbol{\theta}) \in [\mathbf{y}]\} = \mathbf{f}^{-1}([\mathbf{y}]), \tag{3.49}$$

which casts parameter estimation in the context of set inversion. The algorithm SIVIA (for set inversion via interval analysis) [25] can then be used to compute two unions of boxes $\underline{\mathscr{S}}$ and $\overline{\mathscr{S}}$ in parameter space, such that $\underline{\mathscr{S}} \subset \mathscr{S} \subset \overline{\mathscr{S}}$. As previously, search will be carried out within a prior box $[\boldsymbol{\theta}_0]$, assumed to be large enough to contain $\mathscr{S}$, or at least the part of it we are interested in. As with Hansen's algorithm, this prior box is bisected into sub-boxes $[\boldsymbol{\theta}]$ that fall into three categories

- those such that $[\mathbf{f}]([\boldsymbol{\theta}]) \in [\mathbf{y}]$ are proven to be inside $\mathscr{S}$; they are included in $\underline{\mathscr{S}}$ and $\overline{\mathscr{S}}$;
- those such that $[\mathbf{f}]([\boldsymbol{\theta}]) \cap [\mathbf{y}] = \emptyset$ are proven to be outside $\mathscr{S}$; they are discarded;
- all others are bisected into sub-boxes to be further tested, unless their width is smaller than some user-defined threshold δ (in which case they are included in the outer approximation $\overline{\mathscr{S}}$).

Because of the threshold δ, this algorithm stops after a finite number of steps. Upon completion, it produces inner and outer approximations of $\mathscr{S}$, and the distance between these approximations is indicative of the quality of the characterization of $\mathscr{S}$ achieved. This quality can be increased by decreasing δ, at the cost of more computation.

An improved version of this algorithm can be found in [27], where it is also applied to the example of Sect. 3.1. Again, for small enough bounds on the acceptable errors, $\overline{\mathscr{S}}$ turns out to consist of two disconnected subsets with a symmetry around the plane $\theta_1 = \theta_3$, consistent with the identifiability analysis carried out in Example 3.1 but obtained without taking this identifiability analysis into account.

3.7 Conclusions and Perspectives

Structural identifiability is a critical issue when one is interested in estimating the physically meaningful parameters of knowledge-based models. It is also important when physically meaningful state variables have to be estimated using filters based on these models.

Methods of test have been presented for models that may be linear or not in the input-output sense, but are always nonlinear in their parameters. This is the rule for knowledge based models.

Having proved that a model is structurally identifiable does not guarantee that it can actually be estimated satisfactorily. The quality of the estimates crucially depends on that of the data, and optimal experiment design for parameter estimation may be viewed as maximizing a measure of practical identifiability (just as optimal experiment design for model discrimination may be viewed as maximizing a measure of practical distinguishability). The main difficulty with experiment design for parameter estimation in the context of models that are nonlinear in their parameters is that the usual approaches yield experiments that depend on the parameters to be estimated. Several ways of addressing this difficulty have been recalled.

Interval analysis makes it possible to characterize the set of all solutions of the estimation problem while bypassing the study of structural identifiability altogether. This does not make this study obsolete as regards the understanding of the mathematical properties of the model. Optimal experiment design also remains a much useful concept. Despite the meaning of the Latin *data*, it is important to realize that the data may not be a given of the problem, and that much may be gained by collecting them in an optimal manner.

Although the examples considered in this chapter were toys, designed to illustrate specific aspects, the methodology that has been presented already has been used on a number of real-life problems (see, e.g., [7, 21, 49, 59]). The main limitations are due to the curse of dimensionality. Even with computer algebra, it is often impossible to reach a conclusion on the structural identifiability of models of practical interest, and the characterization of the set of all optimal or acceptable parameter vectors is an NP-complete problem, so we need approximations to be able to get solutions without loosing the guaranteed nature of the conclusions. The use of appropriate contractors, which makes it possible to limit the numbers of bisections to be carried out in parameter space, is an important tool in this respect, on which much remains to be done, as on other methods to struggle against the curse of dimensionality.

References

1. Bellman, R., Aström, K.J.: On structural identifiability. Math. Biosci. **7**, 329–339 (1970)
2. Bellu, G., Saccomani, M.P., Audoly, S., D'Angio, L.: DAISY: a new software tool to test global identifiability of biological and physiological systems. Comput. Methods Programs Biomed. **88**, 52–61 (2007)
3. Berman, M., Schoenfeld, R.: Invariants in experimental data on linear kinetics and the formulation of models. J. Appl. Phys. **27**(11), 1361–1370 (1956)
4. Berthier, F., Diard, J.P., Pronzato, L., Walter, E.: Identifiability and distinguishability concepts in electrochemistry. Automatica **32**(7), 973–984 (1996)
5. Berz, M., Makino, K.: Verified integration of ODEs and flows using differential algebraic methods on high-order Taylor models. Reliab. Comput. **4**(4), 361–369 (1998)
6. Boulier, F., Lazard, D., Ollivier, F., Petitot, M.: Computing representations for radicals of finitely generated differential ideals. Tech. Rep. IT306, LIFL (1997)
7. Braems, I., Berthier, F., Jaulin, L., Kieffer, M., Walter, E.: Guaranteed estimation of electrochemical parameters by set inversion. J. Electroanal. Chem. **495**(1), 1–9 (2001)
8. Buchberger, B.: Ein algorithmisches Kriterium für die Lösbarkeit eines algebraischen Gleischungssystems. Aequ. Math. **4**, 374–383 (1970)
9. Chappell, M.J., Godfrey, K.R., Vajda, S.: Global identifiability of the parameters of nonlinear systems with specified inputs: a comparison of methods. Math. Biosci. **102**, 41–73 (1990)
10. Chernoff, H.: Approaches in sequential design of experiments. In: Srivasta, J. (ed.) Survey of Statistical Design and Linear Models, pp. 67–90. North-Holland, Amsterdam (1975)
11. D'Argenio, D.Z.: Incorporating prior parameter uncertainty in the design of sampling schedules for pharmacokinetic parameter estimation experiments. Math. Biosci. **99**, 105–118 (1990)
12. Fedorov, V.V.: Theory of Optimal Experiments. Academic Press, New York (1972)
13. Fedorov, V.V., Atkinson, A.C.: The optimum design of experiments in the presence of uncontrolled variability and prior information. In: Dodge, Y., Fedorov, V.V., Wynn, H.P. (eds.) Optimal Design and Analysis of Experiments. North-Holland, Amsterdam (1988)
14. Fisher, F.M.: Generalization of the rank and order conditions for identifiability. Econometrica **27**(3), 431–447 (1959)
15. Fisher, F.M.: Identifiability criteria in nonlinear systems. Econometrica **29**(4), 574–590 (1961)
16. Ford, I., Silvey, S.D.: A sequentially constructed design for estimating a nonlinear parametric function. Biometrika **67**(2), 381–388 (1980)
17. Glover, K., Willems, J.C.: Parametrizations of linear dynamical systems: canonical forms and identifiability. IEEE Trans. Autom. Control **19**, 640–644 (1974)
18. Godfrey, K.: Compartmental Models and Their Application. Academic Press, London (1983)
19. Goodwin, G.C., Payne, R.: Dynamic System Identification: Experiment Design and Data Analysis. Academic Press, New York (1977)

20. Hansen, E.: Global Optimization Using Interval Analysis. Marcel Dekker, New York (1992)
21. Happel, J., Walter, E., Lecourtier, Y.: Isotopic assessment of fundamental catalytic mechanisms by kinetic modeling. I&EC Fund. **25**, 704–712 (1986)
22. Hoekens, J., Berz, M., Makino, K.: Efficient high-order methods for ODEs and DAEs. In: Corliss, G., Faure, C., Griewank, A. (eds.) Automatic Differentiation: From Simulation to Optimization, pp. 341–351. Springer, New York (2001)
23. Jacquez, J.A.: Compartmental Analysis in Biology and Medicine. University of Michigan Press, Ann Arbor (1985)
24. Jaulin, L., Kieffer, M., Didrit, O., Walter, E.: Applied Interval Analysis. Springer, London (2001)
25. Jaulin, L., Walter, E.: Set inversion via interval analysis for nonlinear bounded-error estimation. Automatica **29**, 1053–1064 (1993)
26. Kieffer, M., Walter, E.: Interval analysis for guaranteed nonlinear parameter and state estimation. Math. Comput. Model. Dyn. Syst. **11**(2), 171–181 (2005)
27. Kieffer, M., Walter, E.: Guaranteed estimation of the parameters of nonlinear continuous-time models: contributions of interval analysis. Int. J. Adapt. Control Signal Process. **25**(3), 191–207 (2011)
28. Kolchin, E.: Differential Algebra and Algebraic Groups. Academic Press, San Diego (1973)
29. Koopmans, T.C., Reiersøl, O.: The identification of structural characteristics. Ann. Math. Stat. **21**(2), 165–181 (1950)
30. Ljung, L., Glad, T.: On global identifiability of arbitrary model parametrizations. Automatica **30**(2), 265–276 (1994)
31. Lohner, R.: Computation of guaranteed enclosures for the solutions of ordinary initial and boundary value problems. In: Cash, J., Gladwell, I. (eds.) Computational Ordinary Differential Equations, pp. 425–435. Clarendon Press, Oxford (1992)
32. Margaria, G., Riccomagno, E., Chappell, M.J., Wynn, H.P.: Differential algebra methods for the study of the structural identifiability of rational function state-space models in the biosciences. Math. Biosci. **174** (2001)
33. Milanese, M., Norton, J., Piet-Lahanier, H., Walter, E. (eds.): Bounding Approaches to System Identification. Plenum, New York (1996)
34. Mitchell, T.J.: An algorithm for the construction of "D-optimal" experimental designs. Technometrics **16**, 203–210 (1974)
35. Moore, R.: Methods and Applications of Interval Analysis. SIAM, Philadelphia (1979)
36. Müller, M.: Uber das Fundamentaltheorem in der Theory der gewöhlichen Differentielgleichungen. Math. Z. **26**, 619–645 (1926)
37. Nedialkov, N.S., Jackson, K.R.: Methods for initial value problems for ordinary differential equations. In: Kulisch, U., Lohner, R., Facius, A. (eds.) Perspectives on Enclosure Methods, pp. 219–264. Springer, Vienna (2001)
38. Neumaier, A.: Interval Methods for Systems of Equations. Cambridge University Press, Cambridge (1990)
39. Norton, J.: Special issue on bouded-error estimation: Issue 1. Int. J. Adapt. Control Signal Process. **8**(1), 1–118 (1994)
40. Norton, J.: Special issue on bouded-error estimation: Issue 2. Int. J. Adapt. Control Signal Process. **9**(1), 1–132 (1995)
41. Ollivier, F.: Le problème de l'identifiabilité structurelle globale: approche théorique, méthodes effectives et bornes de complexité. Ph.D. thesis, Ecole Polytechnique, Palaiseau (1990)
42. Pronzato, L., Walter, E.: Robust experiment design via stochastic approximation. Math. Biosci. **75**, 103–120 (1985)
43. Pronzato, L., Walter, E.: Robust experiment design via maximin optimization. Math. Biosci. **89**, 161–176 (1988)
44. Raissi, T., Ramdani, N., Candau, Y.: Set membership state and parameter estimation for systems described by nonlinear differential equations. Automatica **40**(10), 1771–1777 (2004)
45. Raksanyi, A., Lecourtier, Y., Walter, E., Venot, A.: Identifiability and distinguishability testing via computer algebra. Math. Biosci. **77**(1–2), 245–266 (1985)

46. Ritt, J.F.: Differential Algebra. American Mathematical Society, Providence (1950)
47. Shimizu, K., Aiyoshi, E.: Necessary conditions for min-max problems and algorithm by a relaxation procedure. IEEE Trans. Autom. Control **25**, 62–66 (1980)
48. Vajda, S., Rabitz, H.: State isomorphism approach to global identifiability of nonlinear systems. IEEE Trans. Autom. Control **34**, 220–223 (1989)
49. Venot, A., Walter, E., Lecourtier, Y., Raksanyi, A., Chauvelot-Moachon, L.: Structural identifiability of first-pass models. J. Pharmacokinet. Biopharm. **15**, 179–189 (1987)
50. Walter, E.: Identification de paramètres en cinétique chimique non linéaire à l'aide du modèle de compartiments associé à un indicateur. Ph.D. thesis, Université Paris-Sud, Orsay (1975)
51. Walter, E.: Identifiability of State Space Models. Springer, Berlin (1982)
52. Walter, E. (ed.): Special issue on parameter identification with error bounds. Math. Comput. Simul. **32**(5–6), 447–607 (1990)
53. Walter, E., Kieffer, M.: Interval analysis for guaranteed nonlinear parameter estimation. In: Proc. 13th IFAC Symposium on System Identification (SYSID), Rotterdam, pp. 259–270 (2003)
54. Walter, E., Kieffer, M.: Guaranteed optimisation of the parameters of continuous-time knowledge-based models. In: Commault, C., Marchand, N. (eds.) Positive Systems, pp. 137–144. Springer, Heildelberg (2006)
55. Walter, E., Kieffer, M.: Guaranteed nonlinear parameter estimation in knowledge-based models. J. of Comput. and Applied Math. **199**(2) (2007)
56. Walter, E., Lecourtier, Y.: Unidenfiable compartmental models: What to do? Math. Biosci. **56**, 1–25 (1981)
57. Walter, E., Lecourtier, Y.: Global approaches to identifiability testing for linear and nonlinear state space models. Math. Comput. Simul. **24**, 472–482 (1982)
58. Walter, E., Lecourtier, Y., Happel, J.: On the structural output distinguishability of parametric models, and its relation with structural identifiability. IEEE Trans. Autom. Control **29**, 56–57 (1984)
59. Walter, E., Lecourtier, Y., Happel, J., Kao, J.Y.: Identifiability and distinguishability of fundamental parameters in catalytic methanation. AIChE J. **32**(8), 1360–1366 (1986)
60. Walter, E., Pronzato, L.: Identification of Parametric Models from Experimental Data. Springer, London (1997)
61. Zarrop, M.B.: Optimal Experiment Design for Dynamic System Identification. Springer, Heidelberg (1979)

Chapter 4
Model Structure Identification and the Growth of Knowledge

M.B. Beck, Z. Lin, and J.D. Stigter

When contemplating the interpretation of some time-series data, often have I thought: "Why bother with the struggle, when I could simply pass it all over to the virtuoso—Peter".

By the time I arrived in Cambridge to begin my doctoral studies in October, 1970, I had convinced myself I was fascinated by systems, dynamics, and control. In fact, I rather fancy I wanted simply to continue being a student. In my first term, an acquaintance gathered me up on one of those waves of "environment and conservation" that wash over us from decade to decade. I put two words together: pollution and control. Could I do my PhD on the resulting topic? After all, Peter and I were in what was then called a Control Engineering Group, within the University Engineering Department. From amongst the puzzled faces, Peter emerged to take me on board. And the rest, as *I* say, is *my* history.

At that time (1970), Peter was chipping away at the orthodoxy of the book [13] just published by Box and Jenkins (*Time Series Analysis, Forecasting, and Control*), in what I now recognize as his own inimitable contrarian way. He gave me a passion

Personal Tribute to Professor Peter C. Young from the First Author (MBB).

M.B. Beck (✉)
University of Georgia, Athens, GA, USA
e-mail: mbbeck@uga.edu

Z. Lin
North Dakota State University, Fargo, ND, USA
e-mail: zhulu.lin@ndsu.edu

J.D. Stigter
Wageningen University, Wageningen, The Netherlands
e-mail: hans.stigter@wur.nl

L. Wang, H. Garnier (eds.), *System Identification, Environmental Modelling, and Control System Design*,
DOI 10.1007/978-0-85729-974-1_4,

for recursive estimation, from which I subsequently derived my own career-long commitment to solving the problem of model structure identification. These ideas of recursive estimation were so powerful. They have shaped the way I think, about so many problems, including those having more to do with the social sciences than with engineering. I gave them due recognition by entitling a whole sub-section—"Living in a Recursive Predictive World"—in one of the chapters of the book *Environmental Foresight and Models: A Manifesto* [7].

Somewhere in the 1980s, I came to view the extended Kalman filter (EKF) as akin to a Model T Ford, when I knew I wanted a contemporary BMW 700 Series. I could conceive of the overall design, but I needed first Hans Stigter and then Zhulu Lin to realize my conceptual blueprint. Needless to say, this has since drawn in no small measure upon the contributions Peter has been continuing to make.

This chapter honors Peter, then, in recounting my career-long experience (1970–2010) of staring down the devilishly difficult: the problem of model structure identification—of using models for discovery. I still regard this matter as one of *the* grand challenges of environmental modeling [12]. If I appear modest about our progress in the presence of such enormity, so I am. But let no-one presume that I am therefore not greatly enthused by the progress I believe I and my students (now colleagues) have made over these four decades. It has been a privilege to be allowed the time to work on such a most attractive and engaging topic.

4.1 Introduction

The summer of 1972 in Cambridge was unusually sunny, warm, and dry—an event that one would have noticed, in the light of the then popular image of the English summer. While there had been a decade preceding of river water quality modeling, no-one by 1970 had embarked on collecting appropriate field data for such model calibration and verification. This was especially true of the unsteady-state, dynamic models of particular interest to control theory and control engineers.

The goal of the first author's doctoral research [2] was accordingly to develop a dynamic model of a stretch of the River Cam, just downstream of Cambridge and, more to the point, just downstream of a suitably "exciting" input signal, i.e., the discharge from the Cambridge Sewage Works. The model had been chosen and constructed. It described the dynamic interaction between a measure of gross organic pollution (the concentration of biochemical oxygen demand; BOD) and a measure of the healthy status or otherwise of the river, its concentration of dissolved oxygen (DO). Data collection was a matter of sample retrieval in a pack of bottles transported by bicycle, followed by manual titration. Model calibration was implemented using the extended Kalman filter (EKF). The fact that the model failed to match both the observed DO and BOD behavior during the periods of "good" weather, the struggle to diagnose the failure, and then the much greater struggle to rectify its possible causes, came to be known as *model structure identification* [10]. Its procedure was facilitated by interpreting the temporal variations in the recursive estimates of the model's parameters, conventionally, the supposed "constants", i.e., coefficients $\boldsymbol{\alpha}$.

Such failure was not the intention. For we all want our models to approximate the real thing in some demonstrable manner, for reasons of scientific enquiry or for some other purpose, such as making a prediction in association with determining a course of future actions of environmental stewardship. Indeed, the extent to which the model can be reconciled with past observed behavior is a measure of the extent to which we might judge the primary science to be provisionally corroborated. At the same time, the map of uncertainty attaching to the posterior model's conceptual structure and its constituent mechanisms, after this process of system identification, will have significant consequences for any exercises in forecasting and investigating possible future patterns of behavior [4, 6].

In modern times, in the present decade (2001–2010), the US National Science Foundation (NSF) has committed significant amounts of financial support to the initiation of Environmental Observatories (EOs), in the Ocean Sciences, Ecology, and Hydrology-*cum*-Environmental Engineering (see, for example, [12]). Together with a commitment to radical enhancement of the environmental cyber-infrastructure, the EOs extend the promise of vast streams of high-volume, high-quality (HVHQ) data regarding the behavior of environmental systems. This, in principle, should be "transformative" for model structure identification.

To appreciate the contemporary significance of model structure identification, we first introduce the bare bones of some philosophy and some shorthand definitions of the problems being addressed. They are not matters merely of model calibration. This establishes the ground on which to express the challenge of model structure identification, as seen today. Our chapter will not enter into any algorithmic detail, for this has been recorded and reported fully elsewhere [17–19, 34, 35]. Instead, we focus on a scheme of scientific visualization, inspired by the software of biomolecular graphics. We judge its eventual mechanization and implementation to be one of the key goals in realizing more effective procedures for addressing the challenge of model structure identification. And it *will* continue to be an enduring challenge, we argue (see also [12]).

In the decades hitherto, the significance of model structure identification has consistently been underplayed, even trivialized through its association with model calibration, itself viewed by some as somehow disreputable. A "good, physics-based model", after all, should not need calibration! If this chapter can help put a stop to such historic oversight, it will have served its purpose.

4.2 Model Structure Identification: Problem in Contemporary Context

4.2.1 Models and the Growth of Knowledge

We know that models can be used as succinct archives of knowledge, as instruments of prediction in support of making decisions and stewardship of the environment, or as devices for communicating scientific knowledge to a scientifically lay audience (and each such task may have different obligations for model evaluation; [8, 24]). But how, we must ask, might the development and application of models serve the purposes of basic scientific discovery and, therefore, the growth of knowledge?

In an article on interactive computing as a teaching aid, MacFarlane [22] presented a three-element characterization of knowledge. According to the American philosopher Lewis these three elements are (as reported by MacFarlane):

(E1) the given data;
(E2) a set of concepts; and
(E3) acts which interpret data in terms of concepts.

These three pillars, and their inter-relationships, will help to organize our thinking about the role of model structure identification in core discovery. It would be difficult to assert that any one of these pillars was supremely important. Yet elevation of (E3) to rank on a par with the status of (E1) and (E2) is significant.

Given Lewis's schema, we can see that the impact of NSF's EOs on the "*given data*" (E1) should be substantial and profound. Excellence in modeling can in any case not be achieved in the absence of first-class data for rigorous model testing and evaluation.

Just as profound, if not more so, will be the impact of the environmental cyber-infrastructure on mechanizing the "*set of concepts*" (E2) in computable form—although we should take care not to confuse the notion of a computational model entirely with the "set of concepts" or a theory. For models are a secondary science, in the sense of enabling organized assembly and encoding of the distilled knowledge emerging from the primary field sciences. But that distilled knowledge is not indisputable fact. It is a composite assembly of a host of constituent "atomistic" theoretical elements, each themselves reflecting individual hypotheses quarried from laboratory science or a particular field science, often crafted in disciplinary compartments without the benefit of the entire picture of the whole system necessarily in mind. The environmental systems we observe and study behave as indivisible wholes, so that a basic question becomes: when placed together in the organized structure of a computational model, which of the constituent hypotheses are adequate/inadequate, in terms of determining the performance of the whole; and how should the inadequate constituents be removed, modified, and re-introduced in more adequate form?

This too is model structure identification. The urgency of this matter can only but grow as mounts the number of constituent hypotheses upon which one wishes to draw (for a description of the real system's behavior).

What will be the implications of these profoundly important advances—in the sensing technologies of the EOs and in the environmental cyber-infrastructure—for Lewis's "*acts which interpret data in terms of concepts*" (E3)? What will be their implications, in other words, for system identification and for model structure identification? Indeed, how does this "interpretation" actually come about? How does one, for example, reconcile a large-scale geophysical model of global deglaciation with (reconstructed) relative sea level observations at 392 sites spanning a period of some 15,000 years [37]? More specifically, which constituents of the very large and very complex assembly of micro-scale theory is at fault when the model fails—as inevitably it does—to match the relatively macroscopic historical observations? "Interpretation" is a result of juggling with, and sifting through, a unique assortment of disparate facts and figures assembled by the individual, upon which some kind of order is eventually imposed. It is a subjective mental process. That process, moreover, is sore in need of some technical support, not least from all the innovations in computing over the past forty or so years, as the NSF now recognizes [27].

In short, that there will be significant developments in the technical support necessary for engaging the model in a meaningful interpretation of the data, is by no means assured. News of advances in computational capacity is abundant (witness [26]); news of advances in the technology of instrumentation and remote sensing is commonplace (witness [25]); news of the *increasing* capacity of the brain to juggle with disparate facts and concepts is non-existent. In this resides arguably the greatest of opportunities to flow from the EOs and the oncoming environmental cyber-infrastructure for the future of environmental modeling—in what has therefore been recorded as perhaps *the* core, grand challenge of environmental modeling: model structure identification [12].

4.2.2 In the Gap Between the Model and the "Truth of the Matter"

Let us assume the scope of model building can be succinctly defined by the triplet of the observed inputs ($\boldsymbol{u}$), model ($\boldsymbol{M}$), and observed outputs ($\boldsymbol{y}$), and that the attaching tasks are those of the mathematical textbook: given two out of the three unknowns, find the third. The three principal computational and algorithmic questions are thus:

(Q1) Given $\boldsymbol{u}$ and $\boldsymbol{y}$, find $\boldsymbol{M}$. This we shall refer to as system identification, i.e., largely pillar (E3) above in Lewis's pragmatic school of thought on the growth of knowledge, under which falls the task of choosing the contents of $\boldsymbol{u}$ and $\boldsymbol{y}$ so as to maximize the "identifiability" of $\boldsymbol{M}$, i.e., the design of experiments and sensor networks.

(Q2) Given $\boldsymbol{M}$ and $\boldsymbol{u}$, find $\boldsymbol{y}$. The problems of forecasting, and scenario and foresight generation.

(Q3) Given $\boldsymbol{M}$ and desired, feared, and/or threatened $\boldsymbol{y}$, find $\boldsymbol{u}$. The problems of control, management, decision-support, and policy formulation.

From (Q1) emerges a fourth question, which is, of course:

(Q4) How well does $\boldsymbol{M}$ approximate the real thing, and what are we going to do in respect of the other two questions ((Q2) and (Q3)) given there is never such a match, i.e., that there is more or less substantial uncertainty to be dealt with?

Many models ($\boldsymbol{M}$) of the behavior of environmental systems can be defined according to the following representation of the state variable dynamics of classical mechanics,

$$dx(t)/dt = \boldsymbol{f}\{\boldsymbol{x}, \boldsymbol{u}, \boldsymbol{\alpha}; t\} + \boldsymbol{\xi}(t) \tag{4.1a}$$

with observed outputs being defined as follows,

$$\boldsymbol{y}(t) = \boldsymbol{h}\{\boldsymbol{x}, \boldsymbol{\alpha}; t\} + \boldsymbol{\eta}(t) \tag{4.1b}$$

in which $\boldsymbol{f}$ and $\boldsymbol{h}$ are vectors of nonlinear functions, $\boldsymbol{u}$, $\boldsymbol{x}$, and $\boldsymbol{y}$ are the input, state, and output vectors, respectively, $\boldsymbol{\alpha}$ is the vector of model parameters, $\boldsymbol{\xi}$ and $\boldsymbol{\eta}$ are notional representations respectively of those attributes of behavior and output observation that are not to be included in the model in specific form, and t is continuous time. Should it be necessary, spatial variability of the system's state can be assumed to be accounted for by, for example, the use of several state variables of the same attribute of interest at the several defined locations.

For any system, the choices of $[\boldsymbol{u}, \boldsymbol{y}]$ determine the (observable) *external description* of its behavior. Those aspects of the science base mobilized into the computational encoding of the model—the hypothetical mechanisms considered significant to the manner in which input, causative disturbances ($\boldsymbol{u}$) are transcribed into output effects ($\boldsymbol{y}$)—are signaled by the choices of $[\boldsymbol{f}, \boldsymbol{h}; \boldsymbol{x}, \boldsymbol{\alpha}]$. In short, the *structure* of the model ($\boldsymbol{M}$) is most succinctly conveyed in terms of $[\boldsymbol{f}, \boldsymbol{h}]$, which denote the logical inter-connections among $\boldsymbol{u}$, $\boldsymbol{x}$, and $\boldsymbol{y}$, while $\boldsymbol{\alpha}$ signifies parameterization of the particular mathematical expressions of all the hypothetical mechanisms believed to underpin these interactions. We may call $[\boldsymbol{x}, \boldsymbol{\alpha}]$ the *internal description* of the system's behavior, as the complement of $[\boldsymbol{u}, \boldsymbol{y}]$.

If the "truth" of the matter could be represented in a model, which it cannot (by definition), the structure of the system's behavior could be supposed to be of infinitely high order. Let us denote this as $[\boldsymbol{f}^{\infty}, \boldsymbol{h}^{\infty}]$. We, with our models in the realm of the finite, $[\boldsymbol{f}^{0}, \boldsymbol{h}^{0}]$ say, work on a much more macroscopic plane. Our models have a crude resolving power, even for those of a very high order ($+N$), with structure $[\boldsymbol{f}^{+N}, \boldsymbol{h}^{+N}]$. What exactly, however, should we suppose is the content of the *gap* between $[\boldsymbol{f}^{0}, \boldsymbol{h}^{0}]$ and $[\boldsymbol{f}^{\infty}, \boldsymbol{h}^{\infty}]$, the structural error and structural uncertainty in the model, that is? For the anomalies observed during the period of good weather over the River Cam in 1972 derived from within that gap.

Put simply, this inadequacy, or error and uncertainty of approximation, may enter into (4.1a), (4.1b) through $\boldsymbol{\alpha}$, $\boldsymbol{\xi}$, and $\boldsymbol{\eta}$, although these points of entry differ in their interpretation and significance. The principal distinction is between $\boldsymbol{\alpha}$, embedded within the choices for $[\boldsymbol{x}, \boldsymbol{\alpha}, \boldsymbol{f}, \boldsymbol{h}]$, which signify that which we presume (or wish) to know of the system's behavior, relative to the purpose of the model—to be denoted as {presumed known}—and $[\boldsymbol{\xi}, \boldsymbol{\eta}]$, which acknowledge in some form that which falls outside the scope of what we believe we know, to be denoted the {acknowledged unknown}. Much, of course, must be subsumed under the latter, that is, under the definitions of $\boldsymbol{\xi}$ and $\boldsymbol{\eta}$. We may have chosen to exclude from the model some of that which was known beforehand, but which was judged not to be significant. There may be features for which there are no clear hypotheses (and therefore no clear mathematical expressions), other than that these may in part be stochastic processes with presumably quantifiable statistical characteristics. There may be yet other features of conceivable relevance, but of which we are simply ignorant. And, as is most familiar, there may be factors affecting the processes of observation such that we are unable to have uncorrupted, perfect access to knowledge of the values of the inputs, states, or outputs.

Essentially, the model is all that we have to work with to cope with the gap between $[\boldsymbol{f}^0, \boldsymbol{h}^0]$ and $[\boldsymbol{f}^\infty, \boldsymbol{h}^\infty]$, where this gap will constitute the whole of the {acknowledged unknown} *and* of the {presumed known} being wrongly presumed known. It is all we have to apprehend something of significance, to our understanding and actions, within the gap. In particular, in this process of apprehension, the model—being the vessel containing all the relevant hypothetical knowledge from the science base—is to be pitted against all the relevant experience of the observed past behavior.

Given that there will always be such a gap between the model ($\boldsymbol{M}$) and the "truth of the matter", we must ask:

> Can we identify the nature of what lies within it, through model structure identification?
> Can we estimate the magnitude of the inevitable error and uncertainty still remaining (after such identification)?
> How might we best approximate the consequences of this structural error and uncertainty in accounting for its propagation in predictions of future behavior?

In this sense of accounting for uncertainty, questions (Q1) (system identification), (Q2) (forecasting), and (Q3) (policy formation) of our foregoing threesome of textbook problems, are intimately inter-related [4, 6]. In particular, the same line of algorithmic framing of this accountancy, i.e., recursive estimation, is presently being used to make in-roads into quantifying structural error and uncertainty in a model [20].

Given NSF's Environmental Observatory initiatives, which are designed to provide access to unprecedented streams of data $[\boldsymbol{u}, \boldsymbol{y}]$, there is arguably no greater challenge than that of responding to the novelty unleashed thereby in those "acts" of Lewis, "which interpret data in terms of concepts", i.e., system identification, at the core of which resides model structure identification. This is model calibration writ immensely more richly.

4.2.3 Neither Model Calibration Nor Trivial

Because model calibration is so familiar and routine to implement (if *not* to succeed)—and because the richer, more philosophical facets of system identification can so often be obscured by the straightforward pragmatism of model calibration—there is considerable intricacy and deeper subtlety now to be conveyed.

For calibration, the structure of the model ($[\boldsymbol{f}^0, \boldsymbol{h}^0]$) is routinely presumed known. Whatever resides in the gap between this structure of the model $\boldsymbol{M}$ and the truth of the matter ($[\boldsymbol{f}^\infty, \boldsymbol{h}^\infty]$), the algorithm of calibration has not been designed to assist in seeking it out. It is not usually the intent of the calibration exercise to do so, anyway. For calibration, the parameters ($\boldsymbol{\alpha}$) are routinely presumed everywhere invariant in three-dimensional space ($\boldsymbol{s}$) and time (t), i.e., they are random variables but invariant in truth. They are, in any case, *not* usually acknowledged as $\boldsymbol{\alpha}(\boldsymbol{s}, t)$, let alone estimated as such. In a Popperian sense, these presumptions about the model's structure and its parameters are all as they should be. They are bold conjectures, made all the more readily falsifiable by their very boldness, *if* there were a more deliberate and determined over-arching intent to employ calibration and its routine presumptions to root out and explain any errors and uncertainties in the model's structure—which, in general, there is not.

Our model has been cast at some level of inevitably macroscopic resolving power (0), i.e., $[\boldsymbol{f}^0, \boldsymbol{h}^0]$, relative to the "real thing" resolved down to some infinitesimally fine degree, i.e., $[\boldsymbol{f}^\infty, \boldsymbol{h}^\infty]$. The parameters of the model, denoted more precisely as $\boldsymbol{\alpha}^0$, which we would very much prefer to understand as (temporally) *invariant* quantities, must in practice subsume a bundle of states and parameters $[\boldsymbol{x}^q, \boldsymbol{\alpha}^q]$ that would be present in a more refined model, had it been possible or desirable to cast the model at that more refined level (q) of description. Since state variables are by definition quantities that *vary* with time, things contained in the gap between $[\boldsymbol{f}^0, \boldsymbol{h}^0]$ and $[\boldsymbol{f}^\infty, \boldsymbol{h}^\infty]$, i.e., within the structural error/uncertainty, imply that, given an invariant form for $[\boldsymbol{f}^0, \boldsymbol{h}^0]$, the model's parameters must necessarily, and in principle, be capable of *variation with time* t, if the complex of $[\boldsymbol{f}^0, \boldsymbol{h}^0, \boldsymbol{\alpha}^0]$ is required to mimic the behavior of $[\boldsymbol{f}^\infty, \boldsymbol{h}^\infty]$.

One only has to conceive of a chemical kinetic rate *constant*, α^0—presumed to account for a biochemical transformation enacted by a population of bacteria, $x^1(t)$, whose population numbers *change* with time, according to a set of parameters of growth and death ($\boldsymbol{\alpha}^1$)—to appreciate the significance of this. Indeed, even this collection of state and parameters at the more refined resolving power (of $+1$), will be well known to be "in truth" a function of yet more refined states, such as intracellular concentrations of enzymes ($\boldsymbol{x}^q$) and the parameters ($\boldsymbol{\alpha}^q$) appearing in the web of cellular biochemistry considered to be taking place at a yet more refined level of description (q) (see, for example, [1]). In this sense, then, the occurrence of apparent temporal change in the structure of the model, manifest in terms of $\boldsymbol{\alpha}^0(t)$, is a universal possibility. We might better conceive of the nature of our model's parameters, $\boldsymbol{\alpha}^0$, therefore, *not* as random variables, i.e., as uncertain *constants*, but as stochastic processes, whose variation through time is largely systematic—and capable, in principle, of interpretation—but also random and accordingly ascribable only to the actions of pure chance [9].

When the River Cam studies were begun in 1970, it was an unquestioned commonplace to talk of a BOD decay-rate *constant*. In the model (4.1a), this would simply account for a host of bacterial species metabolizing (and thereby degrading) a multitude of complex organic substances—in truth, something approaching $[\boldsymbol{f}^{+N}, \boldsymbol{h}^{+N}; \boldsymbol{x}^{+N}(t), \boldsymbol{\alpha}^{+N}]$, yet described as though a single invariant α^0. The hope was, presumably, that all of this non-linearity would in the end obey some "law of large systems" and add up to nothing more than a rudimentary linearity of chemical kinetics.

4.2.4 *It Matters: Both Philosophically and Pragmatically*

The previous decade of the 1960s had been a time of "youthful exuberance" in the development of environmental systems simulation. With this technocratic optimism, fueled by man's landing on the moon, we entered the 1970s. For calibration, the prior, rudimentary practice of trial and error—of trying out different values for the model's parameters ($\boldsymbol{\alpha}$) until the "curve" of the estimated outputs would match satisfactorily (in some sense) the "dots" of the observed output data—was to be supplanted by the more systematic, objective procedures of mathematical programming, optimization, mathematical filtering theory, and the like. The modernism of "automatic calibration", detached from subjective manipulation, was to supercede the craft-skill of "calibration by hand".[1] To the consternation of all, automatic calibration turned out to be supremely successful in revealing the very considerable difficulty in locating the *uniquely best* set of parameter estimates. And this is what we know (only too well) as the problem of a lack of model identifiability, hence too all the artful ways of trying to constrain automated calibration routines not to deliver nonsensical parameter estimates.

At the time, one might have argued that this problem arose from inappropriate choices for the contents and forms of $\boldsymbol{u}$ and $\boldsymbol{y}$, including—of great concern in control theory—a choice of $\boldsymbol{u}$ that is not "persistently exciting". The freedom of such choice remains remote when studying the behavior of environmental systems *in situ*. Choosing to observe $\boldsymbol{u}$ just downstream of Cambridge Sewage Works was as good as that part of the experimental design was going to be, in 1972. Moreover, the abundant lack of model identifiability was manifest even where there were reasonable approximations of naturally perturbing signals, i.e., precipitation events in hydrological modeling. In the early 1970s, input perturbation (experimental) design was itself being studied as a subject of optimization, to serve the needs of the then burgeoning schemes of adaptive, real-time (on-line) control in engineered systems. Today, wherever the nonlinear Michelis-Menten or Monod kinetics for the growth of microbial organisms appears in a model—and it is ubiquitous, in wastewater treatment [15, 31, 36], river water quality and lake ecology [14, 29] or oceanography [33]—some detailed account of the problem of (a lack of) model identifiability is given.

The consequences of such nonlinear kinetics, albeit not necessarily realized as Monod kinetics, were very probably present in the observed behavior of the Cam in 1972. What might lie beneath their macroscopic approximations exemplifies the foregoing conceptual discussion of what might reside in the gap between $[\boldsymbol{f}^0, \boldsymbol{h}^0]$ and $[\boldsymbol{f}^\infty, \boldsymbol{h}^\infty]$. And the consequences of similar nonlinear biochemical kinetics will be just as apparent and dominant in the case study to follow below.

Attempting to overcome a lack of model identifiability matters philosophically—in the growth of secure knowledge—because this implies a determined attempt at expunging ambiguity in competing interpretations of Lewis's "data" (E1) and at reducing to a singularity an otherwise plurality in his plausible "sets of concepts" (E2).

[1] It did not, as it happens. The two co-exist fruitfully today, notwithstanding the supposed academic inferiority of the latter.

It matters in practice to the public too, since Mooney fully intends scientifically-lay members thereof to read his (2007) popular account of "*Storm World—Hurricanes, Politics, and the Battle Over Global Warming*" [23]. His account (literally) personifies what we consider the challenge of model structure identification (Lewis's (E3)).

4.3 Scientific Visualization: Towards a Contemporary "Solution"

What resides in the gap between the model and the truth of the matter has two parts to it: error in the {presumed known} and uncertainty in the {acknowledged unknown}. Yet in the representation of (4.1a), (4.1b), the burden of discovery is tilted towards just the former, as revealed through temporal variability in estimates of what we must henceforth understand herein as the (conventional) model parameters ($\boldsymbol{\alpha}$). But little can be discerned of the latter, i.e., the presence/absence of systematic, non-random features appearing in the {acknowledged unknown}.

4.3.1 The Algorithm: Special Role of Innovations Representation

We can re-phrase our model in the naturally recursive format of what is called an innovations representation of the system's behavior, as follows [5, 21, 34, 35]:

$$d\boldsymbol{x}(t|t_{k-1})/dt = \boldsymbol{f}\{\boldsymbol{x}(t|t_{k-1}), \boldsymbol{u}(t), \boldsymbol{\alpha}\} + \boldsymbol{K}\boldsymbol{\epsilon}(t|t_{k-1}), \tag{4.2a}$$

$$\boldsymbol{y}(t_k) = \boldsymbol{h}\{\boldsymbol{x}(t_k|t_{k-1}), \boldsymbol{\alpha}\} + \boldsymbol{\epsilon}(t_k|t_{k-1}). \tag{4.2b}$$

Here, attention has been restricted merely to the conventional intervals of time passing from one observing instant t_{k-1} to the next, t_k. In spite of the formalities, what is of importance will prove to be of conceptual, as much as algorithmic, significance. Thus, formally, the argument $(t|t_{k-1})$ signals a predicted value of the associated quantity at some (future) time t utilizing the model and all observed information, in particular, in respect of the observed output $\boldsymbol{y}$, up to and including that available at the most recent sampling instant, t_{k-1}. $\boldsymbol{\epsilon}(t_k|t_{k-1})$ is the innovation, i.e., the mismatch between the predicted and observed values of the output at the next sampling instant in discrete time, t_k, in (4.2b); $\boldsymbol{\epsilon}(t|t_{k-1})$ in (4.2a) is the value of this quantity at times not coincident with the sampling instant. $\boldsymbol{K}$ is a weighting matrix and can be thought of as a device—a throttle or valve—for distributing the impacts of the innovations among the constituent representations of the various state variable dynamics, i.e., the representations $f_i(\cdot)$ for each state x_i [5].

$\boldsymbol{K}$ is central to the conceptual argument we now present.

First, comparing (4.1a), (4.1b) and (4.2a), (4.2b), it is evident that the cleavage in the one, between the {presumed known} and the {acknowledged unknown}, is as that between $[\boldsymbol{f}, \boldsymbol{h}]$ and $[\boldsymbol{\xi}, \boldsymbol{\eta}]$ (in (4.1a), (4.1b)), while in the other (4.2a), (4.2b) it is as that between $[\boldsymbol{f}, \boldsymbol{h}]$ and $[\boldsymbol{K}\boldsymbol{\epsilon}, \boldsymbol{\epsilon}]$.

Second, unlike ξ and η, $\epsilon(t_k|t_{k-1})$ is a computable quantity, being the mismatch (in (4.2b)) between the forecast value of the output and the observed output—albeit *not* the truth of the matter (hence some of the necessary approximation in our argument). In this way, ϵ is a kind of gauge of the foregoing gap between $[\boldsymbol{f}^0, \boldsymbol{h}^0]$ and $[\boldsymbol{f}^\infty, \boldsymbol{h}^\infty]$.

Third, just as we are familiar with the notion of reconciling the model's behavior with that observed of the real thing, in order to adjust and estimate the values of the model's *conventional* parameters ($\boldsymbol{\alpha}$), so this same process of reconciliation can be employed to reconstruct values for the elements ($\boldsymbol{\kappa}$, say) of the matrix $\boldsymbol{K}$. Indeed, the motivation for using the algorithmic form of (4.2a), (4.2b) was precisely this: to reconstruct $\boldsymbol{\kappa}$, instead of setting their values by prior assumption, in order to reconstruct estimates of $\boldsymbol{\alpha}$ [21]. If then the estimates of $\boldsymbol{\kappa}$ remain essentially the same as their prior, presumed values of 0.0, none of the empirical mismatches between the model and the data—in effect the innovations ϵ—are fed back into the predictions made of future behavior. Our predictive instrument is operating essentially on the basis of the {presumed known} alone. Should the elements of $\boldsymbol{\kappa}$ come to be reconstructed in the course of events as substantially non-zero, our predictive instrument is beginning to rely on the {acknowledged unknown}, and perhaps predominantly so.

Armed with these three conceptual interpretations of the formalities of (4.2a), (4.2b), we can proceed to our vital insight into the role of $\boldsymbol{K}$. Given the association of $[\boldsymbol{K}\epsilon, \epsilon]$ with the {acknowledged unknown}, $\boldsymbol{\kappa}$ can be attached to the parameterization of this entity in the same manner as $\boldsymbol{\alpha}$ has been the device for parameterizing the {presumed known}. We have thus the {presumed known ($\boldsymbol{\alpha}$)} and {acknowledged unknown ($\boldsymbol{\kappa}$)}, where now, given the computability of ϵ, we have an empirical means of both identifying the inadequacies of what has been included in the model and apprehending something of significance in what has been excluded from it. Further, as with all the individual elements of $\boldsymbol{\alpha}$, what transpires in reconstructing the individual elements of $\boldsymbol{\kappa}$ can provide pointers to the specific consequences of this "something of significance"—something of substance in guiding the search for the reasons underlying the deformation and/or change of structure.

Alternatively, think of this as follows. Randomness in the gap between $[\boldsymbol{f}^0, \boldsymbol{h}^0]$ and $[\boldsymbol{f}^\infty, \boldsymbol{h}^\infty]$ should cause flutter in ϵ, possibly even of high amplitude (for example, from the spurious corruption of observing errors). Persistent mismatches of significance in ϵ should eventually cause adaptation and change within $\boldsymbol{\alpha}$ and $\boldsymbol{\kappa}$, the one pointing to structural errors in the expression of the {presumed known}, the other to something of significance being apprehended in the {acknowledged unknown}.

4.3.2 The Visual Metaphor

Figuratively (and approximately), the structure of the model has been parameterized by the branches of the network of Fig. 4.1 [7]. The $[\alpha_{ij}]$ are included within

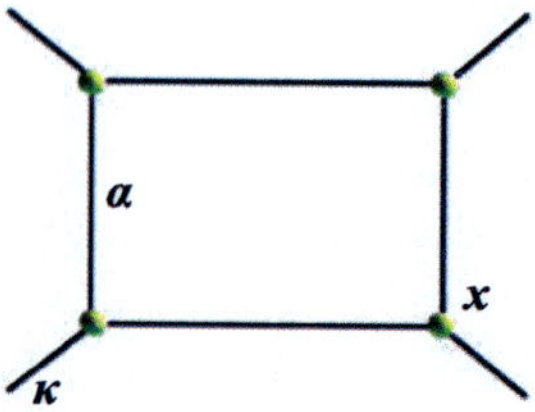

Fig. 4.1 Archetypal link-node network visualization of model structure. Nodes represent state variables ($\boldsymbol{x}$). Branches represent interactions among the state variables and are parameterized according to elements of vector $\boldsymbol{\alpha}$ for the {presumed known} and elements of vector $\boldsymbol{\kappa}$ for the {acknowledged unknown}

the basic rectangular frame connecting the states ($\boldsymbol{x}$) with each other, while the $[\kappa_{ij}]$ attach to the frame but point outwards symbolically into the space surrounding the structure of the frame, visually suggestive of probing the gap between the model $[\boldsymbol{f}^0, \boldsymbol{h}^0]$ and the truth of the matter $[\boldsymbol{f}^\infty, \boldsymbol{h}^\infty]$. Figuratively, oscillation and/or deformation of the branches of Fig. 4.1 should alert us to something being amiss in our understanding. The template of the model structure ($[\boldsymbol{f}^0, \boldsymbol{h}^0]$) (Lewis's (E2)) has caught on something of significance in the space of all possibilities around it, as it is being navigated through the given, observed behavior (E1) of the real system. In particular, the indications from engagement between the two—Lewis's (E3)—should direct our attention into specific avenues for discovery of the source of the anomalies, through the tagging devices of $\boldsymbol{\alpha}$ and $\boldsymbol{\kappa}$.

Reduced to its essence, the challenge of model structure identification obliges us in turn:

(S1) To demonstrate unequivocally (*a posteriori*) the inadequacy of the model's structure ($[\boldsymbol{f}^0, \boldsymbol{h}^0]$)—with yet the bold intent (*a priori*) not to succeed in this;
(S2) To diagnose the sources of this failure; and then
(S3) To reason our way through rectification of the causes of inadequacy and failure.

The metaphor of Fig. 4.1 was introduced as early as 1975. It was indispensable to grasping better the nature of the problem, hence to fathom what kind of recursive estimation algorithm—or (at that time) what better way of running the EKF—might enable the then dimly perceived steps (S1) through (S3) to be realized computationally. The EKF required many (arbitrary) assumptions about the nature of the variance-covariance properties of the system ($\boldsymbol{\xi}$) and (less so) the observation ($\boldsymbol{\eta}$) noise processes. It would not be until 1979, however, that Ljung would publish his seminal paper on the Recursive Prediction Error (RPE) algorithm for circumventing some of this arbitrariness of the EKF (for the purpose of parameter estimation). And it was not until the early 1990s that Stigter would implement the RPE for the purposes of model structure identification, starting with the (by then) most familiar test-bed of the 1972 Cam data [34, 35].

4.4 Case Study: Mechanization of the Visual Metaphor

Shortly after its publication, the book "*Environmental Foresight and Models: A Manifesto*" [7] came somehow to the attention of a bio-pharmaceutical scientist, whose interest lay in optimizing patient treatments for cancer, for example, of the liver (now expressed fully in [16]). From this chance encounter in 2002—so unexpected that it very narrowly escaped being pre-emptively deleted as junk email—has come the central burden of the present chapter: visualizing model structure identification (four decades on). The encounter has also come to epitomize the nature of inter-disciplinary research in applied systems analysis [12]. Indeed, what is now being written would probably not have emerged without yet another related serendipitous spark of insight across the endless inter-disciplinary gaps: the visual matching of a diagram such as Fig. 4.1 with images from a "Gallery of Biomolecular Simulations", for changes over time in the spatial structure of complex biological molecules (also see [12]).

4.4.1 Plethora of Numbers

Everything we wish to know about the performance of recursive estimation algorithms for the purposes of model structure identification, cast according to the innovations representation of (4.2a), (4.2b), is associated with prodigious volumes of numbers. There are numbers to record the propagation through the discretized time-space continuum of various high-dimensional estimation error variance-covariance (and other) matrices; and there are numbers about the like propagation of input, state, parameter, and output vectors. So why should visualization be highlighted in this manner for solving, in particular, the problems of model structure identification? Our response is this: because learning, discovery, and the forensic science of model structure identification in the growth of knowledge, are all about the highly condensed visual apprehension of the myriad diagnostic facets of the comparisons and juxtapositions entailed therein. This is especially the case in complex multivariable situations of HVHQ data and very high order models (or VHOMs).

We need hardly be reminded of the startling expansion over the past few decades in our capacity to simulate the behavior of systems, in theory, in ever more detail and completeness on the computer. Equally obvious is the substantial impact of the EOs and environmental cyber-infrastructure in expanding our technical capacity for observation, i.e., the volume and quality of data streams. By comparison, there has been no advance in the capacity of the human brain to juggle with a huge entanglement of computational estimates and observed facts—no advance in our capacities for lateral thinking, as we have already said—in order to reconcile bundles of obscurely and obliquely discerned anomalies, where data and theory seem to diverge, and not through the action of spurious chance occurrences. Imagine what is to be supported: reconstruction in a computational world of a complex assembly of experimental tests of multiple, constituent hypotheses; which hypotheses are of varying prior strengths and impossible to isolate clinically from the whole for examination one by one as singlets; and whose observable causes and consequences all interact with each other.

What is called for, above all, is a succinct visual representation of the structure of the model: along the lines of animating the branch-node network of Fig. 4.1, thereby achieving compression of the plethora of numbers through the rich visual complexity of color, movement, and animation of the model's structure. Visualization is necessary just as much for the "acts" (E3) of system identification as it is (already) for the "data" (E1) and for the "set of concepts" (E2). It may be as familiar as the computer graphics of games, films, and the scientific reconstruction of history and the imagination of future threats (for the television programs of the *History* and *National Geographic* channels, for instance).

The need has been long-standing: for the kind of software environment enabling rewiring of the constituents within the whole of the model, almost as quickly and easily as the serendipitous thought surfaces in the brain; and for support of the kinds of scientific visualization that will enable the serendipitous thought to occur sooner rather than later. Much of what is called for in wrestling with model structure identification, especially in respect of (S2) and (S3), is likely to depend on an essential element of such serendipity, something which by definition defies full automation and systematization in any form of environmental cyber-infrastructure.

Our task is to demonstrate how these issues may be addressed using a case study.

4.4.2 Visual Demonstration of Structural Inadequacy

Suppose we have access to HVHQ data for the nutrient-biological dynamics of a manipulated aquaculture pond, a posterior conceptual model of which is shown in Fig. 4.2 [18]. Figure 4.3 demonstrates the performance of this model. The result can be thought of as but a "snapshot" in the ongoing process of reconciling a succession of evolving candidate model structures with a portion of the HVHQ data. At this particular juncture, the most significant element of the posterior structure of Fig. 4.2 is its incorporation of an account of the dynamics of duckweed and alkalinity-related features. Both had been omitted from the immediately previous prior model structure, but had been emerging from the joint experience of modeling and working with the field system as prime candidates for inclusion in the next (trial) model.

When reconciliation of the prior candidate model structure ($\boldsymbol{M}_{prior}$) with the field data was attempted—*en route* subsequently to the posterior structure ($\boldsymbol{M}_{posterior}$) of Fig. 4.2—that act (*sensu* Lewis) yielded the recursively generated trajectories of parameter estimates of Figs. 4.4 and 4.5. These attach respectively to the {presumed known ($\boldsymbol{\alpha}$)} and {acknowledged unknown ($\boldsymbol{\kappa}$)} divisions of the relevant (prior) knowledge base. The estimates derive from a Recursive Prediction Error (RPE) algorithm, but with the specific modification [18] of being cast in the parameter space of $\boldsymbol{\alpha}$ and $\boldsymbol{\kappa}$, where all the elements of $\boldsymbol{\alpha}$ and $\boldsymbol{\kappa}$ can be treated as stochastic processes represented by generalized random walk (GRW) models [38, 40, 41]. In respect

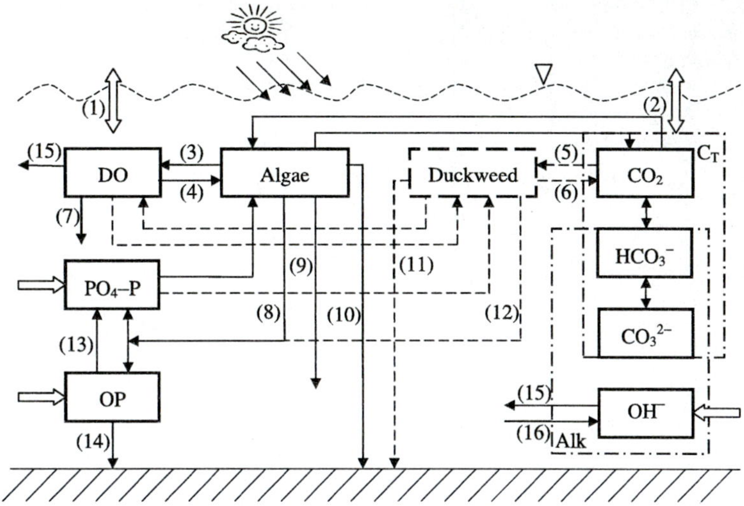

Fig. 4.2 Typical block diagram for an *a posteriori* model structure ($\boldsymbol{M}_{posterior}$) of nutrient, algal, and duckweed dynamics in a manipulated aquaculture pond: blocks represent state variables ($\boldsymbol{x}$); model parameters ($\boldsymbol{\alpha}$) will typically be associated with the mathematical expressions describing interactions among the state variables (*lines/arrows* in the diagram) (originally as in [18])

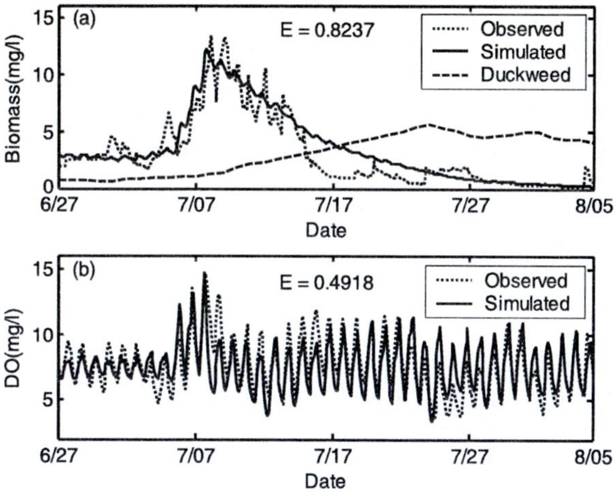

Fig. 4.3 Match of behavior of posterior model structure ($\boldsymbol{M}_{posterior}$) with field observations of state variables ($\boldsymbol{x}$): (**a**) algal biomass (chlorophyl-*a*) concentration and (**b**) dissolved oxygen concentration (DO). The reconstructed (unobserved) state variable for duckweed biomass is shown as the *dashed line* in (**a**) (originally as in [18])

of identifying the successes and failures of the various components of the model's structure, what is of special interest is bound up with the temporal variability in the estimates of $\boldsymbol{\alpha}$ (Fig. 4.4) and $\boldsymbol{\kappa}$ (Fig. 4.5).

In the following, key is the mechanizing of the visual aspect of the numerical results, not their interpretation with respect to the physics, chemistry, and biology of the given case study (more detailed discussion of which can be found in [18]). Also to be borne in mind is the fact that neither the field work nor these studies in modeling were set up and implemented *a priori* with the benefit of the hindsight now being

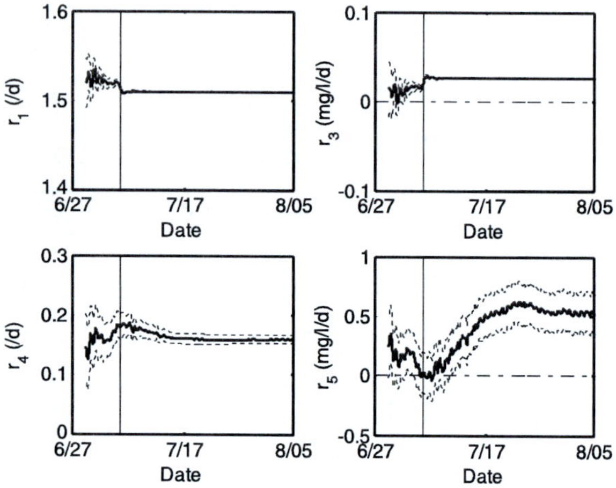

Fig. 4.4 Estimates from a Recursive Prediction Error (RPE) algorithm for parameters (α) logically attaching to the {presumed known} of the prior model structure (M_{prior}) (originally as in [18])

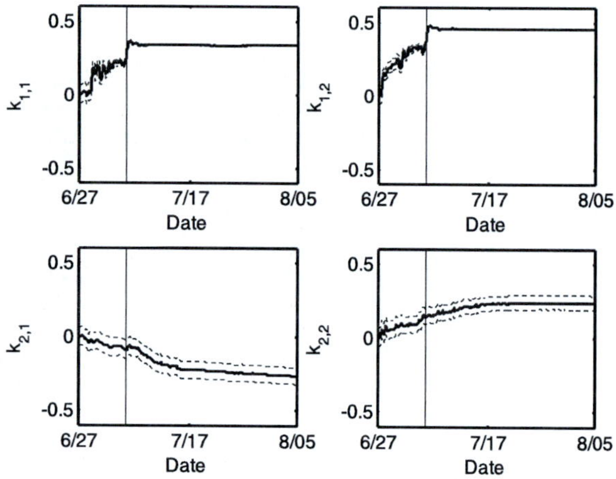

Fig. 4.5 Estimates from a Recursive Prediction Error (RPE) algorithm for parameters (κ) logically attaching to the {acknowledged unknown} of the prior model structure (M_{prior}) (originally as in [18])

applied to this presentation of them. Fertilization of the pond was intended to excite algal growth (which it did), not a duckweed bloom, which it also achieved—seen by all who visited the pond, but not formally observed through any scientifically designed instrument. We note merely in passing, therefore, that the relative conceptual "insecurity" of any conjectures about the behavior of the duckweed is signaled by the dashed box for its biomass and dashed lines for all that might relate to this state variable in the posterior candidate model structure of Fig. 4.2.

4.4.2.1 The {Presumed Known}

The prior structure (M_{prior}) has fourteen elements in the vector α, with the following presumed strengths and weaknesses:

(PK1) Parameters (ten in number): so boldly known as to be constant, certain, hence entirely absent from Fig. 4.4;
(PK2) Parameters r_1, r_3, and r_4: presumed invariant with time, but uncertain;
(PK3) Parameter r_5: presumed variable with time, uncertain, hence behaving as $r_5(t)$.

Inspecting the trajectories of the three elements of (PK2) in Fig. 4.4, all converge to invariant values. In the case of r_1, but not r_3 or r_4, its final estimated value turns out to be close to its initial value, as chosen from the literature. None of these three trajectories is greatly affected by the excitation of the pond through the input of a substantial quantity of fertilizer, whose timing is marked by the vertical bar in Figs. 4.4 and 4.5 and whose purpose was to provoke information-rich responses in the system's behavior. Parameter r_5 of (PK3), permitted to be $r_5(t)$ within the presumed known ($\boldsymbol{\alpha}$) of the model's structure, is included in the dynamics (f_2) of the state variable (x_2) for DO concentration. It purports to represent the lumped consequences of sources of DO in the pond water other than those articulated through the markedly bolder conjectures attaching to (PK1), which are here explicit in the express mathematical forms for the processes of re-aeration and algal photosynthesis. Figure 4.4 shows that r_5 does in fact vary over time and is almost always positively valued. Old habits die hard. For the same kind of parametric ($\boldsymbol{\alpha}$) device was incorporated into revised model structures conjured up to try and pinpoint the anomalies of the 1972 Cam study (and then explain them).

Three decades on (in 2000), the real-time monitoring capabilities of the University of Georgia's Environmental Process Control Laboratory (EPCL; see also [12]) were forearmed for observing the response of algae to the fertilization in Fig. 4.3, but not for anticipating the eventually impressive growth of the duckweed *Lemma*. In 1972, DO and BOD had been observed in the Cam, but not algae. In the 2000 aquaculture pond manipulation, a host of variables were monitored, including algae, but not the duckweed. Life *in situ* is always more complex and expansive than the reach of our experiments and the evolving capacity of our observing systems.[2]

The observing devices themselves may also fit awkwardly into the mathematical formality of (4.1b) of the model. The BOD measurement of 1972 essentially sought to mimic *in situ* microbial behavior in a bottled test of a water sample taken back to the laboratory [3]. The EPCL in 2000 was conversely taken to the shoreline of the aquaculture pond. It withdrew its sample of water from the pond continuously, to expose that flow to various automated sensing devices housed within this mobile laboratory. So significant was the sample flow that independent supplementary measurements of the vertical profile of temperature in the pond strongly suggest destratification of the pond's waters when the EPCL was switched on. The observing device, often taken for granted under the formality of (4.1b), was arguably changing

[2]The EPCL, a platform for real-time monitoring of water quality in a variety of aquatic environments, was operated from 1997 through 2008. All the data bases gathered with it are archived in the Georgia Watershed Information System (GWIS) and are publicly and freely available for downloading and analysis at www.georgiawis.org.

the behavior of the observed entity, generally the exclusive focus of all attention in (4.1a)—and here at a somewhat larger scale than that of quantum physics.

4.4.2.2 The {Acknowledged Unknown}

With two state variables ($\boldsymbol{x}$) in $\boldsymbol{M}_{prior}$ and two observed outputs ($\boldsymbol{y}$), matrix $\boldsymbol{K}$ has four elements. Each, within the {acknowledged unknown ($\boldsymbol{\kappa}$)}, is:

(AU1) Presumed constant but uncertain.

The recursively estimated trajectories of all four (in Fig. 4.5) are perturbed by the observed responses to the act of fertilization of the pond, and more so for $k_{1,1}(t_k)$ and $k_{1,2}(t_k)$ than for $k_{2,1}(t_k)$ and $k_{2,2}(t_k)$. In the first row of $\boldsymbol{K}$, the former pair of elements are the "throttles" regulating the injection of innovations errors into the dynamics of the first model state, algal biomass. The second pair act likewise for the second state, DO. All four elements progress towards values that are estimated to be significantly non-zero.

Crudely speaking, it would appear that much of significance in accounting for the pond's behavior, relative to this candidate prior model structure ($[\boldsymbol{f}^i, \boldsymbol{h}^i]$), must be present in the {acknowledged unknown}. Not readily forthcoming, however, is what might be discerned more incisively from these results regarding the possible nature of the important features omitted from the model and cast under the domain of the {acknowledged unknown}. The results do not point clearly in the direction of the influence of the duckweed, the primary suspected missing feature in the prior candidate model structure.

4.4.2.3 Coping With Complexity and Bewilderment

The evidence of Figs. 4.4 and 4.5 is only a part—yet an important part—of what must be fed into the expression of Fig. 4.2 from diagnosis of the failure of the prior model. Crucially, the availability of such kinds of evidence on parametric variations (or their invariance) should accelerate arrival of the moment at which the serendipitous thought occurs. It is as though the structure underlying the behavior captured in the data might be as that encapsulated broadly in the posterior structure ($[\boldsymbol{f}^{i+1}, \boldsymbol{h}^{i+1}]$), but demonstrably not so in respect of the prior structure ($[\boldsymbol{f}^i, \boldsymbol{h}^i]$). A number of constituent members of the latter—hypotheses, embedded in which are parameters—are shown as failing in the attempt to reconcile that prior structure with the data. This is the outcome of step (S1) of model structure identification.

To assist, even accelerate, the laborious process of proceeding from an obviously inadequate prior model structure ($\boldsymbol{M}_{prior}$) to a less inadequate posterior ($\boldsymbol{M}_{posterior}$), what we should need, *in general*, is something such as that of Fig. 4.6. This substantial advance upon the rudimentary structure of Fig. 4.1 was inspired by what the June (2007) issue of *The MathWorks News & Notes* called the "world's most complex dynamic systems". In fact, Fig. 4.6 is but a small segment of the structure

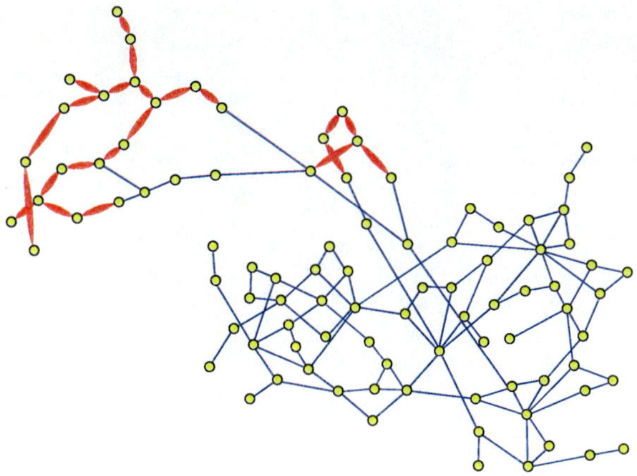

Fig. 4.6 Link-node network diagram representing a model's structure, based (in part) on the schematic representation of a biological/pharmaceutical system: state variables ($\boldsymbol{x}$) are denoted as *yellow* nodes in this structure, while model parameters ($\boldsymbol{\alpha}$; and, in principle, $\boldsymbol{\kappa}$) are associated with the *blue* (or *red*) branches connecting the nodes to each other. *Blue* branches signal those facets (constituent hypotheses) of the model structure associated with model parameters found to be invariant and, therefore, robust and reliable in the face of the given test against field observations. Conversely, red branches indicate significant, non-random variability in what are presumed to be (ideally) constants and, accordingly, failure of the model structure, *in specific*, constituent parts

displayed in the article. While the article trumpets its subject as the "most complex", it is merely a part of one of those kinds of biological *sub*-systems of molecules inside the cells of the organs and organisms that make up the kinds of ecosystems underpinning the behavior of entities such as a BOD decay-rate "constant"!

Coloring of the branches in this visualization of the essential concept of "model structure" is quite deliberate: blue for invariant parameter estimates and therefore provisionally secure constituent parameters; red for deformation over time, as the given constituent members (hypotheses) of the structure buckle (fail). We know in principle how the RPE algorithm could generate these colors and their changes with time, which obviously would require some form of animated scientific visualization.

4.4.3 Animating Flexure and Collapse of Model Structure in Lewis's Acts of System Identification

Figure 4.6 conveys an "artist's impression" of a model's structure, representing a prior candidate model structure ($\boldsymbol{M}_{prior}$) with its apparent failings, as a step *en route* to that of an improved posterior structure ($\boldsymbol{M}_{posterior}$), all in the overall process of model structure identification. Transformed into the three-dimensional representation of Fig. 4.7, we can acquire a yet better visualization of a "real" structure

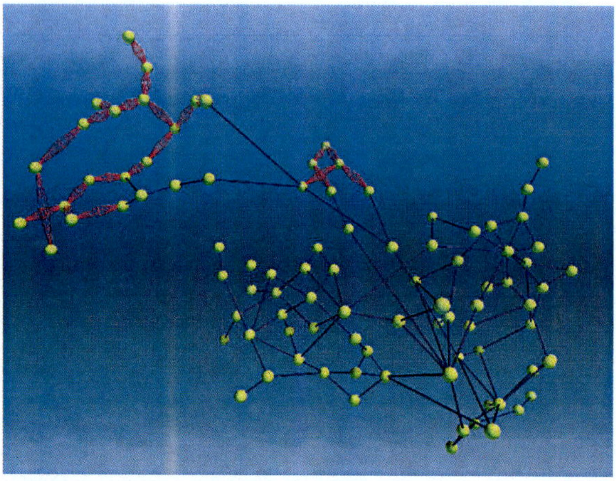

Fig. 4.7 Towards model structure identification: three-dimensional representation of the model structure (previously depicted merely in two dimensions in Fig. 4.6)

(and certainly something indicative of the schematized 3-D structure of a biological molecule). Now (in Fig. 4.7) the insecure members of the model's structure are visualized as the more "diffuse", "multiple" spans of red branches connecting state-variable node to state-variable node. Yet Fig. 4.7, like Fig. 4.6 before it, remains a static snapshot: of either an instant in time t_k; or (but not intended here) the "average" outcome of some attempt at automatic calibration of the model for some entire block of time-series data.

Animation would permit changes of color over time. And to color could be added the dimensions of width and weight of line and flexure, deformation, and oscillation in these branches pinning together the nodes (state variables) of the structure. This, however, can only be inadequately shown on a page of text, in the format of Fig. 4.8. As the current candidate structure of the model is tested and stressed under the conditions of step (S1), the visualization of Fig. 4.8(a)—its nodes fixed in their positions on the screen—will reveal changes of color, changes in shape, and changes in the breadth ("dispersion") of the flexing branches (parameters) of the constituent hypotheses implied in the node-node links. What began as a straight blue branch may evolve into a distorted, flexed red branch, as the mean and variance of the reconstructed parameter estimate (α in Fig. 4.4; κ in Fig. 4.5) increases/decreases and wanes/waxes respectively. At some point (t_k), the analyst visually apprehends failure of a sufficient magnitude and freezes the animated visualization, i.e., stops the RPE algorithm, at Fig. 4.8(a). S/he would then have the facility to rotate the 3-D structure to the perspective of Fig. 4.8(b), and from there extract just the buckled (red) portion of the structure for closer inspection.

In this substantially more complex model of, say, the behavior of the aquaculture pond, did the set of concepts (Lewis's (E2)) fail to be reconciled with the data (E1) because of the characterization of the fluid mechanics of pond stratification, or because of its microbial ecology? If the latter, was this because of the algae or the duckweed or something else—and so on? Failure of the prior model structure has been apprehended, *not* in some aggregate whole sense, of a lumped, non-specific, statistical summary of a "failed $\boldsymbol{M}_{prior}$", but in the specifically targeted sense of α_d

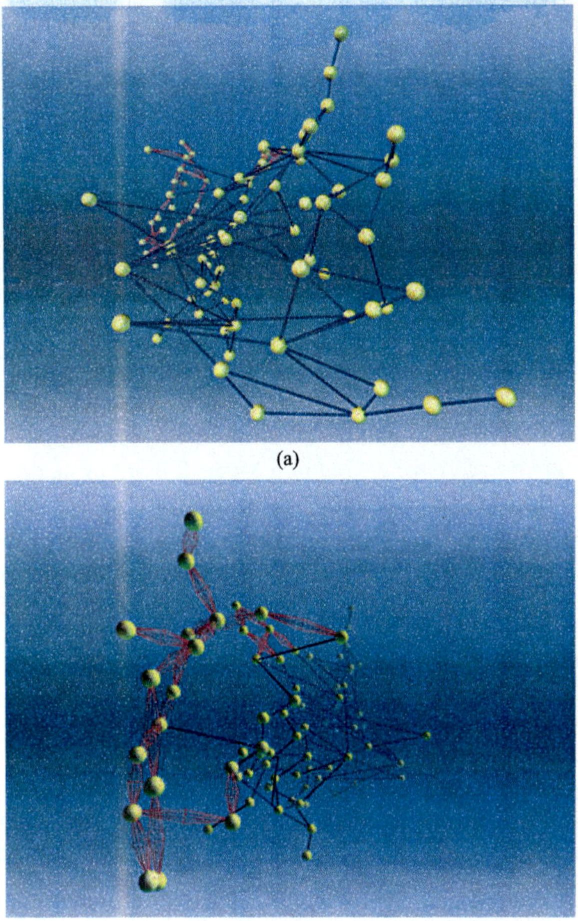

(a)

(b)

Fig. 4.8 Towards model structure identification through animation of flexure and collapse of model structure: (**a**) frozen frame of the animation as the analyst first detects a *red* web of faulty behavior to the rear of the three-dimensional model structure, as the model is in the process of being reconciled with a recorded span of field data; (**b**) same frozen frame as (**a**) but rotated in the three-dimensional space of the visualization of the model's structure in order to reveal more clearly the failing constituents (hypotheses) of the model's structure

and/or α_e and/or α_f of the red, buckled sub-segment of Fig. 4.8(b). In the (idealized) procedure of model structure identification, thus would the analyst move from step (S1) to the diagnoses of step (S2).

4.4.4 Procedure

It is important to keep in mind the fact that the performance of an algorithm such as the RPE is primed—by the choices made about the various initial estimates of quantities and their attaching variance-covariance matrices (such as those of (PK1)-(PK3) and (AU1))—to transcribe significant mismatches between overall observed and conjectured behavior preferentially into changes and fluctuations in those quantities deemed to be time-varying, i.e., $r_5(t)$. This particular RPE algorithmic implementation of a candidate model structure straddles awkwardly steps (S1) and (S2) of the current expression of the procedure of model structure identification. It seeks to demonstrate unequivocally the inadequacy of the prior model structure ($\boldsymbol{M}_{prior}$), i.e., (S1), with yet an equivocal mix of bold-tentative conjectures embedded in the {presumed known ($\boldsymbol{\alpha}$)}, i.e., (PK1)–(PK3). Choosing parameter $r_5(t)$ to represent a host of more specific, putative mechanisms, all of which might amount to a single, lumped entity presumed *a priori* to be time-varying, undermines the power of the test of (S1), while anticipating the beginnings of a diagnosis of the sources of the expected structural failure of ($\boldsymbol{M}_{prior}$), i.e., (S2). Such ambiguity from a muddling of these two procedural steps has long been recognized (as long ago as Beck [5], in fact). How to avoid it should clearly be the subject of further research.

Yet strict adherence to procedure is but an ideal; and one which might miss the unexpectedly beneficial outcome of procedure applied faultily in the messiness of real-life case studies. Our purpose here, moreover, is to attain a blueprint for mechanizing formal, scientific visualizations of the informal pictogram of Fig. 4.1—again, to move from the analog of a Model T Ford to that of a contemporary BMW 700 Series. Implementation of the RPE algorithm appears genuinely to require fewer arbitrary assumptions than did the EKF [35], and it was at least designed primarily for the purpose of parameter estimation, as opposed to state estimation in nonlinear systems. Here we have been relating the performance of an RPE algorithm adapted still further, but well beyond its original remit, for the purposes of model structure identification.

Assumptions have been standardized in the current form of our RPE [18], to those associated with the variance-covariance properties of the white-noise sequences perturbing the GRW models of parametric variation, hitherto for $\boldsymbol{\alpha}$, if not yet $\boldsymbol{\kappa}$. This permits exquisite sophistication in the analyst's specification of the relative strengths and weaknesses of the model's constituent hypotheses. It also permits endless variations on this theme, for there is uncertainty about these enumerations of model structure error and uncertainty (see also [20]). Furthermore, we should not overlook the complexity that will rapidly ensue from the increasing order of vector $\boldsymbol{\kappa}$. This is a function of the *product* of the (fixed) order of the observation vector ($\boldsymbol{y}$) and the almost inevitably increasing order of the state vector ($\boldsymbol{x}$), as the model structure is progressively refined through successive iterations around $\boldsymbol{M}_{prior}$, $\boldsymbol{M}_{posterior}$, $\boldsymbol{M}_{prior}$, $\boldsymbol{M}_{posterior}$, and so on. Sophistication, as so often, is a two-edged sword.

In essence, the trajectories of the reconstructed parameter estimates vary, both in terms of departure from their initial values and over extended intervals (in some cases), yet not in an utterly random manner incapable of sustaining any further interpretation. Such interpretation, through step (S2), is genuinely a struggle. It is not trivial, even for such a simple, initial candidate model structure. But neither is it aimless. There are pointers as to where to seek insight within a rich base of hypothetical knowledge surrounding possible forms of the posterior model, albeit rarely directed at description of the dynamics of duckweed. Nevertheless, it is not hard to imagine the bewilderment of the reader of this chapter, as s/he in turn struggles to follow the deeper complexities behind the words of text employed here in our attempts to relate our own (ever-evolving) appreciation of how to interpret Figs. 4.4 and 4.5 in the scheme of model structure identification—the very acts indeed of Lewis's (E3).

Here now words are failing us, which is why we need a picture—and a rather clever one too.

4.5 Above and Beyond: Diagnosis and Rectification

Our animated visual metaphor has promise. Yet it does not obviate the procedural pitfalls of addressing step (S1) as distinct from (S2), or those of making therefore the various numerical assumptions needed for implementing the RPE algorithm.

4.5.1 Step (S2): Diagnosis

The acts of system identification (E3) have conventionally been articulated within just the space of the system's and model's outputs, $\mathbf{y}$, where the curve should be seen to pass through the dots. In this space, we know that the familiar theory-based models tacitly dominant in this discussion of discovery and learning can readily be found to suffer from a lack of model identifiability. Unambiguous interpretation of the data is not possible. On the other hand, the data-based models of Statistics, the antithesis thereof, are derived directly from the data (E1), deliberately with no prejudices about the set of concepts (E2) that might in due course explain the data. They can be well identified, using presumed objective methods of statistical inference. Yet customarily they are believed incapable of supporting a satisfactory theoretical interpretation of the observed behavior they demonstrably replicate.

That conventional perception is changing, driven on the one side by the ideas of "data-based *mechanistic* modeling" of Young [39], Young and Ratto [43]. The essence of the dynamic behavior of the identified realizations of these models can frequently be encapsulated in simple macro-parameters ($\boldsymbol{\beta}$), such as the system's time-constant and steady-state gain. The essence of the various parts of the dynamic behavior of the theory-based models can similarly be encapsulated in identical terms. We know exactly how this is done: see, for example, Young and Parkinson [42]. Instead of supposing that theory will be entirely successfully confronted with data in the space of $\mathbf{y}$, by way of evaluating the validity of that theory, abstracted features of the macro-parameters of the *theory*-based model can be juxtaposed with those of the *data*-based model, and conclusions drawn from this juxtaposition in the space of $\boldsymbol{\beta}$ (about how theory diverges from observation). Along this continuum of transformations of "information"

$$\text{(E2) Theory} \rightleftharpoons \text{Theory-based model} \rightleftharpoons \text{Macro-parameters } (\boldsymbol{\beta}) \\ \rightleftharpoons \text{Data-based model} \rightleftharpoons \text{Data (E1)}$$

the goal is to deduce useful insights about the relationship between theory and data, as reflected in their shared macro-parameters space [19]. If the recursively estimated trajectories of $\boldsymbol{\beta}_{tbm}$ from the theory-based model match those of $\boldsymbol{\beta}_{dbm}$ from the data-based model (and their variations are not randomly insignificant), directions towards specifically inadequate constituent hypotheses can, in principle, be deduced [17, 19]. Although both at one remove from Lewis's concepts ($\boldsymbol{\beta}_{tbm}$ from (E2)) and data ($\boldsymbol{\beta}_{dbm}$ from (E1)), reconciling "abstracted" $\boldsymbol{\beta}_{tbm}$ with "abstracted" $\boldsymbol{\beta}_{dbm}$ embodies his "acts" (E3).

This continuum of transformations, and its speculated mechanization in serving the growth of core scientific knowledge, will readily and convincingly appear distanced from the immediacy of the (very) public debate over climate change and hurricane intensity [23]. Yet Mooney structures his book around those characters (scientists) promoting empiricism over theory, who plead for "the data to speak for themselves", and those who promote theory over empiricism. Thus he sculpts (with seemingly little literary license) the essential difficulty: of reconciling empiricism with theory—Lewis's (E3)—and the attaching computational complexity of VHOMs, about which such controversy has boiled. Theorists stand resolutely at the point of "Theory-based model" in the above continuum; empiricists are mustered at their "Data" station; and with no apparent meeting of minds somewhere in between.

In less literary terms, that essential difficulty has to do with the vastly different orders of magnitude of the data bases to which we have had access—the orders and samples of $[\boldsymbol{u}, \boldsymbol{y}]$ being customarily small—and these VHOMs with high-dimensional state and parameter vectors $[\boldsymbol{x}, \boldsymbol{\alpha}]$. It is akin to looking at the world and trying to comprehend it through a pair of binoculars, with one eye-piece a microscope, the other a telescope. The device of macro-parameter vector $\boldsymbol{\beta}$ has the appeal of harmonizing the foci of the two eye-pieces, as a part of what could be needed for better realizing step (S2) of model structure identification.

4.5.2 Step (S3): Rectification

How then should we reason our way through rectification (step (S3)) of the causes of inadequacy diagnosed (S2) as being at the root of demonstrable failure (S1)? Words—and now pictures—fail us. For this is the domain of "illogical" serendipity. Being flawed in one's logic, however, is far from being entirely unproductive [5].

4.6 Conclusions

Over four decades, we have made progress in shaping, addressing, and resolving the issues of model structure identification. Punctuation marks in this intellectual journey are discernible:

(P1) A certain frustration with the promise and pain of working with the extended Kalman filter (EKF), for it had not been designed for the purpose of model parameter estimation, let alone model structure identification. Yet some of that frustration doubtless inspired both the visual metaphor of the node($\boldsymbol{x}$)-link($\boldsymbol{\alpha}$) network of Fig. 4.1 and the procedural steps of (S1), (S2), and (S3).

(P2) The most welcome relief upon the introduction of the recursive prediction error (RPE) algorithm of Ljung [21]—as, at least, an algorithm for parameter estimation—and its subsequent re-orientation from its original state-space representation to that of the parameter-space, and to the recognition of how to parameterize the {acknowledged unknown ($\boldsymbol{\kappa}$)} as significantly distinct from the already parameterized {presumed known ($\boldsymbol{\alpha}$)}.

(P3) Recognition of the possibility of diverting the software of molecular graphics into serving the purpose of scientific visualization in supporting the procedural steps (S1) through (S3) of model structure identification, as a direct result of an entirely chance encounter with a bio-pharmaceutical scientist. What is more, we are far from exhausting the repertoire of graphics and visualization in the biomedical sciences, as any leafing through current issues of *Science* (such as that of 29 October, 2010) will reveal.

We have still not had the temerity to begin the design of a recursive estimation algorithm—*de novo*, if necessary—for model structure identification, for its own sake. However, we may explore the scope for improving the performance of further adaptations of the RPE algorithm, specifically in respect of estimating time-varying parameters, for example, through incorporating a Fixed Interval Smoothing (FIS) algorithm [28].

As for metaphor in guiding any such design of novel algorithms, this chapter has entirely overlooked the image Fig. 4.1 evokes of the link-node network representations used to teach undergraduate students of civil engineering how to design engineered structures to resist failure, deformation, buckling, and collapse. As another punctuation mark in this narrative, therefore:

(P4) There is a well known duality (static-kinematic) at the basis of the methods for analyzing the elasto-plastic behavior of engineering structures, which so closely mirrors the duality of setting up an estimation algorithm in the state space ($\boldsymbol{x}$) and its complement in the parameter space ($\boldsymbol{\alpha}$).

This structural metaphor drove thinking about model structure identification through (P1) and (P2) above to its culmination in the contribution of Beck et al. [11] to the book "*Environmental Foresight and Models: A Manifesto*" [7]—which itself unexpectedly prompted (P3). Pursuit of (P4) in respect of pushing the problem-solution couple of model structure identification beyond where it is being left at the end of this chapter, would be quite an ambitious agenda for the future.

Where we stand now evokes then the title of one of philosopher Popper's books: "*Unending Quest*" [32]. This chapter has been about the overlooked role of model structure identification in the core scientific matter of discovery of new knowledge, here according to philosopher Lewis's characterization of the growth of knowledge.

Decisions and policies regarding environmental stewardship cannot be deferred for ever, of course. In that pragmatic world—upon the arrest of the endless attempts at model structure identification (and including in respect of climate change; [30])—the need may increasingly be to quantify the remaining structural error and uncertainty in a model and to account for their impacts on predictions of future behavior.

Meeting that need, barely begun, strikes one as an equally massive agenda for future research [20].

Good, generic problems will keep manifesting themselves in the specifics of case study after case study, until they demand dedicated, unrelenting attention. Model structure identification has become just this kind of a challenge, for some of us. Such was never intended in 1970.

Acknowledgements Support for this work has been provided over the decades by the University of Cambridge, the International Institute for Applied Systems Analysis, Imperial College London, and the University of Georgia (UGA). In particular, funding for the Environmental Process Control Laboratory of UGA, together with support for graduate assistantships for ZL and JDS, has come from the Wheatley-Georgia Research Alliance endowed Chair in Water Quality and Environmental Systems. The freedom of enquiry enabled through this form of financial support has simply been invaluable. We are also indebted to J P Bond, for his visualization and graphic design of Figs. 4.6, 4.7 and 4.8.

References

1. Alvarez-Vasquez, F., Sims, K.L., Cowart, L.A., Okamoto, Y., Voit, E.O., Hannun, Y.A.: Simulation and validation of modelled sphingolipid metabolism in *Saccharomyces Cerevisiae*. Nature **433**, 425–430 (2005)
2. Beck, M.B.: The application of control and systems theory to problems of river pollution. Ph.D. dissertation, University of Cambridge, UK (1973)
3. Beck, M.B.: Model structure identification from experimental data. In: Halfon, E. (ed.) Theoretical Systems Ecology: Advances and Case Studies, pp. 259–289. Academic, New York (1979)
4. Beck, M.B.: Uncertainty, system identification and the prediction of water quality. In: Beck, M.B., van Straten, G. (eds.) Uncertainty and Forecasting of Water Quality, pp. 3–68. Springer, Berlin (1983)
5. Beck, M.B.: Structures, failure, inference, and prediction. In: Barker, H.A., Young, P.C. (eds.) Identification and System Parameter Estimation, pp. 1443–1448. Pergamon, Oxford (1985)
6. Beck, M.B.: Water quality modeling: a review of the analysis of uncertainty. Water Resour. Res. **23**(8), 1393–1442 (1987)
7. Beck, M.B. (ed.): Environmental Foresight and Models: A Manifesto. Elsevier, Oxford (2002), 473 pp.
8. Beck, M.B.: Model evaluation and performance. In: El-Shaarawi, A.H., Piegorsch, W.W. (eds.) Encyclopedia of Environmetrics, vol. 3, pp. 1275–1279. Wiley, Chichester (2002)
9. Beck, M.B.: Structural change: a definition. In: Beck, M.B. (ed.) Environmental Foresight and Models: A Manifesto, pp. 51–60. Elsevier, Amsterdam (2002)
10. Beck, M.B., Young, P.C.: Systematic identification of DO-BOD model structure. J. Environ. Eng. Div. **102**(5), 909–927 (1976). Proceedings American Society of Civil Engineers
11. Beck, M.B., Stigter, J.D., Lloyd Smith, D.: D: Elasto-plastic deformation of the structure. In: Beck, M.B. (ed.) Environmental Foresight and Models: A Manifesto, pp. 323–350. Elsevier, Oxford (2002)
12. Beck, M.B., Gupta, H., Rastetter, E., Shoemaker, C., Tarboton, D., Butler, R., Edelson, D., Graber, H., Gross, L., Harmon, T., McLaughlin, D., Paola, C., Peters, D., Scavia, D., Schnoor, J.L., Weber, L.: Grand challenges of the future for environmental modeling. White Paper, National Science Foundation, Arlington, Virginia (2009) (ISBN: 978-1-61584-248-3)
13. Box, G.E.P., Jenkins, G.M.: Time Series Analysis, Forecasting and Control. Holden Day, San Francisco (1970)

14. Brun, R., Reichert, P., Künsch, H.R.: Practical identifiability analysis of large environmental simulation models. Water Resour. Res. **37**(4), 1015–1030 (2001)
15. Brun, R., Kühni, M., Siegrist, H., Gujer, W., Reichert, P.: Practical identifiability of ASM2d parameters—systematic selection and tuning of parameter subsets. Water Res. **36**(16), 4113–4127 (2002)
16. Hunt, C.A., Ropella, G.E.P., Lam, T.N., Tang, J., Kim, S.H.J., Engelberg, J.A., Sheikh-Bahaei, S.: At the biological modeling and simulation frontier. Pharm. Res. **26**(11), 2369–2400 (2009). doi:10.1007/s11095-009-9958-3
17. Lin, Z.: Modeling environmental systems under uncertainty: towards a synthesis of data-based and theory-based models. Ph.D. dissertation, University of Georgia, Athens, Georgia (2003)
18. Lin, Z., Beck, M.B.: On the identification of model structure in hydrological and environmental systems. Water Resour. Res. **43**, W02402 (2007a). doi:10.1029/2005WR004796
19. Lin, Z., Beck, M.B.: Understanding complex environmental systems: a dual approach. Environmetrics **18**(1), 11–26 (2007b)
20. Lin, Z., Beck, M.B.: Accounting for structural error and uncertainty in a model: An approach based on model parameters as stochastic processes. Environ. Model. Softw. (2010, in press)
21. Ljung, L.: Asymptotic behaviour of the extended Kalman filter as a parameter estimator. IEEE Trans. Autom. Control **24**, 36–50 (1979)
22. MacFarlane, A.G.J.: Interactive computing: a revolutionary medium for teaching and design. Comput. Control J. **1**(4), 149–158 (1990)
23. Mooney, C.: Storm World—Hurricanes, Politics, and the Battle Over Global Warming. Harcourt, Orlando (2007)
24. NRC: Models in Environmental Regulatory Decision Making. National Research Council, National Academy Press, Washington (2007), 267 pp
25. NSF: Sensors for environmental observatories. Report of the NSF-sponsored Workshop, December 2004, National Science Foundation (2005), 64 pp
26. NSF: Simulation-based engineering science: revolutionizing engineering science through simulation. Report of the National Science Foundation Blue Ribbon Panel, National Science Foundation (2006), 65 pp
27. NSF: Cyber-enabled discovery and innovation (CDI). Program Solicitation NSF 07-603 (2007) (www.nsf.gov)
28. Norton, J.P.: Optimal smoothing in the identification of linear time-varying systems. Proc. Inst. Electr. Eng. **122**, 663–668 (1975)
29. Omlin, M., Brun, R., Reichert, P.: Biogeochemical model of lake Zürich: sensitivity, identifiability and uncertainty analysis. Ecol. Model. **141**(1–3), 105–123 (2001)
30. Oppenheimer, M., O'Neill, B.C., Webster, M., Agrawala, S.: The limits of consensus. Science **317**, 1505–1506 (2007)
31. Petersen, B., Gernaey, K., Vanrolleghem, PA: Practical identifiability of model parameters by combined respirometric-titrimetric measurements. Water Sci. Technol. **43**(7), 347–355 (2001)
32. Popper, K.R.: The Unending Quest: An Intellectual Autobiography. Fontana-Collins, Glasgow (1976)
33. Raick, C., Soetart, K., Grégoire, M.: Model complexity and performance: how far can we simplify? Prog. Oceanogr. **70**, 27–57 (2006)
34. Stigter, J.D.: The development and application of a continuous-discrete recursive prediction error algorithm in environmental systems analysis. Ph.D. dissertation, University of Georgia, Athens, Georgia (1997)
35. Stigter, J.D., Beck, M.B.: On the development and application of a continuous-discrete recursive prediction error algorithm. Math. Biosci. **191**(2), 143–158 (2004)
36. Stigter, J.D., Vries, D., Keesman, K.J.: On adaptive optimal input design: a bioreactor case study. AIChE J. **52**(9), 3290–3296 (2006)
37. Tushingham, AM, Peltier, W.R.: Validation of the ICE-3G model of Würm-Wisconsin deglaciation using a global data base of relative sea level histories. J. Geophys. Res. **97**(B3), 3285–3304 (1992)
38. Young, P.C.: Recursive Estimation and Time Series Analysis: An Introduction. Springer, New York (1984)

39. Young, P.C.: Data-based mechanistic modelling of environmental, ecological, economic and engineering systems. Environ. Model. Softw. **12**, 105–122 (1998)
40. Young, P.C.: Nonstationary time series analysis and forecasting. Prog. Environ. Sci. **1**, 3–48 (1999)
41. Young, P.C.: The identification and estimation of nonlinear stochastic systems. In: Mees, A.I. (ed.) Nonlinear Dynamics and Statistics, pp. 127–166. Birkhäuser, Boston (2001)
42. Young, P.C., Parkinson, S.: Simplicity out of complexity. In: Beck, M.B. (ed.) Environmental Foresight and Models: A Manifesto, pp. 251–301. Elsevier, Oxford (2002)
43. Young, P.C., Ratto, M.: A unified approach to environmental systems modeling. J. Stoch. Environ. Res. Risk Assess. **23**, 1037–1057 (2009)

Chapter 5
Application of Minimum Distortion Filtering to Identification of Linear Systems Having Non-uniform Sampling Period

Graham C. Goodwin and Mauricio G. Cea

5.1 Introduction

Sampling of continuous time systems has been studied for many decades [10, 19]. In the linear case, when the sampling rate is constant, there is a closed form for the sampled data model of a continuous-time linear system. Estimation of the model parameters is then straightforward. However, when the sampling is non-uniform (time variable), the discrete-time model becomes time varying and, in this case, the estimation of model parameters becomes more difficult [15, 34, 35]. The approach we follow in this work, is to express the dependency of the discrete-time parameters on the sampling period explicitly, and then to use this nonlinear parameterized discrete-time model to obtain estimates of the continuous parameters. We use Nonlinear Filtering tools to carry out the associated estimation.

Filtering appears in many areas of Science and Engineering, see for example [25, 28, 48]. When the system of interest is linear, then the Kalman Filter provides an elegant and simple solution to the problem [2, 18, 28]. However, it is often the case that practical problems are inherently nonlinear [27, 31, 43, 44]. Unfortunately, the Kalman Filter is not directly applicable to these processes due to the presence of nonlinearities. Hence, there has been on going interest in various approximate nonlinear filtering algorithms.

Here we use a novel class of algorithms, known as Minimum Distortion Filtering (MDF) [6, 7, 16, 17]. MDF is a class of algorithms based on Vector Quantization [17, 45]. The aim, of Vector Quantization, is to choose a representative set of

G.C. Goodwin (✉) · M.G. Cea
School of Electrical Engineering and Computer Science, University of Newcastle, University Drive NSW 2308, Australia
e-mail: Graham.Goodwin@newcastle.edu.au

M.G. Cea
e-mail: Mauricio.Cea@uon.edu.au

L. Wang, H. Garnier (eds.), *System Identification, Environmental Modelling, and Control System Design*,
DOI 10.1007/978-0-85729-974-1_5,

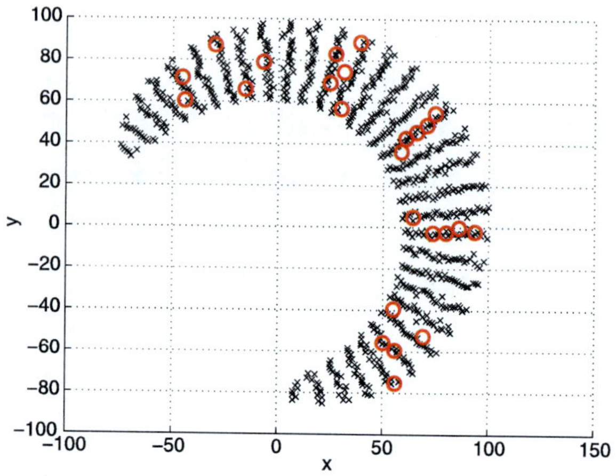

Fig. 5.1 Vector Quantization is a powerful tool. The figure shows a large Vector (*crosses*) which has been approximated by a relatively small one (*red circles*). This allows one to capture high complexity distributions. Note there are several techniques available to perform Vector Quantization. Here we use a Lloyd-based approach

points (Vectors) with N_x elements from a larger Vector with L elements [12, 21, 37], Fig. 5.1 shows an example of quantization of a large vector. The goal of quantization is to reduce computational load.

We also compare the MDF approach with another class of algorithm known as Sequential Monte Carlo or Particle methods. Particle methods are based on using a large number of random points (or particles) to approximate a distribution. Particle methods have been studied for several decades [22, 32, 39, 42]. Various different variants of the basic algorithm have been developed. These strategies provide widely accepted solutions to the general nonlinear filtering problem. However, due to the large number of particles required, the methods are computationally expensive. This motivates the search on more numerically "efficient" algorithms specially when computational resources are limited.

The outline of the remainder of this chapter is as follows: In Sect. 5.2 we review aspects of sampling. In Sect. 5.3 we formulate the non-uniform sampling identification problem. In Sect. 5.4 we explain how to estimate states and parameters using nonlinear filtering methods. In Sect. 5.5 we review nonlinear filtering theory at a conceptual level. Section 5.6 presents, a classification of approximate nonlinear filtering algorithms. In Sect. 5.7 we develop an algorithm based on MDF. In Sect. 5.8 we briefly describe Particle Methods. In Sect. 5.9 we present a simple numerical example. Finally, Sect. 5.10 presents Conclusion.

5.2 From Continuous to Discrete Systems

In this section we briefly discuss the relationship between continuous and discrete time models.

Most physical systems are described by continuous-time models. However, in practice, one needs to interact with these systems in some way. Thus, real implementations are subject to different constraints, such as sensors, actuators and communication channels. These add extra ingredients to the problem. A key issue in almost all problems is that of *sampling*. Sampling provides the link between continuous-time systems and discrete-time models.

Discrete system theory for uniformly sampled linear systems has been studied over many decades using purely discrete methods [10]. However, our interest here is in non-uniform sampling. In this case, it is convenient to model the system in continuous time since the associated parameters are then invariant with respect to different sample periods.

We summarize below some well known results regarding sampling of continuous-time systems in state-space form. Consider the continuous time system:

$$dx = f_c(x(t), u(t))dt + d\omega, \tag{5.1}$$

where $x(t) \in \mathbb{R}^n$, $u(t) \in \mathbb{R}^m$, are the state, input signal and output signal respectively. $f_c(\cdot) : \mathbb{R}^{n+m} \mapsto \mathbb{R}^n$. The process $u(t)$ is a known input and we assume that $\omega(t) \in \mathbb{R}^n$ is a stationary Wiener processes with incremental covariance $Q_c(x(t)) = Qdt$. We also assume that x_0 has Gaussian distribution with mean $\bar{x}_0$ and covariance P_0. The matrices Q and P_0 are symmetric and positive semi-definite.

For the linear case, $f_c(x(t), u(t)) = Ax(t) + Bu(t)$, with $A \in \mathbb{R}^{n\times n}$, $B \in \mathbb{R}^{n\times m}$. We assume that a zero-order hold (ZOH) is used for the input signal $u(t)$. Then, the corresponding discrete-time model with sample period Δ_k takes the form:

$$x_{k+1} = A_d(\Delta_k)x_k + B_d(\Delta_k)u_k + \omega_k, \tag{5.2}$$

where k is the discrete-time index.

We assume the following discrete-time measurement equation:

$$y_{k+1} = C_d x_k + v_k. \tag{5.3}$$

In (5.2), (5.3), ω_k and v_k are discrete-time white noise processes having covariance matrix

$$\Sigma_d(\Delta_k) = \begin{bmatrix} Q_d & 0 \\ 0 & R_d \end{bmatrix} = \begin{bmatrix} \Delta_k(Q + \frac{\Delta_k}{2}(AQ + QA^T) + \cdots) & 0 \\ 0 & R_d(\Delta_k) \end{bmatrix}, \tag{5.4}$$

where Q_d is a symmetric semi-positive definite and R_d is positive definite.

The system matrices in (5.2) are given by

$$A_d(\Delta_k) = e^{A\Delta_k} = I + A\Delta_k + \frac{1}{2}A^2\Delta_k^2 + \cdots, \tag{5.5}$$

$$B_d(\Delta_k) = A^{-1}(e^{A\Delta_k} - I)B = \left(B + \frac{1}{2}AB\Delta_k + \cdots\right)\Delta_k, \tag{5.6}$$

where $A \in \mathbb{R}^{n\times n}$, $B \in \mathbb{R}^n$, $C_d \in \mathbb{R}^{p\times n}$.

The discrete-time model described above use a time-varying sampling sequence $\{\Delta_k\}$, where

$$\Delta_k = t_{k+1} - t_k > 0 \quad \forall k \in \mathbb{N}. \tag{5.7}$$

Here, $\{t_k\}$ denotes the sample times. We assume that data is collected on the time interval $[0; T_f]$, where $T_f = \sum_{k=0}^{N_f-1} \Delta_k$ and N_f is the total number of samples.

5.3 Non-uniform Sampling as an Identification Problem

In this section we formulate the non-uniform sampling system identification problem. When the sampling rate is constant and the underlying continuous time system is time invariant, then the discrete-time model is time invariant. In this case, the parameters can be estimated directly on the discrete-time model. There are several possibilities. For example, one can first estimate the discrete-time model matrices. In a second step, a transformation can be applied to recover the continuous time model matrices. This operation may involve computation of the logarithm of the system matrix or the use of Padé like approximations [38, 47]. Different derivative approximations can also be used to discretize the model. However, the choice of the particular approximation may have a direct impact on the quality of the estimates (see, for example, [35]).

When the sampling rate is non-uniform, the discrete-time model becomes time varying. In this case, estimating the continuous parameters becomes more difficult.

Continuous-time system identification from non-uniform sampled-data has been considered from several different perspectives. For example, in [15] approximate output spectrum reconstruction is performed using B-spline functions. Another identification procedure for the non-uniform sampling was proposed in [9]. In [9] a least squares approach is used, where the states are estimated using a Kalman filter in shift operator form.

Here, we use a different approach. We include the sampling period explicitly in the discrete-time model. We also retain the continuous-time parameters. This leads to the following model which restates (5.2), (5.3):

$$x_{k+1} = e^{A\Delta_k} x_k + A^{-1}(e^{A\Delta_k} - I)Bu_k + \omega_k, \tag{5.8}$$

$$y_{k+1} = Cx_k + \nu_k, \tag{5.9}$$

where A, B and C are the associated continuous time matrices, Δ_k is the sampling period between the kth and $(k+1)$th samples. The process noise ω_k is Gaussian and has a time varying covariance which is proportional to the sampling period $Q_k = \Delta_k Q$. We consider the measurement noise Gaussian and independent of ω_k. For simplicity we assume that the measurement noise has fixed covariance R_d.

Remark 5.1 This formulation leads to a linear discrete-time model with respect to the states, but with significant nonlinearities with respect to the parameters.

5.4 Systems Identification as a Nonlinear Filtering Problem

It is well known that parameter estimation can be combined with state estimation by augmenting the model with additional states for the parameters. Thus, if we begin with the linear model (2) to (8), then we have the following model which explicitly includes the parameters.

$$x_{k+1} = A_d(\Delta_k, \theta_k)x_k + B_d(\Delta_k, \theta_k)u_k + \omega_k, \tag{5.10}$$

$$y_{k+1} = C(\theta_k)x_k + \nu_k, \tag{5.11}$$

where θ_k is the unknown parameter vector. We model the time evolution of this vector by an additional state space model. For example a random-walk process leads to:

$$\theta_{k+1} = \theta_k + {\omega^{(2)}}_k, \tag{5.12}$$

where ${\omega^{(2)}}_k$ is assume to be white Gaussian noise with zero mean and variance Q_θ. The covariance Q_θ models how much we expect the parameter θ to change over time. In turn, this effects the "memory" of the parameter estimator. For example, if Q_θ is large, then the model predicts rapid parameter variations and then the filter will automatically "discard" the data save very recent observations. Conversely, if Q_θ is small, then the model predicts slow parameter variations and then the filter will "retain" the data save for observations that are far removed from the present time.

Remark 5.2 Even if we believe θ is constant, it is usually a good idea to use a value of Q_θ different from zero. The reason is that, otherwise, the parameter estimator will "lock-up" and not use the on-going data. This is acceptable under ideal conditions but can lead to erroneous estimates in practical cases, e.g. due to the influence of outliers.

5.5 Nonlinear Filtering: General Concepts

In this section we introduce general concepts of Nonlinear Filtering. The continuous-time case is described in [4, 33]. Here we focus on the discrete time case. We note that the full state space model (5.10) to (5.12) is linear if considered as a function of x_k only but is nonlinear in the extended state $\bar{x}_k = [x_k^T, \theta_k^T]^T$. We write the augmented model (5.10) to (5.12) in the following general form as:

$$\bar{x}_{k+1} = \bar{f}(\bar{x}_k, u_k) + \bar{\omega}_k, \tag{5.13}$$

$$y_{k+1} = \bar{h}(\bar{x}_k, u_k) + \bar{\nu}_k. \tag{5.14}$$

The associated discrete nonlinear filter can then be expressed by two equations,

The State-update (Chapman-Kolmogorov) equation [4, 28]

$$\mathscr{P}(\bar{x}_{k+1}|\mathscr{Y}_k) = \int \mathscr{P}(\bar{x}_{k+1}|\bar{x}_k)\mathscr{P}(\bar{x}_k|\mathscr{Y}_k)d\bar{x}_k \tag{5.15}$$

and the Observation-update (Bayes Rule) [4, 28]

$$\mathscr{P}(\bar{x}_{k+1}|\mathscr{Y}_{k+1}) = \frac{\mathscr{P}(\bar{x}_{k+1}|\mathscr{Y}_k)\mathscr{P}(y_{k+1}|\bar{x}_{k+1})}{\int \mathscr{P}(\bar{x}_{k+1}|\mathscr{Y}_k)\mathscr{P}(y_{k+1}|\bar{x}_{k+1})d\bar{x}_{k+1}}, \tag{5.16}$$

where $\mathscr{Y}_k$ denotes the set of measurements up to the kth sample.

The above equations provide a complete conceptual solution to the sampled data nonlinear filtering problem. However, these equations are infinite dimensional and then can only be solved in very special cases; e.g. linear-Gaussian problems, in which case, the solutions reduce to the Kalman Filter. In the general nonlinear case, various approximations are used to generate state estimates. A review of some of the existing algorithms is presented below.

5.6 Review of Approximate Algorithms for Discrete Nonlinear Filtering

There exists a huge volume of research on approximate algorithms for discrete time nonlinear filtering. Useful reviews can be found in [3, 4, 8]. The existing algorithms can be broadly classified into 5 categories:

1. Linearization algorithms:

   - Here one linearizes about the current estimate $\hat{x}$. This leads to the Extended Kalman Filter (EKF) [2, 28]. Various embellishments are possible, e.g. re-linearizing about updated estimates, leading to the Iterated Extended Kalman Filter (IEKF) [11, 28].

   The advantage of these algorithms is that they are very simple. The disadvantage is that they will frequently fail when the nonlinearities are far from linear.
2. Mixed Algorithms:

   - Here one uses a Gaussian approximation, but then chooses several representative points to pass through the nonlinearities. These are re-averaged after passing though the nonlinearity: An example of this class of algorithms is the Unscented Kalman Filter (UKF) [29, 30]. A more recent algorithm from the same general class is the algorithm described in [3] which uses Gauss-Hermite Quadrature.

   Again these algorithms are very simple. However, the disadvantage is that they only work for simple nonlinearities. Also they focus on estimating the mean of the posterior distribution. This can be acceptable in some case but is, in general, an inadequate description of the posterior distribution as is clear form the distribution shown in Fig. 5.1.

3. Deterministic Griding Algorithms:

   - One can obtain an approximate filtering algorithm by simply representing the distribution of the states on a finite grid. One choice would be a uniform grid. However, this is often infeasible since a very large number of grid points are typically needed.
   - Another related idea is to, a-priori, choose a grid that is more focused on the "likely" areas of the state space where the states might lie. For example [40] uses vector quantization to choose a grid based on the prior distribution for the state $\bar{x}$.

   Unfortunately, these methods do not account well for disturbances or uncertainty in the state trajectories.
4. Monte Carlo/Particle Filtering:

   - This technique accounts for disturbances by drawing a set of random samples from the disturbance distribution. Thus, a discrete approximation to the posterior distribution is generated which is based on a set of randomly chosen points. The approximation converges in probability with order $1/\sqrt{N}$, where N is the number of chosen samples. The main disadvantages of this class of algorithm is that a very large number of points may be needed and also these points need to be, in some sense, related to the distribution of interest. Also, the number of points grows exponentially with time unless some form of reduction is used. Thus, there are many ad-hoc fixes needed to get this type of algorithm to work in practice. Such fixes include the use of proposal distributions, resampling methods, etc. see [8, 42].

5. Minimum Distortion Filtering [MDF]: This is a new class of algorithm. It was first described in [16, 17]. More details of the computational details are given in [6, 7]. We provide a summary of the algorithm in the next section.

5.7 Minimum Distortion Filtering Algorithm

The key idea underlying this class of algorithm is to utilize Vector Quantization to generate, on-line, a finite approximation to the a-posteriori distribution of the states.

Say that one begins with a discrete approximation to the distribution of $\bar{x}_0$ on N_x grid points. Also assume that one has a finite approximation to the distribution of the process noise on N_w grid points. Then utilizing the discretized version of (5.15), one obtains a finite approximation to $\mathscr{P}(\bar{x}_1)$ on $N_x \times N_w$ grid points. Then, one uses the discrete equivalent of (5.16) to obtain a finite approximation to $\mathscr{P}(\bar{x}_1|y_1)$ on $N_x \times N_w$ points. Finally, one uses vector quantization ideas to re-approximate $\mathscr{P}(\bar{x}_1|y_1)$ back to N_x points. One iterates from the beginning to obtain a discrete approximation to $\mathscr{P}(\bar{x}_2|y_1)$ on $N_x \times N_w$ points and so on. The algorithm is summarized in Table 5.1.

The key step in the MDF algorithm is the vector quantization step (Step 5 in Table 5.1). We give details of this step below.

Table 5.1 MDF algorithm

Step	Description
1	Initialization: Quantize $\mathscr{P}(\bar{x}_0)$ to N_x points by x_i, p_i; $i = 1, \dots, N_x$. Quantize $\mathscr{P}(\bar{\omega})$ to N_ω points by w_j, q_j; $j = 1, \dots, N_w$
2	Begin with $\mathscr{P}(\bar{x}_k \mid \mathscr{Y}_k)$ represented by $x_i, p_i; i = 1, \dots, N_x$
3	Approximate $\mathscr{P}(\bar{x}_{k+1} \mid \mathscr{Y}_k)$ via (5.15) on $N_x * N_\omega$ points
4	Evaluate $\mathscr{P}(\bar{x}_{k+1} \mid \mathscr{Y}_{k+1})$ on $N_x * N_w$ points via (5.16)
5	Quantize back to N_x points
6	Iterate from step 2

Assume we have a vector discrete distribution for some distribution $\mathscr{P}(\bar{x})$, where $\bar{x} \in \mathbb{R}^n$, quantized to a very large (but finite) set of points. Our goal is to quantize $\mathscr{P}(\bar{x})$ to a *smaller* finite set of points $x_i, p_i, i = 1, \dots, N$. The first step in Vector Quantization is to define a measure to quantify the "Distortion" of a given discrete representation. This measure is then optimized to find the optimal representation which minimizes the distortion. In summary, we look for a finite set $\mathscr{W}_x = \{x_1, \dots, x_N\}$ and an associated collection of sets $\mathscr{S} = \{S_1, \dots, S_N\}$ such that $\bigcup_1^N S_i = \mathbb{R}^n$ and $S_i \cap S_j = 0; i \neq j$. We choose $\mathscr{W}_x, \mathscr{S}_x$ by minimizing a cost function of the form:

$$\mathscr{J}(\mathscr{W}_x, \mathscr{S}_x) = \sum_{i=1}^{N} E\big\{(\bar{x} - x_i)^T W(\bar{x} - x_i) \mid \bar{x} \in S_i\big\}, \tag{5.17}$$

where $W = \operatorname{diag}(W_1, \dots, W_N)$. Other choices of the distance measure, can also be used; e.g. Manhattan, $\mathscr{L}_1$, Jaccard, etc., see [45].

If we fix $x_1, \dots, x_N$ (the set of grid points), then the optimal choice of the sets S_i is the, so called, Voronoi cells [12, 21]

$$S_i = \big\{\bar{x} \mid (\bar{x} - x_i)^T W(\bar{x} - x_i) \leq (x - x_j)^T W(x - x_j); \forall j \neq i\big\}. \tag{5.18}$$

Similarly, if we fix the sets $S_1, \dots, S_N$, then the optimal choice for x_i is the centroid of the sets $\mathscr{S}_i$, i.e.

$$x_i = E(\bar{x} \mid \bar{x} \in S_i). \tag{5.19}$$

Many algorithms exist for minimizing functions of the form (5.17) to produce a discrete approximation. One class of algorithm (known as Lloyd's algorithm [12, 21, 37]) iterates between the two conditions (5.18) and (5.19).

Thus Lloyd's algorithm begins with an initial set of grid points $\mathscr{W}_x = \{x_i; i = 1, \dots, N_x\}$. Then one calculates the Voronoi cells $\mathscr{S}_x$ of $\mathscr{W}_x$ using (5.18). Next, one computes the centroids of the Voronoi cells $\mathscr{S}_x$ via (5.19). One then returns to the calculation of the associated Voronoi cells and so on. Lloyd's algorithm iterates these steps until the distortion measure (5.17) reaches a local minimum, or until the change in the distortion measure falls below a given threshold, i.e.

$$\frac{\mathscr{J}(\mathscr{W}_x^{k+1}, \mathscr{S}_x^{k+1}) - \mathscr{J}(\mathscr{W}_x^k, \mathscr{S}_x^k)}{\mathscr{J}(\mathscr{W}_x^k, \mathscr{S}_x)^k} \leq \varepsilon, \tag{5.20}$$

where $\mathscr{W}_x^k$ and $\mathscr{S}_x^k$, is the codebook and Voronoi cells at iteration k respectively.
For further details, we refer the reader to [6, 7].

5.8 Particle Methods

Here we describe an alternative scheme for nonlinear filtering based on Particle filtering. This is actually one of the most commonly used schemes in practical nonlinear filtering problems.

Particle methods deal with the problem of recursively estimating the probability density function $\mathscr{P}(\bar{x}_k|\mathscr{Y}_k)$ by using Monte Carlo ideas. The key idea is to represent the probability density function by a set of random samples having associated weights.

$$\mathscr{P}(\bar{x}_k|\mathscr{Y}_k) = \sum_{i=1}^{M}(q_k^{(i)}\delta(\bar{x}_k - \bar{x}_k^{(i)})), \quad \sum_{i=1}^{M} q_k^i = 1, \quad q_k^i \geq 0, \tag{5.21}$$

where $\delta(\cdot)$ is the Dirac delta function and $q_k^{(i)}$ denotes the weight associated with the particle $\bar{x}_k^{(i)}$. The subscript k indicates the discrete-time index and the superscript (i) denotes a particular particle.

In obtaining this approximation, one has to be able to draw random numbers from complicated distributions. The approximation (5.21) can also be obtained using stochastic integration ideas, see e.g., [5, 13] for related, slightly different, approaches. In practice. one needs to use a relatively large number of random samples to adequately represent a given distribution. It is important to note that the problem of generating random numbers from complicated distributions has previously been assessed in a non-recursive setting using Markov Chain Monte Carlo methods (MCMC).

The generation of the random samples presents a major problem. In the literature one can find various ideas on how to handle the fact that we cannot generate samples directly from the target density. One option is to use a marginalized particle filter. This method can be employed when there is a linear, Gaussian sub-structure available in the model equations. For further details on this topic see [42] and the references therein. Further details used in our implementation are outlined below:

5.8.1 Random Number Generation

The problem of interest is to generate samples from some known probability density function, referred to as the target density $t(x)$. However, since we cannot generate samples from $t(x)$ directly, the idea is to employ an alternate density that is simple to draw samples from, referred to as the sampling density $s(x)$. When a sample $\bar{x} \sim s(x)$ is drawn the probability that it was in fact generated from the target density can be calculated. This probability can then be used to decide whether $\bar{x}$ should be considered as a sample from $t(x)$ or not. This probability is referred to as the acceptance probability, and it is typically expressed as a function of another variable $q(\bar{x})$, defined by the following relationship,

$$t(\bar{x}) \propto q(\bar{x})s(\bar{x}). \tag{5.22}$$

Depending on the exact details of how the acceptance probability is computed different methods are obtained. Some of the best known methods are *Sampling Importance Resampling*, *Acceptance-Rejection Sampling* and *Metropolis-Hastings Independence Sampling*. For a more detailed explanation, see, e.g., [14, 41, 42, 46] comparison of different methods is provided in [36].

5.8.2 Particle Filter

In the Bayesian framework used in the current chapter, we note that the Observation update equation given by (5.16) resembles (5.22). Then we can use this to define the target and sampling density as follows:

$$\underbrace{\mathscr{P}(x_{k+1}|\mathscr{Y}_{k+1})}_{t(\bar{x})} = \frac{\mathscr{P}(x_{k+1}|\mathscr{Y}_k)\mathscr{P}(y_{k+1}|x_{k+1})}{\int \mathscr{P}(x_{k+1}|\mathscr{Y}_k)\mathscr{P}(y_{k+1}|x_{k+1})dx_{k+1}}$$

$$\propto \underbrace{\mathscr{P}(x_{k+1}|\mathscr{Y}_k)}_{s(\bar{x})}\underbrace{\mathscr{P}(y_{k+1}|x_{k+1})}_{q(\bar{x})}. \tag{5.23}$$

The typical Particle Filter is derived using the *Sampling Importance Resampling* technique and many derivations can be found in the literature, see [20, 23, 42]. Later it was independently rediscovered by [26, 32]. Some early ideas relating to the particle filter are given in [1, 22, 24, 39].

The Particle Filter used in this chapter is described in Table 5.2. It is based on ideas presented in [42] (see Algorithm 4.4).

5.9 Simulation Example

Here we use a simple example to illustrate the ideas discussed above. We assume an underlying continuous-time system (5.1) of the form:

$$dx(t) = ax(t)dt + bu(t) + d\omega, \tag{5.24}$$

where $d\omega$ is defined as in Sect. 5.2 and has variance $0.1dt$.

Table 5.2 MDF algorithm

Step	Description
1	Initialization: Quantize $\mathscr{P}(\bar{x}_0)$ to M particles $\{\bar{x}^{(i)}\}_{i=1}^{M}$;
2	Use the State Update equation to propagate the Particles to produce $\mathscr{P}(\bar{x}_{k+1}^{(i)}\vert\mathscr{Y}_k)$
3	Measurement Update: Calculate the importance weights $\{q_k^{(i)}\}_{i=1}^{M}$ according to $q_k^{(i)} = \mathscr{P}(y_{k+1}\vert\bar{x}_{k+1}^{(i)})$
4	Resampling: Draw M particles, with replacement, according to $\mathscr{P}(x_{k+1}^{i} = x_k^{j}) = q_k^{j}$
6	Iterate from step 2

Non-uniform sampling is used with a total of $K = 100$ samples. We use an upper bound $\Delta_{MAX} = 0.5$ on the sample period. For estimation purposes we consider the true parameters as $a = -1$ and $b = 5$.

The model used for parameter estimation is an extension of (5.8), (5.9) using the technique described in Sect. 5.4, i.e., the augmented state equation is

$$x_{k+1} = e^{a_k\Delta_k}x_k + a_k^{-1}(e^{a_k\Delta_k} - 1)b_k u_k + \omega_k^{(1)}, \tag{5.25}$$

$$a_{k+1} = a_k + \omega_k^{(2)}, \tag{5.26}$$

$$b_{k+1} = b_k + \omega_k^{(3)}. \tag{5.27}$$

The measurement equation is taken to be

$$y_k = x_k + v_k. \tag{5.28}$$

Here the extended state vector is $\bar{x}_k = [x_k; a_k; b_k]$; $\omega_k^{(1)}$, $\omega_k^{(2)}$, $\omega_k^{(3)}$ and v_k are assumed to have variance $0.1\Delta_k$, $0.005\Delta_k$, $0.0.005\Delta_k$ and 0.1 respectively. Note that when we sample a continuous-time system, the variance of the process noise typically grows proportional to the sampling period, see (5.4).

Figure 5.2 shows one particular realization of the sampling period Δ_k. Note that this realization changes for each experiment.

5.9.1 MDF

The MDF algorithm is used to obtain an estimate of the continuous-time parameters. We quantize the density function of the augmented state on $N_x = 27$ points. Note that we are quantizing a 3 dimensional state vector. Another $N_w = 27$ points are used to quantize the process noise.

The filter is initialized with $\mathscr{P}(\bar{x}_0)$ having a Gaussian distribution with random initial value and variance $Q_0 = \text{diag}(3, 3, 3)$.

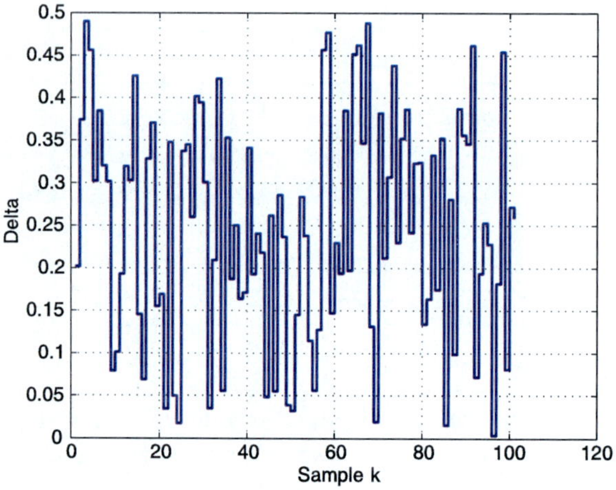

Fig. 5.2 Δ_k vs sample

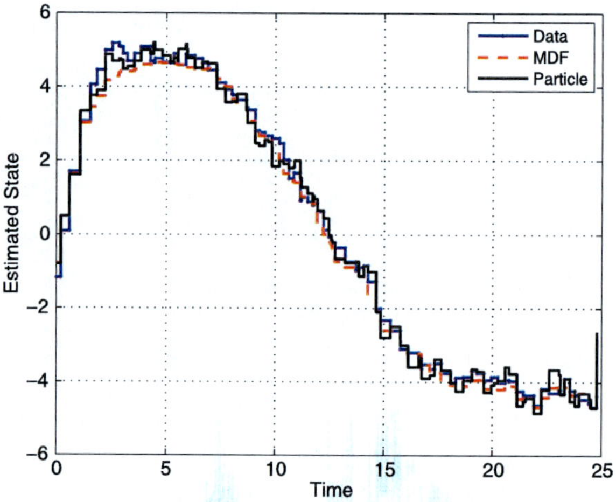

Fig. 5.3 State x_k and the estimates (mean value of $\mathscr{P}(x_k|\mathscr{Y}_k)$) at each sample

5.9.2 Particle Filter

The Particle Filter parameters chosen for this simulation are as described in Sect. 5.8.2 with 100,000 particles.

5.9.3 Results

Here we explain the results obtained using the MDF and Particle Filtering schemes. Figure 5.3 shows the true state evolution and the estimate obtained by both algorithms. Clearly, both algorithms provide good estimates for the state, although the Particle Filter achieves a slightly better result. (This is perhaps not surprising given

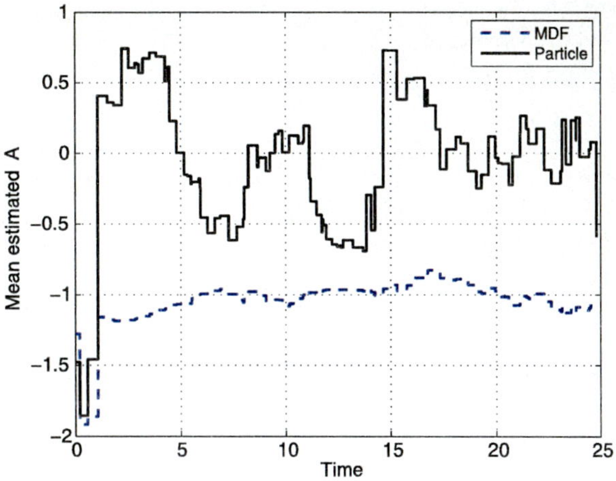

Fig. 5.4 Mean value of $\mathscr{P}(a_k|\mathscr{Y}_k)$ at each sample k. Particle Filter (*black: solid*) MDF (*blue: dashed*)

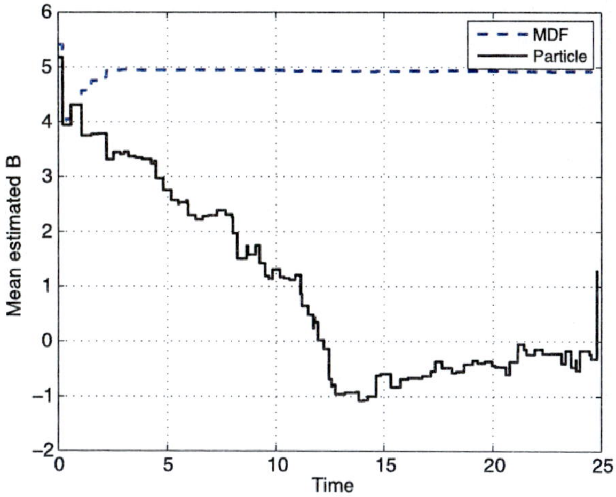

Fig. 5.5 Mean value of $\mathscr{P}(b_k|\mathscr{Y}_k)$ at each sample k. Particle Filter (*black: solid*) MDF (*blue: dashed*)

that the MDF uses 27 points in the approximation of the posterior distribution for $\bar{x}$ whereas the Particle Filter uses 100,000 points.)

Figures 5.4 and 5.5 show the estimates for the parameters obtained using the MDF algorithm and Particle Filter for a particular data set. It is clear from the figures, that the MDF algorithm provides much better performance than the Particle Filter. Of course it is possible that better results could be obtained if the Particle Filter were to be refined. Thus the authors make no general claim from this example.

Remark 5.3 It is worth noticing that the MDF algorithm uses only 27 points to quantize the distribution, yet provides excellent results.

Figures 5.6 and 5.7 shows the cumulative distribution at a particular sample time for the MDF algorithm on a particular data set. The distribution for the parameters can clearly be seen to be non Gaussian which confirms the nonlinear nature of the

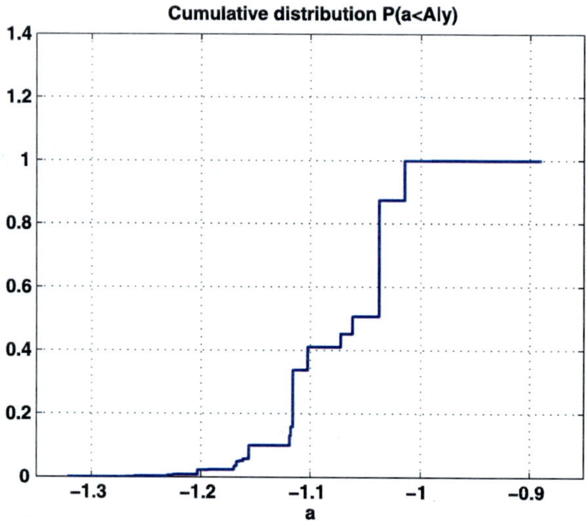

Fig. 5.6 Cumulative marginal distribution $\mathscr{P}(a_k < A_k | y_k)$ at $k = 17$

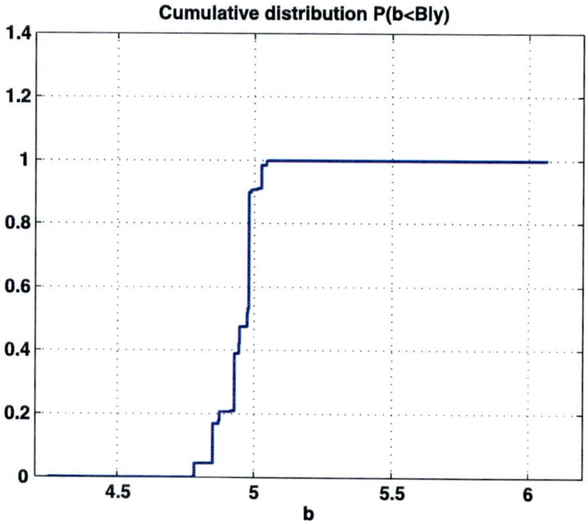

Fig. 5.7 Cumulative marginal distribution $\mathscr{P}(b_k < B_k | y_k)$ at $k = 17$

continuous-discrete time mapping. To reinforce this idea we also show the discrete probability density function at the same sample time in Figs. 5.8 and 5.9.

5.9.4 Robustness

To illustrate the robustness of the MDF algorithm, we run the simulation 20 times with a different seed. In particular, we have different initial conditions for the real state, $\mathscr{P}(\bar{x}_0)$ and the sampling period sequence $\{\Delta_k\}$. The average of the estimates

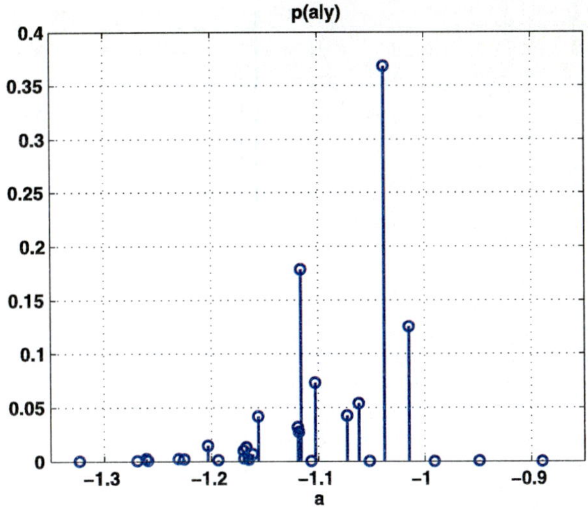

Fig. 5.8 Density function a at $k = 17$

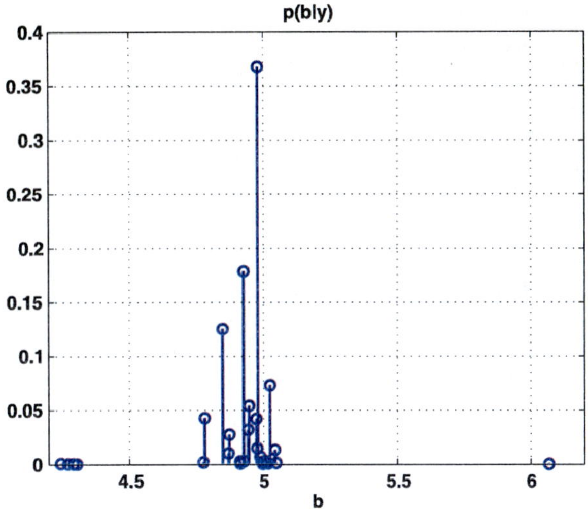

Fig. 5.9 Density function b at $k = 17$

after a fixed number of samples are compared in Table 5.3. Note that the averages are after a fixed number of samples which will correspond to different time periods depending on the realization of the sampling period sequence.

It can be seen that the MDF algorithm provides consistent estimates. On the other hand, the Particle Filter has trouble finding the continuous-time parameters.

5.10 Conclusions

We have described how Minimum Distortion Filtering can be applied to the problem of system identification when the data is collected with non-uniform sampling pe-

Table 5.3 Average Estimate 20 Experiments

Sample	A-MDF	B-MDF	A-Particle	B-Particle	True-A	True-B
10	−1.17	5.07	0.28	3.69	−1	5
30	−1.10	5.04	0.22	2.63	−1	5
50	−1.0003	5.02	−0.15	0.677	−1	5

riod. A comparison with an alternative scheme based on Particle Filtering has also been given. The results appear encouraging and show that the Minimum Distortion Filtering Algorithm is capable of providing excellent estimates despite the presence of highly variable sampling periods.

References

1. Akashi, H., Kumamoto, H.: Random sampling approach to state estimation in switching environments. Automatica **13**, 429–434 (1977)
2. Anderson, B., Moore, J.B.: Optimal Filtering. Prentice Hall, Englewood Cliffs (1979)
3. Arasaratnam, I., Haykin, S., Elliott, R.: Discrete-time nonlinear filtering algorithms using gauss n Hermite quadrature. Proc. IEEE **95**(5), 953–977 (2007)
4. Bain, A., Crisan, D.: Fundamental of Stochastic Filtering, vol. 60. Springer, Berlin (2009). ISBN 978-0-387-76895-3
5. Bergman, N.: Recursive Bayesian estimation navigation and tracking applications. PhD thesis, Linkĺoping University (1999)
6. Cea, M.G., Goodwin, G.C.: A new paradigm for state estimation in nonlinear systems using Minimum Distortion Filtering. In: 18th IFAC World Congress, Milan (2011)
7. Cea, M.G., Goodwin, G.C.: A novel technique based on up-sampling for addressing modeling issues in sampled data nonlinear filtering. In: 18th IFAC World Congress, Milan (2011)
8. Chen, Z.: Bayesian filtering: From Kalman filters to particle filters, and beyond. Available at: http://users.isr.ist.utl.pt/~jpg/tfc0607/chen_bayesian.pdf (2003)
9. Ding, F., Qiu, L., Chen, T.: Reconstruction of continuous-time systems from their non-uniformly sampled discrete-time systems. Automatica **45**(2), 324–332 (2009)
10. Feuer, A., Goodwin, G.: Sampling in Digital Signal Processing and Control. Birkhäuser Boston, Cambridge (1996)
11. Gelb, A.: Applied Optimal Estimation. MIT Press, Cambridge (1974)
12. Gersho, A., Gray, R.M.: Vector Quantization and Signal Compression. Springer International Series in Engineering and Computer Science (1992)
13. Geweke, J.: Monte Carlo Simulation and Numerical Integration. Elsevier, Amsterdam (1996). Chap. 15, pp. 731–800
14. Gilks, W.R., Richardson, S., Spiegelhalter, D.: Markov Chain Monte Carlo in Practice: Interdisciplinary Statistics, 1st edn. Chapman & Hall/CRC Interdisciplinary Statistics. Chapman & Hall/CRC, London (1995)
15. Gillberg, J., Ljung, L.: Frequency domain identification of continuous-time output error models, Part II: Non-uniformly sampled data and b-spline output approximation. Automatica **46**(1), 11–18 (2010)
16. Goodwin, G., Cea, M.G.: State and parameter estimation via minimum distortion filtering with application to chemical processes control. In: 4th International Symposium on Advanced Control of Industrial Processes (2011)

17. Goodwin, G.C., Feuer, A., Müller, C.: Sequential Bayesian Filtering via Minimum Distorion Filtering, in Three Decades of Progress in Control Sciences, 1st edn. Springer, Berlin (2010)
18. Goodwin, G.C., Graebe, S., Salgado, M.E.: Control System Design. Prentice Hall, Upper Saddle River (2001)
19. Goodwin, G.C., Yuz, J.I., Agüero, J.C., Cea, M.G.: Sampling and sampled-data models. In: Proceedings of American Control Conference, Baltimore, Maryland, USA (2010)
20. Gordon, N.J., Salmond, D.J., Smith, A.F.M.: Novel approach to nonlinear/non-Gaussian Bayesian state estimation. IEE Proc., F, Radar Signal Process. **140**(2), 107–113 (2002)
21. Graf, S., Lushgy, H.: Foundations of Quantization for Probability Distributions. Lecture Notes in Mathematics, no: 1730. Springer, Berlin (2000)
22. Hammersley, J.M., Morton, K.W.: Poor man's HEX:92s Monte Carlo. J. R. Stat. Soc. B **16**(1), 23–28 (1954)
23. Handschin, J.: Monte Carlo techniques for prediction and filtering of non-linear stochastic processes. Automatica **6**(4), 555–563 (1970)
24. Handschin, J.E., Mayne, D.Q.: Monte Carlo techniques to estimate the conditional expectation in multi-stage non-linear filtering. Int. J. Control **9**, 547–559 (1969)
25. Haykin, S.: Kalman Filtering and Neural Networks. Wiley-Interscience, New York (2001)
26. Isard, M., Blake, A.: A Smoothing Filter for Condensation. Lecture Notes in Computer Science, vol. 1406 (1998)
27. Isidori, A.: Nonlinear Control Systems. Springer, New York (1995)
28. Jazwinski, A.: Stochastic Processes and Filtering Theory. Academic Press, San Diego (1970)
29. Julier, S., Uhlmann, J.: Unscented filtering and nonlinear estimation. Proc. IEEE **92**(3), 401–422 (2004)
30. Julier, S., Uhlmann, J., Durrant-Whyte, H.: A new method for the nonlinear transformation of means and covariances in filters and estimators. IEEE Trans. Autom. Control **45**(3), 477–482 (2000)
31. Khalil, H.: Nonlinear Systems, 2nd edn. Prentice-Hall, Upper Saddle River (1996)
32. Kitagawa, G.: Monte Carlo filter and smoother for non-Gaussian nonlinear state space models. J. Comput. Graph. Stat. **5**(1), 1–25 (1996)
33. Kushner, H.J.: On the differential equations satisfied by conditional probability densities of Markov processes, with applications. J. SIAM Control **2**(1) (1962)
34. Larsson, E.K., Mossberg, M., Söderström, T.: Identification of continuous-time ARX models from irregularly sampled data. IEEE Trans. Autom. Control **52**(3), 417–427 (2007)
35. Larsson, E.K., Söderström, T.: Identification of continuous-time AR processes from unevenly sampled data. Automatica **38**(4), 709–718 (2002)
36. Liu, J.S.: Metropolized independent sampling with comparisons to rejection sampling and importance sampling. Stat. Comput. **6**(2), 113–119 (1996)
37. Lloyd, S.: Least squares quantization in pcm. IEEE Trans. Inf. Theory **IT-28**, 127–135 (1982)
38. McKelvey, T., Helmersson, A.: State-Space Parametrizations of Multivariable Linear Systems Using Tridiagonal Matrix Forms, vol. 4, pp. 3654–3659 (1996)
39. Metropolis, N.: The Monte Carlo method. Journal of the American Statistical Association **44**(247) (1949)
40. Pagès, G., Pham, H.: Optimal quantization methods for nonlinear filtering with discrete-time observations. Bernoulli **11**(5), 893–932 (2005)
41. Robert, C.P., Casella, G.: Monte Carlo Statistical Methods. Springer Texts in Statistics. Springer, Berlin (2005)
42. Schön, T.B.: Estimation of nonlinear dynamic systems—theory and applications. PhD thesis, Linköping Studies in Science and Technology (2006). http://www.control.isy.liu.se/research/~reports/Ph.D.Thesis/PhD998.pdf
43. Seborg, D.E., Edgar, T.F., Mellichamp, D.A.: Process Dynamics and Control, 2nd edn. Wiley, New York (2003)
44. Strogatz, S.: Nonlinear Dynamics and Chaos: With Applications to Physics, Biology, Chemistry, and Engineering. Studies in Nonlinearity. Westview Press, Boulder (2001)
45. Tan, P.N., Steinbach, M., Kumar, V.: Introduction to Data Mining. Addison Wesley, Reading (2005)

46. Tanizaki, H.: Nonlinear and non-Gaussian state space modeling using sampling techniques. Ann. Inst. Stat. Math. **53**, 63–81 (2001). doi:10.1023/A:1017916420893
47. Wills, A., Ninness, B., Gibson, S.: Maximum likelihood estimation of state space models from frequency domain data. IEEE Trans. Autom. Control **54**(1), 19–33 (2009)
48. Zarchan, P., Musoff, H.: Fundamentals of Kalman Filtering: A Practical Approach, 2nd edn. Progress in Astronautics and Aeronautics. AIAA, Washington (2005)

Chapter 6
Averaging Analysis of Adaptive Algorithms Made Simple

Victor Solo

6.1 Introduction

Adaptive signal processing and adaptive control developed slowly and independently until the 1970s. Early workers in the signal processing areas were Widrow, creator of the LMS algorithm and his colleagues [2]. In the control and system identification community Peter Young was one of the pioneers [3].

From the 1970s there followed a very rapid theoretical and computational development fed partly by the computing revolution but also by demand from control and communications. By now there a number of books emphasizing methods of algorithm stability analysis and performance analysis, [1, 2, 4–8] and more particularly for adaptive control [9]. Peter Young's book [3] is one of the few discussing time varying parameter estimation in an offline setting; see also [10].

But even now the diffusion of analysis tools across the subdiscipline boundaries remains slow. Thus e.g. the powerful tools of averaging analysis remain under applied in adaptive signal processing while performance measures such as excess lag remain almost unknown in adaptive control. This is at least partly because a deterministic setting dominates in control while a stochastic setting is standard in adaptive signal processing. Use of averaging to analyse adaptive (or learning) algorithms outside signal processing and control e.g. in machine learning is essentially unknown. We hasten to add however that averaging analysis itself is widely used in other areas of applied mathematics, e.g. [11, 12].

Although numerous methods have been applied to the analysis of adaptive algorithm it began to become clear in the late 1970s that averaging methods were capable of providing, under reasonably realistic conditions, an analysis of the behaviour of just about any adaptive algorithm no matter how complicated. No other methods could do this or even come near.

V. Solo (✉)
School of Electrical Engineering, University of New South Wales, Sydney, Australia
e-mail: v.solo@unsw.edu.au

L. Wang, H. Garnier (eds.), *System Identification, Environmental Modelling, and Control System Design*,
DOI 10.1007/978-0-85729-974-1_6,

In applied mathematics averaging analysis has a long history being closely related to perturbation analysis [12]. It first emerged in a deterministic setting and then later stochastic versions were developed. There are three approaches to deterministic averaging; Bogoliubov's transformation technique; Gikhman's direct averaging technique; and finally the perturbed test function method of Papanicolau. The first method is very cumbersome and does not extend to a stochastic setting. The other two methods are relatively straightforward and extend easily to the stochastic setting. Both these methods have undergone considerable development since their initial introduction [1, 5, 7, 8]. The latter method may well be however ultimately the more powerful and we use it here (and in [1]). There seems to be no good reason for using Bogoliubov's method. Further discussion can be found in [1, Sects. 7.10, 8.7, 9.9, 10.3].

Stochastic averaging has been viewed as an advanced technique requiring a considerable level of mathematical sophistication. This is partly because it has been closely associated with weak convergence methods [7, 8] and the ode (ordinary differential equation) method [4]. But in fact averaging is a distinct method from these approaches and can be developed without them. This was indeed the agenda in [1] where new stochastic averaging methods were developed to parallel the deterministic approach. Further partly because of the significant success of the ode method it has often been assumed that discrete time algorithms can only be analysed by converting them into continuous time. Again this is simply not the case and was again part of the message of [1].

In this chapter we provide a simple heuristic approach to (discrete time) averaging analysis and illustrate the method on a non-trivial example. While the approach described here has been previously developed in the author's book [1] we expand on that discussion. While averaging has a long history as a method of perturbation analysis in applied mathematics, development has continued. And this brief review is timely if only due to recent advances in stochastic averaging made by the author [13, 14].

The remainder of the chapter is organised as follows. In Sect. 6.2 we briefly discuss adaptive algorithms in general and review the typology from [1]. In Sect. 6.3 we introduce adaptive algorithms and averaging via the best known example, the least mean square (LMS) algorithm. In Sect. 6.4 we illustrate averaging on a non-trivial example. In Sect. 6.5 we compare averaging briefly with other related approaches such as weak convergence and the ode method. Conclusions are in Sect. 6.5.

Acronyms AR(p) = autoregression of order p; LMS = least mean square; wp1 = with probability 1.

6.2 Adaptive Algorithms

In this section we introduce a basic classification of adaptive algorithms that points up crucial aspects of their behaviour. We then discuss some of the consequent stability analysis issues.

6.2.1 Algorithm Classification

A typical adaptive algorithm has the form,

$$\hat{\theta}_{new} = \hat{\theta}_{old} + \text{gain} \times \text{gradient} \times \text{error}.$$

This admits the following classification [1, Sect. 4.1],

(a) Long memory or Short memory algorithm.
For a long memory algorithm, gain $\to 0$ as time $\to \infty$. For a short memory algorithm, gain $>$ constant > 0 for all time. Only short memory algorithms can track time varying parameters. A long memory algorithm loses its ability to adapt as time increases. In practice then, with few exceptions, only short memory algorithms are used.

(b) Single Time Scale or Mixed Time Scale.
The gradient consists of external signals for a single time scale algorithm. Whereas for a mixed time scale algorithm the gradient is generated by an associated (fast) state equation driven by an external signal. Mixed time scale algorithms are usually much harder to analyse than single time scale algorithms.

(c) Gradient Construction.
'Instantaneous' steepest descent algorithms use only first order gradient information. 'Instantaneous' Newton algorithms use second order information.

There are other classifications but these are the main ones from an analysis point of view.

6.2.2 Stability Analysis Issues

Firstly we emphasize again that for practical use *short memory* or fixed gain algorithms are by far the most important. However long memory algorithms can sometimes occur in an auxiliary role. Indeed this is the case in our main illustration later. Unfortunately however much of the algorithm analysis in machine learning has concentrated on the long memory case; see [15] and comments therein.

A crucial feature of short memory adaptive algorithms is their ability to *track* time varying parameters. But by far the bulk of stability analysis is silent on this issue. This feature is however easily handled by averaging analysis and is discussed in [1, 5].

From the point of view of stability analysis, single time scale systems are much easier to analyze than *mixed time scale* systems. Although the averaging approach works just as well in each case. For this reason we consider only single time scale algorithms in this expose.

With mixed time scale systems a crucial issue is the stability of the associated fast state equation. Particularly with stochastic algorithms this has proved to be a major problem known as the *boundedness problem* [16]. In fact the boundedness problem is more general but particularly severe for mixed time scale problems. Numerous monitoring schemes have been introduced but until recently with limited success. Recently the author has introduced a new approach which resolves this problem for both single and mixed time scale algorithms [13, 14].

Since the algorithms of interest operate in real time then the stability analysis of most import is a *realization wise analysis* i.e. analysis with probability one (wp1). That is precisely what averaging analysis is able to deliver. Unfortunately much of the current analysis in the signal processing literature focuses on convergence in probability i.e. analysis across realizations and so is not directly relevant to adaptive algorithm stability.

In this discussion we consider only *first order analysis* i.e. stability analysis. For second order analysis i.e. analysis of fluctuations about the first order trajectory the reader is referred to [1, 5]. We simply note that the nature of the gradient term in the adaptive algorithm controls the size of these fluctuations.

Finally we consider the nature of the analysis. Since adaptive algorithms operate in real time we have only one realization to deal with. This mean the appropriate analysis mode is realization-wise and not averaged across realizations. This means one needs to consider convergence wp1 and not convergence in probability. This fundamental point is appreciated to a certain extent in the adaptive control literature and almost completely unknown in adaptive signal processing. Indeed in adaptive signal processing the standard mode of analysis uses convergence in probability. As we illustrate below it is usually the case that fixed gain algorithms do not converge, rather they hover in the vicinity of the equilibrium points of the averaged system. These two fundamental points are poorly appreciated if at all.

6.3 The LMS Algorithm

We begin by recalling perhaps the best known adaptive algorithm; namely the least mean squares algorithm (LMS) due originally to [2, 17]. We use a discussion of its stability as a means of motivating and introducing averaging analysis.

6.3.1 LMS Defined

Consider on-line estimation of a finite impulse response (FIR) filter relating two observed signals y_k, u_k. If the filter has p taps then the relation can be described as a regression

$$y_k = x_k^T w + \varepsilon_k, \tag{6.1}$$

where w is a p-dimensional *weight* vector of filter taps; $x_k = (u_{k-1}, \dots, u_{k-p})^T$; and ε_k is a noise sequence independent of x_k. The LMS algorithm attempts to minimize the squared error $e_k^2(w)$, $e_k(w) = y_k - x_k^T w$ by an instantaneous steepest descent. Since $\frac{\partial e_k^2(w)}{\partial w} = -x_k e_k(w)$ we get an update,

$$w_{k+1} = w_k + \mu x_k e_k,$$
$$e_k = y_k - x_k^T w_k,$$

where μ is a step size or *gain*.

A fundamental question for any adaptive algorithm is its stability. To pursue that we always need first to convert the algorithm into so-called *error form*. Introduce the deviation $\tilde{w}_{k+1} = w_{k+1} - w_k$ and use the regression equation from the LMS update to find,

$$\tilde{w}_{k+1} = \tilde{w}_k - \mu x_k x_k^T \tilde{w}_k + \mu x_k \varepsilon_k. \tag{6.2}$$

This is a linear time-varying difference equation and we need to analyse its stability. We call this the *primary* system. We note that it has the general form,

$$\delta\tilde{w}_{k+1} = \tilde{w}_{k+1} - \tilde{w}_k = \mu f(k, \tilde{w}_k), \tag{6.3}$$

where in the LMS case, $f(k, \tilde{w}) = -x_k x_k^T \tilde{w} + x_k \varepsilon_k$.

Before proceeding we introduce an assumption on the signals,

S1 x_k, ε_k are jointly strictly stationary and independent.

From S1 and the ergodic theorem,

$$\frac{1}{M}\sum_{s=k}^{k+M} x_s x_s^T \to R_x \quad \text{wp1 as } M \to \infty,$$

$$\frac{1}{M}\sum_{1}^{M} x_s \varepsilon_s \to 0 \quad \text{wp1 as } M \to \infty.$$

We now consider stability analysis.

6.3.2 LMS Equilibrium Points

The most fundamental aspect of stability analysis is to find the equilibrium points $\tilde{w}_e$ of the primary system if any. We find equilibrium points by setting $\delta\tilde{w}_{k+1} = 0$. This yields the requirement,

$$x_k x_k^T \tilde{w}_e + x_k \varepsilon_k = 0 \quad \text{for all } k.$$

Multiplying through by $\tilde{w}_e^T$ gives,

$$(x_k^T \tilde{w}_e)^2 + x_k^T \tilde{w}_e \varepsilon_k = 0 \quad \text{for all } k.$$

This implies $x_k^T \tilde{w}_e = 0$ or $x_k^T \tilde{w}_e + \varepsilon_k = 0$ for all k wp1. Since ε_k, x_k are independent the latter cannot hold. Thus $x_k^T \tilde{w}_e = 0$ whereupon $x_k \varepsilon_k = 0$ wp1 which again cannot hold.

We conclude the equilibrium point condition has no solution. This delivers,

Result 6.1 Under condition S1 the LMS algorithm has no equilibrium points!

In particular this means the LMS algorithm cannot converge! Although this result is understood to some extent in parts of the adaptive signal processing community it is hard to find a reference clearly stating this. Indeed some references give the contrary impression. This is partly because most discussions do not treat convergence wp1 rather only convergence in probability.

6.3.3 *Averaged LMS System*

To find out what then happens, we approximate the primary system as follows. Change the time index to s and then sum over a time interval $s = k$ to $s = k + N$, N to be chosen,

$$\tilde{w}_{k+N+1} = \tilde{w}_k - \mu \sum_k^{N+k} x_s x_s^T \tilde{w}_s.$$

Now since μ is small, then in the sum if N is not too large then $\tilde{w}_s$ will not differ too much from $\tilde{w}_k$ so we can approximate the equation by,

$$\bar{w}_{k+N+1} = \bar{w}_k - \mu \sum_k^{N+k} x_s s_s^T \bar{w}_k.$$

On the other hand, in view of S1 and the ergodic theorem, if N is large enough, then $\sum_k^{N+k} x_s x_s^T \approx N R_x$. Thus we get, $\bar{w}_{k+N+1} = \bar{w}_k - \mu N R_x \bar{w}_k$; or differencing, we get the *averaged* system,

$$\bar{w}_{k+1} = \bar{w}_k - \mu R_x \bar{w}_k = (I - \mu R_x)\bar{w}_k. \tag{6.4}$$

This is a linear time-invariant difference equation and its stability analysis is straightforward. In particular,

Result 6.2 (LMS Averaged System) For the averaged system (6.4), let $\lambda_{\max}$ be the largest eigenvalue of R_x and $\lambda_{\min}$ the smallest. Then provided, $0 < \mu\lambda_{\min} < \mu\lambda_{\max} < 2$ we have $\bar{w}_k \to 0$ as $k \to \infty$.

Proof Elementary and omitted. □

The question now is; what is the relation between the primary and averaged systems? We postpone a formal treatment of this question. But let us say informally that under certain regularity conditions the primary system trajectory hovers in the vicinity of the averaged system trajectories.

6.3.4 *Averaging Analysis*

Now let us apply the same approximation method to the general system (6.3). This leads to the averaged system,

$$\delta \bar{w}_{k+1} = \mu f_{\mathrm{av}}(\bar{w}_k), \tag{6.5}$$

where we assume,

A1 $f_{\mathrm{av}}(\bar{w}) = \lim_{M\to\infty} \frac{1}{M} \sum_{s=k}^{k+M} f(k, \bar{w})$ exists.

The connexion between the primary system and averaged system is addressed via,

Result 6.3 (Finite Time Hovering Theorem [1, Sect. 7.2]) Consider the primary system (6.3) and its averaged counterpart (6.5). Assume they start from the same initial condition. And let $T > 0$ be fixed. Then under certain technical conditions,

$$\max_{1\le k\le T/\mu} \|\tilde{w}_k - \bar{w}_k\| \le c_T(\mu) b_T,$$

where $b_T > 0$ is a constant and $c_T(\mu) \to 0$ as $\mu \to 0$.

The result says that the primary system trajectory hovers in the vicinity of the averaged system trajectory uniformly for all time on any finite time interval. Since by result II the averaged system converges to 0, then the primary system hovers around 0.

To develop an infinite time version of the result where T/μ is replaced by ∞ requires some extra technical work for which the reader is referred to [1].

To state the technical conditions we introduce the perturbation function,

$$p(k, \tilde{w}) = \sum_{1}^{k} [f(s, \tilde{w}) - f_{\mathrm{av}}(\tilde{w})].$$

There are five technical conditions [1, Sect. 9.2: conditions 9.2A1–9.2A5]. The main two are,

A2 $f(k, w)$ obeys a stochastic Lipschitz condition.

$$\|f(k, w) - f(k, w')\| \le L_k \|w - w'\|,$$

where L_k obeys a strong law of large numbers, $\lim_{M\to\infty} \frac{1}{M} \sum_{1}^{M} L_k \to L$ wp1.

A3 The perturbation function $p(k,w)$ obeys a stochastic Lipschitz condition

$$\|p(k,w)-p(k,w')\| \le M_k\|w-w'\|,$$

where $M_k/k \to 0$ wp1 as $k\to\infty$. Also $\|p(k,0)\| \le p_k$ where $p_k/k \to 0$ as $k\to\infty$.

In [1, Condition 9.2A1] it is assumed that $f_{\text{av}}(w)=E(f(k,w))$ which is time invariant. But a perusal of the proof shows that we can dispense with the time invariance (it is not actually used in the proof) and we can replace the definition of $f_{\text{av}}(w)$ with that in A1. There are two other technical conditions. The first [1, Condition 9.2A4] requires the trajectory of the averaged system to be bounded. This is trivially satisfied for LMS and the example in Sect. 6.4 below. The second requires that $\|f_{\text{av}}(w)\| \le B$ when $\|w\| \le h$. Again for LMS and the example below this is trivially satisfied.

Turning to the main conditions, in the case of the LMS algorithm we have firstly

$$\|f(k,w)-f(k,w')\| = \|x_k x_k^T(w-w')\| \le \|x_k\|^2\|w-w'\|$$

and $\lim_{M\to\infty}\frac{1}{M}\sum_1^M\|x_k\|^2 \to \text{tr}(R_x)$ wp1 by S1 and the ergodic theorem.

Secondly, $p(k,w)=\sum_1^k[x_s x_s^T - R_x]w$ and so,

$$M_k/k = \left\|\frac{1}{k}\sum_1^k[x_s x_s^T - R_x]\right\| \to 0 \quad \text{as } k\to\infty \text{ by S1 and the ergodic theorem.}$$

Also in this case $\|\frac{1}{k}p(k,0)\| = \|\frac{1}{k}\sum_1^k x_s\varepsilon_s\| \to 0$ by S1 and the ergodic theorem. So A2, A3 hold for the LMS algorithm and we obtain result (6.3).

6.4 A More Difficult Illustration

Here we consider a more difficult example based on a modification of the recent work of [18]. We assume the linear model (6.1) but now the regressors are lagged values of an observed stationary autoregressive time series AR(p),

$$\begin{aligned} u_k &= \sum_1^q u_{k-r}a_r + \nu_{o,k} \\ &= \zeta_k^T a + \nu_{o,k}, \\ \zeta_k^T &= (u_{k-1},\dots,u_{k-q}), \end{aligned}$$

where $\nu_{o,k}$ is a white noise sequence independent of ε_k. We can then write the regressor vector x_k as,

$$x_k = Z_k a + v_k,$$

$$Z_k = \begin{pmatrix} \zeta_{k-2}^T \\ \vdots \\ \zeta_{k-p}^T \end{pmatrix},$$

$$v_k = (\nu_{o,k-1}, \ldots, \nu_{o,k-p})^T.$$

The algorithm has two components. An update for w_k and an auxiliary update for $\hat{a}_k$ which is a long memory estimator of a. To keep the analysis manageable we have replaced the short memory estimator of a used by [18]. The long memory estimator of a is the least squares estimator

$$\hat{a}_k = \left(\sum_1^k \zeta_s \zeta_s^T\right)^{-1} \left(\sum_1^k \zeta_s u_s\right). \tag{6.6}$$

The update for w_k is,

$$w_{k+1} = w_k + \mu \frac{\hat{v}_k}{\|\hat{v}_k\|^2} e_k, \tag{6.7}$$

$$e_k = y_k - x_k^T w_k,$$

$$\hat{v}_k = x_k - Z_k \hat{a}_k.$$

We call this the LMSAR algorithm. The error system is,

$$\tilde{w}_{k+1} = \tilde{w}_k - \mu \frac{\hat{v}_k x_k^T}{\|\hat{v}_k\|^2} \tilde{w}_k + \mu \frac{\hat{v}_k}{\|\hat{v}_k\|^2} \nu_{o,k}. \tag{6.8}$$

We now introduce some assumptions on the signals.

S2 $x_k, \nu_{o,k}, \varepsilon_k$ are jointly strictly stationary independent and each has finite variance; further for some $\delta > 0$ $E(|\nu_{o,k}|^{4+2\delta}) < \infty$.

S3 $E[\frac{1}{\|v_k\|^{4+2\delta}}] < \infty$ for some $\delta > 0$.

S3 holds e.g. if v_k is multivariate Gaussian.

6.4.1 Equilibrium Points of LMSAR

As with LMS we seek equilibrium points by setting $\delta \tilde{w}_{k+1} = 0$. This leads to

$$\frac{\hat{v}_k x_k^T}{\|\hat{v}_k\|^2} \tilde{w}_k + \frac{\hat{v}_k}{\|\hat{v}_k\|^2} \nu_{o,k} = 0 \quad \Rightarrow \quad \hat{v}_k = 0 \quad \text{or} \quad x_k^T \tilde{w}_k + \nu_{o,k} = 0.$$

And repeating the LMS type analysis we find the latter condition cannot hold for any choice of $\tilde{w}_e$. The former condition also cannot hold since $\hat{v}_k$ is stochastic. We conclude,

Result 6.4 Under condition S2 the LMSAR algorithm (6.7), (6.6) has no equilibrium points.

Again this means LMSAR cannot converge.

6.4.2 Averaged LMSAR System

From (6.8) the averaged system is (6.5) where,

$$f_{\mathrm{av}}(\bar{w}) = \lim_{N\to\infty} \frac{1}{N}\sum_{1}^{N} f(k,\bar{w}), \quad f(k,\bar{w}) = -\frac{\hat{v}_k x_k^T}{\|\hat{v}_k\|^2}\bar{w} + \frac{\hat{v}_k}{\|\hat{v}_k\|^2}\nu_{o,k}.$$

We need to compute $f_{\mathrm{av}}(\bar{w})$. Note that $\hat{v}_k = v_k - Z_k\tilde{a}_k$, $\tilde{a}_k = \hat{a}_k - a$. We then use the following result.

Result 6.5 Under conditions S2, S3

$$\tilde{a}_k \to 0 \quad \text{wp1 as } k \to \infty.$$

$$\|\hat{v}_k - v_k\|/\|v_k\| \to 0 \quad \text{as } k \to 0 \text{ wp1}.$$

Proof See Appendix A.1. □

To aid further analysis we introduce the idealized signal

$$f_o(k,\bar{w}) = -\frac{v_k x_k^T}{\|v_k\|^2}\bar{w} + \frac{v_k}{\|v_k\|^2}\nu_{o,k}.$$

The ergodic theorem delivers, $\lim_{N\to\infty}\frac{1}{N}\sum_1^N f_o(k,\bar{w}) = -G\bar{w}$, where $G = E(\frac{v_k x_k^T}{\|v_k\|^2})$ and we have used the fact that, $E(\frac{v_k}{\|v_k\|^2}\nu_{o,k}) = E(\frac{v_k}{\|v_k\|^2})E(\nu_{o,k}) = 0$. In Appendix A.4 we show $\lim_{N\to\infty}\frac{1}{N}\sum_1^N (f(k,\bar{w}) - f_o(k,\bar{w})) = 0$ wp1 and so the averaged system is,

Result 6.6 Under conditions S2, S3 the averaged LMSAR system is $\delta\bar{w}_{k+1} = -\mu G\bar{w}_k$.

Now we need to analyse the behaviour of the averaged system. For this we need to investigate the eigenvalues of A. Since the AR polynomial is stable we have a Wold representation $u_k = \sum_0^\infty c_r \nu_{o,k-r}$ where $\frac{1}{A(z^{-1})} = \sum_0^\infty c_r z^{-r}$. Using this in the definition of G we find,

$$G = E\left(\frac{1}{\|v_k\|^2}\right)\begin{pmatrix}\nu_{o,k-1}\\ \vdots \\ \nu_{o,k-p}\end{pmatrix}(\nu_{o,k-1},\ldots,\nu_{o,k-p})^T L + \Delta,$$

$$L = \begin{pmatrix} 1 & 0 & \dots & 0 \\ c_1 & 1 & \dots & 0 \\ \vdots & \vdots & \ddots & 0 \\ c_{p-1} & c_{p-2} & \dots & 1 \end{pmatrix},$$

$$\Delta_r = E(m_k \xi_{k-p-r}), \quad r \geq 1,$$

where ξ_{k-r-p} depends on $H_{k-p-r} = (v_{o,k-r-p}, v_{o,k-r-p-1}, \dots)$ and m_k is a vector depending on $(v_{o,k-1}, \dots, v_{o,k-p})$. Then by iterated conditional expectation

$$\Delta_r = E(E(m_k | H_{k-r-p})\xi_{k-r-p}) = m_{o,k} E(\xi_{k-r-p}) = 0.$$

We deduce that $G = RL$, $R = E(v_k v_k^T / \|v_k\|^2)$.

Result 6.7 For the averaged system of result (6.6), let $\lambda_{\max}$ be the largest eigenvalue of RL and $\lambda_{\min}$ the smallest. Then provided, $0 < \mu\lambda_{\min} < \mu\lambda_{\max} < 2$ we have $\bar{w}_k \to 0$ as $k \to \infty$.

Proof Elementary and omitted. □

We can get some information about $\lambda_{\max}$ as follows. We have,

$$|\lambda_{\max}| \leq \|G\| = \|RL\| \leq \|R\| \|L\|.$$

Now for any fixed α,

$$\alpha^T R \alpha = E((v_k^T \alpha)^2 / \|v_k\|^2) \leq \alpha^T \alpha \quad \Rightarrow \quad \|R\| \leq 1.$$

Next to find $\|L\|$ we have to consider $\alpha^T L^T L \alpha$. However $L^T L$ is very close to the Toeplitz matrix Ω of autocovariances for the MA process $y_t = \eta_t + \sum_1^p c_r \eta_{t-r}$ where η_t is a unit variance white noise. Except for the first diagonal entry the remaining diagonal entries are too small. Now let $F(\omega)$ be the corresponding MA spectrum. Thus,

$$\begin{aligned} \alpha^T L^T L \alpha \leq \alpha^T \Omega \alpha &= \sum \sum \alpha_t \alpha_s \int_{-\pi}^{\pi} e^{j\omega(t-s)} F(\omega) \frac{d\omega}{2\pi} \\ &= \int_{-\pi}^{\pi} \left| \sum \alpha_t e^{j\omega t} \right|^2 F(\omega) \frac{d\omega}{2\pi} \\ &\leq F_{\max} \int_{-\pi}^{\pi} \left| \sum \alpha_t e^{j\omega t} \right|^2 \frac{d\omega}{2\pi} = \alpha^T \alpha F_{\max}, \end{aligned}$$

where $F_{\max} = \max_\omega F(\omega)$. Thus the upper bound of result (6.7) holds if $\mu F_{\max} < 2$.

6.4.3 Averaging Analysis of LMSAR

We now turn to checking conditions A2, A3; as indicated earlier the other technical conditions are trivially satisfied.

For A3, we find easily that $M_k/k = \|m_{a,k} + m_{b,k}\|$ where,

$$m_{a,k} = \frac{1}{k}\sum_1^k\left(\frac{v_s x_s^T}{\|v_s\|^2} - G\right),$$

$$m_{b,k} = \frac{1}{k}\sum_1^k\left(\frac{\hat{v}_s}{\|\hat{v}_s\|^2} - \frac{v_s}{\|v_s\|^2}\right)x_s^T.$$

By S2 and the ergodic theorem, $m_{a,k} \to 0$ wp1. In Appendix A.2 we show $m_{b,k} \to 0$ wp1. Next $\|\frac{1}{k}p(k,0)\| \le \|c_k\| + \|b_k\|$ where,

$$c_k = \frac{1}{k}\sum_1^k \frac{v_k}{\|v_k\|}v_{o,k} \quad \text{and} \quad b_k = \frac{1}{k}\sum_1^k\left(\frac{\hat{v}_s}{\|\hat{v}_s\|^2} - \frac{v_s}{\|v_s\|^2}\right)v_{o,s}^T.$$

Now $\|b_k\| \to 0$ by the same argument as used for $m_{b,k}$. And by S2, S3 and the ergodic theorem, $c_k \to E(\frac{v_k}{\|v_k\|}v_{o,k}) = 0$ wp1.

For A2, we also find easily that $L_k = \frac{\|x_k\|}{\|\hat{v}_k\|}$. Introduce $L_{o,k} = \frac{\|x_k\|}{\|v_k\|}$. By S2 and the ergodic theorem $\frac{1}{N}\sum_1^N L_{o,k} \to E(\frac{\|x_k\|}{\|v_k\|})$ wp1. However,

$$E\left(\frac{\|x_k\|}{\|v_k\|}\right) \le \sqrt{E(\|x_k\|^2)E\left(\frac{1}{\|v_k\|^2}\right)} < \infty \quad \text{by S2, S3.}$$

In Appendix A.3 we show $\frac{1}{N}\sum_1^N(L_{o,k} - L_k) \to 0$ wp1.

Thus A2, A3 are established and (6.3) follows.

6.5 Comparison with Other Approaches

Our approach may be compared with the weak convergence method. Firstly our approach is much simpler since no weak convergence framework is needed. We make repeated use of the ergodic theorem, elementary wp1 properties of stochastic sequences and elementary bounding arguments. In fact our approach mimics the kind of arguments used in deterministic averaging analysis and thus shows the strong connexions between the two ([1] does both). Secondly the weak convergence approach is not capable of delivering results valid for fixed μ like the Hovering theorem. Thirdly the weak convergence method can only deliver continuous time approximations. It cannot provide the type of stability conditions obtained in Results 2 and 7. These last two criticisms apply also to the ode method.

On a deeper level we note that in both single time scale and mixed time scale cases, the weak convergence method can in principle produce results when the Lipschitz condition A2 is required of $f_{av}(\bar{w})$ and not of $f(k,w)$ [5, 7]. This is significant since it allows discontinuous $f(k,w)$ but the corresponding $f_{av}(\bar{w})$ is nevertheless usually continuous. We say in principle because there remain doubts about the boundedness condition in these works. However the corresponding result in the deterministic case is already given in [1] and the stochastic version is developed in more recent work [13, 14] where in both single time scale and mixed time scale cases the boundedness problem has been overcome by a novel monitoring procedure.

Appendix

The following lemmas are used repeatedly below.

Lemma 6.1 *If ξ_s is a strictly stationary sequence of random vectors with $E\|\xi_s\| < \infty$ then, $\xi_s/s \to 0$ as $s \to \infty$.*

Proof By the ergodic theorem, $\bar{\xi}_n = \frac{1}{n}\sum_1^n \|\xi_s\| \to E\|\xi_0\|$ wp1. Now apply the update rule for the sample mean,

$$\bar{\xi}_n = \bar{\xi}_{n-1} + \frac{1}{n}(\|\xi_n\| - \bar{\xi}_{n-1})$$

$$\Rightarrow \quad \frac{1}{n}\|\xi_n\| = \bar{\xi}_n - \left(1 - \frac{1}{n}\right)\bar{\xi}_{n-1} \to E\|\xi_0\| - E\|\xi_0\| = 0 \quad \text{as } n \to \infty. \qquad \square$$

Remark Suppose $\delta > 0$ and $E\|\xi_0\|^{2+\delta} < \infty$ then as $s \to \infty$, $\|\xi_s\|^{2+\delta}/s \to 0 \equiv \|\xi_s\|/s^{\frac{1}{2}-\varepsilon} \to 0$ where $\varepsilon = \delta/(2(2+\delta))$.

Lemma 6.2 *If d_s, ξ_s are sequences of positive random variables with $d_s \to 0$ wp1 as $s \to \infty$ and $\frac{1}{n}\sum_1^n \xi_s \to c < \infty$ wp1 as $n \to \infty$, then,*

$$T_n = \frac{1}{n}\sum_1^n d_s\xi_s \to 0 \quad \textit{wp1 as } n \to \infty.$$

Proof We write,

$$T_n = \frac{1}{n}\sum_1^n \xi_s \sum_1^n d_s a_{n,s},$$

$$a_{n,s} = \xi_s \bigg/ \left[\sum_1^n \xi_s\right], \quad \sum_1^n a_{n,s} = 1.$$

The first term in $T_n \to c$. By the Toeplitz lemma [1] the second term $\to 0$ if $a_{n,s} \to 0$ wp1. But $a_{n,s} = \frac{\xi_s}{n}/[\frac{1}{n}\sum_1^n \xi_s] \to 0$ as $n \to \infty$ since s is fixed. □

Lemma 6.3

$$\big|\|a\| - \|b\|\big| \leq \|a - b\|.$$

Proof Square both sides to obtain

$$\begin{aligned} a^T a + b^T b - 2\|a\|\|b\| &\leq a^T a + b^T b - 2a^T b \\ &\equiv 2a^T b \leq \|a\|\|b\| \end{aligned}$$

which holds by the Cauchy-Schwarz inequality.

Returning to the main topic of this appendix we have,

$$m_{b,k} \leq \frac{1}{k}\sum_1^k \|\Delta_s\|\|x_s\|,$$

$$\Delta_s = \hat{v}_s/\|\hat{v}_s\| - v_s/\|v_s\|.$$ □

Lemma 6.4 *If* $\|\hat{v}_s - v_s\|/\|v_s\| \to 0$ *as* $s \to \infty$ *then,*

(a) $\|\hat{v}_s - v_s\|/\|\hat{v}_s\| \to 0$,
(b) $\big|\|\hat{v}_s\|/\|v_s\| - 1\big| \to 0$,
(c) $\big|\|v_s\|/\|\hat{v}_s\| - 1\big| \to 0$,
(d) $|\hat{v}_s^T v_s/[\|v_s\|\|\hat{v}_s\|] - 1| \to 0$.

Proof (a) Set $d_s = \|\hat{v}_s - v_s\|$. Then

$$\begin{aligned} \|\hat{v}_s - v_s\|/\|\hat{v}_s\| &= d_s/\|v_s + \hat{v}_s - v_s\| \\ &\leq d_s/\big|\|v_s\| - d_s\big| \quad \text{by Lemma 6.3} \\ &= [d_s/\|v_s\|]/[1 - d_s/\|v_s\|] \to 0. \end{aligned}$$

(b) $\big|\|\hat{v}_s\|/\|v_s\| - 1\big| = \big|\|\hat{v}_s\| - \|v_s\|\big|/\|v_s\| \leq \|\hat{v}_s - v_s\|/\|v_s\| \to 0$ by Lemma 6.3.

(c) $\big|\|v_s\|/\|\hat{v}_s\| - 1\big| = \big|\|v_s\| - \|\hat{v}_s\|\big|/\|\hat{v}_s\| \leq \|\hat{v}_s - v_s\|/\|\hat{v}_s\| \to 0$ by (a).

(d) $$\begin{aligned} |\hat{v}_s^T v_s/[\|v_s\|\|\hat{v}_s\|] - 1| &\leq |(\hat{v}_s - v_s)^T v_s|/[\|v_s\|\|\hat{v}_s\|] + \|v_s\|/[\|\hat{v}_s\|] - 1| \\ &\leq \|\hat{v}_s - v_s\|/\|\hat{v}_s\| + |\|v_s\|/[\|\hat{v}_s\|] - 1| \to 0 \text{ by (a), (c).} \end{aligned}$$

□

A.1

We have $\tilde{a}_k = (\sum_1^k \zeta_s \zeta_s^T)^{-1}(\sum_1^k \zeta_s \nu_{o,s})$. By the ergodic theorem,

$$\frac{1}{k}\sum_1^k \zeta_s \zeta_s^T \to E(\zeta_0 \zeta_0^T) = R_\zeta \quad \text{wp1},$$

$$\frac{1}{k}\sum_1^k \zeta_s \nu_{o,s} \to E(\zeta_0 \nu_{o,0}) = 0$$

and the first part is established. But we need something stronger provided in the second part.

We claim $k^{\frac{1}{2}-\varepsilon}\tilde{a}_k \to 0$ wp1. This follows if $\sum_1^k \zeta_s \nu_{o,s}/k^{\frac{1}{2}+\varepsilon} \to 0$. By Kronecker's lemma [1] this follows if,

$$\sum_1^\infty \zeta_k \nu_{o,k}/k^{\frac{1}{2}+\varepsilon} < \infty \quad \text{wp1}.$$

By the martingale convergence theorem [1] this follows if,

$$\sum_1^\infty \|\zeta_s\|^2 \nu_{o,s}^2/s^{1+2\varepsilon} < \infty \quad \text{wp1}$$

which follows if

$$\sum_1^\infty E(\|\zeta_s\|^2 \nu_{o,s}^2)/s^{1+2\varepsilon} < \infty \quad \text{wp1}.$$

This holds by S2 since,

$$E(\|\zeta_s\|^2 \nu_{o,s}^2) = E(\|\zeta_s\|^2)E(\nu_{o,s}^2) < \infty.$$

Now consider that,

$$\begin{aligned}\|\hat{v}_k - v_k\|/\|v_k\| &\le \|Z_k\|\|\tilde{a}_k\|/\|v_k\| \\ &= \left(k^{-\frac{1}{2}+\varepsilon}\frac{\|Z_k\|}{\|v_k\|}(\|\tilde{a}_k\|k^{\frac{1}{2}-\varepsilon})\right).\end{aligned}$$

Now the second term $\to 0$ and the first does also by the remark following Lemma 6.1, if, $E(\|Z_k\|/\|v_k\|)^{2+\delta} < \infty$. By the Cauchy-Schwarz inequality this follows from S2, S3.

A.2

Proof That $m_{b,k} = \frac{1}{k}\sum_1^k (\frac{\hat{v}_s}{\|\hat{v}_s\|^2} - \frac{v_s}{\|v_s\|^2})x_s^T \to 0$.

By Lemma 6.2 this will hold if

$$\Delta_s = \left\| \frac{\hat{v}_s}{\|\hat{v}_s\|^2} - \frac{v_s}{\|v_s\|^2} \right\| \|v_s\| \to 0,$$

$$\frac{1}{n}\sum_1^n \|x_s\|/\|v_s\| \to c < \infty \quad \text{wp1}.$$

The second follows from S2, S3. For the first

$$\Delta_s^2 = \|v_s\|^2[1/\|\hat{v}_s\|^2 + 1/\|v_s\|^2 - 2\hat{v}_s^T v_s/(\|\hat{v}_s\|^2\|v_s\|^2)]$$
$$= \frac{\|v_s\|}{\|\hat{v}_s\|}[\|v_s\|/\|\hat{v}_s\| + \|\hat{v}_s\|/\|v_s\| - 2\hat{v}_s^T v_s/(\|\hat{v}_s\|\|v_s\|)].$$

By Result 6.5 and Lemmas 6.3(b), 6.3(c), 6.3(d) the term in square rackets, denoted $[\cdot] \to 0$. Then by Lemma 6.3(c)

$$\Delta_s^2 = (\|v_s\|/\|\hat{v}_s\| - 1)[\cdot] + [\cdot] \to 0.$$

□

A.3

Proof That $\frac{1}{N}\sum_1^N (L_{o,k} - L_k) \to 0$ wp1.

We have,

$$|L_{o,k} - L_k| = \|x_k\||1/\|\hat{v}_k\| - 1/\|v_k\||$$
$$= \frac{\|x_k\|}{\|v_k\|}|\|\hat{v}_k\| - \|v_k\||/\|\hat{v}_k\|$$
$$\le \frac{\|x_k\|}{\|v_k\|}\|\hat{v}_k - v_k\|/\|\hat{v}_k\| \quad \text{by Result 6.5 and Lemma 6.3.}$$

And the result follows from Lemma 6.4a, S3 and Lemma 6.2. □

A.4

Proof That $\lim_{N\to\infty} T_N = 0$ wp1; $T_N = \frac{1}{N}\sum_1^N (f(k,\bar{w}) - f_o(k,\bar{w}))$.

In fact T_N is a sum of two terms. The first is just $m_{b,N}$ of Appendix A.2. The second has the same form as $m_{b,N}$ except with x_s^T replaced by $\nu_{o,s}$. So the result follows from Appendix A.2. □

References

1. Solo, V., Kong, X.: Adaptive Signal Processing Algorithms. Prentice Hall, Upper Saddle River (1995)
2. Widrow, B., Stearns, S.: Adaptive Signal Processing. Prentice Hall, London (1985)
3. Young, P.: Recursive Estimation and Time Series Analysis. Springer, Berlin (1984)
4. Ljung, L.: Theory and Practice of Recursive Identification. MIT Press, Cambridge (1983)
5. Benveniste, A., Metivier, M., Priouret, P.: Adaptive Algorithms and Stochastic Approximations. Springer, New York (1990)
6. Kushner, H., Clark, D.: Stochastic Approximation Methods for Constrained and Unconstrained Systems. Springer, New York (1978)
7. Kushner, H.: Approximation and Weak Convergence Methods for Random Processes with Application to Stochastic System Theory. MIT Press, Cambridge (1984)
8. Kushner, H.J., Yin, G.: Stochastic Approximation Algorithms and Applications. Springer, New York (1997)
9. Sastry, S., Bodson, M.: Adaptive Control. Prentice Hall, New York (1989)
10. Kitagawa, G., Gersch, W.: Smoothness Priors Analysis of Time Series. Springer, Berlin (1996)
11. Sanders, J., Verhulst, F.: Averaging Methods in Nonlinear Dynamical Systems. Springer, New York (1985)
12. Nayfeh, A.: Introduction to Perturbation Techniques. Wiley, New York (1981)
13. Solo, V.: Averaging analysis of a point process adaptive algorithm. J. Appl. Probab. A **41**, 361–372 (2004)
14. Solo, V.: On the boundedness problem of stochastic adaptive algorithms. In: Proc. IEEE CDC, Beijing China, pp. 3472–3476. IEEE, New York (2009)
15. Tadic, V.: Asymptotic analysis of temporal-difference learning algorithms with constant stepsizes. Mach. Learn. **63**, 107–133 (2006)
16. Andrieu, C., Moulines, E., Priouret, P.: Stability of stochastic approximation under verifiable conditions. In: 44th IEEE Conference on Decision and Control and the European Control Conference 2005, Seville, Spain, December 12–15, 2005, pp. 6656–6661. IEEE, New York (2005)
17. Widrow, B., Hoff, M.: Adaptive switching circuits. In: IRE Wescon Convention Record Part IV, pp. 96–104 (1960)
18. de Almeida, S., Bermudez, J., Bershad, N.: A stochastic model for a pseudo affine projection algorithm. IEEE Trans. Signal Process. **57**, 107–118 (2009)

Chapter 7
Graphs for Dependence and Causality in Multivariate Time Series

Christoph Flamm, Ulrike Kalliauer, Manfred Deistler, Markus Waser, and Andreas Graef

7.1 Introduction

In this paper we describe and discuss measures of dependence between single time series in the context of a multivariate stationary process.

We consider an n-dimensional stochastic process $(x(t))_{t\in\mathbb{Z}}$, $x(t) : \Omega \to \mathbb{R}^n$, which is weakly stationary with mean zero. Its covariance function is given as $\gamma(s) = \mathbb{E}\,x(t+s)x(t)'$. Although the covariance function in general does not contain the full information about the underlying stochastic process, the analysis presented in this paper is based on the covariance only.

As is well known, see [36] and [23], a stationary process has a representation of the form

$$x(t) = \int_{-\pi}^{\pi} e^{it\lambda}\, dz(\lambda), \tag{7.1}$$

C. Flamm (✉) · M. Deistler · M. Waser · A. Graef
Institute for Mathematical Methods in Economics, Vienna University of Technology, Vienna, Austria
e-mail: christoph.flamm@tuwien.ac.at

M. Deistler
e-mail: manfred.deistler@tuwien.ac.at

M. Waser
e-mail: markus.waser@tuwien.ac.at

A. Graef
e-mail: andreas.graef@tuwien.ac.at

U. Kalliauer
VERBUND Trading AG, Vienna, Austria
e-mail: ulrike.kalliauer@verbund.com

L. Wang, H. Garnier (eds.), *System Identification, Environmental Modelling, and Control System Design*,
DOI 10.1007/978-0-85729-974-1_7,

where $(z(\lambda)|\lambda \in [-\pi,\pi])$, $z(\lambda) : [-\pi,\pi] \to \mathbb{C}^n$ is a random process with orthogonal increments, which is uniquely defined by $x(t)$.

The *spectral distribution function* $F(\lambda)$ of $x(t)$ is defined by $F(\lambda) = \mathbb{E} z(\lambda)z(\lambda)^*$, where $\cdot^*$ denotes the conjugate transpose. For convenience we will use the notation $dF(\lambda) = \mathbb{E} dz(\lambda)dz(\lambda)^*$. Note that $dF(\lambda)$ describes the importance of a frequency band in terms of its contribution to the overall variance.

Under the assumption $\sum_{s=-\infty}^{\infty} \|\gamma(s)\| < \infty$ the spectral distribution function is absolutely continuous, and the *spectral density function* is defined as $f(\lambda) = dF(\lambda)/d\lambda$ in the Radon Nykodym sense. In this case, there is a one-to-one relation between the covariance function and the spectral density:

$$\gamma(s) = \int_{\lambda \in [-\pi,\pi]} f(\lambda)e^{i\lambda s} d\lambda, \tag{7.2}$$

$$f(\lambda) = \frac{1}{2\pi} \sum_{s=-\infty}^{\infty} \gamma(s)e^{-i\lambda s}. \tag{7.3}$$

In this paper we only consider linearly regular processes, see [36] and [23], i.e. processes where the best linear least squares forecasts tend to zero if the forecast horizon tends to infinity. Linearly regular processes admit a Wold representation

$$x(t) = \sum_{j=0}^{\infty} K(j)\varepsilon(t-j), \tag{7.4}$$

where $\varepsilon(t)$ is n-dimensional white noise, i.e. $\mathbb{E}\varepsilon(t) = 0$, $\mathbb{E}\varepsilon(s)\varepsilon(t)^* = \delta_{st}\Sigma$ and $K(j) \in \mathbb{R}^{n\times n}$, $\sum_{j=0}^{\infty} \|K(j)\|^2 < \infty$. Furthermore $\varepsilon(t)$ are the innovations of $x(t)$, i.e. the one step ahead prediction errors of the best linear least squares forecast of $x(t)$ given its past $x(t-1), x(t-2), \ldots$. In addition we assume Σ is non-singular.

There are two important special cases for linearly regular processes: ARMA processes and AR(∞) processes. ARMA processes are important, because every linearly regular process can be approximated with arbitrary accuracy by an ARMA process, see [24] for further details. In general these two model classes are not the same, but there are overlappings.

For the remainder of this paper we will consider AR(∞), i.e. infinite autoregressive, systems and processes only. An AR(∞) *system* is a linear system of the form

$$\sum_{j=0}^{\infty} A(j)x(t-j) = \varepsilon(t), \tag{7.5}$$

where $A(j) \in \mathbb{R}^{n\times n}$, $\sum_{j=0}^{\infty} \|A(j)\| < \infty$ holds and $\varepsilon(t)$ is white noise. We use z to denote the backshift operator on $\mathbb{Z}$: $z(x(t)|t \in \mathbb{Z}) = (x(t-1)|t \in \mathbb{Z})$, as well as a complex variable. We rewrite (7.5) as

$$a(z)x(t) = \varepsilon(t), \tag{7.6}$$

where $a(z) = \sum_{j=0}^{\infty} A(j)z^j$ exists inside and on the unit circle. We also assume the *stability condition*:

$$\det a(z) \neq 0 \quad \text{for } |z| \leq 1. \tag{7.7}$$

With this assumption, the *transfer function* $k(z) = a^{-1}(z) = \sum_{j=0}^{\infty} K(j)z^j$ exists inside and on the unit circle, see [5]. There is a unique weakly stationary solution of (7.5) of the form

$$x(t) = \sum_{j=0}^{\infty} K(j)\varepsilon(t-j) = k(z)\varepsilon(t). \tag{7.8}$$

This solution (7.8) of the system (7.5) is called an *autoregressive* (∞) *process*. It corresponds to the Wold representation, and here even $\sum_{j=0}^{\infty} \|K(j)\| < \infty$ holds.

For the sake of simplicity of notation we will skip the (∞) sign henceforth.

Every linearly regular (and hence every autoregressive) process has a spectral density of the form, see e.g. [36]

$$f(\lambda) = k(\lambda)\Sigma k(\lambda)^*, \tag{7.9}$$

where we write $k(\lambda)$ for $k(e^{-i\lambda})$. Analogously we use $a(\lambda)$ for $a(e^{-i\lambda})$.

Conversely the transfer function $k(z)$ can be uniquely determined from a spectral density $f(\lambda)$ under the assumptions: $\det k(z) \neq 0$ for $|z| \leq 1$, $k(z)$ has a Taylor series expansion in $|z| \leq 1$, $k(0) = I$ and $\Sigma > 0$. For the remainder of this paper, we will impose all these assumptions, and so for every linearly regular process we can find the transfer function $k(z)$ in a unique way from its spectral density. By inverting the transfer function $a(z) = k^{-1}(z)$ we get an AR representation (7.5), where $a(0) = I$ holds.

By using $\tilde{a}(z) = -\sum_{j=1}^{\infty} A(j)z^j$ we rewrite (7.6) as follows

$$x(t) = \tilde{a}(z)x(t) + \varepsilon(t). \tag{7.10}$$

This paper is concerned with measures of dependence and causality between two univariate component processes $(x_i(t))_{t\in\mathbb{Z}}$ and $(x_j(t))_{t\in\mathbb{Z}}$ $(i \neq j)$ of the n-dimensional process $x(t) = (x_1(t), \ldots, x_i(t), \ldots, x_j(t), \ldots, x_n(t))'$. These measures are based on the second moments only and due to weak stationarity they are invariant under time translations.

Dependencies in a multivariate process can be described by a graph. In this paper we will only consider graphs, where the vertices correspond to the one-dimensional subprocesses and the edges are defined by a dependency or causality measure. Other kinds of graphs are possible, see [9]. In particular we distinguish between directed and undirected graphs respectively, depending whether a directed or an undirected measure is considered. The construction of these graphs is described in Sect. 7.5 of this paper.

Using this methodology we obtain information concerning the complexity and interaction in a multivariate time series. We will discuss in particular Granger causality and the use of graphical modeling. At the end of the paper we will present an example where these measures are used to detect the focus, and to track the propagation, of an epileptic seizure.

The emphasis of this introductory treatment is on description and discussion of the most common measures.

7.2 Undirected Measures of Dependence

7.2.1 Coherence

From the spectral representation of a stationary process (7.1) we obtain a measure of the strength of linear dependence in frequency domain. Let $x_i(t)$ and $x_j(t)$ be univariate subprocesses of $x(t)$ as mentioned in Sect. 7.1, with the corresponding orthogonal increment processes $z_i(\lambda)$ and $z_j(\lambda)$ respectively. The idea of *coherence* is to measure the squared coefficient of correlation between $dz_i(\lambda)$ and $dz_j(\lambda)$

$$C_{ij}^2(\lambda) = \frac{|\mathbb{E}\{dz_i(\lambda)\overline{dz_j(\lambda)}\}|^2}{\mathbb{E}\,|dz_i(\lambda)|^2\mathbb{E}\,|dz_j(\lambda)|^2} = \frac{|f_{ij}(\lambda)|^2}{f_{ii}(\lambda)f_{jj}(\lambda)}, \tag{7.11}$$

where f_{ij} is the (i, j)-element of f. Thus, the coherence is a frequency specific measure for dependence between $x_i(t)$ and $x_j(t)$. It is a measure of the strength of dependence between the frequency weights $dz_i(\lambda)$ and $dz_j(\lambda)$. Since $C_{ij}^2(\lambda)$ is obviously symmetric, it is not possible to detect a direction of influence from $C_{ij}^2(\lambda)$.

7.2.2 Partial Spectral Coherence (PSC)

Of course, one could calculate all pair-wise coherences in an n-dimensional process. However, in such a case, it is impossible to distinguish between direct and indirect influences. This leads to the partial spectral coherence, see [5] and [8].

The idea of PSC is as follows: In order to measure the dependence between $x_i(t)$ and $x_j(t)$ $(i \neq j)$ after removing the influence of all other variables $Y_{ij}(t) = (x_k(t)|k \neq i, j)$, we project $x_i(t)$ as well as $x_j(t)$ onto the Hilbertspace spanned by all Y_{ij} in the L^2 over the underlying probability space. This projection leads to the residuals $\eta_i(t)$ and $\eta_j(t)$

$$\eta_i(t) = x_i(t) - \sum_{k=-\infty}^{\infty} D_i(k)Y_{ij}(t-k) = x_i(t) - d_i(z)Y_{ij}(t),$$

$$\eta_j(t) = x_j(t) - \sum_{k=-\infty}^{\infty} D_j(k)Y_{ij}(t-k) = x_j(t) - d_j(z)Y_{ij}(t),$$

where the filters $d_i(z)$ and $d_j(z)$ minimize the variance of the residuals. Now we look at the spectrum of the process $(\eta_i, \eta_j)'$. Let $f_{\eta_i\eta_j}$ denote the corresponding cross-spectrum. This cross spectrum is a frequency specific measure for the dependence between $x_i(t)$ and $x_j(t)$ given all Y_{ij}. Rescaling leads to the definition of the *partial spectral coherence* (PSC)

$$R^2_{ij|ij^c}(\lambda) = \frac{|f_{\eta_i\eta_j}(\lambda)|^2}{f_{\eta_i\eta_i}(\lambda) f_{\eta_j\eta_j}(\lambda)}. \tag{7.12}$$

As has been shown, see e.g. [8], there exists a more convenient way to compute the partial spectral coherence using the inverse of the spectral density $f^{-1}(\lambda)$ of the original process $x(t)$:

$$R^2_{ij|ij^c}(\lambda) = \frac{|(f^{-1}(\lambda))_{ij}|^2}{(f^{-1}(\lambda))_{ii}(f^{-1}(\lambda))_{jj}},$$

where $(f^{-1}(\lambda))_{ij}$ is the (i, j)-element of $f^{-1}(\lambda)$. Given actual data, the spectral density can be estimated by fitting an finite AR model or by using non-parametric spectral estimators. For dealing with actual data, a test with the zero hypothesis $\mathscr{H}_0 : R^2_{ij|ij^c}(\lambda) \equiv 0$ has been described in [5] and [10].

7.3 Directed Measures of Dependence

In this section we present important directed measures based on the (infinite) AR representation (7.5), such as the directed transfer function and the partial directed coherence.

7.3.1 Directed Transfer Function (DTF)

The first directed dependency measure described, is the *directed transfer function* (DTF), as proposed in [27]. This measure is often used in the neuroscience literature and is defined as

$$\gamma^2_{ij}(\lambda) = \frac{|k_{ij}(\lambda)|^2}{\sum_{m=1}^{n} |k_{im}(\lambda)|^2}, \tag{7.13}$$

where k_{ij} is the (i, j)-th component of the transfer function k. The denominator in (7.13) provides a normalization. The nominator of the directed transfer function measures the total information flow from x_j to x_i. This can be seen by expanding $k(z) = a(z)^{-1}$ as a geometric series as seen below, and using $\tilde{a}(z)$ as introduced in (7.10)

$$k(z) = a(z)^{-1} = (I - \tilde{a}(z))^{-1} = \sum_{m=0}^{\infty} \tilde{a}(z)^m = I + \tilde{a}(z) + \tilde{a}(z)^2 + \cdots.$$

Considering the off diagonal elements (i.e. $i \neq j$) we obtain

$$k(z)_{ij} = (a(z)^{-1})_{ij} = \tilde{a}(z)_{ij} + \sum_m \tilde{a}(z)_{im}\tilde{a}(z)_{mj} + \sum_{m,\ell} \tilde{a}(z)_{im}\tilde{a}(z)_{m\ell}\tilde{a}(z)_{\ell j} + \cdots .$$

This shows that the nominator is the sum of the direct and all indirect information flows from the j-th to the i-th component.

Obviously the DTF is bounded by 0 and 1. Also, obviously the directed transfer function is a directed measure.

As the DTF measures the total information flow between two components in a multivariate system, the direct and all indirect ones, no conclusions may be drawn concerning the pathways of information propagation. So the DTF is not very useful in cases when we want to find the causal structure of a multivariate system.

7.3.2 *Direct Directed Transfer Function (dDTF)*

In order to overcome the problem of indirect information flows, Korzeniewska et al. proposed a combination measure of the DTF and the PSC, see [28], the so called *direct directed transfer function* (dDTF) defined by

$$\delta_{ij}^2(\lambda) = \gamma_{ij}^2(\lambda) R_{ij|ij^c}^2(\lambda). \tag{7.14}$$

The DTF is used to identify the direction of the information flow, and the PSC is used to filter out the indirect flows, so the direct information flows are the only remaining ones. As has been pointed out in [13] the statistical properties of the dDTF have not been investigated so far and an analysis of actual data based on the dDTF could detect wrong relationships.

7.3.3 *Partial Directed Coherence (PDC)*

Now we look at another frequency specific measure, which was introduced by Baccala and Sameshima in [1], the *partial directed coherence* (PDC), which is defined as

$$\pi_{ij}^2(\lambda) = \frac{|\tilde{a}_{ij}(\lambda)|^2}{\sum_{m=1}^n |\tilde{a}_{mj}(\lambda)|^2}. \tag{7.15}$$

The PDC can be seen as the ratio of the direct information flow from x_j to x_i normalized by all outflows of x_j. The careful reader may note, that this normalization is different to the one of the DTF. Of course, other normalizations would also be possible. Obviously, the PDC is bounded by 0 and 1.

Note that "partial" as part of this measure's name does not relate to removing the influences of all other variables $Y_{ij}(t)$. It stems from the derivation of the measure, where Baccala and Sameshima factorized the partial spectral coherence, and skipped some components. Obviously the PDC is a directed measure (and thus not symmetric).

The advantage of this measure is the clear interpretation as the direct information flow. Furthermore it has a connection to the causal interpretation of the multivariate process, as discussed in Sect. 7.4.

7.3.4 *Generalized Partial Directed Coherence (GPDC)*

A disadvantage of the PDC is, that it is not scale invariant, meaning that it is not invariant under different choices of the unit of measurement. To overcome this problem, Baccala et al. introduced an extension of the PDC in [2], the *generalized partial directed coherence* (GPDC)

$$\tilde{\pi}_{ij}^2(\lambda) = \frac{\Sigma_{ii}^{-1} |\tilde{a}_{ij}(\lambda)|^2}{\sum_{m=1}^{n} \Sigma_{mm}^{-1} |\tilde{a}_{mj}(\lambda)|^2}, \tag{7.16}$$

where Σ_{ii} is the (i,i)-component of the error variance-covariance matrix. This modification turns out to be more robust than the PDC when processing actual data.

7.3.5 *Other Directed Measures*

There exists a broad range of directed measures between two signals in the literature, not necessarily based on a linear model. There exist measures based on information theory, see [19, 38, 40], as well as measures based on time continuous models, see [18]. A not exhaustive list of other surveys on this topic is [7, 32, 33, 37].

7.4 Granger Causality

There have been long and thorough discussions about causality, both philosophical and mathematical in nature, a brief summary can be found in [34]. This paper focuses on time series and their causal investigation, using the information contained in their second moments. As opposed to the iid case, the temporal ordering of the time series contains causal information.

The concept we will use here is Granger causality (introduced by Granger in [22]). The basic idea is simple: If the knowledge of the past of one time series improves the predictability of another time series, it is Granger causal for the second series. This is a plausible definition of causality, but there may occur problems arising from this definition, which we will address in this section. The interested reader may be referred to [4] and [30] for additional information.

First we introduce some notation. We write $x_I(t) = (x_{i_1}(t), \ldots, x_{i_k}(t))'$ where $i_j \in I \subseteq \{1, 2, \ldots, n\}$ and $i_1 < i_2 < \cdots < i_k$ to denote the subprocess of $x(t)$ corresponding to the index set I. We use $\Omega_I(t) = \overline{\mathrm{span}(\{\{x_i(s)|s \leq t\}|i \in I\})}$ with $I \subseteq \{1, \ldots, n\}$ to denote the space spanned by the past and present of x_I in the Hilbertspace of all square integrable random variables.

For simplification we will use: $\Omega_i(t) = \Omega_{\{i\}}(t)$ and $\Omega_{IJ}(t) = \Omega_{I \cup J}(t)$.

By $\hat{x}_J(t+1|\Omega_I(t))$, $J \subset \{1, \ldots, n\}$ we denote the best linear least squares predictor of $x_J(t+1)$ based on the knowledge of $\Omega_I(t)$, e.g. the past and present values of x_I. We denote the prediction error by $\varepsilon_J(t+1|\Omega_I(t)) = x_J(t+1) - \hat{x}_J(t+1|\Omega_I(t))$ and its covariance matrix is $\Sigma_J(t+1|\Omega_I(t)) = \mathbb{E}\varepsilon_J(t+1|\Omega_I(t))\varepsilon_J(t+1|\Omega_I(t))'$.

7.4.1 Granger Causality in the Bivariate Case

The original definition of Granger has been given for the bivariate case:

$$x(t) = \begin{pmatrix} x_1(t) \\ x_2(t) \end{pmatrix}.$$

If the knowledge of the past of $x_2(t+1)$ improves the prediction of $x_1(t+1)$ compared to the prediction from its own past only, then x_2 is said to be *Granger causal* for x_1. The criterion for this improvement of prediction is the variance of the errors: If

$$\Sigma_1(t+1|\Omega_{12}(t)) < \Sigma_1(t+1|\Omega_1(t)) \tag{7.17}$$

holds x_2 is Granger causal for x_1. Of course other measures for the errors are possible. We consider the linear least squares predictor in this definition, which in general does not have to be the best predictor, therefore we could speak of linear Granger causality.

In the original paper [22], Granger discussed the relations between two one-dimensional time series, given all other information in the universe. As he explains, the assumption of this ultimate knowledge does not hold in general with actual data. Unknown influences, so called latent variables, can cause spurious causalities, as shown in [26].

In the following we will give extensions of the two dimensional definition of Granger causality.

7.4.2 Granger Causality in the Multivariate Case

The concept of Granger causality can be extended to the multivariate case: We partition $x(t)$ into $x_I(t)$ and $x_J(t)$ such that $I \cup J = \{1, \ldots, n\}$ and $I \cap J = \emptyset$. Then x_J

is said to be *Granger causal* for x_I if

$$\Sigma_I(t+1|\Omega_{IJ}(t)) < \Sigma_I(t+1|\Omega_I(t)). \tag{7.18}$$

Here $B < A$ means that $A - B$ is positive definite.

7.4.3 Conditional Granger Causality

Now we extend this framework to the case where $I \cup J \subseteq \{1, 2, \dots, n\}$. We say x_J is *Granger causal* for x_I *conditional* on the rest x_K, where I and J are disjoint and $K = \{1, \dots, n\} \backslash (I \cup J)$ if

$$\Sigma_I(t+1|\Omega_{IJK}(t)) < \Sigma_I(t+1|\Omega_{IK}(t)). \tag{7.19}$$

That means that we look at the causality of $x_I(t)$ and $x_J(t)$ given the "remaining" process $x_K(t)$. Together these three processes give $x(t)$. (Note for $I = \{i\}$ and $J = \{j\}$ we have $x_K(t) = Y_{ij}(t)$.)

If the variances in (7.19) are equal, then $x_J(t)$ is Granger non-causal for $x_I(t)$, and, as Eichler mentions in [13], this is equal to the condition that

$$A_{IJ}(k) = 0 \quad \text{for all } k \in \mathbb{N} \tag{7.20}$$

where the $A_{IJ}(k)$ are the coefficients in the series representation of $\tilde{a}(z)$ (introduced in (7.10)) corresponding to the sets I and J. This equality is important since it directly links the autoregressive coefficients with (conditional) Granger non-causality.

7.4.4 General Granger Causality

Of course the most general type of so called Granger causality would be, if we look at three disjoint index sets I, J and K with $I \cup J \cup K \subseteq \{1, \dots, n\}$. So we may only look at subprocesses of $x(t)$.

Calculating all possible Granger causalities in an n-dimensional process would be a lot of work. To cope with this problem, Eichler used a graph theoretical approach to explore the non-causalities rather than the causalities of all subsystems, see [13, 15, 16]. In Sect. 7.6 we will investigate this approach more closely and will also introduce Granger Causality Graphs.

7.4.5 Granger Causality Index

When we look at the definitions of different kinds of Granger causality, we can only say whether one series is causal for the other or not. In order to measure the strength of dependence, which in a certain sense all aforementioned measures do, Geweke introduced the Granger causality index in [20]. Although the original definition was based on Granger causality in the multivariate case (i.e. $I \cup J = \{1, 2, \ldots, n\}$), we present an adapted version for the conditional Granger causality. The *conditional Granger causality index* from x_J to x_I, where I and J are disjoint and $K = \{1, \ldots, n\} \backslash (I \cup J)$, is defined as

$$cGCI_{x_J \to x_I} = \ln \frac{\det \Sigma_I(t+1|\Omega_{IK}(t))}{\det \Sigma_I(t+1|\Omega_{IJK}(t))}. \tag{7.21}$$

Instead of the determinant, in the definition, also the trace could be used as a measure, see [3]. The cGCI is a directed measure and indicates conditional Granger causality.

7.4.6 Connection to other Dependency Measures

The computation of Granger dependencies in a high dimensional time series is computationally intensive. It would be desirable to have a frequency based measure that indicates Granger causality.

The PSC is an undirected measure, so it will not be usable for the detection of Granger Causality.

In the bivariate case the DTF indicates (bivariate) Granger causality, but as Eichler showed in [14] this does not hold generally in the multivariate case.

Because the PDC is directly based on the coefficients A_{ij}, it indicates conditional Granger causality from $x_j(t)$ to $x_i(t)$ given the remaining processes, as we have seen in (7.20).

7.4.7 Extension to the Non-linear Case

The notion of Granger Causality introduced here is based on a linear model. There exist ideas to extend the definition for non-linear models, a not exhaustive list of surveys on this topic is [17, 31].

7.5 Construction of Directed and Undirected Graphs

So far we have discussed about measures of dependence and causality and now we explain how to construct a graph based on these measures. Normally we distinguish between directed and undirected graphs, depending on the considered measure. In this work we only consider graphs, where the vertices correspond to the one-dimensional subprocesses.

First we draw a node for each component $x_i(t)$. Then, for undirected graphs, we draw an edge j - - - i for all pairs of vertices $(i, j), j \neq i$ unless $m_{ij}(\lambda) \equiv 0$, where $m(\lambda)$ is the considered undirected measure. When a directed measure $\mu(\lambda)$ is considered, we draw an edge $j \longrightarrow i$ for all pairs of vertices $(i, j), j \neq i$ unless $\mu_{ij}(\lambda) \equiv 0$, where $\mu(\lambda)_{ij}$ measures the influence from x_j to x_i.

Of course a lot of other kinds of graphs exist, like graphs with directed and undirected edges, see the next section, or graphs where the vertices correspond to just one random variable, see e.g. [9].

7.6 Graphical Modeling

The ultimate aim here is the analysis of the "inner structure" an n-dimensional process. As we have suggested in the last section, the analysis of the whole process as well as of all subprocesses would be advisable in order to understand the underlying structure. Of course this kind of analysis would be computationally intensive. Here we present an easy way for gaining more insights in the structure of the processes. This approach is called graphical modeling.

In general, graphical modeling refers to the use of graphs and graph theory in order to analyze the causal structure of some variables. In the last decades there has been a substantial interest in graphical modeling and a lot of research has been conducted. Most of this research has been focused on the dependence structure in the iid case. A not exhaustive list of surveys on this topic is [11, 29, 34, 41].

In this paper, we want to focus on the use of graphical modeling for time series. To the best of our knowledge this analysis was introduced by Brillinger in [6] and Dahlhaus in [8]. A good overview can be found in [9, 13]. The interested reader may be referred to Eichler [12, 15, 16].

We will consider two special kinds of graphical models and their use. First the partial correlation graphs, which are undirected, and second Granger causality graphs, which are mixed graphs, because they contain directed and undirected edges. The presented methodology is applicable for the graphs presented in this section, but not for all graphs in general.

Consider a graph $G = (V, E)$, where $V = \{1, \dots, n\}$ is the *vertex set* and E is the *edge set*. According to the types of connections in E we distinguish undirected, directed and mixed graphs. We will use the notation $x_V(t)$ for $x(t)$ in this section to stress the fact, that the elements of V correspond to the one dimensional component processes of $x(t)$.

A *path* in a graph $G = (V, E)$ is a sequence $p = (e_1, e_2, \dots, e_k)$ of edges $e_i \in E$ with an associated sequence of vertices $(v_0, v_1, \dots, v_k)$ such that the edge e_i connects the distinct vertices v_{i-1} and v_i.

In undirected graphs for $I, J, K \subset V$ (pairwise disjoint) we say that K *separates* I and J, if every path from an element of I to an element of J contains at least one element of the separation set K.

7.6.1 Partial Correlation Graphs

As the name suggests, the partial correlation graph is a graph (constructed under the aforementioned rules) based on the partial spectral coherence (PSC), see Sect. 7.2. It was introduced by Dahlhaus in [8]. The idea behind this graph is, that an edge (i, j) is missing, if the components $x_i(t)$ and $x_j(t)$ are uncorrelated conditional on the other components of the process. For the sake of completeness, we give the exact definition.

Definition 7.1 Let $x_V(t)$ be an autoregressive process (7.5). Then the *partial correlation graph* $G_{PC} = (V, E)$ for x_V is a graph with vertex set V and edge set E such that $(i, j) \notin E \Leftrightarrow R^2_{ij|ij^c}(\lambda) \equiv 0$ for $i \neq j$.

With Definition 7.1 and the common definition of separation in undirected graphs we get the following important theorem.

Theorem 7.1 *Suppose x_V is an autoregressive process* (7.5) *and $G_{PC} = (V, E)$ its corresponding partial correlation graph . Let $I, J, K \subset V$ where K separates I and J, then x_I is conditionally uncorrelated from x_J given x_K.*

With the help of the partial correlation graph using this theorem, we are able to compute independence relations in all subsystems. Note that this approach does not necessarily give all independences of the subsystems.

7.6.2 Granger Causality Graphs

We extend the idea given above to Granger causality, or rather Granger non-causality, following Eichler in [13]. For this purpose we reconsider the definition of conditional Granger Causality in Sect. 7.4, or rather the definition of non-causality given in (7.20), which states that x_J is Granger non-causal for x_I conditional on the rest, if the coefficients of the respective components in the AR representation are zero, i.e. $A_{IJ}(k) = 0$ for all k.

Definition 7.2 Let $x_V(t)$ be an autoregressive process (7.5). Then the *path diagram associated with* x_V is a graph $G_{GC} = (V, E)$ with vertex set V and edge set E such that for $i, j \in V$ with $i \neq j$

(i) $j \longrightarrow i \notin E \Leftrightarrow A_{ij}(k) = 0$ for $k \in \mathbb{N}$
(ii) $j \dashrightarrow i \notin E \Leftrightarrow \Sigma_{ij} = 0$ for $k \in \mathbb{N}$.

Now we have a mixed graph (because it contains directed and undirected edges), which contains the Granger non-causality information for the whole process $x_V(t)$. In order to use this graph for gaining information about the subprocesses we have to introduce a notion of separation for this kind of graphs. This is rather technical and we will skip the details, we will use the *m-separation* criterion introduced by Richardson in [35], which is an extension of the normal separation, and the *i-pointing* property introduced by Eichler, see [13].

Theorem 7.2 *Suppose x_V is an autoregressive process (7.5) and let G_{GC} be the path diagram associated with x_V. Additionally suppose that $K \subset V$ and let I, J be two disjoint subsets of K. If every I-pointing path between J and I is m-blocked given $K \backslash J$, then x_J is Granger non-causal for x_I with respect to x_K.*

With the graph and the m-separation criterion, we are able to determine Granger non-causalities for arbitrary subprocesses of $x_V(t)$. Here again we want to stress the fact, that we do not necessarily get all independences of the subprocesses, other independences in the subprocesses could hold additionally.

Because these graphs reveal Granger causal structures, path diagrams associated with a process are also called *Granger causality graphs*.

The aim of using graphs is to better understand the inner structure of a process. When working with real data, we normally have latent variables, that are not known. These latent influences may generate spurious causalities. Graphical modeling provides a tool to cope with the problems associated with latent variables, see [12] and [16].

7.7 Detection of the Focus of Epileptic Seizures

In this section we describe the application of some of the measures discussed above, to invasive EEG data of a (human) patient suffering from epileptic seizures. The aim is to localize the focus of the seizure and to describe its spread, based on the observed data. This analysis was performed in cooperation with the AIT (Austrian Institute of Technology GmbH).

7.7.1 Epilepsy

An epileptic seizure is the clinical manifestation of excessive hyper-synchronous discharges of neurons in the cerebral cortex. In most cases epilepsy can be controlled by a drug therapy, but in some severe cases, surgical intervention is needed. In this intervention the focus of the seizure is removed surgically. Thus it is desirable to localize this focus as precisely as possible.

As part of a presurgical examination the skullcap of the patient is opened and EEG electrodes are directly placed on the cortex of the patient. These invasive EEG electrodes (also called channels) remain there for some time and the electric potential of the brain is measured in this period. Normally some epileptic seizures occur in during this observation.

In order to localize the focus of these epileptic seizures the signals from the channels are analyzed in periods of epileptic activity. Up to now a visual interpretation of the invasive EEG by an experienced doctor is state of the art.

In the subsequent surgery the focus is removed. The effect of this surgical intervention is successful in 50 to 70% of the cases.

With our methods we want to help clinical doctors in localizing the focus of the epileptic seizures. In this paper we will present some results from [39], where we analyzed an epileptic seizure from a patient suffering from temporal lobe epilepsy.

7.7.2 Processing of the Data

In the brain 28 electrodes are implanted and the EEG signals are recorded at a frequency of 256 Hz and the line interference (50 Hz) is filtered out. As a reference value a non-affected channel is chosen. In the next step the data are averaged and used as a new reference.

Considering the data, we observe that within an epileptic seizure the variances of the invasive EEG signals of channels showing epileptic activity are significantly larger than the variances of non-affected ones. Thus such time series cannot be stationary over the whole time period. For finding the focus only the beginning and the propagation of the seizure seems to be important. In this section we analyze only the first seconds of the signal after the onset of the seizure. The data based identification of the (temporal) onset of the seizure will not be considered, but one possibility to do this would be to look at the variances.

In order to be able to work with methods based on stationarity, the sample has to be segmented into stationary parts. Here, for the sake of brevity, we do not discuss methods for doing so.

To achieve a precise localization of the focus, the distances between electrodes are quite small. This leads to strong correlations between neighboring channels and thus to ill-conditioned variance matrices. In order to avoid the problem associated with this ill-conditioning, we include only a part of the 28 recorded channels to our analysis. The selection of the included channels is done empirically, channels which are not affected by the epileptic seizure and which are far away from the epileptic area are removed. In this way approximately half of the channels are removed.

7.7.3 Clinical Description

According to the description (see Table 7.1) given by the medical doctors, the seizure considered had its onset at channels 15 to 21 and at channels 24 to 27. A propagation of the epileptic activity to the other half of the hemisphere is noticed thirteen seconds later, where channels 10 and 11 are infected. The seizure ends after about one minute.

We have tried different measures (see [21] and [39]) and the best results will be presented here.

Table 7.1 Clinical description of the seizure

Time	Activity	Electrodes
Second 0	Start	Channels 15–21, 24–27
Second 13	Propagation	Channels 10, 11
Second 69	End	Channels 10, 11

7.7.4 Graphs Based on Generalized Partial Directed Coherence

As mentioned in Sect. 7.3 the generalized partial directed coherence (GPDC) (7.16) measures the information flow from time series x_j to time series x_i.

As mentioned we try to identify the propagation of one epileptic seizure. The idea of using a coherence measure is based on the observed synchronization effects of epileptic channels. If we are able to find one electrode which influences many others at the onset, we have already found the focus. Hence we try to identify an information flow from the measurement of the focus channel x_j to other channels x_i.

Because this measure depends on the chosen frequency, we derive a binary measure by integrating the modulus of the GPDC over all frequencies. To define the significance of the interaction between two signals we need a threshold, which may be defined by a test, but for simplicity in our application it was chosen manually.

Figure 7.1 shows the location of the implemented electrodes in the brain and the generalized partial directed coherence. To be able to see the propagation of the seizure, four snapshots are used to represent the first four time windows. Each time window has a duration of 4 seconds.

The electrodes (channels) are represented by circles and crosses. Whereas circles represent the channels included in the analysis and crosses represent the others.

An arrow is drawn from channel j to channel i, if the GPDC from j to i is significantly high (it exceeds the threshold). Our hope is, that the identified information flow shows the epileptic activity.

In this figure we can see that an interaction between channels 15 and 25 and between channels 26 and 27 is identified. Comparing these results with the description of the medical doctors, they found that these channels to show epileptic activity. In a short time window (from second 8 to second 12) the information flow between channels 26 and 27 seems to be too low to be measured, but some seconds later (from second 12 to 16) it gets strong enough to be detected again.

According to the medical doctors a propagation of the seizure to the other side of the hemisphere (to channels 10 and 11) took place after 13 seconds. In comparison with the result obtained by us, we can indeed identify a high value of the GPDC there.

Summarizing it can be stated that the GPDC seems to be a good measure to identify epileptic activity.

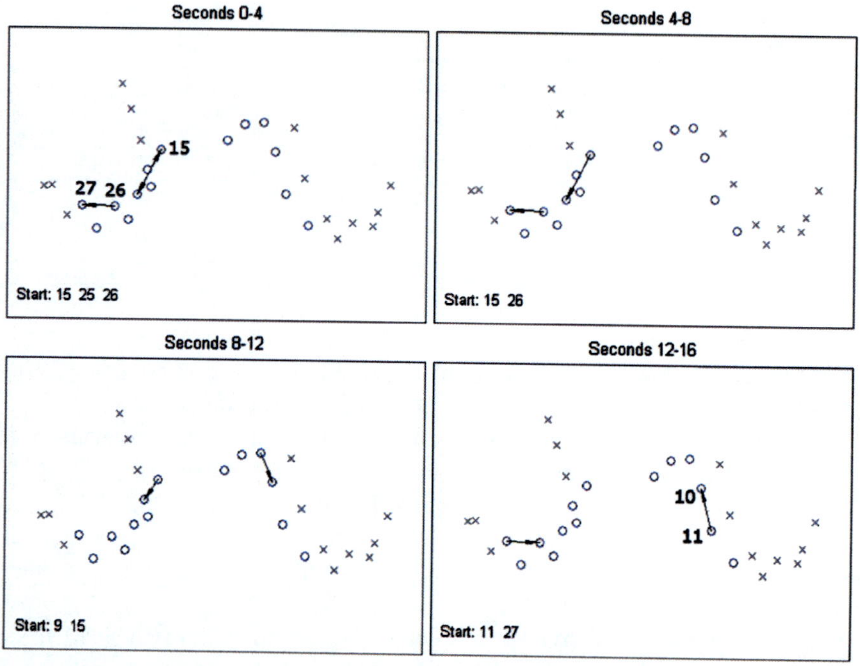

Fig. 7.1 Results of the generalized partial directed coherence: These four snapshots represent the brain with the implanted electrodes and the GPDC in four time windows (each of them is 4 seconds long). Circles represent selected channels (electrodes), whereas *crosses* represent non-selected ones. The *small arrows* between two points represent the influence of one channel to another. The generalized PDC is chosen to measure the influence

7.7.5 Graphs Based on Conditional Granger Causality

As opposed to the previous case, where we performed a calculation over stationary segments, here we use adaptive estimation.

To be able to cope with the non-stationarity of the time series, we use a finite AR model with time varying coefficients, estimated by a recursive least squares algorithm. This Recursive Least Squares Algorithm (see [25]) is equivalent to the minimization of the weighted sum of residuals

$$C(t) = \sum_{k=1}^{t} \lambda^{t-k} \varepsilon_x(k)\, \varepsilon_x^*(k)$$

in every time step t, where λ is the so called forgetting factor (here it was chosen to be 0.995)

As defined in Sect. 7.4, x_j is said to be Granger causal for x_i conditional on all other channels $Y_{ij}(t) = (x_k(t) | k \neq i, j)$, if the knowledge of the past of x_j improves the one step predictor of x_i.

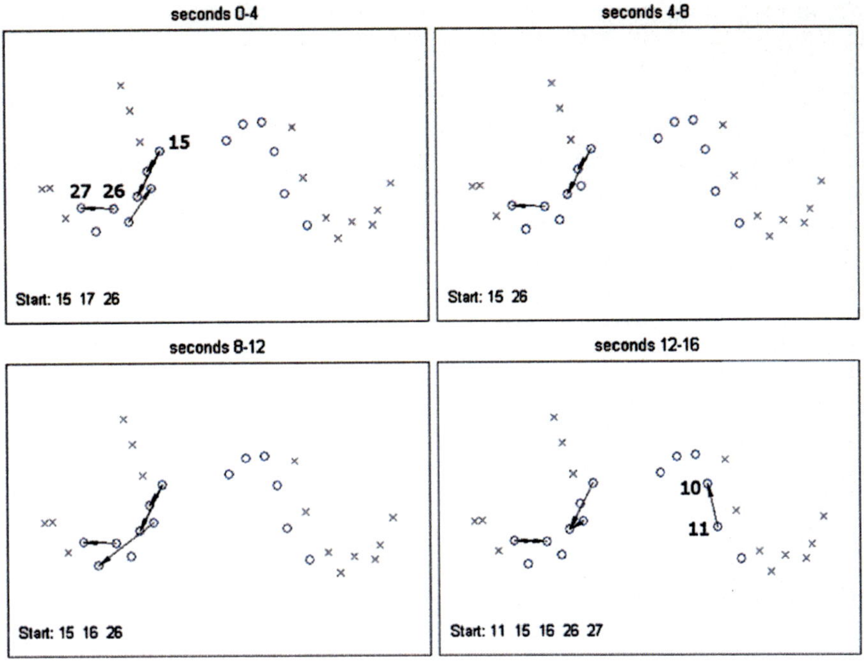

Fig. 7.2 Analysis with conditional Granger causality: The results of using conditional Granger causality as a measure for epileptic activity are shown in four plots, representing four time windows (each with a duration of 4 seconds). Selected channels (electrodes) are represented with *circles*, the others with *crosses*. An *arrow* from channel j to channel i represents a high conditional Granger causality index $cGCI_{x_j \to x_i}$ and hence a causal influence form j to i

In order to be able to differentiate between normal brain activity and epileptic activity, we calculate the conditional Granger causality index (7.21) for each time t.

In Fig. 7.2 we see the location of the electrodes in the brain. Again we had to select a smaller set of channels and therefore used the same channels as in the preceding analysis. To explore the propagation of the seizure, four snapshots are shown. Each of them represents a duration of 4 seconds.

The conditional Granger causality is illustrated with arrows. An arrow from channel j to channel i is drawn if the corresponding index exceeds a manually chosen threshold for more than a quarter of a second. Again we assume that this measure is able to identify epileptic activity.

According to the medical doctors, the onset should be in channels 15 to 21 and in channels 24 to 26. Considering our result, the cGCI is significant at these channels.

In the last snapshot (seconds 12 to 16) we may identify an information flow between channels 10 and 11. According to the clinical description there should be an epileptic activity after 13 seconds. Again we have found a perfect matching between this measure and the epileptic activity as described by the medical doctors.

7.7.6 Conclusion

Both methods, the generalized partial directed coherence, as well as the Granger causality, have delivered a result which is in good accordance with the findings of medical experts. In general, conditional Granger causality leads to slightly better results than the GPDC. We explain this fact, however, not by the use of a different measure, but by the capability of the RLS algorithm to better cope with the non-stationary invasive EEG data.

References

1. Baccala, L.A., Sameshima, K.: Partial directed coherence: a new concept in neural structure determination. Biol. Cybern. **84**, 463–474 (2001)
2. Baccala, L.A., Takahashi, D.Y., Sameshima, K.: Generalized partial directed coherence. In: Proc. of the 15th International Conference on Digital Signal Processing (2007)
3. Barrett, A.B., Barnett, L., Seth, A.K.: Multivariate granger causality and generalized variance (2010). arXiv:1002.0299v2
4. Bressler, S.L., Seth, K.S.: Wiener-granger causality: a well established methodology. NeuroImage (2010). doi:10.1016/j.neuroimage.2010.02.059
5. Brillinger, D.R.: Time Series, Data and Analysis. Holden Day, Oakland (1981)
6. Brillinger, D.R.: Remarks concerning graphical models for time series and point processes. Rev. Econom. **16**, 1–23 (1996)
7. Chavez, M., Martinerie, J., Le Van Quyen, M.: Statistical assessment of nonlinear causality: application to epileptic EEG signals. J. Neurosci. Methods **124**, 113–128 (2003)
8. Dahlhaus, R.: Graphical interaction models for multivariate time series. Metrika **51**, 157–172 (2000)
9. Dahlhaus, R., Eichler, M.: Causality and graphical models in time series analysis. In: Green, P., Hjort, N., Richardson, S. (eds.) Highly Structured Stochastic Systems, pp. 115–137. Oxford University Press, London (2003)
10. Dahlhaus, R., Eichler, M., Sandkuehler, J.: Identification of synaptic connections in neural ensembles by graphical models. J. Neurosci. Methods **77**, 93–107 (1997)
11. Edwards, J.: Introduction to Graphical Modelling, 2nd edn. Springer, Berlin (2000)
12. Eichler, M.: A graphical approach for evaluating effective connectivity in neural systems. Philos. Trans. R. Soc. B **360**, 953–967 (2005)
13. Eichler, M.: Graphical modeling of dynamic relationships in multivariate time series. In: Schelter, B., Winterhalder, M., Timmer, J. (eds.) Handbook of Time Series Analysis, pp. 335–372. Wiley-VCH, New York (2006)
14. Eichler, M.: On the evaluation of information flow in multivariate systems by the directed transfer function. Biol. Cybern. **94**, 469–482 (2006)
15. Eichler, M.: Granger-causality and path diagrams for multivariate time series. J. Econom. **137**, 334–353 (2007)
16. Eichler, M.: Causal inference from multivariate time series: What can be learned from granger causality. In: Glymour, C., Wang, W., Westerstahl, D. (eds.) Proceedings from the 13th International Congress of Logic, Methodology and Philosophy of Science. King's College Publications, London (2009)
17. Freiwald, W.A., Valdes, P., Bosch, J., Biscay, R., Jimenez, J.C., Rodriguez, L.M., Rodriguez, V., Kreiter, A.K., Singer, W.: Testing non-linearity and directedness of interactions between neural groups in the macaque inferotemporal cortex. J. Neurosci. Methods **94**, 105–119 (1999)

18. Friston, K.J., Harrison, L., Penny, W.: Dynamic causal modelling. NeuroImage **19**, 1273–1302 (2003)
19. Gabor, D.: Theory of communication. J. IEEE **93**(26), 429–457 (1946)
20. Geweke, J.: Measurement of linear dependence and feedback between multiple time series. J. Am. Stat. Assoc. **77**(378), 304–313 (1982)
21. Graef, A.: Nonstationary autoregressive modeling for epileptic seizure propagation analysis. Master's thesis, Vienna University of Technology (2008)
22. Granger, C.: Investigating causal relations by econometric models and cross-spectral methods. Econometrica **37**, 424–438 (1969)
23. Hannan, E.J.: Multiple Time Series. Wiley, New York (1970)
24. Hannan, E.J., Deistler, M.: The Statistical Theory of Linear Systems. Wiley, New York (1988)
25. Haykin, S.: Adaptive Filter Theory, 4th edn. Prentice Hall, Upper Saddle River (2002)
26. Hsiao, C.: Autoregressive modeling and causal ordering of econometric variables. J. Econom. Dyn. Control **4**, 243–259 (1982)
27. Kaminski, M., Blinowska, K.: A new method of the description of the information flow in the brain structures. Biol. Cybern. **65**, 203–210 (1991)
28. Korzeniewska, A., Manczak, M., Kaminski, M., Blinowska, K., Kasicki, S.: Determination of information flow direction among brain structures by a modified directed transfer function (dDTF) method. J. Neurosci. Methods **125**, 195–207 (2003)
29. Lauritzen, S.L.: Graphical Models. Oxford University Press, London (1996)
30. Luetkepohl, H.: Introduction to Multiple Time Series Analysis, 2nd edn. Springer, Berlin (1993)
31. Matrinazzo, D., Pellicoro, M., Stramaglia, S.: Kernel method for nonlinear granger causality. Phys. Rev. Lett. **100**, 144103 (2008)
32. Matysiak, A., Durka, P., Montes, E., Barwinski, M., Zwolinski, P., Roszkowski, M., Blinowska, K.: Time-frequency space localization of epileptic EEG oscillations. Acta Neurobiol. Exp. **65**, 435–442 (2005)
33. Osterhage, H., Mormann, F., Wagner, T., Lehnertz, K.: Measuring the directionality of coupling: Phase versus state space dynamics and application to EEG time series. Int. J. Neural Syst. **17**, 139–148 (2007)
34. Pearl, J.: Causality. Cambridge University Press, Cambridge (2000)
35. Richardson, T.: Markov properties for acyclic directed mixed graphs. Scand. J. Stat. **30**, 145–157 (2003)
36. Rozanov, Y.A.: Stationary Random Processes. Holden Day, Oakland (1967)
37. Schelter, B., Winterhalder, M., Dahlhaus, R., Kurths, J., Timmer, J.: Partial phase synchronization for multivariate synchronizing systems. Phys. Rev. Lett. **96**, 208103 (2006)
38. Schreiber, T.: Measuring information transfer. Phys. Rev. Lett. **85**(2), 461–464 (2000)
39. Schuster, T., Kalliauer, U.: Localizing the focus of epileptic seizures using modern measures from multivariate time series analysis. Master's thesis, Vienna University of Technology (2009)
40. Shannon, C.E., Weaver, W.: The Mathematical Theory of Information. University of Illinois Press, Urbana (1949)
41. Whittaker, J.: Graphical Models in Applied Multivariate Statistics. Wiley, New York (2000)

Chapter 8
Box-Jenkins Seasonal Models

Granville Tunnicliffe Wilson and Peter Armitage

8.1 Seasonality in Time Series

There is a pattern in many time series that reflects the cycles of our calendar. It is characterized by an approximate repetition of the pattern after each successive period s of the season, though it may be superimposed on other features such as trends and other irregular cycles, not directly linked to the calendar, that arise from the dynamics of economic and business activity. The period of the cycle is usually an integer, such as the twelve months of the year and seven days of the week, but it may not be an exact integer, for example if measurements are made every four weeks giving rise to a period slightly greater than thirteen. There may also be more than one seasonal period in a series, and we shall give an example of half-hourly electricity demand with a daily period of 24 hours and weekly period of 168 hours.

Some seasonal patterns are very regular, but more usually the pattern will be modified over time, eventually becoming noticeably different in shape. However, the pattern may be modulated (rather than modified) around a persistent long term shape. Our calendar also imposes some irregular, though predictable, modifications of seasonal patterns, particularly in monthly data. For example monthly sales of seasonal clothing might be recorded from weekly figures, with the weekly accounting period ending on Saturday. The series will then be strongly affected by the number of Saturdays in the month. This is not always the same from year to year and may appear to be just part of the seasonal irregularity. Programs such as X12-ARIMA (http://www.census.gov/srd/www/x12a/), designed to adjust time series for their

G. Tunnicliffe Wilson (✉)
Lancaster University, Lancaster, England, UK
e-mail: g.tunnicliffe-wilson@lancaster.ac.uk

P. Armitage
The Civil Service College, London, England, UK

L. Wang, H. Garnier (eds.), *System Identification, Environmental Modelling, and Control System Design*,
DOI 10.1007/978-0-85729-974-1_8,

seasonal variations, recognize such *calendar effects*, which may also include the variable dates of Easter and other holidays. Some seasonal patterns may not be visually apparent in a series because they are relatively minor and masked by other sources of variation; nevertheless it may be important to detect and model them.

The aim of this chapter is to review the approach to seasonal time series modelling and forecasting introduced by Box and Jenkins. This is widely used, not least in the X12-ARIMA program for reducing the magnitude of revisions in seasonal adjustment. We shall re-examine the data widely know as the *Airline Series* used by Box and Jenkins [1] to introduce their seasonal model, and suggest some alternatives to the methodology for identifying this model. In particular we will make use of the sample spectrum of the series as a supplement to the display of sample autocorrelations. The spectrum is a particularly appropriate statistical summary for seasonal data because of its cyclical nature.

8.2 The Airline Model and Related Predictors

The first series shown in Fig. 8.1 is the logarithms of the airline passenger totals from Box and Jenkins. The argument of Box and Jenkins is that it might be appropriate to predict the sub-series of, say, January, figures, by an exponentially weighted moving average (EWMA) with trend. Moreover, the same might be applied to each month of the year, using the same discount parameter across all months. This is equivalent to fitting to the whole series the seasonal integrated moving average (SIMA) model

$$x_t - x_{t-12} = c + f_t - \Theta f_{t-12},$$

or $\nabla_{12} x_t = c + (1 - \Theta B^{12}) f_t$ in their notation, where c is the mean annual increase in x_t. The series f_t is the set of 12 month ahead forecast errors from this model and is shown as the second plot in Fig. 8.1. Note that in predicting one particular January, say, the information in the previous eleven months has been ignored, so it is unsurprising that successive values of this error series are strongly dependent. Box and Jenkins then proposed that corrections to this first stage, of year on year forecasting, be made by a second stage, of month on month EWMA forecasting of the error series f_t. This is equivalent to fitting to f_t the non-seasonal integrated moving average (IMA) model

$$f_t - f_{t-1} = e_t - \theta e_{t-1},$$

or $\nabla f_t = (1 - \theta B) e_t$, where e_t is the error in predicting f_t, and hence also the final error in the corrected forecast of x_t. Putting together the two steps results in their famous *Airline* model

$$\nabla \nabla_{12} x_t = (1 - \theta B)(1 - \Theta B^{12}) e_t. \tag{8.1}$$

The third plot in Fig. 8.1 shows the final error series which is not far from random, which implies that the resulting predictor is close to optimal. The parameters

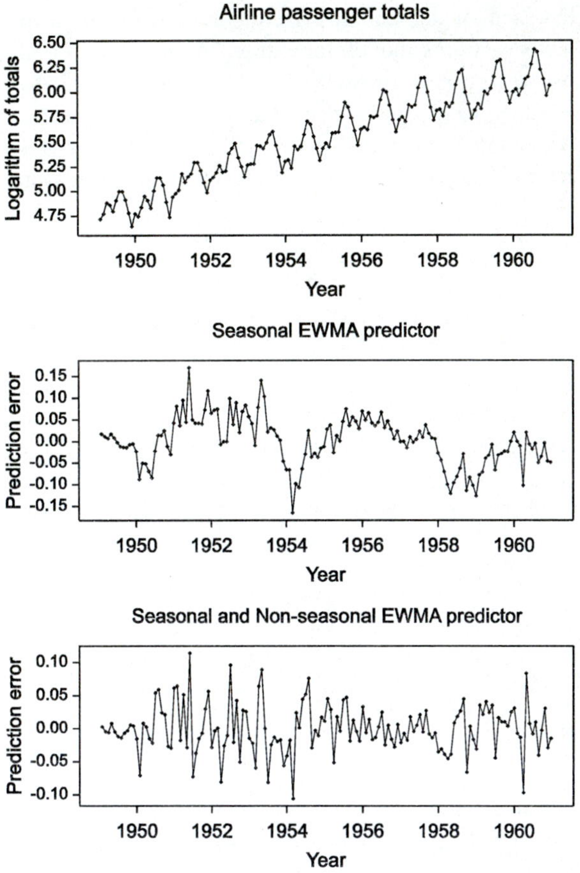

Fig. 8.1 The *first plot* is of the logarithms of the Airline Passenger Total series. The *second plot* shows the errors from predicting the series using a year on year EWMA predictor. The *third plot* shows the final errors from predicting the seasonal errors by a standard EWMA

estimated by this two stage process are $\Theta = 0.3404$ and $\theta = 0.4114$, and the residuals e_t have sample autocorrelation of -0.24 at lag 12. The parameters estimated simultaneously to minimize the final residual sum of squares are $\Theta = 0.5571$ and $\theta = 0.4018$, and the corresponding lag 12 autocorrelation is then -0.085. The difference in the two sets of parameters is explained by the fact that the first stage EWMA is trying to track both the trend and the seasonal variation. The Airline model is described by Box and Jenkins as a multiplicative model because the operators in the model are products of non-seasonal and seasonal terms; in this case nonseasonal and seasonal IMA(1,1) models. This must not be confused with models for multiplicative trend and seasonality—these effects are additive in this model.

The Airline model (8.1) is one of three well known adaptive predictors for seasonal time series models which track changes in the level, trend and seasonality. Of the other two, one is the additive seasonal Holt-Winter predictor [13], and the second is the seasonal state space model, the *Basic Structural Model* (BSM) of Harvey and Todd [4]. All three can be expressed in the predictor-corrector form. As each new observation is made of the series, the prediction error is used to update the states which characterize the level α_t, slope of trend, β_t and current seasonal factor γ_t of the model, for extrapolation of future values. A thorough exposition of this view point is presented by Newbold [8]. The Holt-Winter predictor is defined as generalization of exponential smoothing in terms of three smoothing parameters A, B, and C. Newbold [8, p. 115] shows that the weights applied to the prediction error e_t in updating the current level, trend and seasonal factor are

$$\begin{aligned} A^* &= A + C(1 - A)/12 = 0.391, \\ B^* &= A\,B = 0.009, \\ C^* &= 11C(1 - A)/12 = 0.733. \end{aligned} \tag{8.2}$$

The numerical values shown in (8.2) are those obtained by tuning the weights to minimize the prediction error sum of squares, which Haywood and Tunnicliffe Wilson [5] show for the Airline series leads to $A = 0.35$, $B = 0.01$, $C = 0.75$. Newbold [8, p. 117] also shows how the corresponding state updating form of the Airline model can be derived, leading to corresponding weights, given in terms of $\lambda = 1 - \theta$ and $\Lambda = 1 - \Theta$, of

$$\begin{aligned} \lambda\left(1 - \frac{13}{24}\Lambda\right) + \frac{\Lambda}{12} &= 0.49 \\ \frac{\lambda\Lambda}{12} &= 0.022 \\ \frac{11}{12}\Lambda - \frac{11}{24}\lambda\Lambda &= 0.2823. \end{aligned} \tag{8.3}$$

The numerical values used here are those obtained for the Airline model. The state space model of Harvey and Todd contains three variance ratio parameters controlling the independent evolution of the level, trend and seasonal factors. The corresponding updating weights can be derived from these by application of the Kalman Filter. Tunnicliffe-Wilson [12, p. 70] shows that the corresponding weights obtained from fitting the BSM to the Airline series are

$$\begin{aligned} K_L &= 0.6380, \\ K_T &= 0.0255, \\ K_S &= 0.1510. \end{aligned} \tag{8.4}$$

We have shown only the weight used to update the factor for the current month. What we have not shown are the weights for updating the seasonal factors for the previous eleven months; it is these weights that distinguish the three models, all of which may also be represented in a seasonal ARIMA form with $\nabla\nabla_{12}x_t$ expressed as a moving average of order 13. Both the Holt-Winter predictor and Harvey and Todd's BSM allow three free parameters to model the 13 weights, or equivalently 13 moving average parameters, but Box and Jenkins's Airline Model allows only two. This seems overly restrictive since it constrains the three rates at which the predictor adapts to changes in level, trend and seasonality. This criticism can be countered by the fact that precise estimation of the variance ratio parameter for the trend evolution in the BSM is difficult for many practically occurring series, with an estimate at the boundary value of zero not being unusual. Diagnostic checks on residual correlations also indicate that the Airline model adequately captures the moving average structure. Moreover the updating weights for the trend term are not very dissimilar in (8.2), (8.3) and (8.4).

For series with the same general appearance as the logarithms of the Airline data, the Airline model is recommended as a first trial model. This is based on our wide experience. The model is quickly fitted and little is lost if diagnostic checks reveal inadequacies, but very often the checks are satisfied. The model is robust and we have even applied it to generate credible forecasts from as little as 18 months of data, using the same parameters as for the Airline data. This is not to say that the model cannot be improved upon, even for the Airline data. In the next section we shall look at this more closely. We shall then perform a comparative exercise on a time series with similar appearance, a record of monthly atmospheric CO_2 measurements.

8.3 A Closer Look at the Airline Data

Figure 8.2 shows, on the left, the forecasts of the last two years of the logarithms of Airline totals obtained from the first 12 years of the series using the standard Airline model. Figure 8.3 shows a plot of the estimated one-step prediction errors, or residuals, from this model, the sample autocorrelations (sacf) of these residuals and also their sample spectrum. These last two plots are designed to reveal any remaining linear predictability in the residuals, by departures of the sacf from zero and of the spectrum from the general appearance of uniformity. Formal tests for these conditions have been devised, and the error limits on the sacf give some guidance as to values that might be significantly non-zero. The residuals appear to be somewhat more variable in the first two thirds of the series which contains several more extreme values, leading to some excess kurtosis in the whole series. The sacf value at lag 23 lies just outside the limits but on inspecting the residual scatter plot at that lag, it is accounted for mostly by a pair of positive large residuals at time indices 29 and 52 and a pair of negative large residuals at lags 39 and 62. It might therefore be concluded that there is no way to improve the predictive ability of the model, but the sample spectrum contains a clue which does open the way to this.

The sample spectrum does exhibit several large peaks, but it is well known that attaching meaningful interpretation to the frequencies at which these peaks happen to occur, is generally misleading. The distribution of the maximum of the sample spectrum over the whole frequency range is much more extreme than the (exponential) distribution of its value at any specified frequency. A large peak at a prior specified frequency will, however, be significant. The sample spectrum shown is of the normalized residuals, so has mean two, and there is 5% probability that the spectrum exceeds 6.0 at any specified frequency. The frequency of prior importance is 0.348 at which there is a spectrum peak of height 8.0. The importance of this

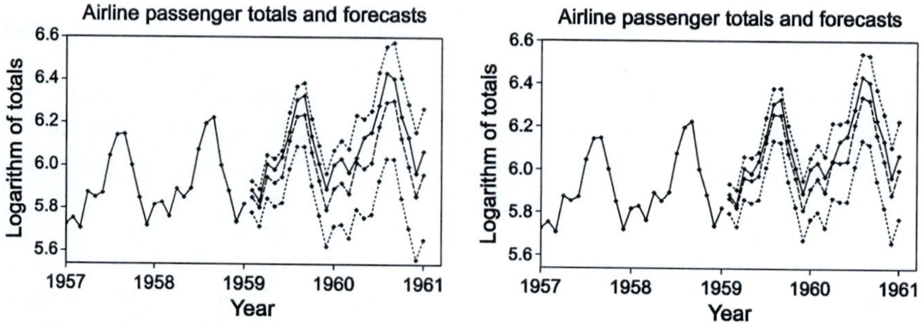

Fig. 8.2 *On the left* are forecasts of the last two years of the logarithms of the Airline Passenger Totals, using the Airline Model fitted to the previous series values. The *solid lines* are the true series values, the *dashed lines* are the forecasts and the *dotted lines* are 95% error limits. *On the right*, the forecasts are constructed using a regression on a cycle for calendar effects, and with the Airline model extended to include a non-seasonal moving average term at lag 3

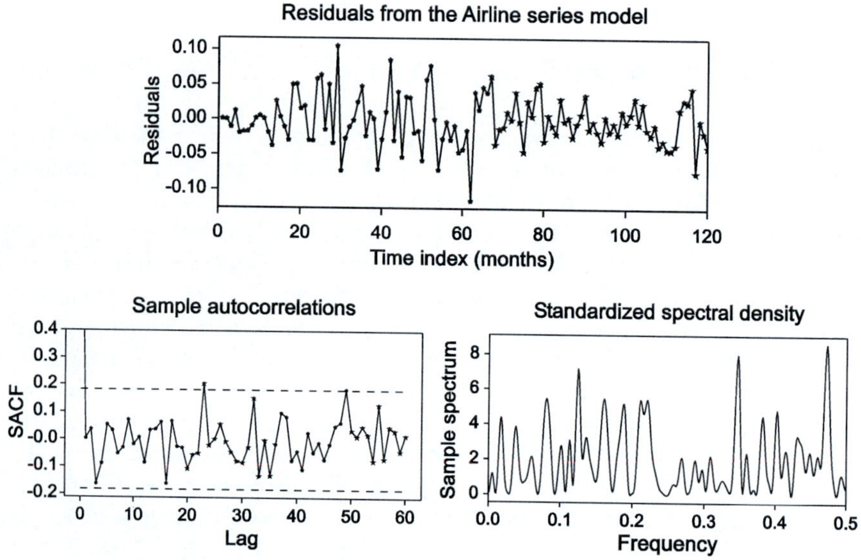

Fig. 8.3 Above are the residuals from the Airline series model of Box and Jenkins. *Below* are the sacf (*on the left*), with nominal two standard error limits shown by the *dashed lines*, and the standardized sample spectrum (*on the right*) of the residuals

frequency, as explained by Cleveland and Devlin [3], is that it is characteristic of a calendar effect in monthly data. This is not surprising in airline totals and investigation of explanatory regressors such as the number of weekends in each month, would repay effort in a serious development of the model. An ad-hoc, but highly recommended alternative is simply to include a cycle at frequency 0.348 as a regressor. We find that the significance of this regression corresponds to a p-value of 0.017. The forecasts are only slightly improved, but, most noticeably, the residual

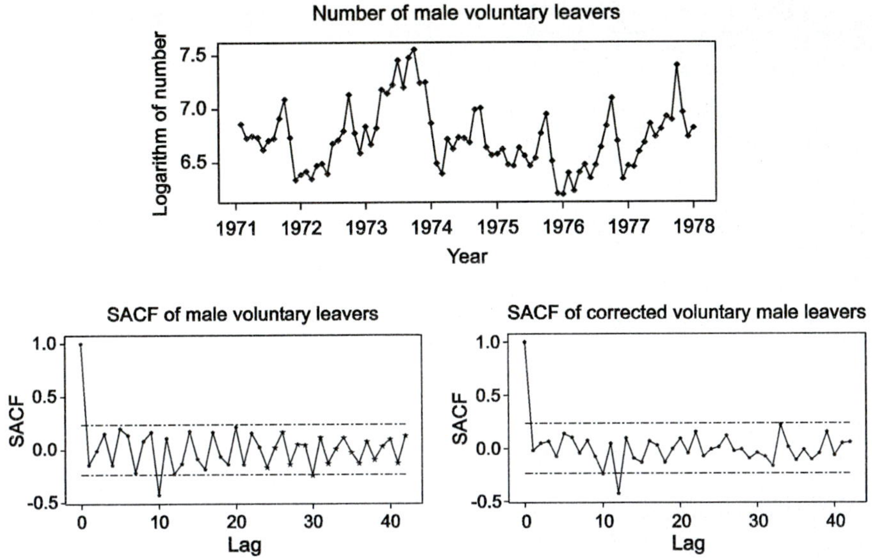

Fig. 8.4 *Above* is the logarithms of numbers of male voluntary leavers from the British Civil Service in each month from 1971 to 1978. The *lower left plot* shows the sacf after application of seasonal and non-seasonal differencing. The *lower right plot* shows the same sacf but using the series corrected by regression on the number of Fridays in the corresponding months

sacf at lag 3 now has the highly significant value of −0.256. The model is therefore extended to include a non-seasonal moving average term at lag three, which on estimation has a significant t-value of 2.67. The p-value of the cyclical regression is also reduced to 0.0012. Figure 8.2 shows, on the right, the forecasts from this new model. There is a noticeable improvement in the forecast accuracy, including narrower forecast error limits.

This last illustration emphasizes the importance of modeling the variability arising from calendar effects on a time series and we conclude this section with an example in which making the correct allowance for such effects is essential for Box-Jenkins seasonal modeling. The series shown in the upper plot of Fig. 8.4 is the logarithms of the number (in thousands) of male voluntary leavers from the British Civil Service in each month from 1971 to 1977. The lower left plot in the figure is the sacf of the series following application of seasonal and non-seasonal differencing. Curiously, this has a large negative value at lag 10, rather than lag 12 which would characterize a Box-Jenkins seasonal model. There is also a strong oscillation in the sacf values with a period close to three. However, the sample spectrum (not shown) of the differenced series clearly reveals the tell-tale spike associated with calendar effects. In this example we surmise that the number of leavers counted in each month will be strongly affected by the number of Fridays in that month, so we corrected the series by a simple regression on a dummy indicator of that number. The sacf of the corrected series, after applying the same differencing, is shown in the lower right plot of Fig. 8.4. The strong negative value is now at lag 12.

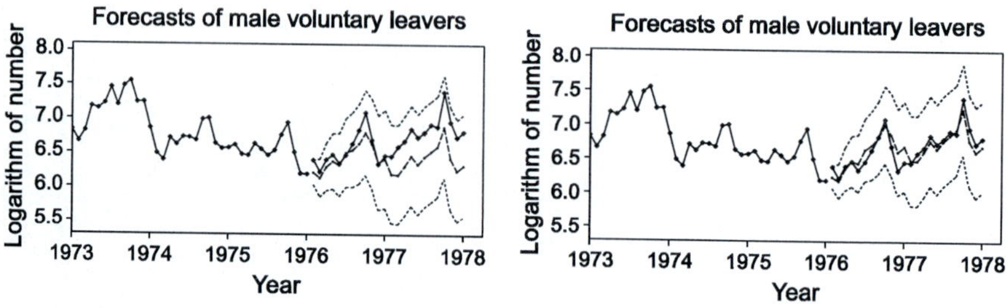

Fig. 8.5 The *left plot* shows forecasts using the Airline model extended to a second order seasonal moving average. The *right plot* shows forecasts using seasonal regressors with IMA(1,1) non-seasonal part and MA(1) seasonal part of a multiplicative model. True series value are shown with a *solid line*, forecasts with a *dashed line* and error limits with *dotted lines*

We proceeded to fit the Airline model to this series with the indicator for Fridays as a regressor. The estimate of θ was small but was retained. Unusually, a second order seasonal moving average was indicated and both associated parameters were strongly significant. The forecasts of the last two years of the series, using this model fitted to the previous values, is shown on the left in Fig. 8.5. However, the seasonal moving average operator had a root of unity, indicating that the seasonal pattern was fixed. Seasonal sinusoidal regressors (described in the next section) were therefore added to the model and the seasonal part of the Airline model reduced to MA(1) with no seasonal differencing. The forecasts from this second model are shown on the right in Fig. 8.5. They are surprisingly accurate, but the error limits are wide, indicating that on the past behavior of the series the accuracy is somewhat fortuitous. An important comment must be made on the difference between these two models. The (extended) Airline model implicitly includes a linear trend term in the forecast function; the regression model does not—we omitted it as being inappropriate to a series that gave no indication of any long term trend. If we had included it, the fit and forecasts would have been the same for both models. We shall shortly see this with another example of monthly atmospheric CO_2 concentrations. Here we have a distinction between using a unit root seasonal model for fixed seasonality, and explicitly using seasonal regressors. However, the main point of this example is being aware of calendar effects.

8.4 Atmospheric CO_2 Concentration

Figure 8.6 shows the logarithms of the series of monthly atmospheric CO_2 concentrations recorded at Mauna Loa from 1974 to 1986 by Keeling et al. [6]. These appear superficially similar to the Airline data and to make a start we simply fit the airline model. We keep back the last two years of the series for comparing with the forecasts obtained from fitting the model to the earlier data. The forecasts are shown on the left in Fig. 8.7. They are quite accurate because the trend and seasonal pattern are so regular. However, the estimated seasonal moving average parameter

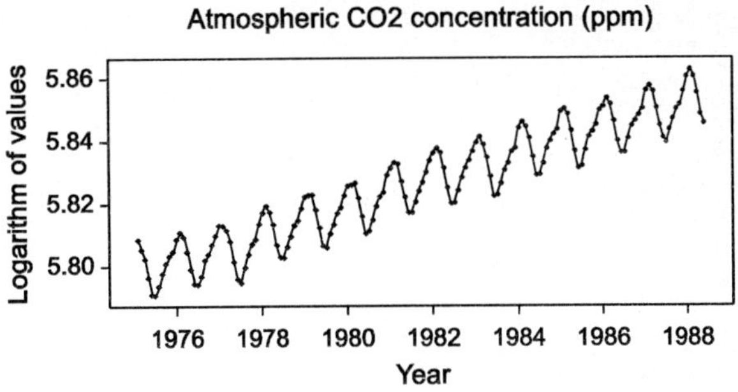

Fig. 8.6 Logarithms of the atmospheric concentration of CO_2 in ppm

is $\Theta = 0.9943$. A value so close to unity suggests that the seasonal pattern is not evolving—it is fixed. In that case the seasonal pattern can be represented by fixed regressors and the seasonal EWMA part of the model can be removed. One possibility for these regressors is a set of indicator variables, one for each month, with the regressor for a particular month taking the value zero in every month except at that particular month, when it takes the value one. As a set, these are collinear with a constant, so it is usual to omit one of them, usually the indicator for December. An alternative is a set of eleven sinusoidal seasonal regressors at the harmonic frequencies $f_j = j/12$:

$$v_{j,t} = \cos(2\pi f_j t) \quad \text{for } j = 1 \ldots 6, \tag{8.5}$$

$$w_{j,t} = \sin(2\pi f_j t) \quad \text{for } j = 1 \ldots 5. \tag{8.6}$$

These regressors were then fitted with just a non-seasonal IMA(1,1) error model, the role of the seasonal part of the model now being taken over by the regressors. A constant term can be included in the error model to allow for the trend or a trend term can be included (and the constant omitted) as a further regressor; they are equivalent. The fit of this model was identical to that of the airline model; the non-seasonal moving average parameter was the same and the forecasts were identical. More generally, regressors for seasonality should be introduced whenever seasonal differencing is removed from the model. Stationary variations of a seasonal nature may, however, remain, and may be evident as significant residual sacf values at seasonal lags—multiples of the seasonal period. There is no such evidence in this example, but for confirmation, both first order seasonal autoregressive and moving average terms were introduced into the model and estimated. Neither of them were significant.

The IMA(1,1) error model includes adaptability to level but not trend. But the series may indeed follow a trend with fixed level. This can be investigated by replacing the integrated IMA(1,1) error model by the stationary ARMA(1,1) model. The regressors will now include a constant term besides the trend and seasonal regressors. The stationary error model encompasses the IMA(1,1) model, which is got by constraining the autoregressive parameter ϕ to one. When estimated, $\phi = 0.948$ with standard error 0.042. A likelihood ratio test shows that the improvement in fit is far short of significant. Even so, as shown on the right in Fig. 8.7, the forecast error limits are slightly reduced though the forecasts are hardly changed. At a lead

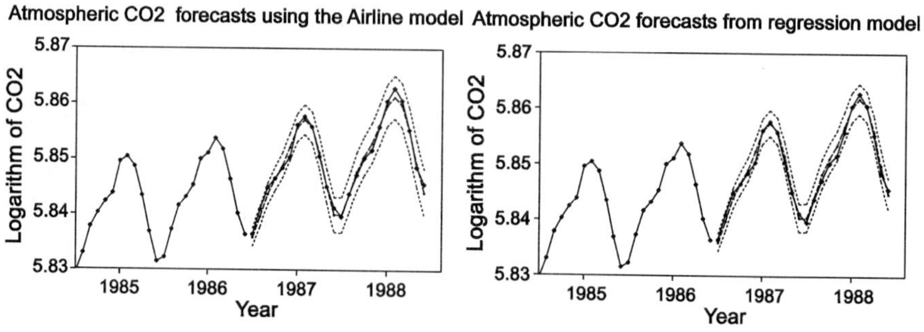

Fig. 8.7 *On the left* are forecasts of the last two years of the logarithms of the atmospheric CO_2 concentration, using the Airline Model fitted to the previous series values. The *solid lines* are the true series values, the *dashed lines* are the forecasts and the *dotted lines* are 95% error limits. *On the right*, the forecasts are constructed using regression on a trend and seasonal sinusoids with an ARMA(1,1) error model

time of 12 the error limits using the IMA(1,1) error model are less than 25% greater than those of the ARMA(1,1) model, but at lead time 24 they are over 50% greater. It may be argued that there is no reason to constrain $\phi = 1$, and the ARMA(1,1) model should be used. On the other hand the IMA(1,1) error model is robust to future level changes. Neither is there is any necessity to introduce regressors for seasonality; the Airline model as initially fitted to this series gives precisely the same forecasts: it is both acceptable and appropriate for this series. The residuals from both models were very similar and showed no evidence of model inadequacy.

8.5 Identification of the Box-Jenkins Seasonal Model

Model identification of Box and Jenkins involves selection of the orders of non-seasonal and seasonal differencing (d, D), autoregressive (p, P) and moving average (q, Q) operators. Their general approach is to inspect the sample autocorrelation function (sacf) and sample partial autocorrelation function (spacf) of the differences of the series for a limited range of orders of non-seasonal and seasonal differences. For a seasonal time series this is usually limited to the combinations of $d = 0, 1$ and $D = 0, 1$. The difficulty with this strategy is that the application of seasonal differencing in particular, can distort the lower order values of the sacf that are needed to identify the non-seasonal orders. It is true that for the Airline model there is an attractive separation of the sacf for $\nabla\nabla_{12}x_t$: the low lag values characterize the non-seasonal moving average part of the model and the values at seasonal lags (multiples of 12) characterize the seasonal moving average part. There is a problem, though, when the non-seasonal part of the model is best modeled as a stationary ARMA, with significant terms of the sacf extending well beyond lag 12, so that information relating to the seasonal part of the model interferes with that of the non-seasonal part. We first discuss questions relating to whether seasonality is present, and whether it is fixed or evolving. We then describe and illustrate a procedure which can help in seasonal model identification.

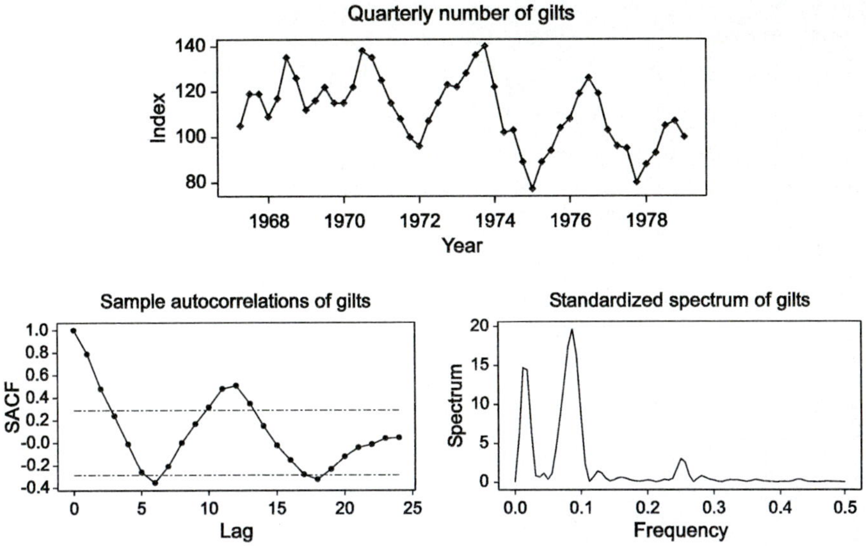

Fig. 8.8 *Above* is the quarterly index of gilt numbers. *Below* are the sacf (*on the left*), with nominal two standard error limits shown by the *dashed lines*, and the standardized sample spectrum of the series (*on the right*)

8.5.1 Checking if Seasonality is Present

One question to be asked, and answered, is whether a given time series contains any seasonality. Some caution is advisable here, because even if a time series is purely non-seasonal, the application of seasonal differencing will induce some, usually negative, seasonal autocorrelation at or around the seasonal lag. This could result in the unnecessary inclusion of a seasonal IMA component in the model. If seasonality is present it will usually be visually evident as a pattern in the original series. It is, however, possible that this is modest in amplitude and masked by noise and other features such as a business cycle. A good check for the presence of seasonality is then to inspect the sacf of the (non-seasonally) differenced series. Differencing reduces the amplitude of trends and other cycles and a pattern of significant values at seasonal lags is then usually quite evident if indeed seasonality is present. The sample spectrum of the differenced series will, in that case, reveal clear peaks at the seasonal fundamental and harmonic frequencies.

As an illustration, the upper plot in Fig. 8.8 shows a UK quarterly index of the number of gilts from 1967 to 1978, which are sows in pig for the first time. This is a measure of investment in pig production, which was quite volatile at the time. The series and its sacf, shown in the lower left plot of the figure, reveal a strong market cycle of approximately 3 years (12 quarters), but no obvious sign of a quarterly pattern, except possibly just at the start of the series. The sample spectrum in the lower right plot of the figure does indicate some seasonality of period 4 by a small peak at frequency 0.25, but this is masked in the other two plots by the strong cycle.

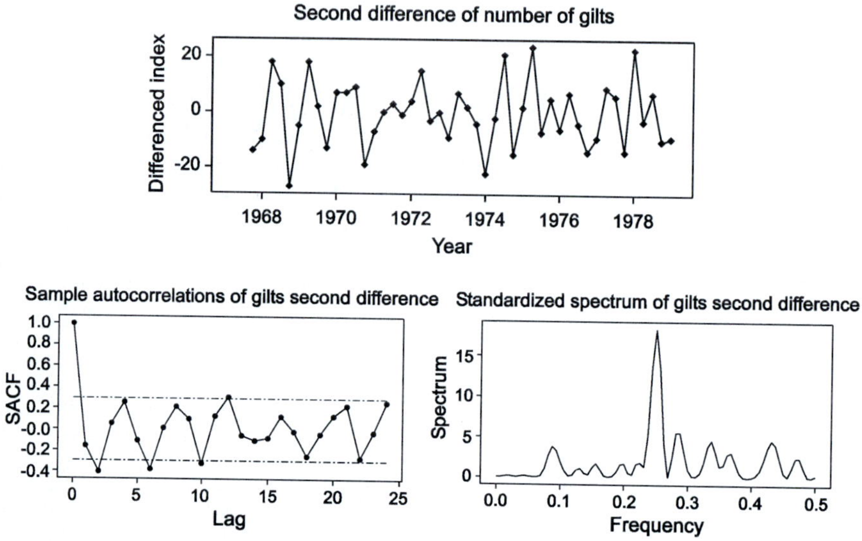

Fig. 8.9 *Above* is the second difference of the gilt numbers. *Below* are the sacf (*on the left*), with nominal two standard error limits shown by the *dashed lines*, and the standardized sample spectrum of the differenced series (*on the right*)

The upper plot in Fig. 8.9 shows the result of applying second differencing to the series which reduces the amplitude of the market cycle. A quarterly pattern is not so evident to the eye in this series, but it is now clear in its sacf, shown in the lower left plot of the figure, which peaks at lags 4, 8, 12 and 16. The sample spectrum in the lower right plot of the figure also reveals the quarterly nature of the series with a very strong peak at frequency 0.25. Although seasonality is a relatively small component of this series, its omission could distort a modeling analysis.

8.5.2 Checking if Seasonality is Fixed or Evolving

A further question to be asked, when seasonality is quite evident in the series, is whether it is fixed or evolving. In the former case the seasonal pattern may be represented by the fixed regressors that we described earlier, either seasonal indicators or seasonal sinusoids. In the latter case a seasonal ARIMA model is advocated. As we have explained with the example of atmospheric CO_2 concentrations, a seasonal IMA(1,1) model with Θ very close to one can give the same fit and forecasts as a model with trend and seasonal regressors, but estimation problems can arise with Θ so close to the invertibility boundary. A quick response to the question of whether seasonality is fixed or evolving, is to fit a simple model with fixed trend and seasonal regressors and look for evidence of any remaining seasonality in the residuals. A simple ARMA(1,1) error model is usually sufficient to reduce, if not eliminate, low lag residual sample autocorrelation. Evidence of seasonality that is not removed

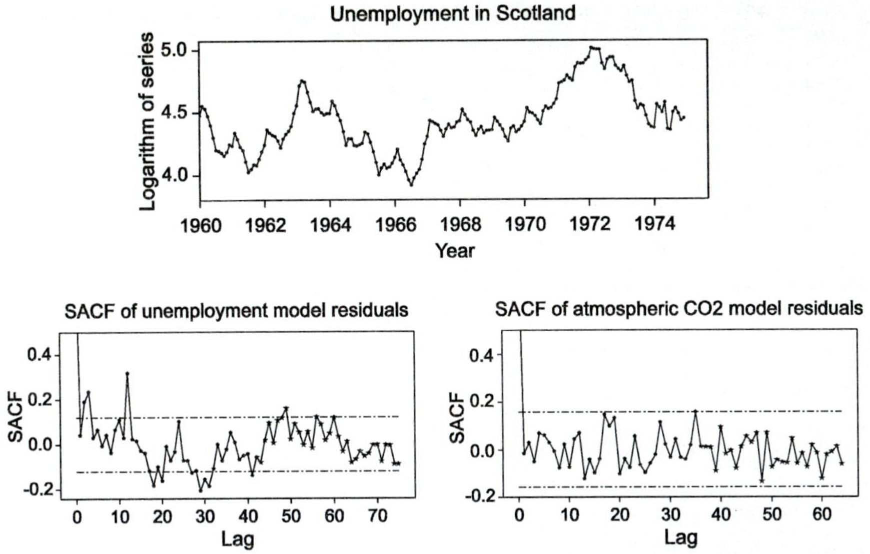

Fig. 8.10 The *upper plot* shows the logarithms of the level of unemployment in Scotland from January 1952 to November 1977. *On the left below* is the sacf after fitting to this series a fixed trend and seasonal regressors with an ARMA(1,1) error model. *On the right below* is the sacf for the same model fitted to the series of atmospheric CO_2 concentrations

by fitting the fixed regressors should then be seen as peaks in the residual sacf at multiples of the seasonal period. We illustrate this with a series of monthly unemployment level for Scotland over the period January 1952 to November 1977. The logarithms of this series are shown in the upper plot of Fig. 8.10. The lower left plot in this figure shows the residual sacf from fitting the fixed seasonal regressors as described. There is still some low lag correlation, but the peaks in the sacf at lags 12, 24 and possibly 36, show clear evidence that the seasonality is not fixed but changing. Although we do not show it here, the residual sample spectrum has a broad peak around frequency $1/12$ which also confirms the presence of remaining seasonality. In contrast, the lower right plot in Fig. 8.10 shows the corresponding residual sacf for the atmospheric CO_2 series. This shows no evidence of seasonality that has not been accounted for by the fixed regressors.

8.5.3 Identification of Non-seasonal Structure

If seasonality is present in a series, whether it is fixed or evolving, it will be substantially removed by application of seasonal differencing. Because of this, Box and Jenkins advocate that inspection of the sacf of the seasonally differenced series should be one of the steps in identifying an appropriate model. In particular, it better reveals the non-seasonal structure in the series. However, although it removes

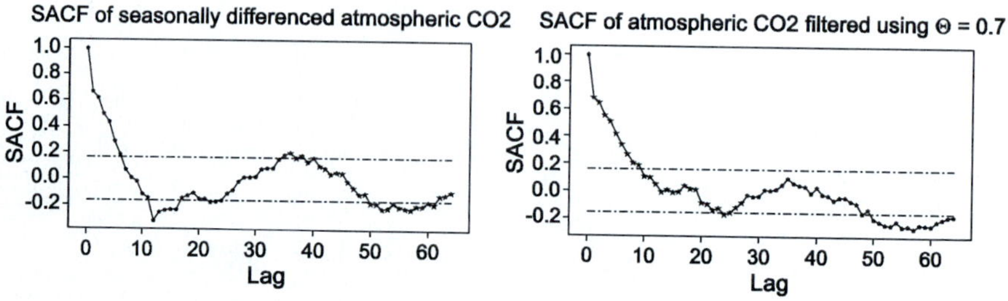

Fig. 8.11 *On the left* is the sacf of the seasonal differences of the atmospheric CO_2 series. *On the right* is the sacf after applying the filter in 8.7, with $\Theta = 0.7$, to the atmospheric CO_2 series

the strong pattern of seasonality in the series, seasonal differencing also induces negative correlation at lag 12 in the sacf of the differenced series. The plot on the left of Fig. 8.11 illustrates this for the atmospheric CO_2 series. If, as experience has shown, the seasonal structure of the series is well modeled by a seasonal IMA(1,1), this induced correlation can be countered by applying, instead, the operator associated with this part of the model:

$$\frac{\nabla_{12}}{1 - \Theta B^{12}}. \tag{8.7}$$

This will leave a filtered series with only the non-seasonal structure to be identified from its sacf. The question, though, is how to determine Θ. For many series this is answered by reference to the motivating argument that we initially presented for the Airline model: simply fit the seasonal IMA(1,1) model. The residuals from this are then precisely the series obtained by applying the operator (8.7). This works well for the Airline series, but we did note that the value of $\Theta = 0.340$ obtained by this strategy was somewhat smaller than the value $\Theta = 0.557$ estimated for the final model. For the series of unemployment in Scotland, the value of $\Theta = -0.415$ estimated by fitting the simple seasonal IMA(1,1) model, is very far from the value $\Theta = 0.528$ which we eventually find in the best model for this series. Applying the operator (8.7) with $\Theta = -0.415$ leads to greater distortion of the sacf, shown on the left in Fig. 8.12, than applying differencing alone!

This leads us to the proposal not to estimate Θ, but to apply the operator (8.7) with the specified value $\Theta = 0.7$ for all series. The sacf of the resulting series is shown for the atmospheric CO_2 series on the right in Fig. 8.11 and for the unemployment in Scotland on the right in Fig. 8.12. These lead us correctly to identify ARMA(1,1) and ARMA(2,1) models respectively for these series (using also the pattern of the associated partial sacfs which are not shown). The point is that using $\Theta = 0.7$ for applying (8.7) is an acceptable compromise which leads to minimal distortion of the filtered series for a wide range of true values of Θ. In our experience its use is certainly to be advocated as strongly as simple seasonal differencing, as a means of revealing the non-seasonal structure of the series.

The same procedure applied to the series of number of gilts also identified an ARMA(2,1) model for the seasonal part. Figure 8.13 shows forecasts derived for

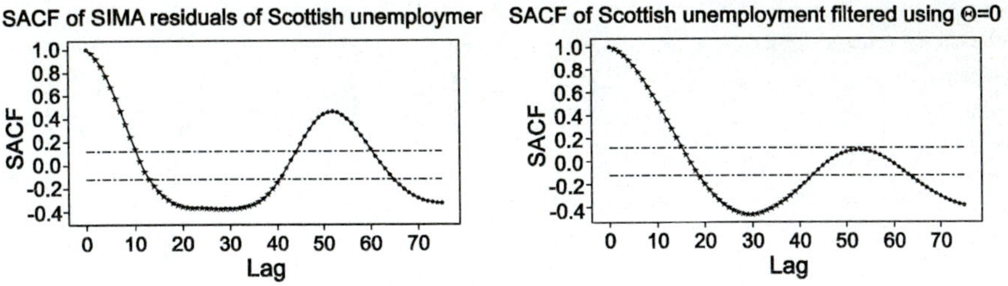

Fig. 8.12 *On the left* is the sacf of the residuals from modelling the unemployment in Scotland with a SIMA model. *On the right* is the sacf after applying the filter in 8.7, with $\Theta = 0.7$, to the series of unemployment in Scotland

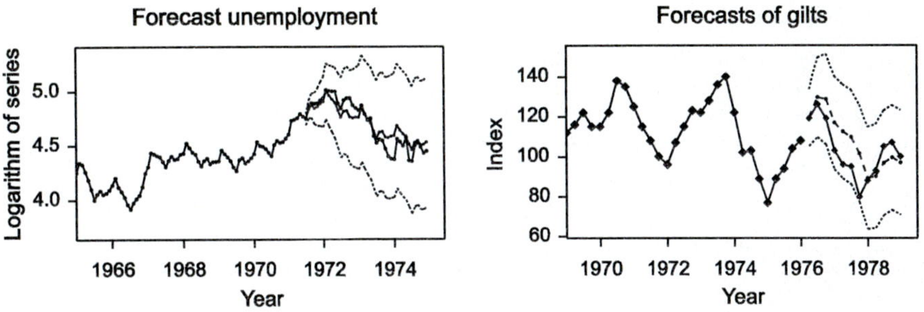

Fig. 8.13 Forecasts of the series of unemployment in Scotland *on the left*, and the number of gilts *on the right*, generated from models fitted to the previous series values

both the unemployment and gilts series using the models identified in this way. Redefinition of official statistics is always a challenge to forecast construction, and affects the forecast period shown for the unemployment series. From August 1972 adult students were included in the total, but were not included before this time. Another group, of temporary stopped workers, was also excluded from November 1972.

8.6 Series with Two Seasonal Periods

The Box-Jenkins Airline model has been successfully extended to time series of electricity demand, for which there are two natural seasonal periods of one day and one week. Brubacher and Tunnicliffe-Wilson [2] modeled hourly observations for which the period lengths are 24 and 168. Taylor et al. [10] compared forecasts from Box-Jenkins models with those from other schemes, using half-hourly observations for which the period lengths are 48 and 336, and Taylor [9] has more recently considered models with three natural seasonal periods. Matteson [7] describes a different approach to modeling a similar data set of hourly call arrival rates for emergency medical services. We are grateful to Taylor for providing us with data to which we

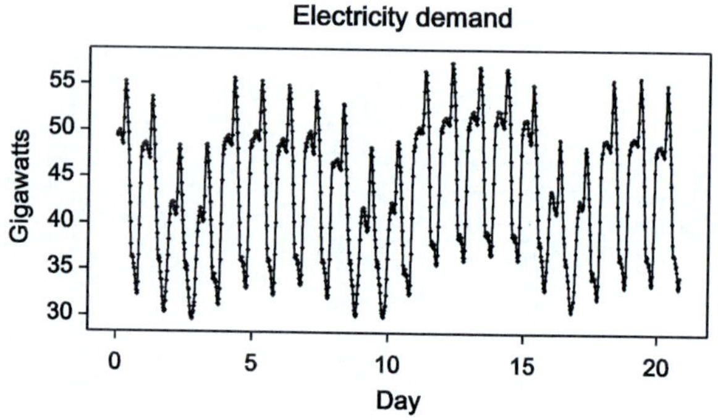

Fig. 8.14 Half hourly electricity demand for 21 days

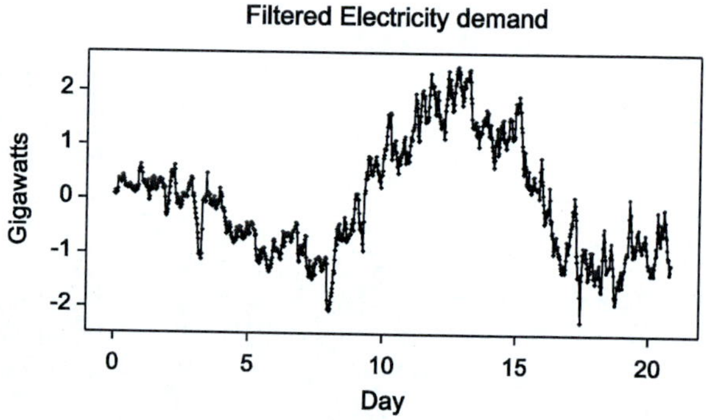

Fig. 8.15 Filtered series of electricity demand

apply our identification strategy. Figure 8.14 shows a plot of this data over a period of 21 days, with three weekend periods of lower demand quite evident. To this data we applied the operator (8.7) twice, but with the seasonal period replaced by first 336 and then 48. The value of $\Theta = 0.7$ was used in both operations. The series resulting from this operation is shown in Fig. 8.15 and is very similar to a random walk. It's first difference has significant, though small, sacf values at lags 1,5 and 9. Consequently a model of the form (8.8) was identified for the original series:

$$\nabla\nabla_{48}\nabla_{336}x_t = (1-\theta_1 B - \theta_5 B^5 - \theta_9 B^9)(1-\Theta_{48}B^{48})(1-\Theta_{336}B^{336})e_t. \quad (8.8)$$

Figure 8.16 shows forecasts of the last three days of the series, based on fitting model (8.8) to the previous observations. The forecasts are very good, even though the forecast origin is in the closing hours of a weekend. The forecast error limits are rather wide which suggest that this accuracy cannot always be expected, but overall this extension of the Airline model appears to provide a highly successful forecasting procedure.

8.7 Conclusion

We have shown how the seasonal models of Box and Jenkins, and the procedures for identifying these models, can be extended in various ways. Not least, we have

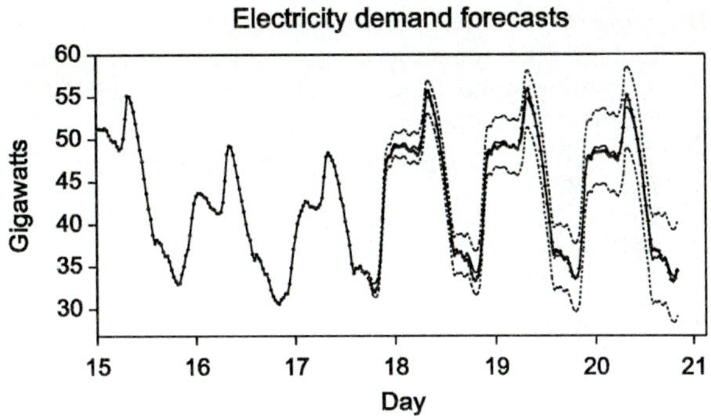

Fig. 8.16 Forecasts of electricity demand up to three days ahead using an extended Airline model fitted to previous series values

demonstrated how successful these models can be at constructing forecasts of a wide range of seasonal times series. All the computations used for the illustrations in this chapter were carried out in the Genstat package (http://www.vsni.co.uk/) and take at the most two or three seconds to execute on a modern computer. A particular feature of Genstat time series estimation software is the ability rigorously to compare models, which use either trend and seasonal regressors or seasonal ARIMA operators, by the marginal likelihood criterion of Tunnicliffe Wilson [11].

Several of the series used to illustrate the methods of this chapter were taken from a set of case studies developed by the second author. These time series were initially brought along by participants in a series of courses organized at the Civil Service College, and later at Lancaster University, by Peter Armitage. Peter Young and the first author were invited lecturers on these courses and enjoyed close collaboration with Peter Armitage, who sadly died early in 2010. This chapter is therefore written partly in memory of Peter Armitage as well as a joint tribute to Peter Young.

References

1. Box, G.E.P., Jenkins, G.M.: Time Series Analysis, Forecasting and Control. Holden-Day, San Francisco (1970)
2. Brubacher, S.R., Tunnicliffe Wilson, G.: Interpolating time series with application to the estimation of holiday effects on electricity demand. J. R. Stat. Soc. C **25**, 107–116 (1976)
3. Cleveland, S., Devlin, S.J.: Calendar effects in monthly time series: detection by spectrum analysis and graphical methods. J. Am. Stat. Assoc. **75**, 487–496 (1980)
4. Harvey, A.C., Todd, P.H.J.: Forecasting economic time series with structural and Box-Jenkins models: a case study. J. Bus. Econ. Stat. **1**, 299–307 (1983)
5. Haywood, J., Tunnicliffe Wilson, G.: Selection and estimation of component models for seasonal time series. J. Forecast. **19**, 393–417 (2000)
6. Keeling, R.F., Piper, S.C., Bollenbacher, A.F., Walker, J.S.: Atmospheric carbon dioxide record from Mauna Loa (2000). doi:10.3334/CDIAC/atg.035
7. Matteson, D.S., Mathew, W., McLean, D., Woodard, S., Henderson, S.G.: Forecasting emergency medical service call arrival rates. Ann. Appl. Stat. **5**, 1379–1406 (2011)
8. Newbold, P.: Predictors projecting linear trends plus seasonal dummies. J. R. Stat. Soc. D **37**, 111–127 (1988)
9. Taylor, J.: Triple seasonal methods for short-term electricity demand forecasting. Eur. J. Oper. Res. **204**, 139–152 (2010)

10. Taylor, J.W., Menezes, L.M., McSharry, P.E.: A comparison of univariate methods for forecasting electricity demand up to a day ahead. Int. J. Forecast. **22**, 1–16 (2006)
11. Tunnicliffe Wilson, G.: On the use of marginal likelihood in time series estimation. J. R. Stat. Soc. **51**, 15–27 (1989)
12. Tunnicliffe Wilson, G.: Structural models for structural change. Quad. Stat. Econom., **14**, 63–77 (1992)
13. Winters, P.R.: Forecasting sales by exponentially weighted moving averages. Manag. Sci. **6**, 324–342 (1960)

Chapter 9
State Dependent Regressions: From Sensitivity Analysis to Meta-modeling

Marco Ratto and Andrea Pagano

9.1 Introduction

The general concept of State Dependent Parameter (SDP) models for nonlinear, stochastic dynamic time series was suggested by Priestley [8] and it has been extensively developed by P.C. Young and co-workers in the last two decades: readers can refer to [20, 23, 24, 26] for a description of the method and a full list of references on the background to its development. SDP modeling is a very useful and efficient tool in signal processing and time series analysis; it has been successfully applied for many years in non-stationary and non-linear signal processing, e.g. to identify non-linearities in the context of dynamic transfer function models and in the framework of Young's Data-Based Mechanistic modeling [21]. The SDP model takes the simplified form of a the State Dependent Regression (SDR) when it is used to identify 'static' (non-dynamic) non-linear regression models, i.e. in the non-parametric regression context [24]. As such, SDR could well be considered for applications like sensitivity analysis and meta-modelling. Applying the SDR approach, Ratto et al. [11] have first developed a nonparametric approach for the efficient estimation of sensitivity indices in the framework Global Sensitivity Analysis (GSA, [13]). Subsequently, the main goal of our work has been to exploit SDR as an efficient identification tool for building emulators or meta-models with tensor product smoothing splines ANOVA models [10].

M. Ratto (✉) · A. Pagano
JRC, Joint Research Centre, The European Commission, TP 361, 21027 Ispra (VA), Italy
e-mail: marco.ratto@jrc.ec.europa.eu

A. Pagano
e-mail: andrea.pagano@jrc.ec.europa.eu

L. Wang, H. Garnier (eds.), *System Identification, Environmental Modelling, and Control System Design*,
DOI 10.1007/978-0-85729-974-1_9,

In our framework we move from a mathematical (or computational) model

$$Y = f(\mathbf{X}) = f(X_1, \ldots, X_p), \tag{9.1}$$

where Y is the model output that depends on $\mathbf{X}$, a vector of p model parameters (the 'input factors' in GSA terminology). In GSA and meta-modeling it is very important to consider the ANOVA decomposition of f into terms of increasing dimensionality:

$$f(X_1, X_2, \ldots, X_p) = f_0 + \sum_i f_i + \sum_i \sum_{j>i} f_{ij} + \cdots + f_{12\ldots p}, \tag{9.2}$$

where each term is a function only of the factors in its index, i.e. $f_i = f(X_i)$, $f_{ij} = f(X_i, X_j)$ and so on.

The input factors X_i have a domain of variability U, linked to the uncertainty about their precise value. We interpret the term 'factor' in a very broad sense: namely, a factor is anything that can be subject to some degree of uncertainty in the model. As such, input factors are treated as random variables characterised by specified distributions. Therefore also the Y is a random variable with a probability distribution, whose characterization is the main goal of our work.

The various terms are defined as follows:

$$\begin{aligned} f_0 &= E(Y), \\ f_i &= E(Y|X_i) - f_0, \\ f_{ij} &= E(Y|X_i, X_j) - E(Y|X_i) - E(Y|X_j) - f_0, \\ &\vdots \end{aligned} \tag{9.3}$$

and they are as many as $2^p - 1$. This ANOVA decomposition is strictly linked to sensitivity analysis: the so-called variance-based sensitivity indices directly derive from (9.2)–(9.3), as shown in [13]:

$$\begin{aligned} S_i &= V(f_i)/V(Y), \\ S_{ij} &= V(f_{i,j})/V(Y), \\ &\vdots \end{aligned} \tag{9.4}$$

Moreover, as discussed in [11], $f(X_I)$'s (I being a multi index $I = i_1 < i_2 < \cdots < i_s$ where $1 \le s \le p$) provide the best approximation to $f()$ in a least squares sense. If the input factors are independent, all the terms of the decomposition are orthogonal and the decomposition in (9.2) is unique. Therefore, estimating $f(X_I)$

provides a route for model approximation, as done by all statistical methods estimating ANOVA models, like tensor product cubic splines [4]. In this direction, SDR modelling is another class of non-parametric smoothing methods which are available to obtain estimations of the $f(X_I)$ terms. Later in this paper we will show that SDR may be effective in detecting important functional components (i.e. the $f(X_I)$'s), providing added value for the smoothing splines techniques (see [10]).

9.2 Estimating Truncated ANOVA Representations with SDR

9.2.1 Additive Models

In this Section we will describe some of the key features of the SDR technique applied to a first order (additive) ANOVA representation of model (9.1), expressed as:

$$f(\mathbf{X}) = f_0 + \sum_i f_i(X_i) + \varepsilon, \tag{9.5}$$

where ε is a Gaussian white noise accounting for all neglected ANOVA terms of order higher than one.

The estimation of f_i's is usually performed on a Monte Carlo sample of computer experiments $\{Y_k, \mathbf{X}_k\}$ of dimension N (see Sect. 3.1 in [11] for a discussion about possible sampling strategies).

The typical form of an SDR model for this kind of computer experiments can be expressed as [24]

$$\begin{aligned} Y_k - f_0 &= \mathbf{X}_k^T \mathbf{s}_k + e_k \\ &= s_{1,k} X_{1,k} + s_{2,k} X_{2,k} + \cdots + s_{p,k} X_{p,k} + e_k, \\ e_k &= N(0, \sigma^2), \end{aligned} \tag{9.6}$$

where as in (9.5), we may assume that all ANOVA terms of order higher than one can be approximated by a Gaussian white noise process with zero mean and variance σ^2, and the index k spans over the entire Monte Carlo dimension, i.e. $k = 1, \ldots, N$.

Comparing (9.5) and (9.6), we have that $f_i(X_{i,k}) = s_{i,k} X_{i,k}$, provided that each $s_{i,k}$ is a function of the corresponding input factor $X_{i,k}$. Hence, as noted in [11], estimating the terms $s_{i,k} X_{i,k}$ provides an estimate of the first order terms f_i. We should notice that whenever the support of some X_i contains zero we may face some singularity problems. Though it would be possible to overcome this issue by shifting the parameter by a constant value, it is preferable to reformulate (9.6) as

$$\begin{aligned} Y_k - f_0 &= \mathbf{1}_k^T \mathbf{s}_k + e_k \\ &= s_{1,k} + s_{2,k} + \cdots + s_{p,k} + e_k, \\ e_k &= N(0, \sigma^2), \end{aligned} \tag{9.7}$$

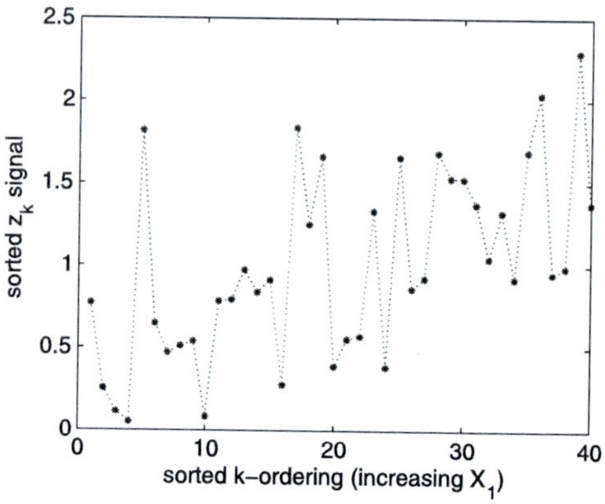

Fig. 9.1 The k-ordering for recursive estimation in SDR

having introduced constant unity regressors. In this case we have directly $f_i(X_{i,k}) = s_{i,k}$.

The next step will be to translate (9.7) in terms of State Space formulation. Each state dependent parameter $s_{i,k}$ needs to be characterized in some stochastic manner. As reported in [11] this is preferably accomplished by employing the Integrated Random Walk (IRW) process that is, in fact, characterized by similar smoothing properties as cubic splines. Assuming that the variability of $s_{i,k}$ follows an IRW process we write the State Space equations as:

$$\begin{aligned} &\text{Observation equation:} \quad && Y_k = \mathbf{s}_k + e_k, \\ &\text{State equations:} && s_{i,k} = s_{i,k-1} + d_{i,k-1}, \\ & && d_{i,k} = d_{i,k-1} + \eta_{i,k}, \end{aligned} \tag{9.8}$$

where e_k and $\eta_{i,k}$, $i = 1, 2, \ldots, p$ are zero mean white noise inputs with variance σ^2 and $\sigma^2_{\eta_{i,\cdot}}$, respectively. Here, the $\eta_{i,k}$ ('system disturbances' in systems terminology) provide the stochastic stimulus for parametric change in the model and they are assumed to be independent of each other and independent of the observation noise e_k.

It is important to recall here that the SDR estimation are based on the recursive Kalman filter (KF) and associated fixed interval smoothing (FIS) algorithms. As such, SDR needs to be applied to variables having some kind of meaningful ordering (think of time series where time gives a natural ordering) and this is not certainly the case of Monte Carlo samples. Hence, for each variable X_i, the k sorting index spans the Monte Carlo in the ascending order $X_{i,1} < X_{i,2} < \cdots < X_{i,k} < \cdots < X_{i,N}$ (Fig. 9.1). Clearly, for each input factor, a different order is used. Hence, a backfitting procedure is employed (see [23, 24] for details about the backfitting). Given the ascending ordering of the Monte Carlo sample, s_k can be estimated by using the KF and FIS recursive algorithms (see e.g. [5, 22] for details).

It seems necessary to discuss the term e_k in (9.8). Normality and independence is strictly appropriate when there is observational error in the data but can be reasonable for smoothing observed data even in computer experiments. First, because there can be applications where the 'computed' value is produced with some error or variability, due to e.g. convergence of numerical algorithms. Furthermore, these residuals also reflect the *truncated* ANOVA expansion that is used to approximate $Y(\mathbf{X})$. In practice, this is done by including a 'small' subset of q ANOVA terms (e.g. main effects and low order interactions) that are statistically identifiable from the available Monte Carlo sample of computer experiments. Thus, e_k can be seen as the sum of all the terms that are not included in this process of model complexity reduction. This set of dropped ANOVA terms usually includes a very large number of elements (namely $2^p - q$, where $q \ll 2^p$), which are orthogonal (independent) by definition. It does not seem out of place to model the sum of a large number of independent variables in statistical terms (Central Limit Theorem). As shown in [10] the inclusion of this 'error' term, rather than being a drawback of this method, turns out to be an advantage (see Examples), since it implies that the ANOVA model approximation (and therefore 'prediction' at untried X values) is performed only using statistically significant ANOVA terms, enhancing the robustness in out-of-sample performances.

In order to obtain the proper estimate of $s_{i,k}$, it is first necessary to optimize the hyper-parameters associated with the state space model (9.8), namely the white noise variances σ^2 and $\sigma^2_{\eta_i}$. As discussed for example in [24], by a simple reformulation of the KF and FIS algorithms, the IRW model can be entirely characterized by one Noise Variance Ratio (NVR) hyper-parameter, where $\mathrm{NVR}_i = \sigma^2_{\eta_i}/\sigma^2$. This NVR value is, of course, unknown *a priori* and needs to be optimized: for example, in all SDP references, this is accomplished by maximum likelihood optimization (ML) using prediction error decomposition (see [15]). The NVR_i plays the inverse role of a smoothing parameter: the smaller the NVR_i, the smoother the estimate of $s_{i,k}$ (and at the limit $\mathrm{NVR}_i = 0$, $s_{i,k}$ will be a straight line). Given the NVR_i, the FIS algorithm then yields an estimate $\hat{s}_{i,k|N}$ of $s_{i,k}$ at each data sample and it can be seen that the $\hat{s}_{k|N}$ from the IRW process is the equivalent of $f(X_k)$ in the cubic smoothing splines model. At the same time, the recursive procedures provide, in a natural way, standard errors of the estimated $\hat{s}_{k|N}$, that allow for the testing of their relative significance.

9.2.2 Extension to 2nd Order Interactions

In order to extend the pure recursive SDR approach to the estimation of 2nd order interaction terms we need to define a sorting strategy for points on a surface. It is well known that it is not possible to define a total ordering on $\mathbb{R}^2$ and in [11] the following approach was proposed. Assume we want to estimate $f_{12}(X_1, X_2)$: on the $\{X_1\, X_2\}$ plane, the k sorting index is defined ordering the pairs $(X_{1,k}, X_{2,k})$ according to their position with respect to the closed trajectory as in Fig. 9.2.

Similarly to the one dimensional case, this special 2-dimensional k-sorting provides a low frequency characteristics for the pair $\{X_1\, X_2\}$ of input factors, while all other factors are still characterized by a high frequency noisy spectrum (Fig. 9.3).

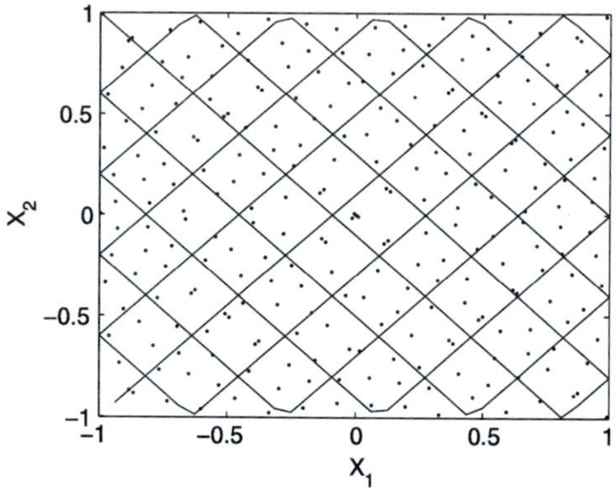

Fig. 9.2 The 2-dimensional k-ordering for recursive estimation of interactions

The corresponding sorted output signal Y_k can therefore be analyzed by the SDR algorithms and the 2nd order interaction term associated to the pair $X_{1,k}\,X_{2,k}$ will be identified (Fig. 9.4). Since each interaction effect will have a different sorting, the backfitting procedure has to be exploited for the interaction effects as well.

Considering the above procedure within the SDR formalism, the ANOVA terms of second order can be expressed as:

$$\sum_{j>i} s_{ij,k}(\chi_{ij}) + \text{h.o.t.}, \tag{9.9}$$

where each state-dependent parameter $s_{ij,k}(\chi_{ij})$, $j > i = 1,\ldots,p$, follows a stochastic IRW process and depends on a state variable χ_{ij} that moves according to the 2-dimensional sorting strategy along the co-ordinates of the pair of factors indexed by ij.

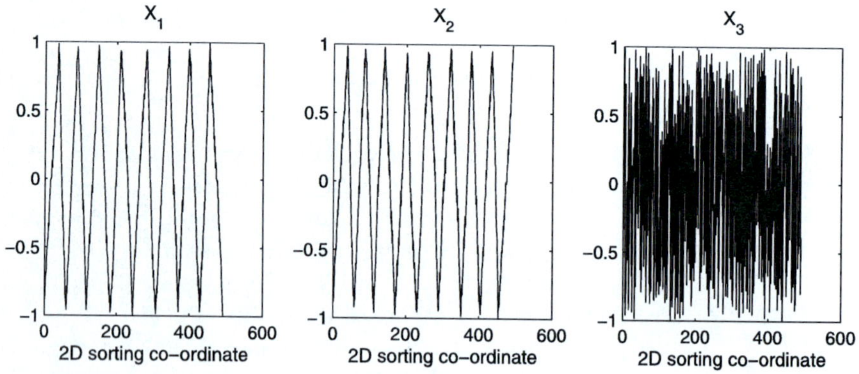

Fig. 9.3 Patterns of input factors produced by the 2-dimensional ordering for the couple $[X_1, X_2]$

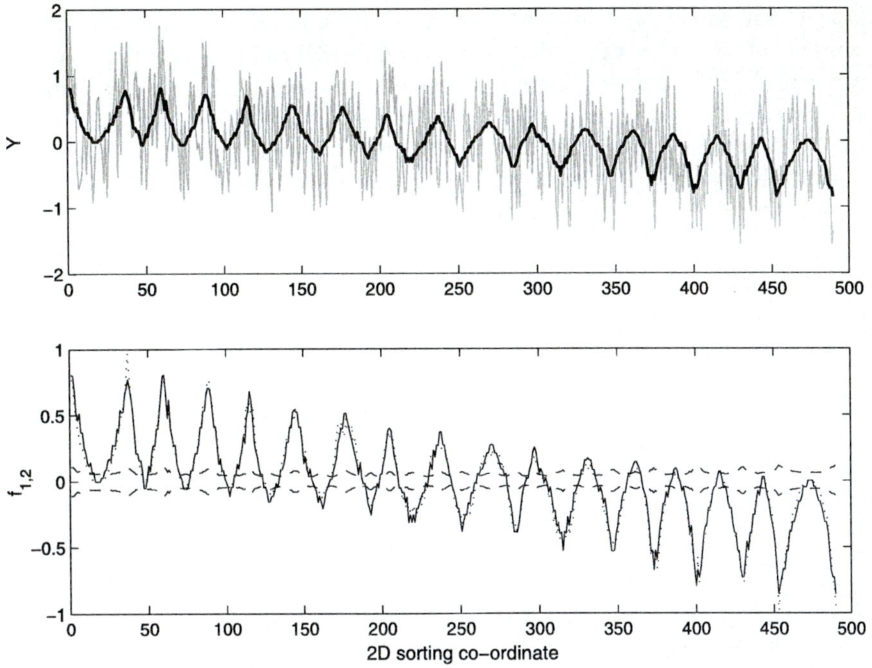

Fig. 9.4 Example of Y model output along the 2-dimensional ordering. *Lower panel* shows the estimated interaction term $f_{1,2}$ (*solid line*) compared to the true sorted interaction signal (*dotted*). *Dashed lines* indicate the width of the uncertainty estimate of the interaction term

9.3 Estimation of Higher Order Moments with SDR

The SDP approach is very flexible and can be adapted to a wide variety of smoothing problems. For example, random walk (RW) or smoothed random walk (SRW) models for the SDP's might be preferable, in some circumstances, because they yield less smooth estimates than the IRW model. Moreover, if any sharp changes or jumps are present, then these can be handled using 'variance intervention' (see [7]). Within the sensitivity analysis framework, in [9] we considered such effects that cannot be attributed to shifts in the mean and are not accounted for by $E(Y|X_i)$ as in (9.3)–(9.4). For example, this happens when the 'observation noise' $e_i = Y - E(Y|X_i)$ has a heteroscedastic nature. This can be modeled by assuming that the variance of $e_{i,k}$ is modulated by X_i in some state-dependent manner. This can be achieved by introducing the state-dependent decomposition [27]:

$$e_{i,t}^2 = m_{2,i,t}(X_i) + n_{2,i,t}, \tag{9.10}$$

where now the state dependent parameter $m_{2,i,t}(X_i)$ accounts for the heteroscedasticity in $e_{i,t}$. Hence, remembering the conditional variance expression

$$\mathrm{Var}(Y|X_i) = E(Y^2|X_i) - E^2(Y|X_i) = E[(Y - E(Y|X_i))^2|X_i] = E(e_i^2|X_i) \tag{9.11}$$

we see that $\mathrm{Var}(Y|X_i) = m_{2,i}$. Moreover, as shown in [27], feeding back the estimated pattern in the observation error within the KF and FIS algorithms used to estimate $E(Y|X_i)$, one can obtain a much more accurate and efficient estimation of $E(Y|X_i)$ itself.

This procedure can be further expanded to identify the presence of patterns in third and fourth order moments and, in particular, to detect changes in the skewness γ_1 or in the kurtosis γ_2 of the distribution of Y which may be driven by some input factor.

For the third moment and skewness, first the state dependent third moment is defined as:

$$e_{i,t}^3 = m_{3,i,t}(X_i) + n_{3,i,t}, \tag{9.12}$$

which then yields the 'local' skewness $\gamma_1(X_i) = m_{3,i}/m_{2,i}^{3/2}$. Alternatively, this local skewness can be estimated directly using the standardized residuals $\tilde{e}_i = (Y - E(Y|X_i))/\sqrt{\mathrm{Var}(Y|X_i)}$:

$$\tilde{e}_{i,t}^3 = \tilde{m}_{3,i,t}(X_i) + \tilde{n}_{3,i,t} \tag{9.13}$$

providing directly $\gamma_1(X_i) = \tilde{m}_{3,i}$.

Similarly for the fourth moment and kurtosis: first, the state dependent fourth moment is defined as

$$e_{i,t}^4 = m_{4,i,t}(X_i) + n_{4,i,t}, \tag{9.14}$$

which then yields the local kurtosis $\gamma_2(X_i) = m_{4,i}/m_{2,i}^2$. As for skewness, the local kurtosis can be directly estimated using the standardized residuals,

$$\tilde{e}_{i,t}^4 = \tilde{m}_{4,i,t}(X_i) + \tilde{n}_{4,i,t} \tag{9.15}$$

yielding directly $\gamma_2(X_i) = \tilde{m}_{4,i}$.

All of these smoothing procedures for second, third and fourth order moments are performed on the same MC sample used for the 'standard' smoothing estimation of $E(Y|X_i)$. As a result, they have no additional cost in terms of model evaluation. At the same time, they provide extremely useful information about sensitivity patterns that are not detectable using standard variance-based techniques; or about parameters that drive shifts in the distribution of Y, like a change in variance, asymmetry and fat tails. As discussed in [9], this additional information complements variance-based sensitivity analysis in a very similar manner to other techniques, such as entropy-based measures and moment-independent measures [1, 2], but with much smaller computational requirements.

9.4 SDR and Smoothing Splines ANOVA Models

Here we summarize the application of SDR within the framework of smoothing splines ANOVA models. Moving from the early work of Wahba (see [17]) and Gu (see [4]), recently, Storlie et al. (see [14]) presented the ACOSSO, 'a new regularization method for simultaneous model fitting and variable selection in nonparametric regression models in the framework of smoothing splines ANOVA'. This method is an improvement of the COSSO (see [6]), penalizing the sum of component norms, instead of the squared norm employed in the traditional smoothing splines method. In ACOSSO, an adaptive weight is used in the COSSO penalty which allows for more flexibility in estimating important functional components while giving a heavier penalty to unimportant functional components.

We will summarize here the main results of [10], where we propose the use of SDR as the identification step in detecting the important ANOVA functional components. This turns out to be a very effective approach, adding valuable information in the ACOSSO framework.

The use of recursive algorithms in smoothing splines is not new in statistical literature: the works of Weinert et al. [19] and of Wecker and Ansley [18] demonstrated the applicability of a stochastic framework for recursive computation of smoothing splines. However, such works were limited to the univariate case, while the subsequent history of tensor product smoothing splines developed in the 'standard' non-recursive form. The SDR recursive approach of Young [24, 26] provides an extension to such seminal papers, which is applicable to the multivariate case, as well as for interaction terms.

9.4.1 Additive Models

As before, we begin our analysis studying an additive model

$$f(\mathbf{X}) = f_0 + \sum_{j=1}^{p} f_j(X_j). \tag{9.16}$$

To estimate f we can use a multivariate (cubic) smoothing splines minimization problem, that is, given $\lambda = (\lambda_1, \ldots, \lambda_p)$, find the minimizer $f(\mathbf{X})$ of:

$$\frac{1}{N}\sum_{k=1}^{N}(Y_k - f(\mathbf{X}_k))^2 + \sum_{j=1}^{p}\lambda_j \int_0^1 [f_j''(X_j)]^2 dX_j, \tag{9.17}$$

where a Monte Carlo (MC) sample of dimension N is assumed, as usual. This statistical problem requires the estimation of the p hyper-parameters λ_j (also denoted as smoothing parameters). There exist various ways of doing that: by applying generalized cross-validation (GCV), generalized Maximum Likelihood procedures (GML) and so on (see e.g. [4, 17]). Note that in the cubic splines situation higher values of the smoothing parameters λ_j correspond to smoother estimates, while in the SDR recursive approach the NVR plays the inverse role of a smoothing parameter: smaller NVR's correspond to smoother estimates.

In the estimation of the hyper-parameters λ_j within the classical statistical framework , we observe that, to avoid a perfect fit solution, a penalty term is necessary. To fix ideas, let us consider the GCV optimization. In the cubic splines context, with GCV we look for λ minimizing

$$\mathrm{GCV}_\lambda = 1/N \cdot \frac{\sum_k (Y_k - f_\lambda(X_k))^2}{(1 - df(\lambda)/N)^2}, \tag{9.18}$$

where $df \in [0, N]$ denotes the 'degrees of freedom' as function of λ. The smaller λ, the larger df, i.e. the smoothing splines model will possibly tend to over-fit the data.

As discussed in previous sections, using the SDR recursive estimation approach the additive model is formalized as in (9.7). Expressing GCV in the SDR notation, we look for NVR minimizing

$$\mathrm{GCV}_{\mathrm{NVR}} = 1/N \cdot \frac{\sum_k (Y_k - \hat{s}_{k|N})^2}{(1 - df(\mathrm{NVR})/N)^2}, \tag{9.19}$$

where, in this case, the 'degrees of freedom' df depend on the NVR: the smaller NVR, the larger df.

Using classical statistical optimization procedures for the smoothing parameters like GCV and GML, without the penalty term, the optimum would always be attained at $\lambda = 0$ (or NVR$\rightarrow \infty$), i.e. perfect fit.

One key issue of the SDR methodology is that, applying ML optimization within the recursive framework, a perfect fit solution is impossible. The reason is that the prediction error (based on Maximal likelihood) estimate uses the *filtered estimates* $\hat{s}_{k|k-1}$ and *not the smoothed estimate* $\hat{s}_{k|N}$ as in (9.18)–(9.19).

In other words, using ML, we are looking for NVR maximizing the log-likelihood function L, where:

$$-2 \cdot \log(L) = \mathrm{const} + \sum_{k=3}^{N} \log(1 + P_{k|k-1}) + (N-2) \cdot \log(\hat{\sigma^2}),$$

$$\hat{\sigma^2} = \frac{1}{N-2} \sum_{k=3}^{N} \frac{(Y_k - \hat{s}_{k|k-1})^2}{(1 + P_{k|k-1})} \tag{9.20}$$

and where $\hat{\sigma^2}$ is the 'weighted average' of the squared innovations (i.e. the prediction error of the IRW model), $P_{k|k-1}$ is the one step ahead forecast error of the state $\hat{s}_{k|k-1}$ provided by the KF (both $P_{k|k-1}$ and $\hat{s}_{k|k-1}$ are functions of NVR). Since $\hat{s}_{k|k-1}$ is based only on the information contained in the sample values $[1, \ldots, k-1]$ (while smoothed estimates use the entire information set $[1, \ldots, N]$), it can be easily seen that the limit NVR $\rightarrow \infty$ is not a 'perfect fit' situation, since a zero variance for

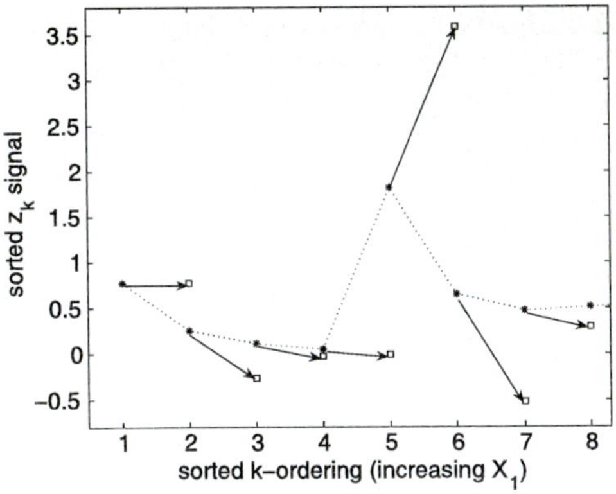

Fig. 9.5 Example of one-step-ahead predictions of the IRW model for NVR = 0

e_k implies $\hat{s}_{k|k-1} = s_{k-1} + d_{k-1} = Y_{k-1} + d_{k-1}$, i.e. the one step ahead prediction of Y_k is given by the linear extrapolation from the adjacent value Y_{k-1}, so implying a non-zero prediction error in this limit case.

This is further exemplified in Fig. 9.5: the squares in the plots denote the one step ahead prediction $\hat{s}_{k|k-1}$ and the arrows show the linear extrapolation mechanism of the IRW process when NVR $\to \infty$. Such a prediction departs considerably not only from a 'perfect fit' situation but also from a 'reasonable fit', implying that the ML estimate will automatically penalize this kind of situation and provide the 'right' value for the NVR.

These properties of the ML optimization makes it appealing to properly identify smoothing parameters in smoothing splines. This is made possible by an equivalence between the hyper-parameters (λ's for the cubic splines and NVR for SDR). It can be easily verified that by setting $\lambda = 1/(\text{NVR} \cdot N^4)$, and with evenly spaced X_k values, the $f(X_k)$ estimate in the cubic smoothing splines model equals the $\hat{s}_{k|N}$ estimate from the IRW process.

9.4.2 Second Order Models

The additive model concept (9.16) can be generalized to include 2-way (and higher) interaction functions via the functional ANOVA decomposition [4, 17]. For example, we can let

$$f(\mathbf{X}) = f_0 + \sum_{j=1}^{p} f_j(X_j) + \sum_{j<i}^{p} f_{j,i}(X_j, X_i). \tag{9.21}$$

In the ANOVA smoothing splines context, corresponding optimization problems with interaction functions and their solutions can be obtained conveniently with the reproducing kernel Hilbert space (RKHS) approach (see [17]).

As discussed in Sect. 9.2.2, in [11] a first attempt to extend the SDR for interaction terms was done, applying the 2-dimensional sorting strategy. We then tried to extend the equivalences between NVR and λ's also for this case and exploit them in the second order tensor product cubic splines models. The results of these trials highlighted limitations of this approach, with the resulting ANOVA model performing worse than using more classical statistical approaches. Therefore, in [10] we proposed to formulate an interaction function as the product of two state variables $s_1 \cdot s_2$, each of them characterized by an IRW stochastic process. Hence the estimation of a single interaction term $Y^*(\mathbf{X}_k) = f(X_{1,k}, X_{2,k}) + e_k$ is expressed as:

$$\begin{aligned} &\text{Observation equation:} \quad && Y_k^* = s_{1,k}^I \cdot s_{2,k}^I + e_k, \\ &\text{State equations: } (j=1,2) \quad && s_{j,k}^I = s_{j,k-1}^I + d_{j,k-1}^I, \\ & && d_{j,k}^I = d_{j,k-1}^I + \eta_{j,k}^I, \end{aligned} \tag{9.22}$$

where Y^* is the model output after having taken out the main effects, $I = 1,2$ is the multi-index denoting the interaction term under estimation and $\eta_{j,k}^I \sim N(0, \sigma^2_{\eta_j^I})$. The two terms $s_{j,k}^I$ are estimated iteratively by running the recursive procedure in turn, i.e.

- take an initial estimate of $s_{1,k}^I$ and $s_{2,k}^I$ by regressing Y^* with the product of simple linear or quadratic polynomials $P_1(X_1) \cdot P_2(X_2)$ and set $s_{j,k}^{I,0} = P_j(X_{j,k})$;
- iterate $i = 1,2$:
  - fix $s_{2,k}^{I,i-1}$ and estimate NVR_1^I and $s_{1,k}^{I,i}$ using the recursive procedure;
  - fix $s_{1,k}^{I,i}$ and estimate NVR_2^I and $s_{2,k}^{I,i}$ using the recursive procedure;
- the product $s_{1,k}^{I,2} \cdot s_{2,k}^{I,2}$ obtained after the second iteration provides the recursive SDR estimate of the interaction function.

The latter stopping criterion is a convenient choice to limit the computation time, and is due to the observation that the estimate of the interaction term never changed too much in any subsequent iteration. We also observe that the recursive form for this kind of estimation of second order interactions uses a standard sorting along each co-ordinate X_1 and X_2. Therefore it does not make any use of the 2-dimensional partial ordering discussed in Sect. 9.2.2.

Unfortunately, as for the generalized 2-dimensional sorting strategy, we could not derive an explicit and full equivalence between SDR and cubic splines of the type mentioned for first order ANOVA terms. Therefore, in order to be able to exploit the SDR estimation results in the context of a smoothing spline ANOVA model, we proposed in [10] to take a different approach, similar to the ACOSSO case.

9.4.3 Short Summary of ACOSSO

We make the usual assumption that $f \in \mathscr{F}$, where $\mathscr{F}$ is a RKHS. The space $\mathscr{F}$ can be written as an orthogonal decomposition $\mathscr{F} = \{1\} \oplus \{\bigoplus_{j=1}^q \mathscr{F}_j\}$, where each $\mathscr{F}_j$ is itself a RKHS and $j = 1, \ldots, q$ spans ANOVA terms of various orders. Typically q includes the main effects plus relevant interaction terms.

We re-formulate (9.17) for the general case with interactions using the function f that minimizes:

$$\frac{1}{N}\sum_{k=1}^{N}(Y_k - f(\mathbf{X}_k))^2 + \lambda_0 \sum_{j=1}^{q} \frac{1}{\theta_j} \|P^j f\|^2_{\mathscr{F}}, \tag{9.23}$$

where $P^j f$ is the orthogonal projection of f onto $\mathscr{F}_j$ and the q-dimensional vector θ_j of smoothing parameters needs to be optimized somehow. This is typically a formidable problem and in the simplest case θ_j is set to one, with the single λ_0 estimated by GCV or GML.

Problem (9.23) also poses the issue of selection of $\mathscr{F}_j$ terms: this is tackled rather effectively within the COSSO/ACOSSO framework.

The COSSO [6] penalizes the sum of norms, using a LASSO type penalty [16] for the ANOVA model, which allows us to identify the informative predictor terms $\mathscr{F}_j$ with an estimate of f that minimizes

$$\frac{1}{N}\sum_{k=1}^{N}(Y_k - f(\mathbf{X}_k))^2 + \lambda \sum_{j=1}^{Q} \|P^j f\|_{\mathscr{F}} \tag{9.24}$$

using a single smoothing parameter λ, and where Q includes *all* ANOVA terms to be potentially included in f, e.g. with a truncation up to 2nd or 3rd order interactions.

It can be shown that the COSSO estimate is also the minimizer of

$$\frac{1}{N}\sum_{k=1}^{N}(Y_k - f(\mathbf{X}_k))^2 + \sum_{j=1}^{Q} \frac{1}{\theta_j} \|P^j f\|^2_{\mathscr{F}} \tag{9.25}$$

subject to $\sum_{j=1}^{Q} 1/\theta_j < M$ (where there is a 1–1 mapping between M and λ). So we can think of the COSSO penalty as the traditional smoothing splines penalty plus a penalty on the Q smoothing parameters used for each component. The LASSO type penalty has the effect of setting some of the functional components ($\mathscr{F}_j$'s) equal to zero (e.g. the variable X_j or the interaction (X_j, X_i) is not in the model), thus it 'automatically' selects the appropriate subset q of terms out of the Q 'candidates'. The key property of COSSO is that with one single smoothing parameter (λ or M) it provides proper estimates of all θ_j parameters: therefore it improves considerably the problem (9.23) with $\theta_j = 1$ (still with one single smoothing parameter λ_0) and is much more computationally efficient than the full problem (9.23) with optimized θ_j's.

In the adaptive COSSO (ACOSSO) of [14], $f \in \mathscr{F}$ minimizes

$$\frac{1}{N}\sum_{k=1}^{N}(Y_k - f(\mathbf{X}_k))^2 + \lambda \sum_{j=1}^{q} w_j \|P^j f\|_{\mathscr{F}}, \tag{9.26}$$

where $0 < w_j \leq \infty$ are weights that depend on an initial estimate of $\tilde{f}$, either using (9.23) with $\theta_j = 1$ or the COSSO estimate (9.24). The adaptive weights are obtained as $w_j = \|P^j \tilde{f}\|_{L_2}^{-\gamma}$,typically with $\gamma = 2$ and the L_2 norm $\|P^j \tilde{f}\|_{L_2} = (\int (P^j \tilde{f}(\mathbf{X}))^2 d\mathbf{X})^{1/2}$. The use of adaptive weights improves the predictive capability of ANOVA models with respect to the COSSO case.

9.4.4 Combining SDR and ACOSSO for Interaction Functions

As discussed in [10], there is an obvious way of exploiting the SDR identification and estimation steps in the ACOSSO framework: namely, the SDR estimates of additive and interaction function terms can be taken as the initial $\tilde{f}$ used to compute the weights in the ACOSSO. However, this would be a minimal approach, whereas the SDR identification and estimation provides more detailed information about ANOVA terms that is worth exploiting. We define $\mathcal{K}_{\langle j \rangle}$ to be the reproducing kernel (r.k.) of an additive term $\mathcal{F}_j$ of the ANOVA decomposition of the space $\mathcal{F}$. In the cubic splines case, this is constructed as the sum of two terms $\mathcal{K}_{\langle j \rangle} = \mathcal{K}_{01\langle j \rangle} \oplus \mathcal{K}_{1\langle j \rangle}$ where $\mathcal{K}_{01\langle j \rangle}$ is the r.k. of the parametric (linear) part and $\mathcal{K}_{1\langle j \rangle}$ is the r.k. of the purely non-parametric part. The second order interaction terms are constructed as the tensor product of the first order terms, for a total of four elements, i.e.

$$
\begin{aligned}
\mathcal{K}_{\langle i,j \rangle} &= (\mathcal{K}_{01\langle i \rangle} \oplus \mathcal{K}_{1\langle i \rangle}) \otimes (\mathcal{K}_{01\langle j \rangle} \oplus \mathcal{K}_{1\langle j \rangle}) \\
&= (\mathcal{K}_{01\langle i \rangle} \otimes \mathcal{K}_{01\langle j \rangle}) \oplus (\mathcal{K}_{01\langle i \rangle} \otimes \mathcal{K}_{1\langle j \rangle}) \oplus (\mathcal{K}_{1\langle i \rangle} \otimes \mathcal{K}_{01\langle j \rangle}) \\
&\quad \oplus (\mathcal{K}_{1\langle i \rangle} \otimes \mathcal{K}_{1\langle j \rangle}).
\end{aligned} \tag{9.27}
$$

In general, considering the problem (9.23), one should attribute a specific coefficient $\theta_{\langle \cdot \rangle}$ to each single element of the r.k. of $\mathcal{F}_j$ (see e.g. [4], Chap. 3), i.e. two θ's for each main effect, four θ's for each two-way interaction, and so on. In fact, each $\mathcal{F}_j$ would be optimally fitted by opportunely choosing weights in the sum of $\mathcal{K}_{\langle \cdot,\cdot \rangle}$ elements. This, however, makes the estimation problem rather complex, so, usually, the tensor product (9.27) is directly used, without tuning the weights of each element of the sum. This strategy is also applied in ACOSSO.

Instead, we propose to use SDR estimates of interaction to set the weights.

In particular, we can see that the SDR estimate of the interaction (9.22) is given by the product of two univariate cubic splines. So, one can easily decompose each estimated $\hat{s}_j^I$ into the sum of a linear ($\hat{s}_{01\langle j \rangle}^I$) and non-parametric term ($\hat{s}_{1\langle j \rangle}^I$). This provides a decomposition of the SDR interaction of the form

$$
\hat{s}_i^I \cdot \hat{s}_j^I = \hat{s}_{01\langle i \rangle}^I \hat{s}_{01\langle j \rangle}^I + \hat{s}_{01\langle i \rangle}^I \hat{s}_{1\langle j \rangle}^I + \hat{s}_{1\langle i \rangle}^I \hat{s}_{01\langle j \rangle}^I + \hat{s}_{1\langle i \rangle}^I \hat{s}_{1\langle j \rangle}^I, \tag{9.28}
$$

which can be thought as a proxy of the four elements of the r.k. of the second order tensor product cubic splines.

This suggests that a natural use of the SDR identification and estimation in the ACOSSO framework is to apply specific weights to each element of the r.k. $\mathcal{K}_{\langle \cdot,\cdot \rangle}$ in (9.27). In particular the weights are the L_2 norms of each of the four elements estimated in (9.28). As shown in the examples, this choice can lead to significant improvement in the accuracy of ANOVA models with respect to the original ACOSSO approach.

9.5 Examples

In this section we use an analytical example to put in practice what we have discussed so far. For further details and other examples we refer to [11] where SDR techniques were first applied for computing sensitivity indices and to [10] where SDR is used together with ACOSSO. To help readers to follow the whole history, we consider an analytical example, based on the Sobol' g-function, as in [11]. In the Sobol' g-function a set of parameters can be modulated in order to achieve different degrees of complexity, as follows:

$$Y = \prod_{i=1}^{p} g_i(X_i), \quad \text{where } g_i(X_i) = \frac{|4X_i - 2| + a_i}{1 + a_i}, \tag{9.29}$$

with $a_i \geq 0$ and each factor X_i are uniformly distributed in the interval [0, 1]. This is a strongly non-linear and non-additive model used in the past to test Global Sensitivity Analysis (GSA) methods [13]. The value for p can be chosen to analyze the dependence of the method on the number of input factors. Moreover, by tuning the spectrum of parameters a_i, the relative importance of the X_i's can be modified. The importance of an input factor is higher when a_i is small; while high values of a_i ($a_i \geq 99$) corresponds to almost null significance of the corresponding factor.

9.5.1 Estimating First Order Sensitivity Indices

As we discussed previously, one may apply SDR to compute sensitivity indices as well as to identify the additive terms as function of each single input factor. In [11] sensitivity indices for several g-functions were computed. We consider the results obtained for a particular g-function, with $p = 15$, whose a_i spectrum is shown in Table 9.1.

According to the theoretical properties of the g-functions, we have:

- four very significant factors: X_7, X_9, X_{11} and X_{13} which definitely should be included in the emulator;
- four medium factors: X_3, X_4, X_{12} and X_{14} which may or may not be included in the emulator;
- seven insignificant factors: X_1, X_2, X_5 X_6 X_8 X_{10} and X_{15} which do not add any significant information.

Using only SDR, as discussed in [11], we show in Table 9.2 the results of the estimation of the first order ANOVA model, using a training sample of 1024 Monte

Table 9.1 g-function spectrum of a_i coefficients

X_1	X_2	X_3	X_4	X_5	X_6	X_7	X_8	X_9	X_{10}	X_{11}	X_{12}	X_{13}	X_{14}	X_{15}
99	99	4.5	4.5	99	99	1	99	0	99	1	9	0	9	99

Table 9.2 First order

Parameter	99	99	4.5	4.5	99	99	1	99	0	99	1	9	0	9	99
Analytical	0	0	0.0096	0.0096	0	0	0.0726	0	0.2905	0	0.0726	0.029	0.2905	0.029	0
Estimated	0	0	0.0102	0.0112	0	0	0.0723	0	0.2856	0	0.0670	0.0046	0.3146	0.0025	0
In sample R^2	0.7576														

Carlo runs. The additive part of the g-function is very well identified by the SDR procedure: this is indicated by the 75.76% in-sample R^2, which corresponds to the true analytical 75.15% of the additive part for this test function.

It is worth mentioning that the estimates of first order sensitivity indices depend on the model approximation degree. For example, adding interactions we will improve not only the whole approximated model, but also the additive part itself.

For the g-function one can see that the S_i mean absolute error (MAE) for the pure additive model is equal to 0.059 while for the second order model is equal to 0.029.

9.5.2 Estimating Second Order Sensitivity Indices

We now consider the estimation of second order sensitivity indices. According to the analytical properties of the g-function we have:

- one very significant interaction: $\langle X_9, X_{13}\rangle$ (9.68%);
- four significant interactions: $\langle X_7, X_9\rangle, \langle X_7, X_{13}\rangle, \langle X_{11}, X_9\rangle$ and $\langle X_{11}, X_{13}\rangle$ (2.42%);
- one very mild interaction: $\langle X_7, X_{11}\rangle$ (0.61%);
- all other interactions not significant (their contribution being $\ll$1%).

We report here the estimation of the five largest interactions with the combined SRD-ACOSSO approach, using the same training sample of 1024 Monte Carlo run used for first order indices. Results are shown in Fig. 9.6.

It is interesting to consider four different estimations at increasing Monte Carlo samples size: 128, 256, 512 and 1024 (see Table 9.3). Adding second order ANOVA terms, the fit obviously improves. Moreover, analyzing the results, one may notice that, as Monte Carlo dimension increases, we are able to better identify how parameters really interact, as indicated by the increasing R^2 portion attributed to non-additive terms, while at the same time the estimated amount of the additive component properly converges to the true theoretical value (75%).

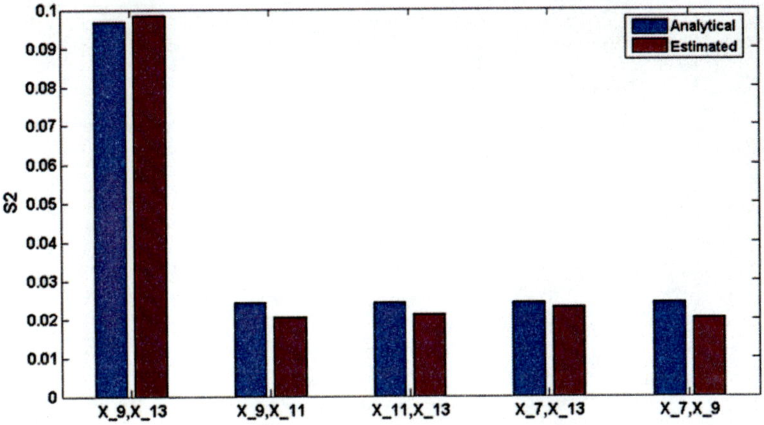

Fig. 9.6 Analytical and estimated second order sensitivity indices for the g-function

Table 9.3 In sample fit

Monte Carlo length	128	256	512	1024
First order ANOVA model R^2	0.7171	0.7814	0.7755	0.7576
Second order ANOVA model R^2	0.9013	0.9441	0.9793	0.9802

9.5.3 Building a Meta-model

As we have pointed out earlier, one of the major strengths of SDR relies in its robustness when out of sample performance is considered.

As reported in [10], SDR-ACOSSO performs quite well in term of out-of-sample fit. Using the same analytical model as before, we compare the results obtained by using SDR-ACOSSO with those given by DACE.[1]

In the following exercise we build a meta-model with SDR-ACOSSO and DACE using a training set of 512 Monte Carlo runs. Then we validate these meta-models using an out-of-sample set of another 256 MC runs. We repeated 25 random replicas of this exercise and computed the validation R^2 of the two meta-models in predicting the out-of-sample values of the model output. Results are summarized in the boxplots in Fig. 9.7, where we can see, for this example, the excellent predictive capability of the SDR-ACOSSO meta-model compared to DACE. We also show the results obtained by using pure ACOSSO without SDR, to highlight the improvements one may achieve by using SDR to identify the ANOVA terms.

[1]DACE is a Matlab toolbox used to construct a kriging approximation models on the basis of data coming from computer experiments (see [3]).

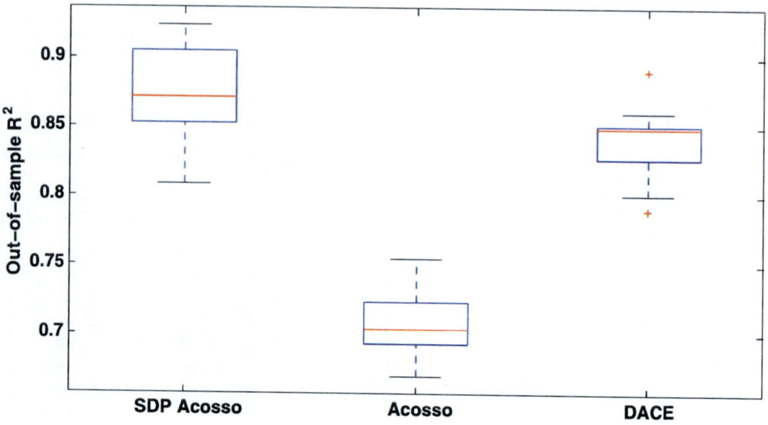

Fig. 9.7 Boxplots of out-of-sample R^2 of SDR-ACOSSO, ACOSSO and DACE

9.6 Conclusion

We have been glad to contribute a Chapter to this book, having the occasion to show how Peter Young's work contributed significantly also in the framework of sensitivity analysis and emulation. There is nowadays a vast stream of research on meta-modeling, which is becoming a key ingredient to handle complex computational models as well as in the DBM framework, as underlined also in the present book by Young [25]. Although kriging and Gaussian Process emulation is most widely applied in the statistical literature on this subject, non-parametric methods like SDR demonstrated to be very useful and efficient tools in improving the efficiency and robustness of emulators. As clearly demonstrated in [10], in fact, it is very difficult to identify a method which outperforms the others in all applications. According to our experience, SDR is extremely efficient and accurate in identifying *additive models*, at a quite small computational cost, due to its full recursive form. In the case of ANOVA models with interaction components, ACOSSO provides very good performances in terms of efficiency and low computational cost. When the model includes interactions, SDR combined with ACOSSO improves ACOSSO in many cases, although at the price of a higher computational cost. SDR-ACOSSO also compares very favorably with respect to DACE in many cases, even if there are cases where DACE outperforms SDR-ACOSSO in out-of-sample prediction. The main drawback of DACE seems to be the occurrence of very bad outliers in out-of-sample forecasting, implying some lack of robustness. The computational cost of DACE can be very sensitive to the underlying model. In terms of computational burden, we found that SDR (for additive models) and ACOSSO (for models with interactions) should be taken as the first choice for a *rapid and reliable* emulation exercise. Whenever ACOSSO is unable to explain large part of the mapping, SDR-ACOSSO or DACE may be considered. Another very important issue that we discussed in the present Chapter concerns the multivariate extension of SDP modeling. We have summarized here two approaches that have been developed in the context of sensitivity analysis [11] and emulation [10]. In the latter case [10], in particular, the SDR recursive identification is combined with a final en bloc estimation (the ACOSSO) to produce the full emulator to be used in forecasting. This approach, albeit providing very good performances, breaks the appeal and elegance of the pure recursive methods. In this context, ongoing research at Lancaster University on Multi-state dependent parameter modeling [12] is providing useful and promising contributions in the direction of a full recursive formulation.

References

1. Borgonovo, E.: Measuring uncertainty importance: investigation and comparison of alternative approaches. Risk Anal. **26**, 1349–1361 (2006)
2. Borgonovo, E.: A new uncertainty importance measure. Reliab. Eng. Syst. Saf. **92**, 771–784 (2007)
3. Lophaven, S., Nielsen, H., Sondergaard, J.: DACE a Matlab kriging toolbox, version 2.0. Technical Report IMM-TR-2002-12, Informatics and Mathematical Modelling, Technical University of Denmark (2002). http://www.immm.dtu.dk/~hbn/dace
4. Gu, C.: Smoothing Spline ANOVA Models. Springer, Berlin (2002)
5. Kalman, R.: A new approach to linear filtering and prediction problems. J. Basic Eng. D **82**, 35–45 (1960)
6. Lin, Y., Zhang, H.: Component selection and smoothing in smoothing spline analysis of variance models. Ann. Stat. **34**, 2272–2297 (2006)
7. Ng, C., Young, P.C.: Recursive estimation and forecasting of non-stationary time series. J. Forecast. **9**, 173–204 (1990)
8. Priestley, M.B.: Nonlinear and Nonstationary Time Series Analysis. Academic Press, New York (1988)
9. Ratto, M., Pagano, A., Young, P.C.: Non-parametric estimation of conditional moments for sensitivity analysis. Reliab. Eng. Syst. Saf. **94**, 237–243 (2009)
10. Ratto, M., Pagano, A.: Using recursive algorithms for the efficient identification of smoothing spline ANOVA models. AStA Adv. Stat. Anal. **94**(4), 367–388 (2010)
11. Ratto, M., Pagano, A., Young, P.C.: State dependent parameter metamodelling and sensitivity analysis. Comput. Phys. Commun. **177**, 863–876 (2007)
12. Sadeghi, J., Tych, W., Chotai, A., Young, P.C.: Multi-state dependent parameter model identification and estimation for nonlinear dynamic systems. Electron. Lett. **46**(18), 1265–1266 (2011)
13. Saltelli, A., Chan, K., Scott, M. (eds.): Sensitivity Analysis. Wiley, New York (2000)
14. Storlie, C., Bondell, H., Reich, B., Zhang, H.: Surface estimation, variable selection, and the nonparametric oracle property. Stat. Sin. **21**(2), 679–705 (2011)
15. Schweppe, F.: Evaluation of likelihood functions for Gaussian signals. IEEE Trans. Inf. Theory **11**, 61–70 (1965)
16. Tibshirani, R.: Regression shrinkage and selection via the LASSO. J. R. Stat. Soc. B **58**(1), 267–288 (1996)
17. Wahba, G.: Spline Models for Observational Data. CBMS-NSF Regional Conference Series in Applied Mathematics (1990)
18. Wecker, W.E., Ansley, C.F.: The signal extraction approach to non linear regression and spline smoothing. J. Am. Stat. Assoc. **78**, 81–89 (1983)
19. Weinert, H., Byrd, R., Sidhu, G.: A stochastic framework for recursive computation of spline functions: Part II, smoothing splines. J. Optim. Theory Appl. **30**, 255–268 (1983)
20. Young, P.C.: Time variable and state dependent modelling of nonstationary and nonlinear time series. In: Rao, T.S. (ed.) Developments in Time Series Analysis, pp. 374–413. Chapman and Hall, London (1993)

21. Young, P.C.: Data-based mechanistic modeling of environmental, ecological, economic and engineering systems. Environ. Model. Softw. **13**, 105–122 (1998)
22. Young, P.C.: Nonstationary time series analysis and forecasting. Progr. Environ. Sci. **1**, 3–48 (1999)
23. Young, P.C.: Stochastic, dynamic modelling and signal processing: Time variable and state dependent parameter estimation. In: Fitzgerald, W.J., Smith, R.L., Walden, A.T., Young, P.C. (eds.) Nonlinear and Nonstationary Signal Processing, pp. 74–114. Cambridge University Press, Cambridge (2000)
24. Young, P.C.: The identification and estimation of nonlinear stochastic systems. In: Mees, F.A.I. (ed.) Nonlinear Dynamics and Statistics. Birkhäuser, Boston (2001)
25. Young, P.C.: Data-based mechanistic modelling: natural philosophy revisited? (in this book)
26. Young, P.C., McKenna, P., Bruun, J.: The identification and estimation of nonlinear stochastic systems. Int. J. Control **74**, 1837–1857 (2001)
27. Young, P.C., Pedregal, D.J.: Recursive fixed interval smoothing and the evaluation of Lidar measurements. Environmetrics **7**, 417–427 (1996)

Chapter 10
Multi-state Dependent Parameter Model Identification and Estimation

Włodek Tych, Jafar Sadeghi, Paul J. Smith, Arun Chotai, and C. James Taylor

10.1 Introduction

It was what was later termed the Data Based Mechanistic (DBM) modelling approach (Young [17] and references therein) with its underlying notion that we can only build a model as good as the data we have, and that data along with the mechanistic interpretation of the resulting model should be the main driver dictating the model structure, that led to the numerous developments sparked by Peter Young's thinking.

From the personal perspective of the first author, this approach also led to the life-changing and never regretted decision to learn more from Peter Young, taken after attending his seminar at IIASA about 30 years ago and implemented some eight years later.

The inductive framework of Data Based Mechanistic Modelling is best explained in Peter Young's own chapter in the present volume [21], so we will not provide any further context here.

Time Varying Parameters (TVP) model estimation based on general approach of Kalman [7], has been explored by Young [13–16, 26] as well as others, e.g. Harvey [4]. What makes Young's approach stand out is that it remains firmly within the DBM framework through the use of general model structures and minimal assumptions made about the model structure prior to its identification. A simple (with the

W. Tych (✉) · J. Sadeghi · P.J. Smith · A. Chotai
Lancaster Environment Centre, Lancaster University, Lancaster, UK
e-mail: w.tych@lancaster.ac.uk

A. Chotai
e-mail: a.chotai@lancaster.ac.uk

C.J. Taylor
Engineering Department, Lancaster University, Lancaster, UK
e-mail: c.taylor@lancaster.ac.uk

L. Wang, H. Garnier (eds.), *System Identification, Environmental Modelling, and Control System Design*,
DOI 10.1007/978-0-85729-974-1_10,

benefit of after-sight) shift in the focus of Kalman Filter made a large difference in its applicability. Parameters of a general linear model became states within the stochastic state-space framework, which led to such developments as GRW smoothers and Dynamic Harmonic Regression to name but two.

It is easy to see the attractiveness of linear dynamic systems theory, leading to well defined control system design and time series methods. The growth of interest in non-linear systems led to numerous attempts of building a bridge between the well established linear theory and the non-linear systems, creating an arguably less general, but instead manageable non-linear systems theory. One of the approaches was making the coefficients of the linear dynamic system model functions of other variables. While this brings in some control-theoretical complications within the linear-made-nonlinear paradigm, it remains a pragmatic and very powerful technique (e.g. Taylor et al. and references therein chapter in the present volume). One of the difficulties with what is often termed functional coefficients models (see below) is that they may well lead to poorly defined, over-parameterised models due to the inevitable arbitrary steps in model formulation. This quite fundamental issue can be overcome by using DBM methodology.

It was in the context of DBM combined with that of TVP estimation that Young introduced State Dependent Parameter (SDP) models. He noticed [13] that the recursive TVP estimation can be used not just in the time domain, but in the state and parameter space, which later led to the now well established SDP, based on ordering of the varying parameter estimates according to a specific parameter-driving state of the system. This simple concept or moving from time domain to state domain and performing filtering and smoothing in this new domain, allows to establish, and to statistically assess any dependencies between the two (Young et al. [27]).

While this approach is very general and has been widely recognised and applied in numerous developments, one acknowledged shortcoming of the so far implemented SDP estimation procedures within the DBM framework was that each of the parameters of the system could only depend on a single state. While this is very often sufficient, and there is usually a single dominant influence on each system parameter, the need for generalisation into multi-state dependency remained.

State and multi-state dependencies of linear dynamic system parameters have been present in the literature for a while. It was originally suggested by Priestley [8], but some variations of it were also described by Hastie and Tibshirani [6]. We should also mention NARMAX and related models of Chen and Billings [2], functional coefficients of Chen and Tsay [1], wavelets of Truong and Wang [12] to name but a few. Until SDP however they have not been formulated within the objective, top-down DBM framework of Young [20] and references therein. Instead arguably all have been based on multi-variable functional surface approximations. Non-parametric SDP estimation has some similarity to Generalised Additive Modelling (GAM) [5]. However, GAM utilises conventional methods of scatter plot smoothing, rather than the recursive KF/FIS approach. This is important in the present dynamic context since it allows for maximum likelihood optimisation of hyper-parameters in the stochastic dynamic model.

While functional approximation is a step within DBM approach, it is a step following the crucial non-parametric identification process based on the TVP estimates

(Young [17]). The importance of the DBM identification stage has been shown in numerous papers of Young and others [16, 18, 25] and hence there was a need for a DBM generalisation of the SDP implementation involving more than one state per model parameter.

This Multi-State Dependent Parameter (MSDP) presented in this Chapter fits in seamlessly where SDP is now used, as it operates within the same DBM and state-dependency paradigms. Therefore all that is said about SDP applications in the associated chapter (Taylor et al in the present volume [11]) can be transparently ported into the MSDP context. This is particularly significant in the context of the new theoretical results in SDP-NMSS control methods (Taylor et al. [9]) where the main streams of methodologies of System Identification and Control introduced by Peter Young and colleagues (model identification: [26] and later works; control: [9, 22, 24] and many others) come together within a powerful methodological framework.

The multi-state algorithmic extensions of the SDP concept, which naturally does not exclude multiple driving states, have been developed by building up on the Generalised Random Walk models, used extensively within the TVP model framework [14] and set within the DBM [20] framework of modelling uncertain dynamic systems.

In the sequel we show how the univariate SDP algorithm with its associated DBM conceptual base can be non-trivially extended into multi-state dependency using recent algorithmic developments. Two documented examples written in Matlab will be presented in a tutorial manner, showing the DBM context of the approach, its wide applications and consequences.

10.2 Generalisation of the Univariate SDP Algorithm

Non-parametric SDP estimation produces a graphical estimate of the SDP as a function of the variables on which it depends. This helps to identify the structural form of the SDP model and the location of the principal nonlinearities within this structure. It is a prelude to the parameterisation of the model and its eventual estimation in this parametric form using specific functional bases (e.g. [12, 19]). This two-stage procedure of identification and estimation helps to ensure that the final parametric model is parsimonious and so can be contrasted with the direct estimation of more general parametric models cited above.

Filtering and smoothing require defining a sequence in which the points in the state sub-space (the subset of states which the given parameter depends upon). A similar definition is also required to define the samples preceding the current sample as well as those which are successors. It is worth noting here that the three terms (current, preceding and succession) all normally refer to temporal sequences, inheriting our univariate concept of time. These concepts of a sequence are easily ported into state spaces with one dimension; more thought is required to define them in a multi-variable state space.

To focus attention we shall introduce the general concepts of sequence in two dimensions, this can be easily generalised to higher dimensional state-spaces. We need to define three concepts and pose them within the Stochastic State-Space framework:

(i) The sequence of samples in the multivariable parameter space
(ii) The selection for preceding samples (precursors) to allow filtering
(iii) The selection for following samples (successors) to allow smoothing

All three need to be defined to allow for the problem to be cast in the Stochastic State Space framework. We shall first introduce the terms of reference, and then follow with the definitions, the proposed algorithm and examples.

10.2.1 Terms of Reference

To focus the attention we shall investigate a specific class of models: State Dependent Parameter ARX (SDARX) models. The DARX model relating a single input variable to an output variable, can be written in the following form:

$$y_t = -\sum_{i=1}^{n} \alpha_{i,t} y_{t-i} + \sum_{j=0}^{m} \beta_{j,t} u_{t-\delta-j} + e_t. \tag{10.1}$$

The term δ is a pure time delay, measured in sampling intervals, which is introduced to allow for any time delay that may occur between the incidence of a change in the input u_t and its first effect on the output y_t. α and β are time varying parameters (index t). e_t is a zero mean, white noise input with a Gaussian amplitude distribution and variance σ^2 (although this assumption is not essential to the practical application of the resulting estimation algorithms).

In this simplest Single-Input-Single-Output (SISO) form, the more complex, nonlinear SDARX model equation can be written conveniently for estimation purposes in the following form:

$$y_t = \sum_{i=1}^{N_z} a_i \left(\mathbf{X}_t^{(i)}, t\right) z_{i,t} + e_t, \tag{10.2}$$

with

$$\mathbf{X}_t^{(i)} = \left[x_{1,t}^{(i)}\, x_{2,t}^{(i)} \ldots x_{j,t}^{(i)} \ldots x_{ns_i,t}^{(i)}, \right], \tag{10.3}$$

$N_z = n + m + 1$ being the number of regressor terms, and $e_t = N(0, \sigma^2)$ Here $a_i(\cdot)$ is the ith SDP

$$a_i = \begin{cases} -\alpha_i & i \leq n, \\ \beta_{i-(n+1)} & i > n \end{cases} \tag{10.4}$$

while y_t and $x_{j,t}^{(i)}$ are the observed output and the jth state corresponding to $a_i(\cdot)$ is the ith at temporal sample t respectively. In addition, $n+m+1$ is the number of parameters of this model, ns_i is the number of states which parameter $a_i(\cdot)$ depends on and $z_{i,t}$ is the ith regressor:

$$z_{i,t} = \begin{cases} y_{t-i} & i \leq n, \\ u_{t-\delta+n+1-i} & i > n. \end{cases} \tag{10.5}$$

A very simple example of an SDARX model which can be used to clarify this notation is a version of the logistic growth equation:

$$\begin{aligned} y_t &= \alpha_1(y_{t-1}) \cdot y_{t-1} + \beta_0 \cdot u_{t-1} + e_t, \\ \alpha_1(y_{t-1}) &= \eta - \rho \cdot y_{t-1}^2, \end{aligned} \tag{10.6}$$

where $e_t \sim N(0, \sigma^2)$ and (η, ρ) satisfy the conditions

$$0 < \eta < 3, \qquad \rho > 0.$$

In this case $a_1(y_{t-1})$ depends on y_{t-1}, and a_2 is a constant, so we only have one single-state dependency of one of the parameters in this model.

Casting (10.6) in terms of (10.4) and noting that: $x_{1,t}^{(1)} = y_{t-1}$ we have:

$$\begin{cases} \alpha_1(y_{t-1}) = a_1(x_{1,t}^{(1)}) = \eta - \rho y_{t-1}^2 & \forall t. \\ \beta_0 = a_2 \end{cases} \tag{10.7}$$

It is easy to see how this simple example can be expanded to multi-state dependency, and we shall look at such more complex example in the sequel.

10.2.1.1 Modelling Variation of the Parameters

Denoting the value of the i-th parameter a and its derivatives (if the latter exist) at the kth sample, where ξ_k denote the states affecting this parameter, and ignoring index i temporarily for clarity, as $\mathbf{A}_k = [a(\xi_k)\ a'(\xi_k)\ a''(\xi_k)\ \ldots\ a^{(q)}(\xi_k)]^T$ and $a^{(q+1)}(\xi_k) = \nu_k$ where the noise term ν_k is defined as $\nu_k \sim N(0, \sigma_\nu^2)$ or other serially uncorrelated process fulfilling the assumptions of the Kalman Filter.

From the qth order Taylor expansion of derivatives of parameter $a(\cdot)$:

$$\mathbf{A}_{k+1} = \mathbf{F}_k \mathbf{A}_k + \mathbf{G}_k \nu_k, \tag{10.8}$$

where:

$$\mathbf{F}_k = \begin{bmatrix} 1 & \Delta_k & \frac{\Delta_k^2}{2!} & \cdots & \frac{\Delta_k^q}{q!} \\ 0 & 1 & \Delta & \cdots & \frac{\Delta_k^{q-1}}{(q-1)!} \\ 0 & 0 & 1 & \cdots & \frac{\Delta_k^{q-2}}{(q-2)!} \\ \vdots & \vdots & \vdots & \ddots & \vdots \\ 0 & 0 & 0 & \cdots & 1 \end{bmatrix}, \tag{10.9}$$

$$\mathbf{G}_k = \begin{bmatrix} \frac{\Delta_k^{q+1}}{(q+1)!} & \frac{\Delta_k^q}{q!} & \frac{\Delta_k^{q-1}}{(q-1)!} & \cdots & \Delta_k \end{bmatrix}^T$$

and $\Delta_k = \|\boldsymbol{\xi}_{k+1} - \xi_k\|$ is the distance in $\boldsymbol{\xi}$ space.

Note here that in a uniformly sampled series case of $\Delta_k \equiv 1$ this definition leads to the widely used Stochastic State Space time series model, with (reintroducing parameter index i):

$$\mathbf{F}_k = \begin{bmatrix} 1 & 1 & \frac{1}{2} & \cdots & \frac{1}{q_i!} \\ 0 & 1 & 1 & \cdots & \frac{1}{(q_i-1)!} \\ 0 & 0 & 1 & \cdots & \frac{1}{(q_i-2)!} \\ \vdots & \vdots & \vdots & \ddots & \vdots \\ 0 & 0 & 0 & \cdots & 1 \end{bmatrix}, \tag{10.10}$$

$$\mathbf{G}_k = \begin{bmatrix} \frac{1}{(q_i+1)!} & \frac{1}{q_i!} & \frac{1}{(q_i-1)!} & \cdots & 1 \end{bmatrix}^T$$

and when additionally $q = 1$ (10.10) describes an Integrated Random Walk model for a regularly sampled time series. It is worth noting at this point that the **G** matrix can take a different form depending on the assumptions about the way the system noise affects the states.

The problem is now cast in the Stochastic State-Space terms, where the parameters $\mathbf{A}_k$ can be estimated along the spatial sequence of their respective driving states.

It is worth to remind here the difference between both concepts of states used here: parameters $\mathbf{A}_k$, $k = 1 \ldots N$ are estimated as states in State equation (10.8), while other arbitrary states of the investigated system our *parameter driving states* $\{\boldsymbol{\xi}_k, k = 1 \ldots N\}$—are influencing parameters $\mathbf{A}_k$, $k = 1 \ldots N$.

10.2.1.2 The Sequence of Samples in the Parametric State Space

In order to use the Kalman Filter and Fixed Interval Smoother in the state-space to estimate the states-parameter dependency it is required to define a sequence of samples, which no longer are defined as a well ordered, uniformly sampled time series, but as a cloud of points in the state space. There are two components to this definition.

Firstly, it is a matter of an unambiguous ordering of samples in n-dimensional space in order to define the order of processing the samples. This is the equivalent of the temporal sequence k, $k = 1 \ldots N$.

Secondly, it should be noted here that there is no reason why in the ns_i-dimensional state space for parameter i we should be using a single preceding sample to calculate the equivalent of an estimate, which would be expressed as $\hat{\mathbf{A}}(k|k-1)$ within the usual temporal framework. In fact this could be simply incorrect in a multi-variable case, which is why we need to define the *predecessor* (or *parent*) and the *successor* (or *child*) sample sets so that instead of $\hat{\mathbf{A}}(k|k-1)$ we use $\hat{\mathbf{A}}(k|\{l_k^p\})$ where $\{l_k^p\}$ is the set (list) of samples in the state-space which *precede* the current sample k. And by analogy, the successor set $\{l_k^s\}$ needs to be defined in order to use the Fixed Interval Smoothing algorithm for the backward pass through the sequence in the state-space. Each sample (apart from the first and the last in the defined spatial sequence) will have its own sets of *predecessors* and *successors*. This definition also alleviates the potential (albeit highly unlikely) issue when the same values of states are associated with different values of parameters, these will be dealt with in the same way as multiple base sets $\{l_k^p\}$ and $\{l_k^s\}$.

10.2.2 Proposed Algorithm

It should be noted that sorting in a multi-variable state space is non-unique. The solution we propose is simple and based on the Euclidean norms of the states (distances from the origin), which have been normalised to lie within a unit hyper-cube prior to building the sequence. This approach is clearly consistent with single state dependency (SDP) and its sorting along the single-states values.

10.2.2.1 Definition of the Sequence

We shall normalise the sequence of the original ns_i dimensional state variables driving parameter i: $\{\xi_k, k = 1 \ldots N\}$, so that they lie within an ns_i dimensional unit hyper-cube and call it $\{\xi_{k^*}^n\}$. We can then define the sequence $\{k^*, k^* = 1 \ldots N\}$ such that $\|\xi_{k^*+1}^n\| \geq \|\xi_{k^*}^n\|$, where in this implementation of the algorithm norm $\|\cdot\|$ is the standard Euclidean norm taken in the normalised co-ordinates $\{\xi_{k^*}^n\}$.

10.2.2.2 Definition of the Predecessor and Successor Sets

In agreement with the ordering criteria above, we can define the predecessor set $\{l_k^{\mathrm{P}}\}$ for each ξ_{k^*} as the nearest p points (p is a parameter of the algorithm see below) which are closest to ξ_{k^*} in terms of the same Euclidean metric in the normalised state-space *and* which precede ξ_{k*} in the sequence $\{k^*, k^* = 1 \ldots N\}$ defined by their norm for any specific $k_0 > 1$. Obviously there will be fewer such points for the first p points in the sequence, which does not influence the algorithm. We can thus define:

$$\left\{l^P_{k_0} : \xi^n\left(l^P_{k_0}\right) = \underset{k^*}{\arg\min} \left\| \xi^n_{k^*} - \xi^n_{k_0} \right\|,\ k^* < k_0 \right\}. \tag{10.11}$$

In the univariate case this is clearly consistent with simple sorting of the parameter driving state.

The successor set $\{l^S_k\}$ for ξ_{k*} is defined using the predecessor set definition: the set includes p points in the normalised state space which are further from the origin than ξ_{k*}, for which ξ_{k*} belongs to their predecessor set $\{l^P_k\}$. This simple requirement provides the required continuity of the sequence.

The number of base points p is set in this algorithm as $(ns_i)^2$, so it is 4 for the simplified 2D example given below. Note that at the ends of the sequence this number will shrink, and also that in this case we have only 3 successor points due to the duplication of points 7 and 8 state values. The reason for taking the square of the dimension as the numbers of base points arose in the context of solving the problem using stochastic splines method. This will be the subject of another publication and will not be discussed here. There are of course other possibilities of defining the sequences, all with their own robustness and other characteristics, but we focus here on the specified definitions.

A simple example using only 20 samples for clarity of the presentation is shown in Fig. 10.1.

The illustration shows all the positions of $\xi_i, i = 1:20$ ordered according to the Euclidean distance from the origin in normalised space (which the presented plot does not have to be), with the current $\xi_{k*}, k^* = 9$ where ξ_{k*} is defined by two states. The numbers show the ordering sequence according to the distance from the lower left-hand corner of the bounding box shown in Fig. 10.1 with dotted lines (zero in the normalised space). The star marker shows the state values for the current sample, the upward pointing triangular markers show the preceding samples set, the downward pointing triangular markers show the successors set. It is necessary to note here, as it would be hard to show it in the figure, that the current point ξ_9 belongs to the predecessor sets of the points in the successor set according to the definition above. A careful reader will also notice that points 7 and 8 overlap and are both zeroes in the 2D space. This is the result of the artificial character of the data set having been simulated, it would be highly unlikely to happen with real data, and it is also gracefully handled by the algorithm.

The larger than 1 number of predecessor and successor points forces an interesting variation of the filtering and smoothing procedures. It can arguably be compared to an ensemble Kalman Filter. Although no formal claim of this kind can be fully substantiated at this stage, averaging between several possible state trajectories reaching the current point in the state space appears to have a moderating effect on the resulting estimates.

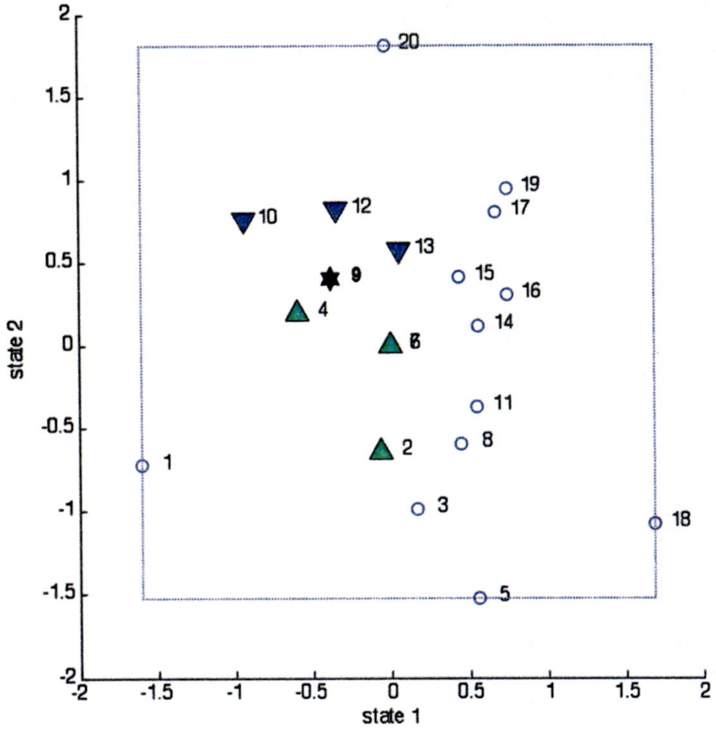

Fig. 10.1 Illustration of predecessor △ and successor ▽ (parent and child) points of temporal sample number 9 (*highlighted*) in an short example multi-variable (2D) sequence of just 20 samples. Note how there is no reflection of temporal order (*numbers shown*) in the sequence within the 2D state space

10.2.2.3 Detailed Estimation Algorithm

The estimation algorithm remains largely unchanged compared to the standard SDP, except for the enlarged predecessor and successor sets, for which simple group averages of filtered and smoothed parameters are calculated, and for the sequence defined earlier.

For completeness we provide the outline of the algorithm here, although it has been well described already in earlier works by Young and co-workers e.g. [27].

At the highest level, the Backfitting Algorithm (BA) is used to isolate the individual contributions from the additive terms. Starting from the original model (10.2)

$$y_t = \sum_{i=1}^{Nz} a_i\left(\mathbf{X}_t^{(i)}, t\right) z_{i,t} + e_t, \tag{10.12}$$

with

$$e_t \sim N\left(0, \sigma^2\right), \tag{10.13}$$

$$\mathbf{X}_t^{(i)} = \left[x_{1,t}^{(i)} \; x_{2,t}^{(i)} \; \ldots \; x_{j,t}^{(i)} \; \ldots \; x_{ns_i,t}^{(i)} \right] \tag{10.14}$$

by defining: $f_t^{(i)} = a_i(\mathbf{X}_t^{(i)}, t) z_{i,t}$ we can cast the problem in terms of an Additive Model

$$y_t = \sum_{i=1}^{Nz} f_t^{(i)} + e_t,$$

where $f_t^{(1)}, f_t^{(2)}, \ldots, f_t^{(Nz)}$ can be estimated by using Backfitting Algorithm (BA). By defining $y_t^{(i)} = f_t^{(i)} + e_t$, the BA can be used to estimate $f_t^{(1)}, f_t^{(2)}, \ldots, f_t^{(Nz)}$ through the following iterations (Young, [27], also in the Additive Models context Hastie and Tibshirani [6]):

1. Initialise $f_t^{(i)} = a_i(\mathbf{X}_t^{(i)}, t) z_{i,t}$
2. Cycle for $i = 1, 2, \ldots, Nz$
   (a) Compute: $y_t^{(i)} = f_t^{(i)} + e_t = y_t - \sum_{\kappa=1,\kappa\neq i}^{n+m+1} \hat{f}_t^{(\kappa)}$
   (b) Estimate $\hat{a}_i(\mathbf{X}_k^{(i)})$—the parameter value at the kth sample—from
   $$y_t^{(i)} = a_i(\mathbf{X}_t^{(i)}) z_{i,t} + e_t$$
   (c) Update $\hat{f}_t^{(i)} = \hat{a}_i(\mathbf{X}_t^{(i)}, t) z_{i,t}$
3. Continue step 2 until convergence.

Estimates in the step (2.b.) of the BA above are obtained by running a Kalman Filter and Fixed Interval Smoother using the Stochastic State Space form (10.8) of parameter variation, with the filtering and FIS in the form as described for instance in Young [26], with the sole assumption of each parameter behaving as a Generalised Random Walk of order q along the defined sequence.

The additional characteristic of the algorithm is that at each step along the trajectory in the state-space, up to p recursive estimates are calculated and pooled to provide a single summary of the distribution of the estimate in the form of its mean and variance at the current point in the state-space. The details of this pooling are provided in the Appendix.

The hyper-parameters for the KF/FIS stage are optimised exactly as they are in the original SDP approach using Maximum Likelihood for this model.

It is worth noting here that no parametric form is used at all in the algorithm, thus providing a perfect non-parametric DBM base for further efficient parameterisation of the investigated relationship.

10.2.3 Method Limitations

So far we have discovered two properties, affecting both MSDP and SDP. As it will become apparent, both can be derived from common-sense practical considerations and do not detract from the attractiveness of the SDP approach within the DBM framework.

10.2.3.1 Multiple Co-linearity Considerations

This can be illustrated using a simple SDP DARX example of:

$$y_t = a_t y_{t-1} + b_t u_{t-1} + e_t, \tag{10.15}$$

where $a_t = f(u_{t-1,}) = \sin(u_{t-1,})$ and $b_t = g(y_{t-1}) = \tan(y_{t-1})$. A simple second order Taylor Series approximation of a_t and b_t around zero (for simplicity and without loss of generality) produce: $a_t \approx u_{t-1} + \frac{(u_{t-1})^3}{3!}$ and $b_t \approx y_{t-1} + \frac{(y_{t-1})^3}{3!}$.

By substituting these approximations into the model equation (10.15) it is easy to see how identical expressions ($u_{t-1}y_{t-1}$) enter the equation and produce linear dependencies:

$$y_t \approx \left(u_{t-1} + \frac{(u_{t-1})^3}{3!}\right)y_{t-1} + \left(y_{t-1} + \frac{(y_{t-1})^3}{3!}\right)u_{t-1} + e_t.$$

While this is a simple and deliberately exaggerated example, it illustrates the importance of correct formulation of the model.

10.2.3.2 Singularity Considerations

Using an even simpler AR model:

$$y_t = a_t y_{t-1} + e_t$$

it is quite apparent that any estimation of a_t will be equivalent to dividing the observation y_t by its preceding sample value y_{t-1} which will be creating singularities when y_{t-1} is close to zero.

It is clear that both of these properties would affect any estimation method and will usually result from a badly defined model in the first place. What is interesting is that when the multiple co-linearity is introduced into a model such as in the example below, the resulting non-parametric estimate is not unstable, but merely showing no significant relationship.

10.3 Examples

We shall illustrate the effectiveness of the method with two examples. The first one is a direct comparison of the original SDP results with the result from MSDP for a single-state dependency where both methods can be applied. The results should be very close in both cases. This first example uses real hydrological data and has been used in an earlier publication by Young [28]. The second example includes a two-dimensional state-dependency in a simulated system. A simulated example has been used to provide the reference for the obtained estimate.

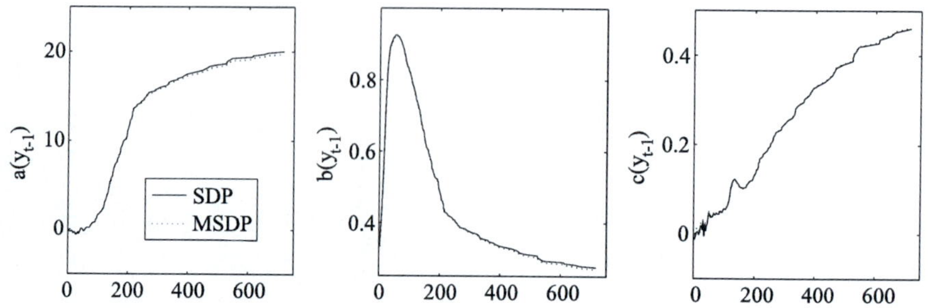

Fig. 10.2 Comparison between SDP and MSDP results (no uncertainty estimates shown but these are equally consistent). The *plots* are all showing the model coefficients (*from the left*): $a(Q_t)$, $b(Q_t)$ and $c(Q_t)$ as shown in model (10.16)

10.3.1 Example 1: Single State Dependency Comparison Between SDP and MSDP

This example with three single-state-dependent parameters was chosen to provide a comparison between the original SDP and MSDP. In the example a simple SDP DARX model of river discharge Q_t is analysed, with the regressors being: lagged river discharge Q_{t-1}, lagged rainfall R_{t-1} and lagged temperature T_{t-1}

$$Q_t = a(Q)Q_{t-1} + b(Q)R_{t-1} + c(Q)T_{t-1} + e_t. \tag{10.16}$$

The model rationale and its physical interpretation are given in Young [23] and are outside the scope of this paper. The example is only used here to verify the consistency of MSDP with the original SDP in a case where they both apply.

The results shown in Fig. 10.2 indicate very good agreement between the two estimation techniques where they are compatible. Although the uncertainty estimates are not shown for reasons of clarity of the presentation, they are similarly consistent. There are no differences between the model fit, and the run times are of similar order of magnitude, although MSDP is about a third faster in this case, which can be attributed to different convergence criteria handling. In Fig. 10.2 the lines showing the estimated non-parametric relationship between the states and model parameters are nearly indistinguishable.

This example shows full compatibility of the more general MSDP with the original SDP in a single-nonlinearity case.

10.3.2 Example 2: Simulated DARX Model with a Two-State and a Single-State Parameter Dependencies

This model is simulated, so that it is possible to assess the quality of the SDP estimation by comparison of the estimated with simulated relationships. A two-state bounded but significant non-linearity puts the method to test.

$$\begin{cases} y_t^* = a_t y_{t-1}^* + b_t u_{t-1} + v_t, \\ y_t = y_t^* + e_t. \end{cases} \tag{10.17}$$

Input u_t, system noise v_t, and observation noise e_t, are Gaussian distributed, serially uncorrelated signals with variances of $\sigma_u^2 = 1$, $\sigma_v^2 = 1$ and $\sigma_e^2 = 1$ respectively. As is usually assumed y_t^* is the observation noise free output. System is simulated for 1000 sampling times in this example.

The first SDP is a saddle-shaped function of two state variables ($x_{1,t}^{(1)} = y_{t-2}^*$ and $x_{1,t}^{(1)} = u_{t-2}^*$) while the second SDP is a sine function of only one state variable ($x_{1,t}^{(2)} = u_{t-2}^*$).

$$\begin{aligned} a_t &= 0.5 \tan^{-1}\left(y_{t-2}^* \cdot u_{t-2}\right), \\ b_t &= \sin(2u_{t-2}). \end{aligned} \tag{10.18}$$

To make the problem more realistic, the noise-free output y_t^* values are not available, so the observed output values y_t are used in the estimation making the system used for estimation:

$$y_t = \hat{a}_t y_{t-1} + \hat{b}_t u_{t-1} + \hat{e}_t, \tag{10.19}$$

with its state-dependent parameters:

$$\begin{aligned} \hat{a}_t(y_{t-2}, u_{t-1}) &= 0.5 \tan^{-1}(y_{t-2} \cdot u_{t-1}), \\ \hat{b}_t(u_{t-2}) &= \sin(2u_{t-2}). \end{aligned} \tag{10.20}$$

For the purpose of estimation the parameters are both assumed to vary as Integrated Random Walk (IRW in the state-space cf. (10.10) with $q = 1$). Estimation uses the backfitting algorithm with standard Maximum Likelihood estimation of Noise Variance Ratios for each of the IRW parameters, as in Sect. 10.2.2.3.

Overall model fit is characterised by $R^2 = 0.96$ comparing the model output to the noise-free data. It is interesting to note that the fit of the first parameter (with dependency on 2 states) estimate to the simulated surface (as in Fig. 10.3) is $R^2 = 0.87$, and of the second—single-state dependency parameter—is $R^2 = 0.99$. The run-time was 20.2 sec (on a standard Windows laptop computer).

The main illustration of the MSDP result is shown in Fig. 10.3 where 3D surfaces of simulated (offset for clarity) and estimated parameter: $a_t = 0.5 \tan^{-1}(y_{t-2}^* \cdot u_{t-2})$ and $\hat{a}_t(y_{t-2} \cdot u_{t-2})$ respectively, are shown along with the scatter of simulated observations of the two driving states. The presented surface of the estimate $\hat{a}_t(y_{t-2}, u_{t-2})$ is created from the non-parametric estimates of the values of a at the marked data points, with the illustrated interpolated values obtained using Delaunay triangulation. It is offset vertically by 2 in order to show the shape of both surfaces at the same time.

It is visible that the general shape of the saddle function is well recovered, which is further shown in another visualisation in Fig. 10.3. In this figure the simulated

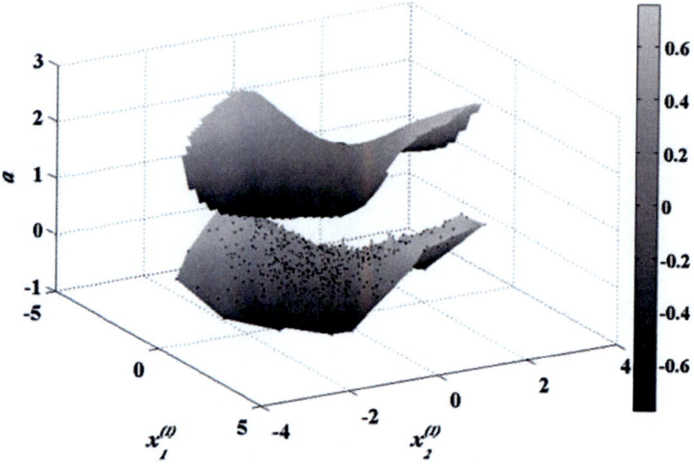

Fig. 10.3 Dual state-dependence of parameter $a_t(y_{t-2}, u_{t-2})$: simulated above (*offset vertically* for clarity of the presentation) and estimated below, including the data points

parametric surface is shown as semi-transparent, with grid. The estimated triangulated surface is shown as solid. The data points are shown as small circles. Although it is difficult to see much detail, the figure is provided to show the generally good agreement of the original and recovered 3D surfaces.

Around the centre of the surface there is a visible roughness of the estimated surface. This is related to the singularity issue (see above in Sect. 10.2.3.2) and arises as small estimation errors in the back-fitting algorithm are amplified through implicit division by the value of the driving state being close to zero. The same effect arises in the single state SDP modelling where near-zero values of the driving state are taken into account. It should be noted however, that this non-parametric estimation is the first step in the DBM process serving to identify the shape and complexity of the state-dependency relationship. It is normally followed by an efficient parameterisation of the state dependency surface. Therefore such "noise-like", non-systematic errors will not be a serious influence upon the modelling process, a the parameterise surface will be smooth in the middle. It should be noted that as these errors are far smaller away from the singularity point and the shape of the surface is revealed well, MSDP estimate serves its purpose well.

In Fig. 10.4 it is visible how the estimation procedure gives a good approximation of the original simulated single-variable highly non-linear dependency, with discrepancies appearing only at the ends of the state range where a smaller number of state values is available. These results are consistent with published single state SDP results for similar non-linearities.

The two presented examples indicate that the proposed MSDP method of extending Young's SDP to multi-state parametric dependency is consistent with the original where both are applicable, and that it produces informative results that can be used further, aiding in generating efficient parameterisation of the multi-state-dependencies for use in control design and other modelling applications.

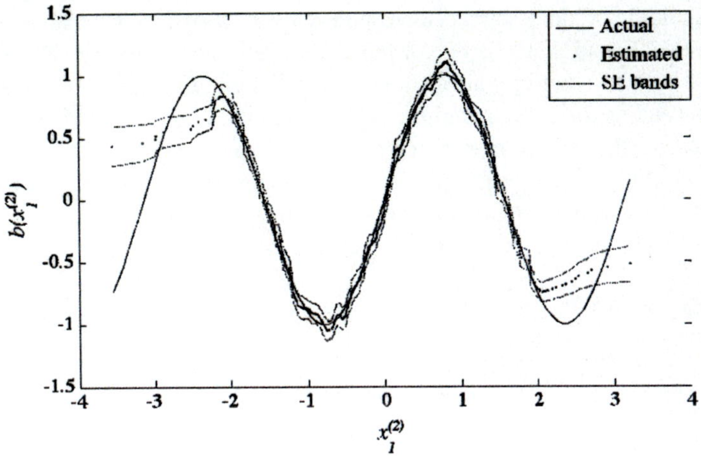

Fig. 10.4 State-dependence of parameter $b_t(u_{t-2})$ shown with data points and uncertainty band

10.4 Summary and Future Developments

The presented approach provides a natural extension to the SDP as developed by Young, by using a well defined sequence in the multivariable state-space, which builds on the original idea of smoothing not in the time-domain, but in the state-space driving each of the systems parameters.

The algorithm seems to work well, resulting in well defined and fully non-parametric hyper-surfaces with acceptable uncertainty, as shown in the examples provided. For single-state dependencies the results are close to those obtained using the well tested original single-state SDP.

Since the method is based on the well proven Kalman Filter and Fixed Interval Smoother, it seems reasonable to assume that there is no need to prove its statistical properties, which it inherits from the applied algorithms. There is a number of possible extensions and variations of the algorithm, which are briefly introduced below.

1. The sorting sequence can use a different norm and different reference point. For example a different normalisation of the states can be used, for instance reducing the states data to zero mean and unit covariance.
2. The numbers of points in predecessor and successor sets at this stage these are still somewhat arbitrary, even if the implemented quadratic dependence is based on other considerations.
3. The distances $\Delta\xi$ between the points in the state space (see (10.8)) are not included in the current simplified algorithm, just as it is implemented in the original SDP. So far this has not proved to be problematic. Further work is underway to create a general version of the algorithm with variable **F** and **G** matrices incorporating all the distances between the samples in the state-space of each parameter, as in the formulation (10.8) of the stochastic state-space model.

4. The multiple predecessor/successor points are a "poor man's version of an Ensemble Kalman Filter", as one of the colleagues commented recently. This is a fair point, and other options of filtering and smoothing will be considered, going beyond the basic Kalman Filter/Fixed Optimal Smoother approach. In order to improve the algorithm, the approximation used in averaging the covariance between the contributing sample points will be replaced by the exact calculation.
5. Finally, development of other approaches to the MSDP problem is the subject of current work and will be presented in forthcoming publications.

Acknowledgements Peter Young's help and advice was invaluable in the development of the project, built largely upon his ideas and philosophy. It is a great satisfaction and an honour to contribute to this volume.

This generalisation of SDP has been developed at Lancaster University during Jafar Sadeghi's Ph.D. studentship funded by the Iranian Government between 2003–2006.

Thanks are also due to Katarzyna M. Tych of Leeds University for comments and suggestions on data and results visualisation.

Appendix: Kalman Filter and Fixed Interval Smoother Modification for Multiple Predecessor and Successor Points

Starting from the basic TVP model:

$$y_t = \mathbf{z}_t^T \mathbf{a}_t + e_t \tag{10.21}$$

as introduced in the main text, with $\mathbf{a}_t$ being the vector of time varying parameters and $\mathbf{z}_t$ being the regressors (inputs), we assume that the parameters $\mathbf{a}_t$ vary according to the Generalised Random Walk (GRW) model of the form:

$$\mathbf{A}_{i,t} = \mathbf{F}_i \mathbf{A}_{i,t-1} + \mathbf{G}_i \nu_{i,t-1}, \quad i = 1, 2, \ldots, Nz \tag{10.22}$$

as explained in the main text (Sect. 10.2.1, in (10.3)) as well as in numerous references (e.g. Young [27]). We shall bring the initial definitions from these references here in order to define the multiple point version.

In the present version of the algorithm, matrices **F** and **G** defined in (10.9) are not time- (or state-)varying, but are fixed. This simplification is consistent with the original SDP and appears to be non-critical.

Note here that according to the defined sequence of filtering, index t in the usual definitions of filtering and smoothing now has a different meaning, and it belongs not to the original time sequence, but to the k^* sequence defined in the main text.

Because we are not assuming interdependency between the additive components $i = 1, \ldots, N_z$ the estimation can be implemented either using a large block-diagonal system, or individually, which works well with the Backfitting Algorithm. Therefore in the sequel for simplicity of notation we shall not refer to i-th equation referring to $\mathbf{A}_{i,t}$, but simply to $\mathbf{A}_t$—the block-diagonal sub-system.

We now have the following overall State Space model defined by the observation and state equations:

$$\begin{aligned} y_t &= \mathbf{H}_t \underline{\mathbf{A}}_t + e_t, \\ \underline{\mathbf{A}}_t &= \mathbf{F}\underline{\mathbf{A}}_{t-1} + \mathbf{G}\boldsymbol{v}_{t-1}, \end{aligned} \tag{10.23}$$

where: $\underline{\mathbf{A}}_t = [\mathbf{A}_{1,t}^T \ \mathbf{A}_{2,t}^T \ \dots \ \mathbf{A}_{Nz,t}^T]^T$; $\mathbf{F}_i$ is a block diagonal matrix with blocks defined by the $\mathbf{A}_i$ matrices in (10.10); $\mathbf{G}$ is a block diagonal matrix with blocks $\mathbf{G}_i$ defined by the corresponding subsystem matrices in (10.10); and $\boldsymbol{v}$ is an $Nz \times 1$ vector containing the white noise input v_i to each of the GRW models in individual equations (10.10). These white noise inputs are assumed to be independent of the observation noise e_t and have a covariance in the form of diagonal $Nz \times Nz$ matrix $\mathbf{Q}$ with diagonal elements Q_i corresponding to $v_{i,t}$. Finally, $\mathbf{H_t}$ is a row vector of the following form

$$\mathbf{H}_t = \left[z_{1,t} \ \mathbf{0}_1 \ z_{2,t} \ \mathbf{0}_2 \dots z_{Nz,t} \ \mathbf{0}_{Nz} \right] \tag{10.24}$$

with $\mathbf{0}_i$, $i = 1, 2, \dots, Nz$ defined as $1 \times q_i$ vectors of zeroes, empty when $q_i = 0$.

Within this formulation the basic prediction step of the Kalman Filter (in the usual notation) is:

$$\begin{aligned} \mathbf{A}_{t|t-1} &= \mathbf{F}\hat{\mathbf{A}}_{t-1}, \\ \mathbf{P}_{t|t-1} &= \mathbf{F}\hat{\mathbf{P}}_{t-1}\mathbf{F}^T + \mathbf{G}\mathbf{Q}_r\mathbf{G}^T \end{aligned} \tag{10.25}$$

and the basic correction step:

$$\begin{aligned} \hat{\mathbf{A}}_t &= \mathbf{A}_{t|t-1} + \hat{\mathbf{P}}_t\mathbf{H}_t^T\{y_t - \mathbf{H}_t\mathbf{A}_{t|t-1}\}, \\ \hat{\mathbf{P}}_t &= \mathbf{P}_{t|t-1} - \mathbf{P}_{t|t-1}\mathbf{H}_t^T\left[1 + \mathbf{H}_t\mathbf{P}_{t|t-1}\mathbf{H}_t^T\right]^{-1}\mathbf{H}_t\mathbf{P}_{t|t-1}. \end{aligned} \tag{10.26}$$

Let us consider the pooling process of the multiple predecessor filtering results. The same reasoning will apply to smoothing. There are two ways that pooling can be combined with filtering. (1) pool the predecessors to estimate the first two moments of the distribution of the predecessors, then perform a single prediction-correction step of the Kalman Filter; (2) perform the prediction step for each predecessor, thus forming an ensemble of forecasts, which are then pooled before the correction step; presuming that the predecessors are IID (independent and identically distributed), this is analogous to Ensemble Kalman Filter [3].

This work uses an approximation of the second method, using the assumption that (means of) the predecessor points are close together in terms of their variances—in other words that their distributions are similar. If this assumption holds we can make the following approximations: (1) that the pooling can occur after the correction step, and (2) that the estimate of the pooled covariance is given by (10.30).

Both KF steps can be written in this simple functional form for brevity:

$$\begin{aligned} \hat{\mathbf{A}}_t &= g_1(\hat{\mathbf{A}}_{t-1}, \hat{\mathbf{P}}_{t-1}, y_t, \mathbf{H}_t), \\ \hat{\mathbf{P}}_t &= h_1(\hat{\mathbf{P}}_{t-1}, \mathbf{H}_t). \end{aligned} \tag{10.27}$$

Similarly, a single step of an implementation of Fixed Interval Smoother (Q algorithm):

$$
\begin{aligned}
\hat{\mathbf{A}}_{t|N} &= \mathbf{F}^{-1}\big[\hat{\mathbf{A}}_{t+1|N} + \mathbf{G}\mathbf{Q}_r\mathbf{G}^T\mathbf{L}_t\big], \\
\mathbf{L}_t &= \begin{cases} [\mathbf{I} - \mathbf{H}_{t+1}^T\mathbf{H}_{t+1}\hat{\mathbf{P}}_{t+1}]\{\mathbf{F}^T\mathbf{L}_{t+1} - \mathbf{H}_{t+1}^T(y_{t+1} - \mathbf{H}_{t+1}\mathbf{A}_{t+1|t})\}, \\ \quad t < N, \\ 0, \quad t = N, \end{cases} \\
\mathbf{P}_{t|N} &= \begin{cases} \hat{\mathbf{P}}_t + \hat{\mathbf{P}}_t\mathbf{F}^T\mathbf{P}_{t+1|t}^{-1}[\mathbf{P}_{t+1|N} - \mathbf{P}_{t+1|t}]\mathbf{P}_{t+1|t}^{-1}\mathbf{F}\hat{\mathbf{P}}_t, & t < N, \\ \hat{\mathbf{P}}_t, & t = N. \end{cases}
\end{aligned}
\tag{10.28}
$$

Can be written as:

$$
\begin{aligned}
\mathbf{L}_t &= \begin{cases} g_2(\hat{\mathbf{A}}_t, \hat{\mathbf{P}}_t, \mathbf{L}_{t+1}, y_{t+1}, \mathbf{H}_{t+1}), & t < N, \\ 0, & t = N, \end{cases} \\
\mathbf{P}_{t|N} &= \begin{cases} h_2(\hat{\mathbf{P}}_t, \mathbf{P}_{t+1|N}), & t < N, \\ \hat{\mathbf{P}}_t, & t = N. \end{cases}
\end{aligned}
\tag{10.29}
$$

In this simplified notation the Kalman Filter for the groups of p predecessor points becomes:

$$
\begin{aligned}
\hat{\mathbf{A}}_k &= \frac{1}{p}\sum_{j=1}^{p} g_1(\hat{\mathbf{A}}_{k'_j}, \hat{\mathbf{P}}_{k'_j}, y_k, \mathbf{H}_k), \\
\hat{\mathbf{P}}_k &= \frac{1}{p}\sum_{j=1}^{p} h_1(\hat{\mathbf{P}}_{k'_j}, \mathbf{H}_k).
\end{aligned}
\tag{10.30}
$$

Importantly, this is now done in the previously defined sequence (see Sect. 10.2.2.1 and Sect. 10.2.2.2) with k'_j, $j = 1, 2, \dots, p$—the predecessor points taking on the role of $t - 1$, and k—the current point—the role of t in (10.22) and subsequent expressions.

The applied approximation and averaging make intuitive sense when the estimated mapping from state-space to parameter space is smooth, and its values do not change much between the state values included in estimating $\hat{\mathbf{A}}_{k'_j}$, in other words, when the coverage of the parameter space by data is good.

Similarly, in the smoothing pass we have:

$$
\begin{aligned}
\hat{\mathbf{A}}_{k|N} &= \hat{\mathbf{A}}_k - \hat{\mathbf{P}}_k\mathbf{F}^T\mathbf{L}_k, \\
\mathbf{L}_k &= \begin{cases} \frac{1}{f}\sum_{j=1}^{f} g_2(\hat{\mathbf{A}}_k, \hat{\mathbf{P}}_k, \mathbf{L}_{k''_j}, y_{k''_j}, \mathbf{H}_{k''_j}), & f > 0, \\ 0, & f = 0, \end{cases} \\
\mathbf{P}_{k|N} &= \begin{cases} h_2(\hat{\mathbf{P}}_k, \frac{1}{f}\sum_{j=1}^{f} \mathbf{P}_{k''_j|N}), & f > 0, \\ \hat{\mathbf{P}}_k, & f = 0. \end{cases}
\end{aligned}
\tag{10.31}
$$

This is now done in the defined sequence, where k''_j, $j = 1, 2, \ldots, f$ are the successor points taking on the role of $t + 1$, and k the current point - the role of t in 10.29. Note that in general $f \leq p$, as there may be fewer than p successor points, and the start of the smoother from the "far" end of the data requires careful programming.

Again, the same approximation is being made with regard to averaging the covariance matrices, as for the Kalman Filter sequence.

It is worth noting that the averaging has to take the varying trajectories for multiple state-dependencies into account, hence the index forms of k'_j, $j = 1, 2, \ldots, p$ and k''_j, $j = 1, 2, \ldots, f$.

We should note again, that in the extended case where distances between state-space points are taken into account, the **F** and **G** matrices become dependent upon these distances as defined in (10.9), and will vary from sample to sample, depending on the sequence of points.

The above extensions are now the subject of further evaluation. It needs to be established to what degree the numerical complexity that they add to the present algorithms can be justified by the improved statistical rigour and possible robustness to poor sample coverage. Following the evaluation and possible improvements the MSDP algorithm will be included in a forthcoming release of the Captain Toolbox for Matlab [10].

References

1. Chen, R., Tsay, R.: Functional-coefficient autoregressive models. J. Am. Stat. Assoc. **88**, 298–308 (1993)
2. Chen, S., Billings, S.A.: Representations of non-linear systems: the Narmax model. Int. J. Control **49**(3), 1013–1032 (1989)
3. Evensen, G.: The ensemble Kalman filter: theoretical formulation and practical implementation. Ocean Dyn. **53**, 343–367 (2003)
4. Harvey, A.: Forecasting, Structural Time Series Models and the Kalman Filter. Cambridge University Press, Cambridge (1989)
5. Hastie, T., Tibshirani, R.: Generalized Additive Models. Chapman and Hall, London (1993)
6. Hastie, T., Tibshirani, R.: Varying-coefficient models. J. R. Stat. Soc. B **55**(4), 757–796 (1993)
7. Kalman, R.: A new approach to linear filtering and prediction problems. J. Basic Eng. **82**(2), 35–45 (1960)
8. Priestley, M.: Spectral Analysis and Time Series: Probability and Mathematical Statistics. Academic Press, New York (1994)
9. Taylor, C.J., Chotai, A., Young, P.C.: Non-linear control by input-output state variable feedback pole assignment. Int. J. Control **82**(6), 1029–1044 (2009)
10. Taylor, C.J., Pedregal, D.J., Young, P.C., Tych, W.: Environmental time series analysis and forecasting with the captain toolbox. Environ. Model. Softw. **22**(6), 797–814 (2007)
11. Taylor, J., Chotai, A., Tych, W.: Linear and nonlinear non-minimal state space control system design. In: Wang, L. (ed.) System Identification, Environmental Modelling and Control. Springer, Berlin (2011)
12. Truong, N.V., Wang, L., Wong, P.K.: Modelling and short-term forecasting of daily peak power demand in Victoria using two-dimensional wavelet based sdp models. Int. J. Electr. Power Energy Syst. **30**(9), 511–518 (2008)

13. Young, P.C.: A general theory of modelling for badly defined dynamic systems. In: Modeling, Identification and Control in Environmental Systems. North Holland, Amsterdam (1978)
14. Young, P.C.: Recursive extrapolation, interpolation and smoothing of non-stationary time-series. In: Chen, C.F. (ed.) Identification and System Parameter Estimation, pp. 33–44. Pergamon Press, Oxford (1988)
15. Young, P.C.: Time variable and state dependent modelling of nonstationary and nonlinear time series. In: Subba Rao, T. (ed.) Developments in Time Series, Volume in Honour of Maurice Priestley, pp. 374–413. Chapman and Hall, London (1993). Chap. 26
16. Young, P.C.: Data-based mechanistic modelling of environmental, ecological, economic and engineering systems. Environ. Model. Softw. **13**(2), 105–122 (1998)
17. Young, P.C.: Data-based mechanistic modelling generalised sensitivity and dominant mode analysis. Comput. Phys. Commun. **117**(1–2), 113–129 (1999)
18. Young, P.C.: Stochastic, dynamic modelling and signal processing: time variable and state dependent parameter estimation. In: Fitzgerald, W.J. et al. (eds.) Nonlinear and Nonstationary Signal Processing, pp. 74–114. Cambridge University Press, Cambridge (2000)
19. Young, P.C.: Comment on: "A Quasi-Armax approach to modelling nonlinear systems", by J. Hu et al. Int. J. Control **74**, 1767–1771 (2001)
20. Young, P.C.: Top-down and data-based mechanistic modelling of rainfall-flow dynamics at the catchment scale. Hydrol. Process. **17**(11), 2195–2217 (2003)
21. Young, P.C.: Data-based mechanistic modelling: natural philosophy revisited. In: Wang, L.P. (ed.) System Identification, Environmental Modelling and Control. Springer, Berlin (2011)
22. Young, P.C., Behzadi, M.A., Wang, C.L., Chotai, A.: Direct digital and adaptive control by input-output state variable feedback pole assignment. Int. J. Control **46**(6), 1867–1881 (1987)
23. Young, P.C., Casteletti, A., Pianosi, F.: The data-based mechanistic approach in hydrological modelling. In: Topics on System Analysis and Integrated Water Resources Management, pp. 27–48. Elsevier, Oxford (2007)
24. Young, P.C., Chotai, A., McKenna, P., Tych, W.: Proportional-integral-plus (pip) design for delta (delta) operator systems. Part 1. Siso systems. Int. J. Control **70**(1), 123–147 (1998)
25. Young, P.C., Parkinson, S., Lees, M.: Simplicity out of complexity in environmental modelling: Occam's razor revisited. J. Appl. Stat. **23**(2), 165–210 (1996)
26. Young, P.C.: Recursive Estimation and Time Series Analysis: An Introduction. Springer, Berlin (1984)
27. Young, P.C., McKenna, P., Bruun, J.: Identification of non-linear stochastic systems by state dependent parameter estimation. Int. J. Control **74**(18), 1837–1857 (2001)
28. Young, P.C., Castelletti, A., Pianosi, F.: The data-based mechanistic approach in hydrological modelling. In: Castelletti, A., Sessa, R.S. (eds.) Topics on System Analysis and Integrated Water Resources Management. Elsevier, Amsterdam (2007). ISBN: 13:978-0-0080-44967-8

Chapter 11
On Application of State Dependent Parameter Models in Electrical Demand Forecast

Nguyen-Vu Truong and Liuping Wang

11.1 Introduction

Privatization and deregulation of power system industries in many countries (i.e. UK, Japan, USA, Australia, etc.) in recent years have lead to the rise of competitive energy markets. This turns electricity into a commodity and trading article which can be sold and bought at market prices. Nevertheless, unlike other commodities, electricity can not be stored; and its transmission is limited by physical and reliability constraints. As a result, in order to effectively manage and plan the production for economical electricity utilities as well as to gain competitiveness in the market, an accurate forecast of future demands at regular time intervals[1] is of great importance for the management and planning of power production, operation and distribution as well as customer services.

Electrical demand modeling and forecast is, however, quite challenging due to the complex dynamics and behaviours exhibited from the load series. That is the electrical demand pattern is not only dependent on historical demands but as well influenced by a number of external variables (i.e. weather related variables, household number, etc.). As a result, from a system's point of view, this can be interpreted as a *complex nonlinear dynamic system*.

[1] Which can change from hour or day for short-term forecasts to week or year for medium and long-term forecasts respectively.

N.-V. Truong
Institute of Applied Mechanics and Informatics, Vietnam Academy of Science and Technology, Hanoi, Vietnam

L. Wang (✉)
School of Electrical and Computer Engineering, RMIT University, Melbourne, Australia
e-mail: liuping.wang@rmit.edu.au

L. Wang, H. Garnier (eds.), *System Identification, Environmental Modelling, and Control System Design*,

DOI 10.1007/978-0-85729-974-1_11,

To address this problem, various approaches have been reported in the open literature. Traditional approaches include regression methods, exponential smoothing, Kalman filtering as well as non-parametric methods (i.e. [23–29], etc.). More recent and most common methods in the area of electricity demand/price forecast rely on artificial intelligence (AI) techniques, for example, Expert Systems (i.e. [9, 10]), Fuzzy logic (i.e. [11, 12]) and especially Artificial Neural Network (i.e. [13–22], etc.) which have received quite considerable research interests in the past 2 decades.

The major advantage of such artificial intelligence based approaches is obvious. That is *no complex mathematical formulation or quantitative correlations between inputs and outputs is required*. Nevertheless, they suffer from a number of shortcomings, for example:

- Expert System based approach (i.e. [9, 10]) exploits human expert knowledge to develop set of rules for the purpose of forecasting by utilizing a comprehensive database. Nevertheless, the major disadvantage of this approach lies on the difficulties to transform this expert knowledge into a set of mathematical rules.
- Fuzzy logic based approach (i.e. [11, 12]) has similar problems, as it maps input variables to outputs using a set of logic statements (fuzzy rules) which could be developed solely from expert knowledge. In addition, when the problem becomes more complicated, it might lead to a significant increase in the number of fuzzy rules used for model building, which is as well another common disadvantage of fuzzy based approaches.
- Although ANN based approach can overcome some of the shortcomings of expert systems as it can directly acquire experience from training data, it suffers from a number of limitations including (1) the need of an excessively large number of parameters used in the model which can subsequently lead to the danger of *over-fitting*, (2) *difficulties in determining optimum network topology as well as training parameters* (i.e. number and size of the hidden layers, type of neuron transfer functions for various layers, training rate, etc.) and so on. Another limitation of this approach is the *'black box'* nature of ANN models. These models give little insight into the modelled relationships and the relative significance of various variables, thus providing *poor explanation* about the system under study.

To tackle such shortcomings, the State Dependent Parameter (SDP) model structure provides a natural way to express nonlinear systems [1–8]. Model of this type is quasi-linear ARX structured, but with State Dependent Parameters which are nonlinear functions of the respective state variables (i.e. derivatives or lagged values of the input and output variables) to describe the system's nonlinearities. It means that at a particular sampling instance, such a model is, in turn, a *"frozen", locally valid linear system,* providing useful property of the local dynamics. Such a representation enables internal connection between various state variables to be descriptively exploited, thus the dynamic system's nonlinearities can be represented in a very interpretable way in comparison to the conventional '*black-box*' nonlinear system identification approaches.

This chapter presents an improved methodology to electrical demand forecast using wavelet based SDP (WSDP) models [5] in which the associated State Dependent Parameters are compactly parameterized in the form of wavelets (i.e. [1–4]). In the present study, the essential of the multi-variable state dependencies in the load dynamics[2] is effectively captured and parsimoniously realized using 2-D wavelet series expansions. This formulates the so called 2-dimensional wavelet based SDP modelling (2-DWSDP) approach [4]. Here, PRESS statistics in conjunction with a forward regression are applied to detect efficient nonlinear model structures. Model obtained in such a manner is *parsimonious and descriptive*, enhancing its generalization capability which is very useful for this particular forecasting application.

This study considers one day ahead forecast of daily peak electrical demands in the state of Victoria, Australia. Using 2-DWSDP model, various dependencies among historical demand and weather related variables (in this situation, daily peak temperature is likely among the most influential) can be exploited and realized through a very *compact and descriptive mathematical formulation.*

Section 11.2 of this chapter reviews the 2-DWSDP modeling approach. A feasible model structure for the daily peak electrical demand forecast under study is discussed in Sect. 11.3. The modeling results are presented in Sect. 11.4 which illustrates the merits and efficiency of the proposed approach. Finally, Sect. 11.5 concludes the chapter.

11.2 2-DWSDP Model

It is assumed that a nonlinear system can be represented by the following 2-D *State Dependent Parameter* (SDP) model:

$$y(k) = \sum_{q=1}^{n_y} f_q(x_{m_q,n_q})y(k-q) + \sum_{q=0}^{n_u} g_q(x_{l_q,p_q})u(k-q) + e(k) \tag{11.1}$$

where f_q, g_q (regarded as 2-D SDPs to carry the nonlinearities) are dependent on $x_{m_q,n_q} = (x_{m_q}, x_{n_q} \in x)$ and $x_{l_q,p_q} = (x_{l_q}, x_{p_q} \in x)$ in which $x = \{y(k-1), \ldots, y(k-n_y), u(k), \ldots, u(k-n_u)\}$; $u(k)$ and $y(k)$ are, respectively, the sampled input-output sequences; while $\{n_u\ n_y\}$ refer to the maximum number of lagged inputs and outputs. Finally, $e(k)$ refers to the noise variable, assumed initially to be a zero mean, white noise process that is uncorrelated with the input $u(k)$ and its past values.

For example, a first order 2-D SDP model representation of a nonlinear system can take the following form:

$$y(k) = f_1[y(k-1), u(k)]y(k-1) + g_0[u(k), u(k-1)]u(k). \tag{11.2}$$

Let $x = \{x_1, x_2, x_3\} = \{y(k-1), u(k), u(k-1)\}$. In this case, the 2-D State Dependent Parameters f_1 and g_0 are dependent on $x_{1,2} = \{x_1, x_2\} = \{y(k-1), u(k)\}$ and $x_{2,3} = \{x_2, x_3\} = \{u(k), u(k-1)\}$ respectively.

[2]There are a number of variables which directly and indirectly influence the electrical demand at a particular point of time, such as historical demand, weather related variables (i.e. humidity, wind and cloud conditions, minimum temperature, etc.), household number and so on.

Using the 2-D wavelet series expansion (i.e. [4]), the 2-D State Dependent Parameters $f_q(x_{m_q,n_q})$ and $g_q(x_{l_q,p_q})$ can be approximated as

$$f_q(x_{m_q,n_q}) = \sum_{i_{\min}}^{i_{\max}} \sum_{k_1 \in L_{ixmq}} \sum_{k_2 \in L_{ixnq}} a_{fq,i,k_1,k_2} \Psi^{[2]}_{i,k_1,k_2}(x_{m_q,n_q}), \tag{11.3}$$

$$g_q(x_{l_q,p_q}) = \sum_{i_{\min}}^{i_{\max}} \sum_{k_1 \in L_{ixlq}} \sum_{k_2 \in L_{ixpq}} b_{gq,i,k_1,k_2} \Psi^{[2]}_{i,k_1,k_2}(x_{l_q,p_q}) \tag{11.4}$$

in which, $\Psi^{[2]}_{i,k_1,k_2}(x, y)$ regards the scaled and translated version of a 2-Dimensional wavelet function $\Psi^{[2]}(x, y)$, i.e.

$$\Psi^{[2]}_{i,k_1,k_2}(x_{m_q,n_q}) = \Psi^{[2]}(2^{-i}x_{m_q} - k_1, 2^{-i}x_{n_q} - k_2),$$
$$\Psi^{[2]}_{i,k_1,k_2}(x_{l_q,p_q}) = \Psi^{[2]}(2^{-i}x_{l_q} - k_1, 2^{-i}x_{p_q} - k_2).$$

To formulate a 2-D wavelet basis function $\Psi^{[2]}(x, y)$, a natural approach is based on the tensor product of 2 1-D wavelet functions $\Psi(x)$ and $\Psi(y)$, i.e.

$$\Psi^{[2]}(x, y) = \Psi(x)\Psi(y). \tag{11.5}$$

For example, if $\Psi(x)$ is chosen to be a 1-D Mexican hat wavelet, i.e.

$$\Psi(x) = \left\{ \begin{matrix} (1 - x^2)e^{-0.5x^2} & \text{if } x \in (-4, 4) \\ 0 & \text{otherwise} \end{matrix} \right\}. \tag{11.6}$$

Then its 2-D version[3] (shown in Fig. 11.1) will take the following form:

$$\Psi^{[2]}(x, y) = \left\{ \begin{matrix} (1 - x^2)(1 - y^2)e^{-0.5(x^2+y^2)} & \text{if } x, y \in (-4, 4) \\ 0 & \text{otherwise} \end{matrix} \right\}. \tag{11.7}$$

$i_{\min}$ and $i_{\max}$ refer to the minimum and maximum scaling parameters (finest and coarsest) which determine the set of terms used for the approximation, and finally, $\{L_{ixmq}, L_{ixnq}\}$ and $\{L_{ixlq}, L_{ixpq}\}$ correspond to the translation libraries with respect to $\{\Psi(x_{mq}), \Psi(x_{nq})\}$ and $\{\Psi(x_{lq}), \Psi(x_{pq})\}$ at scale i.

Substituting (11.3) and (11.4) into (11.1), we obtain

$$y(k) = \sum_{q=1}^{n_y} \left[\sum_{i_{\min}}^{i_{\max}} \sum_{k_1 \in L_{ixmq}} \sum_{k_2 \in L_{ixnq}} a_{fq,i,k_1,k_2} \Psi^{[2]}_{i,k_1,k_2}(x_{m_q,n_q}) \right] y(k - q)$$

[3]For simplicity, through-out this paper, this form of 2-D wavelet functions is used.

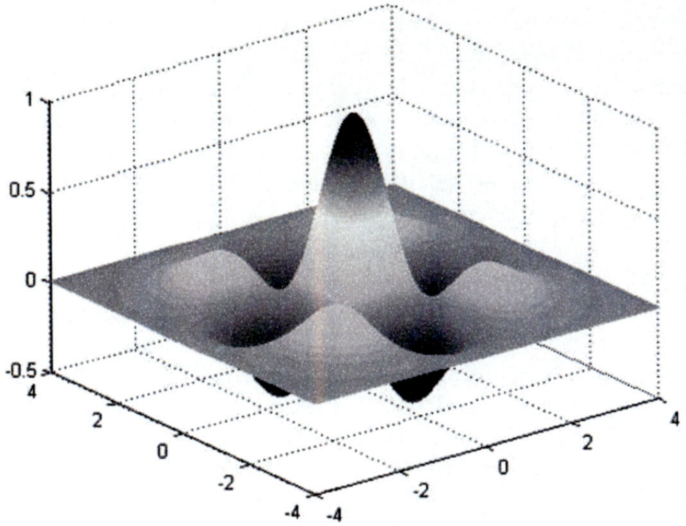

Fig. 11.1 2-D Mexican hat wavelet function

$$+\sum_{q=0}^{n_u}\left[\sum_{i_{\min}}^{i_{\max}}\sum_{k_1\in L_{ixlq}}\sum_{k_2\in L_{ixpq}} b_{gq,i,k_1,k_2}\Psi^{[2]}_{i,k_1,k_2}(x_{l_q},p_q)\right]u(k-q)+e(k). \tag{11.8}$$

Equation (11.8) is regarded as *2-D Wavelet based SDP model* (2-DWSDP).

At this point, (11.8) can be formulated as in the *linear-in-the-parameter regression equation*, i.e.

$$Y = P\theta + \Xi \tag{11.9}$$

in which,

$$\begin{aligned} Y &= \left[y(0),\ldots,y(N-1)\right]^T, \\ U &= \left[u(0),\ldots,u(N-1)\right]^T, \\ \Xi &= \left[e(0),\ldots,e(N-1)\right]^T. \end{aligned} \tag{11.10}$$

11.2.1 Model Structure Selection and Parameter Estimation

One of the keys in nonlinear system identification is to effectively select candidate structures. This is among the most challenging tasks due to infinite possible combinations of nonlinear regression terms. Therefore, it is critical, at the first step, to reduce the set of candidate structures to a manage-able size based on some known characteristics about the system under study. In the situation of 2-D SDP models as considered in this chapter, the finest and coarsest scaling parameters $i_{\min}$, $i_{\max}$ determine the set of terms as well as their associated characteristics[4] used for the approximation of the respective 2-D SDP relationship. As a result, they play an important roles in the selection of candidate model structures for the nonlinear system identification. If $i_{\min}$ and $i_{\max}$ are properly selected and a compactly supported mother wavelet is chosen, the set of candidate structures is now limited and deterministic.[5] This reduces the computational load and improves the efficiency of the optimized model structure selection algorithm.

The 2-DWSDP model as formulated in (11.8) includes all the possible combinations of the parameters. Thus, it is regarded as an over-parameterized model. As a result, in order to obtain a compact representation of nonlinear systems, an efficient model structure determination approach based on the PRESS statistics and forward regression is implemented (see [1–4]). This procedure uses the incremental value of PRESS[6] ($\Delta PRESS$) as criterion to detect the significance of each terms within the model in which the maximum $\Delta PRESS$ signifies the most significant term, while its minimum reflects the least significant term. Based on this, the algorithm initializes with the initial subset being the most significant term. It then starts to grow to include the subsequent significant terms in a forward regression manner, until a specified performance is achieved. Here, the incorporation of *Orthogonal Decomposition* (OD) into the model structure selection algorithm helps to avoid any *ill-conditioning* problems associated with the parameter estimation.

PRESS based selection algorithm For the ease of representation, let us denote ϕ_i be the $(i+1)$th column of Φ: $\phi_i = \Phi(:, i+1)$, and $P^{(-i)}$ denotes the matrix which is resulted from excluding the ith column from the original matrix P.

1. Initialize $\Phi = P$, $[N, m] = size(P)$
2. Orthogonal Decomposition
   a. $[N, m_1] = size(\Phi)$. Initialize $\omega_0 = \phi_0$, $g_0 = \frac{\omega_0^T Y}{\omega_0^T \omega_0}$.
   b. For $1 \le i \le m_1 - 1$, compute

$$\alpha_{j,i} = \frac{\omega_j^T \phi_i}{\omega_j^T \omega_j}, \quad j = 0, 1, \ldots, i-1,$$

$$\omega_i = \phi_i - \sum_{j=0}^{i-1} \alpha_{j,i} \omega_j,$$

[4]A small value of $i_{\min}$ results in a large number of wavelet elements with higher frequency characteristics to be contained in the function's library. And vice versa, with a large value of $i_{\max}$, the function's library will consist of a large number of wavelet elements that are at lower frequency features.

[5]Criteria to guide the selection of the scaling parameters $i_{\min}$ and $i_{\max}$ is described in [4].

[6]The difference between the overparameterized (original) model's PRESS value and the one calculated by excluding a term from the original model.

$$g_i = \frac{\omega_i^T Y}{\omega_i^T \omega_i}.$$

3. PRESS computation

$$\xi_{-k}(k) = \frac{y(k) - \sum_{i=0}^{m_1-1} \omega_i(k) g_i}{1 - \sum_{i=0}^{m_1-1} \frac{\omega_i(k)^2}{\|\omega_i\|^2}},$$

$$PRESS = \sum_{k=0}^{N-1} \xi_{-k}^2(k).$$

4. $PRESS(m) = PRESS$. For $1 \le i_1 \le m$,
   a. Set $\Phi = P^{(-i_1)}$. Repeat steps 2 and 3.
   b. $PRESS_m^{-i_1}(m-1) = PRESS$. Calculate

$$\Delta PRESS_{i_1} = PRESS_m^{-i_1}(m-1) - PRESS(m).$$

5. Based on the largest $\Delta PRESS_{i_1}$ value, select the most significant term to be added to the regressor matrix.
6. Solve for the intermediate parameter estimate in a least squares manner.
7. Calculate the approximation accuracy, and compare it to the desired value:
   - If satisfactory performance is achieved, stop the algorithm;
   - Otherwise, add extra terms into the regressor matrix based on the next largest $\Delta PRESS_{i_1}$ values, and repeat from step 6 to 7.

Upon determining the optimized nonlinear model structure for the over-parameterized representation as in (11.9), the final identified model structure is generally found to be

$$y(k) = \sum_{q=1}^{n_y} \left[\sum_{j=1}^{nf_q} a_{q,j} \varphi_{q,j}^{[2]}(x_{m_q,n_q}) \right] y(k-q) + \sum_{q=0}^{n_u} \left[\sum_{j=1}^{ng_q} b_{q,j} \phi_{q,j}^{[2]}(x_{l_q,p_q}) \right] u(k-q) + e(k). \tag{11.11}$$

11.2.2 Identification Procedure

The overall nonlinear system identification using the proposed approach can be summarized into the following steps:

1. *Determining the 2-D SDP model's initial conditions*. This includes the following:

a. Select the initial values[7] of n_y and n_u.
b. Based on the available *a priori* knowledge, select the significant variables from all the candidate lagged output and input terms (i.e. $y(k-1), \ldots, y(k-n_y), u(k), \ldots, u(k-n_u)$) and the significant 2-D state dependencies (i.e. $f_q(x_{m_q,n_q}), g_q(x_{l_q,p_q})$) formulated by the selected significant variables. Note that these *a priori knowledge* can be some known structural characteristics, or based on some hypothesis and assumption made about the system under study.
c. Otherwise, if there is no *a priori* knowledge available, all the possible variables as well as their associated possible 2-D dependencies for the selected model order (n_y and n_u) need to be considered. For example, if $n_y = 1$ and $n_u = 1$, the possible variables are $y(k-1), u(k), u(k-1)$, leading to the possible 2-D dependencies between: $\{y(k-1), u(k)\}, \{y(k-1), u(k-1)\}, \{u(k), u(k-1)\}$.

2. *2-DWSDP's optimized model structure selection.* This involves the following steps:
   a. Select the associated scaling parameters $[i_{\min}, i_{\max}]$ to be used for the 2-D SDP parameterization.
   b. Formulate an over-parameterized 2-DWSDP model by expanding all the 2-D SDPs (i.e. $f_q(x_{m_q,n_q}), g_q(x_{l_q,p_q})$) via 2-D wavelet series expansion using the selected scaling parameters $[i_{\min}, i_{\max}]$.
   c. Using the *PRESS* based selection algorithm, determine an optimized model structure from the candidate model terms.
3. *Final parametric optimization.*
   - Using the measured data, estimate the associated parameters via a Least Squares algorithm.
4. *Model validation.*
   - If the identified values of n_y and n_u as selected in step 1 provides a satisfactory performance over the considered data, terminates the procedure.
   - Otherwise, increase the model's order, i.e. $n_y = n_y + 1$ and/or $n_u = n_u + 1$, and repeat Steps 1b, 2 through 4.

11.3 Model Structure Development

The development of a model for daily peak electrical demand forecast in this study relies on the hypothesis that the daily peak demand for a certain day in a week is dependent on the following factors:

- The historical peak demands of the previous days (i.e. peak demands of the previous two days)

[7] Which normally start with lower values.

- The peak demand at the same day of the previous week. This looks after the *weekly trend* in the power demand behaviour. It is most likely the fact that the power consumption during working days (i.e. from Monday to Friday) is higher than that during the Weekends (Saturday and Sunday).
- The weather related variables associated with these days, particularly in this study, the peak temperature is used due to its strong link with the power consumption. During a hot day (i.e. summer days), the electrical demand is significantly increased due to the power consumption for cooling. Similarly, due to the power utilization for heating, the demand for a cold day (i.e. winter days) increases for that day as well.

Let $y(k)$ and $u(k)$ respectively denote the peak electrical demand and temperature at the day index k, a feasible model for daily peak electrical demand forecast can be realized in the following form using a 2-DWSDP model, i.e.

$$\begin{aligned}
y(k) = {} & f_1^{[2]}\big[y(k-1), u(k-1)\big]\, y(k-1) + g_1^{[2]}\big[y(k-1), u(k-1)\big]\, u(k-1) \\
& + f_2^{[2]}\big[y(k-2), u(k-2)\big]\, y(k-2) + g_2^{[2]}\big[y(k-2), u(k-2)\big]\, u(k-2) \\
& + f_7^{[2]}\big[y(k-1), y(k-2)\big]\, y(k-7) + g_7^{[2]}\,[u(k-1), u(k-2)]\, u(k-7) \\
& + g_0\,[u(k)]\, u(k)
\end{aligned} \tag{11.12}$$

in which, the components associated with the functions $f_1^{[2]}(\cdot,\cdot)$, $g_1^{[2]}(\cdot,\cdot)$, $f_2^{[2]}(\cdot,\cdot)$ and $g_2^{[2]}(\cdot,\cdot)$ represent the contribution of the peak electrical demand and temperatures in the last 2 days to that of the current day; these associated with the functions $f_7^{[2]}(\cdot,\cdot)$ and $g_7^{[2]}(\cdot,\cdot)$ realize the nonlinear interactions as well as the contribution of the peak demands and temperatures in the previous 2 days and the previous week to that of the current day. Finally, the component associated with $g_0(\cdot)$ represents the relationship between the peak temperature and electrical demand of the current day.

11.4 Results

In this study, the peak temperature and daily electrical demand (from the 1st January to the 24th August of 2007) of the state of Victoria, Australia are used. This data (Fig. 11.2) was obtained from the Australian National Electricity Market Management Company (NEMMCO)[8] and the Australian Government Bureau of Meteorology.[9] For the purpose of model building exercise, the data set was standardized (still designated as $\{y(k), u(k)\}$) and separated into (1) estimation set for the model building (from 1st January 2007 to 8th August 2007) and (2) validation set used for the evaluation of the model forecast capability (from 9th August 2007 to 24th August 2007).

[8]National Electricity Market Management Company, http://www.nemweb.com.au/.

[9]Australian Government Bureau of Meteorology, http://www.bom.gov.au/weather/vic/.

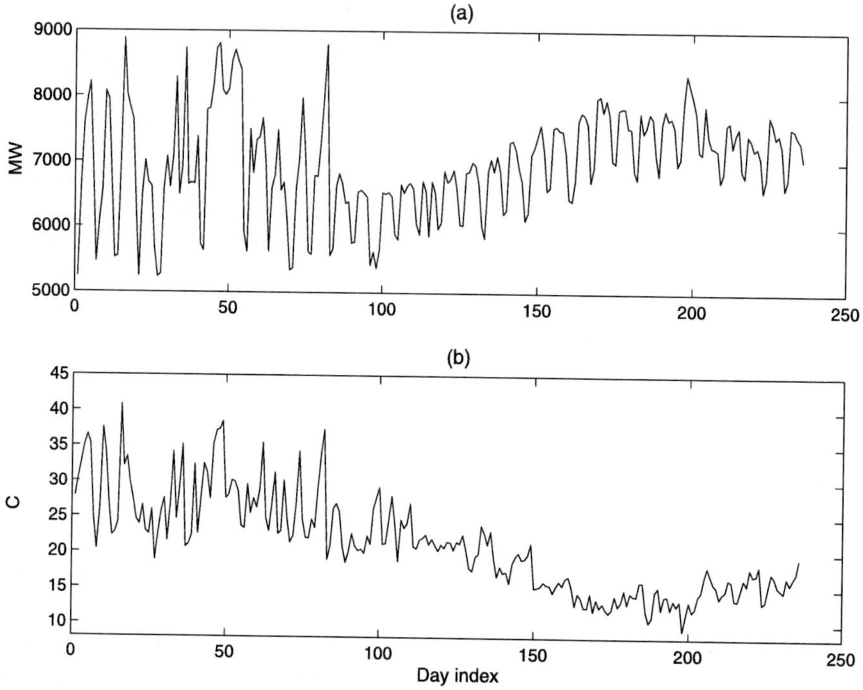

Fig. 11.2 (**a**) Daily peak power demand (**b**) peak temperature data in the time period under study

With $i_{\min}$ and $i_{\max}$ chosen to be -3 and 3, the final identified model is found to be:

$$\begin{aligned} y(k) = {} & \hat{f}_1^{[2]}\left[y(k-1), u(k-1)\right] y(k-1) + \hat{g}_1^{[2]}\left[y(k-1), u(k-1)\right] u(k-1) \\ & + \hat{f}_2^{[2]}\left[y(k-2), u(k-2)\right] y(k-2) + \hat{g}_2^{[2]}\left[y(k-2), u(k-2)\right] u(k-2) \\ & + \hat{f}_7^{[2]}\left[y(k-1), y(k-2)\right] y(k-7) + \hat{g}_7^{[2]}\left[u(k-1), u(k-2)\right] u(k-7) \\ & + \hat{g}_0\left[u(k)\right] u(k) \end{aligned} \tag{11.13}$$

in which,

$$\begin{aligned} \hat{f}_1^{[2]}(x_1, x_2) = {} & 0.3007\Psi_{3,0,0}^{[2]}(x_1, x_2) + 1.0539\Psi_{0,0,0}^{[2]}(x_1, x_2) \\ & + 0.5396\Psi_{1,-1,0}^{[2]}(x_1, x_2), \end{aligned} \tag{11.14}$$

$$\hat{g}_1^{[2]}(x_1, x_2) = 0.9056\Psi_{3,0,2}^{[2]}(x_1, x_2) + 0.0503\Psi_{-1,3,5}^{[2]}(x_1, x_2), \tag{11.15}$$

$$\begin{aligned} \hat{f}_2^{[2]}(x_1, x_2) = {} & 1.0751\Psi_{3,0,2}^{[2]}(x_1, x_2) + 0.3806\Psi_{1,1,1}^{[2]}(x_1, x_2) \\ & + 0.1523\Psi_{1,-1,0}^{[2]}(x_1, x_2) + 0.1993\Psi_{1,0,0}^{[2]}(x_1, x_2), \end{aligned} \tag{11.16}$$

$$\begin{aligned}\hat{g}_2^{[2]}(x_1,x_2) &= 0.0699\Psi_{3,0,0}^{[2]}(x_1,x_2)+0.2716\Psi_{1,0,1}^{[2]}(x_1,x_2)\\ &\quad +0.8007\Psi_{0,2,1}^{[2]}(x_1,x_2)+0.3904\Psi_{-1,3,5}^{[2]}(x_1,x_2)\\ &\quad +0.5806\Psi_{-1,2,3}^{[2]}(x_1,x_2)+0.5206\Psi_{-1,4,4}^{[2]}(x_1,x_2)\\ &\quad +0.4516\Psi_{0,0,0}^{[2]}(x_1,x_2), \end{aligned} \tag{11.17}$$

$$\begin{aligned}\hat{f}_7^{[2]}(x_1,x_2) &= 0.4354\Psi_{3,0,0}^{[2]}(x_1,x_2)+0.6399\Psi_{1,1,1}^{[2]}(x_1,x_2)\\ &\quad +0.3224\Psi_{0,0,1}^{[2]}(x_1,x_2)+0.0702\Psi_{1,1,0}^{[2]}(x_1,x_2), \end{aligned} \tag{11.18}$$

$$\begin{aligned}\hat{g}_7^{[2]}(x_1,x_2) &= 0.5860\Psi_{3,0,2}^{[2]}(x_1,x_2)+0.6567\Psi_{1,0,-1}^{[2]}(x_1,x_2)\\ &\quad +0.8944\Psi_{1,-1,0}^{[2]}(x_1,x_2)+0.2043\Psi_{0,1,1}^{[2]}(x_1,x_2)\\ &\quad +1.5331\Psi_{-1,4,3}^{[2]}(x_1,x_2), \end{aligned} \tag{11.19}$$

$$\begin{aligned}\hat{g}_0(x) &= 1.0146\Psi_{1,1}(x)-0.0754\Psi_{3,0}(x)+0.4196\Psi_{0,3}(x)\\ &\quad +0.4023\Psi_{-3,21}(x)+0.0662\Psi_{-1,4}(x), \end{aligned} \tag{11.20}$$

where

$$\Psi_{i,k1,k2}^{[2]}(x_1,x_2) = \Psi^{[2]}(2^{-i}x_1-k_1, 2^{-i}x_2-k_2), \tag{11.21}$$

$$\Psi^{[2]}(x_1,x_2) = (1-x_1^2)(1-x_2^2)e^{-0.5(x_1^2+x_2^2)}, \tag{11.22}$$

$$\Psi_{i,k}(x) = \Psi(2^{-i}x-k), \tag{11.23}$$

$$\Psi(x) = (1-x^2)e^{-0.5x^2}. \tag{11.24}$$

Here, Relative Error of forecasting-**REF** and Mean Absolute Prediction Error-**MAPE** are used to measure, thus quantify the model's forecasting performance:

$$\mathbf{REF}_k = \frac{y^p(k)-y(k)}{y(k)} \times 100\%, \tag{11.25}$$

$$\mathbf{MAPE} = \text{Mean}\,[|\mathbf{REF}_k|], \tag{11.26}$$

in which, $y^p(k)$ denotes the forecasted value of the peak demand of the day index k.

Figure 11.3 compares the prediction (which is recovered to its original amplitude by de-standardization) of the model (see (11.13)) versus the actual daily peak demand over the estimation set (from 1st January 2007 to 8th August 2007), in which the identified model fits 95.71% of the data. Figure 11.4 demonstrates the model performance in the forecasting of daily peak demands over the validation set (from 9th January 2007 to 24th August 2007). These forecasted values are tabulated in Table 11.1 in comparison with the actual values. The **MAPE** over the forecasted period is 1.9%, while the standard deviation of $|\mathbf{REF}_k|$ is calculated to be 1.6%.

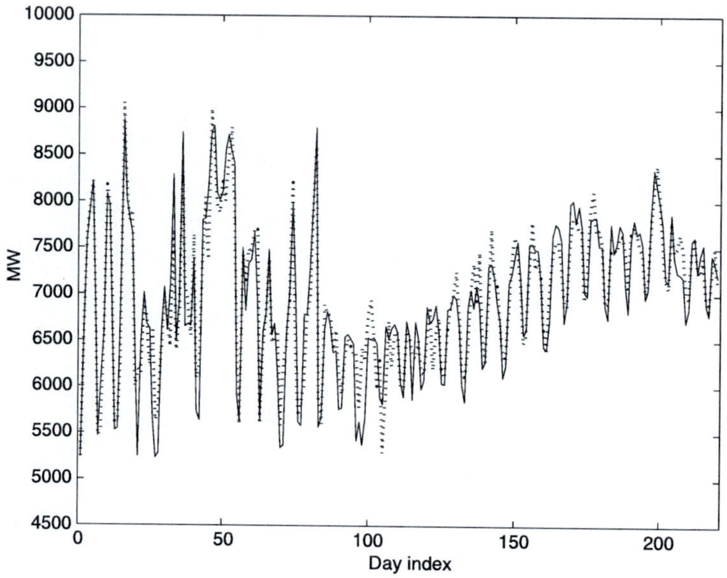

Fig. 11.3 Model (11.13) prediction (*dot-dot*) versus the actual daily peak demand (*solid*) over the estimation set

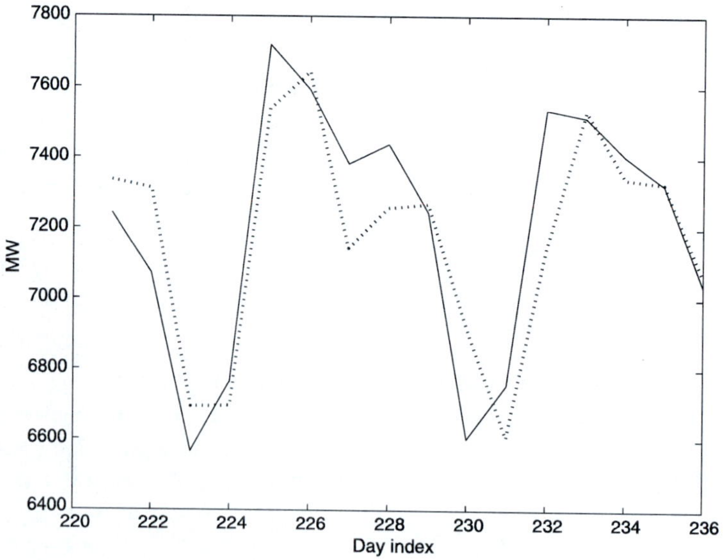

Fig. 11.4 Model (11.13) performance in forecasting the daily peak power demands over the validation set: forecasted values (*dot-dot*) versus actual values (*solid*)

Table 11.1 Forecasted daily peak electrical demands versus the actual values

Date	Actual values (MW)	Forecasted values (MW)	*REF*
09/08/2007	7242.1	7336.3	1.3%
10/08/2007	7071.7	7312.8	3.41%
11/08/2007	6565.4	6692.8	1.94%
12/08/2007	6764.2	6695.9	−1.01%
13/08/2007	7718.8	7538.1	−2.34%
14/08/2007	7592.6	7644.5	0.68%
15/08/2007	7382.2	7143.8	−3.23%
16/08/2007	7437.7	7259.0	−2.40%
17/08/2007	7245.5	7269.3	0.33%
18/08/2007	6602.5	6917.6	4.77%
19/08/2007	6755.0	6606.8	−2.19%
20/08/2007	7536.1	7141.7	−5.23%
21/08/2007	7514.3	7532.7	0.25%
22/08/2007	7405.1	7341.2	−0.86%
23/08/2007	7323.6	7327.4	0.05%
24/08/2007	7037.8	7062.1	0.35%
MAPE			1.9%

From Table 11.1, it can be observed that the forecasted peak demands are very close to the actual daily peak power demands. These results indicate the model's excellent performance in the sense that it can capture the significant essentials about this complex nonlinear dynamic system through a very compact mathematical realization (30 terms). This model provides a descriptive representation to the system under study. The respective State Dependent Parameters (SDPs) are shown in Figs. 11.5 and 11.6, demonstrating very clear views about the interaction and relationships between various components used in building the model.

As shown in Fig. 11.5(a) and Fig. 11.6(a), it demonstrates strong multi-variable dependencies in the relationship between the electrical demand and its historical data. Since, there is a strong link between the peak electrical demands and peak temperatures, this implies that there exists significant multi-variable dependencies in the relationship between the historical temperature data and the respective peak demands (as shown in Figs. 11.5(b), (c), (d), and Fig. 11.6(b)).

Figure 11.6(c) demonstrates the direct relationship between the peak demand and temperature in a certain day. A clearer view can be explored by plotting $[\hat{g}_0(x)]x$ against the actual temperature range under study. Figure 11.7 indicates that power consumption at cold temperature (9°C) is significantly higher than that at normal temperature (i.e. 22°C). The power consumption trend decreases as the temperature increases from 9°C to 22°C, and reaches its minimum value at 22°C (thermal comfort). When the temperature goes higher than 22°C, the power consumption increases, and reaches its maximum value at 39°C. Note that, the rate of change in

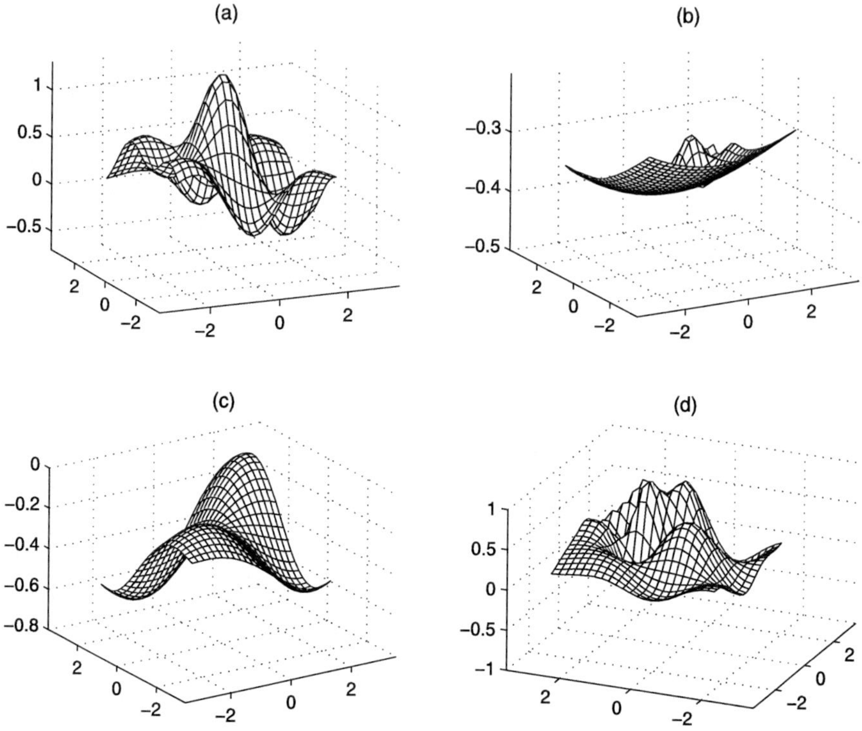

Fig. 11.5 SDPs' plots: **(a)** $\hat{f}_1^{[2]}(x_1,x_2)$, **(b)** $\hat{g}_1^{[2]}(x_1,x_2)$, **(c)** $\hat{f}_2^{[2]}(x_1,x_2)$ and **(d)** $\hat{g}_2^{[2]}(x_1,x_2)$

the power consumption at the temperature above 25°C (especially, above 36°C, the power consumption is dramatically increased) is quicker than that at the temperature below 22°C. It demonstrates a common fact that the power consumption during hot weather (i.e. summer) is higher than that during cold weather (i.e. winter). This phenomenon has been adequately explained by the model.

11.5 Conclusion

This chapter has described an application of a particular class of State Dependent Parameter models, 2-DWSDP model, to electrical demand modeling and forecast. In the present study, forecast of daily peak demand in the state of Victoria, Australia was considered. The obtained mathematical model is parsimonious, yet descriptive, enhancing its generalization capability while providing reasonable insights about the interactions and relationships between several variables within this highly complex, nonlinear system. Excellent performance in forecasting daily peak power demand as demonstrated in the modelling results illustrates the merit of this approach, in which the identified model efficiently captures the essentials of the system's dynamics. In addition, this approach could be generally applicable to some other ap-

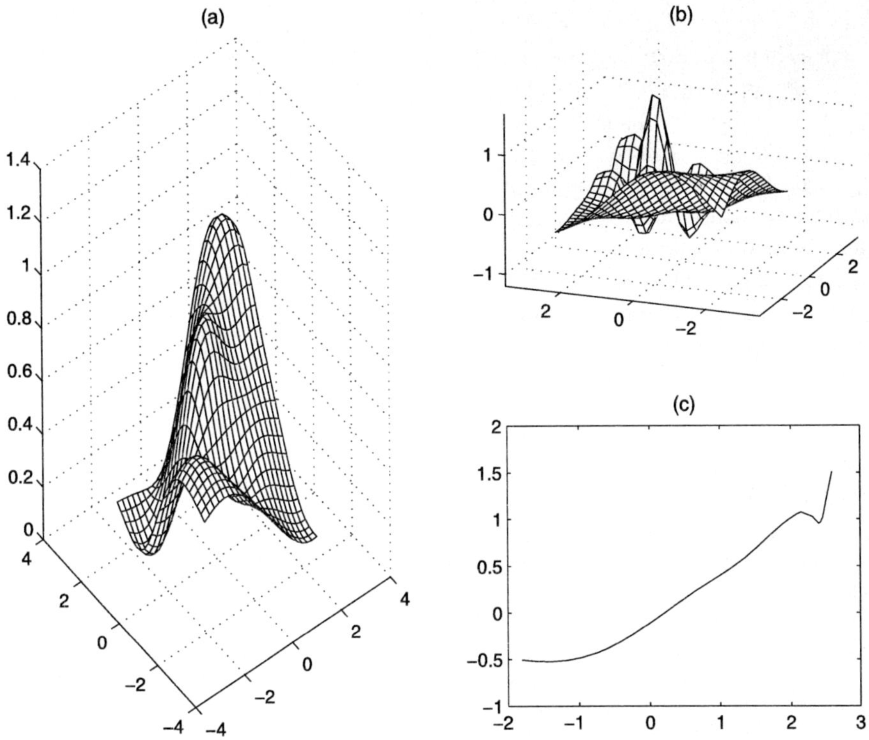

Fig. 11.6 SDPs' plots: **(a)** $\hat{f}_7^{[2]}(x_1, x_2)$, **(b)** $\hat{g}_7^{[2]}(x_1, x_2)$ and **(c)** $\hat{g}_0(x)$

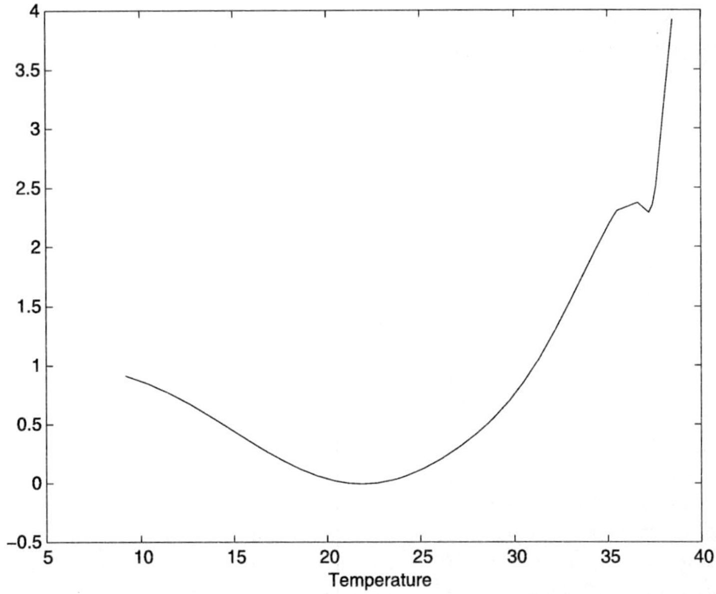

Fig. 11.7 $[\hat{g}_0(x)]x$ versus temperature

plications in power systems research area which concerns the needs of modelling, such as in power distribution line modelling, load behaviour study, and so on.

There are several other weather related variables which directly and indirectly affect the power demand, such as humidity, wind and cloud conditions, minimum temperature, etc. However, the obtained model's performance suggests that for the daily peak power demand modelling and forecasting problem in the present study, daily peak temperature is the most influential weather-related variable. In the future work, all the relevant variables will be incorporated into the model to further enhance the model's performance, particularly (1) the development of a composite variable (i.e. apparent temperature, chill factor, etc.) which looks after all the relevant weather variables as well as (2) the incorporation of some special variables such as customer's variables, holiday, etc.

References

1. Truong, N.V., Wang, L., Young, P.C.: Nonlinear system modeling using nonparametric identification and linear wavelet estimation of SDP models. In: Proc. 45th IEEE Conf. on Decision and Control, San Diego, USA, pp. 2523–2528 (2006)
2. Truong, N.V., Wang, L., Young, P.C.: Nonlinear system modeling using nonparametric identification and linear wavelet estimation of SDP models. Int. J. Control **80**(5), 774–788 (2007)
3. Truong, N.V., Wang, L., Huang, J.M.: Nonlinear modeling of a magnetic bearing using SDP model and linear wavelet parameterization. In: Proc. 2007 American Control Conference, New York, USA, pp. 2254–2259 (2007)
4. Truong, N.V., Wang, L.: Nonlinear system identification using two dimensional wavelet based SDP models. Int. J. Syst. Sci. **40**(11), 1161–1180 (2009)
5. Truong, N.V., Wang, L., Wong, K.C.: Modelling and short-term forecasting of daily peak power demand in Victoria using two-dimensional wavelet based SDP models. Int. J. Electr. Power Energy Syst. **30**(9), 511–518 (2008)
6. Young, P.C.: Stochastic, dynamic modelling and signal processing: time variable and state dependent parameter estimation. In: Fitzgerald, W.J., Walden, A., Smith, R., Young, P.C. (eds.) Nonlinear and Nonstationary Signal Processing, pp. 74–114. Cambridge University Press, Cambridge (2000)
7. Young, P.C.: The identification and estimation of nonlinear stochastic systems. In: Mees, A.I. (ed.) Nonlinear Dynamics and Statistics, pp. 127–166. Birkhäuser, Boston (2001a)
8. Young, P.C., McKenna, P., Bruun, J.: Identification of nonlinear stochastic systems by state dependent parameter estimation. Int. J. Control **74**, 1837–1857 (2001)
9. Ho, K.L.: Short-term load forecasting of Taiwan power system using a knowledge based expert system. IEEE Trans. Power Syst. **5**(4), 1214–1219 (1990)
10. Rahman, S., Hazim, O.: Load forecasting for multiple sites: development of an expert system based technique. Electr. Power Syst. Res. **39**(3), 161–169 (1996)
11. Huang, S., Shih, K.: Application of a fuzzy model for short-term load forecast with group method of data handling enhancement. Int. J. Electr. Power Energy Syst. **24**, 631–638 (2002)
12. Al-Kandari, A.M., Soliman, S.A., El-Hawary, M.E.: Fuzzy short-term electric forecasting. Int. J. Electr. Power Energy Syst. **26**, 111–122 (2004)
13. Senjyu, T., Mandal, P., Uezato, K., Funabashi, T.: Next day load curve forecasting using recurrent neural network structure. IEE Proc., Gener. Transm. Distrib. **151**, 388–394 (2004)
14. Beccali, M., Cellura, M., Lo Brano, V., Marvuglia, A.: Forecasting daily urban electric load profiles using artificial neural networks. Energy Convers. Manag. **45**, 2879–2900 (2004)

15. Baczynski, D., Parol, M.: Influence of artificial neural network structure on quality of short-term electric energy consumption forecast. IEE Proc., Gener. Transm. Distrib. **151**, 241–245 (2004)
16. Saini, L.M., Soni, M.K.: Artificial neural network based peak load forecasting using Levenberg-Marquardt and quasi-Newton methods. IEE Proc., Gener. Transm. Distrib. **149**, 578–584 (2002)
17. Kumluca, A., Erkmen, I.: A hybrid learning for neural networks applied to short term load forecasting. Neurocomputing **51**, 495–500 (2003)
18. Niebur, D.: Artificial Neural Networks in the power industry, survey and applications. Neural Netw. World **6**, 945–950 (1995)
19. Desouky, A.A., Elkateb, M.M.: Hybrid adaptive techniques for electric-load forecast using ANN and ARIMA. IEE Proc., Gener. Transm. Distrib. **147**, 213–217 (2000)
20. Liang, R., Cheng, C.: Short-term forecasting by a neuro-fuzzy based approach. Int. J. Electr. Power Energy Syst. **24**(2), 103–111 (2002)
21. Gao, R., Tsoukalas, L.H.: Neural-wavelet methodology for load forecasting. J. Intell. Robot. Syst. **31**, 149–157 (2001)
22. Kermanshashi, B., Iwamiya, H.: Up to year 2020 load forecasting using neural nets. Int. J. Electr. Power Energy Syst. **24**, 789–797 (2002)
23. Bunn, D.W., Farmer, E.D.: Comparative Models for Electrical Load Forecasting. Wiley, New York (1985)
24. Moghram, I., Rahman, S.: Analysis and Evaluation of five short-term load forecasting techniques. IEEE Trans. Power Syst. **4**(4), 1484–1491 (1989)
25. Bunn, D.W., Vassilopoulos, A.I.: Comparison of seasonal estimation methods in multi-item short-term forecasting. Int. J. Forecast. **15**, 431–443 (1999)
26. Al-Hamadi, H.M., Soliman, S.A.: Short-term electric load forecasting based on Kalman filtering algorithm with moving window weather and load model. Electr. Power Syst. Res. **68**(1), 47–59 (2004)
27. Taylor, J.W.: Short-term electricity demand forecasting using double seasonal exponential smoothing. J. Oper. Res. Soc. **54**, 799–804 (2003)
28. Taylor, J.W., Buizza, R.: Using weather ensemble predictions in electricity demand forecasting. Int. J. Forecast. **19**, 57–70 (2003)
29. Asber, D., Lefebvre, S., Asber, J., Saad, M., Desbiens, C.: Non-parametric short-term load forecasting. Int. J. Electr. Power Energy Syst. **29**, 630–635 (2007)

Chapter 12
Automatic Selection for Non-linear Models

Jennifer L. Castle and David F. Hendry

12.1 Introduction

It is a pleasure to contribute a chapter on non-linear model selection to a volume in honor of Peter C. Young, who has himself contributed so much to modeling, understanding and capturing key aspects of non-linearity, and to data basing the choice of which models work in a wide range of important areas in statistics, environmental studies and economics. While we do not also address his interests in forecasting, we share them strongly and have tried to advance that subject in other publications—and as a further objective, trying to establish the general approach adopted here for dynamic, non-stationary processes. We congratulate Peter on his successes to date and look forward to many more.

Economic processes are complicated entities, which are often modeled by linear approximations, leading to possible mis-specification when non-linearity matters. This chapter develops a strategy for selecting non-linear-in-variables models for cross-section data, following the automatic general-to-specific (*Gets*) multi-path

We thank participants of the Royal Economic Society Conference 2006, Econometric Society European and Australasian Meetings, 2006, *Journal of Econometrics* Conference, 2007, and the *Arne Ryde* Lectures 2007 for helpful comments and suggestions on an earlier version. Financial support from the ESRC under grant RES 051 270035 and from the Open Society Institute and the Oxford Martin School is gratefully acknowledged.

J.L. Castle (✉)
Magdalen College & Institute for New Economic Thinking at the Oxford Martin School, University of Oxford, Oxford, UK
e-mail: jennifer.castle@magd.ox.ac.uk

D.F. Hendry
Economics Department & Institute for New Economic Thinking at the Oxford Martin School, University of Oxford, Oxford, UK
e-mail: david.hendry@nuffield.ox.ac.uk

L. Wang, H. Garnier (eds.), *System Identification, Environmental Modelling, and Control System Design*,
DOI 10.1007/978-0-85729-974-1_12,

search algorithms of *PcGets* (see [28], which built on [35]), and *Autometrics* within *PcGive* (see [17], and [27]). The general properties of *Autometrics* model selection are established in [9], multiple breaks are investigated by [10], and an empirical application is provided in [30]. These properties of *Autometrics* can be summarized as follows for a linear static model. When there are K candidate variables, and k of these are relevant, then $\alpha(K-k)$ irrelevant variables will be retained on average, where α is the chosen significance level. Because it selects variables (K), rather than models (2^K), that result continues to hold even when K is greater than the sample size, N, provided $N > k$. Also, the k relevant variables will be retained with a probability close to the theoretical t-test powers determined by the non-centralities of their parameters. For example, if $K-k=100$ and $\alpha=0.01$, then one irrelevant variable will be retained on average by chance sampling, despite the plethora of candidate variables. Moreover, coefficients with |t|-values greater than about $c_\alpha = 2.6$ will be retained on average. Next, although selection only retains variables whose estimated coefficients have $|t| \geq c_\alpha$, the resulting selection bias is easily corrected, which greatly reduces the mean-square errors (MSEs) of retained irrelevant variables: see [29]. Finally, the terminal models found by *Autometrics* will be congruent (well specified), undominated reductions of the initial general unrestricted model (GUM). We will not discuss the details of the multi-path search algorithms that have made such developments feasible, as these are well covered elsewhere (see e.g., [17, 27, 28], and [19]): the reader is referred to those publications for bibliographic perspective on this exciting and burgeoning new field. The latest version of the model selection algorithm *Autometrics* is likelihood based, so can accommodate discrete variable models such as logit and probit, along with many other econometric specifications, but we focus on non-linear regression analysis here.

Thus, we investigate non-linear modeling as part of a general strategy of empirical model discovery. Commencing with a low-dimensional portmanteau test for non-linearity (see [8]), non-rejection entails remaining with a linear specification, whereas rejection leads to specifying a general non-linear, identified and congruent approximation. Next, the multi-path search procedure seeks a parsimonious, still congruent, non-linear model, and that in turn can be tested against specific non-linear functional forms using encompassing tests (see, e.g., [40], and [32]), and simplified to them if appropriate.

Since the class is one of non-linear in variables, but linear in parameters, the most obvious approach is to redefine non-linear functions as new variables (e.g., $x_i^2 = z_i$ say), so the model becomes linear but larger, and standard selection theory applies. However, non-linearity *per se* introduces five specific additional problems even in cross sections, solutions to which need to be implemented as follows.

First, determining whether there is non-linearity. The low-dimensional portmanteau test for non-linearity in [8] is applied to the unrestricted linear regression to check whether any non-linear extension is needed. Their test is related to the test for heteroskedasticity proposed by [49], but by using squares and cubics of the principal components of the linear variables, the test circumvents problems of high-dimensionality and collinearity, and is not restricted to quadratic departures. Providing there are fewer linear variables, K, than about a quarter of the sample size,

N, the test can accommodate large numbers, M_K, of potential non-linear terms, including more than N, where for a cubic polynomial:

$$M_K = K(K+1)(K+5)/6.$$

If the test does not reject, the usual *Gets* approach is applied to the linear model. Otherwise, a non-linear, or indeed non-constant, model is needed to characterize the evidence, so these possibilities must be handled jointly, as we do below.

Second, including both the linear and non-linear transformations of a variable can generate substantial collinearity, similar to slowly-varying regressors (as in [42]). Such collinearity can be problematic for estimation and selection procedures, as the information content of the extra collinear variables is small, yet disrupts existing information attribution. When the additional transformed variables are in fact irrelevant, model selection algorithms may select poorly between the relevant and irrelevant variables, depending on chance sampling. In a sense, automatic algorithms still perform adequately, as they usually keep a 'representative' of the relevant effect. Nevertheless, orthogonality is beneficial for model selection in general, both for that reason, and because deleting small, insignificant coefficients leaves the retained estimates almost unaltered. We use a simple operational de-meaning rule to eliminate one important non-orthogonality prior to undertaking model selection.

Third, non-linear functions can generate extreme outcomes, and the resulting 'fat tails' are problematic for inference and model selection, as the assumption of normality is in-built into most procedures' critical values. Non-linear functions can also 'align' with outliers, causing the functions to be retained spuriously, which can be detrimental for forecasting and policy. Thus, data contamination, outliers and non-linearity interact, so need to be treated together. To do so, we use impulse-indicator saturation (denoted IIS), which adds an indicator for every observation to the candidate regressor set (see [34], and [36]) to remove the impact of breaks and extreme observations in both regressors and regressand, and ensure near normality. Johansen and Nielsen [36] show that IIS is a robust estimation method, and that despite adding greatly to the number of variables in the search, there is little efficiency loss under the null of no contamination. In the present context, there is also a potentially large gain by avoiding non-linear terms that chance to capture unmodeled outliers, but there are always bound to be more candidate variables for selection than the sample size.

General non-linear functional approximations alone can create more variables than observations. However, building on [29], *Autometrics* already handles such situations by a combination of expanding and contracting searches (see [15]). Nevertheless, the number of potential regressors, M_K, grows rapidly as K increases:

K	1	2	3	4	5	10	15	20	30	40
M_K	3	9	19	30	55	285	679	1539	5455	12300

(12.1)

An additional exponential component adds K more to M_K, and impulse-indicator saturation (IIS) adds N more dummies for a sample of size N (below, we use more than 5000 observations). Selections of such a magnitude are now feasible, but lead to the next problem.

The fourth is the related problem of excess retention of linear and non-linear functions and indicators due to a highly over-parameterized GUM. This is controlled by implementing a 'super-conservative' strategy for the non-linear functions, where selection is undertaken at stringent significance levels to control the null rejection frequency. For example, when $M_K + K + N = 8000$ and no variables actually matter, a significance level of $\alpha = 0.001$ would lead on average to 8 irrelevant retentions, of which 5 would simply be indicators, which just dummy out their respective observations (so is 99.9% efficient). As discussed in [29] and [10], post-selection bias correction will drive the estimated coefficients of adventitiously retained variables towards the origin, leading to small mean square errors, so is not a problematic outcome from learning that 7992 of the candidate variables do not in fact matter. Thus the distribution under the null is established as retaining $\alpha(M_K + K + N - k)$ chance significant effects when k variables matter.

Finally, non-linearity comprises everything other than the linear terms, so some functional form class needs to be assigned to search across, and that is almost bound to be an approximation in practice. In a cross-section context, polynomials often make sense, so we use that as the basis class. To then implement any economic-theory based information, encompassing tests of the entailed non-linear form against the selected model can be undertaken, and this order of proceeding avoids the potential identification problems that can arise when starting with non-linear-in-parameters models (see [22]). However, we do not focus on that aspect here.

We undertake an empirical study of returns to education for US males, using 1980 census data, applying the proposed non-linear algorithm after finding strong evidence for non-linearity using the [8] test. The log-wage data are non-normal, but we use IIS to obtain an approximation to normality, adding the indicators to a general non-linear GUM, which controls for a wide range of covariates such as education, experience, ability, usual hours worked, marital status, race, etc. The non-linear selection algorithm finds a congruent model in which non-linear functions play a key role in explaining the data.

The structure of the chapter is as follows. Section 12.2 outlines the non-linear specification procedure to which a model selection algorithm such as *Autometrics* is applied, and details the non-linear functions used, related to the RETINA algorithm in [41]. Section 12.3 addresses the five intrinsic problems of selecting models that are non-linear in the regressors. First, Sect. 12.3.1 sketches the non-linearity test, then Sect. 12.3.2 demonstrates the collinearity between linear and non-linear functions, and proposes a solution by simply de-meaning all functions of variables. Third, Sect. 12.3.3 outlines the issue of non-normality, with a Monte Carlo study that highlights the problem of extreme observations for model selection, and explains the application of IIS jointly with selecting variables. Finally, Sect. 12.3.5 discusses the super-conservative strategy to ensure non-linear functions are retained only when there is definite evidence of non-linearity in the data. Section 12.4 applies the non-linear selection algorithm to a cross section of log wages, modeling the returns to education: there is strong evidence both for non-linearity and outliers that are captured by the algorithm. Finally, Sect. 12.5 concludes.

12.2 The Non-linear Algorithm

Finding a unique non-linear representation of an economic process can be formidable given the complexity of possible local data generating processes (LDGPs, namely the DGP in the space of the variables under analysis). As there are an infinite number of potential functional forms that the LDGP may take, specifying a GUM that nests the unknown LDGP is problematic. Here, we assume the LDGP is given by:

$$y_i = f(x_{1,i}, \ldots, x_{k,i}; \boldsymbol{\theta}) + \varepsilon_i \quad \text{where } \varepsilon_i \sim \mathsf{IN}\left[0, \sigma_\varepsilon^2\right], \tag{12.2}$$

for $i = 1, \ldots, N$, with $\boldsymbol{\theta} \in \mathscr{T} \subseteq \mathsf{R}^n$. Three key concerns for the econometrician are the specification of the functional form, $f(\cdot)$, the identification of $\boldsymbol{\theta}$, and the selection of the potentially relevant variables, $\mathbf{x}_i' = (x_{1,i}, \ldots, x_{k,i})$ from an available set of candidates $(x_{1,i}, \ldots, x_{K,i})$ where $K \geq k$.

The initial GUM includes all K candidates, in some non-linear form $g(\cdot)$:

$$y_i = g(x_{1,i}, \ldots, x_{K,i}; \boldsymbol{\phi}) + v_i \quad \text{where } v_i \sim \mathsf{IN}\left[0, \sigma_v^2\right]. \tag{12.3}$$

Economic theory, past empirical and historical evidence, and institutional knowledge all inform the specification of the variables in the GUM and their functional form. If the initial specification is too parsimonious, relevant variables may be omitted leading to a mis-specified final model. Theory often has little to say regarding the functional-form specification, so an approximating class is required from the infinite possibilities of non-linear functions. Many non-linear models—including smooth-transition regressions, regime-switching models, neural networks and non-linear equations—can be approximated by Taylor expansions, so polynomials form a flexible approximating class for a range of possible LDGPs.

A Taylor-series expansion of (12.3) around zero results in (see e.g., [43]):

$$\begin{aligned} g(x_{1,i}, \ldots, x_{K,i}; \boldsymbol{\phi}) = \phi_0 + \sum_{j=1}^{K} \phi_{1,j} x_{j,i} + \sum_{j=1}^{K} \sum_{l=1}^{j} \phi_{2,j,l} x_{j,i} x_{l,i} \\ + \sum_{j=1}^{K} \sum_{l=1}^{j} \sum_{m=1}^{l} \phi_{3,j,l,m} x_{j,i} x_{l,i} x_{m,i} + \cdots . \end{aligned} \tag{12.4}$$

While motivating the use of polynomial functions, (12.4) demonstrates how quickly the number of parameters increases as (12.1), shows, exacerbated when N impulse indicators are added. Polynomial functions are often used in economics because of Weierstrass's approximation theorem whereby any continuous function on a closed and bounded interval can be approximated as closely as one wishes by a polynomial, so if $x \in [a, b]$, for any $\eta > 0$ there exists a polynomial $p(x) \in [a, b]$ such that $|f(x) - p(x)| < \eta \; \forall x \in [a, b]$. However, the goodness of the approximation is unknown *a priori* in any given application, although it can be evaluated by testing against a higher-order formulation and by mis-specification tests.

A wide range of non-linear functions has been considered to approximate (12.2), including various orthogonal polynomials, such as Hermite, Fourier series, asymptotic series (see e.g., [13]), squashing functions (see [50]), and confluent hypergeometric functions (see [1]). Here, we include cubic functions, as these are sign-preserving (so could represent, say, non-linear demand or price responses), and add to the flexibility of the transformations, potentially approximating ogives. We do not include exponential components, although the most general test in [8] does. If the LDGP contains an inverse polynomial function, the polynomial will detect this form of non-linearity due to the high correlation between the variable and its inverse. Although the selected model might then be prone to misinterpretation, we consider the polynomial approximation to be an intermediate stage before testing parsimonious encompassing by a specific functional form.

Many other functional forms have been proposed in the literature: for example, RETINA (see [41]) uses the transformations (see [7]):

$$\sum_{j=1}^{K}\sum_{l=1}^{K}\beta_{j,l}x_{j,i}^{\lambda_1}x_{l,i}^{\lambda_2} \quad \text{for } \lambda_1, \lambda_2 = -1, 0, 1. \tag{12.5}$$

Although we exclude inverses, squared inverses, and ratios due to their unstable behavior potentially creating outliers, and adequate correlations with levels (12.4) includes the remaining terms. Also, for example, logistic smooth transition models (LSTAR: see e.g., [48]) will be approximated by the third-order Taylor expansion given by (12.4). Thus, (12.4) approximates or nests many non-linear specifications.

While (12.4) already looks almost intractable, the inclusion of more variables than observations does not in fact make it infeasible for an automatic algorithm, enabling considerable flexibility when examining non-linear models despite the number of potential regressors being large. When $N > K$, the *Gets* approach is to specify a GUM that nests the LDGP in (12.2), to ensure the initial formulation is congruent. As $K > N$, both expanding and contracting searches are required, and congruence can only be established after some initial simplification to make it feasible to estimate the remaining model. Here, we propose using the general formulation:

$$\begin{aligned} y_i = \phi_0 &+ \sum_{j=1}^{K}\phi_{1,j}x_{j,i} + \sum_{j=1}^{K}\sum_{l=1}^{j}\phi_{2,j,l}x_{j,i}x_{l,i} \\ &+ \sum_{j=1}^{K}\sum_{l=1}^{j}\sum_{m=1}^{l}\phi_{3,j,l,m}x_{j,i}x_{l,i}x_{m,i} \\ &+ \sum_{j=1}^{N}\delta_j 1_{\{j=i\}} + u_i \end{aligned} \tag{12.6}$$

with K potential linear regressors, $\mathbf{x}_i$, where $1_{\{j=i\}}$ is an indicator for the ith observation.

12.3 Problems When Selecting Non-linear Models

There are five problems that arise when selecting from a GUM that consists of a large set of polynomial regressors as in (12.6). These problems include first detecting non-linearity (Sect. 12.3.1), reducing collinearity (Sect. 12.3.2), handling non-normality (Sect. 12.3.3) leading to more variables than observations (Sect. 12.3.4), and avoiding potential excess retention of irrelevant regressors (Sect. 12.3.5). Solutions to all of these problems are now proposed, confirming the feasibility of our non-linear model selection strategy.

12.3.1 Testing for Non-linearity

The LDGP in (12.2) has k relevant and $K - k$ irrelevant variables when $f(\cdot)$ is linear. The first stage is to apply the test for non-linearity in [8] to see if it is viable to reduce (12.6) directly to:

$$y_i = \sum_{j=1}^{K} \beta_j x_{j,i} + \sum_{j=1}^{N} \delta_j 1_{\{j=i\}} + e_i. \tag{12.7}$$

If outliers are likely to be problematic, IIS could first be applied to (12.7) to ascertain any major discrepancies, leading to say r indicators being retained (see Sect. 12.3.4):

$$y_i = \sum_{j=1}^{K} \beta_j x_{j,i} + \sum_{j=1}^{r} \delta_j 1_{\{j=i\}} + e_i. \tag{12.8}$$

When $\mathbf{x}_i$ denotes the set of linear candidate regressor variables, to calculate their principal components, denoted $\mathbf{z}_i$, define $\mathbf{H}$ and $\mathbf{B}$ as the eigenvectors and eigenvalues of $N^{-1}\mathbf{X}'\mathbf{X}$, such that:

$$\mathbf{z}_i = \mathbf{B}^{-\frac{1}{2}}\left[(\mathbf{H}'\mathbf{x}_i) - \overline{(\mathbf{H}'\mathbf{x}_i)}\right]. \tag{12.9}$$

Let $z_{j,i}^2 = w_{j,i}$ and $z_{j,i}^3 = s_{j,i}$, then the test for non-linearity is the F-test of H_0: $\boldsymbol{\beta}_2 = \boldsymbol{\beta}_3 = 0$ in:

$$y_i = \beta_0 + \boldsymbol{\beta}_1'\mathbf{x}_i + \boldsymbol{\beta}_2'\mathbf{w}_i + \boldsymbol{\beta}_3'\mathbf{s}_i + \sum_{j=1}^{r} \delta_j 1_{\{j=i\}} + \varepsilon_i, \tag{12.10}$$

where $r = 0$ if IIS is not first applied. If the F-test does not reject, the GUM is taken to be linear, and the usual selection algorithm is applied to select the relevant regressors. Conversely, if the test rejects, non-linearity is established at the selected significance level, so the remaining four problems need resolving for a viable approach. If IIS was not applied, non-linearity is only contingently established, as it may be proxying outliers as Sect. 12.3.3 shows.

12.3.2 Collinearity

Multicollinearity was first outlined by [20] within the context of static general-equilibrium linear relations. Confluence analysis was developed to address the problem, although that method is not in common practice now (see [31]). The definition of collinearity has shifted over the years, but for an $N \times K$ regressor matrix $\mathbf{X}$, we can define perfect collinearity as $|\mathbf{X}'\mathbf{X}| = 0$, and perfect orthogonality as a diagonal $(\mathbf{X}'\mathbf{X})$ matrix. Since collinearity is not invariant under linear transformations, it is difficult to define a 'degree of collinearity', as a linear model is equivariant under linear transformations, and so the same model could be defined by various isomorphic representations, which nevertheless deliver very different inter-correlations. Hence, collinearity is a property of the parametrization of the model, and not the variables *per se*. Moreover, $|\mathbf{X}'\mathbf{X}| = 0$ whenever $N > K$ anyway.

Nevertheless non-linear transformations can generate substantial collinearity between the linear and non-linear functions. We consider a simple case in which we add the irrelevant transformation $f(w_i) = w_i^2$ to a linear model in w_i. This polynomial transform is common in economics: see Sect. 12.4 for an empirical application. The degree of collinearity varies as the statistical properties of the process vary: collinearity between w_i and w_i^2 is zero when $\mathsf{E}[w_i] = 0$, but dramatically increases to almost perfect collinearity as $\mathsf{E}[w_i] = \mu$ increases. To see that, consider the DGP given by the linear conditional relation:

$$y_i = \beta w_i + e_i = 0 + \beta w_i + 0 w_i^2 + \varepsilon_i, \tag{12.11}$$

where $\varepsilon_i \sim \mathsf{IN}[0, \sigma_\varepsilon^2]$ with $i = 1, \ldots, N$, and:

$$w_i \sim \mathsf{IN}\left[0, \sigma_w^2\right]. \tag{12.12}$$

Since (12.11) is equivariant under linear transformations, in that both the dependent variable and the error process are unaffected, it can also be written for $z_i = w_i + \mu$ as:

$$\begin{aligned} y_i &= -\beta\mu + \beta(w_i + \mu) + 0(w_i + \mu)^2 + \varepsilon_i \\ &= -\beta\bar{z} + \beta z_i + 0 z_i^2 + \varepsilon_i \\ &= 0 + \beta(z_i - \bar{z}) + 0(z_i - \bar{z})^2 + \varepsilon_i. \end{aligned} \tag{12.13}$$

Correspondingly, there are three models, namely, the original zero-mean case:

$$y_i = \beta_0 + \beta_1 w_i + \beta_2 w_i^2 + u_i \tag{12.14}$$

the non-zero-mean case:

$$y_i = \gamma_0 + \gamma_1 z_i + \gamma_2 z_i^2 + u_i \tag{12.15}$$

and the transformed zero-mean case:

$$y_i = \lambda_0 + \lambda_1 z_i + \lambda_2\left(z_i^2 - \overline{z^2}\right) + u_i, \tag{12.16}$$

where $\overline{z^2}$ is the sample mean of z_i^2.

First, letting $\mathbf{X}$ denote the general regressor matrix, for (12.15) with a non-zero mean:

$$\mathsf{E}\left[N^{-1}\mathbf{X}'\mathbf{X}_{(\mu)}\right] = \mathsf{E}\left[\begin{pmatrix} 1.0 & \overline{z} & \overline{z^2} \\ \overline{z} & N^{-1}\sum z_i^2 & N^{-1}\sum z_i^3 \\ \overline{z^2} & N^{-1}\sum z_i^3 & N^{-1}\sum z_i^4 \end{pmatrix}\right]$$

$$= \begin{pmatrix} 1.0 & \mu & \mu^2+\sigma_w^2 \\ \mu & \mu^2+\sigma_w^2 & \mu^3+3\mu\sigma_w^2 \\ \mu^2+\sigma_w^2 & \mu^3+3\mu\sigma_w^2 & 3\sigma_w^4+\mu^4+6\mu^2\sigma_w^2 \end{pmatrix} \tag{12.17}$$

with the inverse:

$$\left(\mathsf{E}\left[N^{-1}\mathbf{X}'\mathbf{X}_{(\mu)}\right]\right)^{-1} = \frac{1}{2\sigma_w^6}\begin{pmatrix} \mu^4\sigma_w^2+3\sigma_w^6 & -2\mu^3\sigma_w^2 & \mu^2\sigma_w^2-\sigma_w^4 \\ -2\mu^3\sigma_w^2 & 2\sigma_w^4+4\mu^2\sigma_w^2 & -2\mu\sigma_w^2 \\ \mu^2\sigma_w^2-\sigma_w^4 & -2\mu\sigma_w^2 & \sigma_w^2 \end{pmatrix}. \tag{12.18}$$

There is substantial collinearity between the variables, except for the squared term, which is irrelevant in the DGP. As μ—an incidental parameter here—increases, $\mathsf{E}[N^{-1}\mathbf{X}'\mathbf{X}_{(\mu)}]$ tends towards singularity, and for $\sigma_w^2 = 1$, the ratio R of the largest to the smallest eigenvalues in (12.18) grows dramatically from $R = 5.83$ when $\mu = 0$ through $R = 60223$ for $\mu = 4$ to $R = 5.6 \times 10^7$ when $\mu = 10$. Note that age enters some regressions below, often with a mean above 20.

Next, in the zero-mean model in (12.14):

$$\mathsf{E}\left[N^{-1}\mathbf{X}'\mathbf{X}_{(0)}\right] = \mathsf{E}\left[\begin{pmatrix} 1.0 & \overline{w} & \overline{w^2} \\ \overline{w} & N^{-1}\sum w_i^2 & N^{-1}\sum w_i^3 \\ \overline{w^2} & N^{-1}\sum w_i^3 & N^{-1}\sum w_i^4 \end{pmatrix}\right] = \begin{pmatrix} 1.0 & 0.0 & \sigma_w^2 \\ 0.0 & \sigma_w^2 & 0.0 \\ \sigma_w^2 & 0.0 & 3\sigma_w^4 \end{pmatrix} \tag{12.19}$$

so the inverse is:

$$\left(\mathsf{E}\left[N^{-1}\mathbf{X}'\mathbf{X}_{(0)}\right]\right)^{-1} = \frac{1}{2\sigma_w^6}\begin{pmatrix} 3\sigma_w^6 & 0 & -\sigma_w^4 \\ 0 & 2\sigma_w^4 & 0 \\ -\sigma_w^4 & 0 & \sigma_w^2 \end{pmatrix}. \tag{12.20}$$

There is no collinearity between w_i and w_i^2 although there is an effect on the intercept, but this does not cause a problem for either estimation or a selection algorithm.

Finally, in the transformed zero-mean model in (12.16):

$$\left(\mathsf{E}\left[N^{-1}\mathbf{X}'\mathbf{X}_{(0,0)}\right]\right)^{-1} = \frac{1}{3\sigma_w^6}\begin{pmatrix} 3\sigma_w^6 & 0.0 & 0.0 \\ 0.0 & 3\sigma_w^4 & 0.0 \\ 0.0 & 0.0 & \sigma_w^2 \end{pmatrix}. \tag{12.21}$$

Thus, a near orthogonal representation can be achieved simply by taking deviations from means, which re-creates the specification in terms of the original variables w_i and w_i^2 as $z_i = w_i + \mu$ where $\mathsf{E}[\overline{z}] = \mu$ and $\mathsf{E}[\overline{z^2}] = \mu^2 + \sigma_w^2$:

$$\mathsf{E}\left[N^{-1}\mathbf{X}'\mathbf{X}_{(\overline{\mu})}\right] = \begin{pmatrix} 1.0 & 0.0 & 0.0 \\ 0.0 & \sigma_w^2 & 2\mu\sigma_w^2 \\ 0.0 & 2\mu\sigma_w^2 & 3\sigma_w^6 - \sigma_w^4 + 4\mu^2\sigma_w^2 \end{pmatrix} \tag{12.22}$$

with the inverse:

$$\left(\mathsf{E}\left[N^{-1}\mathbf{X}'\mathbf{X}_{(\overline{\mu})}\right]\right)^{-1} = \frac{1}{\sigma_w^6(3\sigma_w^2 - 1)} \times \begin{pmatrix} 3\sigma_w^8 - \sigma_w^6 & 0.0 & 0.0 \\ 0.0 & 3\sigma_w^6 - \sigma_w^4 + 4\mu^2\sigma_w^2 & -2\mu\sigma_w^2 \\ 0.0 & -2\mu\sigma_w^2 & \sigma_w^2 \end{pmatrix}. \tag{12.23}$$

Taking deviations from sample means delivers a reduction in collinearity, which is particularly marked for the intercept, but worse for the linear term $(z_i - \overline{z})$. Again the irrelevant squared term 'benefits'. To remove the collinearity, first de-mean z_i, then also de-mean z_i^2. The linear term remains $(z_i - \overline{z})$, but the squared term becomes $(z_i - \overline{z})^2 - [\mathsf{E}(z_i - \overline{z})]^2$ which will result in a model that is identical to (12.16). Double de-meaning thus removes the collinearity generated by the non-zero mean, and Monte Carlo evidence confirms this is an effective solution to mean-induced collinearity.

A non-linear selection strategy should automatically double de-mean the generated polynomial functions prior to formulating the GUM. Two caveats apply. First, the orthogonalizing rules will not remove all collinearity between higher-order polynomials. We considered orthogonalizing using the Choleski method (see [45]), but double de-meaning removed enough collinearity to ensure the *Autometrics* selection had the appropriate properties. Second, any information contained in the intercepts of the explanatory variables will be removed, although there is rarely a theory of the intercept when developing econometric models, especially for cross-section data.

12.3.3 Non-normality

Normality is a central assumption for inference, as conventional critical values tend to be used, so null rejection frequencies would be incorrect for non-normality. Normality tends to be even more vital for selection, when many decisions are made. In non-linear models, normality is essential, as problems arise when fat-tailed distributions result in extreme observations, as there is an increased probability that non-linear functions will align with extreme observations, effectively acting as indicators and therefore being retained too often (see e.g., [11]).

We now show by a Monte Carlo example that non-normal variables pose similar problems. Consider these DGPs for four variables:

$$x_{i,t} = \varepsilon_{i,t}, \quad \varepsilon_{i,t} \sim \mathsf{IN}[0,1] \quad \text{for } i = 1, \ldots, 4. \tag{12.24}$$

We generate non-linear functions given by the inverses of these normal distributions (as in RETINA):

$$x_{i,t}^{-1} = \frac{1}{x_{i,t}}. \tag{12.25}$$

The GUM contains twenty irrelevant variables given by:

$$x_{1,t}^{-1} = \rho_0 + \sum_{i=1}^{4} \rho_i x_{1,t-i}^{-1} + \sum_{j=2}^{4} \sum_{m=0}^{4} \rho_{j,m} x_{j,t-m}^{-1} + \varepsilon_t. \tag{12.26}$$

Then selecting from (12.26) leads to |t|-values as large as 19 for variables with zero non-centralities. Such a variable would unequivocally, but incorrectly, be retained as a DGP variable. On average, two of the twenty irrelevant regressors are retained at the 1% significance level. This implies that a fat-tailed distribution would have a null rejection frequency of 10% at the 1% significance level. If the dependent variable is $x_{i,t}$ rather than $x_{i,t}^{-1}$, the retention probabilities are correct as normality results. Non-normal errors can also pose a similar problem (see [10]). Hence, the problem of model selection is exacerbated by the inclusion of non-linear functions, such as inverses, which generate extreme observations.

12.3.4 Impulse-Indicator Saturation

[34] propose the use of impulse-indicator saturation to detect and remove outliers and breaks, utilizing the fact that *Autometrics* can handle more variables than observations. Here the aim is to ensure that the selection process will not overly favor non-linear functions that chance to capture outliers. The modeling procedure generates impulse indicators for every observation, $1_{\{i=s\}}\ \forall s$. The indicators are divided into J subsets, which form the initial GUMs (including an intercept) and *Autometrics* selects the significant indicators from each subset, which are then stored as terminal models. The joint model is formulated as the union of the terminal models and *Autometrics* re-selects the indicators. Under the null that there are no outliers, αN indicators will be retained on average for a significance level α. Johansen and Nielsen [36] show that the cost of testing for the significance of N indicators under the null is low for small α: for example, when $\alpha = 1/N$, only one observation is 'removed' on average. Also, [10] show that IIS alleviates fat-tailed draws, and allows near-normal inference, important both during search and for the post-selection bias correction which assume normality.

Impulse-indicator saturation also overcomes the problem of 'undetectable' outliers. One concern with non-linearity is that it is difficult to distinguish between extreme observations that are outliers or data contamination and extreme observations that are due to the non-linearity in the data. Non-linear functions can 'hide' outliers by fitting to the extreme values, or conversely, methods that remove extreme observations could be in danger of removing the underlying non-linearity that should be modeled. IIS avoids this problem by including all potentially relevant variables as well as indicators for all observations in the initial GUM, effectively applying IIS to the residuals of the model as opposed to the dependent variable itself. Removing the extreme observations in conjunction with selecting the non-linear functions avoids both problems of removing observations that generate the non-linearity and finding spurious non-linearity that merely captures outliers.

In fact the empirical example does not carry out the strategy precisely as proposed here because the distributions transpired to be so highly non-normal, specifically very badly skewed. Since there were more variables (including indicators) than observations, initial selection inferences based on subsets of variables could be distorted by that skewness. Thus, we added a stage of pre-selecting indicators to 'normalize' the dependent variable. Johansen and Nielsen [36] show the close relationship of IIS to robust statistics: both can handle data contamination and outliers, and IIS appears to be a low cost way of doing so. Thus, in the spirit of robust statistics, we sought the sub-sample that would be near normal, representing the most discrepant observations by indicators rather than dropping them, so this was only a transient stage. Those indicators are then retained as if they were additional regressors. If the indicators are essential, then better initial selection inferences will ensue, and if they really are not needed, as there were no outliers after the non-linear terms were included, then they should drop out during selection.

12.3.5 Super-conservative Strategy

Irrelevant non-linear functions that are adventitiously retained are likely to be detrimental to modeling and forecasting, making such models less robust than linear models, by 'amplifying' changes in collinearity between regressors (see e.g., [12]), and location shifts within the equation or in any retained irrelevant variables. Hence, non-linear functions should only be retained if there is strong evidence. Given the possible excess retention of irrelevant functions due to the large number of potential non-linear functions in the candidate set, much more stringent critical values must be used for the non-linear, than linear, functions during multi-path searches. Critical values should also increase with the number of functions included in the model, and with the sample size, although as with all significance levels, the choice can also depend on the preferences of the econometrician and the likely uses of the resulting model. Parsimonious encompassing of the feasible GUM by the final selected model helps control the performance of the selection algorithm: see [16].

A potential problem could arise if the selection procedure eliminated all non-linear functions, contradicting the results of the non-linearity test: it is feasible that the ellipsoid for a joint test at a looser significance level does not include the origin, whereas the p-value hyper-square from individual tests at a tighter significance level does. This can be avoided by then repeating the multi-stage strategy with tests undertaken at consecutively looser significance levels. Rules for the super-conservative strategy could be similar to those implemented for the Schwarz information criterion (see [5]), so the selection strategy should deliver an undominated, congruent, specific non-linear model that parsimoniously encompasses the feasible GUM.

We have now resolved the main problems likely to distort selection for a non-linear model, relative to what is known about its performance in linear settings, so now apply the approach in *Autometrics* to empirically modeling the returns to education.

12.4 Empirical Application: Returns to Education

A natural application of the non-linear algorithm is returns to education. The literature is replete with empirical studies: see, *inter alia*, [21, 24] and [3]. We focus on a one-factor model, where education is summarized as a single measure defined by years of schooling, in keeping with the homogeneous returns literature of [23] and [6]. We do not allow for unobserved heterogeneity, capturing heterogeneity through the conditioning variables, following [14]. There are a range of estimation procedures commonly used, including instrumental variables, control functions and matching methods (see [4], for an overview), all of which have been developed to mitigate the biases induced by least-squares estimation. There are 3 sources of biases in a least-squares regression of wage on schooling:

(i) the ability bias, where there is a correlation between the length of schooling and an individual's inherent, but unobserved, ability;
(ii) the returns bias, where the marginal return is correlated with the length of schooling; and
(iii) measurement-error bias due to incorrect measurement of the schooling variable.

In our simple one-factor model, these biases are likely to be small, and [6] argues that there is some evidence that the biases balance out, resulting in near consistent OLS estimates of the returns' coefficient. In order to reduce the biases it is important to include many control variables that can capture omitted factors. Since the functional forms cannot be deduced from theory in this context, a non-linear model must be postulated and so an automatic selection algorithm is a natural tool to use.

We use data from the 1980 US census, based on a random draw of 0.01% of the population of US males in employment, resulting in 5173 observations. Wage income has been top coded at $75,000, resulting in 204 observations that are truncated. Figure 12.1 records the density and distribution of log wages (w_i) with their Gaussian reference counterparts. Normality is strongly rejected for w as $\chi^2(2) = 1018.0^{**}$, with substantial skewness in the left tail. Many studies have

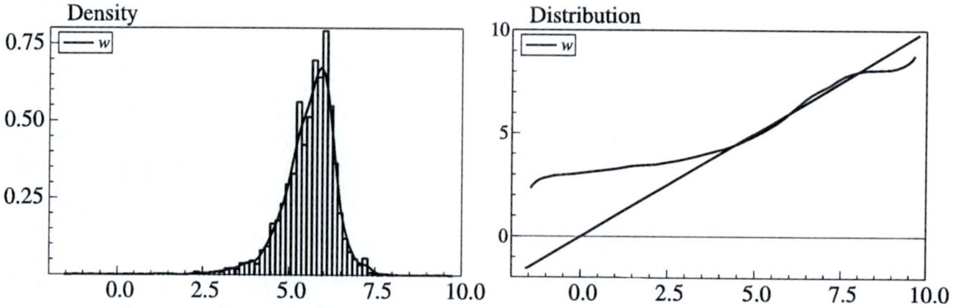

Fig. 12.1 Distribution of log wages

Table 12.1 Potential explanatory variables

Variable	Label	Definition	Mean	Variance	Min	Max
Wage	*w*	logs	5.58	0.59	−1.24	9.47
Experience	*exp*	Age-years education-6 (÷10)	1.82	1.81	−0.3	5.7
Education	*edu*	Grade completed (21 categories) (÷10)	1.26	0.01	0	2
Usual hours worked	*hrs*	Log ave. hours worked in 1979	3.70	0.11	0	4.6
Metropolitan status	*met*	City/rural (5 categories)	–	–	0	1
Race	*race*	(9 categories)	–	–	0	1
State	*state*	FIPS code (62 categories)	–	–	0	1
No. of own children	*child*	in household	1.01	1.69	0	9
Marital status	*mar*	(6 categories)	–	–	0	1
Educational attainment	*attain*	(9 categories)	6.97	3.12	1	9

considered alternative distributions to the log-normal including the Pareto, Champernowne and inverse Gaussian: see [25, 37, 47] and [2]. Instead, we apply IIS as outlined in Sect. 12.3.4. Table 12.1 records summary statistics for wages and the covariates.

12.4.1 Fitting the Theory Model

The standard reduced-form model of returns to education is the Mincer regression [38, 39]:

$$w_i = \beta_0 + \beta_1 edu_i + \beta_2 exp_i + \beta_3 exp_i^2 + u_i, \tag{12.27}$$

where β_1 measures the 'rate of return to education' which is assumed to be the same for all education levels, and $\mathsf{E}[u_i|edu_i, exp_i] = 0$. In practice, conditioning on additional covariates reduces the impact of omitted variable bias. Here, the results for the augmented Mincer regression are:

$$\widehat{w}_i = \underset{(0.25)}{2.74} + \underset{(0.081)}{0.624} edu_i + \underset{(0.029)}{0.436} exp_i - \underset{(0.006)}{0.066} exp_i^2 + \underset{(0.014)}{0.002} attain_i$$
$$+ \underset{(0.028)}{0.404} hrs_i + \underset{(0.009)}{0.019} child_i + 47 \text{ dummies}, \tag{12.28}$$
$$\mathsf{R}^2 = 0.308, \quad \widehat{\sigma} = 0.645, \quad \chi^2(2) = 2065.7^{**}, \quad SC = 1.997,$$
$$N = 5173, \quad \mathsf{F}_{\mathsf{het}}(58, 5114) = 2.7^{**}, \quad \mathsf{F}_{\mathsf{reset}}(2, 5117) = 0.074.$$

In (12.28), R^2 is the squared multiple correlation, $\widehat{\sigma}$ is the residual standard deviation, SC is the Schwarz criterion (see [46]), and coefficient standard errors are shown in parentheses. The diagnostic tests are of the form $\mathsf{F_j}(k, T - l)$ which denotes an approximate F-test against the alternative hypothesis j for: heteroskedasticity ($\mathsf{F_{het}}$: see [49]) and the RESET test ($\mathsf{F_{reset}}$: see [44]); and a chi-square test for normality ($\chi^2_{\mathsf{nd}}(2)$: see [18]). * and ** denote rejection at 5% and 1% respectively.

The model shows a positive *ex post* average rate of return to education of 6% which is broadly in line with the Mincer regression results in [26, Table 2] although these are slightly higher at 10–13% as they consider separate regressions for blacks and whites, whereas we take a random sample of the population and condition on 3 continuous dummies (*hrs*, *child*, *attain*) and 47 0/1 dummies (*met*, *race*, *state*, *mar*) to control for omitted variable bias. The economic theory leads to a relatively poor fit ($R^2 = 31\%$), and does not capture well the behavior of the observed data as the model fails mis-specification tests for normality and heteroskedasticity. Despite poor model specification, the elasticity signs are 'correct', with positive returns to education and experience and an earnings profile that is concave with a significant negative estimated coefficient for experience squared ($\mathsf{t}_{exp^2} = -10.5$).

12.4.2 Theory Equation with IIS

Given the poor performance of the theory model, and the highly significant non-normality test statistic, we next introduce IIS into the specification, using a 0.001 significance level. The resulting model is:

$$\widehat{w}_i = \underset{(0.095)}{2.44} + \underset{(0.062)}{0.641} edu_i + \underset{(0.022)}{0.470} exp_i - \underset{(0.005)}{0.072} exp_i^2 - \underset{(0.011)}{0.006} attain_i$$
$$+ \underset{(0.023)}{0.473} hrs_i + \underset{(0.007)}{0.0055} child_i + 10 \text{ dummies} + 114 \text{ indicators}, \tag{12.29}$$
$$\mathsf{R}^2 = 0.597, \quad \widehat{\sigma} = 0.496, \quad \chi^2(2) = 249.8^{**}, \quad SC = 1.725,$$
$$N = 5173, \quad \mathsf{F}_{\mathsf{het}}(21, 5037)^{**} = 9.60, \quad \mathsf{F}_{\mathsf{reset}}(2, 5040) = 3.34^{*}.$$

IIS does not remove the heteroskedasticity found in (12.28) (note that the test for heteroskedasticity excludes the indicators from the variable set), which suggests that

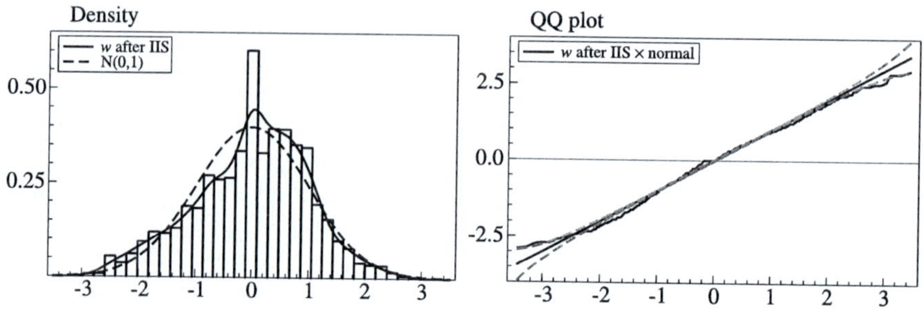

Fig. 12.2 Log wages adjusted for extreme observations

an alternative functional form should be sought. The RESET test indicates that there is functional form mis-specification at the 5% significance level; we will see if we can improve on the functional-form specification in Sect. 12.4.3.[1] The normality test still fails, but the statistic value is vastly reduced. At a significance level of 0.1%, with 5173 observations, 5 variables will be retained on average under the null, and t-statistics of approximately 3.3 or greater would be retained under normality. Autometrics finds 114 indicators (less than 2% of observations) and this greatly reduces non-normality. The test is only an indication, as there is a mass at zero due to the indicators, although [33] show that forming indexes of the indicators can avoid this problem. Figure 12.2 records the density and QQ plot of log wages once the indicators have been included: there is some deviation from the normal distribution in the tails with the distribution falling outside the pointwise asymptotic 95% standard error bands.

We also applied IIS at $\alpha = 0.05\%$ and $\alpha = 0.01\%$, which would imply that under the null of no outliers we would retain 2.5 and 0.5 of an indicator on average. The resulting Mincer regressions are similar to (12.29) with 58 and 17 indicators retained.

12.4.3 *Non-linear Models*

In this section, we extend the Mincer regression in (12.27) to allow for non-linearities that may enter other than through the experience squared term. We apply the non-linear algorithm presented in Sect. 12.2, first without IIS and then with IIS to assess the importance of removing outliers.

12.4.3.1 Testing Non-linearity

The first stage of the algorithm is to test for non-linearity using the test proposed by [8]. Here $\mathbf{x}_i' = \{exp_i, edu_i, hrs_i, child_i, attain_i\}$, so the regressors are a combi-

[1] p-values shown in brackets.

nation of variables and continuous dummies with very different ranges, but principal components standardize the linear combinations. We apply IIS to the linear model in which we fix the linear regressors in the model, i.e. do not select over them, and apply model selection to the impulse-indicators (including discrete dummies), which is equivalent to applying IIS to the residuals after conditioning on the linear regressors. We retain $r = 114$ indicators ($\mathsf{F}(114, 5042) = 32.55[0.00]^{**}$). We then compute the non-linearity test (12.10) based on (12.9). The test statistic, $\mathsf{F}(10, 5049) = 1293.2[0.00]^{**}$, strongly rejects the null hypothesis of linearity. Given the strong evidence for a squared experience term in (12.28) and (12.29), the test may seem redundant, but we wish to illustrate the general approach in action. In many applications, theory does not provide such a direct non-linear functional-form specification, so there is value in confirming the need for a non-linear specification in advance of model selection to avoid over-parameterizing the GUM with non-linear functions when they are not required.

12.4.3.2 Modeling Non-linearity Without IIS

We form the non-linear GUM given by (12.6), but excluding the impulse indicators, which results in 132 regressors (we exclude non-linear functions of the discrete dummy variables, computing 2nd and 3rd powers of *edu*, *exp*, *hrs*, *child* and *attain*). The resulting model nests the Mincer regression (12.28). All functions are double de-meaned as in Sect. 12.3.2. The GUM equation standard error is $\widehat{\sigma}_{GUM} = 0.640$. Selection is undertaken using *Autometrics* at the 0.1% significance level, and (12.30) reports the selected model.

$$\begin{aligned}\widehat{w}_i = {} & \underset{(0.21)}{3.43} + \underset{(0.129)}{0.817} edu_i^2 + \underset{(0.010)}{0.190} exp_i - \underset{(0.007)}{0.071} exp_i^2 + \underset{(0.053)}{0.26}\, hrs_i \\ & - \underset{(0.051)}{0.17}\, hrs_i^2 - \underset{(0.021)}{0.115} exp_i \times hrs_i + \underset{(0.012)}{0.053} child_i - \underset{(0.004)}{0.013} child_i^2 \\ & + \underset{(0.013)}{0.099} attain_i + 9 \text{ dummies} + 6 \text{ cubics}, \end{aligned} \tag{12.30}$$

$$\mathsf{R}^2 = 0.314, \quad \widehat{\sigma} = 0.641, \quad \chi^2(2) = 2218.1^{**}, \quad SC = 1.983,$$

$$N = 5173, \quad \mathsf{F}_{\mathsf{het}}(36, 5136) = 4.27^{*}, \quad \mathsf{F}_{\mathsf{reset}}(2, 5146) = 2.34.$$

Experience and experience squared are retained with the correct signs and are highly significant. Education is no longer significant, but its square is, as are hours squared and child squared, and an interaction of experience and hours, as well as 6 cubic terms, possibly representing a problem of over-fitting when outliers are not accounted for. There is little improvement in fit compared to (12.28), but again the model fails the diagnostic tests apart from RESET and selection using critical values based on the normal distribution is clearly violated. We next consider a model that includes both the non-linear functions and IIS.

12.4.3.3 Modeling Non-linearity with IIS

The previous regressions demonstrate that both augmenting the Mincer regression with additional non-linear functions and applying IIS to account for outliers are necessary but insufficient steps on their own in developing a theory-consistent model that also captures the key characteristics of the data. Instead of applying both jointly, we add a preliminary step in which IIS is first applied by itself to the linear model (12.7) to eliminate the most extreme observations: from Sect. 12.4.3.1 we find $r = 114$ indicators. Johansen and Nielsen [36] show that under the null, impulse-indicator saturation can be applied to any asymmetric distribution as long as the first four moments exist, and the distribution satisfies some smoothness properties. The reason for this preliminary stage, as opposed to the simultaneous application of IIS and selection of non-linear functions (as recommended above to overcome the problem of extreme observations), is that by obtaining a reasonable first approximation to normality, conventional critical values are then applicable throughout the selection process, which perforce includes both expanding as well as the usual contracting searches as all variables cannot be included in the GUM from the outset. By selecting over the indicators again in the non-linear GUM, the problem of extreme observations is overcome, and this second stage can be undertaken at looser significance levels as the procedure will involve fewer variables than observations.

Augmenting the GUM in Sect. 12.4.3.2 with the 114 impulse indicators results in 217 regressors in the initial GUM. The GUM equation standard error is $\widehat{\sigma}_{GUM} = 0.487$, which is only slightly smaller than (12.29), although an F-test of the reduction to (12.29) (excluding indicators) is rejected ($\mathsf{F}(87, 4956) = 5.78[0.00]^{**}$). Selection is undertaken using *Autometrics* at the 0.1% significance level, and equation (12.31) reports the selected model, with Fig. 12.3 recording the residual density and residual QQ plot.

$$\begin{aligned}
\widehat{w}_i = {} & \underset{(0.17)}{3.12} + \underset{(0.106)}{0.705} edu_i^2 + \underset{(1.83)}{5.375} edu_i^3 + \underset{(0.007)}{0.196} exp_i - \underset{(0.005)}{0.081} exp_i^2 - \underset{(0.010)}{0.049} edu_i \times exp_i^2 \\
& + \underset{(0.014)}{0.076} exp_i^2 \times hrs_i + \underset{(0.041)}{0.372} hrs_i - \underset{(0.041)}{0.12}\, hrs_i^2 - \underset{(0.016)}{0.11}\, hrs_i^3 \\
& + \underset{(0.014)}{0.048} hrs_i^2 \times child_i + \underset{(0.010)}{0.097} attain_i - \underset{(0.013)}{0.068} attain_i^3 - \underset{(1.11)}{4.141} edu_i^2 \times attain_i \\
& + \underset{(0.216)}{0.986} edu_i \times attain_i^2 + \underset{(0.015)}{0.063} exp_i \times hrs_i \times child_i \\
& + \underset{(0.009)}{0.032} exp_i \times hrs_i \times attain_i + 112 \text{ indicators} + 11 \text{ dummies},
\end{aligned}$$

$$\begin{aligned}
& \mathsf{R}^2 = 0.609, \quad \widehat{\sigma} = 0.489, \quad \chi^2_{\mathsf{nd}}(2) = 251.2^{**}, \quad SC = 1.646, \\
& N = 5173, \quad \mathsf{F}_{\mathsf{het}}(42, 5019) = 5.85^{**}, \quad \mathsf{F}_{\mathsf{reset}}(2, 5031) = 1.54.
\end{aligned} \tag{12.31}$$

17 explanatory variables are retained from the 103. Also, 11 dummies an 112 indicators are retained, picking up most of the left-tail skewness. The model still does not pass all diagnostics partly due to the large number of indicators putting a

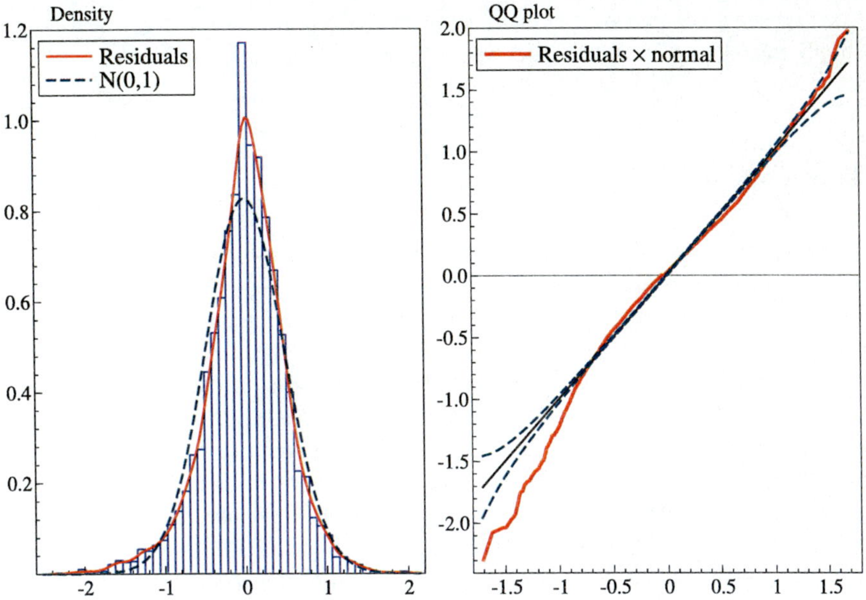

Fig. 12.3 Non-linear wage model with IIS: residual density and residual QQ plot

mass at the origin, and partly due to some residual skewness in the tails: Fig. 12.3 records the QQ plot with 95% pointwise standard error bands around the normal and there are significant deviations in the tails. Education enters as a quadratic and cubic, and experience as a level and quadratic, indicating strong non-linearity, as many authors have found when including age and age-squared terms. Characteristics such as usual hours worked, attainments, and the number of children also help explain wages, with some strong interactions and non-linear terms. Some effects enter with opposite signs on the level and quadratic term suggesting concave functions. The equation standard error is similar to the GUM: the parsimonious encompassing test of the specific model against the GUM is $\mathsf{F}(77, 4956) = 0.879$, so a valid reduction has been undertaken.

Double de-meaning was important: the correlation between *exp* and exp^2 was 0.974, but after double de-meaning the correlation was reduced to -0.327. Impulse-indicator saturation was also needed to obtain near-normality for selection and inference. Finally, tight significance levels were vital to prevent excess retention of irrelevant variables.

Although we do not have a substantive functional form specification deduced from a prior theory to test as an encompassing reduction here, the logic thereof is fairly clear. Adding such a functional form to (12.31) should eliminate many of the selected non-linear terms in favor of the theory-based form, thereby delivering a more robust, identified, interpretable and parsimonious form that does not impugn the congruence of the model or its parsimonious encompassing of the initial GUM, and indeed could even improve the fit while reducing the number of parameters.

Equally, such a theory-based function might not remove all the non-linearity, so simply imposing it from the outset would have led to a poorer final model.

12.5 Conclusion

This chapter develops a strategy for the selection of non-linear models, designed to be embedded within the automatic model selection algorithm of *Autometrics*. First, a general unrestricted model (GUM), including all potential variables that are thought to explain the phenomenon of interest, is formulated and estimated. Next, a test of linearity is applied to that initial approximation. If the null is accepted, standard selection procedures are applied to the linear GUM. If the null is rejected, a non-linear functional form is generated using polynomial transformations of the regressors in which all functions are double de-meaned prior to inclusion in the GUM to remove one potential collinearity. A set of N impulse indicators is also generated for a sample of size N, and included in the GUM to remove outliers and data contamination concurrently with selection of the specific model. Above, because normality was so strongly rejected, a preliminary stage was applied with impulse-indicator saturation alone, to ensure more appropriate initial inferences. Selection is then performed using the techniques developed to handle more variables than observations.

The chapter has shown that in order to achieve a successful algorithm, it is important to jointly implement all the developments discussed above, namely:

testing for the need to select a non-linear model when there are many candidates;
transformations to a near-orthogonal representation;
impulse-indicator saturation to remove extreme observations;
tight significance levels to avoid excess retention of irrelevant non-linear functions;
handling more variables than observations.

Removing any one of these ingredients would be deleterious to selection, and hence to the quality of the resulting model.

An empirical study of returns to education demonstrated the applicability of the approach. Fitting theory-based models such as the Mincer equation without paying attention to the data characteristics by addressing evidence of mis-specification and outliers, can result in poor models. Further, many previous empirical studies did not address the implications of induced collinearity by including age and age squared (or experience) without prior de-meaning. The empirical application is large in dimension, with over 5000 observations and many linear covariates, leading to a large number of candidate non-linear functions as well as indicators. Fortunately, advances in automatic model selection make problems of this scale tractable; and the analyses and simulations in recent research demonstrate the high success rates of such an approach.

References

1. Abadir, K.M.: An introduction to hypergeometric functions for economists. Econom. Rev. **18**, 287–330 (1999)
2. Ahmed, S.: Econometric issues on the return to education. MPhil thesis. University of Oxford (2007)
3. Altonji, J., Dunn, T.: Using siblings to estimate the effect of schooling quality on wages. Rev. Econ. Stat. **78**, 665–671 (1996)
4. Blundell, R., Dearden, L., Sianesi, B.: Evaluating the effect of education on earnings: models, methods and results from the National Child Development Survey. J. R. Stat. Soc. A **168**, 473–512 (2005)
5. Campos, J., Hendry, D.F., Krolzig, H.M.: Consistent model selection by an automatic *Gets* approach. Oxf. Bull. Econ. Stat. **65**, 803–819 (2003)
6. Card, D.: The causal effect of education on earnings. In: Ashenfelter, O., Card, D. (eds.) Handbook of Labor Economics, vol. 3A, pp. 1801–1863. North-Holland, Amsterdam (1999)
7. Castle, J.L.: Evaluating PcGets and RETINA as automatic model selection algorithms. Oxf. Bull. Econ. Stat. **67**, 837–880 (2005)
8. Castle, J.L., Hendry, D.F.: A low-dimension, portmanteau test for non-linearity. J. Econom. (2010)
9. Castle, J.L., Doornik, J.A., Hendry, D.F.: Evaluating automatic model selection. J. Time Ser. Econom. 3(1), Article 8 (2011)
10. Castle, J.L., Doornik, J.A., Hendry, D.F.: Model selection when there are multiple breaks. Working Paper 472, Economics Department, University of Oxford (2009)
11. Castle, J.L., Fawcett, N.W.P., Hendry, D.F.: Forecasting breaks and during breaks. In: Clements, M.P., Hendry, D.F. (eds.) Oxford Handbook of Economic Forecasting, Chap. 11, pp. 315–354. Oxford University Press, London (2011)
12. Clements, M.P., Hendry, D.F.: Forecasting Economic Time Series. Cambridge University Press, Cambridge (1998)
13. Copson, E.T.: Asymptotic Expansions. Cambridge University Press, Cambridge (1965)
14. Dearden, L.: The effects of families and ability on men's education and earnings in Britain. Labour Econ. **6**, 551–567 (1999)
15. Doornik, J.A.: Econometric model selection with more variables than observations. Working Paper, Economics Department, University of Oxford (2007)
16. Doornik, J.A.: Encompassing and automatic model selection. Oxf. Bull. Econ. Stat. **70**, 915–925 (2008)
17. Doornik, J.A.: Autometrics. In: Castle, J.L., Shephard, N. (eds.) The Methodology and Practice of Econometrics, pp. 88–121. Oxford University Press, Oxford (2009)
18. Doornik, J.A., Hansen, H.: An omnibus test for univariate and multivariate normality. Oxf. Bull. Econ. Stat. **70**, 927–939 (2008)
19. Doornik, J.A., Hendry, D.F.: Empirical model discovery. Working paper, Economics Department, University of Oxford (2009)
20. Frisch, R.: Statistical Confluence Analysis by Means of Complete Regression Systems. University Institute of Economics, Oslo (1934)
21. Garen, J.: The returns to schooling: a selectivity bias approach with a continuous choice variable. Econometrica **52**(5), 1199–1218 (1984)
22. Granger, C.W.J., Teräsvirta, T.: Modelling Nonlinear Economic Relationships. Oxford University Press, Oxford (1993)
23. Griliches, Z.: Estimating the returns to schooling: some econometric problems. Econometrica **45**, 1–22 (1977)
24. Harmon, C., Walker, I.: Estimates of the economic return to schooling for the UK. Am. Econ. Rev. **85**, 1278–1286 (1995)
25. Harrison, A.: Earnings by size: a tale of two distributions. Rev. Econ. Stud. **48**, 621–631 (1981)

26. Heckman, J.J., Lochner, L.J., Todd, P.E.: Earnings functions, rates of return and treatment effects: the Mincer equation and beyond. In: Hanushek, E., Welch, F. (eds.) Handbook of the Economics of Education, vol. 1. North Holland, Amsterdam (2006). Chap. 7
27. Hendry, D.F., Doornik, J.A.: Empirical Econometric Modelling using PcGive, vol. I. Timberlake Consultants Press, London (2009)
28. Hendry, D.F., Krolzig, H.M.: Automatic Econometric Model Selection. Timberlake Consultants Press, London (2001)
29. Hendry, D.F., Krolzig, H.M.: The properties of automatic Gets modelling. Econ. J. **115**, C32–C61 (2005)
30. Hendry, D.F., Mizon, G.E.: Econometric modelling of changing time series. Working Paper 475, Economics Department, Oxford University (2009)
31. Hendry, D.F., Morgan, M.S.: A re-analysis of confluence analysis. Oxf. Econ. Pap. **41**, 35–52 (1989)
32. Hendry, D.F., Richard, J.F.: Recent developments in the theory of encompassing. In: Cornet, B., Tulkens, H. (eds.) Contributions to Operations Research and Economics. The XXth Aniversary of CORE, pp. 393–440. MIT Press, Cambridge (1989)
33. Hendry, D.F., Santos, C.: Regression models with data-based indicator variables. Oxf. Bull. Econ. Stat. **67**, 571–595 (2005)
34. Hendry, D.F., Johansen, S., Santos, C.: Automatic selection of indicators in a fully saturated regression. Comput. Stat. **33**, 317–335 (2008). Erratum, 337–339
35. Hoover, K.D., Perez, S.J.: Data mining reconsidered: encompassing and the general-to-specific approach to specification search. Econom. J. **2**, 167–191 (1999)
36. Johansen, S., Nielsen, B.: An analysis of the indicator saturation estimator as a robust regression estimator. In: Castle, J.L., Shephard, N. (eds.) The Methodology and Practice of Econometrics, pp. 1–36. Oxford University Press, Oxford (2009)
37. Lehergott, S.: The shape of the income distribution. Am. Econ. Rev. **49**, 328–347 (1959)
38. Mincer, J.: Investment in human capital and personal income distribution. J. Polit. Econ. **66**(4), 281–302 (1958)
39. Mincer, J.: Schooling, Experience and Earnings. Natonal Bureau of Economic Research, New York (1974)
40. Mizon, G.E., Richard, J.F.: The encompassing principle and its application to non-nested hypothesis tests. Econometrica **54**, 657–678 (1986)
41. Perez-Amaral, T., Gallo, G.M., White, H.: A flexible tool for model building: the relevant transformation of the inputs network approach (RETINA). Oxf. Bull. Econ. Stat. **65**, 821–838 (2003)
42. Phillips, P.C.B.: Regression with slowly varying regressors and nonlinear trends. Econom. Theory **23**, 557–614 (2007)
43. Priestley, M.B.: Spectral Analysis and Time Series. Academic Press, New York (1981)
44. Ramsey, J.B.: Tests for specification errors in classical linear least squares regression analysis. J. R. Stat. Soc. B **31**, 350–371 (1969)
45. Rushton, S.: On least squares fitting by orthogonal polynomials using the Choleski method. J. R. Stat. Soc. B **13**, 92–99 (1951)
46. Schwarz, G.: Estimating the dimension of a model. Ann. Stat. **6**, 461–464 (1978)
47. Staehle, H.: Ability, wages and income. Rev. Econ. Stat. **25**, 77–87 (1943)
48. Teräsvirta, T.: Specification, estimation and evaluation of smooth transition autoregressive models. J. Am. Stat. Assoc. **89**, 208–218 (1994)
49. White, H.: A heteroskedastic-consistent covariance matrix estimator and a direct test for heteroskedasticity. Econometrica **48**, 817–838 (1980)
50. White, H.: Artificial Neural Networks: Approximation and Learning Theory. Oxford University Press, Oxford (1992)

Chapter 13
Construction of Radial Basis Function Networks with Diversified Topologies

X. Hong, S. Chen, and C.J. Harris

13.1 Introduction

The identification of nonlinear systems using only observed finite data sets has become a mature research area over the last two decades [1]. A large class of nonlinear models and neural networks can be classified as a linear-in-the-parameters model [2, 3]. These are well structured for adaptive learning, have provable learning and convergence conditions, have the capability of parallel processing and have clear applications in many engineering applications [4–6]. In particular, the radial basis function (RBF) network is a popular type of linear-in-the-parameters model and has been widely applied in diverse fields of engineering [7–10]. The ultimate objective of model construction from observed data sets should be to produce a model which captures the true underlying dynamics and predicts accurately the output for unseen data. This translates into the practical principle in nonlinear modelling of finding the smallest model that generalizes well. Sparse models are preferable in engineering applications since a models' computational complexity scales with its model complexity. Furthermore, a sparse model is easier to interpret from the angle of knowledge extraction from observed data sets.

A fundamental concept in the evaluation of model generalization capability is that of cross validation [11] which is often used to derive the information theoretic metrics, e.g. the leave-one-out (LOO) cross validation has been used to derive

X. Hong (✉)
School of Systems Engineering, University of Reading, Reading, UK
e-mail: x.hong@reading.ac.uk

S. Chen · C.J. Harris
School of Electronics and Computer Science, University of Southampton, Southampton, UK

S. Chen
e-mail: sqc@ecs.soton.ac.uk

L. Wang, H. Garnier (eds.), *System Identification, Environmental Modelling, and Control System Design*,

DOI 10.1007/978-0-85729-974-1_13,

model selective criteria such as the Akaike information criterion (AIC) [12]. Model selective criteria can be used for predicting a model's performance on unseen data and evaluating a model's quality amongst other competitive models. The forward orthogonal least squares (OLS) algorithm is an efficient nonlinear system identification algorithm [13, 14] which selects regressors in a forward manner by virtue of their contribution to the maximization of the model error reduction ratio (ERR). In order to produce a model with good generalization capabilities, the AIC [12] is usually incorporated into the forward orthogonal least squares (OLS) algorithm to determinate the model construction process. The OLS algorithm has become a popular modelling tool in a wide range of applications [15–18]. Note that most of model selective criteria are formula of approximating the LOO mean-square error (mse), and due to the approximation, have lost discriminate power in selecting terms if being used in the forward OLS algorithm. The LOO mean-square error (MSE) criterion, which directly measures the model generalization capability, has been introduced into the framework of forward OLS algorithm [19] in which the LOO mean-square error (MSE) criterion is calculated efficiently (as outlined in Sect. 13.2). An additional advantage is that the process is fully automatic, so that there is no need for the user to specify a termination criterion of the model construction process.

In this review we bring together some of our recent work from the angle of the diversified RBF topologies, including three different topologies; (i) the RBF network with tunable nodes [20]; (ii) the Box-Cox RBF [21]; and (iii) the BVC-RBF network [22]. The RBF network with tunable nodes is initially described in Sect. 13.3. Note that the parameters of the RBF network include its center vectors and variance or the covariance matrices of the basis function as well as the connecting weights from the RBF nodes to the network output. In [19] and many other RBF modelling paradigms [23–26], the RBF centers are restricted to be selected from the input data sets and a common variance is employed for every RBF node. The common variance should be treated as a hyperparameter and determined via cross-validation, which may be computationally costly. The recent work [20] has introduced a construction algorithm for the tunable RBF network, where each RBF node has a tunable center and an adjustable diagonal covariance matrix. An OFS procedure is developed to append the RBF units one by one by minimizing the LOO mse. Because the extra flexibility for the basis functions is allowed in the tunable RBF topology and all the parameters are optimized via minimizing the LOO mean-square error (MSE) criterion, the algorithm is computationally efficient and the resultant models have sparser representations with excellent generalization capability, in comparison with the existing sparse kernel modeling methods.

In Sect. 13.4, the Box-Cox RBF topology and its fast model construction algorithm [21] is described. It is a common practice to construct the RBF network in order to represent a systems' input/output mapping. For the network training the system output observations are used as the direct target of the model output. Least squares algorithm is often used as the parameter estimator, which is equivalent to the maximum likelihood estimator (MLE) under the assumption that the noise is additive and independent identically distributed (i.i.d.) Gaussian with zero mean and constant variance. In practice the variance of process noise may vary with the output, e.g. the variance of noise may increase as the system output increases. For some dynamical processes in which the model residuals exhibit heteroscedasticity, e.g. with nonconstant variance or skewed distribution, or being multiplicative to the model, using conventional RBF models to construct a direct systems' input/output mapping based on the least squares estimator is no longer appropriate. The work [21] has modified RBF topology based on Box-Cox transformation. The fast identification algorithm [21] is developed based on a maximum likelihood estimator (MLE) to find the required Box-Cox transformation. It is shown the OFS-LOO algorithm is readily applicable to construct a sparse Box-Cox RBF model with good generalisation [19, 21, 27].

Finally Sect. 13.5 describes the topology of the BVC-RBF network [22]. Note that most of RBF modelling algorithms fit into the statistical learning framework, i.e. the model is determined based on the observational data only. In many modelling tasks, there are more or less prior knowledge available. Although any prior knowledge about the system should help to improve the model generalization, in general incorporating the deterministic prior knowledge into a statistically learning paradigm would make the development of modelling algorithms more difficult if not impossible. The work [22] has introduced the idea of modifying RBF topology in order to enhance its capability of automatic constraints satisfaction. We considered a special type of prior knowledge given by a type of boundary value constraints (BVC), and introduced the BVC-RBF as a new topology of RBF neural network that has the capability of satisfying the BVC automatically. The BVC-RBF network [22] is constructed and parameterized based on the given BVC. It is shown that the BVC-RBF remains as a linear-in-the-parameter structure just as the conventional RBF does. Therefore many of the existing modelling algorithms for a conventional RBF are almost directly applicable to the new BVC-RBF without added algorithmic complexity nor computational cost. Consequently the topology of the BVC-RBF effectively lends itself as a single framework in which both the deterministic prior knowledge and stochastic data are fused with ease.

13.2 Orthogonal Forward Selection (OFS) Algorithm Based on Leave-One-Out (LOO) Criteria

Consider the regression problem of approximating the N pairs of training data $D_N = \{\mathbf{x}_k, y_k\}_{k=1}^N$ with a linear-in-the-parameter model defined in

$$y_k = \hat{y}_k + e_k = \sum_{i=1}^{M} w_i g_i(\mathbf{x}_k) + e_k = \mathbf{g}^T(k)\mathbf{w} + e_k, \tag{13.1}$$

where the input $\mathbf{x}_k \in \Re^m$, the desired output $y_k \in \Re$, $\hat{y}_k$ denotes the model output, $e_k = y_k - \hat{y}_k$ is the modelling error, $g_i(\cdot)$ for $1 \le i \le M$ is a known nonlinear basis function mapping, such as RBF, polynomial or B-spline functions, and $\mathbf{g}(k) = [g_1(\mathbf{x}_k)\ g_2(\mathbf{x}_k)\ \ldots\ g_M(\mathbf{x}_k)]^T$, $\mathbf{w} = [w_1\ w_2\ \ldots\ w_M] \in \Re^M$ is the weight vector, M is the number of basis functions. By defining $\mathbf{y} = [y_1\ y_2\ \dots\ y_N]^T$, $\mathbf{e} = [e_1\ e_2\ \dots\ e_N]^T$, and $\mathbf{G} = [\mathbf{g}_1\ \mathbf{g}_2\ \dots\ \mathbf{g}_M]$ with $\mathbf{g}_l = [g_l(\mathbf{x}_1)\ g_l(\mathbf{x}_2)\ \dots\ g_l(\mathbf{x}_N)]^T$, $1 \leq l \leq M$. The regression model (13.1) over the training data set can be written in the matrix form

$$\mathbf{y} = \mathbf{G}\mathbf{w} + \mathbf{e}. \tag{13.2}$$

Here $\mathbf{g}_l$ is the lth column of while $\mathbf{g}^T(k)$ the kth row of $\mathbf{G}$.

Let an orthogonal decomposition of $\mathbf{G}$ be $\mathbf{G} = \mathbf{P}\mathbf{A}$, where $\mathbf{A} = \{\alpha_{ij}\}$ is an $M \times M$ unit upper triangular matrix and $\mathbf{P} = [\mathbf{p}_1\ \mathbf{p}_2\ \cdots\ \mathbf{p}_M]$ is an $N \times M$ matrix with orthogonal columns that satisfy

$$\mathbf{P}^T\mathbf{P} = \text{diag}\{\kappa_1, \dots, \kappa_M\}, \tag{13.3}$$

where $\kappa_l = \mathbf{p}_l^T\mathbf{p}_l$ for $1 \leq l \leq M$. The regression model (13.2) can be alternatively expressed as

$$\mathbf{y} = \mathbf{P}\boldsymbol{\theta} + \mathbf{e}, \tag{13.4}$$

where $\boldsymbol{\theta} = [\theta_1\ \theta_2\ \dots\ \theta_M]^T$ satisfies the triangular system $\mathbf{A}\mathbf{w} = \boldsymbol{\theta}$. The model output $\hat{y}_k$ can be equivalently expressed by

$$\hat{y}_k = \mathbf{p}^T(k)\boldsymbol{\theta}, \tag{13.5}$$

where $\mathbf{p}^T(k) = [p_1(\mathbf{x}_k)\ p_2(\mathbf{x}_k)\ \cdots\ p_M(\mathbf{x}_k)]$ is the kth row of $\mathbf{P}$.

Consider the modeling process that has produced the n-unit model. Let us denote the constructed n model columns as $\mathbf{P}_n = [\mathbf{p}_1, \dots, \mathbf{p}_n]$, the kth model output of this n unit model identified using the entire training data set as $\hat{y}_k^{(n)} = \sum_{i=1}^{n} \theta_i p_i(k)$, and the corresponding kth modeling error $e_k^{(n)} = y_k - \hat{y}_k^{(n)}$.

Definition 13.1 The leave-one-out (LOO) mse: If we "remove" the kth data point from the trying data set and use the remaining $(N-1)$ data points to identify the n-unit model instead, the "test" error of the resulting model can be calculated on the data point removed from training. This LOO modeling error (this corresponds to the LOO pseudo-modeling error in the context of Box-Cox RBF network (see Sect. 13.4)), denoted as $e_k^{(n,-k)}$, is given by [28]

$$e_k^{(n,-k)} = e_k^{(n)} / \eta_k^{(n)}, \tag{13.6}$$

where $\eta_k^{(n)}$ is the LOO error weighting [28]. The LOO mse (this corresponds to the LOO pseudo-mse in the context of Box-Cox RBF network (see Sect. 13.4)) for the n-unit model is then defined by

$$J_n = \frac{1}{N}\sum_{k=1}^{N}\left(e_k^{(n,-k)}\right)^2 \tag{13.7}$$

which is a measure of the model generalisation capability [11, 28].

For model (13.5) the computation of the LOO criterion J_n is very efficient, because $e_k^{(n)}$ and $\eta_k^{(n)}$ can be computed recursively using [19, 27]

$$e_k^{(n)} = e_k^{(n-1)} - \theta_n p_n(k), \tag{13.8}$$

$$\eta_k^{(n)} = \eta_k^{(n-1)} - \frac{p_n^2(k)}{\kappa_n + \nu}, \tag{13.9}$$

where $\nu \geq 0$ is a small regularization parameter.

The orthogonal forward selection (OFS) algorithm based leave-one-out (LOO) criteria was proposed [19, 27], in which the LOO mse J_n was minimized by searching a set of candidate regressors at each forward orthogonal regression stage. It is shown [19] that J_n is concave with respect to the number of model terms, and this means that the model construction process becomes fully automatic without using additional termination criterion. Furthermore note that J_n directly measures the model generalization capability so that there is no need to use a separate validation data set. Other advantages for using LOO mse criteria are that LOO mse J_n has not lost discriminative power in selecting terms as happens with AIC, and that there is no extra tuning parameters in the model selective criterion.

13.3 RBF Network with Tunable Nodes

A popular approach is to construct the RBF models with the Gaussian basis functions, in which the candidate regressors $g_i(\cdot)$ are formed using the training data set, and a *given* common variance is employed for every RBF node. In order to find a satisfactory value of the common variance, the algorithms in [19, 27] need to be repeated, e.g. via grid search based cross validation. Clearly the true cost of modeling must take into account the cost of determining all the parameters, e.g. optimizing the value the common variance. This is because most of the complexity for many existing learning algorithms is due to the need to tune parameters that have nonlinear relationship to the system output via cross validation. Therefore a model with less parameters that are tuned via cross validation could potentially lead to the significant reduction to the true cost of modeling.

Alternatively if the regressors $g_i(\cdot)$ are viewed as the building blocks of the RBF network, then it is intuitive to make these more flexible by relaxing the constraint that each regressor has the same shape, because this allows the model generalization capability to be maximized for a model with the smallest size. The tunable RBF network was recently introduced [20], in which each node of the network has a tunable center and an adjustable diagonal covariance matrix. Clearly the tunable RBF topology has more parameters that are nonlinear to the system output, and nonlinear optimization is necessary, leading to the additional computation costs. Note that it would be computationally prohibitive to tune a large number of extra parameters via cross validation. Significantly the OFS-LOO algorithm, the construction algorithm developed for the tunable RBF network in [20], optimizes all the associated parameters in order to achieve model generalization without cross validation. This potentially leads to considerable saving in terms of the true cost of modeling, despite the fact that more parameters that have nonlinear relationship to the system output are introduced in the tunable RBF topology.

Consider the general RBF regressor of the form [20]

$$g_i(\mathbf{x}) = K\left(\sqrt{(\mathbf{x}-\boldsymbol{\mu}_i)^T \Sigma_i^{-1}(\mathbf{x}-\boldsymbol{\mu}_i)}\right) \tag{13.10}$$

where $\boldsymbol{\mu}_i$ is the center vector of the ith RBF unit, the diagonal covariance matrix has the form $\Sigma_i = \mathrm{diag}\{\sigma_{i,1}, \dots, \sigma_{i,m}\}$, and $K(\cdot)$ is the chosen basis or kernel function. The proposed algorithm constructs the RBF units one by one by positioning and shaping the RBF nodes while minimizing the LOO mse J_n. Specifically, at the nth stage of the constructing procedure, the nth RBF unit is determined by minimizing J_n with respect to the node's center vector $\boldsymbol{\mu}_n$ and the diagonal covariance matrix Σ_n

$$\min_{\boldsymbol{\mu}_n, \Sigma_n} J_n(\boldsymbol{\mu}_n, \Sigma_n) \tag{13.11}$$

and the construction procedure is automatically terminated when $J_M \leq J_{M+1}$, yielding an M-term RBF network. Intuitively the extra number of tunable parameters in each RBF node can enhance the modeling capability such that the final model size M could be much smaller than that of fixed RBF with each unit having a common variance, leading to another part of saving in computation cost, and this is often confirmed in simulation studies.

In [20], a simple yet efficient global search algorithm called the repeating weighted boosting search (RWBS) algorithm [29] was proposed to solve the task of the nonconvex optimization problem (13.11). The procedure is summarized here. Let $\mathbf{u}$ be the vector that contains $\boldsymbol{\mu}_n$ and Σ_n. Giving the following initial conditions:

$$\left.\begin{array}{l} e_k^{(0)} = y_k \quad \text{and} \quad \eta_k^{(0)} = 1, \quad 1 \leq k \leq N \\ J_0 = \dfrac{1}{N}\mathbf{y}^T\mathbf{y} = \dfrac{1}{N}\displaystyle\sum_{k=1}^{N} y_k^2 \end{array}\right\}$$

Specify the RWBS algorithmic parameters, namely, the population size P_s, the number of generations in the repeated search N_G, and the number of weighted boosting search iterations M_I.

Outer loop: generations For ($l = 1; l \leq N_G; l = l + 1$) {

Generation Initialization: Initialize the population by setting $\mathbf{u}_1^{[l]} = \mathbf{u}_{best}^{[l-1]}$ and randomly generating the rest of the population members $\mathbf{u}_i^{[l]}$, $2 \leq i \leq P_S$, where $\mathbf{u}_{best}^{[l-1]}$ denotes the solution found in the previous generation. If $l = 1$, $\mathbf{u}_1^{[l]}$ is also randomly chosen.

Weighted boosting search initialization: Assign the initial distribution weighting factors $\delta_i(0) = 1/P_s$, $1 \leq i \leq P_s$, for the population. Then

(1) For $1 \le i \le P_S$, generate $\mathbf{g}_n^{i)}$ from $\mathbf{u}_i^{[l]}$, the candidates for the nth model column, and orthogonalize them

$$\alpha_{j,n}^{i)} = \mathbf{p}_j^T \mathbf{g}_n^{i)} / \mathbf{p}_j^T \mathbf{p}_j, \quad 1 \le j < n, \tag{13.12}$$

$$\mathbf{p}_n^{i)} = \mathbf{g}_n^{i)} - \sum_{j=1}^{n-1} \alpha_{j,n}^{i)} \mathbf{p}_j, \tag{13.13}$$

$$\theta_n^{i)} = (\mathbf{p}_n^{i)})^T \mathbf{y} / \left((\mathbf{p}_n^{i)})^T \mathbf{p}_n^{i)} + \nu \right). \tag{13.14}$$

(2) For $1 \le i \le P_s$, calculate the LOO cost for each $\mathbf{u}_i^{[l]}$

$$e_k^{(n)}(i) = e_k^{(n-1)} - p_n^{i)}(k)\theta_n^{i)}, \quad 1 \le k < N, \tag{13.15}$$

$$\eta_k^{(n)}(i) = \eta_k^{(n-1)} - \left(p_n^{i)}(k)\right)^2 / \left((\mathbf{p}_n^{i)})^T \mathbf{p}_n^{i)} + \nu \right), \quad 1 \le k < N, \tag{13.16}$$

$$J_n^{i)} = \frac{1}{N} \sum_{k=1}^{N} \left(\frac{e_k^{(n)}(i)}{\eta_k^{(n)}(i)} \right)^2, \tag{13.17}$$

where $p_n^{i)}(k)$ is the kth element of $\mathbf{p}_n^{i)}$.

Inner loop: weighted boosting search For $(t = 1; t \le M_I; t = t + 1)$ {
Step 1: Boosting

1. Find

$$i_{best} = \arg \min_{1 \le i \le P_s} J_n^{i)}, \tag{13.18}$$

$$i_{worst} = \arg \max_{1 \le i \le P_s} J_n^{i)}. \tag{13.19}$$

Denote $\mathbf{u}_{best}^{[l]} = \mathbf{u}_{i_{best}}^{[l]}$ and $\mathbf{u}_{worst}^{[l]} = \mathbf{u}_{i_{worst}}^{[l]}$,

2. Normalize the cost function values

$$\bar{J}_n^{i)} = \frac{J_n^{i)}}{\sum_{j=1}^{P_s} J_n^{j)}}, \quad 1 \le i \le P_s. \tag{13.20}$$

3. Compute a weighting factor β_t according to

$$\xi_t = \sum_{i=1}^{P_s} \delta_i(t-1) \bar{J}_n^{i)}, \qquad \beta_t = \frac{\xi_t}{1 - \xi_t}. \tag{13.21}$$

4. Update the distribution weightings for $1 \le i \le P_s$

$$\delta_i(t) = \begin{cases} \delta_i(t-1)\beta_t^{\bar{J}_n^{i)}} & \text{for } \beta \le 1, \\ \delta_i(t-1)\beta_t^{1-\bar{J}_n^{i)}} & \text{for } \beta > 1 \end{cases} \tag{13.22}$$

and normalize them

$$\delta_i(t) = \frac{\delta_i(t)}{\sum_{j=1}^{P_s} \delta_j(t)}, \quad 1 \le i \le P_s. \tag{13.23}$$

Step 2: Parameter Updating

1. Construct the $(Ps+1)$th point using

$$\mathbf{u}_{P_s+1} = \sum_{i=1}^{P_s} \delta_i(t)\mathbf{u}_i^{[l]}. \tag{13.24}$$

2. Construct the $(Ps+2)$th point using

$$\mathbf{u}_{P_s+2} = \mathbf{u}_{best}^{[l)]} + \big(\mathbf{u}_{best}^{[l]} - \mathbf{u}_{P_s+1}\big). \tag{13.25}$$

3. Calculate $\mathbf{g}_n^{P_s+1)}$ and $\mathbf{g}_n^{P_s+2)}$ from $\mathbf{u}_{P_s+1}$ and $\mathbf{u}_{P_s+2}$, orthogonalize these two candidate model columns (as in (13.12)–(13.14), and compute the corresponding LOO cost function values $J_n^{i)}$, $i = P_s+1, P_s+2$ (as in (13.15)–(13.17)). Then find

$$i_* = \arg \min_{i=P_s+1, P_s+2} J_n^{i)}. \tag{13.26}$$

$(\mathbf{u}_{i_*}, J_n^{i_*)})$, which replace $(\mathbf{u}_{worst}^{[l]}, J_n^{i_{worst})})$ in the population.

} **End of inner loop** This solution found in the lth generation is $\mathbf{u} = \mathbf{u}_{best}^{[l]}$.
} **End of outer loop** This yields the solution $\mathbf{u} = \mathbf{u}_{best}^{(N_G)}$, i.e., $\boldsymbol{\mu}_n$, Σ_n of the nth RBF node, the nth model column $\mathbf{g}_n$, the orthogonalization coefficients $\alpha_{j,n}$, $1 \le j < n$, the corresponding orthogonal model column $\mathbf{p}_n$, and the weight θ_n, as well as the n-term modelling errors $e_k^{(n)}$ and the associated LOO modelling error weightings $\eta_k^{(n)}$ for $1 \le k \le N$.

Note that the algorithmic parameters P_s, N_G and M_I are found empirically, and some general rules are discussed in [29].

Example 13.1 (Chen et al. [20] Boston Housing Data) This benchmark data set is available at the University of California, Irvine (UCI) repository [30]. The data set comprises 506 data points with 14 variables. The task of predicting the median house value was performed from the remaining 13 attributes. 456 data points were randomly selected from the data set for training and the remaining 50 data points were used as a test data set. The experiment was repeated and the average results over 100 repetitions were given [20]. Three construction algorithms, the ε-SVM [24], the LROLS-LOO [27] and the OFS-LOO [20] were compared, and the Gaussian basis function was used to form the basis function. Table 13.1 summarize the results for three algorithms over the 100 realizations. The experiments parameters setting can be found [20]. Discussions on the computational complexity comparison can be found [20], in which it is argued that the OFS-LOO algorithm is highly

Table 13.1 Comparative results for Boston housing data set [20]; The results were averaged over 100 realisations and quoted as the mean ± standard deviation

Algorithm	RBF type	Model size	Training MSE	Test MSE
ε-SVM	fixed	243.2 ± 5.3	6.7986 ± 0.4444	23.1750 ± 9.0459
LROLS-LOO	fixed	58.6 ± 11.3	12.9690 ± 2.6628	17.4157 ± 4.6670
OFS-LOO	**tunable**	**34.6 ± 8.4**	**10.0997 ± 3.4047**	**14.0745 ± 3.6178**

competitive in terms of the real cost of modeling. The computation efficiency aspect is further improved by a particle swarm optimization (PSO) aided OFS-LOO method [31].

13.4 Box-Cox Output Transformation Based RBF Network (Box-Cox RBF)

In this section we review a modified RBF topology [21], in which a conventional RBF network was introduced to represent the Box-Cox transformed system output, rather than the actual system output. One of the motivations of [21] is to provide a computationally efficient approach to construct a sparse Box-Cox RBF network for some systems with the heteroscedasticity. Provided that there is a suitable Box-Cox transformation, the pseudo model errors that are the model residuals between the transformed system output and model output can be stabilized so that it follows a normal assumption [32–34]. Provided that the optimal parameter λ used in Box-Cox transformation, the number and location of candidate RBF centers are known, various orthogonal forward regression (OFR) algorithms [13, 35–37] are readily applicable to model structure selection and parameter estimation for the proposed Box-Cox transformed based RBF network.

Consider the problem of approximating the N pairs of training data $\{\mathbf{x}_k, y_k\}_{k=1}^N$, where y_k is positive system output. If the original system output is not negative, then $y_k + c \rightarrow y_k > 0$ is used where c is a chosen positive number just large enough to enable y_k to be positive. The Box-Cox transformation is a transformation to the system output given by

$$h(y,\lambda) = \begin{cases} (y^\lambda - 1)/(\lambda \tilde{y}^{\lambda-1}) & \text{if } \lambda \neq 0, \\ \tilde{y}\log(y) & \text{if } \lambda = 0, \end{cases} \tag{13.27}$$

where $\tilde{y} = \sqrt[N]{\prod_{k=1}^N y_k}$, the geometric mean of the output observations.

The Box-Cox transformation based RBF networks (Box-Cox RBF) [21] is illustrated in Fig. 13.1. For a given λ, the Box-Cox RBF network with a single output

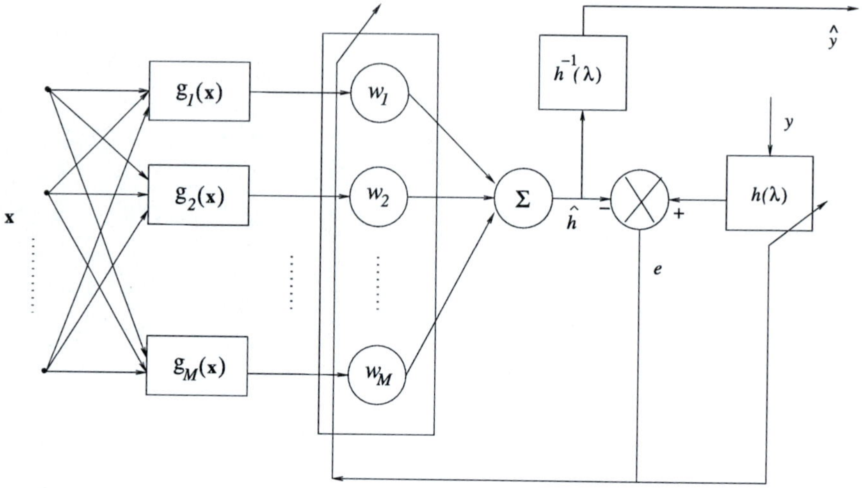

Fig. 13.1 The topology of the Box-Cox RBF network

can be formulated as

$$h(y_k, \lambda) = \hat{h}_k + e_k = \sum_{i=1}^{M} w_i g_i(\mathbf{x}_k) + e_k = \mathbf{g}^T(k)\mathbf{w} + e_k. \tag{13.28}$$

Here $e_k = h(y_k, \lambda) - \hat{h}_k$ is referred as the pseudo error. (In order to reduce the number of notations, e_k is still used here in spite of the difference between (13.1) and (13.28). This allows that the algorithm in Sect. 13.2 to be shared for the different topologies.) The regressors $g_i(\mathbf{x}_k)$ are formed using some known RBF functions (see Sect. 13.2). Note that

$$\lim_{\lambda \to 0} h(y, \lambda) = \lim_{\lambda \to 0} [(y^{\lambda} - 1)/(\lambda \tilde{y}^{\lambda - 1})] = \tilde{y} \log(y) \tag{13.29}$$

and the inverse of Box-Cox transformation upon $\hat{h}_k$ for given $\lambda \neq 0$ is

$$\hat{y}_k = h^{-1}(\hat{h}_k) = \sqrt[\lambda]{1 + \lambda \tilde{y}^{\lambda - 1} \hat{h}_k}. \tag{13.30}$$

If $\lambda = 0$, then $\hat{y}_k = \exp[\hat{h}_k / \tilde{y}]$.

Supposing all the training data were used as centres to construct the candidate regressors $g_i(\mathbf{x}_k)$, (13.28) can be rewritten in a vector form as

$$\mathbf{e} = \mathbf{h}(\lambda) - \mathbf{G}\mathbf{w} \tag{13.31}$$

in which $\mathbf{h}(\lambda) = [h(y_1, \lambda), \ldots, h(y_N, \lambda)]^T \in \Re^N$ is transformed system outputs' vector. $\mathbf{e} = [e_1, \ldots, e_N]^T \in \Re^N$ is the pseudo-error vector.

The parameter estimation for the Box-Cox RBF network is to adapt model parameters based on the fundamentals of feedback learning and weight adjustment found in conventional parametric optimization so that the model produces a good approximation to the true system, e.g. to minimize pseudo errors as shown Fig. 13.1. Compared to the conventional RBF neural networks, there is an additional task of determining the required Box-Cox transformation, i.e. finding the optimal λ. The method introduced in [21] is based on the underlying assumption that there exists a suitable Box-Cox RBF network such that the resultant model residuals, or pseudo errors e_k, become Gaussian with zero mean and constant variance σ^2 [32, 33]. This leads to a fast algorithm for determining λ based on MLE, as described below.

Because the parameter estimators for linear-in-the-parameters models rely on the well-conditioning of the model, yet using the full data set to form RBF regressors usually results in ill-conditioning. Initially we consider the singular value decomposition (SVD) of matrix $\mathbf{G}$ with orthonormal matrix $\mathbf{Q}_N \in \Re^{N\times N}$, such that

$$\mathbf{Q}_N^T\mathbf{G}\mathbf{Q}_N = \boldsymbol{\Sigma}_N = \mathrm{diag}(\sigma_1, \sigma_2, \ldots, \sigma_L, 0, \ldots, 0) \in \Re^{N\times N}, \tag{13.32}$$

where $\sigma_1 \geq \sigma_2 \geq \cdots \geq \sigma_L$ are L nonnegative singular values of $\mathbf{G}$. Denote $\boldsymbol{\Sigma} = \mathrm{diag}(\sigma_1, \sigma_2, \ldots, \sigma_L) \in \Re^{L\times L}$, and the submatrix of the first L columns of $\mathbf{Q}_N$ as $\mathbf{Q} = [\mathbf{q}_1, \ldots, \mathbf{q}_L] \in \Re^{N\times L}$, and $\mathbf{q}_k = [q_k(\mathbf{x}_1), \ldots, q_k(\mathbf{x}_N)]^T$. Equation (13.31) becomes

$$\mathbf{e}(\boldsymbol{\vartheta}_\lambda) = \mathbf{h}(\lambda) - \mathbf{Q}\boldsymbol{\Sigma}\mathbf{Q}^T\mathbf{w} = \mathbf{h}(\lambda) - \mathbf{Q}\boldsymbol{\vartheta} \tag{13.33}$$

in which $\boldsymbol{\vartheta} = [\vartheta_1, \ldots, \vartheta_L]^T \in \Re^L$, $\boldsymbol{\vartheta}_\lambda$ is defined as $\boldsymbol{\vartheta}_\lambda = [\boldsymbol{\vartheta}^T, \lambda]^T$. Denote $\mathbf{e}(\boldsymbol{\vartheta}_\lambda) = [e_1(\boldsymbol{\vartheta}_\lambda), \ldots, e_N(\boldsymbol{\vartheta}_\lambda)]^T$.

Consider the MLE for $\boldsymbol{\vartheta}_\lambda$ under the assumption that the pseudo errors, e_k, is Gaussian with zero mean and constant variance σ^2 [32, 33]. Specifically, suppose that there exists a suitable Box-Cox transformation given by (13.27) such that the transformed output observations $h(y, \lambda)$ satisfy the normal assumption with the probability density function [32, 33] in relation to the original observations y_k, $k = 1, \ldots, N$ proportional to the following function

$$\frac{1}{\sqrt{2\pi}\sigma}\exp\left\{-\frac{e_k^2(\boldsymbol{\vartheta}_\lambda)}{2\sigma^2}\right\}\mathscr{J}(k, \lambda), \tag{13.34}$$

where

$$e_k(\boldsymbol{\vartheta}_\lambda) = h(y_k, \lambda) - \sum_{i=1}^{L} q_i(\mathbf{x}_k)\vartheta_i \tag{13.35}$$

and $\mathscr{J}(k, \lambda)$ is the Jacobian of the Box-Cox transformation given by [32, 33]

$$\mathscr{J}(k, \lambda) = \left.\frac{\partial h(y, \lambda)}{\partial y}\right|_{y=y_k} = \left[\frac{y_k}{\tilde{y}}\right]^{\lambda-1}.$$

Define a loglikelihood function as follows [32, 33]

$$L(\boldsymbol{\theta}_\lambda) = -N\log(\sigma) - \sum_{k=1}^{N}\frac{e_k^2(\boldsymbol{\vartheta}_\lambda)}{2\sigma^2} \tag{13.36}$$

in which (13.36) is applied. Hence MLE of $\boldsymbol{\vartheta}_\lambda$ can be solved by nonlinear least squares algorithm such as Gauss-Newton algorithm to minimize the mean squares pseudo errors $\sum_{k=1}^N e_k^2(\boldsymbol{\vartheta}_\lambda)$.

Consider the minimization of $\sum_{k=1}^N e_k^2(\boldsymbol{\vartheta}_\lambda)$ with respect to $\boldsymbol{\vartheta}_\lambda$ by using Gauss-Newton algorithm [38]. Denote an iteration step variable l by a superscript (l). With an initial $\boldsymbol{\vartheta}_\lambda^{(0)}$, the iteration formula is given by

$$\boldsymbol{\vartheta}_\lambda^{(l)} = \boldsymbol{\vartheta}_\lambda^{(l-1)} + \alpha\{[\underline{\mathbf{Q}}^{(l)}]^T \underline{\mathbf{Q}}^{(l)}\}^{-1}[\underline{\mathbf{Q}}^{(l)}]^T \mathbf{e}(\boldsymbol{\vartheta}_\lambda^{(l-1)}), \tag{13.37}$$

where $\alpha > 0$ is a small positive step size. $\underline{\mathbf{Q}}$ (the superscript (l) is removed here for notational simplicity) is the Jacobian matrix of $e_k(\boldsymbol{\vartheta}_\lambda)$ with respect to $\boldsymbol{\vartheta}_\lambda$, given by

$$\underline{\mathbf{Q}} = \begin{bmatrix} \frac{\partial}{\partial \theta_1} e_1(\boldsymbol{\vartheta}_\lambda) & \frac{\partial}{\partial \vartheta_2} e_1(\boldsymbol{\vartheta}_\lambda) & \dots & \frac{\partial}{\partial \vartheta_L} e_1(\boldsymbol{\vartheta}_\lambda) & \frac{\partial}{\partial \lambda} e_1(\boldsymbol{\vartheta}_\lambda) \\ \frac{\partial}{\partial \theta_1} e_2(\boldsymbol{\vartheta}_\lambda) & \frac{\partial}{\partial \vartheta_2} e_2(\boldsymbol{\theta}_\lambda) & \dots & \frac{\partial}{\partial \vartheta_L} e_2(\boldsymbol{\vartheta}) & \frac{\partial}{\partial \lambda} e_2(\boldsymbol{\vartheta}_\lambda) \\ \vdots & \vdots & \ddots & \vdots & \vdots \\ \frac{\partial}{\partial \vartheta_1} e(\boldsymbol{\vartheta}_\lambda, N) & \frac{\partial}{\partial \vartheta_2} e_N(\boldsymbol{\vartheta}_\lambda) & \dots & \frac{\partial}{\partial \vartheta_L} e_N(\boldsymbol{\vartheta}_\lambda) & \frac{\partial}{\partial \lambda} e_N(\boldsymbol{\vartheta}_\lambda) \end{bmatrix} \in \Re^{N\times(L+1)} \tag{13.38}$$

or equivalently

$$\underline{\mathbf{Q}} = [-\mathbf{Q}, \nabla_\lambda h(k,\lambda)], \tag{13.39}$$

where

$$\nabla_\lambda h(k,\lambda) = \left[\frac{\partial}{\partial\lambda} h(y_1,\lambda), \frac{\partial}{\partial\lambda} h(y_2,\lambda), \dots, \frac{\partial}{\partial\lambda} h(y_N,\lambda)\right]^T \in \Re^N, \tag{13.40}$$

in which,

$$\frac{\partial}{\partial\lambda} h(y_k,\lambda) = \frac{\lambda y_k^\lambda \log[y_k] - (y_k^\lambda - 1)(1 + \lambda \log \tilde{y})}{\lambda^2 \tilde{y}^{\lambda-1}} \tag{13.41}$$

as derived from (13.27). Hence, due to the fact that $\mathbf{Q}$ is orthonormal,

$$\underline{\mathbf{Q}}^T \underline{\mathbf{Q}} = \begin{bmatrix} \mathbf{I} & \mathbf{b}(\lambda) \\ \mathbf{b}^T(\lambda) & d(\lambda) \end{bmatrix} \tag{13.42}$$

in which $\mathbf{I}$ is an unit matrix.

$$\begin{aligned} \mathbf{b}(\lambda) &= -\mathbf{Q}^T \nabla_\lambda h(t,\lambda) = -[\mathbf{q}_1^T \nabla_\lambda h(t,\lambda), \dots, \mathbf{q}_L^T \nabla_\lambda h(t,\lambda)]^T, \\ d(\lambda) &= \{\nabla_\lambda h(\lambda)\}^T \nabla_\lambda h(\lambda). \end{aligned} \tag{13.43}$$

At the lth iteration step with previous parameter estimator as $\boldsymbol{\vartheta}_\lambda^{(l-1)} = [\boldsymbol{\vartheta}^{(l-1)}, \lambda^{(l-1)}]^T$. Denote $\underline{\mathbf{K}}^{(l)} = \{[\underline{\mathbf{Q}}^{(l)}]^T \underline{\mathbf{Q}}^{(l)}\}^{-1}$. Apply the inverse of matrix block decomposition lemma to (13.42), in which $\mathbf{b}(\lambda)$, $d(\lambda)$, $\underline{\mathbf{Q}}$ are replaced by $\mathbf{b}(\lambda^{(l-1)})$,

$d(\lambda^{(l-1)})$ and $\underline{\mathbf{Q}}^{(l)}$, to yield,

$$\underline{\mathbf{K}}^{(l)} = \frac{1}{h(\lambda^{(l-1)})} \begin{bmatrix} \mathbf{I} + \mathbf{b}(\lambda^{(l-1)})\mathbf{b}^T(\lambda^{(l-1)}) & -\mathbf{b}(\lambda^{(l-1)}) \\ -\mathbf{b}^T(\lambda^{(l-1)}) & 1 \end{bmatrix}, \tag{13.44}$$

where

$$h(\lambda^{(l-1)}) = d(\lambda^{(l-1)}) - \mathbf{b}^T(\lambda^{(l-1)})\mathbf{b}(\lambda^{(l-1)}). \tag{13.45}$$

The proposed algorithm is fast and stable, as the update of $\underline{\mathbf{K}}^{(l)}$ over iteration step l is simplified with no need of matrix inversion. Following deriving the MLE for λ by using the above fast Gauss-Newton algorithm, the Box-Cox transformation is readily applied to form the transformed output.

For system modelling and control, it is desirable that the model is represented as (13.28) with a minimal number of M basis functions. Provided that the optimal parameter λ used in Box-Cox transformation, the number and location of candidate RBF centers are known, various orthogonal forward regression(OFR) algorithms [13, 35–37] are readily applicable for model structure selection and parameter estimation for the Box-Cox RBF network, simply by using the transformed system output as target of the RBF networks output. This is based on the assumption that the MLE estimator of λ as derived above can be treated as true parameter of λ. For the complete algorithm to construct a sparse Box-Cox RBF model with good generalisation, see [21], which simply extends the algorithm [19, 27] (see also Sect. 13.2) to Box-Cox RBF model.

Example 13.2 (Hong [21]) Non-stationary time series data: Beveridge wheat price indices from 1500 to 1869 [39]. The comparison study comprises two different topologies, the conventional RBF network and the Box-Cox RBF. For both topologies, all the data ($N = 370$) were used as training data set, and the input vector was set as $\mathbf{x}_k = [y_{k-1}, y_{k-2}, y_{k-3}, y_{k-4}, y_{k-5}]^T$. The thin-plate-spline basis function $g_i(\mathbf{x}_k) = \|\mathbf{x}_k - \mathbf{c}_i\|^2 \log \|\mathbf{x}_k - \mathbf{c}_i\|$ was used as basis function with all data sets initially used as candidate centres $\mathbf{c}_i$'s. The experimental results is given in Fig. 13.2 and the further details can be found in [21].

13.5 The RBF Network with Boundary Value Constraints (BVC-RBF)

In this section we describe a newly introduced RBF topology [22] which aims to handle effectively a special type of prior knowledge given by a type of boundary value constraints (BVC). In many modelling tasks, there are more or less some prior knowledge available. Note that most of the RBF modelling algorithms are conditioned on that the model is determined based on the observational data only, so that these fit into the statistical learning framework. However, despite the fact

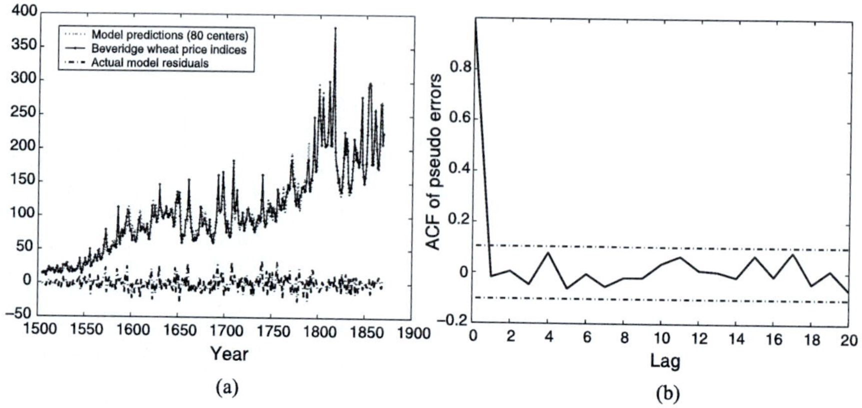

Fig. 13.2 (**a**) Modelling results of the Box-Cox RBF networks (80 centres) for Example 13.2; and (**b**) Autocorrelation function coefficients based on pseudo errors of Box-Cox RBF network (80 centre model) for Example 13.2, where the dotted line is calculated as $\pm\frac{1.96}{\sqrt{N}}$. © 2007 IET

that the availability of prior knowledge about the system could help to improve the model generalization, incorporating the deterministic prior knowledge into a statistically learning paradigm would generally make the development of modelling algorithms more difficult if not impossible.

The new topology of RBF network [22] is referred as the BVC-RBF and as shown in Fig. 13.3. The BVC-RBF is constructed and parameterized based on the given BVC and has the capability of satisfying the BVC automatically. Because the BVC-RBF remains as a linear-in-the-parameter structure just as the conventional RBF does, it is advantageous that many of the existing modelling algorithms for a conventional RBF are almost directly applicable without added algorithmic complexity nor computational cost. Consequently the BVC-RBF effectively lends itself as a single framework in which both the deterministic prior knowledge and stochastic data are fused with ease.

Consider the identification of a semi-unknown system. Given a training data set D_N consisting of N input/output data pairs $\{\mathbf{x}_k, y_k\}_{k=1}^N$, the goal is to find the underlying system dynamics

$$y_k = f(\mathbf{x}_k) + \varepsilon_k. \tag{13.46}$$

The underlying function $f : \Re^m \to \Re$ is unknown. ε_k is the noise, which is often assumed to be independent and identically distributed (i.i.d.) with constant variance. In addition, it is required that the model *strictly* satisfies a set of $\mathscr{L}$ boundary value constraints (BVC) given by

$$f(\mathbf{x}'_j) = d_j, \quad j = 1, \ldots, \mathscr{L}, \tag{13.47}$$

where $\mathbf{x}'_j \in \Re^m$ and $d_j \in \Re$ are known. Note that the information from the given BVC is fundamentally different from that of the observational data set D_N and

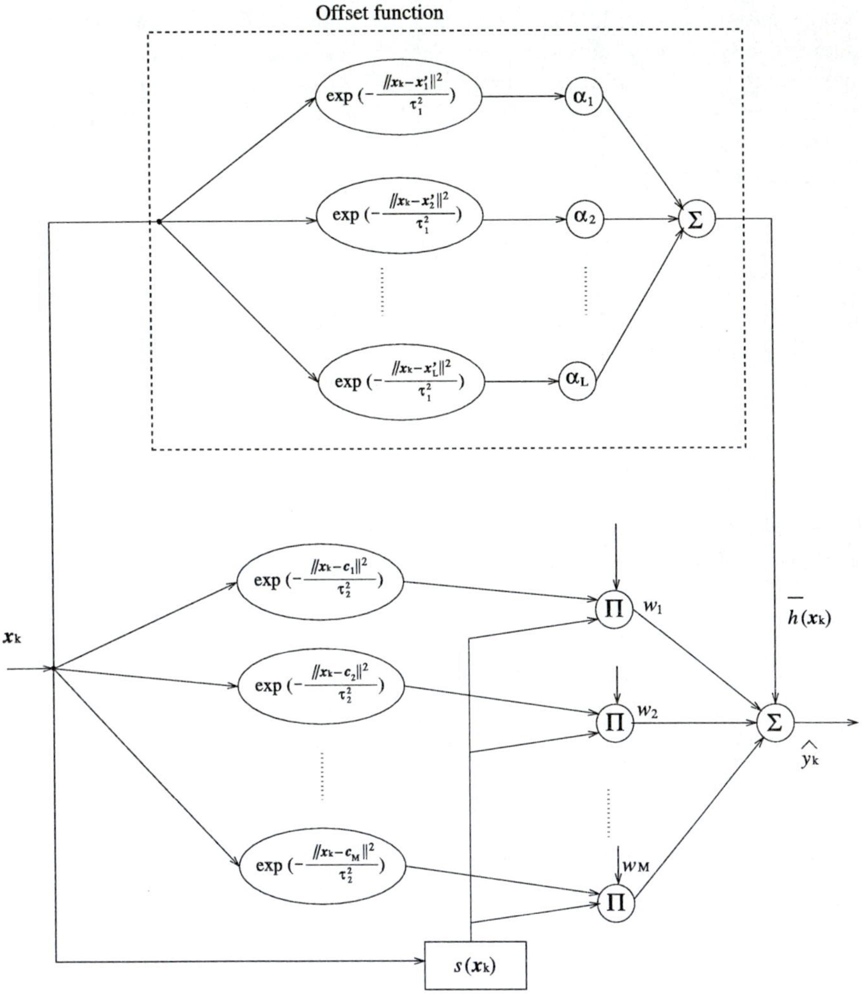

Fig. 13.3 A graphical illustration of the BVC-RBF network

should be treated differently. The BVC is a deterministic condition but D_N is subject to observation noise and possesses stochastic characteristics. The BVC may represent the fact that at some critical regions, there is a complete knowledge about the system.

If the underlying function $f(\cdot)$ is represented by a conventional RBF network (formulated as (13.1)), then resultant RBF network using the conventional modelling procedure, e.g. Sect. 13.2, cannot meet the BVC given by (13.47). Clearly the prior knowledge about the system from BVC should help to improve the model generalization, but equally this makes the modelling process more difficult, since with constraints we are facing a constrained optimization problem. A simple yet effective treatment was introduced to ease the problem [22], as summarized below.

The design goal in [22] is to find a new topology of RBF such that the BVC is automatically satisfied, and as a consequence the system identification can be carried out without added algorithmic complexity nor computational cost compared to any modelling algorithm for a conventional RBF. The BVC-RBF is parameterized and dependent upon the given BVC as shown below. Consider the following BVC-RBF model representation

$$\hat{y}_k = \sum_{i=1}^{M} g_i(\mathbf{x}_k) w_i + \bar{h}(\mathbf{x}_k), \tag{13.48}$$

where the proposed RBF function for BVC-RBF model [22] is given by

$$g_i(\mathbf{x}_k) = s(\mathbf{x}_k) \exp\left(-\frac{\|\mathbf{x}_k - \mathbf{c}_i\|^2}{\tau_1^2}\right), \tag{13.49}$$

where $s(\mathbf{x}_k) = \sqrt[\mathscr{L}]{\prod_{j=1}^{\mathscr{L}} \|\mathbf{x}_k - \mathbf{x}'_j\|}$ is the geometric mean of the data sample $\mathbf{x}_k$ to the set of boundary values $\mathbf{x}'_j$, $j = 1, \ldots, \mathscr{L}$. $\mathbf{c}_i \in \Re^m$ is the RBF centers, τ_1 is a positive scalar.

$$\bar{h}(\mathbf{x}_k) = \sum_{j=1}^{\mathscr{L}} \alpha_j \exp\left(-\frac{\|\mathbf{x}_k - \mathbf{x}'_j\|^2}{\tau_2^2}\right), \tag{13.50}$$

τ_2 is also a positive scalar. α_j is a set of parameters that is obtained by solving a set of linear equations $g(\mathbf{x}_j) = d_j$, $j = 1, \ldots, \mathscr{L}$. That is

$$\boldsymbol{\alpha} = \bar{\mathbf{H}}^{-1}\mathbf{d}, \tag{13.51}$$

where $\boldsymbol{\alpha} = [\alpha_1, \ldots, \alpha_{\mathscr{L}}]^T$, $\mathbf{d} = [d_1, \ldots, d_L]^T$ and $\bar{\mathbf{H}}$ is given by

$$\bar{\mathbf{H}} = \begin{pmatrix} 1 & e^{-\frac{\|\mathbf{x}'_1 - \mathbf{x}'_2\|^2}{\tau_2^2}} & \ldots & e^{-\frac{\|\mathbf{x}'_1 - \mathbf{x}'_{\mathscr{L}}\|^2}{\tau_2^2}} \\ e^{-\frac{\|\mathbf{x}'_2 - \mathbf{x}'_1\|^2}{\tau_2^2}} & 1 & \ldots & e^{-\frac{\|\mathbf{x}'_2 - \mathbf{x}'_{\mathscr{L}}\|^2}{\tau_2^2}} \\ \vdots & \vdots & \ddots & \vdots \\ e^{-\frac{\|\mathbf{x}'_{calL} - \mathbf{x}'_1\|^2}{\tau_2^2}} & e^{-\frac{\|\mathbf{x}'_{\mathscr{L}} - \mathbf{x}'_2\|^2}{\tau_2^2}} & \ldots & 1 \end{pmatrix} \tag{13.52}$$

In the case of the ill-conditioning, the regularization technique is applied to the above solution. It is easy to verify that with the proposed topology of BVC-RBF neural networks, the BVC is automatically satisfied [22]. In general, $g_i(\mathbf{x}_k)$ and $\bar{h}(\mathbf{x}_k)$ act as building blocks of the BVC-RBF networks in (13.48), with a novel feature compared to most of the existent neural networks architecture. That is, by resorting to the given boundary conditions, its topology is designed for the boundary constraints satisfaction, or more generally, for incorporating given prior knowledge. Note that the boundary condition satisfaction via the network topology is an inherent, but often overlooked, feature for any model representation. For example, the autoregressive with exogenous output (ARX) model automatically satisfies the boundary condition of $f(\mathbf{0}) = 0$, and for the conventional RBF with the Gaussian basis functions, $f(\infty) = 0$. The aim of [22] is to introduce and exploit the boundary condition satisfaction via the network topology in a controlled manner, so that the modelling performance may be enhanced by incorporating the a prior knowledge via boundary conditions satisfaction.

Substituting (13.48) into (13.46) and defining an auxiliary output variable $z_k = y_k - \bar{h}(\mathbf{x}_k)$, we have

$$z_k = \sum_{i=1}^{M} g_i(\mathbf{x}_k) w_i + e_k \tag{13.53}$$

conforming to (13.1), except that the auxiliary output variable z_k is used as the target of the first term in (13.48) (the adjustable part of BVC-RBF). Aiming for improved model robustness, the D-optimality in experimental design [40] has been incorporated in the D-optimality based model selective criterion [41] to select M regressors in a forward regression manner. For completeness the combined D-optimality based orthogonal least squares algorithm [41] is used in the following example [22].

Example 13.3 ([22]) The Matlab logo was generated by the first eigenfunction of the L-shaped membrane. A 31×31 meshed data set $f(x_1, x_2)$ is generated by using Matlab command *membrane.m*, which is defined over a unit square input region $x_1 \in [0, 1]$ and $x_2 \in [0, 1]$. The data set $y(x_1, x_2) = f(x_1, x_2) + \varepsilon(x_1, x_2)$ is then generated by adding a noise term $\varepsilon(x_1, x_2) \sim N(0, 0.01^2)$. We use all the data points within the boundary as the training data set D_N consisting of the set of $\{x_1, x_2, y(x_1, x_2)\}$ coordinates ($N = 721$). For comparison, the combined D-optimality based orthogonal least squares algorithm was applied [41] to identify a sparse conventional RBF model. The modeling results are shown in Fig. 13.4 and Table 13.2. It is shown that the BVC-RBF can achieve significant improvement over the RBF in terms of the modeling performance to the true function. In particular we note that the BVC can be satisfied with the proposed BVC-RBF model, but not by the conventional RBF. The detail of the parameters setting for the experiment can be found in [22].

13.6 Conclusions

Our recent work on diversified RBF topologies is reviewed. Three different topologies have been introduced aimed at enhancing the modelling capabilities of RBF network by modifying their topologies for specific problems; (i) the RBF network with tunable nodes is introduced with the aim of flexible basis function shaping for achieving the minimal model and improved model generalisation; (ii) the Box-Cox RBF network is aimed at effectively handling some dynamical processes in which the model residuals exhibit heteroscedasticity; and (iii) the BVC-RBF is introduced

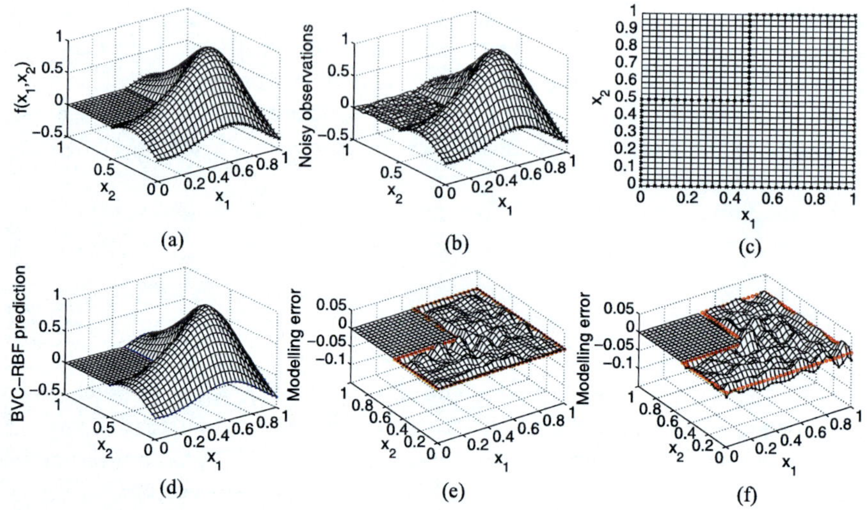

Fig. 13.4 Example 13.3; (**a**) the true function $f(x_1, x_2)$; (**b**) noisy data $y(x_1, x_2)$; (**c**) the boundary points and (**d**) the prediction of the resultant BVC-RBF model; (**e**) the modelling error between the true function and the model prediction ($\hat{y}(x_1, x_2) - f(x_1, x_2)$) for the BVC-RBF model; and (**f**) The modelling error for the RBF model. IEEE©2008 IEEE

Table 13.2 A comparison between the conventional RBF and the BVC-RBF network for Example 13.3

	Model size M	MSE $\frac{1}{N}\sum(\hat{y}-f)^2$	MSE $\frac{1}{N}\sum(\hat{y}-y)^2$	MSE (boundary) $\frac{1}{\mathscr{L}}\sum_j(\hat{y}(\mathbf{x}'_j)-d_j)^2$
BVC-RBF	68	4.3787×10^{-5}	1.0736×10^{-4}	7.2598×10^{-11}
RBF	91	1.0229×10^{-4}	1.6894×10^{-4}	2.1249×10^{-4}

in order to achieve automatic constraints satisfaction and incorporating deterministic *prior* knowledge with ease. It is advantageous that the model construction algorithms for the diversified RBF topologies are either direct application or extension of linear learning algorithms. In each case, an illustrative example is used to demonstrate the efficacy of the proposed topology, together with the application of the modeling construction algorithm.

References

1. Hong, X., Mitchell, R.J., Chen, S., Harris, C.J., Li, K., Irwin, G.W.: Model selection approaches for nonlinear system identification: a review. Int. J. Syst. Sci. **39**(10), 925–946 (2008)
2. Harris, C.J., Hong, X., Gan, Q.: Adaptive Modelling, Estimation and Fusion from Data: A Neurofuzzy Approach. Springer, Berlin (2002)

3. Brown, M., Harris, C.J.: Neurofuzzy Adaptive Modelling and Control. Prentice Hall, Upper Saddle River (1994)
4. Ruano, A.E.: Intelligent Control Systems using Computational Intelligence Techniques. IEE Publishing, New York (2005)
5. Murray-Smith, R., Johansen, T.A.: Multiple Model Approaches to Modelling and Control. Taylor and Francis, London (1997)
6. Fabri, S.G., Kadirkamanathan, V.: Functional Adaptive Control: An Intelligent Systems Approach. Springer, Berlin (2001)
7. Leonard, J.A., Kramer, M.A.: Radial basis function networks for classifying process faults. IEEE Control Syst. Mag. **11**(3), 31–38 (1991)
8. Caiti, A., Parisini, T.: Mapping ocean sediments by RBF networks. IEEE J. Ocean. Eng. **19**(4), 577–582 (1994)
9. Li, Y., Sundararajan, N., Saratchandran, P., Wang, Z.: Robust neuro-H_∞ controller design for aircraft auto-landing. IEEE Trans. Aerosp. Electron. Syst. **40**(1), 158–167 (2004)
10. Ng, S.X., Yee, M.S., Hanzo, L.: Coded modulation assisted radial basis function aided turbo equalization for dispersive Rayleigh-fading channels. IEEE Trans. Wirel. Commun. **3**(6), 2198–2206 (2004)
11. Stone, M.: Cross validatory choice and assessment of statistical predictions. J. R. Stat. Soc. B **36**, 117–147 (1974)
12. Akaike, H.: A new look at the statistical model identification. IEEE Trans. Autom. Control **AC-19**, 716–723 (1974)
13. Chen, S., Billings, S.A., Luo, W.: Orthogonal least squares methods and their applications to non-linear system identification. Int. J. Control **50**, 1873–1896 (1989)
14. Korenberg, M.J.: Identifying nonlinear difference equation and functional expansion representations: the fast orthogonal algorithm. Ann. Biomed. Eng. **16**, 123–142 (1988)
15. Wang, L., Mendel, J.M.: Fuzzy basis functions, universal approximation, and orthogonal least-squares learning. IEEE Trans. Neural Netw. **5**, 807–814 (1992)
16. Hong, X., Harris, C.J.: Neurofuzzy design and model construction of nonlinear dynamical processes from data. IEE Proc., Control Theory Appl. **148**(6), 530–538 (2001)
17. Zhang, Q.: Using wavelets network in nonparametric estimation. IEEE Trans. Neural Netw. **8**(2), 1997 (1993)
18. Billings, S.A., Wei, H.L.: The wavelet-narmax representation: a hybrid model structure combining polynomial models with multiresolution wavelet decompositions. Int. J. Syst. Sci. **36**(3), 137–152 (2005)
19. Hong, X., Sharkey, P.M., Warwick, K.: Automatic nonlinear predictive model construction using forward regression and the PRESS statistic. IEE Proc., Control Theory Appl. **150**(3), 245–254 (2003)
20. Chen, S., Hong, X., Harris, C.J.: Construction of tunable radial basis function networks using orthogonal forward selection. IEEE Trans. Syst. Man Cybern., Part B, Cybern. **39**(2), 457–466 (2009)
21. Hong, X.: Modified radial basisfunction neural networks using output transformation. IEE Proc., Control Theory Appl. **1**(1), 1–8 (2007)
22. Hong, X., Chen, S.: A new RBF neural network with boundary value constraints. IEEE Trans. Syst. Man Cybern., Part B, Cybern. **39**(1), 298–303 (2009)
23. Vapnik, V.: The Nature of Statictical Learning Theory. Springer, New York (1995)
24. Gunn, S.R.: Support vector machine for classification and regression. Technical Report, ISIS Research Group, Dept of Electronics and Computer Science, University of Southampton (1998)
25. Chen, S.S., Donoho, D.L., Saunders, M.A.: Atomic decomposition by basis pursuit. SIAM J. Sci. Comput. **20**(1), 33–61 (1998)
26. Tipping, M.E.: Sparse Bayesian learning and the relevance vector machine. J. Mach. Learn. Res. **1**, 211–244 (2001)
27. Chen, S., Hong, X., Harris, C.J., Sharkey, P.M.: Sparse modelling using orthogonal forward regression with PRESS statistic and regularization. IEEE Trans. Syst. Man Cybern., Part B, Cybern. **34**(2), 898–911 (2004)

28. Myers, R.H.: Classical and Modern Regression with Applications, 2nd edn. PWS-KENT, Boston (1990)
29. Chen, S., Wang, X.X., Harris, C.J.: Experiments with repeating weighted boosting search for optimization signal processing applications. IEEE Trans. Syst. Man Cybern., Part B, Cybern. **35**(4), 682–693 (2005)
30. [online]: Available: http://www.ics.uci.edu/~mlearn/MLRepository.html
31. Chen, S., Hong, X., Luk, B.L., Harris, C.J.: Nonlinear system identification using particle swarm optimization tuned radial basis function models. Int. J. Bio-Inspired Comput. **1**(4), 246–258 (2009)
32. Box, G.E.P., Cox, D.R.: An analysis of transformation. J. R. Stat. Soc. B **26**(2), 211–252 (1964)
33. Carroll, R.J., Ruppert, D.: Transformation and Weighting in Regression. Chapman and Hall, London (1988)
34. Ding, A.A., He, X.: Backpropagation of pseudoerrors: neural networks that are adaptive to heterogeneous noise. IEEE Trans. Neural Netw. **14**(2), 253–262 (2003)
35. Chen, S., Wu, Y., Luk, B.L.: Combined genetic algorithm optimization and regularized orthogonal least squares learning for radial basis function networks. IEEE Trans. Neural Netw. **10**, 1239–1243 (1999)
36. Hong, X., Harris, C.J.: Nonlinear model structure design and construction using orthogonal least squares and D-optimality design. IEEE Trans. Neural Netw. **13**(5), 1245–1250 (2002)
37. Chen, S.: Locally regularised orthogonal least squares algorithm for the construction of sparse kernel regression models. In: Proceedings of 6th Int. Conf. Signal Processing, Beijing, China, pp. 1229–1232 (2002)
38. Powell, M.J.D.: Problems related to unconstrained optimization. In: Murray, W. (ed.) Numerical Methods for Unconstrained Optimization, pp. 29–55. Academic Press, London (1972)
39. Hipel, K.W., Mcleod, A.I.: Time Series Modelling of Water Resources and Environmental Systems. Elsevier, Amsterdam (1994)
40. Atkinson, A.C., Donev, A.N.: Optimum Experimental Designs. Clarendon Press, Oxford (1992)
41. Hong, X., Harris, C.J.: Experimental design and model construction algorithms for radial basis function networks. Int. J. Syst. Sci. **34**(14–15), 733–745 (2003)

Part II
Applications of System Identification

Chapter 14
Application of Filtering Methods for Removal of Resuscitation Artifacts from Human ECG Signals

Ivan Markovsky, Anton Amann, and Sabine Van Huffel

14.1 Introduction

We are dealing with a particular filtering problem that arises in a biomedical signal processing application—removal of resuscitation artifacts from ventricular fibrillation human ECG signals. The measured ECG signal y has two components: the ventricular fibrillation ECG signal v, which is the useful signal, and the resuscitation artifacts c, which is the disturbance. Our goal is to extract the unknown useful signal v from the given signal y. In the application at hand, we are given another signal u that has causal relation with the artifact c.

A method for artifact removal produces an approximation $\widehat{v}$ of v and as an approximation $\widehat{c}$ of c, using the observed signals u and y, see Fig. 14.1. An important requirement for an artifact removal procedure to be practical is that it works in real-time, i.e., it computes an approximation of the ventricular fibrillation ECG signal on-line and at each time instant, it uses only past values of the measurements (causality). Such a procedure is a dynamical system and is called a filter.

In order to specify a well defined mathematical problem for ECG artifacts removal, we need to impose additional assumptions. The main question, addressed in

I. Markovsky (✉)
School of Electronics and Computer Science, University of Southampton,
SO17 1BJ Southampton, UK
e-mail: im@ecs.soton.ac.uk

A. Amann
Innsbruck Medical University and Department of Anesthesia and General Intensive Care,
Anichstr 35, 6020 Innsbruck, Austria
e-mail: Anton.Amann@i-med.ac.at

S. Van Huffel
ESAT-SCD, K.U. Leuven, Kasteelpark Arenberg 10, 3001 Leuven, Belgium
e-mail: Sabine.VanHuffel@esat.kuleuven.be

L. Wang, H. Garnier (eds.), *System Identification, Environmental Modelling, and Control System Design*,
DOI 10.1007/978-0-85729-974-1_14,

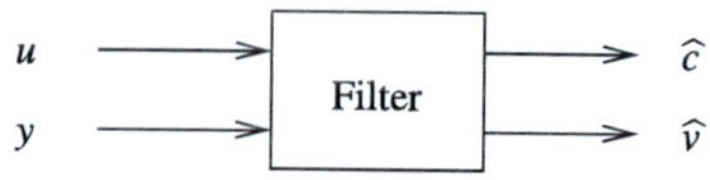

Fig. 14.1 The filter uses as inputs the reference signal u and the artifacts corrupted ECG signal y, and produces as outputs approximations $\widehat{c}$ and $\widehat{v}$ of, respectively, the artifacts and the pure ECG signals

this paper, is what assumptions are relevant for the ECG artifacts removal problem and how they translate to classical filtering problems. Underlying assumptions, used in the literature, are not explicitly spelt out or they are not scrutinised from the point of view of their practical relevance. Next we state three important open question related to design of ECG artifacts removal methods. However, we do not set our objectives too high and address in this paper only one of them.

A global assumption used in the literature is that the observed ECG signal y is the sum of the ventricular fibrillation ECG signal v and the artifact c, i.e.,

$$y = v + c. \tag{14.1}$$

A more general alternative to the linear mixture model (14.1) is $y = f(v, c)$, where f is a possibly nonlinear function. The question

Q1: What class of functions f is "most relevant" for modelling human ECG signals corrupted by artifacts?

is currently unexplored. (The meaning of "most relevant" is related to our question Q3, stated below.) Because of the following reasons, we leave question Q1 for future study and concentrate in this paper on the linear mixture model.

- We do not have the necessary data for an empirical study of question Q1. (For empirical study of question Q1, one needs separately recorded human ECG, human artifacts signals, and corresponding measured human ECG signals corrupted by artifacts.)
- The linear mixture model is the simplest special case of $y = f(v, c)$, however, as discussed in the rest of the paper even in this case there are open problems and room for improvement of the current state-of-the-art.
- Methods developed in the linear setting are a prerequisite for the development of the intrinsically more complicated methods based on nonlinear models.

Even in the linear case, there are infinitely many ways in which y can be decomposed as $y = \widehat{v} + \widehat{c}$ and without extra knowledge they are all equally good. An important question we are going to address in this paper is

Q2: What are the distinguishing properties of the ventricular fibrillation ECG signal v and the resuscitation artifact signal c?

Our study is empirical and is based on a database of separately recorded resuscitation artifacts and reference signals (the arterial blood pressure) from pigs and ventricular fibrillation ECG signals from human. The same data is used in [10] for tuning and evaluation of an adaptive filtering method.

The classical bandpass, Kalman, and adaptive filtering methods rely on different types of prior knowledge that enable the separation of the useful signal from the artifacts. Band-pass filtering relies on the assumption that the useful signal and the artifacts have non-overlapping spectral bands, while the Kalman filter uses a linear model for one of the components. Based on the model the Kalman filter does optimal least squares separation. The adaptive filtering methods are also model based, however, they identify the model in real-time as the data is collected, so that in the adaptive case the model is time-varying.

The methods for ECG artifacts removal, presented in [1, 4, 10] are adaptive. Conceptually these methods are similar: they use a finite impulse response (FIR) model for the relation between the reference signal u and the artifact c and minimise related cost functions in order to identify the model parameters. Apparently the methods differ in their algorithmic implementation: [1] uses directly the least squares solution; [4] uses a recursive algorithm, called matching pursuit; and [10] uses a Kalman filtering algorithm. The methods, however, use different parameters that are empirically chosen for optimal performance on the test data. The main effort in the development of these methods is to choose "suitable" values for these parameters.

In general, different methods for artifacts removal produce different approximations $\widehat{v}$ of the unknown signal v. The quality of approximation of v by $\widehat{v}$ is evaluated in [1, 4, 10] by the signal-to-noise ratio (SNR)

$$\mathrm{SNR}(\widehat{v}, v) := 20 \log_{10}\left(\frac{\|v\|}{\|v - \widehat{v}\|}\right), \tag{14.2}$$

where $\|\cdot\|$ denotes the 2-norm. Of course, $\mathrm{SNR}(\widehat{v}, v)$ can not be computed in real-life applications. It is used for evaluation of the methods on artificially constructed ECG signals y, where the true signal v is known. Although 2-norm is a standard choice in signal processing, its biomedical relevance needs justification. The question

Q3: What norm $\|\cdot\|$ in the definition of (14.2) has medical relevance?

is unexplored. Admittedly the question is fuzzy because it relays on the notion of "medical relevance". This is left as a topic for further study.

The contributions of this paper are:

1. give a nontechnical tutorial exposition of the bandpass, Kalman, and adaptive filtering methods in the context of their application for ECG artifacts removal,
2. clarify the rationale for applying these methods, and
3. perform an empirical study comparing the methods.

Although there are numerous texts describing the theory of the classical filtering methods, there is no single reference describing the bandpass, Kalman, and adaptive methods in the context of ECG artifacts removal. In addition, most of the signal processing literature adopts a stochastic setting that requires certain mathematical sophistication and in our opinion increases the theory–practise gap. For this reason we adopt a simpler and more intuitive deterministic setting.

The rationale for applying the bandpass filtering methods is well known: frequency separation of the signal of interest v and the artifact c. The rationale for applying the more advanced Kalman and adaptive methods, however, is less well known. We point out assumptions on which the Kalman and adaptive filters are based and check to what extent they are satisfied on the ECG signal and the artifacts in the considered database of signals. The prior knowledge for the Kalman filter is the given model for the relation between u and c. The adaptive filters automatically identify such a model, however, for the identification step to be possible, certain conditions have to be satisfied. The tunable parameters of the adaptive methods are related to the complexity of the model class and the identification criterion, so that they influence the identifiability conditions.

The empirical results published in [1, 4, 10] do not compare the methods proposed in the corresponding papers with other methods from the literature. Moreover, the reported empirical results are not reproducible in the sense of [2], i.e., the full computational environment that is used to produce the results is not published. We fulfil this gap in the literature by presenting a comparative study of the methods of [1, 10] and classical filtering methods described in this paper. Matlab implementation of the methods used in our simulation study are available from: ftp://ftp.esat.kuleuven.be/pub/SISTA/markovsky/abstracts/06-212.html The data is available upon request. (Please, direct your requests to Prof. A. Amann (anton.amann@i-med.ac.at).)

14.2 Methods for Artifacts Removal

14.2.1 Band-Pass Filtering

First we consider the application of the classical band-pass filtering method for artifact removal. The key assumption for applying this technique is that the useful signal, the ventricular fibrillation ECG signal v, and the disturbance, the resuscitation artifacts c, have well separated spectral bands. The left plot in Fig. 14.2 shows the power spectral densities of v and c in a particular experiment. (The same data is used for comparison with the other methods in Sects. 14.2.2 and 14.2.3.) The plot shows that up to $f_0 = 0.075$ Hz (normalised frequency, which corresponds to 3 Hz physical frequency) the spectrum of c dominates the spectrum of v and for frequencies above 0.075 Hz the opposite is true—the spectrum of v dominates the spectrum of c. Therefore, low-pass filtering, with a cut-off frequency f_0, can extract from the signal $y = c + v$ an approximation $\widehat{c}$ of the resuscitation artifacts c and then find an approximation of v as $\widehat{v} := y - \widehat{c}$.

The structure of the low-pass filter for artifacts removal is shown in Fig. 14.3. Note that this method does not use the reference signal u, i.e., the method can be applied even in the case when no reference signal is available.

We design a finite impulse response (FIR) low-pass filter with 100 time lags using the window method [9]. The ideal low-pass filter has magnitude one at all frequencies with magnitude less than the cut-off frequency and magnitude zero at all other

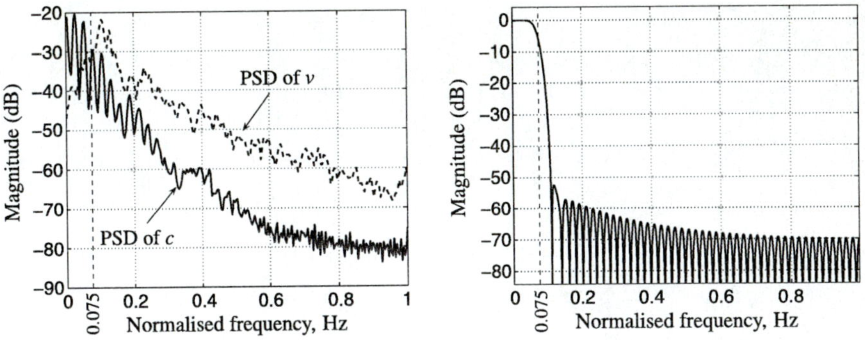

Fig. 14.2 Power spectral densities (PSD) of v and c (*left*) and magnitude response of the low-pass filter (*right*)

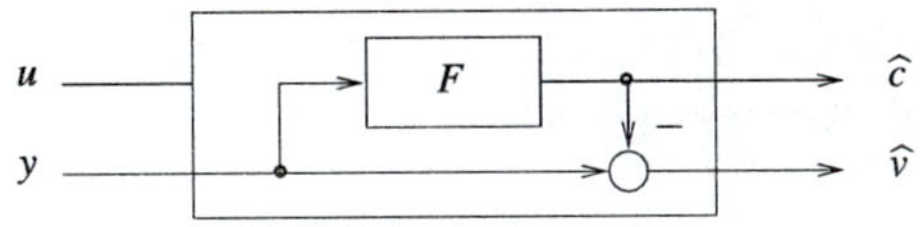

Fig. 14.3 Structure of the low-pass filter for ECG artifacts removal. The filter F is low-pass with a cut-off frequency f_0

frequencies. The impulse response of this ideal filter is infinite and non-causal, so that it is not implementable. The windowing method resolves the implementability problem by truncating the ideal impulse response by a window of a specified length. In the example, we use the Hamming window.

The right plot in Fig. 14.2 shows the filter magnitude response. Due to the overlap in the spectra of v and c, low-pass filtering does not achieve perfect separation. In order to illustrate this we show in Fig. 14.4 the signal $\widehat{c}$ obtained by processing c and v with the low-pass filter. These two cases correspond to the extremes $y = c$ with $\mathrm{SNR}(y, v) = 0$ and $y = v$ with $\mathrm{SNR}(y, v) = \infty$. Ideally, the filter does not modify c and completely rejects v. In practise, it deforms c and attenuates v. Figure 14.5 shows the performance of the low-pass filtering technique for artifact removal on a signal $y = c + v$ with SNR 0 dB. The SNR of the restored signal is 5.7 dB.

In summary, low-pass filtering requires to

1. choose the cut-off frequency f_0,
2. design a low-pass filter F with cut-off frequency f_0, and
3. apply the filter F on the observed data.

Steps 1 and 2 are done *off-line*, so that in the simulations they should use only an identification part of the data, i.e., data available for tuning the methods. Step 3 is done in *real-time* as test data is obtained. Most expensive computationally is step 2. However, there are well developed methods for filter design that are readily available in free and commercial software, e.g., the Signal Processing Toolbox of Matlab, so that from the user point of view, this step is trivial. Once the filter is designed, applying it on the data involves recursive computation that is very fast

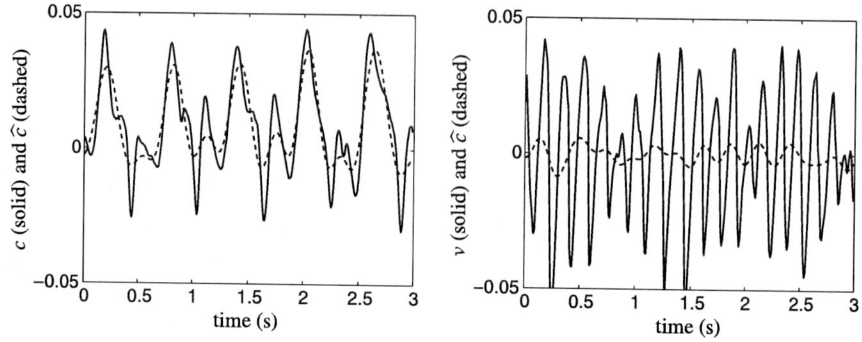

Fig. 14.4 Deformation of the signal c and attenuation of v by the low-pass filter

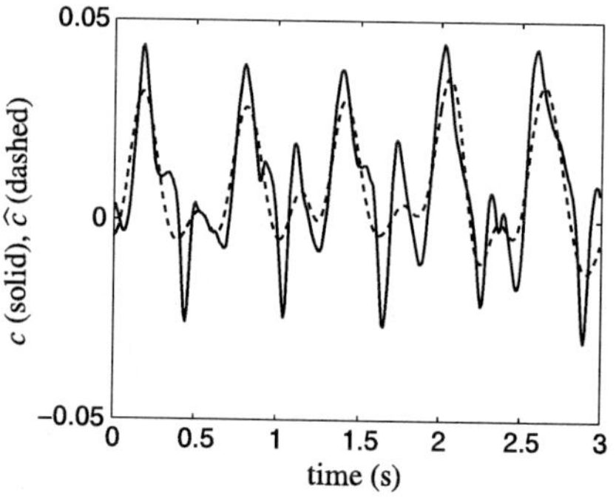

Fig. 14.5 Approximated (*dashed*) and true (*solid*) ventricular artifact signals using low-pass filtering for data signal y with SNR 0 dB

and can be implemented on a digital signal processor for practical implementation of the method.

Step 1 requires human interaction and is ad-hoc. This is a weakness of the method. Note, however, that a *single* design parameter has to be chosen. Simulation results with the considered database show that the choice $f_0 = 0.075$ Hz for the cut-off frequency gives good results on the average for all data sets. This experimental result suggests that the low-pass filtering method is robust with respect to the choice of the f_0 parameter.

14.2.2 Kalman Filtering

By assumption the reference signal u has a causal relation with the artifact signal c. Formally, this means that there is a model M, such that when u is given as an input to M and the initial conditions are properly chosen, the resulting output is c. The model M, however, may be a complicated nonlinear time-varying one, while in this section we consider simple linear time-invariant models. In addition, apart from u other unmeasurable signals may affect c. Finally, there may be measurement

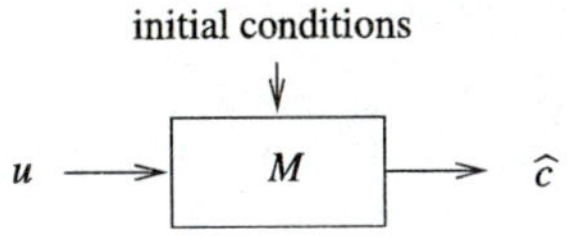

Fig. 14.6 Model for c: $\widehat{c}$ is "close" to c

noise on the reference and the corrupted ECG signals. For these reasons, in practise, we can not obtain an exact relation between u and c. The model M is only an approximation of the unknown exact relation $u \mapsto c$, see Fig. 14.6.

An artifact removal method, called "naive model based method", using a model M for the relation between u and c is shown on the left plot of Fig. 14.7. This method uses the reference signal u and the prior knowledge about the relation between u and c in the form of the model M. Note that the naive method sets the initial conditions for M to zero. This is an arbitrary choice (hence the name of the method).

A proper way of choosing the initial conditions is actually given by the celebrated Kalman filter. The Kalman filter for M accepts as inputs both u and y and recursively updates the initial conditions for optimal prediction $\widehat{c}$ of c. The Kalman filter, however, uses extra prior knowledge: an initial value x_{ini} for the initial conditions and a matrix P_{ini} related to the uncertainty of (or, equivalently, the confidence in) x_{ini}. In addition, the Kalman filter produces optimal prediction of c in certain specified sense, that involves choice of an approximation criterion.

The main question in applying the naive model based method or the optimal (Kalman) filtering method is the selection of the model M. In a practical artifacts removal problem such a model is *not* given as a prior knowledge but has to be deduced from the data. This lead us to the adaptive filtering methods that compute on-line a model M and, based on M, filter the signal. Before explaining the adaptive methods, however, we give some background information on offline identification of a model from data.

We adopt the deterministic identification setting of [8], where the aim is to minimise the fitting error $\|y - \widehat{c}\|$ over the model parameters and the initial conditions. The criterion $\|y - \widehat{c}\|$ corresponds to what is called output error identification [5], however, no stability constraint is imposed on the identified model.

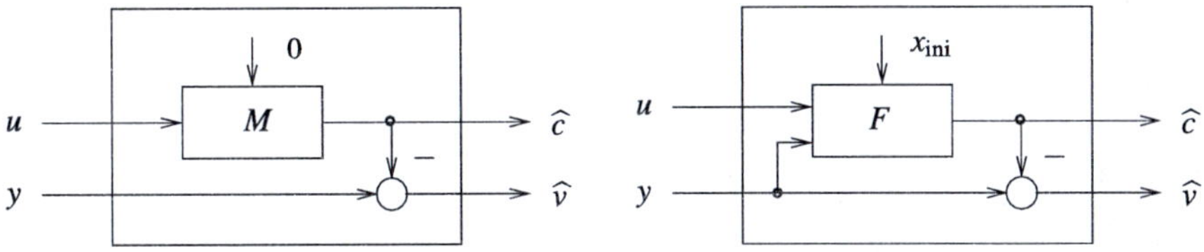

Fig. 14.7 *Left*: naive model based method for artifact removal. The initial conditions of the model M are set to zero. *Right*: method for artifact removal based on the Kalman filter F for M. The initial conditions are recursively updated using the data (u, y), starting from an initial guess x_{ini} with uncertainty P_{ini}

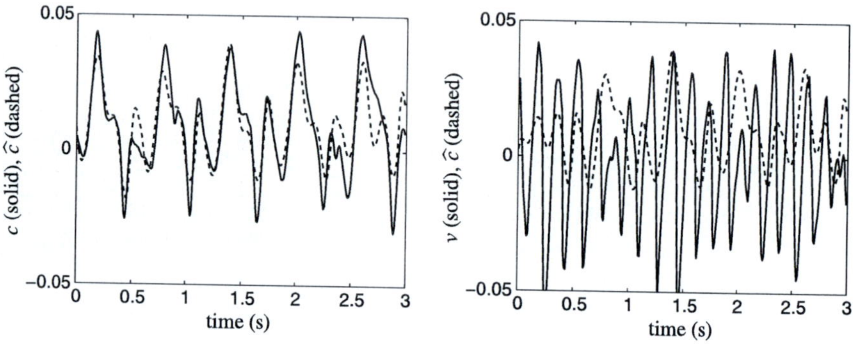

Fig. 14.8 Model validation: fit of the signals c and v (*solid*) by the output $\hat{c}$ (*dashed*) of the model

Fig. 14.9 Approximation (*dashed*) and true (*solid*) artifact signals using Kalman filtering

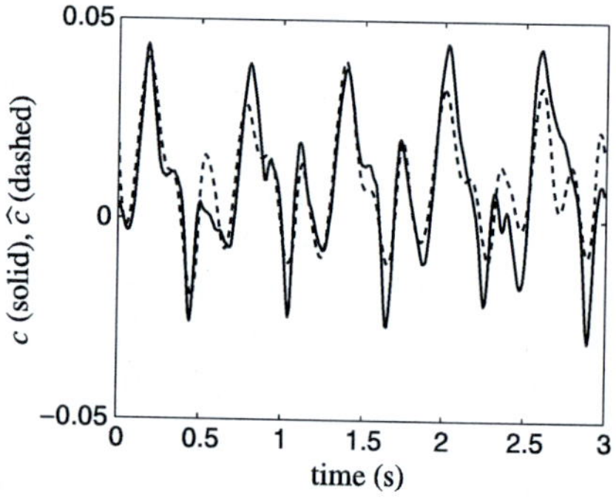

We consider a linear time-invariant model with a state space representation

$$\begin{aligned} x(t+1) &= Ax(t) + Bu(t), \\ \hat{c}(t) &= Cx(t) + Du(t). \end{aligned} \tag{14.3}$$

The Kalman filter is a linear system derived from the identified model, the approximation criterion, and the assumption about the initial conditions $(x_{\mathrm{ini}}, P_{\mathrm{ini}})$. The computational load of finding the approximation $\hat{c}$ by the Kalman filter depends on the order of the model.

Figure 14.8, left, shows the signal c and its best fit $\hat{c}$ by a model of order 5. The match between $\hat{c}$ and c is a measure of the model quality. In the particular example, the identified model (14.3) is a good description of the relation between u and c. Figure 14.8, right, shows the signal v and its best fit $\hat{c}$ by the model. Ideally the Kalman filter should reject v; compare with Fig. 14.4 in the case of low-pass filtering.

Figure 14.9 shows the performance of the Kalman filtering technique for artifact removal on a signal $y = c + v$ with SNR 0 dB. The SNR of the restored signal is 6.08 dB.

In summary, applying the Kalman filtering method requires to

1. identify a model (14.3),
2. design a Kalman filter F for that model, and
3. apply the filter F on the observed data.

As in the case of low-pass filtering, here again, steps 1 and 2 are done off-line and step 3 is done in real-time as the data is observed. Note that identifying the model (14.3) in the Kalman filtering case corresponds to the choice of the cut-off frequency f_0 in the low-pass filtering method. Although there is well developed theory and methods for system identification, step 1 remains the most difficult and ad-hoc one. Once a model is available, the design of the Kalman filter and its application on the data are standard problems that have known solutions and readily available software implementation, e.g., the one in the Control Toolbox of Matlab.

The main questions are what data and what identification method to use. Concerning the identification method, we decide to use the output error criterion without constraining the model to be stable. Software implementation of this method is available in the System Identification Toolbox of Matlab and in [6]. Other identification criteria can also be considered. Currently it is not clear which criterion is most suitable for artifact removal in ECG signals.

Concerning the data used for identifying the model, in a realistic application of the method, one could use only identification data (i.e., data that is not used for testing the method). However, we also apply the Kalman filtering method in an unrealistic test setup: identifying the model (14.3) from the *test* data (u, y). This corresponds to the situation encountered in the next section in the context of the adaptive filtering method, with the difference that the Kalman filtering method is not causal (on the identification step all data is used in batch), while the adaptive filter is causal.

If, in addition to using the test data, the model (14.3) is identified from the true artifact signals c instead of the observed signal y, we refer to the Kalman filtering method as the "reference method". Of course, the reference method is impractical (for a more essential reason than non-causality), but we consider it because it performs Kalman filtering with the "best" possible model for the data. Therefore, it gives the theoretically optimal performance under the assumption that the data (u, c) satisfy a model of the form (14.3) for some "true" parameter values.

14.2.3 Adaptive Filtering

Adaptive filters are conceptually similar to the Kalman filter: they are model-based and perform optimal least squares filtering. The main difference is that in the classical Kalman filter the model is identified off-line or is given as prior knowledge, while in adaptive filters the model is identified from the data in real-time. The recursive identification procedure is the essential part of any adaptive filter.

There is a large variety of adaptive filters depending on the model class and the identification algorithm that are used. A common model class is the set of FIR models with at most l lags

$$\widehat{c}(t) = \widehat{h}_0(t)u(t) + \widehat{h}_1(t)u(t-1) + \cdots + \widehat{h}_l(t)u(t-l). \tag{14.4}$$

For a time-invariant FIR model (i.e., $\widehat{h}$ constant in time), l specifies the model complexity. The vector

$$\widehat{h}(t) := \operatorname{col}\big(\widehat{h}_0(t), \widehat{h}_1(t), \ldots, \widehat{h}_l(t)\big)$$

is the parameter of the FIR model at time t. At each time step, $\widehat{h}(t)$ is determined as a minimum point of a certain cost function depending on $\widehat{h}(t)$ and the data (u, y) (up to and including time instant t, due to the causality requirement). A typical cost function in adaptive filtering is the 2-norm of the error signal $e := y - \widehat{c}$ over the window $t - t_1, t - t_1 + 1, \ldots, t$

$$J\big(\widehat{h}(t)\big) := \sum_{\tau=0}^{t_1} \big(y(t-\tau) - \widehat{c}(t-\tau)\big)^2. \tag{14.5}$$

Common algorithms for adaptive filtering that minimise J are the recursive least squares and gradient descent algorithms. A class of adaptive filters, using gradient descent, is the one of least mean squares (LMS) adaptive filters [3].

Tunable parameters of the FIR adaptive filter are the filter length l and one or more parameters that control the adaptation rate. For example, in the LMS adaptive filters the adaptation rate is determined by the step size λ of the gradient decent algorithm.

14.2.3.1 Overview of the Methods of [1, 4, 10]

The methods of [1, 4, 10] use the FIR model (14.4). The parameter $\widehat{h}$ is selected adaptively from the data, while the parameter l is fixed during the operation of the algorithm and is user defined. In order to distinguish between these two types of parameters, we call the latter ones *hyper-parameters*.

Apart from the model class, another similarity for the methods is that the methods of [1] and [4] minimise the cost function (14.5) and the method [10] minimises a closely related cost function (see (14.11) on p. 286). The window length t_1 is a hyper-parameter that determines the speed of adaptation (larger values of t_1 imply slower adaptation).[1]

The cost function J, forces the approximation $\widehat{c}$ to be as close as possible to the measurement y in a finite horizon 2-norm sense (specified by the hyper-parameter t_1). The adaptive filters determine the model parameters $\widehat{h}(t)$ as a minimum point of J and obtain the filtered signal $\widehat{c}(t)$ at time t from the identified

[1] Actually the methods of [1, 4, 10] relax the causality condition for the filter, which allows for the window defining J to extend in the "future". More specifically, the lower bound for the summation in (14.5) is t_2, where t_2 is a hyper-parameter.

model (14.4). An important implicit assumption, on which these methods are based, is that by minimising J, the filtered signal $\widehat{c}$ becomes closer to the artifact signal c.

Remark 14.1 Without extra assumptions on the data (u, y), there are no reasons for the signal $\widehat{c}$, produced by the FIR adaptive filter, to be close to the signal c. To see this, consider the special case of the algorithm when $l = 0$ and $t_1 = 0$, i.e., consider static model and instantaneous error criterion. The minimisation of J in this case is the trivial problem

$$\min_{h(t),\widehat{c}(t)} \big(y(t) - \widehat{c}(t)\big)^2 \quad \text{subject to} \quad \widehat{c}(t) = h(t)u(t),$$

which solution is $h_0(t) = y(t)/u(t)$ and $\widehat{c}(t) = y(t)$, for all t. Independent of u, $\widehat{c} = y$, so clearly, in this case, the minimisation of J does not lead to the desired result—approximate c by $\widehat{c}$.

The same problem occurs in FIR adaptive filtering with dynamic models ($l > 0$) and non-instantaneous errors ($t_1 > 0$), so that the example is not an artificially constructed nongeneric one. Next we show that, the case $(l + 1)m \leq t_1 + 1$ is not of interest for adaptive filtering. However, the condition $(l + 1)m > t_1 + 1$ is not sufficient to ensure that the filtered signal $\widehat{c}$ approximates the artifact c.

In finding the minimum point of min $J(\widehat{h}(t))$, the following assumption is made: the time-varying FIR filter coefficients $\widehat{h}(\tau)$ for $\tau = t - t_1, t - t_1 + 1, \ldots, t$ are assumed *fixed* and equal to $\widehat{h}(t)$. The assumption implies that the FIR filter is stationary over the window for which J is computed. Such an assumption is justified when t_1 is small with respect to the rate of variation of the system parameters. Under the assumption $\widehat{h}(\tau) = \widehat{h}(t)$ for $\tau = t - t_1, t - t_1 + 1, \ldots, t$, substituting (14.4) into (14.5) gives

$$J\big(\widehat{h}(t)\big) = \left\| \begin{bmatrix} y(t) \\ y(t-1) \\ \vdots \\ y(t-t_1) \end{bmatrix} - \begin{bmatrix} u(t) & u(t-1) & \ldots & u(t-l) \\ u(t-1) & u(t-2) & \ldots & u(t-l-1) \\ \vdots & \vdots & & \vdots \\ u(t-t_1) & u(t-t_1-1) & \ldots & u(t-t_1-l) \end{bmatrix} \begin{bmatrix} \widehat{h}_0(t) \\ \widehat{h}_1(t) \\ \vdots \\ \widehat{h}_l(t) \end{bmatrix} \right\|^2$$

$$=: \|\mathbf{y}(t) - \mathbf{U}(t)\widehat{h}(t)\|^2.$$

In what follows we assume that the matrix $\mathbf{U}(t)$ is full rank. The technical term for this condition is "persistency of excitation" of the input u. It is a necessary condition for identifiability of the model, see [7, Sect. 8.3]. By "identifiability" we mean that if the data (u, y) were generated by a model in the considered model class an, then this model could be recovered back from the data.

There are three cases for the solution of the minimisation of J, depending on the relation between the number of unknowns $(l + 1)m$ and the number of constraints $t_1 + 1$.

- If $(l+1)m > t_1 + 1$, then $\min J(h(t))$ is a linear least squares problem and the unique solution is

$$\widehat{h}(t) = \left(\mathbf{U}(t)\mathbf{U}^\top(t)\right)^{-1}\mathbf{U}^\top(t)\mathbf{y}(t). \tag{14.6}$$

- If $(l+1)m = t_1 + 1$, then $\min J(h(t)) = 0$ and the unique solution is $\widehat{h}(t) = \mathbf{U}^{-1}(t)\mathbf{y}$.
- If $(l+1)m < t_1 + 1$, then $\min J(h(t)) = 0$ and the solution is not unique.

The example shown in Remark 14.1 corresponds to the case $\min J(\widehat{h}(t)) = 0$ with unique solution. From the adaptive filtering point of view, the case $(l+1)m \le t_1 + 1$ is meaningless, because it either leads to a trivial solution or to a nonunique solution. In the former case, $\widehat{c}$ does not depend on u. In the latter case $\widehat{c}$ is not well defined. For this reason next we only consider the case $(l+1)m > t_1 + 1$.

Formula (14.6) gives the solution in closed form and for small number of parameters (i.e., a small value of l) it can be used directly for computing $\widehat{h}(t)$ on each time step. An empirical observation reported in [1] is that best results are obtained by a static model ($l = 0$). This means that $\widehat{h}(t)$ is a scalar, so that (14.6) is cheap to evaluate. With a larger number of parameters, recomputing $\widehat{h}(t)$ from (14.6) may be prohibitive for a real-time application. Computationally, the core of the adaptive filtering method is a recursive algorithm for $\widehat{h}(t)$, i.e., an algorithm that computes the optimal solution $\widehat{h}(t)$ by applying a cheap updating on the optimal solution $\widehat{h}(t-1)$ of the previous time step.

The matching pursuit algorithm of [4] is based on the idea of updating only one component of $\widehat{h}(t)$ at a time. This subproblem can be solved recursively and the resulting algorithm turns out to be computationally cheap. After a finite number of such updates (for any fixed t), the optimal solution is reached. In [4] the number of updates is a hyper-parameter. The reason for this is that fewer updates than necessary to compute the optimal solution result in a suboptimal solution that may be "sufficiently good" for adaptive filtering.

In [4] the authors advocate the method because of its numerical robustness. Ill-conditioning of the matrix $\mathbf{U}(t)\mathbf{U}^\top(t)$ may lead to numerical instability (see (14.6)) and the matching pursuit can handle such a case.[2] The numerical robustness and efficiency make the matching pursuit algorithm the preferred implementation of the method proposed in [1]. In [4], however, different values for the hyper parameters are chosen compared to those in [1], e.g., the model is no longer static but dynamic with $l = 5$ taps and an additional hyper parameter (the number of matching pursuit iterations) is introduced.

[2]Note, that the persistency of excitation assumption ensures that $\mathbf{U}(t)\mathbf{U}^\top(t)$ is invertible. Therefore, ill-conditioning corresponds to input signals that are almost *not* persistently exciting.

The method of [10] is based on more assumptions. It is assumed that there is a "true" FIR model

$$\begin{aligned} c(t) &= \bar{h}_0(t)u(t) + \bar{h}_1(t)u(t-1) + \cdots + \bar{h}_l(t)u(t-l) \\ &=: \underbrace{\big[u(t)\; u(t-1) \cdots u(t-l)\big]}_{\mathbf{u}(t)} \bar{h}(t), \end{aligned} \tag{14.7}$$

i.e., the artifact signal c satisfies an FIR model with l taps for certain "true" parameter $\bar{h}$. It is assumed, moreover, that $\bar{h}$ satisfies the "random walk" equation

$$\bar{h}(t+1) = \bar{h}(t) + w(t), \tag{14.8}$$

where w is a zero-mean stationary white stochastic process. Finally, it is assumed that the ECG signal $v = y - c$ is a zero-mean stationary white stochastic process and is independent of w. Under these assumptions, (14.7) and (14.8) form a state-space representation of a classical linear time-varying stochastic system

$$\begin{aligned} \bar{h}(t+1) &= \bar{h}(t) + w(t), \\ y(t) &= \underbrace{\mathbf{u}(t)\bar{h}(t)}_{c(t)} + v(t). \end{aligned} \tag{14.9}$$

The parameters of this model are *known*, so the adaptive filtering problem reduces to the simpler and easier linear state estimation problem, which optimal (in the minimum variance sense) solution is the Kalman filter. Note that the identification of (14.7) is implicitly done by the Kalman filter for (14.9).

Let W be the covariance matrix of $w(0)$ and let V be the covariance matrix of $v(0)$. Under the above stated assumptions for v and w, the Kalman filter for (14.9) is

$$\begin{aligned} \widehat{h}(t+1) &= \widehat{h}(t) + \Sigma(t)\mathbf{u}^\top(t)\big(\mathbf{u}(t)\Sigma(t)\mathbf{u}^\top(t) + V\big)^{-1}\big(y(t) - \mathbf{u}\widehat{h}(t)\big), \\ \Sigma(t+1) &= \Sigma(t) + W - \Sigma(t)\mathbf{u}^\top(t)\big(\mathbf{u}(t)\Sigma(t)\mathbf{u}^\top(t) + V\big)^{-1}\mathbf{u}(t)\Sigma(t), \end{aligned} \tag{14.10}$$

with initial conditions $\widehat{h}(0) = h_{\text{ini}}$ and $\Sigma(0) = \Sigma_{\text{ini}}$. It defines a minimum variance estimator $\widehat{h}$ for $\bar{h}$ and the corresponding $\widehat{c} = \mathbf{u}(t)\widehat{h}(t)$ is the minimum variance estimate of c. Hyper-parameters in this case are the filter length l, the covariance matrices W and V, and the initial conditions h_{ini}, Σ_{ini}, and $u(0), u(-1), \ldots, u(1-l)$.

The stochastic assumptions imposed on w and v are motivated by the reformulation of the original adaptive filtering problem as a classical linear state estimation problem, of which recursive solution is known. These assumptions, however, are hard to interpret in the context of the original ECG artifacts removal problem and their relevance for the real-life problem is unclear.

A deterministic derivation of the Kalman filter is given in [11]. The main result is that the Kalman filter computes the estimates $\widehat{h}$, $\widehat{v}$, and $\widehat{w}$ by minimising the size of w and v subject to the constraint that y is "explained by the model". More specifically at each time step t it solves the approximation problem

$$\min_{\widehat{w},\widehat{v},\widehat{h}} \sum_{\tau=0}^{t} \widehat{w}^\top(\tau)W^{-1}\widehat{w}(\tau) + \widehat{v}^\top(\tau)V^{-1}\widehat{v}(\tau) + \big(\widehat{h}(0) - h_{\text{ini}}\big)^\top \Sigma_{\text{ini}}^{-1}\big(\widehat{h}(0) - h_{\text{ini}}\big)$$

$$\text{subject to} \quad \begin{array}{l} \widehat{h}(\tau+1) = \widehat{h}(\tau) + \widehat{w}(\tau) \\ y(\tau) = \mathbf{u}(\tau)\widehat{h}(\tau) + \widehat{v}(\tau) \end{array} \quad \text{for } \tau = 0, 1, \ldots, t. \tag{14.11}$$

The deterministic formulation (14.11) shows that the method in [10] derives an optimal approximation of the observed signal y in the sense that

- y satisfies a time-varying FIR model with parameter $\widehat{h}$ up to equation error $\widehat{v}$, which is "small" in the sense that the weighted norm $\sum \widehat{v}^\top(\tau)V^{-1}\widehat{v}(\tau)$ is minimised,
- the FIR model parameter $\widehat{h}$ satisfies a first order auto-regressive equation up to equation error $\widehat{w}$, which is "small" in the sense that the weighted norm $\sum \widehat{w}^\top(\tau)W^{-1}\widehat{w}(\tau)$ is minimised and its initial value $\widehat{h}(0)$ is small in the sense that $(\widehat{h}(0) - h_{\text{ini}})^\top \Sigma_{\text{ini}}^{-1}(\widehat{h}(0) - h_{\text{ini}})$ is minimised.

In [10] the hyper parameters V, W, h_{ini}, Σ_{ini}, and $u(0), u(-1), \ldots, u(1-l)$ are chosen by trial-and-error for optimal performance on the learning part of the data.

14.2.3.2 Numerical Examples

Next, we show the performance of the LMS adaptive filter and the methods of [1] and [10] on the data set used in the simulation examples of Secs. 14.2.1 and 14.2.2. All algorithms were coded in Matlab[3] and tuned by trial-and-error for optimal performance on the specific data set. The selected parameters are used in the following section for evaluation of the methods on the test data.

For the LMS algorithm tunable hyper parameters are the FIR filter length l and the step size λ of the gradient descent algorithm. We chose the following values:

Hyper parameters for the LMS algorithm: $\quad l = 100$ and $\lambda = 0.1$.

For the algorithm of [10] tunable parameters are the covariance matrices V, W, Σ_{ini}, and the initial conditions h_{ini}, $u(0), u(-1), \ldots, u(1-l)$. We chose the following values:

Hyper parameters for the algorithm of [10]:

$$V = 10^{-3}, \qquad W = 2 \times 10^{-8} \times I_6, \qquad \Sigma_{\text{ini}} = 10^{-3} \times I_6,$$

[3]We would like to thank Simon Doclo from K.U. Leuven for a Matlab code for LMS adaptive filtering.

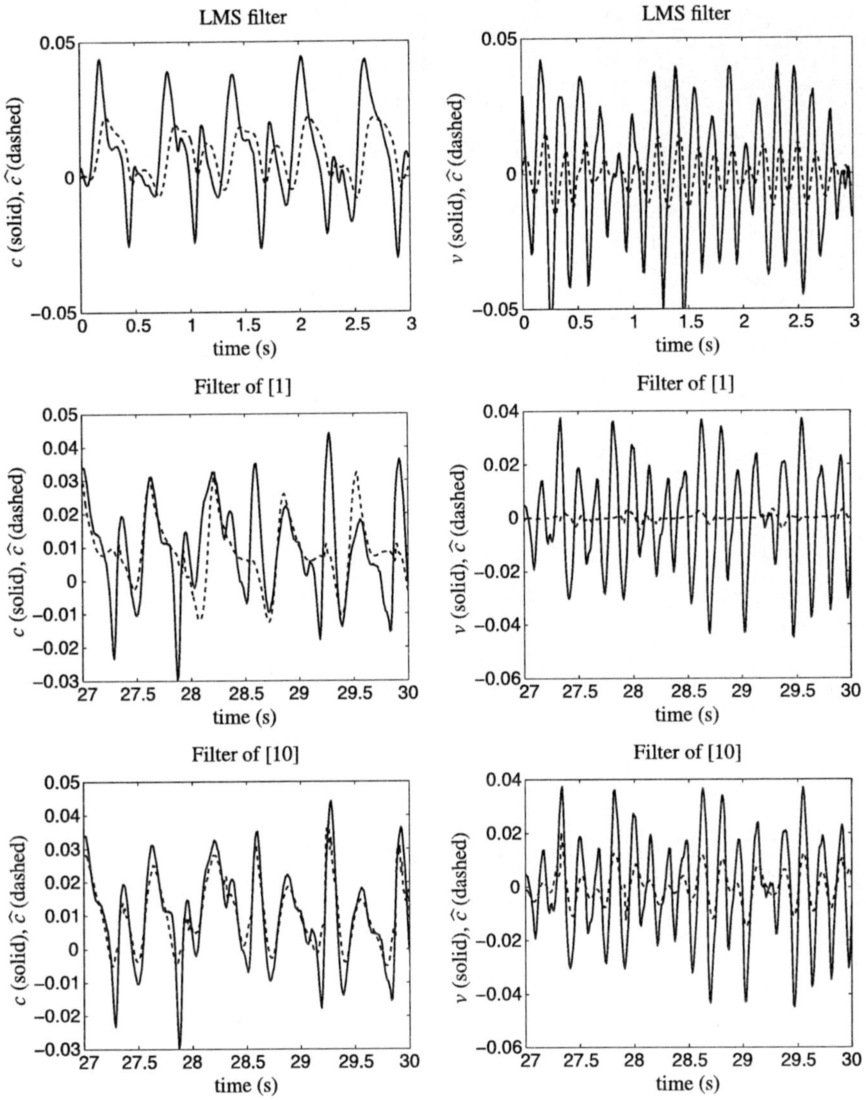

Fig. 14.10 Filtered signals c and v by the adaptive filters

$$h_{\text{ini}} = 10^{-3} \times \text{col}(0.4617, 0.1088, -0.2128, -0.1604, -0.0036, 0.6199),$$

$$u_{\text{ini}} = 0.$$

For the algorithm of [1] the window length t_1 is a tunable parameter. (The FIR filter length is $l = 0$, i.e., the filter is a time-varying static system.) We chose the following value:

$$\text{Hyper parameters for the algorithm of [1]:} \quad t_1 = 10.$$

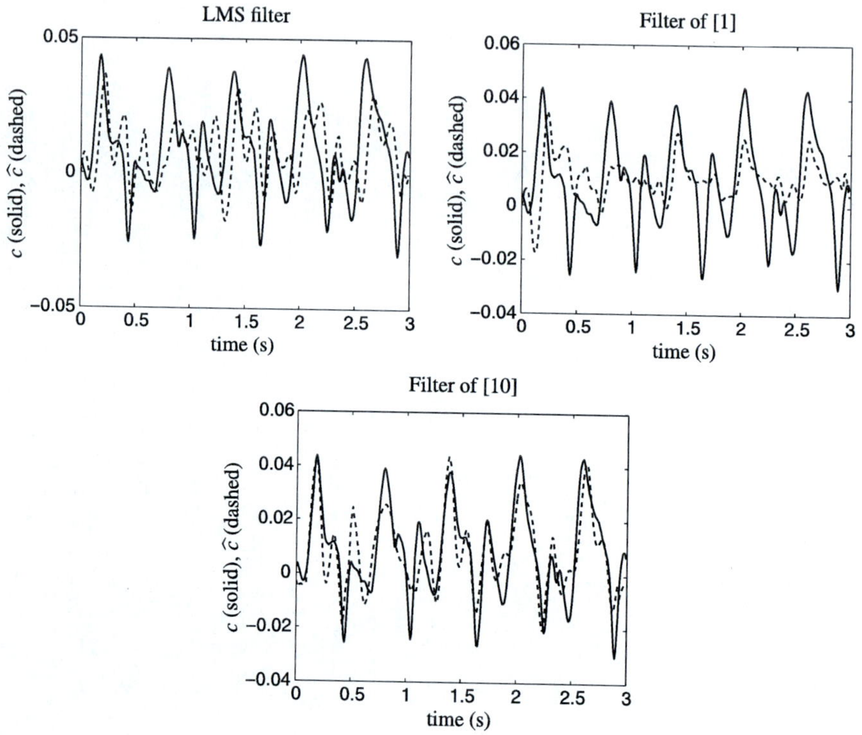

Fig. 14.11 Estimated (*dashed*) and true (*solid*) artifact signals using the adaptive filters

Figure 14.10 shows the signals $\widehat{c}$ and $\widehat{v}$ obtained by the adaptive filters. Ideally c should not be distorted and v should be cancelled. Compare with Figs. 14.4 and 14.8 for similar plots obtained with the low-pass and Kalman filters. Figure 14.11 shows the performance of the adaptive filtering method on a signal $y = c + v$ with SNR 0 dB. The SNR of the restored signal $\widehat{v}$ are 3.4 dB for the LMS method, 5.9 dB for the method of [1], and 3.1 dB for the method of [10].

Note that the performance of the adaptive filter is worse than the one of the Kalman filter. This can be explained as follows: in the case of Kalman filtering, the model is identified off-line using all the data, while in the case of adaptive filtering, the model is identified in real-time, using past data only. Therefore, in the Kalman filter case, more information is used on the identification stage than in the adaptive filter case.

14.3 Results: Performance Evaluation

The seven ventricular fibrillation ECG signals v and the seven resuscitation artifact signals c, available in the test database are combined to form 49 test signals y. The evaluated methods are applied on the 49 signals and the 49 SNRs of the estimated

Table 14.1 Average performance $\text{SNR}(\hat{v}, v)$ in dB of the filtering method on the test data. In brackets is the standard deviation over the 49 experiments

Method	−10	−5	0	5	10
Low-pass	1.5 (2.4)	4.8 (2.0)	6.8 (1.8)	7.8 (1.9)	8.2 (1.9)
Reference	−1.3 (1.8)	3.7 (1.8)	8.7 (1.8)	13.7 (1.8)	18.7 (1.8)
Kalman	−2.5 (1.9)	2.0 (1.8)	6.2 (1.8)	9.2 (1.9)	11.2 (2.9)
LMS	−1.5 (2.0)	2.7 (1.8)	6.0 (1.6)	8.0 (1.6)	9.0 (1.6)
Rheinberger et al.	−5.2 (1.4)	−0.2 (1.4)	4.6 (1.4)	9.2 (1.3)	13.0 (1.3)
Aase et al.	−4.4 (1.6)	0.0 (1.5)	3.4 (1.2)	5.6 (1.1)	6.6 (1.1)

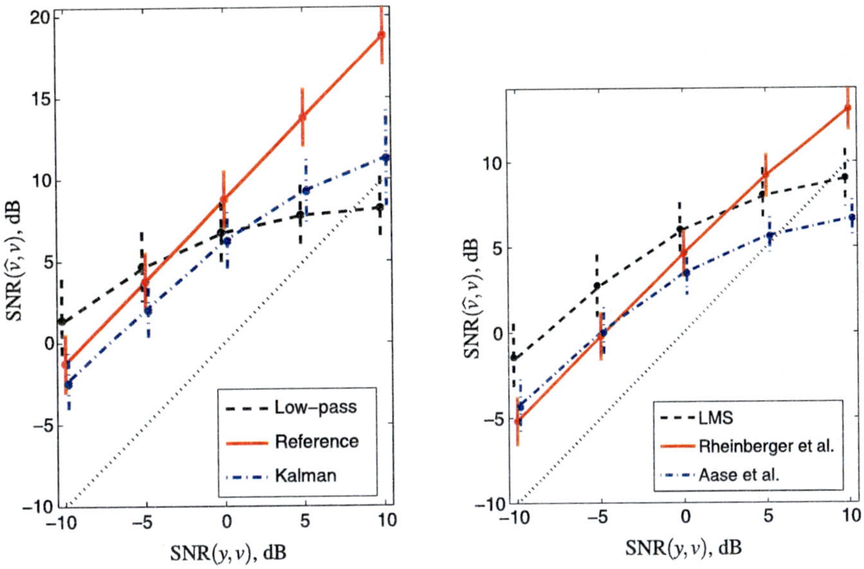

Fig. 14.12 Comparison of the methods in terms of average SNR improvement

signals $\hat{v}$ are computed. Table 14.1 and Fig. 14.12 show the average results. The standard deviations (over the 49 experiments) are visualised with the vertical bars on Fig. 14.12.

For $\text{SNR}(y, v)$ less than −3 dB, the low-pass filter achieves the best performance. A possible explanation is that for low SNR, the system identification methods fail to obtain a sufficiently good model for the observed data and the Kalman filter, which is based on the model, is sensitive to model-data discrepancy. Indeed, obtaining a good model (14.3) for the artifact signal c is a challenging problem even when the data used for identification is (u, c) (reference method). Note that the signals that we use in the simulation study are measured in a real-life environment and therefore they need not satisfy a linear time-invariant model (14.3) of low order. Moreover, the actual identification problem encountered in ECG artifact removal is to derive the model from (u, y) in real-time.

As expected, the reference method outperforms the Kalman filtering method. The explanation is that the reference method is noncausal and in the system identification step uses information (the true artifact signal c) that is not available to the other methods. The Kalman filtering method is also noncausal and therefore has an advantage over the adaptive filtering method. Time-invariant is causal, however, it is based on a time-invariant model, which is a limitation when the data is generated by a time-varying "true" system. For high SNR the best performance is achieved by the adaptive filtering method of [10], which suggests that the true model for v is indeed time-varying.

14.4 Conclusions

The robustness of the filtering methods, i.e., their sensitivity to the models being used, is crucial in the removal of resuscitation artifacts from ventricular fibrillation ECG signals. The low-pass filter uses the simplest model: separation of the spectra of the useful signal and the disturbance. The only tuning parameter in this case is the cut-off frequency. Because of its simplicity the low-pass filtering method is more robust than the Kalman filter, which is based on an accurate model of the data. The experiments show, however, that even in the unrealistic case when the model, used in the synthesis of the Kalman filter, is identified from the unknown true data, the low-pass filter on the average still achieves better performance than the Kalman filter, provided the SNR of the given ECG signal is sufficiently low. For high SNR the best performance achieves the adaptive filtering method of [10].

Acknowledgements The research leading to these results has received funding from the European Research Council under the European Union's Seventh Framework Programme (FP7/2007-2013)/ERC Grant agreement number 258581 "Structured low-rank approximation: Theory, algorithms, and applications"; the Austrian Fonds zur Förderung der Wissenschaftlichen Forschung (FWF, Vienna, grant No. L288), Research Council KUL: GOA-AMBioRICS, GOA-Mefisto 666, Center of Excellence EF/05/006 "Optimization in engineering", several PhD/postdoc & fellow grants; Flemish Government: FWO: PhD/postdoc grants, projects, G.0360.05 (EEG signal processing), G.0321.06 (numerical tensor techniques), research communities (ICCoS, ANMMM); IWT: PhD Grants; Belgian Federal Science Policy Office IUAP P5/22 ('Dynamical Systems and Control: Computation, Identification and Modelling'); EU: BIOPATTERN, ETUMOUR; HEALTHagents.

References

1. Aase, S., Eftestøl, T., Husøy, J., Sunde, K., Steen, P.: CPR artifact removal from human ECG using optimal multichannel filtering. IEEE Trans. Biomed. Eng. **47**, 1440–1449 (2000)
2. Buckheit, J., Donoho, D.: Wavelab and reproducible research. In: Wavelets and Statistics. Springer, Berlin (1995)
3. Haykin, S.: Adaptive Filter Theory. Prentice Hall, Upper Saddle River (1991)

4. Husøy, J., Eilevstrønn, J., Eftestøl, T., Aase, S., Myklebust, H., Steen, P.: Removal of cardiopulmonary resuscitation artifacts from human ECG using an efficient matching pursuit-like algorithm. IEEE Trans. Biomed. Eng. **49**, 1287–1298 (2002)
5. Ljung, L.: System Identification: Theory for the User. Prentice-Hall, Upper Saddle River (1999)
6. Markovsky, I., Van Huffel, S.: High-performance numerical algorithms and software for structured total least squares. J. Comput. Appl. Math. **180**(2), 311–331 (2005)
7. Markovsky, I., Willems, J.C., Van Huffel, S., De Moor, B.: Exact and Approximate Modeling of Linear Systems: A Behavioral Approach. Monographs on Mathematical Modeling and Computation, vol. 11. SIAM, Philadelphia (2006)
8. Markovsky, I., Willems, J.C., Van Huffel, S., De Moor, B., Pintelon, R.: Application of structured total least squares for system identification and model reduction. IEEE Trans. Autom. Control **50**(10), 1490–1500 (2005)
9. Oppenheim, A., Willsky, A.: Signals and Systems. Prentice Hall, Upper Saddle River (1996)
10. Rheinberger, K., Steinberger, T., Unterkofler, K., Baubin, M., Klotz, A., Amann, A.: Removal of CPR artifacts from the ventricular fibrillation ECG by adaptive regression on lagged reference signals. IEEE Trans. Biomed. Eng. **55**(1), 130–137 (2008)
11. Willems, J.C.: Deterministic least squares filtering. J. Econom. **118**, 341–373 (2004)

Chapter 15
Progress and Open Questions in the Identification of Electrically Stimulated Human Muscle for Stroke Rehabilitation

Fengmin Le, Chris T. Freeman, Ivan Markovsky, and Eric Rogers

15.1 Introduction

Almost 85% of the people living in the UK in 2005 with moderate to severe disabilities as a result of a stroke had an initial deficiency in the upper limb [1] and less than 50% recovered useful upper limb function [1, 2]. Moreover, due to an aging population and better acute care, prevalence of stroke is likely to increase. These features are amongst the main drivers for research aimed at providing more effective rehabilitation which is, ideally, available for use outside the hospital, for example, in the patients home. One major area of research in this field is the use of Electrical Stimulation (ES) to improve motor control, which is supported by a growing body of clinical evidence [3–5], and also theoretical support from neurophysiology [6] and motor learning research. There is also strong evidence to support the proposition that functional recovery is enhanced when stimulation is applied coincidently with a patient's voluntary intention whilst performing a task [7]. The need to accurately apply ES to achieve a movement has motivated significant interest in the development and application of techniques that can control upper limb movement to a high level of precision.

A survey of the literature in [8] reveals that a wide range of model-based schemes have been proposed for movement control of paralyzed subjects, including multi-channel Proportional plus Integral plus Derivative (PID) control of the wrist, optimal and H_∞ control, fuzzy control of standing, sliding mode control of shank movement, and data-driven control of the knee joint. A very significant feature to

F. Le · C.T. Freeman · I. Markovsky · E. Rogers (✉)
School of Electronics and Computer Science, University of Southampton,
Southampton SO17 1BJ, UK
e-mail: etar@ecs.soton.ac.uk

I. Markovsky
e-mail: im@ecs.soton.ac.uk

L. Wang, H. Garnier (eds.), *System Identification, Environmental Modelling, and Control System Design*,
DOI 10.1007/978-0-85729-974-1_15,

emerge from this review is that these advanced techniques have not transferred to clinical practice. In particular, the strategies adopted are either open-loop, or the stimulation is triggered using limb position or Electromyographic (EMG) signals to provide a measure of participant's intended movement. Closed-loop control has been achieved using EMG [9] but this has not been incorporated in model-based controllers since EMG does not directly relate to the force or torque generated by the muscle. In the few cases where model-based control approaches have been used clinically, they have enabled a far higher level of tracking accuracy.

A major reason for the lack of model-based methods in a program of patient trials is the difficulty in obtaining reliable biomechanical models of hemiplegic subjects. In the clinical setting there is minimal set-up time, reduced control over environmental constraints and little possibility of repeating any one test in the program of treatment undertaken and consequently controllers are required to perform to a minimum standard across a wide number of subjects and conditions. In particular, the underlying musculoskeletal system is highly sensitive to physiological conditions, such as skin impedance, temperature, moisture and electrode placement, together with time-varying effects such as spasticity and fatigue [10].

Recent work [11–13] has demonstrated that Iterative Learning Control (ILC) is one model-based approach that can be effective clinically. In this work a robotic workstation was designed and constructed for use by stroke patients in order to regain voluntary control of their impaired arm. Here, ES is applied to generate torque about the elbow joint, and ILC is used to update the stimulation level to assist their completion of a planar reaching task. This treatment produced statistically significant improvement for participants across a number of outcome impairment measures [14], but the need for improved modeling of the patient's arm, and the muscle model in particular, was also highlighted by these results.

This chapter begins with a more detailed description of the use of ILC in this setting and then proceeds to give an overview of recent progress on muscle response modeling, which typically employs a Hammerstein structure. Included are results from application to measured data. The chapter concludes by giving some currently open research questions.

15.2 Background

Iterative learning control is a technique for controlling systems operating in a repetitive, or trial-to-trial, mode with the requirement that a reference trajectory $r(p)$ defined over a finite interval $p = 0, 1, \ldots, \alpha - 1$, where α denotes the trial duration, is accurately followed. Examples of such systems include robotic manipulators that are required to repeat a given task to high precision, chemical batch processes or, more generally, the class of tracking systems. Since the original work [15], the general area of ILC has been the subject of intense research effort. Initial sources for the literature here are the survey papers [16, 17].

The robotic workstation which is the starting point for the work covered in this chapter was developed for use by stroke patients in order to regain voluntary control

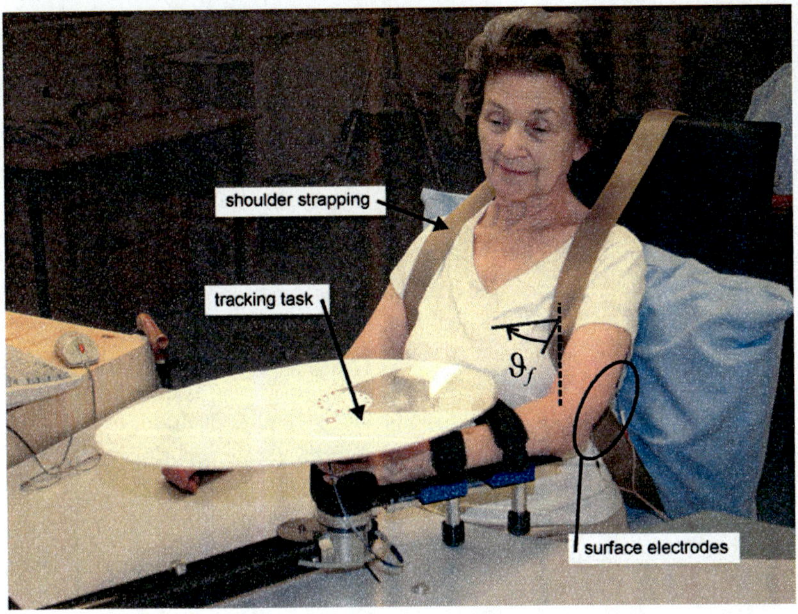

Fig. 15.1 A participant using the robotic workstation

of their impaired arm [11–13]. Here, ES is applied to generate torque about the elbow joint, and ILC is used to update the stimulation level to assist their completion of a planar reaching task.

Figure 15.1 shows a stroke participant using the robotic workstation, where the shoulder strapping is used to prevent trunk movement which would reduce the effectiveness of treatment. In this case, ES is applied to generate torque about the elbow joint, and ILC is used to update the stimulation level to assist their completion of a planar reaching task. In particular, the patient's hand is strapped to the robot and they attempt to follow a point moving along an illuminated elliptical track.

As they complete the task, the error between the required and achieved joint angle, $\vartheta^*(t)$ and $\vartheta(t)$, respectively, is measured and once they reach the end the robot returns their arm to the starting position, and in this resetting time an ILC algorithm is used to update the stimulation to be applied on the next attempt or trial.

Figure 15.2 shows a block diagram of the control scheme, consisting of a feedback controller, a linearizing controller and an ILC feedforward controller. The former block, taken as a proportional plus derivative controller in the clinical tests, acts as a pre-stabilizer and provides satisfactory tracking during initial trials. During the arm resetting time at the end of trial k, the ILC controller uses a biomechanical model of the arm and muscle system, together with the previous tracking error, to produce the feedforward update signal for application on the next trial. A full treatment of the ILC algorithms is given in [11].

The overall performance of this system critically relies on the accuracy of the arm and muscle model, which contains the following components.

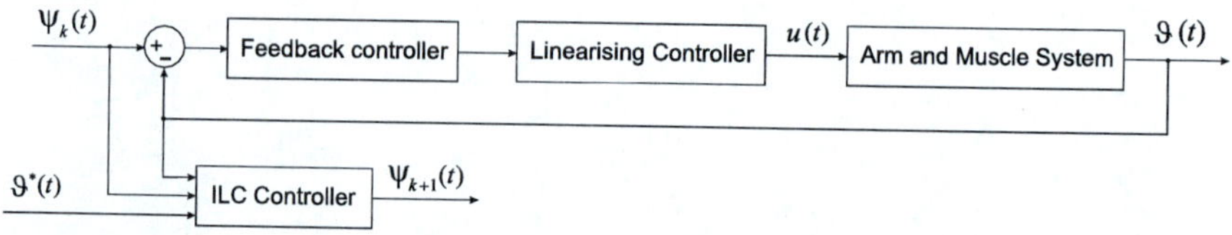

Fig. 15.2 Block diagram representation of the ILC control scheme

- A stimulated muscle structure that accounts for the torque acting about the elbow generated in response to the applied ES, $u(t)$.
- A kinematic model which gives the component of this torque in the horizontal plane of movement.
- A two-link system which provides the resulting angular position, $\vartheta(t)$.

This biomechanical model has been experimentally verified with both unimpaired subjects and stroke patients using a variety of functional parameter forms [13].

Although the model can predict arm movement resulting from applied ES with reasonable accuracy, experimental data confirms that the model of the stimulated muscle used is less accurate than the remaining components of the arm. Hence relatively low ILC learning gains had to be used throughout the clinical trials, but the treatment still resulted in statistically significant improvement for participants across a number of outcome impairment measures [14]. This work established the basic feasibility of the approach, but with a clear requirement to improve the modeling of both the patient's arm and the muscle model.

Muscle models adopted in the wide range of model-based controllers that have been proposed for both the upper and lower limb vary widely in structure, from no explicit form of muscle model in [18–20], linear forms in [21, 22] and a general non-linear form in [23], but the most widely assumed structure is the Hill-type model [24]. This model describes the output force as the product of three independent experimentally measured factors: the force-length property, the force-velocity property and the nonlinear muscle activation dynamics under isometric conditions, respectively, where the latter is termed simply the activation dynamics (AD) of the stimulation input. The form of the first two is typically chosen to correspond with physiological observations, see, for example, [25–27], but they have also been combined in a more general functional form [13, 28].

The activation dynamics are almost uniformly represented by a static nonlinearity in series with linear dynamics, and constitute an important component of the model since controlled motions are typically smooth and slow, so that the effects of inertia, velocity, and series elasticity are small and the isometric behavior of muscle dominates. The non-linearity has been parameterized in a number of ways, taking the form of a simple gain with saturation [29], a piecewise linear function [25, 30] and a predefined functional form [31, 32]. The linear dynamics have been assumed to be first order [25], a series of two first order systems [26, 33], critically damped second order [34–36] or second order with transport delay [30, 37].

The popularity of Hammerstein structure representations of activation dynamics is supported by their correspondence with biophysics. In particular, the static nonlin-

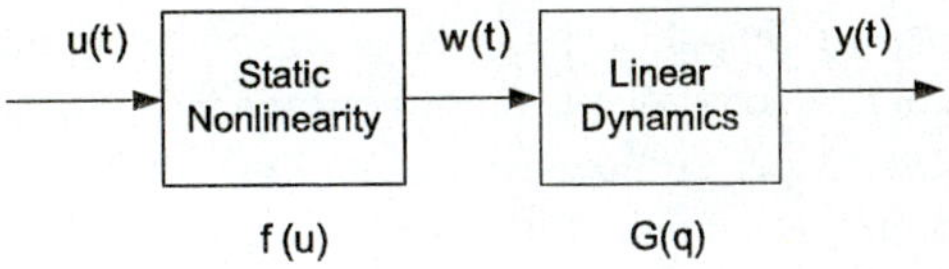

Fig. 15.3 Hammerstein structure representation of activation dynamics

earity, $f(u)$, represents the Isometric Recruitment Curve (IRC), which is the static gain relation between stimulus activation level, $u(t)$, and steady-state output torque, $w(t)$, when the muscle is held at a fixed length. The linear dynamics, $G(q)$, where q denotes the shift operator, represents the muscle contraction dynamics, which combines with the IRC to give the overall torque generated, $y(t)$. These components are shown in the block diagram of Fig. 15.3.

Due to its track record in the modeling of stimulated muscle, and also the subsequent design and implementation of controllers, the Hammerstein structure has been employed in the research discussed in this chapter. This was on the previso that the following limitations restricting its application to upper limb stroke rehabilitation could be overcome.

- ES has been applied to either in vitro or paretic muscles in the vast majority of experimental verification tests. This effectively removes the possibility of an involuntary response to stimulation which may occur when applied to subjects with incomplete paralysis, such as stroke. In addition to motivating the need for experimental validation on such subjects, this also meant that the excitation inputs widely used to identify the Hammerstein structure (Pseudo Random Binary Sequences (PRBS)), white noise and pulses, are not appropriate as they would elicit an involuntary response from the subject.
- The absence of test results from subjects with incomplete paraplegia also meant that physiologically based constraints on the form of the dynamics, such as the assumption of a critically damped system [38, 39], may not be justified.
- Almost all previously reported in vivo studies and control implementations have applied ES to the lower limb, even though upper limb functional tasks require finer control, and are more subject to adverse effects such as sliding electrodes and the activation of adjacent muscles during stimulation.

The work reported in [8] developed a novel identification scheme, and accompanying set of excitation inputs, in order to address these drawbacks through the following attributes.

- The excitation signal must be chosen from a physiological perspective and hence the identification scheme cannot use rapidly changing inputs and must be applicable to an arbitrary choice of signal.
- A general form of linear dynamics represented in transfer-function form is used.
- The use of a smooth function with continuous derivatives is preferable to that of a piecewise linear function in the representation of the static nonlinearity.

The physical realization of the input is a more arbitrary consideration: the stimulus activation level, quantified by the number of muscle fibers activated, can be achieved by varying either the current or voltage amplitude, or the duration (width) of stimulus pulses. The latter method is preferred since it is easier to quantify and control, provides a more consistent response across subjects, requires a smaller charge per stimulus pulse, and allows for greater selectivity of recruitment than amplitude modulation [40].

The modulation by temporal summation (stimulus period modulation, or, inversely, pulse frequency modulation), achieved by varying the time interval between the start of successive pulses, can be represented using a multiplicative function [26]. It was not used in this application, however, because (i) the use of high frequencies (>50 Hz) is known to increase muscle fatigue in stroke patients [10] (frequencies up to 100 Hz are used in the frequency modulation method of [41]), and (ii) frequency modulation alone may not generate the range of torque needed to achieve the wide variety of functional tasks required by a stroke rehabilitation program [42].

Having established the model structure, the task of estimating the model parameters is now considered from a systems identification perspective. There are many methods applicable to Hammerstein models, and in general they can be classified into two categories: iterative methods and non-iterative methods.

In this work, an iterative technique has been applied since this generally leads to improved accuracy. Following the discussion in [8], the separable least squares method using cubic spline nonlinearities [43] appears the most appropriate since it has been successfully used in modeling stretch reflex dynamics. The approach involves dividing the parameters into linear and nonlinear components: the nonlinear parameters start from initial values, which are then updated on each iteration using the Levenberg-Marquardt algorithm [44] to compute the step, and the linear parameters are then similarly updated by linear regression. However, this approach cannot be applied to the present problem because it relies on a finite impulse response representation of dynamics, rather than the infinite impulse transfer-function form that has been adopted in the vast majority of Hammerstein structure applications to muscle modeling, and which also leads to reduced memory requirement and computational work load. Therefore, since [43] is unsuitable, an iterative algorithm has been developed [8] for identification of the form of Hammerstein structure required, which uses a different projection approach to update the nonlinear parameters. The iterative algorithm is now summarized and then experimental results using a human subject will be given in order to evaluate this method with respect to its convergence properties, identification and predictive abilities.

15.3 The Identification Problem

Two discrete-time Hammerstein model structures of the form shown in Fig. 15.4 have been used. The stimulation input u is first scaled by the static nonlinear function f and then passed to a linear time-invariant system described by a transfer-function $G(q) = B(q)/A(q)$. The internal signal w is not measurable and the noise v is zero mean and white.

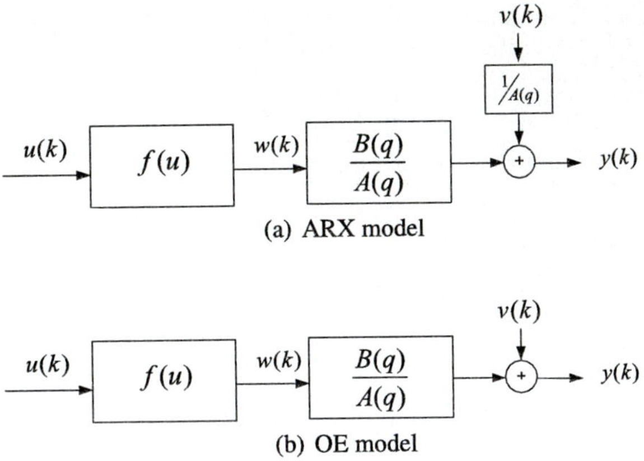

Fig. 15.4 Two discrete-time Hammerstein model structures

The linear system is represented by the transfer-function

$$G(q) = \frac{B(q)}{A(q)} = \frac{b_0 q^{-d} + b_1 q^{-(d+1)} + \cdots + b_n q^{-(n+d)}}{1 + a_1 q^{-1} + \cdots + a_l q^{-l}}, \tag{15.1}$$

where q^{-1} is the delay operator and n, l and d are the number of zeros, poles and the time delay order, respectively, where the last three are assumed to be known.

The nonlinear function $f(u)$ is represented by the cubic spline

$$f(u) = \sum_{i=1}^{m-2} \beta_i |u - u_{i+1}|^3 + \beta_{m-1} + \beta_m u + \beta_{m+1} u^2 + \beta_{m+2} u^3, \tag{15.2}$$

where $u_{\min} = u_1 < u_2 < u_3 < \cdots < u_m = u_{\max}$ are the spline knots,

$$\theta_n = \left[\beta_1\ \beta_2\ \ldots\ \beta_{m+2}\right]^T$$

are the parameters of the nonlinear block and

$$\theta_l = \begin{bmatrix} \theta_a \\ \theta_b \end{bmatrix} = \left[a_1\ \ldots\ a_l\ b_0\ b_1\ \ldots\ b_n\right]^T \tag{15.3}$$

are the parameters of the linear block.

These two Hammerstein models differ in the form of the noise model. In Fig. 15.4(a) an Auto Regressive eXternal (ARX) model is used, in which the noise filter, $H = 1/A(q)$, is coupled to the linear component of the plant model. In Fig. 15.4(b) an Output-Error (OE) model is used and in this case the noise model is $H = 1$.

The identification problem can now be stated as follows: given measured input/output data

$$\left\{\left(u(1), y(1)\right), \ldots, \left(u(N), y(N)\right)\right\}$$

find a parameter vector

$$\theta = \begin{bmatrix} \theta_n \\ \theta_l \end{bmatrix}$$

that minimizes the cost function

$$\|v\|_2^2 = \sum_{k=1}^{N} v^2(k), \tag{15.4}$$

where

$$\frac{1}{\hat{A}(q)} v = y - G(q, \hat{\theta}_l) f(u, \hat{\theta}_n) = y - \frac{\hat{B}(q)}{\hat{A}(q)} f(u, \hat{\theta}_n) \tag{15.5}$$

in the case of ARX noise model and

$$v = y - G(q, \hat{\theta}_l) f(u, \hat{\theta}_n) = y - \frac{\hat{B}(q)}{\hat{A}(q)} f(u, \hat{\theta}_n) \tag{15.6}$$

in the case of OE noise model.

15.4 Identification Algorithm

Assume that an initial estimate of the linear parameter vector, $\hat{\theta}_l$, is available. In which case the nonlinear parameters can be identified as follows.

ARX Model Multiplying (15.5) by $\hat{A}(q)$ and substituting the resulting expression for v in (15.4) yields

$$\hat{\theta}_n v = \arg\min_{\theta_n} \| \hat{A}(q) y - \hat{B}(q) f(u, \theta_n) \|_2. \tag{15.7}$$

From (15.2), it follows that $f(u, \theta_n)$ is linear in θ_n, and hence

$$\begin{aligned} &\big(\hat{B}(q) f(u, \theta_n)\big)(k) \\ &\quad = \sum_{i=1}^{m-2} \beta_i \underbrace{\left(\hat{b}_0 |u(k-d) - u_{i+1}|^3 + \cdots + \hat{b}_n |u(k-d-n) - u_{i+1}|^3\right)}_{f_i(u(k), \hat{\theta}_b)} \\ &\qquad + \beta_{m-1} \underbrace{(\hat{b}_0 + \cdots + \hat{b}_n)}_{f_{m-1}(u(k), \hat{\theta}_b)} \\ &\qquad + \beta_m \underbrace{(\hat{b}_0 u(k-d) + \cdots + \hat{b}_n u(k-d-n))}_{f_m(u(k), \hat{\theta}_b)} \end{aligned}$$

$$+\beta_{m+1}\underbrace{\left(\hat{b}_0 u(k-d)^2+\cdots+\hat{b}_n u(k-d-n)^2\right)}_{f_{m+1}(u(k),\hat{\theta}_b)}$$
$$+\beta_{m+2}\underbrace{\left(\hat{b}_0 u(k-d)^3+\cdots+\hat{b}_n u(k-d-n)^3\right)}_{f_{m+2}(u(k),\hat{\theta}_b)}. \tag{15.8}$$

Consequently (15.7) can be rewritten as an ordinary least squares problem

$$\arg\min_{\theta_n}\left\| Y_n(y,\hat{\theta}_a)-\Phi_n(u,\hat{\theta}_b)\theta_n\right\|_2 \tag{15.9}$$

where, assuming that $l > n+d$,

$$Y_n(y,\hat{\theta}_a)=\begin{bmatrix} y(l+1)+\hat{a}_1 y(l)+\cdots+\hat{a}_l y(1) \\ y(l+2)+\hat{a}_1 y(l+1)+\cdots+\hat{a}_l y(2) \\ \vdots \\ y(N)+\hat{a}_1 y(N-1)+\cdots+\hat{a}_l y(N-l) \end{bmatrix}$$

and

$$\Phi_n(u,\hat{\theta}_b)=\begin{bmatrix} f_1(u(l+1),\hat{\theta}_b) & \cdots & f_{m+2}(u(l+1),\hat{\theta}_b) \\ f_1(u(l+2),\hat{\theta}_b) & \cdots & f_{m+2}(u(l+2),\hat{\theta}_b) \\ \vdots & & \vdots \\ f_1(u(N),\hat{\theta}_b) & \cdots & f_{m+2}(u(N),\hat{\theta}_b) \end{bmatrix}.$$

The solution of (15.7) now is

$$\hat{\theta}_n=\left(\Phi_n(u,\hat{\theta}_b)^T\Phi_n(u,\hat{\theta}_b)\right)^{-1}\Phi_n(u,\hat{\theta}_b)^T Y_n(y,\hat{\theta}_a).$$

OE Model Let $\hat{y}$ be the output of $\hat{G}$ when the input is $f(u,\theta_n)$ or, equivalently,

$$\hat{y}(k)=\frac{\hat{B}(q)}{\hat{A}(q)}f(u,\theta_n) \tag{15.10}$$

or, on multiplying both sides of (15.10) by $\hat{A}(q)$ and expanding $\hat{B}(q)f(u(k),\theta_n)$ as in (15.8),

$$T(\hat{\theta}_a)\hat{Y}=\Phi_n(u,\hat{\theta}_b)\theta_n, \tag{15.11}$$

where

$$T(\hat{\theta}_a)=\begin{bmatrix} \hat{a}_l & \cdots & \hat{a}_1 & 1 & 0 & \cdots & \cdots & 0 \\ 0 & \hat{a}_l & \cdots & \hat{a}_1 & 1 & \cdots & \cdots & 0 \\ \vdots & & & & & & & \vdots \\ 0 & \cdots & \cdots & 0 & \hat{a}_l & \cdots & \hat{a}_1 & 1 \end{bmatrix} \quad\text{and}\quad \hat{Y}=\begin{bmatrix} \hat{y}(1) \\ \hat{y}(2) \\ \vdots \\ \hat{y}(N) \end{bmatrix}$$

in which $T(\hat{\theta}_a)$ is an $(N-l)\times N$ matrix, and hence the solution for $\hat{Y}$ is not unique.

The system theoretic interpretation of this linear algebra fact is that the output cannot be uniquely determined by the given model and input. Indeed, there are additional degrees of freedom in the choice of the initial conditions. In order to enforce a unique solution of (15.11) unique, zero initial conditions are assumed. This choice is justifiable in the context of the muscle identification problem because the experiment starts with the muscle "at rest". The choice of zero initial conditions amounts to extending the data by zeros in the past, which, in turn, means that the matrices $T(\hat{\theta}_a)$ and $\Phi_n(u, \hat{\theta}_b)$ are extended to comprise N columns, and then (15.11) becomes

$$T_{ext}(\hat{\theta}_a)\hat{Y} = \Phi_n(u_{ext}, \hat{\theta}_b)\theta_n, \tag{15.12}$$

with

$$T_{ext}(\hat{\theta}_a) = \begin{bmatrix} 1 & 0 & \cdots & 0 & 0 & \cdots & \cdots & 0 \\ \hat{a}_1 & 1 & 0 & 0 & 0 & \cdots & \cdots & 0 \\ \vdots & \ddots & & & & & & \vdots \\ \hat{a}_l & \cdots & \hat{a}_1 & 1 & 0 & \cdots & \cdots & 0 \\ 0 & \hat{a}_l & \cdots & \hat{a}_1 & 1 & \cdots & \cdots & 0 \\ \vdots & & & & & & & \vdots \\ 0 & \cdots & \cdots & 0 & \hat{a}_l & \cdots & \hat{a}_1 & 1 \end{bmatrix} \quad \text{and}$$

$$\Phi_n(u_{ext}, \hat{\theta}_b) = \begin{bmatrix} f_1(u(1), \hat{\theta}_b) & \cdots & f_{m+2}(u(1), \hat{\theta}_b) \\ f_1(u(2), \hat{\theta}_b) & \cdots & f_{m+2}(u(2), \hat{\theta}_b) \\ \vdots & & \vdots \\ f_1(u(N), \hat{\theta}_b) & \cdots & f_{m+2}(u(N), \hat{\theta}_b) \end{bmatrix}.$$

Now, from (15.12)

$$\hat{Y} = T_{ext}^{-1}(\hat{\theta}_a)\Phi_n(u_{ext}, \hat{\theta}_b)\theta_n$$

and on substituting this expression in (15.6), the cost function (15.4) becomes

$$\hat{\theta}_n = \arg\min_{\theta_n} \|Y - T_{ext}^{-1}(\hat{\theta}_a)\Phi_n(u_{ext}, \hat{\theta}_b)\theta_n\|_2,$$

which can be solved approximately in the least squares sense to obtain the estimate of the nonlinear parameter vector, $\hat{\theta}_n$, as

$$\hat{\theta}_n = \big(\big(T_{ext}^{-1}(\hat{\theta}_a)\Phi_n(u_{ext}, \hat{\theta}_b)\big)^T T_{ext}^{-1}(\hat{\theta}_a)\Phi_n(u_{ext}, \hat{\theta}_b)\big)^{-1} \times \big(T_{ext}^{-1}(\hat{\theta}_a)\Phi_n(u_{ext}, \hat{\theta}_b)\big)^T Y.$$

15.4.1 Identification of the Linear Parameters

Given an estimate $\hat{\theta}_n$ for the nonlinear parameter vector θ_n, the cost function (15.4) can be minimized over the linear parameter vector θ_l. This subproblem is a linear least squares minimization in the ARX case but a difficult nonlinear least squares problem in the OE case.

ARX Model The minimization problem for this model is

$$\hat{\theta}_l = \arg\min_{\theta_l} \| A(q)y - B(q) f(u, \hat{\theta}_n) \|$$

or in matrix form

$$\arg\min_{\theta_l} \| Y' - \Phi_l(u, y, \hat{\theta}_n)\theta_l \|_2, \tag{15.13}$$

where

$$Y' = \begin{bmatrix} y(l+1) \; y(l+2) \; \dots \; y(N) \end{bmatrix}^T$$

and

$$\Phi_l(u, y, \hat{\theta}_n) = \begin{bmatrix} -y(l) & \cdots & -y(1) & f(u(l+1-d), \hat{\theta}_n) & \cdots & f(u(l+1-d-n), \hat{\theta}_n) \\ -y(l+1) & \cdots & -y(2) & f(u(l+2-d), \hat{\theta}_n) & \cdots & f(u(l+2-d-n), \hat{\theta}_n) \\ \vdots & & \vdots & \vdots & & \vdots \\ -y(N-1) & \cdots & -y(N-l) & f(u(N-d), \hat{\theta}_n) & \cdots & f(u(N-d-n), \hat{\theta}_n) \end{bmatrix}.$$

Therefore, the solution of (15.13) is

$$\hat{\theta}_l = \left(\Phi_l(u, y, \hat{\theta}_n)^T \Phi_l(u, y, \hat{\theta}_n)\right)^{-1} \Phi_l(u, y, \hat{\theta}_n)^T Y'.$$

OE Model Recall the partition of the transfer-function linear parameters vector θ_l into parameter θ_a of the denominator A and parameter θ_b of the numerator B. Then the output error can be minimized analytically over θ_b, reducing the number of optimization variables for the minimization problem.

For a given θ_a, (15.10) can be rewritten in a matrix form similar to (15.12) as

$$T_{ext}(\hat{\theta}_a)\hat{Y} = \Phi_l'(u_{ext}, \hat{\theta}_n)\theta_b, \tag{15.14}$$

where

$$\Phi_l'(u_{ext}, \hat{\theta}_n) = \begin{bmatrix} f(u(1-d), \hat{\theta}_n) & \cdots & f(u(1-d-n), \hat{\theta}_n) \\ f(u(2-d), \hat{\theta}_n) & \cdots & f(u(2-d-n), \hat{\theta}_n) \\ \vdots & & \vdots \\ f(u(N-d), \hat{\theta}_n) & \cdots & f(u(N-d-n), \hat{\theta}_n) \end{bmatrix}$$

to give

$$\hat{Y}(\theta_a, \theta_b) = T_{ext}^{-1}(\hat{\theta}_a)\Phi_l'(u_{ext}, \hat{\theta}_n)\theta_b.$$

Thus, for a given $\hat{\theta}_a$, the solution, $\hat{\theta}_b$, for θ_b is given by

$$\hat{\theta}_b = \arg\min_{\theta_b} \|Y - \hat{Y}\|_2$$
$$= \underbrace{\left(\left(T_{ext}^{-1}(\hat{\theta}_a)\Phi_l'(u_{ext}, \hat{\theta}_n)\right)^T T_{ext}^{-1}(\hat{\theta}_a)\Phi_l'(u_{ext}, \hat{\theta}_n)\right)^{-1}\left(T_{ext}^{-1}(\hat{\theta}_a)\Phi_l'(u_{ext}, \hat{\theta}_n)\right)^T Y}_{g(\hat{\theta}_a)}. \tag{15.15}$$

The OE minimization problem has now been reduced to an unconstrained nonlinear least squares problem

$$\hat{\theta}_a = \arg\min_{\theta_a} \|Y - \hat{Y}(\theta_a, g(\theta_a))\|_2$$

with θ_a as the only variable to be optimized. Such a problem can be solved by standard local optimization methods, for example, the Levenberg–Marquardt method [44].

In general it is difficult to impose the requirement that the identified model is stable but in the muscle identification context, a second order system has often been assumed by many authors, and in this case the stability constraint reduces to the following bound constraints on the parameters

$$0 < \hat{a}_2 \leq 1 \quad \text{and} \quad -2 \leq \hat{a}_1 \leq 0.$$

15.4.2 Iterative Algorithms

For both model structures, minimization over the θ_n and θ_l parameters can be executed iteratively, resulting in Algorithms 15.1 and 15.2 given next.

Algorithm 15.1: Iterative algorithm for Hammerstein system identification with ARX model

---

Inputs: an initial value of the linear component, $\hat{\theta}_l^0$, an input/output data set $u(k)$, $y(k)$, $k = 1, 2, \ldots, N$, and a convergence tolerance ε.
$j = 0$
repeat
  $j = j + 1$
  $\hat{\theta}_n^j = \left(\Phi_n(u, \hat{\theta}_b^{j-1})^T \Phi_n(u, \hat{\theta}_b^{j-1})\right)^{-1} \Phi_n(u, \hat{\theta}_b^{j-1})^T Y_n(y, \hat{\theta}_a^{j-1})$
  $\hat{\theta}_l^j = \left(\Phi_l(u, y, \hat{\theta}_n^j)^T \Phi_l(u, y, \hat{\theta}_n^j)\right)^{-1} \Phi_l(u, y, \hat{\theta}_n^j)^T Y'$
until $|V_N(\hat{\theta}_l^j, \hat{\theta}_n^j) - V_N(\hat{\theta}_l^{j-1}, \hat{\theta}_n^{j-1})| < \varepsilon$
Output: $\hat{\theta} = \begin{bmatrix} \hat{\theta}_n^j \\ \hat{\theta}_l^j \end{bmatrix}$

Algorithm 15.2: Iterative algorithm for Hammerstein system identification with OE model

Inputs: an initial value of the linear component, $\hat{\theta}_l^0$, an input/output data set $u(k)$, $y(k)$, $k = 1, 2, \ldots, N$, and a convergence tolerance ε.

$j = 0$

repeat

$j = j + 1$

$$\hat{\theta}_n^j = \left(\left(T_{ext}^{-1}(\hat{\theta}_a^{j-1})\Phi_n(u_{ext}, \hat{\theta}_b^{j-1})\right)^T T_{ext}^{-1}(\hat{\theta}_a^{j-1})\Phi_n(u_{ext}, \hat{\theta}_b^{j-1})\right)^{-1} \times \left(T_{ext}^{-1}(\hat{\theta}_a^{j-1})\Phi_n(u_{ext}, \hat{\theta}_b^{j-1})\right)^T Y$$

$$\hat{\theta}_a^j = \arg\min_{\theta_a} \|Y - \hat{Y}(\theta_a, g(\theta_a))\|_2$$

where $g(\theta_a)$ is defined in (15.15) and $\hat{\theta}_b^j = g(\hat{\theta}_a^j)$

until $|V_N(\hat{\theta}_l^j, \hat{\theta}_n^j) - V_N(\hat{\theta}_l^{j-1}, \hat{\theta}_n^{j-1})| < \varepsilon$

Output: $\hat{\theta} = \begin{bmatrix} \hat{\theta}_n^j \\ \hat{\theta}_l^j \end{bmatrix}$

15.5 Experimental Test Design

The choice of suitable experimental test procedure is a crucial step for any successful model identification. This is especially true in the present case since tests are not applied to a mechanical or physical process, but to a human being; care must be taken to avoid triggering involuntary reflex mechanisms, fatigue, inhibition due to subject discomfort [6] and to operate within physiological constraints and limitations. Whilst the experimental test procedure ensures that the maximum levels of stimulation are within suitable bounds, as discussed in more detail in Sect. 15.6.1 below, it is also necessary to ensure that motor units are recruited gradually rather than abruptly exciting a large number simultaneously [10]. This clearly excludes rapidly increasing input signals with wide amplitude fluctuation, and instead necessitates a more slowly varying signal to ensure the rate of recruitment of nerve fibers is limited above the excitation threshold. The need for a limitation on rapidly decreasing signals is also necessary since the sudden reduction in stimulation is also associated with involuntary reflexes and patient discomfort [10]. A more subject-specific concern is that it has also been observed that certain types of signal elicit a greater degree of involuntary response than others, despite possessing similar characteristics.

Detailed investigation [8] based on the arguments just given, led to four candidate tests for use in the identification of electrically stimulated muscle. Examples of the excitation inputs used in each are given in Fig. 15.5, where it is the pulse duration which is selected as the controlled variable, as discussed in Sect. 15.2. These tests are as follows.

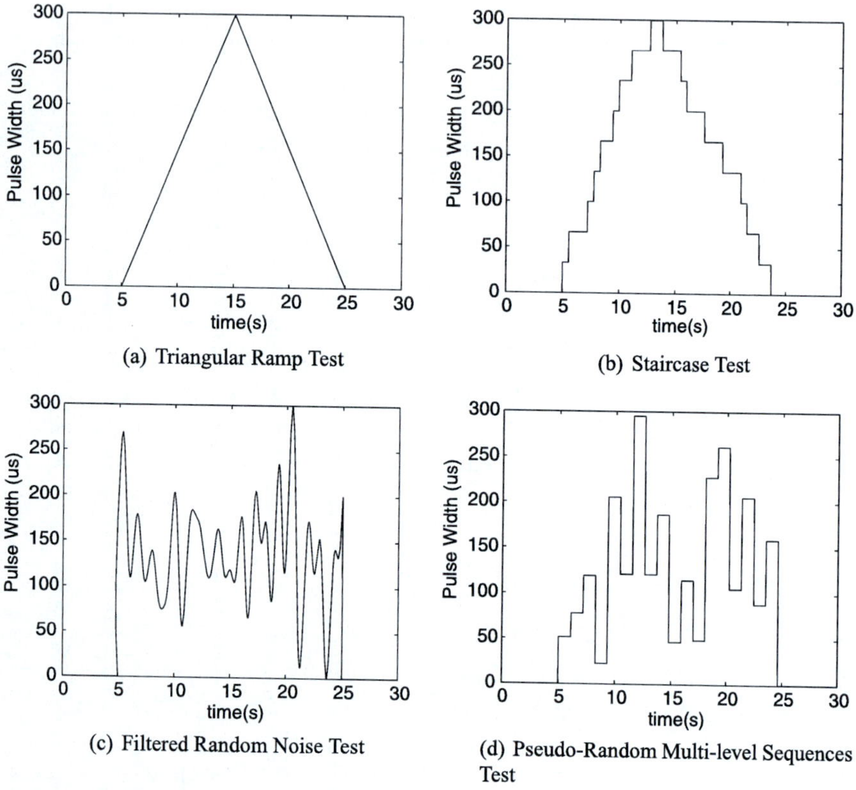

(a) Triangular Ramp Test

(b) Staircase Test

(c) Filtered Random Noise Test

(d) Pseudo-Random Multi-level Sequences Test

Fig. 15.5 Examples of the four candidate tests

- Triangular Ramp (TR) test

  The pulse duration rises linearly from 0 to 300 μs and then returns to 0, its range being uniformly distributed.
- Staircase test

  The duration of each pulse changes step by step. The number of steps should be large enough to identify the nonlinearity and their width chosen carefully. Let $\tau = T_s/4$ (where T_s is the 98% settling time). It is then recommended to use mixed step widths, with step width τ for $1/3$ of the test period, 2τ for another $1/3$ of the test period and 3τ for the remaining $1/3$ of the test period, and to randomize these widths when creating the test signals [45].
- Filtered Random Noise (FRN) test

  The pulse width signal is produced by low-pass filtering white noise, using a suitable cut-off frequency to balance the opposing physiological and identification issues discussed above. Having filtered the signal, an offset and gain are applied to ensure the desired pulsewidth range is spanned.
- Pseudo-Random Multi-level Sequences (RMS) test

The excitation signal is an multi-level pseudo random signal which is a periodic, deterministic signal having an autocorrelation function similar to white noise. The amplitude level is uniformly distributed over the full range.

15.6 Results

15.6.1 Experimental Set-Up

The ILC workstation has been described in Sect. 15.2, and is a platform on which model-based ES has been clinically applied. The system has been used to obtain the experimental results which follow since it

- provides a facility whose software and hardware components, including sensor and stimulation systems, have been experimentally assessed and verified [12]
- ensures that the experimental set-up procedure, as used in clinical trials, is appropriate to the intended application area of stroke rehabilitation.

Tests were performed on a single unimpaired subject, and took place during two sessions conducted over consecutive days. Biometric measurements, including the length of upper arm and forearm, were first made using anatomical landmarks, and then the participant was seated in the workstation. Their right arm was strapped to the extreme link of the five-bar robotic arm which incorporates a six axis force/torque sensor, which provides support and constrains it to lie in a horizontal plane, and straps were also applied about the upper torso to prevent shoulder and trunk movement (as shown in Fig. 15.1). The subject's upper limb was then moved over as large an area as possible and a kinematic model of the arm was produced using the measurements recorded. This kinematic model is the same as that appearing in the arm and muscle system shown in Fig. 15.2, but it is now used to convert the force recorded by the force/torque sensor to a torque acting about the elbow (full details are given in [12]). The electrode was then positioned on the lateral head of triceps and adjusted so that the applied ES generated maximum forearm movement. The stimulation consists of a series of bi-phasic pulses at 40 Hz, whose pulsewidth is variable from 0 to 300 μs with a resolution of 1 μs. The amplitude, which is fixed throughout all subsequent tests, is determined by setting the pulsewidth equal to 300 μs and slowly increasing the applied voltage until a maximum comfortable limit is reached. A sample frequency of 1.6 KHz is used by the real-time hardware, and all calculations are performed using the Matlab/Simulink environment.

The position of the robotic arm was then fixed using a locking pin, at an elbow extension angle of approximately $\pi/2$ rads. This removes the non-isometric components of the biomechanical model, so that the resulting system corresponds to the Hammerstein structure shown in Fig. 15.3. The identification tests that followed were each of 30 sec duration, and used excitation signals in which the first and last 5 sec periods consisted of zero stimulation. Only the middle 20 sec section of input and output data was used for identification, with the adjoining periods used to establish the baseline torque offset (taken as the mean torque value). The identification

Table 15.1 Identification results of Algorithm 15.1 and Algorithm 15.2 for the four candidate tests. The results are in terms of the Best Fit Rate

	Triangular ramp	Filtered random noise	Staircase	PRMS
(a) Algorithm 15.1				
1	85.88	42.54	87.62	50.34
2	88.34	36.39	89.16	52.48
3	91.33	36.09	85.18	52.43
4	89.23	36.58	89.68	36.69
5	92.25	63.38	89.84	
6	90.68	55.23	91.35	
7	89.14	48.12	88.33	
8	91.41	58.09	88.17	
9	94.25	74.74	83.46	
10	89.02	66.84	91.85	
Average	90.15	51.8	88.46	48.00
(b) Algorithm 15.2				
1	92.65	73.03	90.89	66.89
2	92.25	65.19	93.32	78.91
3	93.88	51.69	93.49	63.79
4	93.36	70.92	93.49	65.92
5	93.08	79.94	93.77	
6	91.98	68.46	92.34	
7	95.74	58.48	93.38	
8	92.41	61.50	94.66	
9	95.32	79.74	90.85	
10	92.60	71.32	94.23	
Average	93.33	68.03	93.04	68.88

calculations were carried out immediately following each test in order to establish the efficacy of the data.

For the TR, Staircase and FRN tests, 10 trials were performed, however, in the case of the PRMS test, only 4 trials were carried out as it was evident that the fit rate was poor. Between every two tests there was a rest period of at least 10 min in order to eliminate fatigue [46], and the order of identification tests was also randomized to minimize the effect of subject memory or acclimatization increasing their involuntary response.

15.6.2 Experimental Results

For both algorithms, the identification, validation and cross-validation results are listed in Tables 15.1, 15.2 and 15.3, respectively. Results are in terms of the Best Fit

Table 15.2 Validation results of Algorithm 15.1 and Algorithm 15.2 for the four candidate tests. The model is identified from the listed data set and validated on all the data of the same test. The results are expressed in terms of the average Best Fit Rate

	Triangular ramp	Filtered random noise	Staircase	PRMS
(a) Algorithm 15.1				
1	82.28	28.01	73.99	11.17
2	82.78	45.00	83.63	43.45
3	79.01	40.52	77.77	46.32
4	82.51	8.79	77.27	44.10
5	82.83	37.62	82.72	
6	81.94	28.97	82.28	
7	78.51	40.67	81.20	
8	80.11	30.37	80.68	
9	82.80	44.77	78.09	
10	83.25	−45.65	81.04	
Average	81.60	25.91	79.87	36.26
(b) Algorithm 15.2				
1	76.12	16.90	75.86	46.08
2	80.94	26.34	84.39	32.15
3	76.58	36.32	83.42	29.50
4	81.80	16.31	83.47	50.22
5	81.79	24.37	75.82	
6	75.98	41.43	83.09	
7	68.03	29.49	83.32	
8	79.87	20.76	81.96	
9	80.32	46.80	80.67	
10	78.61	−30.94	83.81	
Average	78.00	22.78	81.58	39.49

rate, defined as the percentage,

$$Best\ Fit = \left(1 - \frac{\|y - \hat{y}\|_2}{\|y - \bar{y}\|_2}\right) \times 100,$$

where y is the measured output, $\hat{y}$ is the simulated model output and $\bar{y}$ is the mean of y. To aid visual comparison of the identification and validation results between Algorithms 15.1 and 15.2, box and whisker plots are given in Fig. 15.6.

The identification results for each individual trial of four candidate tests are given in Table 15.1 together with the average results for all the trials. To obtain the validation results, a model is firstly identified from the data of one trial and then is used to

Table 15.3 Cross Validation results of Algorithms 15.1 and 15.2 for the TR, FRN, and Staircase tests. The model is identified from all the data of one type of test and validated on all the data of the other type of test. The results are the average Best Fit Rate

	Triangular ramp (TR)	Filtered random noise (FRN)	Staircase
(a) Algorithm 15.1			
(TR)	84.83	17.76	47.94
(FRN)	83.96	40.23	71.40
Staircase	79.53	42.67	84.48
(b) Algorithm 15.2			
(TR)	81.80	41.08	64.11
(FRN)	68.75	46.80	67.86
Staircase	80.72	45.00	84.39

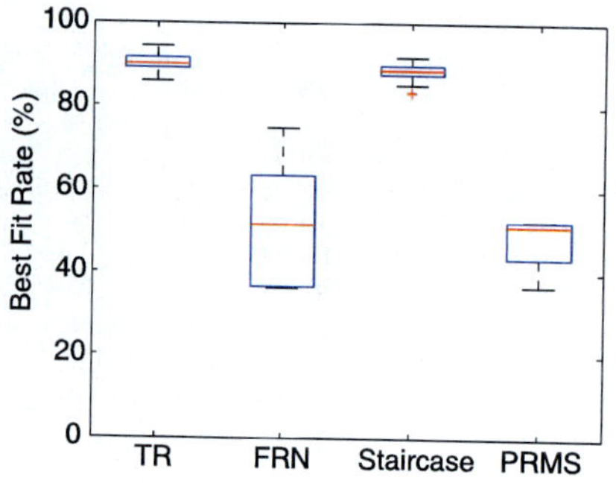

(a) Identification Results for Algorithm 1

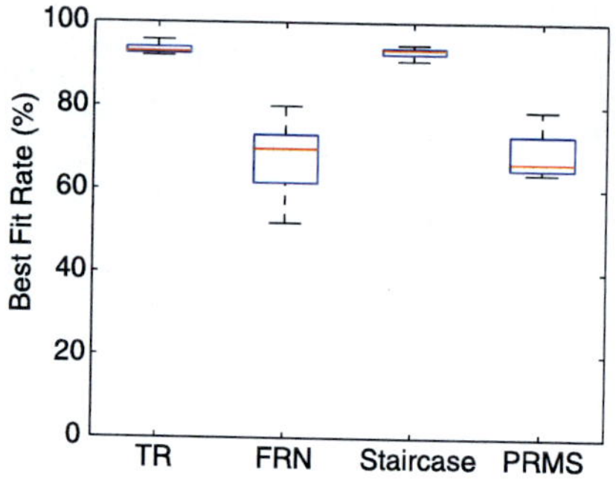

(b) Identification Results for Algorithm 2

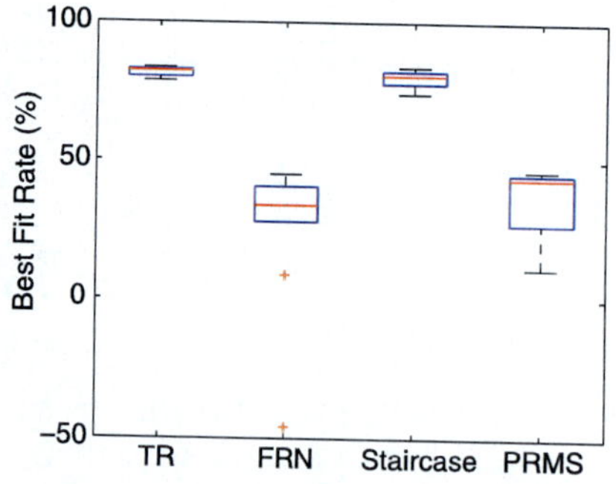

(c) Validation Results for Algorithm 1

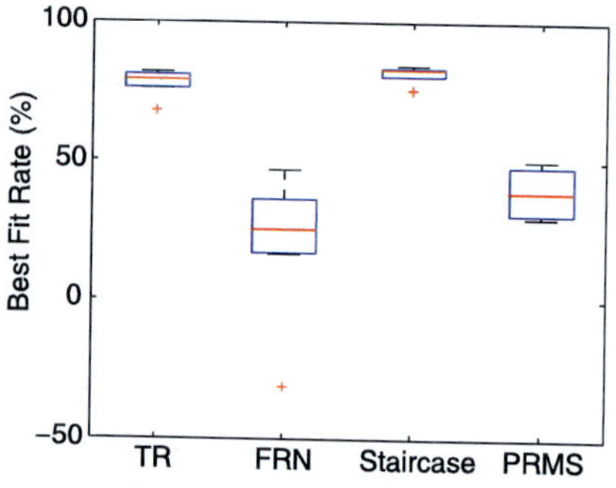

(d) Validation Results for Algorithm 2

Fig. 15.6 Comparison between Algorithm 15.1 and Algorithm 15.2 for the identification and validation results predict the outputs for all the trials in the same type of the tests. The results are the average values of all the prediction results in terms of Best Fit rate. The validation results, in Table 15.2, show the predictive ability within the same type of identification tests. Similarly, in order to show the predictive ability for different stimulation patterns, cross-validation analysis is conducted, see Table 15.3. Firstly, a model is identified from the data of all the trials in one type of test and then is used to predict the outputs for all the trials in one of the other tests. The results are again the average value of the Best Fit rate. Here only the TR, FRN, and Staircase tests are compared, due to the poor performance of the PRMS test in both identification and validation.

15.7 Discussion

15.7.1 Initial Values for Linear Parameters

Both algorithms require the initial values of the linear parameters, which can be obtained using any existing method that applies an input suitable for use with stroke patients. One such technique is the ramp deconvolution method [34], which was used in [13]. By using representative choice of parameters to provide an initial value estimate, both algorithms can achieve convergence after several iterations, illustrated by Fig. 15.7(a) and 15.7(b). However, irrespective of the iteration number, Algorithm 15.2 takes a longer period of time because in each iteration, an iterative search is applied.

In order to expedite the identification procedure of Algorithm 15.2, a better solution to the linear parameter estimate is required. The representative estimate from [13] is obviously not sufficiently accurate and, moreover, the values of the linear parameters vary widely from subject to subject and it is difficult to find a single representative estimate across all subjects. Therefore, the optimal solution of the linear parameters from Algorithm 15.1 has been used to initialize Algorithm 15.2. This thereby unites the two algorithms in a single scheme which combines the speed of the first with the accuracy of the second. The results confirm high accuracy with fewer iterations to converge, as illustrated in Fig. 15.7(c).

15.7.2 Algorithmic Comparison

- *Structure and unknown parameters*

  A Hammerstein structure is used in both algorithms, but with different linear models. Algorithm 15.1 uses the ARX linear model, where the white noise is assumed to pass through the denominator dynamics of the linear block before being added to the output. However, it is perhaps not the most natural form from a physical point of view. Thus, another linear model, the OE model, is assumed in Algorithm 15.2, where white noise is added directly to the output, accounting

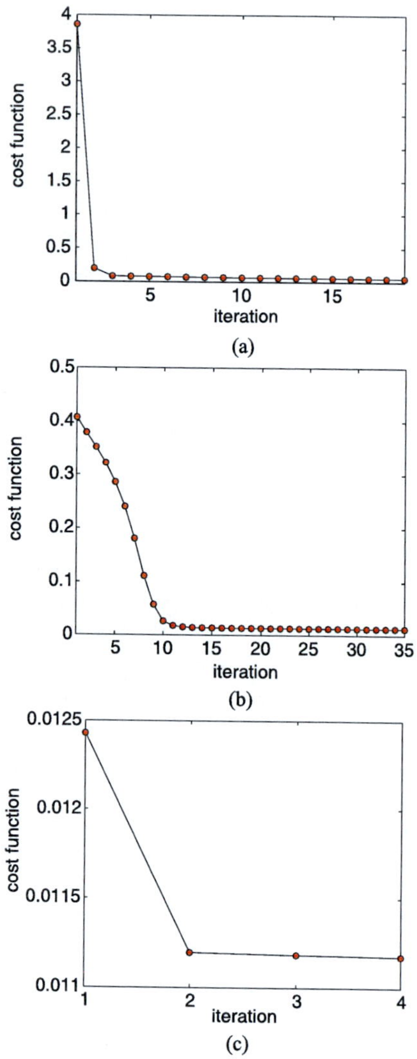

Fig. 15.7 Examples of convergence properties for Algorithm 15.1 and Algorithm15.2: (**a**) A representative estimate from [13] is used to give the initial values and Algorithm 15.1 is applied. Convergence is achieved after 18 trials; (**b**) A representative estimate from [13] is used as the initial values and Algorithm 15.2 is applied. Convergence is achieved after 35 trials; (**c**) The optimal solution from Algorithm 15.1 is used to provide the initial values and Algorithm 15.2 is applied. Convergence is achieved after 4 iterations

for the measured errors from the equipment. The number of unknown parameters is kept the same for both algorithms.

- *Identification procedure*

  The identification procedures for the two algorithms are not the same but they both alternatively optimize the nonlinear and linear parameters at each iteration.

  Algorithm 15.1 is a development of a two stage identification method, see [47], which has been shown to outperform both the ramp deconvolution and Separable Least Squares methods on a simulated muscle system with a range of noise levels. It alternatively solves the least squares problems to optimize the linear and nonlinear parameters. It is computationally easy and is reasonably fast in implementation.

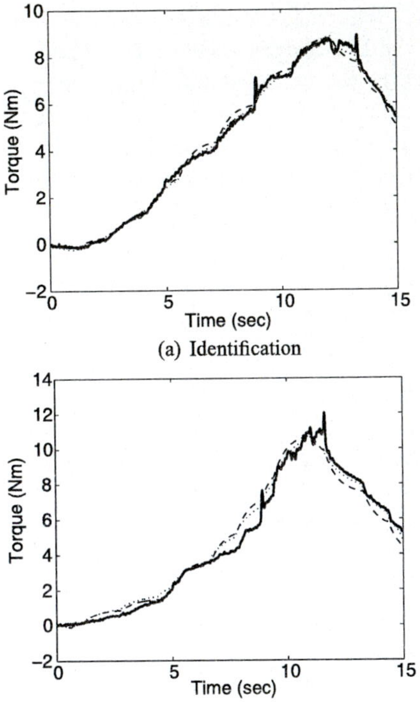

Fig. 15.8 The force outputs of Algorithm 15.1 (*dashed*), Algorithm 15.2 (*dotted*) and the measured force outputs (*solid*) are plotted

The identification procedure of Algorithm 15.2 is more complicated. On each iteration, the nonlinear parameters can be identified in a least squares sense through use of transformations and related assumptions, while the identification of the linear parameters necessitates an iterative search technique to find the local optimal solution. Thus, it is more time consuming than Algorithm 15.1, but, by using the optimal solution from Algorithm 15.1 to provide initial values, the identification procedure of Algorithm 15.2 can be greatly accelerated in terms of speed to the point where it is not a matter of concern.

- *Performance*

  Both algorithms provide good fitting performance and predictive ability. Figure 15.8 shows the fitting performance between the modeled outputs and measured outputs in both identification and validation cases.

  In terms of identification results, Algorithm 15.2 is superior to Algorithm 15.1, as observed directly from Figs. 15.6(a) and 15.6(b). Numerically, Algorithm 15.2 improves the average results by up to 20% compared with Algorithm 15.1, as shown in Table 15.1.

  In validation, both algorithms have similar performance, as shown by Figs. 15.6(c) and 15.6(d). Through examination of Table 15.2, Algorithm 15.1 is seen to be better for TR and FRN test data, while Algorithm 15.2 is better for the staircase and PRMS test data. It is therefore reasonable to conclude that both algorithms have comparable performance in prediction.

The validation and prediction results provide the most direct indication of the models' accuracy when applied to the design of controllers for stroke rehabilitation. Since both algorithms exhibit similar performance in this area, it is Algorithm 15.1's simpler implementation and faster computation that make it the preferable option. Whilst in this application Algorithm 15.2's increased complexity does not translate to improved results in validation and prediction, it is anticipated that applications exist in which it does outperform Algorithm 15.1.

15.7.3 Test Comparisons

Although the TR test is widely used in muscle tests such as [13, 34, 39] and can achieve satisfactory fitting rates (almost the highest values in the identification case and approximately 80% for Algorithm 15.1 and a little lower for Algorithm 15.2 in the validation case), it shows poor capability in predicting other stimulation patterns, such as those in Table 15.3. This is due to its non-persistent excitation property discussed in the previous section, which leads to an unreliable model identified from this test.

For the FRN and PRMS tests, the average identification results for Algorithm 15.1 are 51.8% and 48% respectively, as shown in Table 15.1(a). Although Algorithm 15.2 improves on these by as much as 20%, these tests are still far lower than those of the TR and Staircase tests. The validation results in Table 15.2 are even lower, and this lack of output prediction may be expected to lead to poor results when transferred to model-based control application. There may be two reasons for this: the first is that the model structure and the identification algorithms are not proper, but this is not the case because they perform very well for the other two identification tests. The second reason is that the experimental data is not proper. Considering the effects caused by randomly exciting tests on the human subjects, it is believed that these signals elicit involuntary reflexes and subject discomfort, which results in noisy data.

This is the first time the Staircase test has been used in the identification of electrically stimulated muscles, and it has shown clear advantages over alternatives, that is, it is persistently exciting, gives high fitting rates in the identification case (the second highest one in Table 15.1) and in the validation case (surpassing even the TR test for Algorithm 15.2 in Table 15.2(b)) and shows accurate predictive ability across different stimulation patterns (see Table 15.3). Therefore the Staircase test is highly recommended for the identification of the response of electrically stimulated muscle.

15.8 Conclusions and Open Research Questions

Two identifications schemes have been considered in order to address the limitations in the suitability and effectiveness of existing methods for identification of electrically stimulated muscle models in the case of incomplete paralysis. These limitations relate to the underlying model structure, and also to the associated excitation input required in the identification of its parameters. Experimental results have been used to confirm the efficacy of each approach and assess their performance over a range of identification test inputs with respect to persistence of excitation and modeling accuracy. Limitations in the identification procedure of each were subsequently overcome by combining them into a unified scheme.

The algorithms considered in this chapter represent significant progress in the identification of electrically stimulated muscle, but the resulting models were only verified over a short time interval of 20 sec duration. For application in stroke rehabilitation, however, stimulation must be applied during intensive, goal orientated practice tasks in order to maximize improvement in motor control. In clinical trials this translates to sustained application of stimulation during each treatment session of between 30 minutes and 1 hour duration [4]. In this case, slowly time-varying properties of the muscle system arise due to fatigue, changing physiological conditions or spasticity [46].

One possible route to obtaining more effective algorithms is to use online, or recursive, identification where the model parameters are updated once new data is available. Only a few of the existing identification methods are recursive, and can be divided into three categories. The first category is the recently developed recursive subspace identification method, see, for example, [48], the second is stochastic approximation, see, for example, [49, 50], and the third is recursive least squares or extended recursive least squares. The Recursive Least Squares (RLS) algorithm is a well known method for recursive identification of linear-in-parameter models, and if the data is generated by correlated noise, the parameters describing the model of the correlation can be estimated by Extended Recursive Least Squares (ERLS). Here, a typical way to use these two algorithms is to treat each of the cross-product terms in the Hammerstein system equations as an unknown parameter. This procedure, which results in an increased number of unknowns, is usually referred to as the over-parameterization method [51, 52]. After this step, the RLS or ERLS method can be applied, see, for example, [53].

A review of the limitations of current algorithms leads to the following conclusions with respect to their application the problem area considered in this chapter

- The first two categories have only been applied in simulation and the stochastic approximation has not considered time-varying linear dynamics. The third category is the most promising as it has already been applied to electrically stimulated muscle in [54, 55].
- Most of the test signals used are formed as random noise in order to guarantee persistent excitation, even when applied to the human muscle [55], and use pseudo-random binary sequences. This type of signal, as has been described, excites the motor units abruptly, causing patient discomfort and, and entails eliciting an involuntary response, as reported in [10]. In [54] a test consisting of 25 pulses is used, each of which is of 1 second duration in the form of a noisy triangular wave. This test meets our requirements but is too short to exhibit time-varying properties.

- The most relevant previous work is [54] where the system considered had linear constraints and RLS was developed for constrained systems. However, the results given do not establish that the constraints are achieved. For example, even when considering the prediction error, the posteriori estimated output without constraints is better than the one with constraints. Thus, the idea of adding constraints to RLS, leading to increased computational load, is well worth considering.

Overall, RLS is the most promising technique for recursive identification of electrically stimulated muscle, but the problem of consistent estimation must be resolved [50, 54]. It is, however, possible that unsatisfactory performance will result, especially for noisy measurements, and hence alternative recursive algorithms for Hammerstein systems may be required. Moreover, a long-period test signal needs to be designed for this application area, which is persistently exciting and also gradually recruits the motor units.

References

1. Parker, V.M., Wade, D.T., Langton-Hewer, R.: Loss of arm function after stroke: measurement, frequency and recovery. Int. Rehabil. Med. **8**(4), 69–73 (1986)
2. Broeks, J.G., Lankhorst, G.J., Rumping, K., Previo, A.J.: The long-term outcome of arm function after stroke: results of a follow-up study. Disabil. Rehabil. **21**, 357–364 (1999)
3. de Kroon, J.R., van der Lee, J.H., Ijzerman, M.J., Lankhorst, G.J.: Therapeutic electrical stimulation to improve motor control and functional abilities of the upper extremity after stroke: a systematic review. Clin. Rehabil. **16**(4), 350–360 (2002)
4. De Kroon, J.R., Ijzerman, M.J., Chae, J.J., Lankhorst, G.J., Zilvold, G.: Relation between stimulation characteristics and clinical outcome in studies using electrical stimulation to improve motor control of the upper extremety in stroke. J. Rehabil. Med. **37**(2), 65–74 (2005)
5. Pomeroy, V.M., King, L., Pollack, A., Baily-Hallon, A., Longhorne, P.: Electrostimulation for promoting recovery of movement or functional ability after stroke. The Cochrane Database of Systematic Reviews, Issue 2 (2006)
6. Burridge, J.H., Ladouceur, M.: Clinical and therapeutic applications of neuromuscular stimulation: a review of current use and speculation into future developments. Neuromodulation **4**(4), 147–154 (2001)
7. Rushton, D.N.: Functional electrical stimulation and rehabilitation—an hypothesis. Med. Eng. Phys. **25**(1), 75–78 (2003)
8. Le, F., Markovsky, I., Freeman, C.T., Rogers, E.: Identification of electrically stimulated muscle models of stroke patients. Control Eng. Pract. **18**(4), 396–407 (2010)
9. Thorsen, R., Spadone, R., Ferrarin, M.: A pilot study of myoelectrically controlled FES of upper extremity. IEEE Trans. Neural Syst. Rehabil. Eng. **9**(2), 161–168 (2001)
10. Baker, L.L., McNeal, D.R., Benton, L.A., Bowman, B.R., Waters, R.L.: NeuroMuscular Electrical Stimulation: A Practical Guide, 3rd edn. (1993)
11. Freeman, C.T., Hughes, A.-M., Burridge, J.H., Chappell, P.H., Lewin, P.L., Rogers, E.: Iterative learning control of FES applied to the upper extremity for rehabilitation. Control Eng. Pract. **17**(3), 368–381 (2009)
12. Freeman, C.T., Hughes, A.-M., Burridge, J.H., Chappell, P.H., Lewin, P.L., Rogers, E.: A robotic workstation for stroke rehabilitation of the upper extremity using FES. Med. Eng. Phys. **31**(3), 364–373 (2009)

13. Freeman, C.T., Hughes, A.-M., Burridge, J.H., Chappell, P.H., Lewin, P.L., Rogers, E.: A model of the upper extremity using surface FES for stroke rehabilitation. J. Biomed. Eng. **131**(1), 031011 (2009)
14. Hughes, A.-M., Freeman, C.T., Burridge, J.H., Chappell, P.H., Lewin, P.L., Rogers, E.: Feasibility of iterative learning control mediated by functional electrical stimulation for reaching after stroke. Neurorehabil. Neural Repair **23**(6), 559–568 (2009)
15. Arimoto, S., Kawamura, S., Miyazaki, F.: Bettering operations of robots by learning. J. Robot. Syst. **1**, 123–140 (1984)
16. Bristow, D.A., Tharayil, M., Alleyne, A.G.: A survey of iterative learning control. IEEE Control Syst. Mag. **26**(3), 96–114 (2006)
17. Ahn, H.-S., Chen, Y., Moore, K.L.: Iterative learning control: brief survey and categorization. IEEE Trans. Syst. Man Cybern. **37**(6), 1109–1121 (2007)
18. Popovic, D., Popovic, M.: Tuning of a nonanalytical hierarchical control system for reaching with FES. IEEE Trans. Biomed. Eng. **45**(2), 203–212 (1998)
19. Crago, P.E., Nakai, R.J., Chizeck, H.J.: Feedback regulation of hand grasp opening and contact force during stimulation of paralysed muscle. IEEE Trans. Biomed. Eng. **38**(1), 17–28 (1991)
20. Chizeck, H.J., Lan, N., Palmieri, L.S., Crago, P.L.: Feedback control of electrically stimulated muscle using simultaneous pulse width and stimulus period modulation. IEEE Trans. Biomed. Eng. **38**(12), 1224–1234 (1991)
21. Watanabe, T., Iibuchi, K., Kurosawa, K., Hoshimiya, N.: A method of multichannel PID control of two-degree-of-freedom wrist joint movements by functional electrical stimulation. Syst. Comput. Jpn. **34**(5), 319–328 (2003)
22. Hatwell, M.S., Oderkerk, B.J., Sacher, C.A., Inbar, G.F.: Patient-driven control of FES-supported standing up: a simulation study. IEEE Trans. Rehabil. Eng. **36**(6), 683–691 (1991)
23. Previdi, F., Schauer, T., Savaresi, S.M., Hunt, K.J.: Data-driven control design for neuroprotheses: a virtual reference feedback tuning (VRFT) approach. IEEE Trans. Control Syst. Technol. **12**(1), 176–182 (2004)
24. Hill, A.V.: Then heat of shortening and the dynamic constants of a muscle. Proc. R. Soc. Lond. B, Biol. Sci. **126**, 136–195 (1938)
25. Lan, N.: Stability analysis for postural control in a two-joint limb system. IEEE Trans. Neural Syst. Rehabil. Eng. **10**(4), 249–259 (2002)
26. Riener, R., Fuhr, T.: Patient-driven control of FES-supported standing up: a simulation study. IEEE Trans. Rehabil. Eng. **6**(2), 113–124 (1998)
27. Jezernik, S., Wassink, R.G.V., Keller, T.: Sliding mode closed-loop control of FES: controlling the shank movement. IEEE Trans. Rehabil. Eng. **51**(2), 263–272 (2004)
28. Schauer, T., Negard, N.O., Previdi, F., Hunt, K.J., Fraser, M.H., Ferchland, E., Raisch, J.: Online identification and nonlinear control of the electrically stimulated quadriceps muscle. Control Eng. Pract. **13**(9), 1207–1219 (2005)
29. Ferrarin, M., Palazzo, F., Riener, R., Quintern, J.: Model-based control of FES induced single joint movements. IEEE Trans. Neural Syst. Rehabil. Eng. **9**(3), 245–257 (2001)
30. Hunt, K.J., Munih, M., Donaldson, N.N., Barr, F.M.D.: Investigation of the Hammerstein hypothesis in the modeling of electrically stimulated muscle. IEEE Trans. Rehabil. Eng. **45**(8), 998–1009 (1998)
31. Reiner, R., Quintern, J.: A physiologically based model of muscle activation verified by electrical stimulation. Bioelectrochem. Bioenerg. **43**, 257–264 (1997)
32. Previdi, F., Carpanzano, E.: Design of a gain scheduling controller for knee-joint angle control by using functional electrical stimulation. IEEE Trans. Control Syst. Technol. **11**(3), 310–324 (2003)
33. Happee, R., Van der Helm, F.C.T.V.: The control of shoulder muscles during goal directed movements, an inverse dynamic analysis. J. Biomed. Eng. **28**(10), 1179–1191 (1995)
34. Durfee, W.K., MacLean, K.E.: Methods for estimating isometric recruitment curves of electrically stimulated muscle. IEEE Trans. Biomed. Eng. **36**(7), 654–667 (1989)
35. Baratta, R., Solomonow, M.: The dynamic response model of nine different skeletal muscles. IEEE Trans. Biomed. Eng. **37**(3), 243–251 (1990)

36. Veltink, P.H., Chizeck, H.J., Crago, P.E., El-Bialy, A.: Nonlinear joint angle control for artificially stimulate muscle. IEEE Trans. Biomed. Eng. **39**(4), 368–380 (1992)
37. Chizeck, H.J., Crago, P.E., Kofman, L.S.: Robust closed-Loop control of isometric muscle force using pulsewidth modulation. IEEE Trans. Biomed. Eng. **35**(7), 510–517 (1988)
38. Bernotas, L., Crago, P.E., Chizeck, H.J.: A discrete-time model of electrically stimulated muscle. IEEE Trans. Biomed. Eng. **33**(9), 829–838 (1986)
39. Durfee, W.K., Palmer, K.L.: Estimation of force-activation, force-length, and force-velocity properties in isolated, electrically stimulated muscle. IEEE Trans. Biomed. Eng. **41**(3), 205–216 (1994)
40. Crago, P.E., Peckham, P.H., Thorpe, G.B.: Modulation of muscle force by recruitment during intramuscular stimulation. IEEE Trans. Biomed. Eng. **27**(12), 679–684 (1980)
41. Ding, J., Wexler, A.S., Binder-MacLeod, S.A.: A mathematical model that predicts the force-frequency relationship of human skeletal muscle. Muscle Nerve **26**(2), 477–485 (2002)
42. Carroll, S.G., Triolo, R.J., Chizeck, H.J., Kobetic, R., Marsolias, E.B.: Tetanic responses of electrically stimulated paralyzed muscle at varying interpulse intervals. IEEE Trans. Biomed. Eng. **36**(7), 644–653 (1989)
43. Dempsey, E.J., Westwick, D.T.: Identification of Hammerstein models with cubic spline nonlinearities. IEEE Trans. Biomed. Eng. **51**(2), 237–245 (2004)
44. Marquardt, D.: An algorithm for least-squares estimation from linear parameters. SIAM J. Appl. Math. **11**, 431–441 (1963)
45. Zhu, Y.: Identification of Hammerstein models for control using ASYM. Int. J. Control **73**(18), 1692–1702 (2000)
46. Graham, G.M., Thrasher, T.A., Popovic, M.R.: The effect of random modulation of functional electrical stimulation parameters on muscle fatigue. IEEE Trans. Neural Syst. Rehabil. Eng. **14**(1), 38–45 (2006)
47. Le, F., Markovsky, I., Freeman, C.T., Rogers, E.: Identification of electrically stimulated muscle after stroke. In: Proc. European Control Conference, pp. 3208–3213 (2009)
48. Bako, L., Mercere, G., Lecoeuche, S., Lovera, M.: Recursive subspace identification of Hammerstein models based on least squares support vector machines. IET Proc. D **3**(9), 1209–1216 (2009)
49. Greblicki, W.: Stochastic approximation in nonparametric identification of Hammerstein systems. IEEE Trans. Autom. Control **47**(11), 1800–1810 (2002)
50. Chen, H.F.: Pathwise convergence of recursive identification algorithms for Hammerstein systems. IEEE Trans. Autom. Control **49**(4), 1641–1649 (2004)
51. Bai, E.W.: An optimal two-stage identification algorithm for Hammerstein-Wiener nonlinear systems. Automatica **34**(3), 333–338 (1998)
52. Chang, F.H.I., Luus, R.: A non-iterative method for identification using Hammerstein model. IEEE Trans. Autom. Control **16**(5), 464–468 (1971)
53. Zhao, W.X., Chen, H.F.: Adaptive tracking and recursive identification for Hammerstein systems. Automatica, **45**(12), 2773–2783 (2009)
54. Chia, T.L., Chow, P.C., Chizeck, H.J.: Recursive parameter identification of constrained systems: an application to electrically stimulated muscle. IEEE Trans. Biomed. Eng. **38**(5), 429–442 (1991)
55. Ponikvar, M., Munih, M.: Setup and procedure for online identification of electrically stimulated muscle With Matlab Simulink. IEEE Trans. Neural Syst. Rehabil. Eng. **9**(3), 295–301 (2001)

Part III
Data-Based Mechanistic Modelling and Environmental Systems

Chapter 16
Data-Based Mechanistic Modelling: Natural Philosophy Revisited?

Peter C. Young

> *Hypotheses non fingo*
> Isaac Newton (1713): appears in the concluding 'Scholium Generale' in the revised second edition of the Philosophiae Naturalis Principia Mathematica' and has been translated as 'I frame no hypotheses'.

16.1 Introduction

While there is much debate [5] about exactly what Newton meant by the above phrase, surely there can be little doubt what views he held on the formulation of hypotheses when he says:

> As in Mathematics, so in Natural Philosophy, the Investigation of difficult Things by the Method of Analysis, ought ever to precede the Method of Composition. This Analysis consists in making Experiments and Observations, and in drawing general Conclusions from them by Induction, and admitting of no Objections against the Conclusions, but such as are taken from Experiments, or other certain Truths. For Hypotheses are not to be regarded in experimental Philosophy.
> Isaac Newton (1718). *Opticks, 2nd edition (1718), Book 3, Query 31, 380.*

Here, 'Natural Philosophy' (from the Latin *philosophia naturalis*), is the term used to describe 'science' before the development of 'modern' science in the 19th Century and beyond. Newton was, of course, the greatest 'Natural Philosopher' of his

P.C. Young (✉)
Centre for Research on Environmental Systems and Statistics, University of Lancaster, Lancaster, UK
e-mail: p.young@lancaster.ac.uk

P.C. Young
Fenner School of Environment and Society, Australian National University, Canberra, Australia

L. Wang, H. Garnier (eds.), *System Identification, Environmental Modelling, and Control System Design*,
DOI 10.1007/978-0-85729-974-1_16,

age and, indeed, the English translation of his famous 1687 scientific treatise (see above) is 'The Mathematical Principles of Natural Philosophy'.

Newton clearly equated hypotheses with guesses (particularly, it seems, if they were made by others!). For example, in a famous letter to Edmond Halley[1] concerning his controversy with Robert Hooke over the inverse square law, Newton says:

> ... so Mr Hooke without knowing what I have found out since his letters to me can know no more but that the proportion was duplicate quam proximé at great distances from the centre and only guessed it to be so accurately and guessed amiss in extending that proportion down to the very centre whereas Kepler guessed right at the ellipsis.[2]

Although Newton's use of the word 'guess' is probably meant here to be somewhat derogatory, such an interpretation is perfectly sensible (if rather unfair to Robert Hooke, who was a great scientist in his own right). A dynamic modelling hypotheses, for instance, is a (hopefully inspired) guess based on the current understanding of some physical or other system. But the point made by Newton in the above letter was that this hypothesis must not be a vague statement of belief made in general conversation, but has to be supported by evidence that gives it scientific credence.

The inductive approach to science and mathematical modelling preferred by Newton (1643–1727) was not his invention: it has a long history in philosophy and had been discussed by Francis Bacon (1561–1626) and, almost contemporaneously with Newton, by Robert Boyle (1627–1691). Interestingly, the polymath William Whewell (1794–1866), who was actually born in Lancaster and also came up with the terms 'scientist', and 'physicist', wrote two books on induction: a *History of the Inductive Sciences, from the Earliest to the Present Times* (1837) and *The Philosophy of the Inductive Sciences* (1840).

In the 20th Century, scientific philosophers such as Karl Popper (1959) and Thomas Kuhn (1962), looked at the philosophy of science in a wider context and Popper [13], in particular, was a proponent of an alternative to the inductive approach that he termed the *hypothetico–deductive* method. Here, the model forms a hypothesis that is tested against data, usually obtained from carefully planned experiments. And the aim is not to 'prove' the hypothesis, but rather to attempt its 'falsification' and consider it to be 'conditionally valid' until falsified.

Whether Kuhn [9] subscribed to the hypothetico-deductive concept is not clear. Rather, he viewed science from a 'paradigmatic' standpoint in which most 'ordinary science' worked within, and embroidered, defined paradigms; while the more fundamental achievements of science were those that questioned or even overturned these current paradigms (as Einstein's theories of relativity radically changed the Newtonian view of the World). In this regard, the hypothetico-deductive approach to scientific research used by ordinary scientists often tends to be too constrained by current paradigms: hypotheses are made within the current paradigm and do not often seek to question it.

---

[1] Newton's friend and famous fellow scientist, who convinced him to write the *Principia* and financed its publication.

[2] This extract is taken from the book by Rigaud [15], which is available from Google books and makes very interesting reading.

So, in a modelling context, which is better: the inductive approach that concentrates on inference from experimental or monitored data; or the hypothetico–deductive approach that relies on the creation of a prior hypotheses that are then tested against such data? The answer is, of course, that they are not mutually exclusive methodologies and should be combined in a constructive way to yield an array of models that satisfy different objectives. My own predilection is to concentrate on an inductive approach whenever the availability of suitable data allows for this. Such an approach can often yield a useful, physically meaningful model rather quickly, without being overly constrained by existing hypotheses; and it can also be an aid in the falsification of hypothetico–deductive models (see the later illustrative example). But suitable data are not always available and hypothetico–deductive 'simulation models' provide an obvious alternative in this data-scarce situation that occurs so often in the natural sciences. Moreover, in making inferences about the model structure, inductive analysis is normally guided by existing or new hypotheses concerning the physical interpretation of parsimonious model structures. In other words, inductive and hypothetico–deductive modelling are synergistic activities, the relative contributions of which will depend upon the system being modelled and the information of *all* types, not only time-series data, that are available to the scientist and modeller.

With above caveats in mind, this chapter will present the main aspects of *Data-Based Mechanistic* (DBM) modelling, a predominantly, but not exclusively, inductive approach to modelling that harks back to the era of natural philosophy. It recognises that, in contrast to most man-made dynamic systems, the nature of many natural systems, particularly at the holistic or macro-level (global climate, river catchment, macro-economy), is still not well understood. 'Reductionist' approaches to modelling such systems, based on the aggregation of hypothetico–deductive models at the micro level, or the rather naïve application of micro-scale laws at the macro level, often results in very large simulation models that suffer from 'equifinality' [2, 3] and are not fully identifiable from the available data.

It will be argued that the DBM approach is often a more appropriate method of scientific inference in research on natural systems, where the 'natural laws' at the macro-level, as used in reductionist modelling, are normally untestable by planned experimentation, which is often difficult or impossible in the broad gamut of the natural sciences. In such applications, DBM modelling not only helps to avoid the possibility of false hypotheses and overly-parameterised, poorly identifiable models, but also provides a compelling reason for exploiting the powerful and relatively novel tools of statistical inference that have been developed to service the requirements of DBM modelling. These tools are collected together in the CAPTAIN Toolbox[3] for Matlab and are used to generate the results presented in this chapter.

The latest exposition of the DBM approach [37] is not, however, exclusively inductive in its approach: it recognises the need for, and utility of, hypothetico–deductive simulation modelling and so covers the whole range of model-based scientific inference, from hypothetico–deductive simulation modelling when data are

[3]The CAPTAIN Toolbox can be downloaded from http://www.es.lancs.ac.uk/cres/captain/.

scarce, to inductive modelling when suitable data become available. The present chapter is dominated, therefore, by an illustrative example that considers the modelling in this combined manner, with 'large model emulation' or 'meta-modelling' providing a bridge between the simulation and data-based model forms. This is a particularly suitable example for the present book because it harks back to one of the earliest examples of DBM modelling carried out by the author and his colleagues: the development of the *Aggregated Dead Zone* (ADZ) model [1, 21] for the characterisation of pollutant transport and dispersion in water bodies (see also Chap. 18).

16.2 Data-Based Mechanistic (DBM) Modelling

The term 'data-based mechanistic modelling' was first used in [35] but the basic concepts of this approach to modelling dynamic systems have been developed over many years. For example, they were first applied seriously within a hydrological context in the early 1970s, with application to the modelling of water quality and flow in rivers [22, 34] and set within a more general framework shortly thereafter [23]. Since then, they have been applied to many different systems in diverse areas of application from ecology, through engineering to economics: see e.g. [25, 30].

The seven major phases in the DBM modelling strategy are as follows [37]:

1. The important first stage in any modelling exercise is to define the objectives and to consider the types of model that are most appropriate to meeting these objectives. Since the concept of DBM modelling requires adequate data if it is to be completely successful, this stage also includes considerations of scale and the likely data availability at this scale, particularly as they relate to the defined modelling objectives.
2. In the initial phases of modelling, it may well be that real observational data will be scarce, so that any major modelling effort will have to be centred on simulation modelling, normally based on largely deterministic concepts, such as the conservation laws (mass, energy momentum etc.). In the DBM simulation modelling approach, which is basically Bayesian in concept, these deterministic simulation equations are converted to a stochastic form by assuming that the associated parameters and inputs are inherently uncertain and can only be characterised in some suitable stochastic form, such as a probability distribution function (pdf) for the parameters and a continuous or discrete time-series model for the inputs. The subsequent stochastic analysis uses Monte Carlo Simulation (MCS), to explore the propagation of uncertainty in the resulting stochastic model, and sensitivity analysis of the MCS results to identify the most important parameters which lead to a specified model behaviour: e.g. [12].
3. The initial exploration of the simulation model in stochastic terms is aimed at revealing the relative importance of different parts of the model in explaining the dominant behavioural mechanisms. This understanding of the model is further enhanced by employing *Dominant Mode Analysis* (DMA). This approach to dynamic model order reduction [26] is applied to time-series data obtained from

planned experimentation, not on the system itself, but on the simulation model that, in effect, becomes a surrogate for the real system. In particular, optimal methods of *Refined Instrumental Variable* (RIV) estimation [24, 31] are applied to these experimental data and yield low order approximations to the high order simulation model that are almost always able to explain its dynamic response characteristics to a remarkably accurate degree (e.g. greater than 99.99% of the large model output variance explained by the reduced order model output).

4. A more complete understanding of the links between the high order simulation model and its reduced order representation obtained in stage 3 is obtained by performing multiple DMA analysis over a user-specified range of simulation model parameter values. The mapping between the large and reduced order model parameters or responses then yields a full *Dynamic Emulation (or 'meta') Model* (DEM) that can replace the simulation model over a wide range of parameter values. This approach to high order model emulation is introduced in [37], while [38] describes in detail two methods of emulation: namely, 'stand-alone parameter mapping', which is used in the present chapter, and 'response mapping', with application to the emulation of the Nash-Cascade hydrological model and a large economic model.
5. Once experimental time series data are available, an appropriate model structure and order is identified by a process of statistical inference applied directly to these real time-series data and based on a generic class of dynamic models: in the present chapter, these are simply linear, stochastic models described by continuous-time transfer functions (i.e. lumped parameter differential equations). If such time series data are available at the start of the study, then this analysis will constitute the first stage in DBM modelling. The identification and estimation procedures used in the DBM modelling are the same optimal RIV methods used in dominant mode analysis (see above 3 and 4). Note that statistical terminology is utilised here, so that 'identification' is the process of determining a model structure that is identifiable from the data; and 'estimation' is the statistical estimation of the parameters that characterise this identified structure.
6. If emulation modelling has been carried out prior to the acquisition of data, then the DBM model obtained at the previous stage 5 should be reconciled with the dynamic emulation version of the simulation model considered in stage 4. Otherwise, if time series data are available at the start of the study and a DBM model has been obtained at the previous stage 5, then an emulation model should be considered at this stage and reconciled with the DBM model. Although such reconciliation will depend upon the nature of the application being considered, the DBM model obtained from the real data should have strong similarities with the reduced order dynamic emulation model. If this is not the case, then the differences need to be investigated, with the aim of linking the reduced-order model with the high order simulation model via the parametric mapping of the dynamic emulation model (see later illustrative example).
7. The final stage of model synthesis should always be an attempt at model validation: see e.g. [27]. The word 'attempt' is important since validation is a complex process and even its definition is controversial. Some academics (e.g. Konikow

and Bredehoeft [8], within a ground-water context; and Oreskes et al. [11], in relation to the whole of the earth sciences) question even the possibility of validating models. However, statistical evaluation of the model by confirming that statistical diagnostics are satisfactory (e.g. no significant autocorrelation in the residuals or cross correlation between the residuals and input variables; no evidence of un-modelled nonlinearity etc.) is always possible and can engender greater confidence in the efficacy of the model. Also, one specific, quantitative aspect of validation is widely accepted; namely 'predictive validation' or 'cross-validation', in which the predictive potential of the model is evaluated on data other than that used in the identification and estimation stages of the analysis. When validated in this narrow sense, it can be assumed that the 'conditionally valid' model represents the best theory of behaviour currently available that has not yet been 'falsified' in a Popperian sense.

Although these are the seven major stages in the process of DBM model synthesis, they may not all be required in any specific application: rather, they are 'tools' to be used at the discretion of the modeller. Also, they are not the end of the modelling process. If the model is to be applied in practice (and for what other reason should it be constructed?) then, as additional data are received, they should be used to evaluate further the model's ability to meet its objectives. Then, if possible, both the model parameters and structure can be modified if they are inadequate in any way. This process, sometimes referred to as 'data assimilation', can be achieved in a variety of ways. Since most data assimilation methods attempt to mimic the Kalman Filter, however, it is likely to involve recursive updating of the model parameter and state estimates in some manner, as well as the use of the model in a predictive (forecasting) sense. This process of data assimilation is made simpler in the DBM case because the optimal RIV estimation methods used in DBM modelling (see next Sect. 16.3) are all inherently recursive in form and so can be used directly for on-line, Bayesian data assimilation [16, 24, 29, 32, 33].

16.3 An Illustrative Example: DBM Modelling of Pollution Transport and Dispersion in a Wetland Area

One of the first models to be considered seriously in DBM terms was the *Aggregated Dead Zone* (ADZ) model for the transport and dispersion of solutes in river systems: see e.g. Beer and Young [1]; Wallis et al. [21]; Green et al. [7]. This has also led to related models that describe the imperfect mixing processes that characterise mass and energy flow processes in the wider environment: for instance, Beven and Young [4] use a similar equation for modelling flow through porous media; Young and Lees [35] generalise the concept to an *Active Mixing Volume* (AMV) form and apply it to the DBM modelling of heat flow in soils; while Price et al. [14] and Young et al. [36] show how it can be used very successfully for the DBM modelling of heat flow and the resultant temperature changes in buildings. It is clear therefore that the

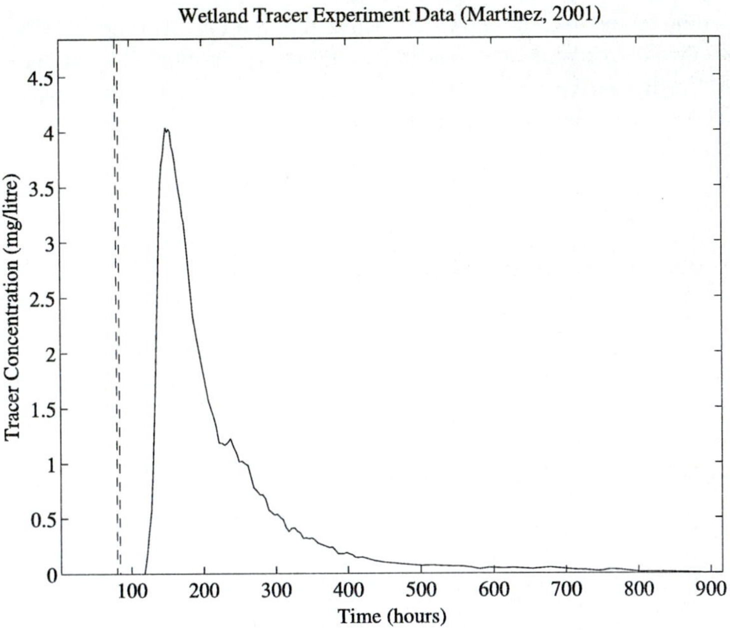

Fig. 16.1 Wetland tracer experiment data: the sampled input $u(t_k)$ is an impulsive (or 'gulp') application of bromide tracer (*dashed line*) and the sampled output $y(t_k)$ is the concentration of bromide measured every two hours at a downstream weir (*full line*)

same ADZ/AMV modelling ideas have fairly wide applicability to flow processes involving mass and energy transfer in both the natural and built environment.

Research in this area of study is aided by the ability to conduct simple planned experiments using conservative tracer materials, such as the fluorescent red dye, Rhodamine WT. Small quantities of such tracers can be injected into the environment, for example into a river system, and then the subsequent low concentrations can be measured using special equipment, such as a fluorometer in the case of Rhodamine WT [21]. The typical results of a tracer experiment in a wetland area are shown in Fig. 16.1, in this case using conservative potassium bromide (KBr) as the tracer material. The wetland area is located in Florida, USA, and it receives treated domestic wastewater which travels slowly through the wetland to allow for further nutrient removal. The tracer experiment was part of a study carried out by Chris Martinez and William R. Wise of the Environmental Engineering Sciences Department, University of Florida for the City of Orlando [10]. The study objective was to determine residence times for each wetland cell in the system and to assess whether the same degree of treatment could be maintained should the wastewater loading be raised from 16 to 20 million gallons per day.

The data shown in Fig. 16.1 are used later as the basis for direct DBM modelling. However, the main objective of this illustrative example is to show how the complete DBM modelling procedure, as outlined in previous sections of this chapter, can develop from an initial, fairly large, simulation model, through emulation modelling, to modelling on the basis of real data, such as those in Fig. 16.1. So, although these data are already available and could be used directly for modelling analysis, let us assume, for illustrative purposes, that the data collection experiments have not yet taken place and, without access to data, we resort to more speculative simulation modelling.

16.3.1 The Large Simulation Model

Although its efficacy as a description of solute transport and dispersion in rivers has been questioned, the *Advection Dispersion Equation* (ADE) has been the basis of many models that have been used in practice over many years. In the case of a conservative (non-decaying or non-reactive) solute, this model takes the form of the following partial differential equation for the solute concentration $c(t,s)$ in time (t) and space (s):

$$\frac{\partial c(t,s)}{\partial t} + U\frac{\partial c(t,s)}{\partial s} = D\frac{\partial^2 c(t,s)}{\partial s^2}, \tag{16.1}$$

where D is the 'dispersion coefficient' and U is the velocity. Its derivation is discussed in many fluid dynamic texts and is outlined in [39].

The ADE derives originally from the work of the great fluid dynamicist G.I. Taylor [18] who used a one dimensional Fickian diffusion equation to describe the random diffusion of a solute in a pipe, once sufficient time had elapsed to allow for full cross sectional mixing. Subsequently, although the same partial differential equation (PDE) model has been applied to solute dispersion in river channels, tracer studies such as that shown in Fig. 16.1 show a longer 'tail' in the response that is redolent of non-Fickian behaviour. Two main solutions to this limitation have been suggested: the ADZ approach mentioned above, which replaces the PDE model by a lumped parameter, ordinary differential equation (ODE) relating solute concentration between spatial locations along the river; and a modified version of the ADE that is now termed the 'transient storage ADE' model [10, 19] and takes the following form:

$$\begin{aligned} \frac{\partial c(t,s)}{\partial t} + U\frac{\partial c(t,s)}{\partial s} &= D\frac{\partial^2 c(t,s)}{\partial s^2} + K\Gamma_c\{u(t,s) - c(t,s)\}, \\ \frac{du(t,s)}{dt} &= K\Gamma_s\{u(t,s) - c(t,s)\}. \end{aligned} \tag{16.2}$$

Here, $u(t,s)$ is the solute concentration in the storage zone; K is the mass exchange coefficient between the storage zone and the main flow; Γ_c is ratio of interfacial area (between the main flow and the dead zone) to the main flow volume; and Γ_s is the ratio of the interfacial area to dead zone volume. In (16.1) and (16.2), it is straightforward to add terms to allow for non-conservative or reactive solutes. Beer and Young [1] suggested, and provided evidence to show, that the effect of the dead zone may be dominant in many rivers and so the transient storage equation can be replaced by the single ADZ equation:

$$\frac{\partial c(t,s)}{\partial t} + U\frac{\partial c(t,s)}{\partial s} = \frac{1}{T}\{u(t) - c(t,s)\}, \tag{16.3}$$

where the two terms on the left hand side represent 'plug' flow, which leads to a pure 'advective' time delay τ between the 'upstream' input $u(t)$ and the 'downstream' output $c(t)$; while the right hand term incorporates the dispersive effect of the, now 'aggregated', dead zone (strictly an 'imperfect mixing' zone).

For our purposes, (16.3) is better written in the ODE form:

$$\frac{dc(t)}{dt} = \frac{1}{T}\{u(t-\tau) - c(t)\}, \tag{16.4}$$

where τ is the pure advective time delay and T is the time constant. This can then be written in the transfer function form:

$$c(t) = \frac{1}{1+Ts}u(t-\tau) = \frac{b_0}{s+a_1}u(t-\tau), \tag{16.5}$$

where $s^r = d^r/dt^r$ is the derivative operator and, in this conservative situation, $b_0 = a_1 = 1/T$. In other words, it is assumed that the length of the river reach is such that the distributed dead zones can be represented by a single dead zone, where T is the 'effective' time constant, or residence time, that characterises the ADZ dynamics.

Equation (16.4) will be recognised as simple mass balance equation, where it is assumed that the mass of solute being lost from the reach, because of the predominantly downstream flow, is proportional to the solute concentration in the reach [35]. Consequently, in a more general, possibly non-conservative situation, the TF equation (16.5) becomes:

$$c(t) = \frac{G}{1+Ts}u(t-\tau), \tag{16.6}$$

where the 'steady state gain' $G = 1.0$ in the conservative case; $G < 1.0$ if mass is being lost; and $G > 1.0$ is there is an accretion of mass for some reason. In addition, the ADZ model can be modified to allow for 'back-flow' caused, for example, by the physical nature of the river channel or, in the extreme case, by tidal effects. In this situation, the ADZ equation takes the following more general form:

$$T\frac{dc(t)}{dt} = -c(t) + Gu(t-\tau) + G_d c_d(t), \tag{16.7}$$

where $c_d(t)$ are the changes in the downstream solute concentration and G_d is the steady state gain for this downstream input that defines its effect on the concentration $c(t)$ of the solute in the reach. In this case, the model is conservative when $G + G_d = 1.0$.

Of course, the above ADZ equations only relate the temporal changes in solute concentration between two spatial locations on a river. However, a *Semi-Distributed*

ADZ (SDADZ) model can be constructed rather simply by a chain of suitably small ADZ elements such as (16.7) connected in series, parallel or even feedback (should this relate to a physically meaningful situation). For example, in this SDADZ model, the $c_d(t)$ in (16.7) would be interpreted as the solute concentration in the immediate downstream reach, which would also be modelled as an ADZ element. In this manner, the equation for transport and dispersion of a conservative solute in the ith ADZ reach, of a uniform river system of n such identical reaches, with all the intermediate pure time delays set to zero, would then take the form:

$$T\frac{dc_i(t)}{dt} = -c_i(t) + Gc_{i-1}(t) + G_d c_{i+1}(t), \quad i = 1, 2, \ldots, n. \tag{16.8}$$

Alternatively, this model can be represented in TF form:

$$c_i(t) = \frac{G}{1+Ts}c_{i-1}(t) + \frac{G_d}{1+Ts}c_{i+1}(t), \quad i = 1, 2, \ldots, n \tag{16.9}$$

or as the following element in a n dimensional state space model:

$$\begin{bmatrix} \frac{dc_{i-1}(t)}{dt} \\ \frac{dc_i(t)}{dt} \\ \frac{dc_{i+1}(t)}{dt} \end{bmatrix} = \begin{bmatrix} -\frac{1}{T} & \frac{G_d}{T} & 0 \\ \frac{G}{T} & -\frac{1}{T} & \frac{G_d}{T} \\ 0 & \frac{G}{T} & -\frac{1}{T} \end{bmatrix} \begin{bmatrix} c_{i-1}(t) \\ c_i(t) \\ c_{i+1}(t) \end{bmatrix}, \tag{16.10}$$

where the output is defined by the state variable $c_i(t)$. The input to the whole system is then the input to the farthest upstream reach, denoted by $u(t) = c_0(t)$, and the output is the output of the farthest downstream reach, denoted by $x(t) = c_n(t)$. If it is assumed that any pure advective time delays are lumped into a single time delay τ at the input, the complete deterministic model can be represented in the following general TF form:

$$x(t) = \frac{B(s)}{A(s)}u(t-\tau) = \frac{b_0 s^{n-1} + b_1 s^{n-2} + \cdots + b_m}{s^n + a_1 s^{n-1} + \cdots + a_n}u(t-\tau). \tag{16.11}$$

If G and G_d are selected so that the system is conservative, then $b_m = a_n$ and the overall steady state gain is unity.

The original idea was to base this example around either the ADE or the transient storage ADE. Although analytical solutions of these models can be obtained for specified inputs [6], the models are normally implemented on a computer using some form of numerical approximation that can, more usefully, apply for any specified input forcing functions. A popular approach is the Crank-Nicolson finite-difference solution developed by Runkle and Chapra [17] for the solution of the transient storage model (16.2), which is the basis for the well known OTIS simulation model: see http://csdms.colorado.edu/wiki/Model:OTIS. However, this model is not available in Matlab and so, as a an interesting alternative, the SDADZ state space model, with elements defined by (16.10), will be used here since it is straightforward to program it in Matlab.

In the present example, the entire system needs to be conservative and uniform, so that all of the reach elements are of the same form (16.10), with $G+G_d=1.0$, and the complete state space model matrix is 3-band diagonal, similar (16.10) but of dimension n. The model is simulated with $n=40$, $\tau=0$ and the $\{G_i\ T_i\}$ parameters defined as appropriate functions of D, V and the reach length dz: e.g. $1/T=-(2D/dz^2+U/dz)$. Note that, since this SDADZ model can be solved explicitly, it can be considered as a possible alternative to the transient storage model (16.2). However, further research is continuing on the model and its relationship with both the ADE and the transient storage model.

16.3.2 Emulation Modelling

Given the artificial assumption that, at this stage, we have no measured data from the wetland area, the simulation modelling is based on its physical characteristics and the measurement locations in the anticipated tracer experiment. This suggests model parameter values in the range of $D:\{0.030-1.5\}\ \text{m}^2/\text{sec}$ and $U:\{0.00045-0.0014\}\ \text{m/sec}$ (although note that these are parameters in the SDADZ model, not the ADE). For the specific measurement location considered later in Sect. 16.3.3, reach 5 appears to be most suitable but, in order to ensure that a single model structure is possible for emulation at all reaches, if this is required, the initial nominal DMA considers both reach 5 and reach 40, with $D=0.318$ and $V=0.0013$ in both cases.

This nominal emulation analysis is carried out using an input signal in the form of two repeated pulses of period 1000 hours and amplitude 180 entering at reach 1. Although this is not an optimal input from a statistical identification and estimation standpoint, it is sufficient to produce good TF emulation models. For instance, the continuous-time RIV (RIVC) identification results (verbatim) for reach 5, as produced by the rivcbjid routine in the CAPTAIN Toolbox, are as follows for reach 5:

den	num	del	AR	MA	YIC	RT2	BIC
5	6	1	0	0	-25.7517	1.000000	-200025
5	6	0	0	0	-18.6377	0.999998	-180477
5	5	0	0	0	-22.4703	0.999998	-180288
4	5	1	0	0	-19.1426	0.999976	-147983
4	5	0	0	0	-11.6862	0.999976	-147983
4	4	0	0	0	-18.1096	0.999976	-147990

In these results, RT2 $=R_T^2$ is the simulation coefficient of determination ($R_T^2=1.0$ indicates a perfect fit), while YIC and BIC are order identification criteria. Although this suggests that the [5 6 1] model is marginally better, the [5 5 0] model is not only almost as good but it has one less parameter and, most importantly, it proves superior in the case of reach 40, where, the best identified model is [5 5 79].

To summarise, therefore, this initial nominal emulation analysis, which examines the extremes of the large simulation model response characteristics, shows that a $[5\ 5\ \tau]$ continuous-time TF model of the TF form;

$$\hat{x}(t) = \frac{\hat{b}_0 s^4 + \hat{b}_1 s^3 + \hat{b}_2 s^2 + \hat{b}_3 s + \hat{b}_4}{s^5 + \hat{a}_1 s^4 + \hat{a}_2 s^3 + \hat{a}_3 s^2 + \hat{a}_4 s + \hat{a}_5} u(t-\tau) \tag{16.12}$$

provides very good emulation at both extremes. For reach 5, $\tau = 0$ and the SRIVC parameter estimates are (note that $a_5 = b_4$ and mass is conserved):

$$\begin{aligned}
&\hat{a}_1 = 0.5003; \quad \hat{a}_2 = 0.0772; \quad \hat{a}_3 = 0.00335; \quad \hat{a}_4 = 4.6033 \times 10^{-5};\\
&\hat{a}_5 = 1.7265 \times 10^{-7}; \quad \hat{b}_0 = 7.0278 \times 10^{-4}; \quad \hat{b}_1 = -4.5845 \times 10^{-4};\\
&\hat{b}_2 = 1.0403 \times 10^{-3}; \quad \hat{b}_3 = 3.1639 \times 10^{-5}; \quad \hat{b}_4 = 1.7265 \times 10^{-7}
\end{aligned} \tag{16.13}$$

while for reach 40, $\tau = 79$ and the estimates are:

$$\begin{aligned}
&\hat{a}_1 = 0.06116; \quad \hat{a}_2 = 0.001475; \quad \hat{a}_3 = 1.6720 \times 10^{-5};\\
&\hat{a}_4 = 8.7469 \times 10^{-8}; \quad \hat{a}_5 = 1.6736 \times 10^{-10}; \quad \hat{b}_0 = 1.2088 \times 10^{-5};\\
&\hat{b}_1 = 1.0735 \times 10^{-6}; \quad \hat{b}_2 = 9.9738 \times 10^{-8}; \quad \hat{b}_3 = 2.8188 \times 10^{-9};\\
&\hat{b}_4 = 1.6736 \times 10^{-10}.
\end{aligned} \tag{16.14}$$

In both cases, the explanation of the large model response, when validated with a single impulse input of 70 for two hours, is almost perfect, as shown in Fig. 16.2, with $R_T^2 = 0.999999$ in both cases. Similarly good validation results are obtained for other inputs but final emulation model validation is considered in more detail later.

The full emulation mapping analysis is carried out using the same input forcing function and it involves 850 separate TF model identification and estimation runs, using 50 equally spaced values of D over the range $\{0.030 - 1.5\}$ $\mathrm{m^2/sec}$ and 17 equally spaced values of V over the range $\{0.00045 - 0.0014\}$ m/sec. Because the computational burden is not too large in this case, it is possible to carry out the mapping over this complete grid of simulation model parameter values and so ensure good mapping coverage without having to resort to MCS randomization (see Sect. 16.2). More specifically, the TF model identification at each combination of D and V values is based on the $[5\ 5\ \tau]$ model with τ considered in the range $\{0-2\}$ sec. These 3 calls to the rivcbj routine in CAPTAIN take about 10 seconds on a quad-core Mac Pro computer, so that the overall computation time for the mapping analysis is about 2.3 hours.

Given the results of these mapping experiments, the mapping relationships are obtained using the interp2 routine in Matlab, with the 'spline' option. Figure 16.3 is a three dimensional plot of the resulting mapping surface for the five TF denominator parameters $a_i, i = 1, 2, \ldots, 5$; while Fig. 16.4 provides a more quantitative idea of this surface by showing how the parameter estimates vary with the dispersion coefficient D for 'slices' across the surface at different values of velocity V.

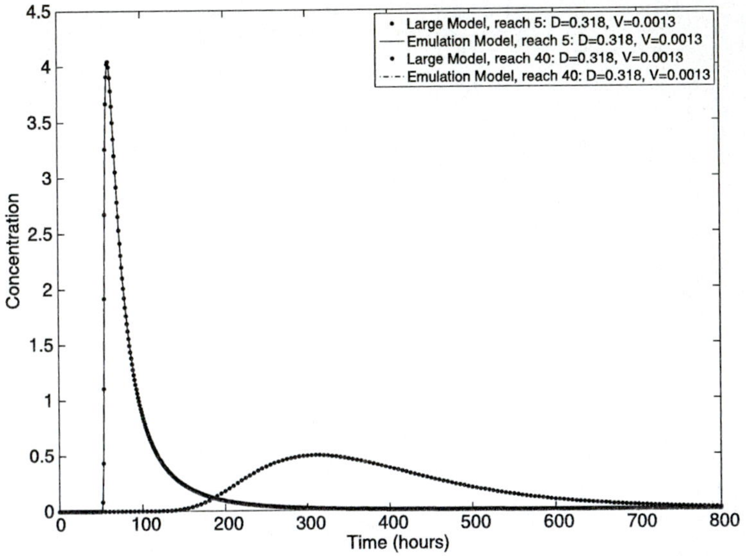

Fig. 16.2 Nominal emulation of the large simulation model at reaches 5 and 40 by continuous-time RIVC estimated 5th order TF models

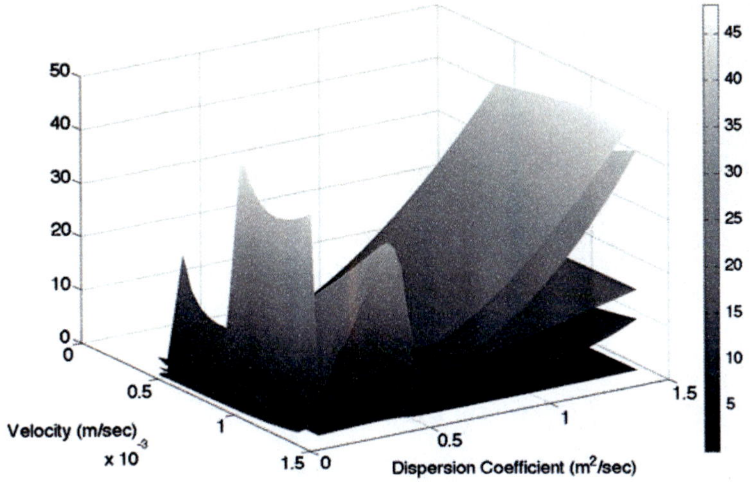

Fig. 16.3 Parameter mapping for reach 5: 3 dimensional plot of the five TF denominator parameters a_i, $i = 1, 2, \ldots, 5$, as functions of the dispersion coefficient, D, and the velocity, V

Note that the mapping surface in Fig. 16.3 is quite smooth for $D > 0.1\ \mathrm{m}^2/\mathrm{sec}$ but there is a quite sharp change at smaller values than this, suggesting that a finer grid might be necessary in this region. However this region is not important in the present case and this has not been investigated further.

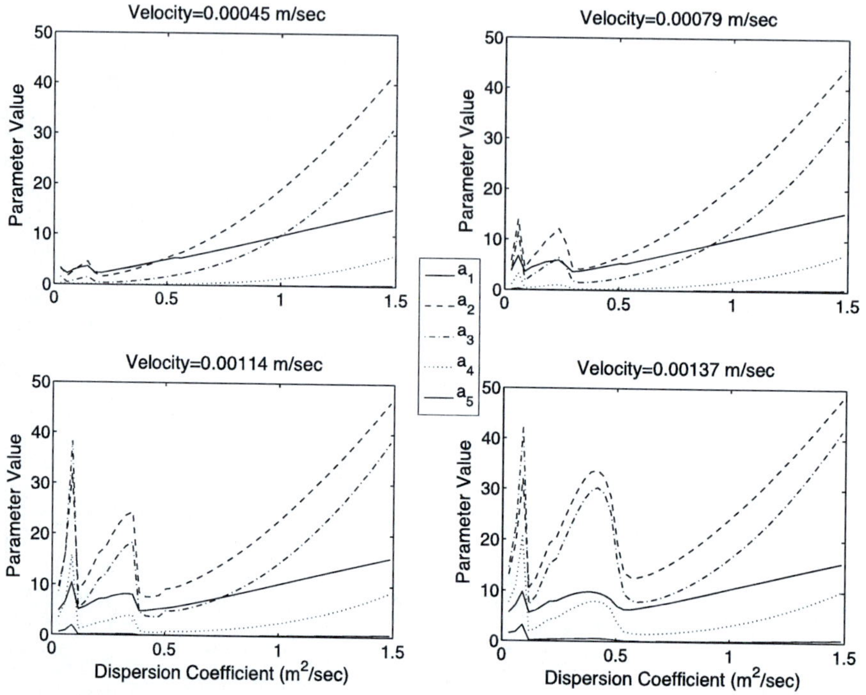

Fig. 16.4 Parameter mapping for reach 5: the TF denominator parameters as changing functions of the dispersion coefficient, D, for four different velocity, V, values ('slices' from Fig. 16.3)

The final stage of the full emulation analysis is validation on interpolated values of parameters and two examples of such validation analysis are shown in Fig. 16.5, for an impulse forcing function, and Fig. 16.6 for a forcing function of the more general type that might be expected if the model was being employed for the evaluation of pollutant transport and dispersion. In both cases, the emulation is exceptional, with R_T^2 values greater than 0.999.

16.3.3 Modelling from Real Data

The standard, inductive DBM approach to the analysis of data such as those shown in Fig. 16.1 is to use them to identify and estimate a TF model in either discrete or continuous-time form. Discrete-time modelling of these data is described in [28], where a [4 2 22] model is identified and estimated in a constrained form to ensure physically interpretable real poles in the estimated TF model. In the present context, it is clear that a continuous time model makes more sense. However, rivcbjid identification in CAPTAIN reveals that a number of models explain the data well in this case, including 5th order models, such as those used in the emulation analysis. Also, as in the discrete-time model case, the poles of the estimated TF are always

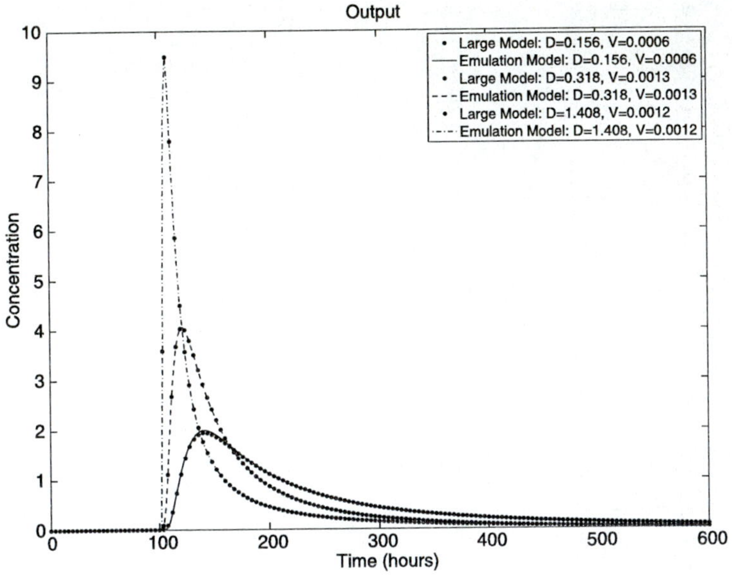

Fig. 16.5 Validation of DBM stand-alone emulation model with an impulsive-type input

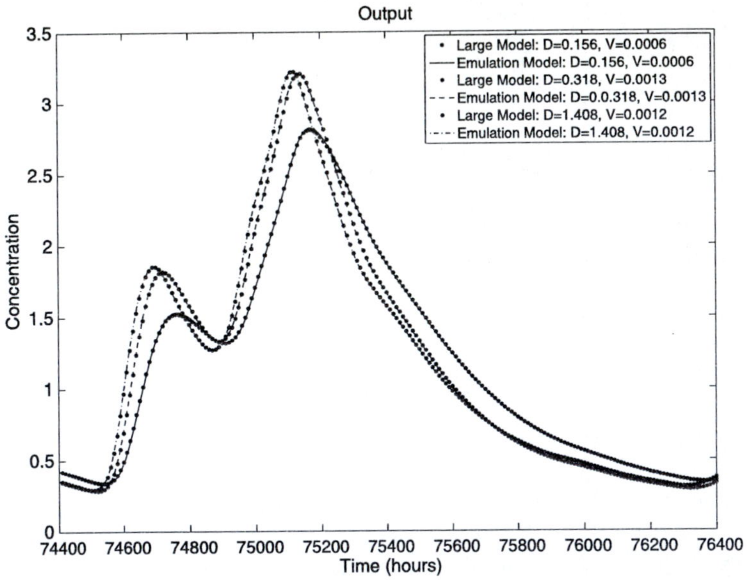

Fig. 16.6 Validation of DBM stand-alone emulation model with a typical input

complex. Using the same arguments as those used in the above reference [28], the continuous-time model, constrained to have real poles, is identified this time to have a [4 2 20] structure with additive coloured noise identified by the *Akaike Information*

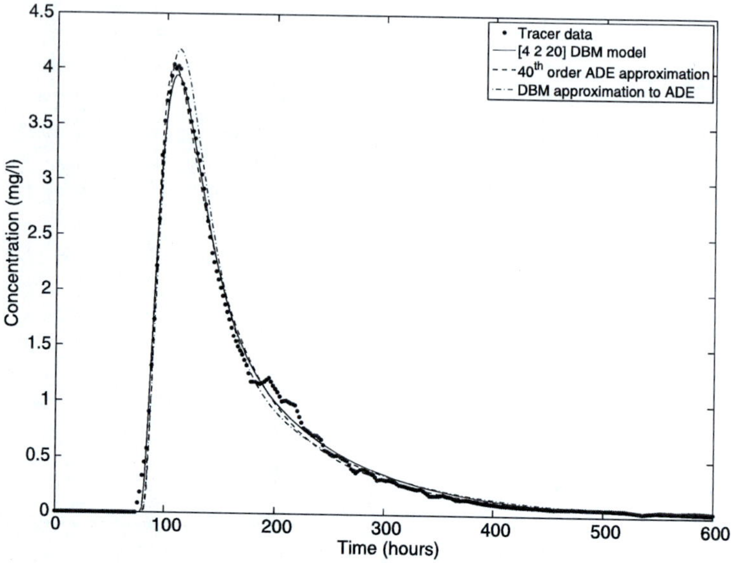

Fig. 16.7 Model outputs compared with the tracer data

Criterion (AIC) as an AR(20) process: i.e.,

$$y(t_k) = \frac{b_0 s + b_1}{(s+\alpha_1)^3(s+\alpha_2)} u(t_k - 20) + \xi(t_k),$$
$$\xi(t_k) = \frac{1}{1 + c_1 z^{-1} + \ldots + c_{20} z^{-20}} e(t_k), \quad e(t_k) = N(0, \sigma^2) \tag{16.15}$$

and the parameter estimates for the system model are as follows, with the standard errors shown in parentheses:

$$\hat{\alpha}_1 = 0.07734(0.0005); \quad \hat{\alpha}_2 = 0.0099(0.00015); \quad \hat{\sigma}^2 = 0.0003;$$
$$\hat{b}_0 = 1.9157 \times 10^{-4}(1.908 \times 10^{-6}); \quad \hat{b}_1 = 4.5415 \times 10^{-6}(3.157 \times 10^{-8}). \tag{16.16}$$

Note the 'hybrid' nature of this TF model, with a continuous-time TF model, a discrete-time noise model, and the changed time argument t_k (denoting 'snapshot' sampled values at time t_k: see [31] for a full explanation). It is characterised by three real, equal modes with short time constants of 12.9 hours and one mode with a long time constant of 101 hours. The tracer data are explained well, as shown in Fig. 16.7, with $R_T^2 = 0.997$ and a standard coefficient of determination for the residual white noise (one-step-ahead prediction errors) of $R^2 = 0.9996$. The autocorrelation function of the estimated residuals $e(t_k)$ shows that they are reasonably white although, as is usual with hydrological systems, they are somewhat heteroscedastic.

But can the model (16.15) be reconciled with the large simulation model? The answer to this question is aided by the nature of this large model which has been chosen specially, for the purposes of this illustrative example, as a particular ODE approximation of the ADE that it is identifiable from the tracer data in Fig. 16.1. It should be noted that this is unusual for large simulation models because they are often over-parameterised. For instance, Wagener et al. [20] have investigated the identifiability of a particular version of the transient storage equation (16.2) and they conclude that:

> It can be seen that very different combinations of the parameters yield identical performances in terms of the selected objective function (the weighted sum of squared differences between simulated and observed concentrations). From this analysis, only one parameter seems identifiable.

This despite the fact that their version of the model has only four parameters.

If, in the present case, the parameters D and V in our 40th order SDADZ model are optimized by simple nonlinear least squares, based on the tracer data in Fig. 16.1 and using the optimisation routines in Matlab, then the estimates are clearly defined at $\hat{D} = 0.3126\ \text{m}^2/\text{sec}$ and $\hat{V} = 0.00126\ \text{m/sec}$. As pointed out previously in Sect. 16.3.1, these numerical values of D and V are not particularly important in the present context and, in any case, should not be interpreted directly in ADE terms: for our purposes, it is only required that the output of this high order SDADZ model is able to explain the tracer data well. This is certainly the case here, as shown in Fig. 16.7, where the simulation coefficient of determination is $R_T^2 = 0.995$.

Now, if a constrained [4 2 0] model, of similar form to (16.15), is estimated *on the basis of deterministic data generated by this SDADZ model*, the estimated parameters are as follows:

$$\begin{aligned} &\hat{\alpha}_1 = 0.07366; \qquad \hat{\alpha}_2 = 0.00888; \qquad \hat{\sigma}_\xi^2 = 0.0021; \\ &\hat{b}_0 = 1.9671 \times 10^{-4}; \qquad \hat{b}_1 = 3.5308 \times 10^{-6}. \end{aligned} \tag{16.17}$$

Note that here $\tau = 0$ since this is based on the SDADZ simulated data (see Sect. 16.3.1); also no AR noise model is estimated and standard errors are omitted, because they are inappropriate in this simple least squares estimation situation, where the SDADZ simulated model output is noise free: i.e. here, $\hat{\sigma}_\xi^2$ is the simply the variance of the fitted residuals, with an associated coefficient of determination of 0.990. Most significantly, the time constants of this model (13.6 and 113 hours) are quite similar to those of the directly estimated DBM model.

Exactly the same results are obtained with a constrained [5 3 0] model because the estimated TF numerator then has a zero that cancels almost exactly with one of the four identical poles, suggesting that any TF model over 4th order is likely to be over-parameterised. This is confirmed if constrained or unconstrained [5 5 0] models are estimated, since there are again clear signs of pole-zero cancellation. These results suggest that, while the large simulation model can be reconciled with the DBM model based on real data, since they can both be fitted to the tracer experiment data and yield similar response characteristics, this reconciliation is only partial. In particular, although the large model has only two parameters, it is a 40th order dynamic system with 40 dynamic modes and so has an enormous surplus capacity when we realise that the data can be explained marginally better ($R_T^2 = 0.997$ vs $R_T^2 = 0.995$) by the 4th order DBM (here an ADZ) model.

With the above results in mind, there is clearly a case for considering how the large SDADZ model that we have emulated here could be simplified in cases where its enormous explanatory potential is not required: for instance, when the detail provided by the 40 short reaches is not essential to the solution of the problem at hand. They also raise the question of how this SDADZ model relates to its 'pure' ADE progenitor. The model is able to explain the rather elevated 'tail' of the tracer response shown in Figs. 16.1 and 16.7 simply because of the 'numerical dispersion' that arises from its lumped parameter ODE approximation. In contrast to this, the pure, partial differential ADE model would not be able to explain this 'tail' at all because the numerical dispersion is not present. Indeed, this was the original reason why the ADZ model was developed. Finally, one might question also whether the additional complication of the transient storage model is required, particularly when the Wagener et al. results mentioned previously suggest that it has severe identifiability problems.

It must be emphasised that this example is intended only to illustrate various aspects of the DBM approach and is in no sense a complete modelling study. For instance, the limited data availability has prevented any predictive validation of the models considered above. It could, however, constitute the first step in a much more comprehensive research project, such as that carried out into the ADZ model by Wallis et al. [21]. This established that, for many natural and man-made channels, the 'dispersive fraction', defined by the ratio $T/(T + \tau)$, is relatively invariant over most of the flow régime so that, once this is established, the ADZ model is able to describe pollution transport and dispersion for any defined flow conditions (which is not possible in the case of the OTIS model: see [10], p. 217). It would be very interesting, therefore, to see whether the limited results presented in this chapter are confirmed in this wider context and with a much larger data base.

16.4 Conclusions

This chapter provides a brief outline of the procedures involved in DBM modelling. Its main aim, however, is to put the DBM approach to modelling in a philosophical context and demonstrate how this is reflected in an illustrative example, where DBM modelling is applied to the investigation of solute transport and dispersion in water bodies. From a philosophical standpoint, DBM modelling stresses the need to rely, whenever possible, on inductive inference from time series data, without over-reliance on pre-conceived notions about the structure of the model that can often lead to over-large simulation models with severe identifiability problems. But, by providing an emulation modelling bridge between such large simulation models, produced in a hypothetico–deductive manner, and parsimonious DBM models that are normally identifiable from the available data, it emphasises the need to utilise both approaches, in an integrated manner, in order to meet multiple modelling objectives.

The chapter is dominated by the illustrative example which is aimed at demonstrating, in as simple a manner possible, how large simulation and DBM models can be reconciled to some extent, so providing useful cross-fertilisation that should lead, in any specific study, to a model which achieves two main aims. First and most importantly, to satisfy the modelling objectives; second, to combine the hypothetico–deductive virtues of good scientific intuition and simulation modelling with the pragmatism of inductive data-based modelling, where more objective inference from data is the primary driving force. In this way, it is hoped to encourage changes in modelling practice away from the fractionisation of modelling activities that has been a feature of much environmental modelling for too long, towards a more integrated and cooperative endeavour that is able to tackle the many problems that remain in the modelling of natural systems from time series data.

Acknowledgements I am extremely grateful to the friends and colleagues I have worked with over the past half century, particularly all of those who have so generously prepared chapters for this book. Thanks also to Chris Martinez and William Wise, for providing the tracer data; and my colleague Keith Beven for his useful comments on the chapter. And, of course, a special thank you to my wife Wendy, who has so selflessly put up with me so well for all this time.

References

1. Beer, T., Young, P.C.: Longitudinal dispersion in natural streams. ASME J. Environ. Eng. **109**, 1049–1067 (1983)
2. von Bertalanffy, K.L.: General System Theory: Foundations, Development, Applications. George Braziller, New York (1968)
3. Beven, K.J.: Prophecy, reality and uncertainty in distributed hydrological modelling. Adv. Water Resour. **16**, 41–51 (1993)
4. Beven, K.J., Young, P.C.: An aggregated mixing zone model of solute transport through porous media. J. Contam. Hydrol. **3**, 129–143 (1988)
5. Cohen, I.B.: The first English version of Newton's *hypotheses non fingo*. ISIS **53**(173), 379–388 (1962)
6. De Smedt, F.: Analytical solutions for transport of decaying solutes in rivers with transient storage. J. Hydrol. **330**, 672–680 (2006)
7. Green, H.M., Beven, K.J., Buckley, K., Young, P.C.: Pollution incident prediction with uncertainty. In: Beven, K.J., Chatwin, P., Millbank, J. (eds.) Mixing and Transport in the Environment, pp. 113–137. Wiley, Chichester (1994)
8. Konikow, L.F., Bredehoeft, J.D.: Ground water models cannot be validated. Adv. Water Resour. **15**, 75–83 (1992)
9. Kuhn, T.: The Structure of Scientific Revolutions. University of Chicago, Chicago (1962)
10. Martinez, C.J., Wise, W.R.: Analysis of constructed treatment wetland hydraulics with the transient storage model OTIS. Ecol. Eng. **20**(3), 211–222 (2003)
11. Oreskes, N., Shrader-Frechette, K., Belitz, K.: Verification, validation, and confirmation of numerical models in the earth sciences. Science **263**, 641–646 (1994)
12. Parkinson, S., Young, P.C.: Uncertainty and sensitivity in global carbon cycle modelling. Clim. Res. **9**, 157–174 (1998)
13. Popper, K.: The Logic of Scientific Discovery. Hutchinson, London (1959)
14. Price, L., Young, P.C., Berckmans, D., Janssens, K., Taylor, J.: Data-based mechanistic modelling and control of mass and energy transfer in agricultural buildings. Annu. Rev. Control **23**, 71–82 (1999)
15. Rigaud, S.P.: Historical Essay on the First Publication of Sir Isaac Newton's Principia. Oxford University Press, Oxford (1838)
16. Romanowicz, R.J., Young, P.C., Beven, K.J.: Data assimilation and adaptive forecasting of water levels in the River Severn catchment. Water Resour. Res. **42**, W06407 (2006). doi:10.1029/2005WR004373

17. Runkel, R.L., Chapra, S.C.: An efficient numerical solution of the transient storage equations for solute transport in small streams. Water Resour. Res. **29**(1), 211–215 (1993). doi:10.1029/92WR02217
18. Taylor, G.I.: The dispersion of matter in turbulent flow through a pipe. Proc. R. Soc. A **223**, 446–468 (1954)
19. Valentine, E.M., Wood, I.R.: Longitudinal dispersion with dead zones. J. Hydraul. Div. **103**(9), 975–990 (1977)
20. Wagener, T., Camacho, L.A., Wheater, H.S.: Dynamic identifiability analysis of the transient storage model for solute transport in rivers. J. Hydroinform. **4**, 199–211 (2002)
21. Wallis, S.G., Young, P.C., Beven, K.J.: Experimental investigation of the aggregated dead zone model for longitudinal solute transport in stream channels. Proc. Inst. Civ. Eng. 2. Res. Theory **87**, 1–22 (1989)
22. Young, P.C.: Recursive approaches to time-series analysis. Bull. Inst. Math. Appl. **10**, 209–224 (1974)
23. Young, P.C.: A general theory of modeling for badly defined dynamic systems. In: Vansteenkiste, G.C. (ed.) Modeling, Identification and Control in Environmental Systems, pp. 103–135. North Holland, Amsterdam (1978)
24. Young, P.C.: Recursive Estimation and Time-Series Analysis. Springer, Berlin (1984) (new revised and enlarged edition published 2011: see http://www.springer.com/engineering/control/book/978-3-642-21980-1)
25. Young, P.C.: Data-based mechanistic modeling of environmental, ecological, economic and engineering systems. Environ. Model. Softw. **13**, 105–122 (1998)
26. Young, P.C.: Data-based mechanistic modelling, generalised sensitivity and dominant mode analysis. Comput. Phys. Commun. **117**, 113–129 (1999)
27. Young, P.C.: Data-based mechanistic modelling and validation of rainfall-flow processes. In: Anderson, M.G., Bates, P.D. (eds.) Model Validation: Perspectives in Hydrological Science, pp. 117–161. Wiley, Chichester (2001)
28. Young, P.C.: Data-based mechanistic modelling of environmental systems. In: Proceedings, IFAC Workshop on Environmental Systems, First Plenary Session Keynote Paper, Yokohama, Japan (2001)
29. Young, P.C.: Advances in real-time flood forecasting. Philos. Trans. R. Soc., Math. Phys. Eng. Sci. **360**(9), 1433–1450 (2002)
30. Young, P.C.: The data-based mechanistic approach to the modelling, forecasting and control of environmental systems. Annu. Rev. Control **30**, 169–182 (2006)
31. Young, P.C.: The refined instrumental variable method: unified estimation of discrete and continuous-time transfer function models. J. Eur. Syst. Autom. **42**, 149–179 (2008)
32. Young, P.C.: Gauss, Kalman and advances in recursive parameter estimation. J. Forecast. **30**, 104–146 (2010) (special issue celebrating 50 years of the Kalman Filter)
33. Young, P.C.: Real-time updating in flood forecasting and warning. In: Pender, G.J., Faulkner, H. (eds.) Flood Risk Science and Management, pp. 163–195. Wiley-Blackwell, Oxford (2010)
34. Young, P.C., Beck, M.B.: The modelling and control of water quality in a river system. Automatica **10**, 455–468 (1974)
35. Young, P.C., Lees, M.J.: The active mixing volume: a new concept in modelling environmental systems. In: Barnett, V., Turkman, K.F. (eds.) Statistics for the Environment, pp. 3–43. Wiley, Chichester (1993)
36. Young, P.C., Price, L., Berckmans, D., Janssens, K.: Recent developments in the modelling of imperfectly mixed airspaces. Comput. Electron. Agric. **26**, 239–254 (2000)
37. Young, P.C., Ratto, M.: A unified approach to environmental systems modeling. Stoch. Environ. Res. Risk Assess. **23**, 1037–1057 (2009)
38. Young, P.C., Ratto, M.: Statistical emulation of large linear dynamic models. Technometrics **53**, 29–43 (2011)
39. Young, P.C., Wallis, S.G.: Solute transport and dispersion in channels. In: Beven, K.J., Kirkby, M.J. (eds.) Channel Network Hydrology, pp. 129–173. Wiley, Chichester (1993)

Chapter 17
Identification and Representation of State Dependent Non-linearities in Flood Forecasting Using the DBM Methodology

Keith J. Beven, David T. Leedal, Paul J. Smith, and Peter C. Young

17.1 Flood Forecasting: Concepts and Issues

There is a wide variety of rainfall-run-off models used in hydrology from simple black-box conceptual structures to distributed models based on process representations of different degrees of complexity (Beven [1]). Most of these different types of models have been used in flood forecasting, from artificial neural networks to complex distributed models. The essential requirements of a flood forecasting model are that it should reflect the non-linear dependence of run-off generation and initial hydrograph rise on the antecedent state of a catchment; that it should then route the generated run-off in time to get the timing of the hydrograph peak right; and that it should minimise the variance of the predictions at the required lead time. This is something that has not always been properly appreciated in the past: flood forecasts have often been made deterministically without any account being taken of the uncertainties inherent in the forecasting process. This is despite the fact that an appreciation of such uncertainties might have an important impact on decisions about flood warnings and post-event evaluations of the success of a warning system (e.g. Pielke [11]).

Minimising the prediction variance at the required lead time is much easier in some catchments and some events than others. This is a result of both the sources

K.J. Beven (✉) · D.T. Leedal · P.J. Smith · P.C. Young
Lancaster Environment Centre, Lancaster University, Lancaster, UK
e-mail: k.beven@lancaster.ac.uk

D.T. Leedal
e-mail: d.t.leedal@lancaster.ac.uk

P.J. Smith
e-mail: p.j.smith@lancaster.ac.uk

P.C. Young
e-mail: p.young@lancaster.ac.uk

L. Wang, H. Garnier (eds.), *System Identification, Environmental Modelling, and Control System Design*,
DOI 10.1007/978-0-85729-974-1_17,

of uncertainty and the natural time delay in a catchment relative to the required lead time. In large catchments, where it may take days to reach a flood peak but there is a standard of service to give 6 hours warning to the public, then the prediction problem (given some upstream flow gauges and raingauges feeding into the warning system) should be much easier. In small catchments, subject to flash floods, where the response time of a catchment above a community at risk might be less than a minimum standard of service for warnings of 2 hours, then the prediction problem will be much more difficult. In this case, to achieve the require lead time will require estimation of the precipitation (or snowmelt) inputs ahead of time which introduces a very significant source of uncertainty, however these estimates are made.

In what follows we will consider the case only of catchments where the natural response time is equal to or greater than the required lead time for flood warnings. In that case, it should be possible to achieve good forecasts at the required lead time using only estimates of the inputs to the catchment up to the time of forecast. The major sources of uncertainty in the flood forecasting process are then due to error in the estimation of rainfall and snowmelt inputs based on the measurements available, and error in the rainfall-run-off model used, including the way in which predicted run-off generation reflects the antecedent state of the catchment. We will not consider the additional error associated with quantitative precipitation forecasts for the flash flood case. Even so, the estimates of past inputs based on available measurements may be subject to significant uncertainty because of lack of spatial coverage of the raingauge network or, if rainfall radar data are available, the many different sources of error in the conversion to rainfall rates. Moreover, estimates of snowmelt (e.g. Young et al. [27]) can introduce special problems.

Thus we should expect that any flood forecast will be in error and, therefore, it should be associated with a representation of that error (for example in the form of a standard error). In very many situations it will be possible to make an assessment of that error, in real-time, by comparing the forecast with the current state of a river as observed at gauging stations sites. Thus, it should be possible to see if the model is under- or over-predicting (for whatever reason) and use this information to both improve the accuracy and minimise the error variance of the next forecast out to the required lead time. This is adaptive forecasting which is now becoming more widely used in forecasting systems world-wide (see Young [28, 29]).

Flood forecasting involves two types of non-linearity. The most important is the non-linearity in run-off generation in relation to the state of the catchment and the temporal and spatial pattern of rainfall (and sometimes snowmelt) inputs in a particular event. Hydrologists have long tried to represent this non-linearity in models, ranging from simple conceptual models to fully distributed representations of infiltration and subsurface flow processes (Beven [1]). Hydraulic theory also then suggests that the routing of that run-off to a point at risk of flooding should also be non-linear, since the celerity of the flood wave will depend non-linearly on the temporal and spatial pattern of run-off generation, the geometry of the channel and flood plain and the magnitude of the flood peak. However, there has been a tradition in hydrology of treating the routing problem as a linear problem, ranging from the unit hydrograph of Sherman [15], to linearisations of the hydrodynamic equations for 1D flow in a channel (see Cunge [3], Dooge et al. [4]). The widely used

Muskingum-Cunge equation is one such linearisation which, because of its lack of an explicit time delay and non-minimum phase characteristics should be used with care if a non-physical, albeit volume preserving, impulse-response is to be avoided (Nash [8]).

The work of Peter Young in developing flood forecasting techniques for both the run-off generation and routing problems goes back a long way (see Young [17, 18, 20]). It makes use of what has come to be known as the Data-Based Mechanistic (DBM) methodology which has the aim of defining a model structure based on the information in the observed response of a system. The foundation of the DBM approach is in the robust estimation of linear transfer functions. Thus, to apply the method to the type of non-linear problems found in flood forecasting, some additional non-linear component is needed. This is usually applied as a non-linear transformation of the input signal, dependent on the current state of the system. This type of state-dependent non-linearity was first introduced to hydrological modelling by Young [19], Young and Beven [23, 24] and first used in flood forecasting in an application to the River Nith to predict flooding in Dumfries, Scotland in Lees et al. [7]. Since then, the methods have been tested on a variety of UK catchments, using a number of different forms of non-linear transform (see for example, Young [20, 21, 25], Young et al. [27], Ratto et al. [12], Pappenberger et al. [9], Romanowicz et al. [13, 14], Leedal et al. [6], Smith et al. [16]). This Chapter presents work carried out using data from the River Eden (Cumbria UK).

17.2 River Eden Study Site Description

DBM models are ideally suited to real-time flood forecasting applications. The flexibility of the approach allows for good model performance over a broad range of catchment characteristics. The parametrically efficient formulation permits rapid computation time. The separation of the model into linear and non-linear components permits straightforward inclusion of a DBM representation into a modified Kalman filter scheme for real-time data assimilation. Two flood forecasting schemes from the Eden catchment were used to illustrate the above properties and the use of the alternate input non-linearity methods. These schemes are described below.

1. A rainfall to level model taking as input the mean value of hourly data from the following UK Environment Agency (EA) gauge sites:
   a. Barras;
   b. Scalebeck;
   c. Wet Sleddale;
   d. Burnbank and
   e. Aisgill.
2. A level to level model taking as input the (EA) gauge site at Kirkby Stephen.

Both models were used to forecast river level at the Appleby level gauging station. The town of Appleby is vulnerable to flooding and relies on an extensive system of demountable defences that require a reasonable forecast lead time to install.

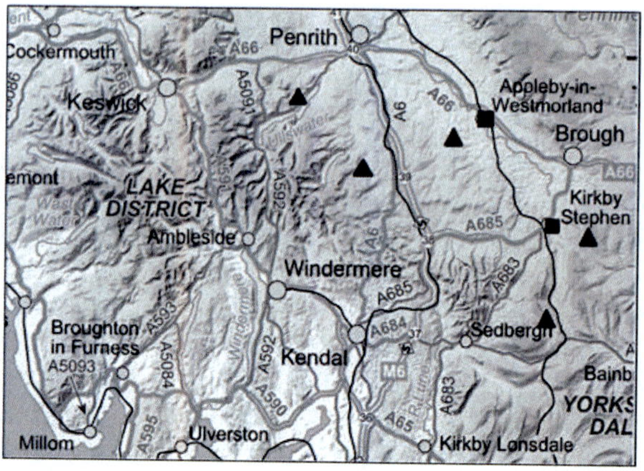

Fig. 17.1 Study site and gauge locations. Key: *rectangles* show river level gauge sites at Kirkby Stephen and Appleby, *triangles* show rain gauge sites at (from *l–r*) Burnbank, Wet Sleddale, Scalebeck, Aisgill and Barras. ©Crown Copyright. O.S. Licence No. A281220

The calibration data extended from the 25th July 2004 to the 10th March 2005. The testing or 'validation' data covered the period from 9th September 2003 to the start of the calibration time series. Figure 17.1 shows the location of gauge sites used by the study.

The various input non-linearity functions together with their associated linear transfer function component were incorporated into a series of rainfall to level and level to level forecast models incorporating a data assimilation scheme designed for real-time flood forecasting as described by Young [22].

17.3 Outline of the DBM Methodology for Real-Time Flood Forecasting

The DBM methodology for real-time flood forecasting aims to produce a physically meaningful model of run-off generation and/or routing processes. The structure and parameterisation of the DBM model is derived from and supported by the information content of input-output time series. The relationship between input and output is split into a non-linear transformation of the inputs and a linear transfer function that distributes the transformed input series through time (a 'Hammerstein' model). In this respect, the DBM approach is similar to the traditional unit hydrograph modelling approach; however, DBM modelling differs by fitting the transfer function in a way that is robust to noisy data. The method for identifying the input non-linearity transformation is also unique in its use of a state dependent parameter estimation algorithm (see later Sect. 17.4). The DBM method for real-time flood forecasting applications then takes the transfer function and input non-linearity function and embeds these within a data assimilation scheme. The model parameters and hyperparameters are optimized so as to minimise the uncertainty in the n-step ahead forecast, where n is a discrete number of observation sample periods.

The DBM model process is described by the following steps.

1. Fit a linear transfer function to the input-output data.
2. Pass through the data again to identify recursively the gain on the inputs required to best fit the output observations.
3. Examine the non-parametric relationship between the gain and an index of the state of the catchment.
4. Find a mathematical function to describe this relationship in a way that can be used as the non-linear transformation of the system input.
5. Apply the transformation and re-identify the transfer function using the transformed inputs in Step 1.
6. Repeat Steps 2 to 5 to convergence if necessary.
7. Embed the transfer function and input non-linearity within a data assimilation scheme and optimize the data assimilation hyperparameters for the chosen n step forecast.

Intrinsic to this method is the estimation of the linear transfer function in a way that is robust to noise in the input-output data using one of the recursive instrumental variable methods developed by Peter Young (see Young [18]) and available in the CAPTAIN toolbox for Matlab. The transfer functions are taken from the general class of linear models described by:

$$y_k = \frac{B(z^{-1})}{A(z^{-1})} u_{k-\delta} + \xi_k, \tag{17.1}$$

where $B(z^{-1})$ and $A(z^{-1})$ are polynomials of order m and n respectively such that $B(z^{-1}) = b_0 + b_1 z^{-1} + \cdots + b_m z^{-m}$ and $A(z^{-1}) = 1 + a_1 z^{-1} + \cdots + a_n z^{-n}$; z^{-1} is the discrete time backwards shift operator such that for example $z^{-i} u_k = u_{k-i}$; ξ_k is a noise input at sample k representing all the stochastic components of the system not represented by the model.

The CAPTAIN algorithms can be used to fit a number of different model structures to find the one best supported by the input-output observations. The choice of model is based on minimising the squared residuals while requiring that parameter estimates have low variance so as to avoid over-fitting. This is achieved using the Young Information Criterion (YIC) which is defined as:

$$\mathbf{YIC} = \log_e \frac{\hat{\sigma}^2}{\sigma_y^2} + \log_e \frac{1}{np} \sum_{i=1}^{np} \frac{\hat{\sigma}^2 \boldsymbol{\Theta}_{i,i}}{\hat{a}_i^2}, \tag{17.2}$$

where $\hat{\sigma}^2$ is the variance of the model residuals; $\hat{\sigma}_y^2$ is the variance of the data; np is the number of model parameters i.e., $n + m + 1$; $\boldsymbol{\Theta}$ is the parameter covariance matrix; $\hat{a}_i^2$ is the square of the ith parameter. The first term on the RHS represents goodness of fit, the second is a joint expression of the variance of the estimated parameter values. Both should be as small as possible (or in log that the YIC be as negative as possible) *while maintaining* a good fit to the data.

Once the linear TF component and a suitable input non-linearity scheme have been identified (see later) the DBM model can be incorporated into a data assimilation algorithm in the following way. Firstly the TF model of (17.1) is recast into the equivalent state space form shown by (17.3).

$$\begin{aligned} \mathbf{x}_k &= \mathbf{F}\mathbf{x}_{k-1} + \mathbf{G}u_{k-\delta} + \boldsymbol{\zeta}_k, \\ y_k &= \mathbf{h}^T\mathbf{x}_k + \xi_k, \end{aligned} \tag{17.3}$$

where $\mathbf{x}_k$ is a vector of model states; the elements of $\mathbf{F}$, $\mathbf{G}$, and $\mathbf{h}$ are determined by the associated linear TF parameters; $u_{k-\delta}$ is a suitably lagged input value (generally rainfall or upstream level) transformed by one of the non-linear functions described in this chapter); δ is the identified advective time delay between input and output; $\boldsymbol{\zeta}_k$ is a vector of process noise $[\zeta_{1,k} \ \dots \ \zeta_{n,k}]^T$ with each element applied to the associated n internal states; ξ_k is the observation noise associated with the measurement. It is possible to transform the noise processes using an Auto Regressive Moving Average (ARMA) filter to account for correlation in time. However, from a pragmatic standpoint, it has often been found acceptable to make the simplifying assumption that the elements of $\boldsymbol{\zeta}_k$ and ξ_k are zero mean, serially uncorrelated and statistically independent, normally distributed random variables with variance at sample k specified by $\zeta_{1,k} \dots \zeta_{n,k}$ and ξ_k. This is the approach taken in the results presented below. The facility to specify variance at each sample period allows for heteroscedasticity within the modelling framework (see later).

The state space form of the model can then be placed within a recursive two-stage filter as shown by (17.4a) and (17.4b).

Forecast:

$$\begin{aligned} \hat{\mathbf{x}}_{k|k-1} &= \mathbf{F}\hat{\mathbf{x}}_{k-1} + \mathbf{G}\mathbf{r}_{k-\delta}, \\ \mathbf{P}_{k|k-1} &= \mathbf{F}\mathbf{P}_{k-1}\mathbf{F}^T + \hat{\sigma}_k^2\mathbf{Q}_r, \\ \hat{y}_{k|k-1} &= \mathbf{h}^T\hat{\mathbf{x}}_{k|k-1}. \end{aligned} \tag{17.4a}$$

Correction:

$$\begin{aligned} \hat{\mathbf{x}}_k &= \hat{\mathbf{x}}_{k|k-1} + \mathbf{P}_{k|k-1}\mathbf{h}[\hat{\sigma}^2 + \mathbf{h}^T\mathbf{P}_{k|k-1}\mathbf{h}]^{-1}\{y_k - \hat{y}_{k|k-1}\}, \\ \mathbf{P}_k &= \mathbf{P}_{k|k-1} - \mathbf{P}_{k|k-1}\mathbf{h}[\hat{\sigma}^2 + \mathbf{h}^T\mathbf{P}_{k|k-1}\mathbf{h}]^{-1}\mathbf{h}^T\mathbf{P}_{k|k-1}, \\ \hat{y}_k &= \mathbf{h}^T\hat{\mathbf{x}}_k. \end{aligned} \tag{17.4b}$$

Here $\mathbf{Q}_r$ is a square, Noise Variance Ratio (NVR) matrix with diagonal entries representing the ratio of variance of the process to observation noise for the model states (off-diagonal entries = 0) i.e., the diagonal elements of $\mathbf{Q}_r$ are $[\frac{\zeta_{1,k}}{\xi_k} \ \dots \ \frac{\zeta_{n,k}}{\xi_k}]$; $\hat{\sigma}_k^2$ is an estimate of the heteroscedastic observation noise variance at sample k calculated using the empirical formula shown by (17.5).

$$\hat{\sigma}_k^2 = \theta_0 + \theta_1\hat{y}_k^2, \tag{17.5}$$

where θ_0 and θ_1 are hyperparameters determining the degree of inflation in observation uncertainty for increasing amplitude of the observation. The n-step forecast (where $n \leq \delta$) is produced by iterating the forecast step of (17.4a) and (17.4b) the required number of times. The estimate of the variance of the forecast output $\hat{y}_{k+n|k}$ is calculated as shown by (17.6).

$$\mathrm{var}(\hat{y}_{k+n|k}) = \hat{q}_{k+n|k} = \hat{\sigma}_k^2 + \mathbf{h}^T \mathbf{P}_{k+n|k}\mathbf{h}. \tag{17.6}$$

17.3.1 Hyperparameter Optimisation

The hyperparameters required to define the data assimilation scheme shown by (17.4a) and (17.4b) cannot be optimized directly due to the multiplicative relationship between θ_0 and $\zeta_{1,k} \ldots \zeta_{n,k}$. However, the scheme can be optimized up to proportionality if θ_0 from (17.5) is set to 1 and the optimized parameters are limited to θ_1 and the diagonal element of $\mathbf{Q}_r$. This provides the ratio of distribution of process noise between the internal state variables (the diagonal elements of $\mathbf{Q}_r$), and the degree of inflation of observation noise variance for increasing observation level (θ_1). Having identified these optimal values, (17.5) is replaced with (17.7):

$$\hat{\sigma}_k^2 = c_{scale}(1 + \theta_1 \hat{y}_k^2). \tag{17.7}$$

If we make the simplifying assumption that the model residuals are distributed normally then c_{scale} can be estimated using (17.8) following an initial optimization.

$$\frac{y_k - \hat{y}_k}{\hat{q}_{k|k-n}} \sim N(0, c_{scale}). \tag{17.8}$$

Further details about hyperparameter optimisation can be found in Smith et al. [16].

The results in this Chapter were produced using the modelling and data assimilation scheme described by (17.1) to (17.7), applied to data collected from the model site described in Sect. 17.2. The remainder of this Chapter focuses on the identification and parameterisation of the input non-linearity function used to transform the observed input (rainfall or upstream level) into the *effective* input term $u_{k-\delta}$ of (17.1), (17.3) and (17.4a) and (17.4b).[1]

[1] A common addition to the scheme described above is an adaptive gain module. This module is not included in the results presented in this chapter but for completeness a brief description follows. An adaptive gain module assumes that the forecast value is scaled by a probabilistic, non-stationary gain whose value is conditioned by the mismatch between observed and forecast output. A NVR hyperparameter for the adaptive gain module determines how quickly the gain reacts to this mismatch. It is usual to set a low NVR value so that in operation the adaptive gain responds sluggishly and can correct for the slow accumulation of model error. This mechanism provides a simple means to correct for scenarios such as seasonal variation in catchment dynamics. Full details of the adaptive gain module can be found in Young [22].

17.4 Methods for Identification of State-Dependent Non-linearity

There are two approaches to identifying the form of the state dependent non-linearity function:

1. A functional representation describing the mechanistic processes assumed to be operating within the catchment
2. An empirical functional form, identified from available data

Here we focus on method two as it appeals to Peter Young's DBM philosophy that emphasises an inductive approach driven by observational data. This allows the data to take precedent over the prior assumptions of the model builder.

The first step in method two is to apply the State Dependent Parameter (SDP) estimation algorithm from the Captain toolbox. The SDP algorithm has been described in detail in several papers (see for example: Young et al. [21]). Here the present authors make some observations relating to the use of the SDP tool that may prove useful to others attempting the DBM modelling approach.

The SDP algorithm is designed as an *identification* tool. The output from the algorithm is not a parametric function and, therefore, is not intended to be used for forecasting or simulation. Instead the shape of the non-parametric curve provides an *indication* of the state dependency identified from the data. This then provides the modeller with a guide for designing an appropriate parameterization scheme (as well as some justification for pursuing this approach).

Because the SDP algorithm is often only required to provide a guide for a subsequently optimized parameterization scheme, it is not usually necessary to include a high order transfer function model structure within the algorithm setup. The method used in this Chapter was to limit the identification to a suitably lagged first order relationship i.e., provide the algorithm with the observed output as the time series; the input lagged by the delay in the system and the observed output lagged by one sample as the regressors; and the observed output is the dependent state for the lagged input regressor. This arrangement produced a good indication of the state dependent relationship linking observed output to the effective input series.

The SDP algorithm requires a NVR hyperparameter (similar to the NVRs defined in $\mathbf{Q}_r$ in (17.4a) and (17.4b)) for estimating the state dependency. This can either be optimized by the algorithm or set manually. If the signal to noise ratio of the state dependency is low the optimisation may result in an overfit to the time series. If automatic optimization of the NVR is not possible then a useful relationship may still be identified by manually limiting the NVR to a low value (for example $1e-7$). However, if the user is required to set the NVR parameter the resulting estimation provides less compelling evidence for the existence of a strong state dependency.

By their nature, extreme flood events are rare. This results in a paucity of data at critical flood levels. As a result, the identification of an input non-linearity function at flood levels is unfortunately often reliant on a small number of data points. It is also challenging to estimate the full uncertainty of the state dependency at high flows. This situation is not critical from an uncertainty representation perspective

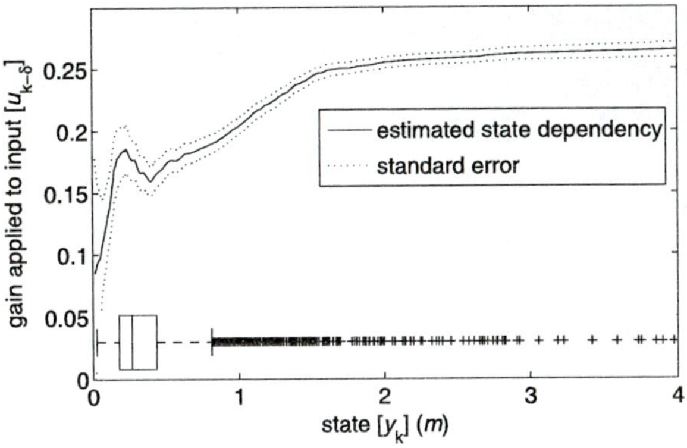

Fig. 17.2 Figure showing the estimated dependency of the model input (level at Kirkby Stephen) to the output state (level at Appleby) for a first order transfer function model produced with the SDP algorithm in the CAPTAIN toolbox for Matlab. The *box* and *whisker plot* shows the density of data points in the time series. *Box* represents lower, median, and upper quartile range; *whiskers* extend to $\times 1.5$ the interquartile range; *crosses mark* data points beyond this range. The relative scarcity of data in the important high level region is clearly shown

as all uncertainty is lumped into the total model error which is conditioned on the performance of the model. Therefore, if the model performs poorly at high levels due to inaccuracy in the representation of the input non-linearity shape at these levels this will still be reflected in the width of the overall forecast uncertainty.

For the Kirkby Stephen to Appleby level to level example the estimated SDP input non-linearity function, determined as described above and with NVR optimized internally by the algorithm, is shown in Fig. 17.2. This non-parametric curve provides evidence to support the inclusion of an input non-linearity scheme in a model representation of the parent system. The box-whisker plot included with Fig. 17.2 provides an indication of the low density of data at high flows.

17.5 Representation of State-Dependent Non-linearity

If an SDP analysis provides compelling evidence to the model builder for the existence of a well defined input non-linearity function, then he or she is faced with the task of choosing a suitable parameterisation scheme. This section considers four methods for generating such a function and provides some commentary on the advantages, disadvantages and issues with each. The methods are:

1. a power law input non-linearity described in Sect. 17.5.1;
2. a radial basis function input non-linearity described in Sect. 17.5.2;
3. a polynomial cubic hermitian spline input non-linearity described in Sect. 17.5.3 and
4. A fuzzy input non-linearity described in Sect. 17.5.4.

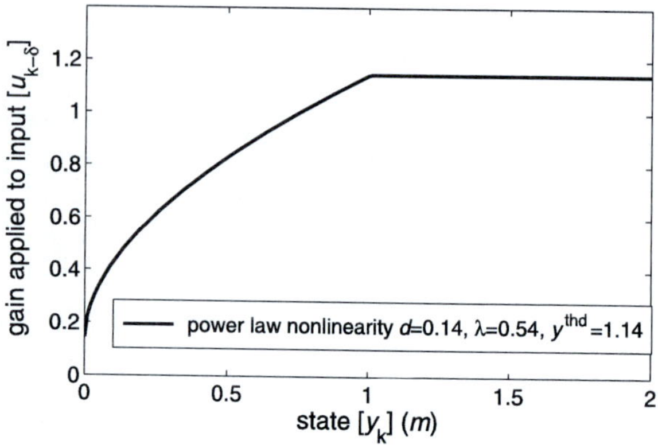

Fig. 17.3 Figure showing the estimated dependency of the model input (rainfall at sites shown in Fig. 17.1) to the output state (level at Appleby) as identified and estimated using a three parameter power law function

17.5.1 Power Law

The power law input transformation is described by (17.9) and (17.10).

$$g_{k-\delta} = d + y_k^{\lambda} \begin{cases} y_k = y_k & \text{if } y_k \leq y^{thd}, \\ y_k = y^{thd} & \text{if } y_k > y^{thd}, \end{cases} \tag{17.9}$$

where d is an offset term to allow for a non-zero gain when y is zero; y is the observed output; k is the present sample period, δ is the pure time delay between input and output measured in an integer number of sample periods; y^{thd} is a threshold level for the output and λ is a power law parameter used to define the non-linear relationship between output and g_k. In many cases both d and y^{thd} are not used. A gain term (g_k) is formed by applying (17.9) to the observed output. This gain is then used to transform the observed input into an effective input r_k using (17.10):

$$r_k = g_k \cdot u_k. \tag{17.10}$$

17.5.1.1 Advantages

Figure 17.3 shows an example power law input non-linearity function including d, λ, and y^{thd}. The key advantage of the power law non-linearity is simplicity. In many cases the optimisation of the function requires only the tuning of a single parameter (λ) with the optimisation expanding to tune d and y^{thd} if necessary. Provided the input non-linearity can be reasonably represented by this family of functions the optimal parameter set is usually well defined and robust to initial conditions. A mechanistic interpretation of the function is straightforward for rainfall as input: low flows represent a 'dry'catchment where a larger proportion of rainfall input is transferred to storage this is represented in the curve as a low gain applied to the input when the output is low.

17.5.1.2 Disadvantages

The power law function is limited to a smooth monotonically increasing or decreasing form. This cannot capture more complex structure in the non-linear relationship that may result from mechanisms such as large changes in channel dynamics at specific points in the level range. In situations where well identified state dependent information is present in the data this information may be lost by choosing the simple power law form for parameterising the input non-linearity function.

17.5.1.3 Comments

In common with the other methods described here it has been found that the most efficient way to achieve optimal parameter estimates is to embed a refined instrument variable (RIV) transfer function optimisation routine within the power law optimisation function. The order of events is then: (1) choose a set of parameters for the power law equations; (2) form the effective input; (3) optimize a linear transfer function between the effective input and the observed output using the CAPTAIN RIV algorithm rivbj form; (4) form the cost function (generally the model residuals). Steps (1) through (4) are generally carried out within a generic non-linear optimisation routine such as Matlab's lsqnonlin.

Additional heuristics can be built into the optimization routine. Options that may be added include a weighting scheme to emphasise specific regions of the hydrograph: for example, placing greater significance on the model fit to the rising limb which is key for flood forecasting applications; and placing constraints on the structure and parameter range of the linear transfer function parameters to guarantee that only models with a sensible mechanistic interpretation (i.e. all real poles, positive flow partitions etc.: see Young [28]) are accepted.

17.5.2 Radial Basis Function Network

The RBF approach described here uses a network of Gaussian basis functions to approximate the input non-linearity function. The operation of the network is described by (17.11).

$$g_{k-\delta} = \sum_{i=1}^{q} \alpha_i \cdot \exp(-\beta_i \| y_k - c_i \|^2), \tag{17.11}$$

where $g_{k-\delta}$ and y_k are defined as previously; q is the chosen number of basis functions; the α_i's, β_i's and c_i's are a set of weights, widths and centre locations respectively describing each of the q basis functions.

RBF methods provide a convenient and flexible function interpolation scheme. The input value is evaluated by each individual RBF, the output from each RBF

Fig. 17.4 Figure showing the estimated dependency of the model input (level at Kirkby Stephen) to the output state (level at Appleby) as identified and estimated using a radial basis function network trained with 6373 hourly data points (*bold line*). The form of the 12 individual RBFs is also shown (*fine lines*)

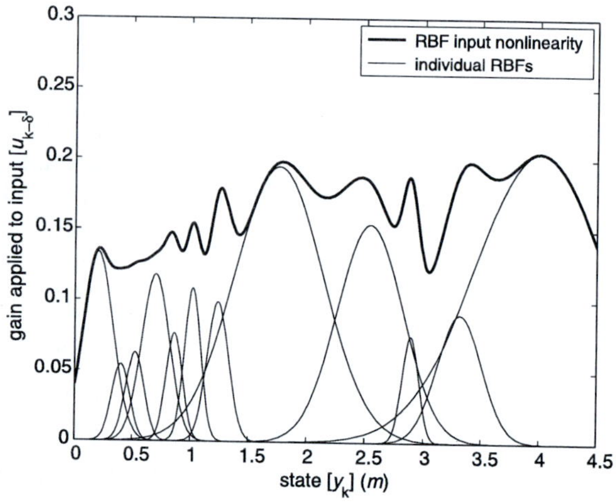

is scaled by the RBF's associated weighting term, finally the output from the RBF group is calculated as the sum of these individual values. It has been shown that given enough individual basis functions this method can be used as a Universal Approximator (Park et al. [10]).

In the examples used here the RBF functions provide an effective way to parameterize a curve in the x, y plane albeit using a relatively large number of parameters. An example RBF input non-linearity function for the Kirkby Stephen to Appleby level to level model is shown in Fig. 17.4.

17.5.2.1 Advantages

As Universal Approximators, an RBF network can replicate any continuous function. This implies that given enough individual basis function components, the RBF network should have the flexibility to approximate any input non-linearity shape provided the shape is well defined by the information content of the data.

17.5.2.2 Disadvantages

The RBF approach requires many parameters. For the example shown in Fig. 17.4, 12 basis functions where used resulting in a total of 36 parameters. This is generally a larger number than would be selected in practice but provides a clear illustration of both the flexibility of the method in terms of curve fitting as well as the drawbacks in terms of over parameterization (see below). The large number of parameters can lead to problems such as long optimisation times, poorly defined parameters or optimization runs that fail to converge on a solution.

The large number of degrees of freedom afforded by the RBF network allows the input non-linearity function to achieve very complex shapes. These shapes may be very specific to the chosen training data i.e., the model can become over-fitted. It could be hypothesised that model over-fit is a particular risk in environmental systems where stochastic fluctuation in system dynamics is the norm in contrast, for example, to engineering systems where system performance is more tightly constrained. A dedicated over-fit protection data set can be included in the optimisation algorithm but this approach requires the user to surrender a portion of data from the calibration or testing set to use for this purpose.

It is impossible to optimize RBFs centred in regions with no observed data i.e., into the region of future unprecedented flood events. The Gaussian basis functions provide a convenient mechanism for the RBF network within the range of the training data; however, these RBFs trail off to zero at the edges of the parameterized function. This may be more desirable than an erratic extrapolation method. However, it seems mechanistically unlikely that the effective input would tend to zero as the output value increases. To counter this it is generally necessary to incorporate an additional function, such as a sigmoid, that maintains an arbitrary constant value beyond the upper range of the RBF network.

17.5.2.3 Comments

Defining the initial parameter vector and setting up the optimisation function for an RBF network can be quite complex. As with the power law input non-linearity described above, the optimisation function incorporates the CAPTAIN rivbj transfer function estimation algorithm placed inside the Matlab lsqnonlin optimisation routine. The optimisation routine is passed a vector of α, β and c parameters that are first used to form a gain time series as a function of the observed output. The gain series is applied to the observed input to form the effective input using (17.11). The rivbj algorithm is then used to estimate the optimal transfer function model. This model is used to form an output estimate and the optimisation cost function is formed from the model deterministic or stochastic residuals. The last weighting term α_q is fixed to prevent the ill-conditioned optimization problem that would result from the interaction of the gain generated by the input non-linearity and the gain of the TF component.

One option for defining the initial parameter values for the optimisation routine is to fit the RBF network to the non-parametric input nonlinearity shape estimated by the CAPTAIN sdp algorithm. This process is made easier by the linear properties of the network. If the c_i's and β_i's are chosen manually, then a curve fit to the SDP results can be performed to estimate the α_i's using ordinary least squares (OLS):

$$\boldsymbol{\alpha} = (\mathbf{X}^T\mathbf{X})^{-1}\mathbf{X}^T\mathbf{g}, \tag{17.12}$$

where $\boldsymbol{\alpha}$ is a column vector of the q individual RBF weights; $\mathbf{X}$ is a $r \times q$ matrix such that $x_{i,j}$ is equal to $\exp(-\beta_j \|y_i - c_j\|^2)$; and $\mathbf{g}$ is a column vector of r input non-linearity gains estimated via SDP i.e., the data pairs (y_i, g_i) form the line shown in Fig. 17.3.

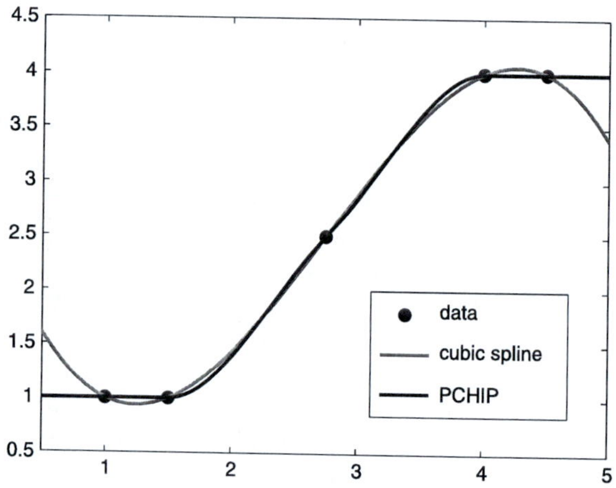

Fig. 17.5 Comparison of PCHIP against cubic polynomial spline for data interpolation. Favourable characteristics of PCHIP include the interpolation performance over flat regions of the data and the sensible endpoint extrapolation

The success or otherwise of an RBF curve fitting scheme can be determined easily from an inspection of the basis functions. When the optimisation *works* the basis functions show a distinct tendency to spread evenly across the input range, and to select weightings and widths that fall within a narrow range. An unsuccessful scheme will tend to group RBFs closely in some regions, include wildly different weights and/or widths, and cancel poorly defined RBFs with a second identical, but negatively weighted partner. These symptoms should prompt the user to repeat the optimisation process using an alternate initial parameter set and/or number of RBFs.

17.5.3 Piecewise Cubic Hermite Data Interpolation (PCHIP)

The PCHIP method for parameterizing curves described in Fritch et al. [5] provides great benefit in situations where the user requires a smooth curve through data pairs, including flat sections and requiring 'sensible'extrapolation characteristics. Figure 17.5 shows a simple example illustrating the advantages of PCHIP over polynomial spline interpolation when a more *natural* and *safe* fit to the data is required.

Interpolation algorithms are a popular method for parameterising functional forms when a set of Cartesian x, y data points are known and the objective is to produce $f(x)$ passing through these points and extended to $f : X \to Y$ where $x \in \mathbb{R}$. For the application described here, an arbitrary set of x, y pairs (or *knots*) are chosen from the estimated non-parametric SDP input nonlinearity function and the Matlab pchip algorithm is used to fit the appropriate function $f(x)$. The set X is held constant and Y is used as initial parameters for an optimisation procedure using Matlab's lsqnonlin. The optimisation then proceeds as follows:

1. an updated Y set is formed
2. the new X, Y set is used to build an updated PCHIP function $f(x)$
3. the updated $f(x)$ is used to form the effective input

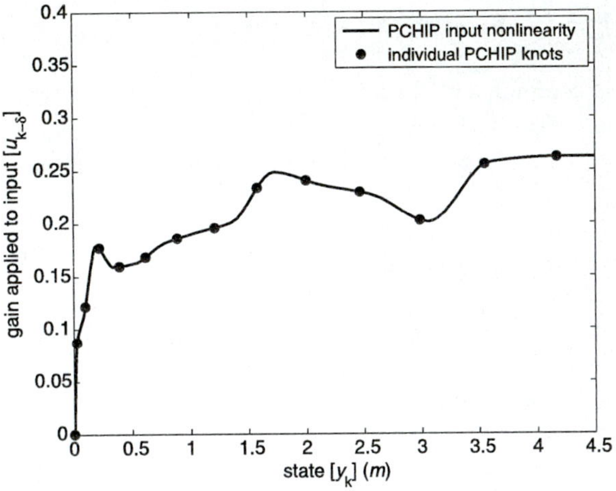

Fig. 17.6 The estimated PCHIP input non-linearity function for the Kirkby Stephen to Appleby level to level model. The *dots* show the location of the chosen X set and corresponding $f(x)$ values (the PCHIP knots). Note the similarity in shape to the RBF network estimation (Fig. 17.4)

4. the CAPTAIN rivbj algorithm is used to estimate the linear transfer function between effective input and observed output
5. the cost function is formed from the model's deterministic or stochastic residuals.

Steps 1 to 5 are repeated until the optimization converges.

Figure 17.6 shows the estimated PCHIP input non-linearity function for the Kirkby Stephen to Appleby level to level model. The PCHIP input non-linearity functions illustrated in this chapter use 13 internal knots plus two end knots. As with the RBF network example, this is more parameters than would generally be used but illustrates well the curve fitting flexibility as well as potential for data over-fit.

17.5.3.1 Advantages

If a human, with understanding of the requirements of a good input non-linearity function, was asked to sketch a line through a set of optimally positioned x, y pairs, the result would look very much like the PCHIP solution. This is the advantage of the PCHIP method; it is able to parameterize a very *natural* looking line through the function knots. Once built, the function can be used to evaluate $f(x)$ for any chosen x. Extrapolated values can be easily controlled by adding a single knot to each end of the domain range holding the extrapolated values constant or applying a gentle linear slope (Fig. 17.6 shows $f(> 4.4)$ held constant at 0.26).

The optimization of each knot of a PCHIP function requires one less parameter than for each individual function from the RBF network method. If the x ordinate of the knots are distributed heuristically across the function domain then the optimisation can be reduced to the corresponding y ordinates. This later method is generally sufficient to produce a successful input non-linearity function and was used to produce the results presented in this chapter.

17.5.3.2 Disadvantages

In common with the RBF network method, the PCHIP function carries the potential to overfit the input non-linearity function to the chosen calibration data. An overfit prevention method could be included as a precaution.

17.5.3.3 Comments

In the results presented in this chapter, a logarithmic spacing for the x ordinates of the PCHIP knots was used. This method was chosen as it allows for more function shape detail to be included towards the origin of the input non-linearity function. In this region there is a higher density of data from which to draw information for function shaping. Other methods for selecting x ordinates are available including a linear spacing, manual grouping of points, fuzzy clustering algorithms and full optimisation of both the x and y component of the PCHIP knots. In practice the authors have found little if any advantage in applying more complex x ordinate locating methods. For the optimisation, the location of the final knot is fixed to prevent the ill-conditioned optimization problem that would result from the interaction of the gain generated by the input non-linearity and the gain of the TF component.

17.5.4 Takagi-Sugeno Fuzzy Inference Method

The methods described thus far have been inductive in that they have relied on numerical optimisation techniques to parameterize the shape of the input non-linearity function based on observed data. This section departs from this approach in-order to explore the possibilities offered by fuzzy inference to translate the expert judgement held by the user into a parametric form of the input non-linearity function. To achieve this a simple Takagi-Sugeno Fuzzy Inference System (T-S FIS) was built using four input membership functions and four first-order output functions together with a rule linking an input to an output function. The input into the T-S FIS is the observed output data y_k, the output is the gain to apply to the observed input at sample $k–\delta$ in order to form the effective input r_k. The form of the T-S FIS used here can be described by (17.13).

$$g_{k-\delta} = \frac{\sum_{i=1}^{N} w_i z_i}{\sum_{i=1}^{N} w_i}. \tag{17.13}$$

Here N is the number of rules; the w_i's are the degree of membership of the input to each of the N membership functions ($w_i \in [0, 1]$); and the z_i's are the first order output functions (in this case simply a set of N constants).

In an attempt to capture the expert judgement of the user, an interactive heuristic procedure is followed whereby the user interacts dynamically with the shape and

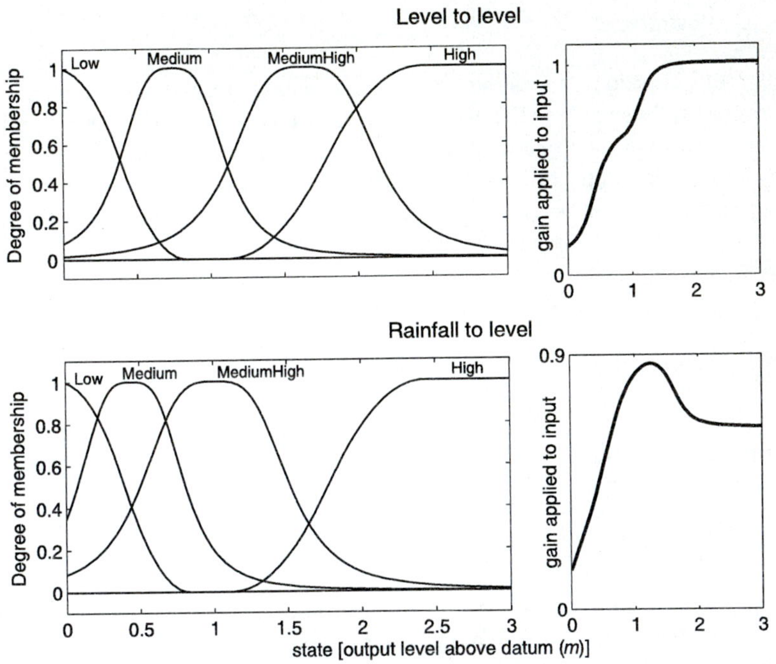

Fig. 17.7 The estimated T-S FIS input non-linearity function for the level to level (*top*) and rainfall to level (*bottom*) modelling schemes. The *left side plots* show the shape and location of the four input membership functions, the *right side plots* show the resulting input non-linearity shape

location of the T-S FIS membership functions in order to shape an overall input non-linearity function. This function is then used to form the effective input series and again the rivbj algorithm from CAPTAIN is used to estimate the optimal linear transfer function component of the model.

The nature of T-S FIS provides a natural set of signifiers in the form of membership functions with which the user can interact. Unlike RBF networks and PCHIP splines, the user can *break down* the problem into a series of decisions about the sub units of the overall function. A set of simple rules form a mechanistically meaningful association between the input and the output (for example: *if* LEVEL *is* LOW *then* OUTPUT *is* LOW). Using the rule base as a start point, the user can then make decisions based on his/her interpretation of the system and any other relevant experience. Example decisions include: where to locate the membership functions, what widths to apply to them and what output gains to associate with each. The combination of a meaningful linguistic naming convention for the T-S FIS membership functions together with a graphical interface such as Matlab's fuzzy tool provide a relatively rapid means to shape and investigate the form of the input non-linearity function.

Figure 17.7 shows example input non-linearity function developed for the models in this chapter.

17.5.4.1 Advantages

The T-S FIS method provides the user with a means to form an input non-linearity function using knowledge of the system or experience of similar systems. By interactively shaping the input non-linearity function, the user gains an intuitive understanding of the impact of the function on the model performance. This type of *human-in-the-loop* computer interaction has been shown to be effective in applications where a straight forward globally optimal parameter set is not available (see for example Colgan et al. [2]).

17.5.4.2 Disadvantages

The resulting model will not be optimal in any statistical sense. The input non-linearity function will be dependent on the skill of the user.

17.5.4.3 Comment

One interesting feature of the T-S FIS process described here is it allows the user to focus the model performance on specific portions of the data. When the model is intended for flood forecasting, the user can *tune* the input non-linearity such that the model performs well on the rising limb of the hydrograph—the most crucial characteristic to capture simply by using a visual inspection of the simulation results. The performance of the model over low flow and recession periods can be discounted to a larger degree. This can also be achieved with the other methods presented but requires a more complex formulation of the optimisation cost function.

17.6 Results

The eight input non-linearity function produced by the methods described above are shown in Fig. 17.8. The figure also shows the input non-linearity shape identified using non-parametric SDP estimation.

17.6.1 Rainfall to Level Forecasting on the River Eden

Figure 17.9 shows the largest event from the data calibration period. The calibration period ran from 17th June 2004 to the 10th March 2005. The detail shows the observed and forecast data for the large event that took place around the 8th January 2005.

Figure 17.10 shows a detail from the validation data period 9th September 2003 to the 25th June 2004 including the largest event in this range that resulted in significant flooding at Appleby on the 2nd February 2004. The forecast lead time was five

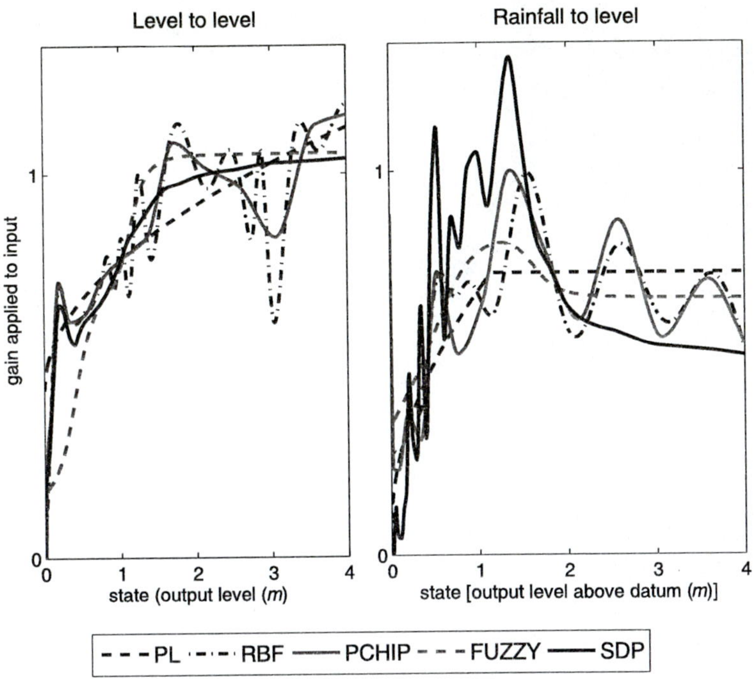

Fig. 17.8 The shape of the input non-linearity functions together with SDP identification for the level to level and rainfall to level model schemes

hours. The model efficiency RT2 (Coefficient of Determination based on the simulation output error) is used here for assessment where an RT2 score of 1 is a perfect fit between modelled and observed data; RT2 $= 0$ occurs when model predictions are as accurate as the mean of the observed data; RT2 < 0 occurs when the residual variance is larger than the data variance. The RT2 scores for the full validation period at each forecast step are shown in Table 17.1.

The results presented in Fig. 17.10 show that all the input non-linearity methods combined with appropriate TF component perform reasonably well. The combination of DBM methods and data assimilation provide a robust forecasting framework. However, within the overall performance range exhibited by the four level to level forecasting schemes, points to note include:

1. The power law method demonstrated an over-estimate of the peak level.
2. The RBF network would have provided the best performance for flood forecasting but by a very slim margin over the other methods.

17.6.2 Level to Level Forecasting on the River Eden

Figures 17.11 and 17.12 show the same subset of the calibration and validation data presented in Figs. 17.9 and 17.10 but here showing the level to level results. The

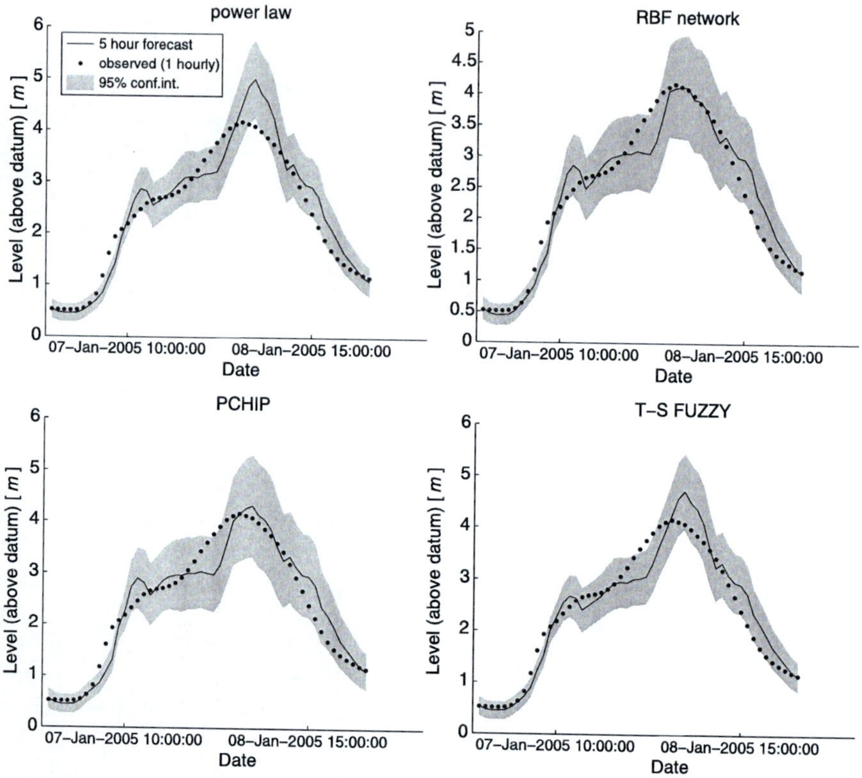

Fig. 17.9 Calibration results for five hour rainfall to level forecast on the River Eden using alternate input non-linearity function parameterisation methods

forecast lead time is three hours. Table 17.2 shows Model Efficiency scores for the full validation period. The results presented in Fig. 17.12 show that all the input non-linearity methods combined with appropriate TF component perform reasonably well. The combination of DBM methods and data assimilation provide a robust forecasting framework. However, within the overall performance range exhibited by the four rainfall to level forecasting schemes points to note include:

1. The RBF network demonstrates some sharp level adjustments which may be a result of the over-fit when this method is applied with many degrees of freedom.
2. The PCHIP method marginally provides the best performance for flood forecasting applications.

For comparison with the results presented above, the optimal linear model (i.e., TF model optimized with no input non-linearity function) results are shown in Fig. 17.13. A visual inspection of the subset of validation data shown in Fig. 17.13 demonstrates that the optimal linear models perform reasonably well but both models show a tendency to lag the rising limb of the storm event. The lag of the rising

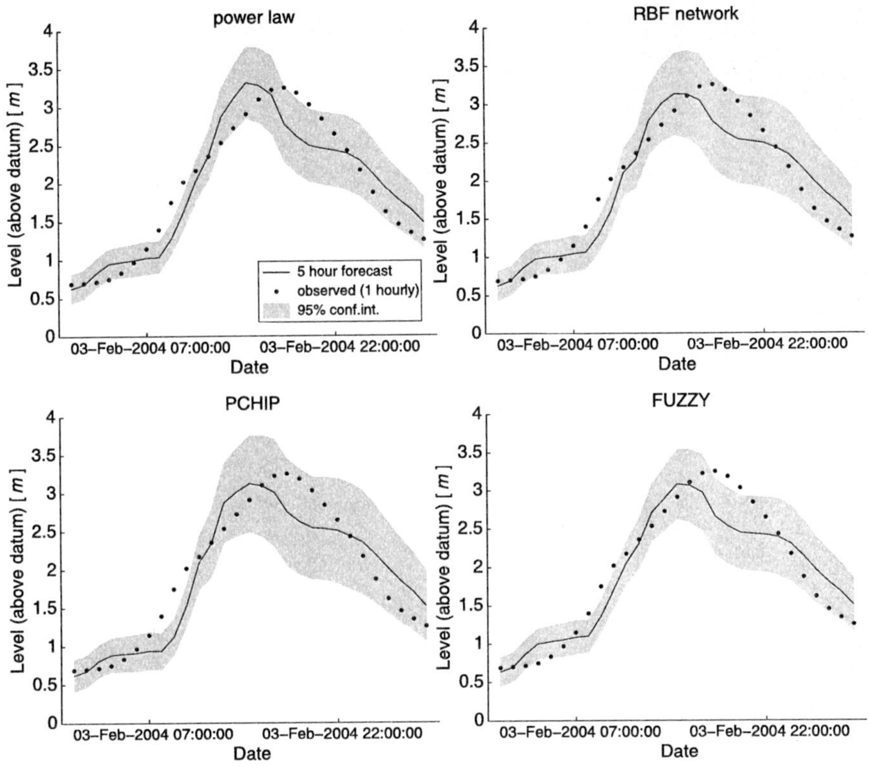

Fig. 17.10 Validation results (detail) for five hour rainfall to level forecast on the River Eden using alternate input non-linearity methods

Table 17.1 Model Efficiency scores (RT2) for rainfall to level forecasting between 1 and 5 hour lead times. Score calculated from validation period data

Non-linearity type	1 hour	2 hour	3 hour	4 hour	5 hour
Power law	0.978	0.963	0.948	0.936	0.926
RBF network	0.979	0.965	0.950	0.938	0.927
PCHIP	0.979	0.965	0.952	0.940	0.931
T-S FIS	0.976	0.961	0.946	0.933	0.923

Table 17.2 Model Efficiency (RT2) scores for level to level forecasting between 1 and 3 hour lead times. Score calculated from validation period data

Non-linearity type	1 hour	2 hour	3 hour
Power law	0.958	0.950	0.946
RBF network	0.956	0.944	0.940
PCHIP	0.954	0.944	0.938
T-S FIS	0.894	0.862	0.838

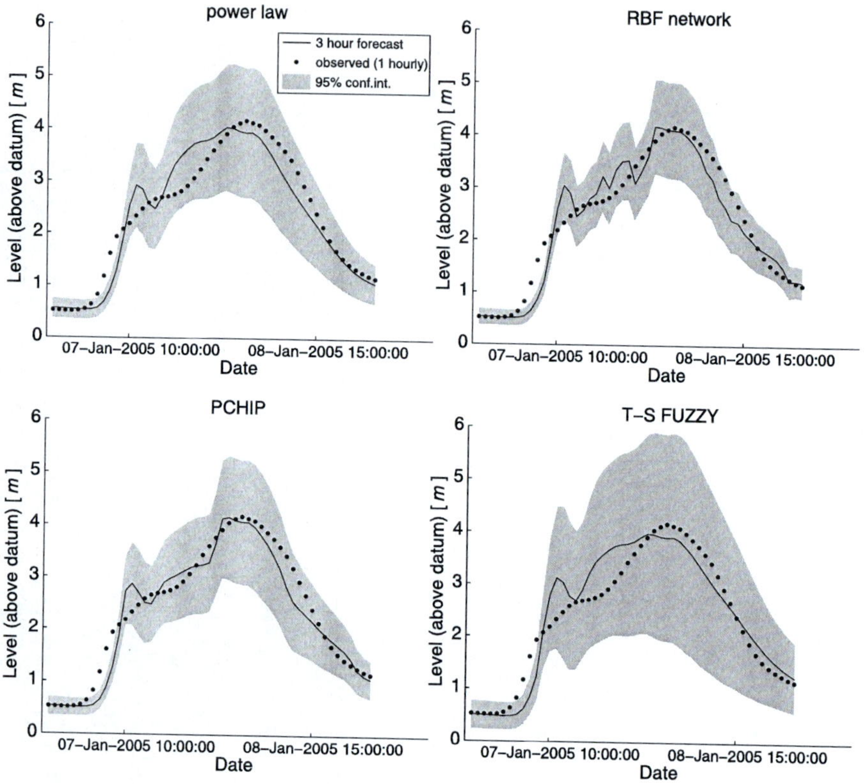

Fig. 17.11 Calibration results for three hour level to level forecast on the river Eden using alternate input non-linearity function parameterisation methods

limb observed in the linear models is more significant than that found in any of the model configurations incorporating an input non-linearity function.

17.7 Conclusions and Comment

Comparing results between Figs. 17.12, 17.10 and 17.13 demonstrates that some form of input non-linearity function is an important component of a flood forecasting system. The optimal linear model produces more significant errors on both the timing and scale of the large event within the calibration data.

The four methods presented here for parameterizing and optimizing the input non-linearity function have associated advantages and disadvantages. The best method to choose for a particular modelling/data assimilation exercise is dependent on the characteristics of the system. The power law method is an appropriate choice when the non-linearity is equally simple. It is also a good choice when there is insufficient data to identify the input non-linearity shape in any detail. The RBF network

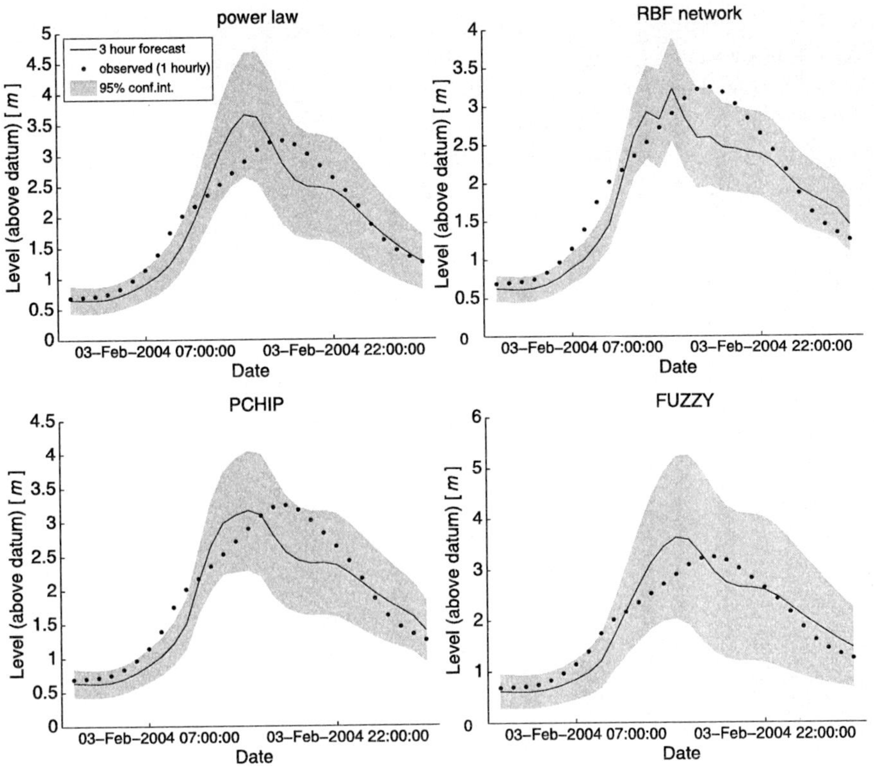

Fig. 17.12 Validation results for three hour level to level forecast on the river Eden using alternate input non-linearity function methods

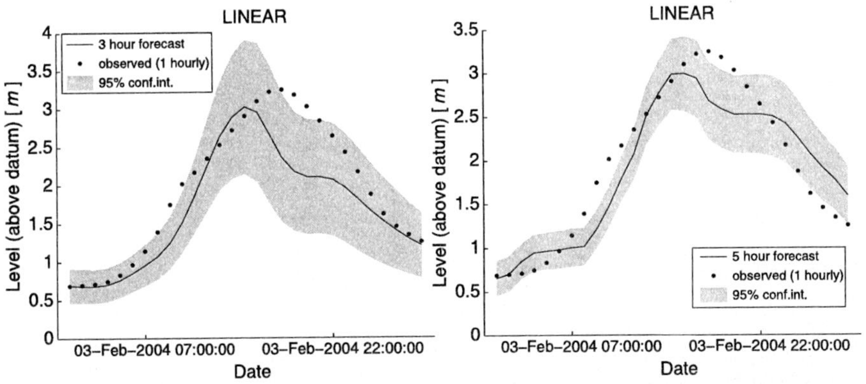

Fig. 17.13 Validation results for three hour level to level (l) and five hour rainfall to level(r) forecast on the River Eden using optimal linear models i.e., models using no input non-linearity function

method is appropriate for systems that exhibit very well defined and complex input non-linearity shapes. However, the flexibility of the RBF network to produce any shape may lead to model over-fit problems. The PCHIP method also provides the flexibility to map complex input non-linearity shapes while providing the ability to maintain a *natural* curve. Overfit to calibration data is also a risk however especially is a large number of knots are used. The T-S FIS method, together with interactive tuning, provides an entirely different approach employing *human-in-the-loop* interaction during the parameter estimation process. This approach can be aided by easy to use visual interfaces such as Matlab's fuzzy toolbox. However, this method is not optimal in any statistical sense.

It is hoped users of the DBM modelling approach will find this chapter provides a solid foundation for producing input non-linearity parameterization schemes. Future users may go on to refine these or develop useful methods of their own.

17.8 Future Work

The DBM modelling approach to real-time flood forecasting and data assimilation is an ongoing field of research. Active areas of research include, but are not limited to, the following.

1. Continuous time transfer function form. With continuous time modelling the system is modelled in differential equation in differential equation terms (or the s operator transfer function equivalent, where $s^n = \frac{d^n}{dt}$: see Young [26] for more detail). This approach will prove advantageous in situations where data are available, and forecasts are required, at unevenly spaced intervals, as well as when the data are sampled rapidly.
2. Single stage optimization. In the examples presented here, the optimisation of the linear transfer function model is embedded within the optimisation of the input non-linearity scheme. Work by Smith et al. in this volume is investigating a single stage recursive optimisation procedure for estimating both parameter sets together. This would be useful for both automating the identification/estimation procedure and also for identifying the covariance structure of the full model parameter set. Also Young [28, 29] presents an adaptive forecasting system where a recursive form of the rivbj algorithm in CAPTAIN is used to continuously update the parameters of the TF model.

Acknowledgements This research was carried out as part of RPA9 and SWP1 of the Flood Risk Management Research Consortium (FRMRC) phases 1 and 2. The principal sponsors of FRMRC are: the Engineering and Physical Sciences Research Council (EPSRC) in collaboration with the Environment Agency (EA), the Northern Ireland Rivers Agency (DARDNI), the United Kingdom Water Industry Research (UKWIR) Organisation, the Scottish Government (via SNIFFER), the Welsh Assembly Government (WAG) through the auspices of the Defra/EA, and the Office of Public Works (OPW) in the Republic of Ireland. For details of the FRMRC, see http://www.floodrisk.org.uk.

References

1. Beven, K.: Rainfall-Runoff Modelling: The Primer. Wiley, New York (2001)
2. Colgan, L., Spence, R., Rankin, P.: The cockpit metaphor. Behav. Inf. Technol. **14**(4), 251–263 (1995)
3. Cunge, J.A.: On the subject of a flood propagation computation method (Muskingum method). J. Hydraul. Res. **7**(2), 205–230 (1969)
4. Dooge, J.C., Strupczewski, W.G., Napiorkowski, J.J.: Hydrodynamic derivation of storage parameters of the Muskingum model. J. Hydrol. **54**(4), 371–387 (1982)
5. Fritsch, F.N., Carlson, R.E.: Monotone piecewise cubic interpolation. SIAM J. Numer. Anal. **17**, 238–246 (1980)
6. Leedal, D., Beven, K.J., Young, P.C., Romanowicz, R.J.: Data assimilation and adaptive real-time forecasting of water levels in the Eden catchment, UK. In: Samuels, P., Huntington, S., Allsop, W., Harrop, J. (eds.) Flood Risk Management Research and Practice. Taylor and Francis, London (2008)
7. Lees, M., Young, P.C., Beven, K.J., Ferguson, S., Burns, J.: An adaptive flood warning system for the river Nith at Dumfries. In: White, W.R., Watts, J. (eds.) River Flood Hydraulics. Institute of Hydrology, Wallingford (1994)
8. Nash, J.E.: A note on the Muskingham flood routing method. J. Geophys. Res. **64**, 1053–1056 (1959)
9. Pappenberger, F., Beven, K.J., Hunter, N., Gouweleeuw, B., Bates, P., de Roo, A.: Cascading model uncertainty from medium range weather forecasts (10 days) through a rainfall-runoff model to flood inundation predictions within the European flood forecasting system (EFFS). Hydrol. Earth Syst. Sci. **9**(4), 1430–1449 (2005)
10. Park, J., Swandberg, I.W.: Universal approximation using radial-basis-function networks. Neural Comput. **3**(2), 246–257 (1991)
11. Pielke, R.A. Jr., Pielke, R.A. Sr.: Hurricanes: Their Nature and Impacts on Society. Wiley, New York (1997)
12. Ratto, M., Young, P.C., Romanowicz, R., Pappenberger, F., Saltelli, Pagano A.: Uncertainty, sensitivity analysis and the role of data based mechanistic modeling in hydrology. Hydrol. Earth Syst. Sci. **11**, 1249–1266 (2007)
13. Romanowicz, R.J., Young, P.C., Beven, K.J.: Data assimilation and adaptive forecasting of water levels in the river Severn catchment, United Kingdom. Water Resour. Res. **42**, W06407 (2006)
14. Romanowicz, R.J., Young, P.C., Beven, K.J., Pappenberger, F.: A data based mechanistic approach to nonlinear flood routing and adaptive flood level forecasting. Adv. Water Resour. **31**(8), 1048–1056 (2008)
15. Sherman, L.K.: Streamflow from rainfall by the unit-hydrograph method. Eng. News-Rec. **108**, 501–505 (1932)
16. Smith, P., Beven, K.J., Tych, W., Hughes, D., Coulson, G., Blair, G.: The provision of site specific flood warnings using wireless sensor networks. In: Samuels, P., Huntington, S., Allsop, W., Harrop, J. (eds.) Flood Risk Management Research and Practice. Taylor and Francis, London (2008)
17. Young, P.C.: Recursive approaches to time-series analysis. Bull. Inst. Math. Appl. **10**, 209–224 (1974)
18. Young, P.C.: Recursive Estimation and Time-Series Analysis. Springer, Berlin (1984)
19. Young, P.C.: Time variable and state dependent modelling of nonstationary and nonlinear time series. In: Subba Rao, T. (ed.) Developments in Time Series Analysis, pp. 374–413. Chapman and Hall, London (1993)
20. Young, P.C.: Data-based mechanistic modelling and validation of rainfall-flow processes. In: Anderson, M.G., Bates, P.D. (eds.) Model Validation: Perspectives in Hydrological Science, pp. 117–161. Wiley, Chichester (2001)
21. Young, P.C.: The identification and estimation of nonlinear stochastic systems. In: Mees, A.I. (ed.) Nonlinear Dynamics and Statistics, pp. 127–166. Birkhäuser, Boston (2001)

22. Young, P.C.: Advances in real-time flood forecasting. Philos. Trans. R. Soc. Lond. A **360**(1796), 1433–1450 (2002)
23. Young, P.C., Beven, K.J.: Computation of the instantaneous unit hydrograph and identifiable component flows with application to two small upland catchments comment. J. Hydrol. **129**(1–4), 389–396 (1991)
24. Young, P.C., Beven, K.J.: Data-based mechanistic (DBM) modelling and the rainfall-flow non-linearity. Environmetrics **5**, 335–363 (1994)
25. Young, P.C.: Top-down and data-based mechanistic modelling of rainfall-flow dynamics at the catchment scale. Hydrol. Process. **17**, 2195–2217 (2003)
26. Young, P.C., Garnier, H.: Identification and estimation of continuous-time data-based mechanistic (DBM) models for environmental systems. Environ. Model. Softw. **21**(8), 1055–1072 (2006)
27. Young, P.C., Castelletti, A., Pianosi, F.: The data-based mechanistic approach in hydrological modelling. In: Castelletti, A., Sessa, R.S. (eds.) Topics on System Analysis and Integrated Water Resource Management, pp. 27–48. Elsevier, Amsterdam (2007)
28. Young, P.C.: Real-time updating in flood forecasting and warning. In: Pender, G.J., Faulkner, H. (eds.) Flood Risk Science and Management, Oxford, UK, pp. 163–195. Wiley-Blackwell, Oxford (2010)
29. Young, P.C.: Gauss, Kalman and advances in recursive parameter estimation. J. Forecast. **30**, 104–146 (2010) (special issue celebrating 50 years of the Kalman Filter)

Chapter 18
Transport and Dispersion in Large Rivers: Application of the Aggregated Dead Zone Model

Sarka D. Blazkova, Keith J. Beven, and Paul J. Smith

18.1 Transport and Dispersion in Large Rivers: Some Issues

The transport and dispersion of solutes in rivers is an important issue for many purposes. Understanding the way in which nutrients are available to phytoplankton and macrophytes; the licensing of effluents; the prediction of pollution incidents and consequent damage to ecological services all depend on the adequate prediction of transport and dispersion (most famously the Sandoz incident on the River Rhine in 1986 [32], but many other incidents have caused significant ecological damage). There is, of course, an extensive body of theory concerned with the transport and dispersion of solutes in rivers (e.g. [9, 30]). There have also been many tracer experiments carried out on both small and large rivers (and laboratory flumes) to try and determine the dispersion characteristics of particular reaches directly.

The advection-dispersion equation (ADE) is the most widely used description of transport and dispersion in rivers. It is based on an assumption of a linear relationship between dispersive flux and concentration gradient, scaled by the dispersion coefficient. This assumption can be justified if the dispersion is assumed to be controlled by the velocity distribution in the vertical [9, 30]. Unfortunately, this theoretical justification does not guarantee that the ADE provides good predictions of tracer observations. The ADE predicts that after an initial mixing length, when the solute becomes "fully mixed" with the flow, the concentration plume develops the shape of a symmetric Gaussian distribution in the downstream direction and a slightly asymmetric as it passes a particular cross-section of the river.

S.D. Blazkova
T G Marsaryk Water Resource Institute, Prague, Czech Republic

K.J. Beven (✉) · P.J. Smith
Lancaster Environment Centre, Lancaster University, Lancaster, UK
e-mail: k.beven@lancaster.ac.uk

L. Wang, H. Garnier (eds.), *System Identification, Environmental Modelling, and Control System Design*,
DOI 10.1007/978-0-85729-974-1_18,

But tracer experiments almost invariably have much heavier tails than predicted by the ADE. This is not a new observation: it has been recognised for decades (e.g. [8, 10, 24, 33–35]). This is the result of the fact that mixing over longer distances is not dominated by shear dispersion (which is actually rather an efficient mixing process). It is rather dominated by the imperfect mixing associated with the lateral shear associated with secondary currents, including secondary circulation cells and the effects of backwaters and other "dead zones" (e.g. [27]). Even with artificially high dispersion coefficients, the ADE cannot adequately mimic these larger scale controls on dispersion. It simply predicts (in many cases) the wrong shape of plume. These larger scale effects also imply that the effective mixing length in real rivers may be much longer than predicted on the basis of vertical velocity shear. This is something that is often observed downstream of junctions between tributaries of similar discharges but with different sediment concentrations. Thus the failure of the ADE should not be a surprise (even if it is still widely used in many water quality models).

The ADE can be modified to predict longer tails by the inclusion of additional exchange terms, such as in the transient storage model of [2] and [29] amongst others. There is however a much simpler alternative based on assuming that dispersion is dominated by dead zone mixing: the Aggregated Dead Zone model.

18.2 The Aggregated Dead Zone Model

The Aggregated Dead Zone (ADZ) model was first introduced by Peter Young and Tom Beer in their 1983 paper [1]. The original motivation for the ADZ was to show that actual dispersion in rivers, as revealed by tracer data, could be predicted quite simply by a first (or higher) order linear transfer function. It followed what Peter Young would later call data-based mechanistic (DBM) principles of letting the data show what form of model might be needed [40, 44]. Further development and testing of the approach was later carried out at Lancaster by Peter Young in collaboration with Steve Wallis, Keith Beven and Hannah Green [13, 14, 31, 36, 37, 40, 42–44] including the collection of many tracer experiments on (mostly small) UK rivers. The Lancaster group also produced software for the analysis of tracer experiments and the prediction of pollution incidents in river networks [3, 5, 6]. An interesting application was made in analysing the sequence of events in a pollution incident in the River Eden in Cumbria, UK [11].

The ADZ concepts have also been used to describe transport in soils [4], in a general water quality mode [22], in bedrock channels [28], and in urban sewer channels [16–18]. The ADZ method did merit a very brief mention in Rutherford's book in 1994 but, perhaps because of its lack of a theoretical link to hydraulic principles and velocity distributions, has not been widely taken up (though see [23] for a method for matching ADZ and transient storage model results through the use of the method of moments). In the UK, it was used as an alternative to the ADE in an analysis of all the UK Environment Agency tracer database [15] but elsewhere there has been less interest (though see [7, 20, 26, 29]). This is unfortunate since it is simple and generally provides excellent predictive capability.

The DBM method is based on the use of general linear transfer functions to identify a model structure from the data by induction, without preconceptions about what form of model should be used. In this case, this has the advantage of producing a representation of the processes that can reproduce the long tails of experimental tracer and pollutant concentration data, which the basic ADE cannot. The ADZ model is best understood in terms of simple "lag and route" concepts. In any transport problem there is a lag between the time at which the solute concentration starts to rise at the start of a reach and when it starts to rise at the end of the reach. This, in the ADZ methodology, is called the advective time delay, τ. It will be expected to decrease with increasing discharge in the reach. The dispersion of the input plume in the reach is modelled as a linear transfer function. In the simplest first-order case, the transfer function is equivalent to a linear store with a mean residence time, T. The mean travel time in the reach is then $(\tau + T)$. Given a steady discharge, Q, the volume of this effective mixing store is then $V_e = Q \cdot T$. The total volume in the reach involved in the transport process is $V \cdot [\tau + T]$. The ratio V_e/V is called the Dispersive Fraction (DF). It can also be defined in terms of the advective travel time and mean residence time as $T/(\tau + T)$.

In some analyses it has been shown that a higher order transfer function can give a slightly better fit to the observed tracer experiments (e.g. [36, 43]. This suggests that there might be different mechanisms affecting the dispersion with rather longer mean residence times than in the bulk flow and its associated dead zones. This might be due to exchanges into larger dead zones or perhaps mixing with hyporheic zone waters in the bed and banks. It is worth noting that, as shown by [27], the lateral velocity shear at the boundary between a dead zone and the main flow can induce locally efficient mixing. However, once solute or pollutant is transferred into such a storage, mixing is much slower and the time scale of pollutant retention much longer. This contributes to the heavy tails of observed tracer concentrations.

18.3 Fitting the ADZ Model to Tracing Experiments in Larger Rivers

Most of the early work with the ADZ model was based on the analysis and prediction of tracer data from small UK river reaches. In smaller rivers the boundary effects (including vegetation, hyporheic zone and other dead zone effects) might be expected to have a greater influence on the transport of solutes. It is therefore of interest to see if the ADZ approach can apply equally to larger rivers.

Fitting the ADZ model to tracer data is equivalent to fitting a linear transfer function to the downstream concentration curve, given the input concentration curve at the entry to a reach. A general linear transfer function model may be written

$$C_t = \frac{b_o + b_1 z^{-1} + \cdots + b_m z^{-m}}{1 - a_1 z^{-1} - a_2 z^{-2} + \cdots + a_n z^{-n}} U_{t-\tau}, \tag{18.1}$$

Table 18.1 Application of the ADZ model to tracer data from the Glen Canyon Dam controlled flood realise in March 1996. Models fitted to input and output data for each reach length with time step of 0.2 hours. Distances in river km from injection point. Advective time delays (τ) in hours

Reach	River km	Reach length	(n, m, τ)	RT2	YIC	DF
Badger Creek Rapid	12.7					
to Little Colorado River	98.3	85.6	(3,2,13.0)	0.9994	−16.58	0.852
to Hance Rapids	123.5	25.2	(1,1,3.2)	0.9988	−16.53	0.758
to Diamond Creek	362.3	238.8	(1,1,33.7)	0.9934	−14.28	0.787

where C_t is the output concentration at the downstream end of a reach at time t, U is the input concentration from upstream and z^{-1} is the backward difference operator such that $U_{t-1} = z^{-1}U_t$. This general model is defined in terms of the triplet (m, n, τ). Here, the CAPTAIN Matlab toolbox routines developed by Peter Young have been used to define the correct order of the model and the associated a and b coefficients. The Young Information Crtierion (YIC) is used to guard against overfitting the data [41].

Three examples demonstrate how well the ADZ model can predict tracer transport over long distances (and the persistence of the long tails in the observed tracer concentrations in such large rivers). The first is the set of tracer data collected in the Colorado River, USA. In this experiment, rhodamine tracer was added during an experimental steady high flow event generated by a controlled flood release from the Glen Canyon Dam in 1996 [12, 21]. A number of input-output concentration curves were available for different reaches. Details of the sites, reach lengths, ADZ model structure and fit are given in Table 18.1. A representative demonstration of the excellent fit by the simplest first order model for the 239 km Hance Rapids to Diamond Creek reach is given in Fig. 18.1. The third order model required to fit the Badger Creek to Little Colorado Reach might be an indication that the initial mixing was incomplete at Badger Creek, only 12.7 km downstream of the injection point.

A second example, from the River Labe/Elbe, crossing the border from the Czech Republic into Germany also illustrates the accuracy of the ADZ model in reproducing the concentration curves from tracing experiments. Table 18.2 provides details of the reaches, ADZ model structures and fit. Figure 18.2 provides a demonstration of the application of the model in calibration. For the first experiment, in June 1997, the first four reaches (to 43 km) were still considered to be in the initial mixing length, as indicated by concentration measurements made at 4 different points in the channel.

The third example makes use of tracer experiments from the River Rhine. Table 18.3 provides details of the reaches, ADZ model structures and fit. Figure 18.3 provides a demonstration of the application of the model in calibration.

Of greater interest, of course, is how well the model can perform in prediction. It is well known that both the mean advective velocity and the dispersive characteristics of rivers can change strongly and nonlinearly with discharge. There is some

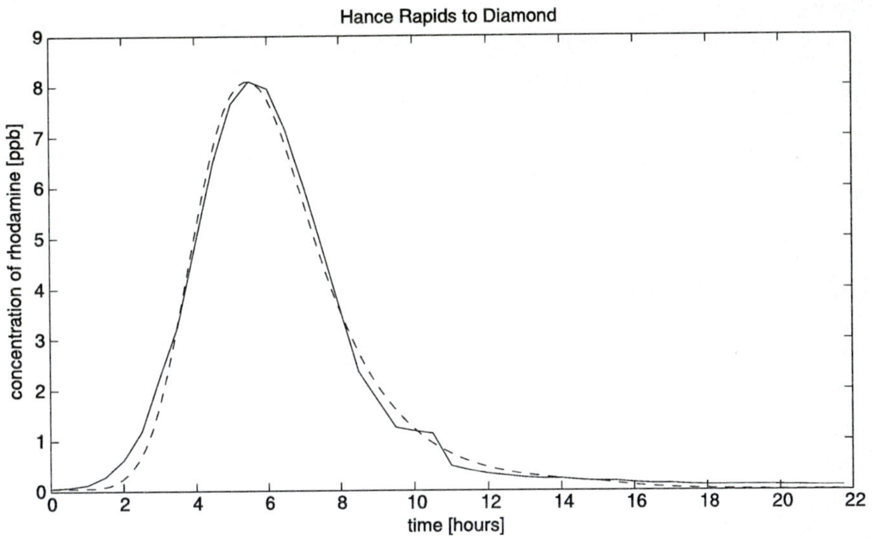

Fig. 18.1 Fit of ADZ model to tracer data during the Glen Canyon flood release, Hance Rapids to Diamond Creek reach, 1996

Fig. 18.2 Fit of ADZ model to tracer data from the Pirna to Dresden reach on the River Elbe

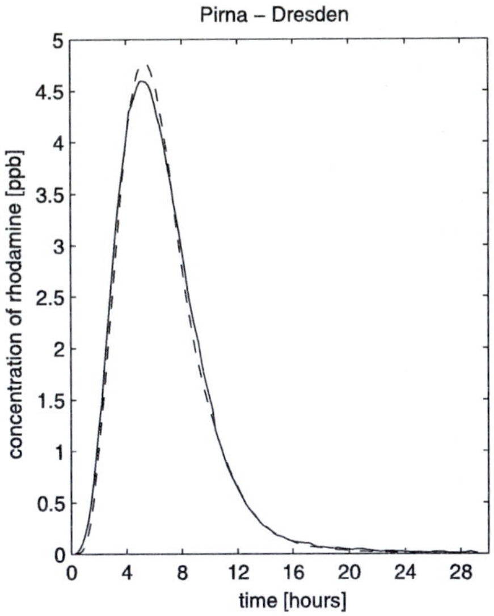

evidence that, in the case of the ADZ model, the Dispersive Fraction is a near constant with discharge in many reaches where multiple tracing experiments have been carried out [13, 36]. There would appear to be no real theoretical reason why this should be the case; it is simply an empirical result. From the definition of the Dispersive Fraction, this means that since the advective time delay is changing with discharge, the mean residence time of the aggregated dead zone is changing with

Table 18.2 Application of the ADZ model to data from two tracer experiments in the River Elbe. Models fitted to input and output data for each reach length. Discharge in June 1997 at the Dresden gauge was 330 m^3/s and in November 1997, 127 m^3/s. Distances in river km from November 1997 injection point at Strekov. Advective time delays (τ) in hours

Reach	River km	Reach length	June 1997				November 1997			
			(n, m, τ)	RT2	YIC	DF	(n, m, τ)	RT2	YIC	DF
Veseli	9									
to Dobkovice	20	11					(1,1,3.3)	0.981	−10.63	0.213
to Loubi	29	9					(1,1,2.2)	0.991	−13.27	0.230
to Hrensko	39	10					(0,1,3.8)	0.976	−12.10	0.185
to Bad Schandau	52	13					(1,1,3.6)	0.997	−14.97	0.177
to Pirna	75	23	(1,1,4.9)	0.994	−14.15	0.235	(1,1,7.3)	0.995	−14.98	0.173
to Dresden	98	23	(1,1,5.0)	0.992	−13.73	0.215	(1,1,7.3)	0.996	−15.03	0.169
to Scharfenberg	116	18	(1,1,4.0)	0.996	−15.66	0.186	(0,1,5.6)	0.994	−15.52	0.151
to Riesa	147	31	(1,1,7.3)	0.996	−15.16	0.180	(1,1,10.0)	0.992	−12.93	0.141
to Muhlberg	167	20	(1,1,4.4)	0.998	−16.02	0.157	(1,1,6.6)	0.998	−15.65	0.134
to Pretzch	224	57	(1,1,13.5)	0.994	−14.74	0.154	(1,1,17.5)	0.990	−13.60	0.126
to Wittenberg	254	30	(1,1,6.9)	0.980	−11.53	0.159	(0,1,10.4)	0.985	−14.26	0.117
to Rosslau	298	44	(1,1,13.3)	0.989	−13.27	0.158	(0,1,15.3)	0.976	−13.43	0.109
to Barby	331	33	(1,1,8.4)	0.992	−13.80	0.160	(1,1,11.6)	0.996	−15.39	0.121
to Magdeburg	358	27	(0,1,8.5)	0.990	−15.33	0.148	(1,1,9.0)	0.997	−15.04	0.119
to Niegripp	385	27	(0,1,5.7)	0.982	−14.26	0.169	(0,1,8.0)	0.985	−14.65	0.116
to Tangermunde	429	44	(1,1,13.9)	0.996	−15.19	0.157	(0,1,17.2)	0.983	−14.54	0.118
to Sandau	456	27	(0,1,9.4)	0.985	−14.78	0.156	(1,1,10.9)	0.994	−14.16	0.119
to Wittenberge	493	37	(0,1,14.5)	0.974	−13.75	0.134	(0,1,15.3)	0.973	−13.77	0.107
to Lenzen	524	31	(1,1,9.8)	0.972	−11.35	0.144	(0,1,11.7)	0.980	−14.46	0.111
to Bleckede	590	66	(0,1,25.0)	0.966	−13.55	0.138	(1,1,27.7)	0.936	−10.53	0.116
to Lauenburg	609	19	(0,1,6.8)	0.986	−15.37	0.135	(0,1,8.0)	0.983	−15.00	0.119
to Geesthacht	625	16	(0,1,9.6)	0.969	−13.90	0.130	(0,1,21.0)	0.970	−14.08	0.103

discharge with a similar scaling. In the ADE model, in contrast, the dispersion coefficient changes approximately with the square of discharge (see e.g. [39]), making it more difficult to extrapolate to different discharges, especially if only one or two tracer experiments are available (as is often the case in larger rivers).

18.4 Making Use of Surrogate Data at Different Flows

Even so, use of the ADZ model in prediction requires information about the variability of the advective time delay and dispersive fraction with discharge. It is more expensive and logistically more difficult to carry out tracing experiments in larger

Table 18.3 Application of the ADZ model to tracer data from the River Rhine. Models fitted to input and output data for each reach length with time step in hours. Distances in river km. Advective time delays (τ) in hours. June 1991 tracer experiment to Mainz analysed separately for data collected at 4 different sites in the cross-section

Reach	River km	Reach length	(n, m, τ)	RT2	YIC	DF
April/May 1989, 1100 m^3/s at Rheinfelden						
Bruecke Neuenburg	199.25					
to Fessenheim	211.1	11.85	(1,1,2.83)	0.986	−11.83	0.151
September 1990, 950 m^3/s at Koblenz						
Koblenz	591					
to Bimmen	865	274	(1,1,71)	0.978	−11.82	0.339
June 1991, 1800 m^3/s at Speyer						
Speyer	400.0					
to Mainz	498.5	98.5				
L1 left bank			(1,1,20)	0.993	−12.77	0.407
L2 right bank			(1,1,22)	0.987	−12.24	0.460
L3 middle left			(1,1,20)	0.992	−12.39	0.412
L4 middle right			(1,1,22)	0.997	−14.43	0.412

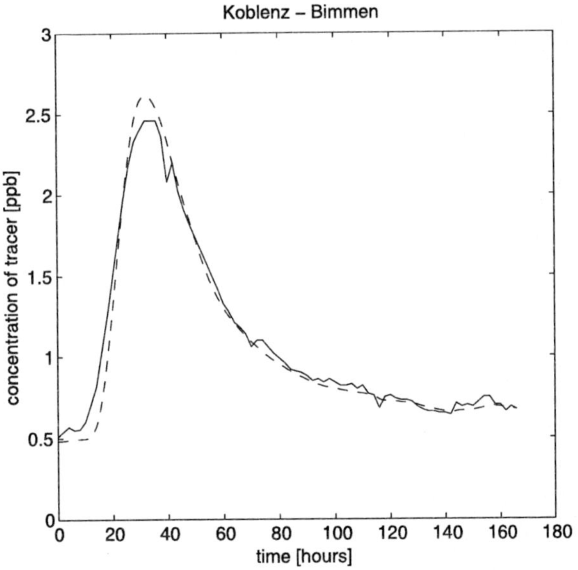

Fig. 18.3 ADZ model fit to tracer data for the Koblenz (discharge 950 m^3/s) to Bimmen (discharge 990 m^3/s) reach of the River Rhine, September 1990

rivers so that (with some exceptions, such as the tracer experiment programme carried out by the United States Geological Survey) the available data on large rivers is relatively sparse. It might also be possible, however, to make use of surrogate data where distinct episodes of changes in water quality can be tracked at successive

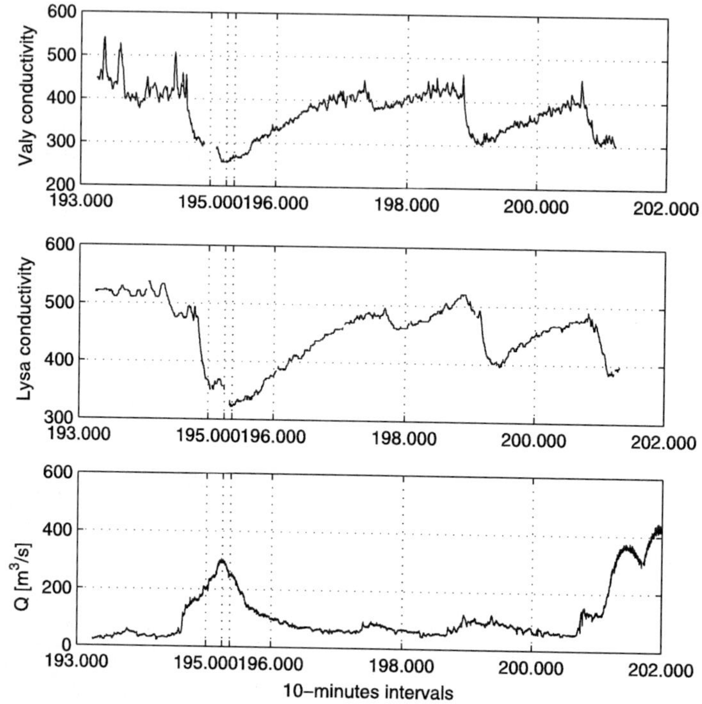

Fig. 18.4 Identification of a transport event using surrogate (electrical conductivity) data: Valy to Lysa reach, River Elbe, Czech Republic

Fig. 18.5 Prediction of advective time delay based on regression of uncertain values from small pollution incidents (*closed circle*), surrogate data at higher non-steady flows (*open circles*), first tracer experiment (*square*), and second tracer experiment used as test (*diamond*): Valy to Lysa reach, River Elbe, Czech Republic

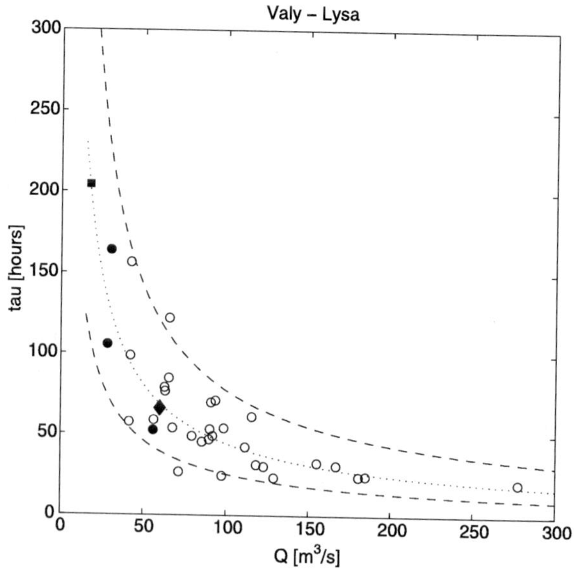

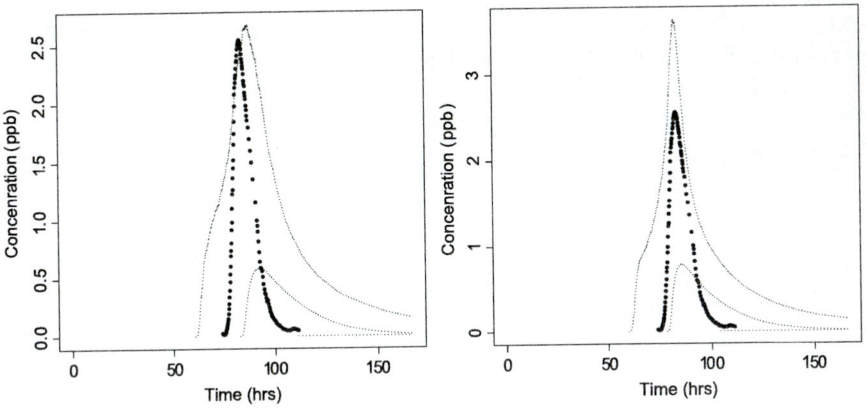

Fig. 18.6 Prediction of a new tracer experiment based on statistical regression of Fig. 18.5: Valy to Lysa reach, River Elbe, Czech Republic. *Left panel*: after excluding first tracer experiment from the regression. *Right panel*: including first tracer experiment in regression

measurement sites downstream. Such surrogate data will be subject to significantly greater uncertainty than planned tracer experiments but Smith et al. [31] have shown that such episodes can still provide a useful constraint on the uncertainty in the prediction of future pollution incidents.

An example of the identification of a time delay from logged electrical conductivity data at the Valy and Lysa sites on the River Elbe is shown in Fig. 18.4. The chosen perturbation in conductivity occurs just following the peak of the hydrograph. Since the ADZ model has an intrinsic assumption that the flow can be treated as steady some account has to be taken of the changing discharge in assessing the change in dispersion characteristics with flow. Smith et al. show how this can be done by allowing for the uncertainty in the observations within a Bayesian statistical regression framework. Figure 18.5 shows the resulting relationship between advective time delay and discharge for this 76.5 km reach, based on different types of data including one tracer experiment, three small pollution incidents and surrogate data of the type shown in Fig. 18.4.

18.5 Predicting Dispersion: Extrapolation to an Arbitrary Discharge

Based on this type of relationship, predictions can be made of dispersion at other discharges. Figure 18.6 shows the results of predicting the concentration curve for a new tracing experiment conditional on the uncertainties in advective time delay shown in Fig. 18.5. Estimating the dispersive fraction, however, requires fitting of the full ADZ model, and was carried out for the three pollution incidents and an earlier tracer experiment. In prediction, the dispersive fraction was then represented by the mean and variance of these estimates for all discharges. The left hand panel of Fig. 18.6 shows the predictions made without taking account of the first tracer experiment; the right hand panel demonstrates the increased accuracy and reduced uncertainty achieved when the information in the first tracer experiment is included. Even so, and despite the uncertainties demonstrated in Fig. 18.5, the dispersive fraction has been overestimated in the mean and the predictions show a longer tail than the observed tracer concentrations.

A further issue in predicting the transport of an arbitrary pollutant is that the gain in a reach might be uncertain due to processes affecting the tracer mass, such as sorption, chemical reactions, or volatilization. This was not an issue with the tracer experiment in Fig. 18.6 since the tracer was chosen so as to be largely conservative. For other pollutants the gain might be less than one, and the estimation of the gain might also be a source of significant uncertainty in predicting transport of that pollutant. This issue arises, of course, with all models of fluvial transport of pollutants, not just the ADZ model.

18.6 Using Fuzzy Regression for the Prediction of Advective Time with Uncertainty

There are limitations to the use of surrogate data in constraining the uncertainty in the prediction of changing ADZ parameters with discharges. On the lower reaches of larger rivers the surrogate data (i.e. consistent changes in water quality parameters which can be recognised in two water quality monitoring stations) can be identified mostly during high flow (smaller flood) discharges. On the Elbe below the monitoring station Lysa, the next downstream station is Obristvi (at a distance of 24 km) which is placed just upstream the confluence with the Vltava River. The Vltava is a major tributary of the Elbe (in fact it usually contributes the greater proportion of the downstream discharge). The last Czech water quality station on the Elbe before the German frontier is then Decin at a distance of 102 km from Obristvi. At these sites surrogate data are mostly masked by other water quality fluctuations while (small) pollution incidents can mostly only be identified at one site, so cannot be used for estimating either the advective time delay or the dispersive fraction.

In such situations, the number of data points that can be used to estimate the change in advective time delay and dispersive fraction with discharge will generally be small. In the stretch Lysa to Obristvi to Decin on the Elbe, there have been 4 cases of flood discharges where drops in conductivity could be recognised at all three stations in addition to the tracer data reported in Table 18.3. There is also data on a pollution incident (P-PO4) that can be tracked downstream, but this has been left for validating the resulting ADZ dispersion estimates. All the data, however, have some problems with uncertainty for various reasons (including problems of changing discharge in space and time at high flows and incomplete concentration curves for the tracer experiments), but there are fewer points than in the study of Smith et al. [31].

In this situation, the assumptions of the statistical regression may not be valid and an alternative approach was taken using the fuzzy regression method HBS2

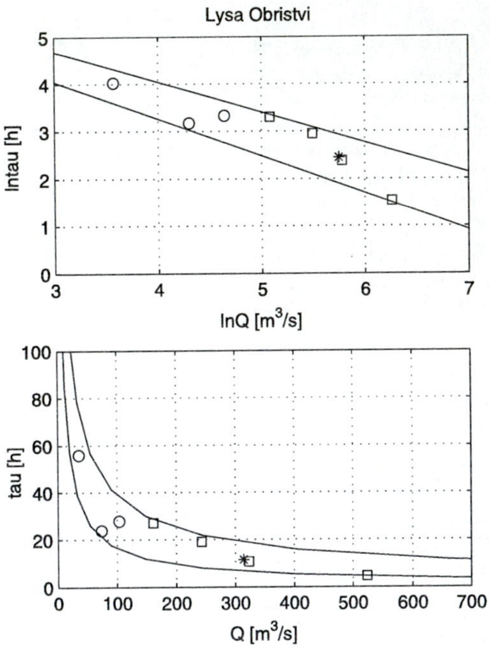

Fig. 18.7 Fuzzy regression of advective time delay on discharge for surrogate data (conductivity during flood discharges), for tracer data and for all data together for the reach Lysa to Obristvi, River Elbe, Czech Republic. The validation P-PO4 pollution incident is also shown but was not included in calculation of regression

(Hojati et al. [19]) which allows both independent and dependent variables to be specified with uncertainty. The fuzzy regression is solved using linear programming in the Excel solver. The results are shown in Fig. 18.7 and Fig. 18.8 for the reaches Lysa to Obristvi and Obristvi to Decin respectively. In each case there were 4 high discharge points from surrogate data and 3 points at lower discharges from the tracing experiments. The advective time delay of the P-PO4 pollution incident is also plotted in the figures for comparison.

The HBS2 regression has been first carried out separately for surrogate data (high discharges) and tracing data (lower and middle discharges) and then all points, except the validation incident, have been evaluated together. For the Lysa to Obristvi reach the regression uncertainty bounds using all the data points are very near to the bounds from tracing alone. It can also be seen from these plots that two of the tracer experiments were rather similar in both discharge and travel time. The real pollution incident being used as validation happened at a higher discharge than any of the tracings but is within the fuzzy uncertainty bounds (and shows almost identical advective time delay to one of the surrogate data points).

The reach Obristvi to Decin is more difficult. Here, the regression has a large uncertainty in the surrogate regression on the intercept and small uncertainty on the slope; the bounds, however, include the validation incident point. The tracer regression bounds and the all-over regression have the uncertainty predominantly on the slope, but they are very different from each other. The tracer regression does not comprise the validation point and the all-over regression misses one surrogate point, albeit by only a small amount in both cases.

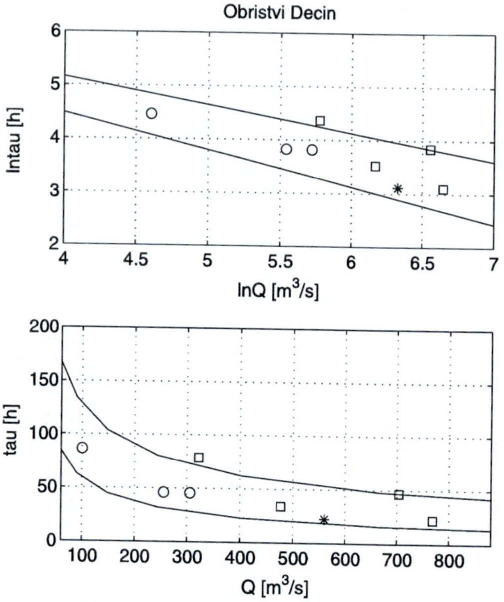

Fig. 18.8 Fuzzy regression of advective time delay on discharge for surrogate data (conductivity and UV absorbance during flood discharges), for tracer data and for all data together for the reach Obristvi to Decin, River Elbe, Czech Republic. The validation P-PO4 pollution incident is also shown but was not included in calculation of regression

Prediction of the validation incident for the two reaches, scaled such that the predicted mass matches the observed, is shown in Fig. 18.9. The ADZ models are sampled from the fuzzy estimates of τ in Fig. 18.7 and Fig. 18.8 and the range of dispersive fraction determined from the available tracer experiments. The uncertainty in the predictions, arising from the uncertainty in predicting both τ and the dispersive fraction is significant, but in both cases spans the observed transport for this incident. In the Obristvi to Decin reach, the observed incident concentrations are only just within the range of the predictions. Figure 18.8 shows, however, that the surrogate data indicates that much slower transport is possible at similar discharges. At the time of the incident the input from the Vltava tributary was adding more than 50% of the total discharge in the main Elbe downstream of Obristvi.

18.7 Conclusions

The results presented in this paper have shown how the ADZ model concepts, originally developed by Peter Young and Tom Beer, can be applied in the analysis and prediction of transport and dispersion in large rivers. In analysis, the model provides much better predictions of concentration curves than the simple ADE model, particularly in reproducing the long tails often seen in experimental tracer data, and is simpler to calibrate and apply than the transient storage form of the ADE.

Tracer experiments in large rivers are, however, expensive, and it has been shown how the information provided from the analysis of tracers can be augmented by the use of pollution incidents and more readily logged surrogate water quality indicators such as electrical conductivity while allowing for the inherent uncertainties in

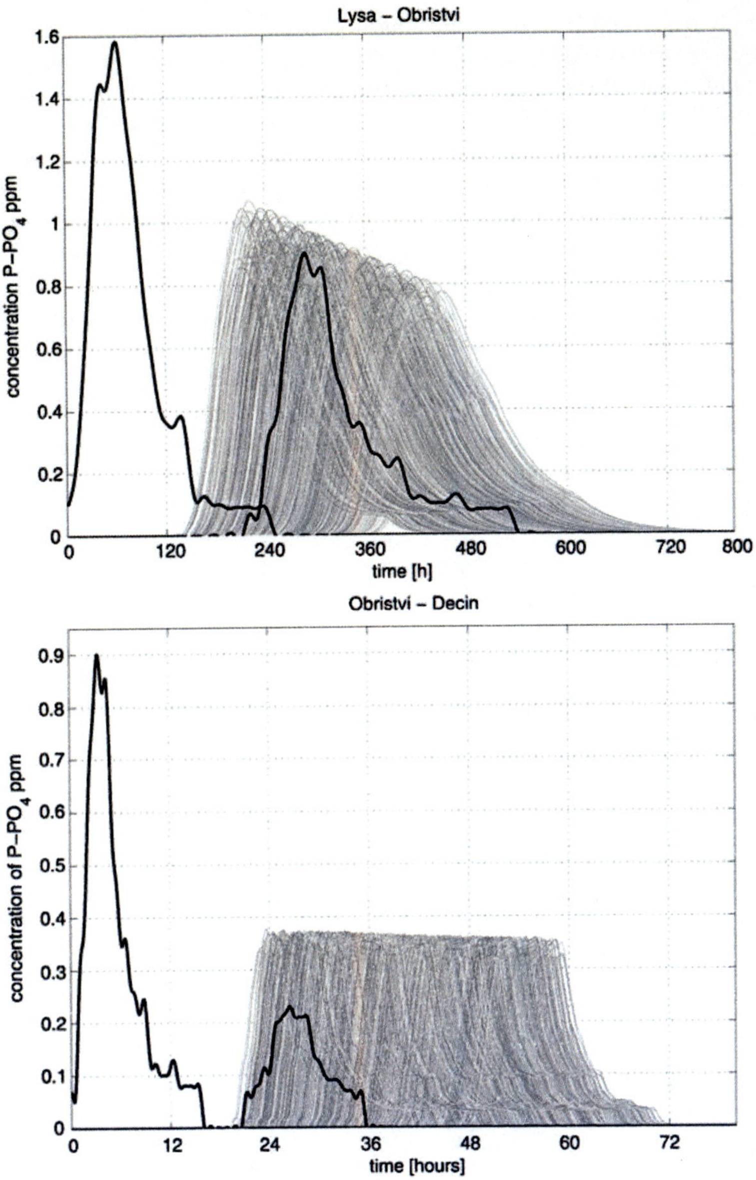

Fig. 18.9 ADZ model prediction of a P-PO4 pollution incident in (*top*) the Lysa—Obristvi and (*bottom*) Obristvi to Decin reaches, River Elbe, Czech Republic, based on the full data sets of Fig. 18.7 and Fig. 18.8

such data and associated (changing) discharges. With large numbers of data points at different discharges, a form of statistical regression has been used to predict the uncertain change of travel times with discharge. The technique allows uncertainty to be considered on both dependent and independent variables. With only small numbers of points, a fuzzy regression has been used. The successful validation prediction of a real P-PO4 pollution incident, given only a small amount of tracer and surrogate data in reaches of the River Elbe has been demonstrated.

There is further research to be done with the ADZ model. In particular, to have a more generally useful pollution incident prediction tool, it would be very useful to relate the advective time delay and dispersive fraction to the physical and hydraulic characteristics of a reach. Neither can be easily linked to the more useful uniform flow estimates of mean velocity in a reach; both will be affected by the full three-dimensional geometry of the reach at different discharges. In addition, it would be useful to develop a database of the likely gains to be expected in predicting the transport of different non-conservative solutes in such rivers to include in prediction, such as that provided by the original ADZ-Protect software [5]. At present, there is very little information on which to base such estimates. Predictions can, of course, be made *as if* the solute is conservative to provide estimates of arrival and duration of an incident at, say, a downstream water intake.

Acknowledgements KB first met Peter Young when he was passing through the University of Virginia on his way back to England to take up the post of Head of Environmental Sciences at Lancaster University in 1981, our talk soon turned to the analysis of tracing experiments. We wrote a grant proposal to start the joint work on the ADZ model when KB also returned to the UK at the Institute of Hydrology, before joining Peter at Lancaster in 1985.

BfG Koblenz, Elbe River Board, TGM WRI and Lancaster University performed the tracer experiments.The surrogate data has been collected by the Elbe River Board. SB has been supported by a grant of Ministry of Environment of the Czech Republic SP/2e7/229/07. Information about tracing experiments on the River Rhine were supplied by A. van Mazijk.

References

1. Beer, T., Young, P.C.: Longitudinal dispersion in natural streams. J. Environ. Eng. **109**, 1049–1067 (1983)
2. Bencala, K.E., Walters, R.A.: Simulation of solute transport in a mountain pool-and-riffle stream: a transient storage model. Water Resour. Res. **19**, 718–724 (1983)
3. Beven, K.J., Buckley, K.M., Young, P.C.: ADZ-analysis manual part I. CRES Technical Report TR/90, Lancaster University (1991)
4. Beven, K.J., Young, P.C.: An aggregated mixing zone model of solute transport through porous media. J. Contam. Hydrol. **3**, 129–143 (1988)
5. Buckley, K.M., Beven, K.J.: ADZ-protect manual. CRES Technical Report TR/92, Lancaster University, UK (1991)
6. Buckley, K.M., Beven, K.J., Young, P.C., Benner, S.: ADZ-analysis manual part II. CRES Technical Report TR/91, Lancaster University, UK (1991)
7. Costa, J.R., Young, P., French, P.: Aggregated dead zone (ADZ) interactive water-quality model for the Ave River. Water Sci. Technol. **19**, 1213–1224 (1987)
8. Day, T.: Longitudinal dispersion in natural channels. Water Resour. Res. **11**, 909–918 (1975)
9. Fischer, H.R., List, E.J., Koh, R.C.Y., Imberger, J., Brooks, N.H.: Mixing in Inland and Coastal Waters. Academic Press, New York (1979)
10. Godfrey, R.G., Frederick, B.J.: Stream dispersion at selected sites. USGS Prof. Paper 433-K, Washington, DC (1970)

11. Green, H.M., Beven, K.J.: Prediction of times of travel for a pollution incident on the River Eden in March 1993. Report for the North West Region, National Rivers Authority, CRES Technical Report TR/99, Lancaster University (1993)
12. Graf, J.B.: Measured and predicted velocity and longitudinal dispersion at steady and unsteady flow, Colorado River, Glen Canyon Dam to Lake Mead. Water Resour. Bull. **31**(2), 265–281 (1995)
13. Green, H.M., Beven, K.J., Buckley, K., Young, P.C.: Pollution incident prediction with uncertainty. In: Beven, K.J., Chatwin, P.C., Millbank, J.H. (eds.) Mixing and Transport in the Environment, pp. 113–140. Wiley, New York (1994)
14. Green, H.M.: Unpublished PhD thesis, Lancaster University, UK (1997)
15. Guymer, I.: A national database of travel time. Dispersion and Methodologies for the Protection of River Abstractions, Environment Agency R & D Technical Report P346, ISBN 1 85705 821 6 (2002)
16. Guymer, I., O'Brien, R.T.: Longitudinal dispersion due to surcharged manhole. J. Hydraul. Eng. **126**, 137–149 (2000)
17. Guymer, I., O'Brien, R., Harrison, C.: Representation of solute transport and mixing within a surcharged benched manhole using an aggregated dead zone (ADZ) technique. Water Sci. Technol. **34**, 95–101 (1996)
18. Guymer, I., O'Brien, R.T.: The effects of surcharged manholes on the travel time and dispersion of solutes in sewer systems. Water Sci. Technol. **31**, 51–59 (1995)
19. Hojati, M., Bector, C.R., Smimou, K.: A simple method for computation of fuzzy linear regression. Eur. J. Oper. Res. **166**, 172–184 (2005). doi:10.1016/j.ejor.2004.01.039
20. Höttges, J., Wallis, S.G., Guymer, I.: Das ATZ-Modell zur Vorhersage des Schadstofftransports in Flüssen'. Wasserwirtschart, pp. 494–497 (October 1992) (in German)
21. Konieczki, A.D., Graf, J.B., Carpenter, M.C.: Streamflow and sediment data collected to determine the effects of a controlled flood in March and April 1996 on the Colorado River between Lees Ferry and Diamond Creek, Arizona: U.S. Geological Survey Open—File Report 97–224, 55 p. (1997)
22. Lees, M.J., Camacho, L.A., Whitehead, P.: Extension of the QUASAR river water quality model to include dead zone mixing. Hydrol. Earth Syst. Sci. **2**, 353–365 (1998)
23. Lees, M., Camacho, L.A., Chapra, S.: On the relationship of transient storage and aggregated dead zone models of longitudinal solute transport in streams. Water Resour. Res. **36**, 213–224 (2000). doi:10.1029/1999WR900265
24. Nordin, C.F., Sabol, G.L.: Empirical data on longitudinal dispersion in rivers. In: USGS Water Resource Investigation, Lakewood, Colorado, pp. 20–74 (1974)
25. Nordin, C.F., Troutman, D.M.: Longitudinal dispersion in rivers: the persistence of skewness in observed data. Water Resour. Res. **16**, 123–128 (1980)
26. Osuch, M., Romanowicz, R., Wallis, S.G.: Uncertainty in the relationship between flow and parameters in models of pollutant transport. Paper presented at 28th International School of Hydraulics, Krag, Poland, 23–26 September and published in Monographic, Volume E-10(406), Institute of Geophysics, Polish Academy of Sciences, pp. 127–138 (2008)
27. Reynolds, C.S., Carling, P.A., Beven, K.J.: Flow in river channels: new insights into hydraulic retention. Arch. Hydrobiol. **121**, 171–179 (1991)
28. Richardson, K., Carling, P.A.: The hydraulics of a straight bedrock channel: insights from solute dispersion studies. Geomorphology **82**, 98–125 (2006). doi:10.1016/j.geomorph.2005.09.022
29. Romanowicz, R.J., Osuch, M., Wallis, S.: Modelling of pollutant transport in rivers under unsteady flow. In: Proceedings of IAHR European Division Conference, Edinburgh, UK, May 2010 (2010). Paper FPIIb
30. Rutherford, J.C.: River Mixing. Wiley, Chichester (1994)
31. Smith, P.J., Beven, K.J., Tawn, J., Blazkova, S., Merta, L.: Discharge dependent pollutant dispersion in rivers: estimation of ADZ parameters with surrogate data. Water Resour. Res. **42**, W04412 (2006). doi:10.1029/2005WR004008

32. Spreafico, M., van Mazijk, A.: Alarmmodell Rhein ein modell für die operationelle Vorhersage des Transportes von Schadstoffen im Rhein. Bericht Nr. I-12, Kommission für die Hydrologie des Rheins, Lelystad (1993)
33. Thackston, E.L., Schnelle, K.B.: Predicting effects of dead zones on stream mixing. J. Sanit. Eng. Div. ASCE **96**, 319–331 (1970)
34. Valentine, E.M., Wood, I.R.: Longitudinal dispersion with dead zones. J. Hydraul. Eng. Div. ASCE **103**, 975–990 (1977)
35. van Mazijk, A., Veling, E.J.M.: Tracer experiments in the Rhine Basin: evaluation of the skewness of observed concentration distributions. J. Hydrol. **307**, 60–78 (2005)
36. Wallis, S.G., Young, P.C., Beven, K.J.: Experimental investigation of the aggregated dead zone model for longitudinal solute transport in stream channels. In: Proceedings of the Institution of Civil Engineers, Part 2, vol. 87, March, pp. 1–22 (1989)
37. Wallis, S.G.: Aggregated mixing zone modelling of solute transport in rivers. In: Proceedings of the Fourth National Hydrology Symposium, Cardiff, 13–16 September, pp. 5.9–5.13 (1993)
38. Wallis, S.G., Clarke, R.F.: Hydrological modelling of solute transport in the river Rhine. In: Proceedings of the 5th National BHS Hydrology Symposium, Edinburgh, 4–7 September, pp. 9.31–9.36 (1995)
39. Whitehead, P.G., Williams, R.J., Hornberger, G.M.: On the identification of pollutant or tracer sources using dispersion theory. J. Hydrol. **84**, 273–286 (1986)
40. Young, P.C.: Parallel processes in hydrology and water quality: a unified time-series approach. J. Inst. Water Eng. Manag. **6**, 598–612 (1992)
41. Young, P.C.: Recursive Estimation and Time-Series Analysis. Springer, Berlin (1984)
42. Young, W.F., Wallis, S.G.: The aggregated dead zone (ADZ) model for dispersion in rivers. In: Int. Conf. on River Quality Modelling in the Inland Natural Environment (Bournemouth) (1986). BHRA Paper LI 421-433
43. Young, P.C., Wallis, S.G.: Solute transport and dispersion in channels. In: Beven, K.J., Kirkby, M.J. (eds.) Channel Network Hydrology, pp. 129–174. Wiley, Chichester (1993)
44. Young, P.C., Lees, M.J.: The active mixing volume: a new concept in modelling environmental systems. In: Barnett, V., Turkman, K. (eds.) Statistics for the Environment, pp. 3–34. Wiley, Chichester (1993)

Chapter 19
Stochastic and Robust Control of Water Resource Systems: Concepts, Methods and Applications

Andrea Castelletti, Francesca Pianosi, and Rodolfo Soncini-Sessa

19.1 Introduction

In order for water resources management to effectively cope with all the key drivers of global change (climate, demographic, economic, social, policy/law/institutional, and technology changes), it is essential that the traditional sectoral management approach to water resources is transformed into a new paradigm, where water is considered as the principal and cross cutting medium for balancing food, energy security, and environmental sustainability. One major technical challenge in expanding the scope of water resources management across sectors and to the river basin level is to develop new methodologies and tools to cope with the increasing complexity of water systems. When dealing with large water resources systems, particularly with water reservoir networks, traditional non-linear, stochastic control approaches, like Stochastic Dynamic Programming, suffer from the curse of dimensionality [5] that makes them essentially unusable.

Stochastic Dynamic Programming (SDP) is by far one of the most studied method to design optimal water reservoir operation (see, e.g., [50] and references therein). SDP is based on the formulation of the control policy design problem as a sequential decision making process. The key idea is to use cost-to-go functions to organize and structure the search for optimal policies. A decision taken now can produce not only an immediate cost, but also affect the next system state and, through that, all the

A. Castelletti (✉) · F. Pianosi · R. Soncini-Sessa
Politecnico di Milano, Milano, Italy
e-mail: castelle@elet.polimi.it

F. Pianosi
e-mail: pianosi@elet.polimi.it

R. Soncini-Sessa
e-mail: soncini@elet.polimi.it

L. Wang, H. Garnier (eds.), *System Identification, Environmental Modelling, and Control System Design*,
DOI 10.1007/978-0-85729-974-1_19,

subsequent rewards. SDP is thus based on looking ahead to future events and computing a backed-up value, which is then used to update the value function. The first application of (deterministic) Dynamic Programming to water systems management is probably owed to [24]. Since then, the method has been systematically applied to reservoir management, particularly for hydropower production (see, e.g., [21, 25, 26, 53]). Beginning in the early 1980s, the interest expands to the stochastic version of dynamic programming for multi-purpose reservoirs operation and networks of reservoirs (see the reviews by [59] and the contributions by [23, 45, 47, 56]). The uncertain version of Dynamic Programming (that we will still call SDP) was proposed by [44].

Despite being studied so extensively in the literature, SDP suffers from a dual curse which, de facto, prevents its practical application to even reasonably complex water systems. (i) The computational complexity grows exponentially with state, control and disturbance dimensions (Bellman's *curse of dimensionality* [5]), so that SDP cannot be used with water systems where the number of reservoirs is greater than a few (2–3) units. (ii) An explicit model of each system component is required (*curse of modeling* [9]) to anticipate the effects of the system transitions. Any information included into the SDP framework can only be either a state variable described by a dynamic model or a stochastic disturbance, independent in time, with the associated pdf. Exogenous information, such as temperature, precipitation, snowpack depth, which could effectively improve reservoir operation [27], cannot be explicitly considered in taking the release decision, unless a dynamic model is identified for each additional information, thus adding to the curse of dimensionality (additional state variables). Further, in large reservoir networks, disturbances are very likely to be spatially and temporally correlated. While including space variability in the identification of the disturbance's pdf can be sometimes rather complicated, it does not add to the computational complexity. Conversely, temporal correlation can be properly accounted for by using a dynamic stochastic model, which could be a cumbersome contribution to the curse of dimensionality.

Attempts to overcome the SDP's curses are ubiquitous in the literature, e.g. Dynamic Programming based on Successive Approximations [6], Incremental Dynamic Programming [32], and Differential Dynamic Programming [28]. However, these methods have been conceived mainly for deterministic problems and are of scarce interest for the optimal operation of reservoirs networks, where the uncertainty associated with the underlying hydro-meteorological processes cannot be neglected. A number of authors propose decomposition/aggregation methods for reducing the system to a smaller, computationally tractable one (see, e.g., [1, 48, 55]). Most of these methods, however, exploit some particular topological feature of the system and are thus problem-specific.

In this chapter we review the most advanced, general approaches available to overcome, or at least effectively mitigate, the SDP limits in dealing with large water reservoir networks. They can be classified in two main classes depending on the strategy they adopt to alleviate the dimensionality burden: methods based on the restriction of the degrees of freedom of the control problem (Approximate Dynamic Programming and Policy Search) and methods based on the simplification of the water system model (on-line suboptimal controllers and Reinforcement Learning). Emphasis is given to the technical implications of the very nature of the water reservoir systems on the problem formulation and solution. Indeed, reservoir networks are high dimension and highly non-linear systems, affected by strong uncertainties, with multiple objective functions usually non-linear, strongly asymmetric and to be defined over an infinite horizon. Further, their operation directly or indirectly involves human beings (decision-makers and stakeholders), whose risk aversion and preference structure influence formulation and solution of the problem. Finally, for each approach a real world numerical application is briefly presented.

19.2 Problem Formulation

We consider a water system composed of reservoirs, natural catchments feeding the reservoirs, diversion dams, water users (e.g. hydropower plants, irrigation districts), and artificial and natural canals that connect all the above components. Even if the physical processes that are involved in the system are time-continuous, the model is time-discrete as decisions are taken at discrete instants of time. The decision time-step is usually one week or one day and, in any case, not smaller than few hours, because of the physical constraints in the implementation of the decision (e.g. dam's gate operation). The system dynamics is thus given by the state transition equation

$$\mathbf{x}_{t+1} = f(\mathbf{x}_t, \mathbf{u}_t, \boldsymbol{\varepsilon}_{t+1}, t), \qquad (19.1)$$

where $\mathbf{x}_t \in \mathbb{R}^{n_x}$ and $\mathbf{u}_t \in U_t \subseteq \mathbb{R}^{n_u}$ are the state and control vectors at time t; and $\boldsymbol{\varepsilon}_{t+1} \in \mathbb{R}^{n_\varepsilon}$ is the disturbance[1] acting in the time interval $[t, t+1)$. The state vector $\mathbf{x}_t$ is composed of the reservoir storages and the state variables of catchments, canals, and water users. The control vector includes the release decisions at the reservoir outlet and the distribution decisions at the regulated diversion dams. The disturbance vector $\boldsymbol{\varepsilon}$ collects the random disturbances acting in the system, e.g. climate or hydrological inputs, and error terms in the model of the system. It can be described either as an uncertain or a stochastic variable and modelled by membership-set Ξ_t or a pdf $\phi_t(\cdot)$ respectively. At each time t, either Ξ_t and $\phi_t(\cdot)$ may be a function of state and control, that is

$$\boldsymbol{\varepsilon}_{t+1} \sim \phi_t(\cdot\,|\mathbf{x}_t, \mathbf{u}_t) \quad \text{or} \quad \boldsymbol{\varepsilon}_{t+1} \in \Xi_t(\mathbf{x}_t, \mathbf{u}_t). \qquad (19.2)$$

For each of the m issues that have to be considered in operating the system (e.g. agricultural and hydropower production, flood control, ecological services) an objective function J^i, with $i = 1, \dots, m$, can be defined to express the cost payed over the time horizon $[0, h]$

$$J^i = \mathop{\Psi}_{\boldsymbol{\varepsilon}_1,\dots,\boldsymbol{\varepsilon}_h} \left[\Phi\left(g_0^i(\mathbf{x}_0, \mathbf{u}_0, \boldsymbol{\varepsilon}_1), \dots, g_{h-1}^i(\mathbf{x}_{h-1}, \mathbf{u}_{h-1}, \boldsymbol{\varepsilon}_h), g_h^i(\mathbf{x}_h)\right)\right], \qquad (19.3)$$

[1] According to the notation adopted, the time subscript of a variable indicates the instant when the its value is deterministically known.

where $g_t^i(\cdot)$, for $t = 1, \dots, h-1$ and with $i = 1, \dots, m$, are step-cost functions associated to the transitions from t to $t+1$, $g_h^i(\cdot)$ is a penalty function over the final state, Φ is an operator for aggregation over time and Ψ a statistic used to filter the disturbance. Common choices for Φ are the sum ($\Phi = \Sigma$) and the maximum ($\Phi = \max$), whereas for Ψ, the expected value is often used ($\Psi = E$), but the maximum ($\Psi = \max$) is preferred when the stakeholders are risk averse [41, 51]. In principle, all the combinations of these operators can be considered; in practice, only two are of interest in real-life applications: $\Psi = E$ and $\Phi = \Sigma$ (Laplace problem), and both Ψ and Φ equal to the maximum operator (Wald problem). The control vector is specified by a time-varying control law

$$\mathbf{u}_t = m_t(\mathbf{x}_t) \tag{19.4}$$

and the aim of the control problem is to define the sequence of control laws $m_t(\cdot)$ over the horizon $[0, h-1]$, i.e. the control policy

$$p = [m_0(\cdot), \dots, m_{h-1}(\cdot)]. \tag{19.5}$$

The optimal control problem is formulated as

$$\min_p [J^1, J^2, \dots, J^m] \tag{19.6}$$

subject to constraints (19.1), (19.2), (19.4), (19.5), and with $\mathbf{x}_0$ given. The pdf formulation in equation (19.2) is used when the expected value is adopted as filtering criterion Ψ in equation (19.3); the membership-set formulation is used when Ψ is the maximum. Note that the control variable is unconstrained because unfeasible decisions are not transformed into feasible ones due to the form of the reservoir's model.

The control problem (19.5) is a multi-objective (MO) optimization problem, whose solution is the set $\mathcal{P}$ of Pareto optimal (efficient) policies (see, e.g., [36]). Each policy in $\mathcal{P}$ can be computed by solving the following single (aggregate) objective (SO) optimal control problem:

$$\min_p J \tag{19.7}$$

subject to constraints (19.1), (19.2), (19.4), (19.5), and with $\mathbf{x}_0$ given, with

$$J = \Psi_{\boldsymbol{\varepsilon}_1, \dots, \boldsymbol{\varepsilon}_h} \big[\Phi\big(g_0(\mathbf{x}_0, \mathbf{u}_0, \boldsymbol{\varepsilon}_1), \dots, g_{h-1}(\mathbf{x}_{h-1}, \mathbf{u}_{h-1}, \boldsymbol{\varepsilon}_h), g_h(\mathbf{x}_h)\big)\big], \tag{19.8}$$

where $g_t(\cdot)$ and $g_h(\cdot)$ are the aggregate step-cost and penalty functions obtained from $g_t^i(\cdot)$ and $g_h^i(\cdot)$ (with $i = 1, \dots, m$) according to the aggregation method (see, e.g., [34]) used to re-conduct the MO problem to a SO problem. The choice of this method is constrained by the formulation adopted for the problem, particularly, by the choice of the filtering operator Ψ.

In the water resources context, the choice of the time horizon and the penalty function $g_h(\mathbf{x}_h)$ might be critical since the life time of the system is infinite. Generally, the adoption of an infinite horizon, which vanquishes the influence of the penalty, is recommended. When the model of the system and all the step-cost functions are cyclostationary with period T, the problem on the infinite horizon is well-posed and the solution is a periodic control policy. The SO problem over an infinite horizon is formulated as

$$\min_{p} \lim_{h\to\infty} J \tag{19.9}$$

subject to (19.1), (19.2), (19.4), given $\mathbf{x}_0$, and

$$p = [m_0(\cdot), \dots, m_{T-1}(\cdot)] \tag{19.10}$$

instead of (19.5). If $\Phi = \Sigma$ in (19.8), the objective function must be adjusted in order to avoid divergence because it is not guaranteed that the controlled system will converge to a stable cycle. To overcome this difficulty, the objective function can be defined as the Total Discounted Cost (TDC)

$$J = \lim_{h\to\infty} \underset{\varepsilon_1,\dots,\varepsilon_h}{\Psi} \left[\sum_{t=0}^{h} \gamma^t g_t(\mathbf{x}_t, \mathbf{u}_t, \varepsilon_{t+1}) \right], \tag{19.11}$$

with $0 < \gamma < 1$, or as the Average Expected Value (AEV)

$$J = \lim_{h\to\infty} \underset{\varepsilon_1,\dots,\varepsilon_h}{\Psi} \left[\frac{1}{h+1} \sum_{t=0}^{h} g_t(\mathbf{x}_t, \mathbf{u}_t, \varepsilon_{t+1}) \right]. \tag{19.12}$$

The TDC form gives more weight to the short-term, transient conditions and it is well suited for expressing economic costs, for which the discount factor γ can be easily estimated. The AEV accounts only for the steady-state conditions and should be preferred over the TDC when social or environmental issues are implicated.

19.3 Traditional Problem Solution: The Dynamic Programming Approach

Stochastic Dynamic Programming (SDP) appears to be the most suitable, and one of the more commonly adopted, method for solving problem (19.7). One pillar of SDP success is its wide applicability. Indeed, the only requirements for its application are: (1) the inputs in the model can only be controls or random disturbances, which means that it is not possible to consider (and condition the policy upon) uncontrolled, exogenous, deterministic variables whose value is known in real time (e.g. rainfall measures), unless these are described by a dynamic model and so are not exogenous inputs anymore; (2) the membership-set or the pdf of the disturbance vector must be in the form as in (19.2), i.e. either the disturbance process is independent in time or, at time t, any dependency on the past could be completely accounted for by the value of the state at the same time; and (3) the step-cost functions $g_t(\cdot)$

only depend upon variables defined for the same time interval. As anticipated in the introduction, the first condition leads to the so-called curse of modelling.

The Bellman equation for the SO finite horizon optimal control problem (19.7) is

$$H_t(\mathbf{x}_t) = \min_{u_t} \underset{\varepsilon_{t+1}}{\Psi} \left[\Phi\left[g_t(\mathbf{x}_t, \mathbf{u}_t, \varepsilon_{t+1}), H_{t+1}(\mathbf{x}_{t+1})\right]\right] \tag{19.13}$$

where $H_t(\cdot)$ is the optimal cost-to-go function for the aggregate objective and only the following combinations of Φ and Ψ are considered

$$\Phi[v, w] = v + w \quad \text{and} \quad \Psi = E,$$
$$\Phi[v, w] = \max\{v, w\} \quad \text{and} \quad \Psi = \max .$$

The solution is obtained by initializing $H_h(\mathbf{x}_h)$ with $g_h(\mathbf{x}_h)$ and recursively computing $H_t(\mathbf{x}_t)$ with (19.13). Once the optimal costs-to-go have been computed for all the time instants $t = h - 1, \dots, 0$, the optimal control law at any time t is derived as

$$m_t(\mathbf{x}_t) = \arg\min_{u_t} \underset{\varepsilon_{t+1}}{\Psi} \left[\Phi\left[g_t(\mathbf{x}_t, \mathbf{u}_t, \varepsilon_{t+1}), H_{t+1}(\mathbf{x}_{t+1})\right]\right]. \tag{19.14}$$

If the system is linear, the step-cost functions quadratic and the random disturbance is stochastic and Gaussian, the analytical solution to the Bellman equation is given as in the well known LQG framework. Unfortunately, this result can not be exploited in most of the water field applications, since none of the assumptions of the LQG framework is satisfied and forcing the system description to fit them can dramatically reduce the solution significance [16].

An approximate solution can be obtained by discretizing sets $\mathcal{S}_{x_t}$, $\mathcal{S}_{u_t}$, and $\mathcal{S}_{\varepsilon_t}$, of state, control and disturbance, and numerically solving the Bellman equation (19.13). Uniform discretization is suitable when no information is available about the form of the optimal cost-to-go function $H_t(\cdot)$. The intuition is confirmed by some numerical results [18], which show that the error in the estimation of $H_t(\cdot)$, given the values that it assumes in P points $(\mathbf{x}_t^i, H_t(\mathbf{x}_{t+1}^i))$ with $\mathbf{x}_t^i \in \mathcal{S}_{x_t}$, is proportional to an index, called discrepancy index, which expresses the minimum density of the points $\mathbf{x}_t^i$ among all the subsets of $\mathcal{S}_{x_t}$. For fixed P, the uniform discretization has a low discrepancy index and thus produces a low estimation error. However, when a uniform grid is adopted, $P = N_{x_t}^{n_x}$ and thus the number of points P can not be increased continuously and the distance between two successive values of P increases exponentially with $n_x - 1$. Methods have been developed (see, e.g. [39]) to iteratively produce non-uniform discretizations whose discrepancy index decreases polynomially with P (low-discrepancy sequences).

When an infinite horizon is considered, the idea is still to recursively solve equation (19.13), however the algorithm is started at time $t = 0$, with a suitable initialization for $H_0(\mathbf{x}_0)$, and proceeds backwards in time until the optimal cost-to-go function converges to a periodic function of period T. The initialization can be arbitrary chosen when $\Psi = E$, while it must be equal to

$$H_0(\mathbf{x}_0) = \inf_{\mathbf{x}_t \in \mathcal{S}_{x_t}, \mathbf{u}_t \in \mathcal{S}_{u_t}, \varepsilon_{t+1} \in \mathcal{S}_{\varepsilon_{t+1}}} g_t(\mathbf{x}_t, \mathbf{u}_t, \varepsilon_{t+1}),$$

when $\Psi = \max$. If the TDC formulation (19.11) is used, the operator $\Phi[\cdot,\cdot]$ in the Bellman equation (19.13) must be defined as $\Phi[v,w] = v + \gamma w$, which guarantees that $H_t(\cdot)$ does not diverge. If instead the AEV formulation (19.12) is used, it is not possible to avoid divergence of $H_t(\cdot)$ if it is recursively computed with (19.13). To overcome this difficulty, the idea is to replace $H_t(\mathbf{x}_t)$ with the difference between $H_t(\mathbf{x}_t)$ and the cost-to-go $H_t(\bar{\mathbf{x}}_t)$ of a reference state $\bar{\mathbf{x}}_t$. Based on this idea, the Successive Approximation Algorithm (ASA) has been proposed for either the stationary [58] and cyclostationary [52] case. Asymptotical convergence of both the algorithms is guaranteed under suitable conditions (see [8] for the stochastic case and [44] for the uncertain one) which are always satisfied by real world water systems.

19.3.1 Curse of Dimensionality

The main limit of SDP is the associated computational complexity. Let N_{x_t}, N_{u_t} and N_{ε_t} be the number of elements in the discretized state, control and disturbance sets $\mathcal{S}_{x_t} \subset \mathbb{R}^{n_x}$, $\mathcal{S}_{u_t} \subset \mathbb{R}^{n_u}$ and $\mathcal{S}_{\varepsilon_t} \subset \mathbb{R}^{n_\varepsilon}$: the recursive resolution of (19.13) for K iteration steps (with $K = h$ if the optimization horizon is finite and $K = kT$ if the horizon is infinite, where T is the period and k is usually lower than ten) requires

$$K \cdot \left(N_{x_t}^{n_x} \cdot N_{u_t}^{n_u} \cdot N_{\varepsilon_t}^{n_\varepsilon}\right) \tag{19.15}$$

evaluations of the operator $\Phi[\cdot,\cdot]$ in (19.13). Equation (19.15) shows the so-called curse of dimensionality: the computational complexity grows exponentially with the state, control and disturbance dimensions. This limits the use of SDP to small water systems where the number of reservoirs is smaller than few units.

19.3.2 Set-Valued Policies

Equation (19.14) might have more than one solution. If this is the case, one can compute the set M_t of all these solutions (equivalent optimal controls). Since this set is a function of the state it can be viewed as a set-valued control law and the sequence $P = [M_0(\cdot), \ldots, M_{h-1}(\cdot)]$ is the optimal set-valued policy. Aufiero et al. [2] prove that P is the 'largest' set-valued policy that solves problem (19.7). Determining the general set-valued policy implies almost the same computing time as determining a point-valued policy and can be much more effective. In fact, not only uniqueness of the solution is not necessary, since the control is supposed to be implemented by a human regulator, but it is not even favourable: leaving the regulator the possibility of choosing a control in M_t is preferable since in this way (s)he can consider other information that are available when the release decision is taken (e.g. down-time periods of some plant) but that have not been included in the model of the system when formulating the control problem. The adoption of a set-valued policy approach turns out to be particularly useful also when some priority among the objectives can be established a priori (e.g. accordingly to national regulations). In this event, the optimal control problem (19.7) can be reformulated decomposing it into a hierarchy of q (with $q \leq m$) single or/and multi-objective subproblems (lexicographic approach), each of which is formulated considering as feasible control set the optimal set-value policies obtained by solving the problem at the higher level in the hierarchy.

19.4 Alternate Problem Solutions

Large water systems comprising more than two/three reservoirs are the rule rather than exception and this is the reason why SDP has had a very limited applicability to real world systems. Since the 1970s, many approaches have been proposed to partially remedy the curse of dimensionality and the topic is still actively investigated [46]. In this section, we present a short survey of alternate approaches we recently explored to overcome the curse of dimensionality in the stochastic or uncertain case, and provide some numerical results of their application to different water systems.

These approaches can be classified in two main classes depending on the strategy they adopt to alleviate the dimensionality burden: methods based on the restriction of the degrees of freedom of the control problem and methods based on the simplification of the water system model. With the first, the problem complexity is reduced by assuming a priori some regularity in the structure either of the optimal cost-to-go function (Sect. 19.4.1) or the control policy (Sect. 19.4.2). With the second, the original model of the system is substituted for a simplified, low order version, in which the state dimension is reduced and the lost information is recovered either by using an on-line suboptimal control scheme (Sect. 19.4.3) or a data-driven learning approach (Sect. 19.4.4).

19.4.1 Approximate Dynamic Programming

The key idea of a broad class of approaches usually categorized as Approximate Dynamic Programming (ADP) is to avert the SDP curse of dimensionality by introducing some hypotheses on the regularity of the optimal cost-to-go function. Since SDP requires discretization of the state and decision spaces, one way to mitigate (but not vanquish) the dimensionality problem is to combine a coarser discretization grid with a continuous approximation of the cost-to-go function.

Instead of computing the exact value of $H_t(\cdot)$ for $N_{x_t}^{n_x}$ state values, the idea is to evaluate it in a smaller number ($\tilde{N}_{x_t}^{n_x} < N_{x_t}^{n_x}$) of points and then interpolate such points with a function belonging to a given class of functions. Thereby (19.13) must be replaced by

$$\hat{H}_t(\mathbf{x}_t) = \min_{u_t} \underset{\boldsymbol{\varepsilon}_{t+1}}{\Psi} \Phi\left[g_t(\mathbf{x}_t, \mathbf{u}_t, \boldsymbol{\varepsilon}_{t+1}), \tilde{H}_{t+1}(\mathbf{x}_{t+1})\right], \tag{19.16}$$

where $\tilde{H}_{t+1}(\cdot)$ is an approximation of the optimal cost-to-go function $H_{t+1}(\cdot)$. The approximation is derived from the $\tilde{N}^{n_x}_{x_{t+1}}$ evaluations of $\hat{H}_{t+1}(\cdot)$ made at previous step, by fitting the approximation function to the points $\{(\mathbf{x}^i_{t+1}, \hat{H}_{t+1}(\mathbf{x}^i_{t+1})); i = 1, \dots, \tilde{N}^{n_x}_{x_{t+1}}\}$. Different classes of approximators have been explored in the literature, including linear polynomials [7, 54], cubic Hermite polynomials [22] and splines [30, 42]. As universal function approximators, artificial neural networks are particularly suited for this purpose, as discussed in [9] and practically demonstrated by [13]. They lead to the so called Neural Stochastic Dynamic Programming (NSDP) approach. NSDP can be used for either finite and infinite horizon except for the AEV formulation, since in this case the convergence of the solution algorithm is not guaranteed. As for the other formulations, [9] proved that under broad hypothesis it is guaranteed that the solution $\tilde{H}_{\cdot}(\cdot)$ lies in a bounded neighbourhood of the exact solution $H_{\cdot}(\cdot)$.

NSDP cuts down the computing time by reducing the term N_{x_t} in (19.15); however, the exponential growth with the state dimension n_x is not avoided and so the curse of dimensionality only mitigated. This is why, with a modern computer, NSDP can be used when n_x is indicatively of the order of ten units at most [49]. Some recent experiments [3, 17] have demonstrated that coupling NSDP and state discretization with low-discrepancy sequences allows for solving problems (on a finite or receding horizon!) with even higher state dimension (30 state variables in [17]).

Application Example A comparison between SDP and NSDP is given by application to the river Piave system, Italy. The system is composed of three main artificial reservoirs (total storage 215 Mm^3) fed by a 4,100 km^2 catchment and operated for hydropower production and irrigation supply. The system description requires at least 3 state variables (reservoir storages), 4 controls (3 release decisions and 1 diversion decision) and 3 disturbances (reservoir inflows). Table 19.1 compares the objective values (average annual revenue from power production and annual irrigation deficit) attained with the optimal policies designed with SDP and NSDP, and different discretization grids. SDP with a dense grid of 10, 6 and 7 discrete values for the first, second and third component of the state (for a total of 420 grid points) and NSDP with a coarse grid of 6, 3, 3 values (54 points) achieve almost the same performance, but NSDP is almost 450% faster. SDP with the coarse discretization grid (6, 3, 3) further reduces the computing time but gives significantly worse (from 5% to 100%, depending on the objective) performance. Finally, NSDP with a slightly

Table 19.1 River Piave system: objective values and computing time of SDP and NSDP with different state space discretization grids

Control scheme	Objectives		Computing time [hours]
	Hydropower [€]	Irrigation [m^3]	
SDP420	30999499	751596	9.1300
NSDP54	30995522	752812	2.2825
SDP54	28336389	1301902	1.3300
NSDP140	31401457	751430	6.1325

denser grid (140 points) overcomes SDP with dense grid, at least for the hydropower objective, while still requiring less computing time. More details on the application can be found in [13].

19.4.2 Policy Search

One way to completely overcome the curse of dimensionality is to directly work in the policy domain (Policy Search) and transfer the ADP's key idea of a regularity of the cost-to go function to the policy structure. Strictly, we can assume that, for any t, the control law (19.4) belongs to a given class of functions $\{m(\cdot; \boldsymbol{\theta}_t)\}$, where $\boldsymbol{\theta}_t$ is a vector of parameters to be estimated. The optimal control problem (over a finite horizon) can be formulated as

$$\min_{\boldsymbol{\theta}_0,\ldots,\boldsymbol{\theta}_{h-1}} \mathop{\Psi}_{\boldsymbol{\varepsilon}_1,\ldots,\boldsymbol{\varepsilon}_h} \left[\Phi(g_0(\mathbf{x}_0, \mathbf{u}_0, \boldsymbol{\varepsilon}_1), \ldots, g_{h-1}(\mathbf{x}_{h-1}, \mathbf{u}_{h-1}, \boldsymbol{\varepsilon}_h), g_h(\mathbf{x}_h))\right]$$

subject to constraints (19.1), (19.2), $\mathbf{x}_0$ given, and

$$\mathbf{u}_t = m(\mathbf{x}_t; \boldsymbol{\theta}_t).$$

The same could be done for an infinite horizon cyclostationary problem, were the unknown would be the sequence $[\boldsymbol{\theta}_0, \ldots, \boldsymbol{\theta}_{T-1}]$.

The clear advantage of Policy Search is that the optimal control problem is reconducted to an optimization problem that can be solved by means of traditional Mathematical Programming techniques, Evolutionary Methods (see, e.g., [37]) or other optimization techniques (e.g. [29]). It does not require any discretization and therefore totally avoids the curse of dimensionality. However, the final result depends on the choice of the class of functions (e.g. linear, piecewise linear, fuzzy rule base, etc.) to which the control law is assumed to belong and optimality can not be guaranteed. Reservoir operation practice often provides indications for this choice, which, however, becomes harder as the complexity of the system increases: a review of Policy Search approaches based on empirical experience can be found in [40]. Alternatively, universal approximators (e.g. Artificial Neural Networks) can be used [3].

Application Example Policy Search is applied to designing the operating policy of the Hoa Binh reservoir on the Da River, Vietnam. The reservoir has a live storage of 5.6 billion m^3 and a total catchment area of 169,000 km^2. The main operation objectives are hydropower production, irrigation supply and flood control in the downstream city of Hanoi. The optimal control problem is reconducted to a nonlinear programming problem by using Artificial Neural Network (ANN) to approximate the unknown optimal control law, which we assume to be time-varying and depending on the storage and the previous day flows. The ANN inputs thus are the current reservoir storage, the inflow to the reservoir in the previous day, and time itself. Multi-Objective Evolutionary Algorithms, namely NSGA II, [19], are

Table 19.2 Hoa Binh reservoir: objective values for the historical operation and different Pareto optimal policies designed by Policy Search (ANN with 1 hidden layer of 4 log-sigmoid neurons and a linear output layer) and DDP, and simulated over the validation period 1989–2004. At each simulation step, physical units are multiplied by a time-varying parameter to account for variations in the value of energy and water during the year

Control scheme		Objectives		
		Hydropower $\sim 10^7 \times$[kWh]	Irrigation $\sim$[m^3/s]^2	Flooding [m]
Historical		2.66	1271	0.063
Policy search	(hydropower only)	3.54	4168	0.056
	(compromise irrigation and flooding)	3.08	66	0.044
	(compromise among all objectives)	3.35	68	0.049
DDP	(hydropower only)	3.56	4907	0.032
	(compromise irrigation and flooding)	2.85	4	0.011
	(compromise among all objectives)	3.52	89	0.022

used to optimize the ANN parameters $\boldsymbol{\theta}$. Time series of measured inflow over the period 1961–1976 are used in the optimization phase, while time series over the period 1989–2004 are used to simulate the performances of the optimized networks (validation) and compare with historical regulation. As a matter of comparison, the performances of Deterministic Dynamic Programming (DDP) are also simulated: they provide the upper bound of performances that could be attained by a manager with perfect knowledge of all future flows. Table 19.2 compares the objective values (average daily power production, irrigation deficit and exceedence of the flooding threshold) of several of these policies: the historical one, the best for hydropower designed by DDP and Policy Search, and two possible compromise solutions. It can be noted that policy search can effectively find policies that Pareto dominate the historical operation and, as far as hydropower production is concerned, almost reach the performances of DDP.

19.4.3 On-line Suboptimal Controllers

The idea is to use a simplified model of the system, where the dynamics of the components that are not influenced by the control action (e.g. uncontrolled catchments) is neglected and their outputs are modelled as disturbances to the reduced model. The loss of information associated with this forced model order reduction is partially compensated by solving the control problem on-line over a finite, receding (or rolling) horizon and using all available information (e.g. the state of the neglected components or other new relevant measurements) to update the disturbance pdfs. The on-line control problem can be formulated as:

1. A deterministic open-loop control problem

$$\min_{\mathbf{u}_t,\ldots,\mathbf{u}_{t+h-1}} \Phi\big(g_t(\tilde{\mathbf{x}}_t, \mathbf{u}_t, \bar{\boldsymbol{\varepsilon}}_{t+1}), \ldots, g_{t+h}(\tilde{\mathbf{x}}_{t+h})\big)$$

subject to

$$\tilde{\mathbf{x}}_{\tau+1} = \tilde{f}_\tau(\tilde{\mathbf{x}}_\tau, \mathbf{u}_\tau, \bar{\boldsymbol{\varepsilon}}_{\tau+1}), \quad \tau = t, \ldots, t+h-1$$
$$\tilde{\mathbf{x}}_t \quad \text{given,}$$

where $\tilde{\mathbf{x}}_\tau$ is the reduced state vector, $\tilde{f}(\cdot)$ is the corresponding state transition function and, for each $\tau = t, \ldots, t+h-1$, $\bar{\boldsymbol{\varepsilon}}_{\tau+1}$ is the expected or maximum value of $\boldsymbol{\varepsilon}_{\tau+1}$ based on $\phi_\tau(\cdot|I_t)$ or $\Xi_\tau(I_t)$, and I_t is a vector including all the real-time information available at time t.

2. A stochastic open-loop control problem

$$\min_{\mathbf{u}_t,\ldots,\mathbf{u}_{t+h-1}} \mathop{\Psi}_{\boldsymbol{\varepsilon}_{t+1},\ldots,\boldsymbol{\varepsilon}_{t+h}} \big[\Phi\big(g_t(\tilde{\mathbf{x}}_t, \mathbf{u}_t, \boldsymbol{\varepsilon}_{t+1}), \ldots, g_{t+h}(\tilde{\mathbf{x}}_{t+h})\big)\big]$$

subject to

$$\tilde{\mathbf{x}}_{\tau+1} = \tilde{f}_\tau(\tilde{\mathbf{x}}_\tau, \mathbf{u}_\tau, \boldsymbol{\varepsilon}_{\tau+1}), \tag{19.17a}$$
$$\boldsymbol{\varepsilon}_{\tau+1} \sim \phi_\tau(\cdot\,|I_t) \quad \text{or} \quad \boldsymbol{\varepsilon}_{\tau+1} \in \Xi_\tau(I_t), \tag{19.17b}$$
$$\tau = t, \ldots, t+h-1, \tag{19.17c}$$
$$\tilde{\mathbf{x}}_t \quad \text{given.} \tag{19.17d}$$

3. A stochastic closed-loop control problem

$$\min_p \mathop{\Psi}_{\boldsymbol{\varepsilon}_{t+1},\ldots,\boldsymbol{\varepsilon}_{t+h}} \big[\Phi\big(g_t(\tilde{\mathbf{x}}_t, \mathbf{u}_t, \boldsymbol{\varepsilon}_{t+1}), \ldots, g_{t+h}(\tilde{\mathbf{x}}_{t+h})\big)\big]$$

subject to (19.17a)–(19.17d) and

$$\mathbf{u}_\tau = m_\tau(\tilde{\mathbf{x}}_\tau), \quad \tau = t, \ldots, t+h-1$$
$$p = [m_t(\cdot), \ldots, m_{t+h-1}(\cdot)].$$

Problem 1 is referred to by [8] as Naive Feedback Control NFC problem, problem 2 as Open-Loop Feedback Control (OLFC) and problem 3 as Partial Open-Loop Feedback Control (POLFC). Problems 1 and 2 can be solved by means of Mathematical Programming techniques, problem 3 is solved by means of SDP.

For all problems, one of the main difficulties is the choice of the penalty function $g_h(\cdot)$, which influences either the performances of the closed loop scheme and its stability [35]. One possibility [38] is to let $g_h(\cdot)$ be equal to the optimal cost-to-go $H_h(\cdot)$ obtained by solving an off-line infinite horizon problem with the reduced model and a trivial predictor, i.e. with a priori pdf or membership-set for the description of the disturbance. However, since the solution of the latter problem requires to use SDP, this approach still suffers from the curse of dimensionality.

Table 19.3 Lake Verbano system: average step-cost (aggregate, flooding and irrigation) with off-line and on-line (POLFC) control scheme

Control scheme		Objectives		
		Aggregate [–]	Flooding [km^2]	Irrigation [m^3/s]^{12}
Off-line		5.43×10^{19}	7.75×10^{-3}	2.04×10^{14}
On-line	$h = 1$ (prediction)	5.20×10^{19}	7.42×10^{-3}	6.21×10^{13}
	$h = 1$ (observation)	5.33×10^{19}	7.61×10^{-3}	5.28×10^{13}
	$h = 2$ (observation)	5.04×10^{19}	7.20×10^{-3}	1.59×10^{14}
	$h = 4$ (observation)	4.73×10^{19}	6.75×10^{-3}	1.38×10^{14}
	$h = 8$ (observation)	4.54×10^{19}	6.48×10^{-3}	1.33×10^{13}

As for optimality, it is well known from the certainty equivalence principle that the solution to problem 1 coincides with the optimal solution of the off-line closed-loop problem with the complete model (19.1)–(19.2), i.e. of the original problem, when the model is linear and the objective function quadratic. [8] proved that, independently of the form of the model, the solution to problem 3 can not be worse than the solution of the off-line open-loop problem with the complete model. As for the other problems, performance generally increases when passing from problem 1 to problem 3, but in some cases the solution to problem 2 is better than that of problem 3.

Application Example As an application of on-line suboptimal control, the performance of the on-line POLFC approach are compared with off-line SDP in the operation of the lake Verbano system, Italy/Switzerland.

The lake is fed by an alpine catchment of about 6560 km^2 and has been operated since from 1942 to increase the water supply for irrigation and hydropower production in the downstream territory. Lake regulation must also consider flood prevention on the lake and the downstream river shores, environmental quality, navigation and other issues (for more details see [15]). In this numerical application, we formulate a simple bi-objective problem (irrigation supply and flood control on the lake shores).

A reduced model of the system is identified with one state variable (the lake storage), one control (the release decision) and one disturbance (the lake inflow). Table 19.3 compares the off-line policy obtained with SDP and the on-line POLFC scheme. Comparison is based on the objective values (aggregated objective and its components: the average daily flooded area around the lake and the daily crop stress, defined as the irrigation deficit at power 12) over a simulation period of 5 years (1998–2002). By comparing the first two lines in the table, it can be noticed that the POLFC scheme with 1-step-ahead (24 hours) inflow prediction is significantly better than the off-line solution on both the objectives. Lines 3–6 report the performances of the on-line POLFC scheme for different length h of the prediction/control horizon. Since the current prediction ability cannot extend over such horizon, the control scheme is simulated using observed inflows in place of inflow forecast. Performances so obtained must thus be regarded as the upper bound that could be attained with perfect prediction ability. The minimum cost is reached when $h = 8$ (days), which is right the time constant of the lake. More details on the application can be found in [43].

19.4.4 Reinforcement Learning

The only way to use the reduced model of the system also in off-line policy design, without resorting to the unrealistic assumption that the outflows from uncontrolled catchments are purely random disturbances as imposed by SDP, is to use a solution approach based on Reinforcement Learning (see, e.g., [4]). With this approach, the control law depends on the reduced state vector $\tilde{\mathbf{x}}_t$ and on an information vector $\mathbf{I}_t$, constituted with the exogenous information (e.g. rainfall, snow cover, snow depth, evapotranspiration) that might have a key role in the outflow formation process and result in an improved control policy. In its original concept, Reinforcement Learning (RL) is based on the idea of designing the control policy through a trial-and-error learning process, in which the model-based estimates of the system transitions are substituted for a learning by experiencing. The learning experience can be acquired on-line, by directly experimenting controls on the real system without any model, or generated off-line, either by using an external simulator or historical observations. While the first option is clearly impracticable on real water systems, off-line learning has been successfully experimented in the operation of water systems, particularly the Q-learning algorithm developed by [57] (see the works by [10, 12, 33]).

Recently, a new approach, called fitted Q-iteration (FQI), which combines the RL concept of off-line learning and functional approximation of the cost-to-go function (here called Q-function) as in ADP, has been proposed [20]. Unlike traditional stochastic approximation algorithms (see, e.g., [54]), which use parametric function approximators and thus require a time consuming parameter estimation process at each iteration step, FQI uses tree-based approximation [11]. The use of tree-based regressors offers a twofold advantage: first, a great modeling flexibility, which is a paramount characteristic in the typical multi-objective context of water reservoir systems with multi-dimensional states, where the Q-functions to be approximated are unpredictable in shape; second, a higher computational efficiency as no optimal parameter estimation is required for the Q-function approximation at each iteration step. Further, while traditional Q-learning has been provably shown to converge only when the Q-function updates are performed incrementally, following the state trajectory produced by the sequence of optimal controls selected at each iteration step, FQI processes the information in a batch mode, by simultaneously using, in making an update of the Q-function, all the learning experience structured as a sample data-set $\mathscr{F}$ of six-tuples $\langle \tilde{\mathbf{x}}_t, \mathbf{I}_t, \mathbf{u}_t, \tilde{\mathbf{x}}_{t+1}, \mathbf{I}_{t+1}, g_{t+1} \rangle$. This has been shown to speed up the convergence rate [31].

The key idea in applying FQI to our problem is to generate the data-set $\mathscr{F}$ using the historical observations (i.e. a time series of $(\mathbf{I}_t, \mathbf{I}_{t+1})$) as a sample of the dynamics of the uncontrolled components. The reduced state sample is obtained via one-step (or multi-step) simulation of the model of the controlled components under different control policies and the historical inflow series (see [14] for more details).

The computational advantages of FQI over SDP are only slightly better than those of ADP methods in alleviating the computational burden associated with the exploration of the reduced state space. However, it provides a remarkable improvement with respect to the exogenous information and the disturbance vector dimension, as the uncontrolled components and the disturbance comes at nearly no computational additional time, and so it is particularly suitable when a large amount of potentially useful exogenous information is available or the exogenous information is strongly temporally correlated, and/or when the disturbance vector as a high dimension.

Application Example As an example of RL to water resources management, FQI is applied to design the optimal operation of a selective withdrawal reservoir in Japan, with the purpose of meeting established water quality/quantity targets both in-reservoir and downstream.

Tono dam (Tottori prefecture, Japan) is being constructed these days and will form an impounded reservoir of 12.4×10^6 m^3 (gross capacity), fed by a 38.1 km^2 catchment. The reservoir is being built for multiple purposes (irrigation, hydropower production, flood control, water supply) and is equipped with a selective withdrawal structure (SWS) through which the water can be released at different heights to preserve the river fish habitat affected by too high or too low water temperature, to prevent in-reservoir algal blooms and the reservoir silting by sedimentation.

A simplified model of the SWS is assumed, with 2 controlled outlets at -3 m and -13 m below the lake surface. The reduced state vector includes 5 components (storage, temperature at -3 m and -13 m, and total suspended solids at -3 m and -13 m), whose dynamics is influenced by more than 15 disturbances (two inflows, different nutrient loads, solar radiation, wind speed, etc.). The sample data set $\mathscr{F}$ is constructed via multi-step simulation of a 1D coupled hydrodynamic-ecological model (DYRESM) under different control policies generated pseudo-randomly and the historical time series of the disturbances. The policies generated with FQI by considering in-reservoir algal bloom and silting reduction, irrigation supply, and preservation of the fish habitat as objectives are compared with the reference operation rule adopted in the dam design and so constructed: the total amount of water to be daily released is computed by solving an open-loop control problem assuming perfect knowledge of the inflow; the allocation of this volume among the SWS outlets is obtained using a scenario analysis. Results (Table 19.4) indicate that a greater control over algal bloom and release temperature can be gained with FQI, which more effectively exploits the operational flexibility provided by the selective structure.

Table 19.4 Tono dam: objective values for the reference rule curve and one of the Pareto optimal policies obtained via simulation over a validation horizon

Control scheme	Objectives			
	Algal-blooms [g (Chl-a)/m^3]	Silting [g (TSS)]	Irrigation [m^3/s]	Fish habitat [C°]
FQI	2.23	2.99×10^6	1.91×10^{-2}	1.31
Reference rule	6.90	2.88×10^6	1.08×10^{-2}	1.6

19.5 Closure

Although the problem of designing Pareto optimal water reservoir operation policies has been extensively studied in the past, it is still a very intriguing and investigated research theme. In this chapter we reviewed some of the recent, and in our opinion, more promising alternatives to SDP in designing (sub-)optimal control policies for large water systems, namely water reservoir networks.

The problem proposed has many other facets that have not been dealt with in the paper, but are very topical, especially in a global change perspective. (1) When new water facilities are being planned, the control problem discussed in this chapter has to be nested into a mathematical programming problem whose arguments are the design parameters (e.g. number and capacity of reservoirs), thus adding to the computational burden. (2) In a globally changing world the control policy should adapt to the underlying variability of hydro-climatic, social and economic processes. New adaptive control approaches have to be developed to link optimality to changes. (3) Water quality is becoming a critical issue in most of reservoirs worldwide. Integrating quality and quantity targets in water resources management will dramatically enlarge the complexity of the problem as biochemical and ecological processes are spatially distributed and intrinsically more complex than the simply storing and moving water volumes in space and time. (4) In a multipurpose and multistakeholder context the choice of the policy to adopt in the set of the Pareto optimal policies is the final step of a complex, often recursive, decision making process that involve many different phases: from the stakeholder analysis, through the system model identification and the very policy design, to comparison and negotiations of the policies. The activities within these phases require full stakeholder involvement and integration among the different and disparate issues. They have to be organized in a procedure [15] and supported by proper computer tools, namely, Multi-Objective Decision Support Systems.

References

1. Archibald, T.W., McKinnon, K.I.M., Thomas, L.C.: An aggregate stochastic dynamic programming model of multireservoir systems. Water Resour. Res. **33**(2), 333–340 (1997)
2. Aufiero, A., Soncini-Sessa, R., Weber, E.: Set-valued control laws in TEV-DC control problem. In: Proceedings of 15th IFAC World Congress on Automatic Control, Barcelona, 21–26 July 2002

3. Baglietto, M., Cervellera, C., Sanguineti, M., Zoppoli, R.: Water reservoirs management under uncertainty by approximating networks and learning from data. In: Topics on System Analysis and Integrated Water Resource Management. Elsevier, Amsterdam (2006)
4. Barto, A., Sutton, R.: Reinforcement Learning: An Introduction. MIT Press, Boston (1998)
5. Bellman, R.E.: Dynamic Programming. Princeton University Press, Princeton (1957)
6. Bellman, R.E., Dreyfus, S.: Applied Dynamic Programming. Princeton University Press, Princeton (1962)
7. Bellman, R.E., Kabala, R., Kotkin, B.: Polynomial approximation—a new computational technique in dynamic programming. Math. Comput. **17**(8), 155–161 (1963)
8. Bertsekas, D.P.: Dynamic Programming and Stochastic Control. Academic Press, New York (1976)
9. Bertsekas, D.P., Tsitsiklis, J.N.: Neuro-Dynamic Programming. Athena Scientific, Boston (1996)
10. Bhattacharya, A., Lobbrecht, A.H., Solomatine, D.P.: Neural networks and reinforcement learning in control of water systems. J. Water Resour. Plan. Manag. **129**(6), 458–465 (2003)
11. Breiman, L., Friedman, J., Olsen, R., Stone, C.: Classification and Regression Trees. Wadsworth & Brooks, Pacific Grove (1984)
12. Castelletti, A., Corani, G., Rizzoli, A.E., Soncini-Sessa, R., Weber, E.: A reinforcement learning approach for the operational management of a water system. In: Proceedings of IFAC Workshop Modelling and Control in Environmental Issues, 22–23 August 2001. Elsevier, Yokohama (2001)
13. Castelletti, A., de Rigo, D., Rizzoli, A.E., Soncini-Sessa, R., Weber, E.: Neuro-dynamic programming for designing water reservoir network management policies. Control Eng. Pract. **15**(8), 1001–1011 (2007)
14. Castelletti, A., Galelli, S., Restelli, M., Soncini-Sessa, R.: Tree-based reinforcement learning for optimal water reservoir operation. Water Resour. Res. (2010)
15. Castelletti, A., Soncini-Sessa, R.: A procedural approach to strengthening integration and participation in water resource planning. Environ. Model. Softw. **21**(10), 1455–1470 (2006)
16. Castelletti, A., Pianosi, F., Soncini-Sessa, R.: Water reservoir control under economic, social and environmental constraints. Automatica **44**(6), 1595–1607 (2008)
17. Cervellera, C., Chen, V.C.P., Wen, A.: Optimization of a large-scale water reservoir network by stochastic dynamic programming with efficient state space discretization. Eur. J. Oper. Res. **171**(3), 1139–1151 (2006)
18. Cervellera, C., Muselli, M.: Deterministic design for neural network learning: an approach based on discrepancy. IEEE Trans. Neural Netw. **15**(3), 533–544 (2004)
19. Deb, K., Pratap, A., Agarwal, S., Meyarivan, T.: A fast and elitist multiobjective genetic algorithm: NSGA-II. IEEE Trans. Evol. Comput. **6**(2), 182–197 (2002)
20. Ernst, D., Geurts, P., Wehenkel, L.: Tree-based batch mode reinforcement learning. J. Mach. Learn. Res. **6**, 503–556 (2005)
21. Esogbue, A.O.: Dynamic programming and water resources: origins and interconnections. In: Dynamic Programming for Optimal Water Resources Systems Analysis. Prentice-Hall, Englewood Cliffs (1989)
22. Foufoula-Georgiou, E., Kitanidis, P.K.: Gradient dynamic programming for stochastic optimal control of multidimensional water resources systems. Water Resour. Res. **24**, 1345–1359 (1988)
23. Gilbert, K.C., Shane, R.M.: TVA hydroscheduling model: theoretical aspects. J. Water Resour. Plan. Manag. **108**(1), 21–36 (1982)
24. Hall, W.A., Buras, N.: The dynamic programming approach to water resources development. J. Geophys. Res. **66**(2), 510–520 (1961)
25. Hall, W.A., Butcher, W.S., Esogbue, A.: Optimization of the operation of a multi-purpose reservoir by dynamic programming. Water Resour. Res. **4**(3), 471–477 (1968)
26. Heidari, M., Chow, V.T., Kokotovic, P.V., Meredith, D.: Discrete differential dynamic programming approach to water resources systems optimisation. Water Resour. Res. **7**(2), 273–282 (1971)

27. Hejazi, M.I., Cai, X., Ruddell, B.L.: The role of hydrologic information in reservoir operation—learning from historical releases. Adv. Water Resour. **31**(12), 1636–1650 (2008)
28. Jacobson, H., Mayne, Q.: Differential Dynamic Programming. American Elsevier, New York (1970)
29. Jalali, M.R., Afshar, A., Marino, M.A.: Reservoir operation by ant colony optimization algorithms. The Iranian Journal of Science **30**, 107–117 (2006)
30. Johnson, S.A., Stedinger, J.R., Shoemaker, C., Li, Y., Tejada-Guibert, J.A.: Numerical solution of continuous-state dynamic programs using linear and spline interpolation. Oper. Res. **41**, 484–500 (1993)
31. Kalyanakrishnan, S., Stone, P.: Batch reinforcement learning in a complex domain. In: The Sixth International Joint Conference on Autonomous Agents and Multiagent Systems, May 2007
32. Larson, R.E.: State Incremental Dynamic Programming. American Elsevier, New York (1968)
33. Lee, J.-H., Labadie, J.W.: Stochastic optimization of multireservoir systems via reinforcement learning. Water Resour. Res. **43**(11), 1–16 (2007)
34. Lotov, A.V., Bushenkov, V.A., Kamenev, G.K.: Interactive Decision Maps Approximation and Visualization of Pareto Frontier. Springer, Heidelberg (2004)
35. Mayne, D.Q., Rawlings, J.B., Rao, C.V., Scokaert, P.O.M.: Constrained model predictive control: stability and optimality. Automatica **36**, 789–814 (2000)
36. Miettinen, K.: Nonlinear Multiobjective Optimization. Kluwer Academic, Dordrecht (1999)
37. Momtahen, Sh., Dariane, A.B.: Direct search approaches using genetic algorithms for optimization of water reservoir operating policies. J. Water Resour. Plan. Manag. **133**(3), 202–209 (2007)
38. Nardini, A., Piccardi, C., Soncini-Sessa, R.: A decomposition approach to suboptimal control of discrete-time systems. Optim. Control Appl. Methods **15**(1), 1–12 (1994)
39. Niederreiter, H.: Random Number Generation and Quasi-Monte Carlo Methods. SIAM, Philadelphia (1992)
40. Oliveira, R., Loucks, D.P.: Operating rules for multireservoir systems. Water Resour. Res. **33**(4), 839–852 (1997)
41. Orlovski, S., Rinaldi, S., Soncini-Sessa, R.: A min max approach to reservoir management. Water Resour. Res. **20**(11), 1506–1514 (1984)
42. Philbrick, C.R., Kitanidis, P.K.: Improved dynamic programming methods for optimal control of lumped-parameter stochastic systems. Oper. Res. **49**, 398–412 (2001)
43. Pianosi, F., Soncini-Sessa, R.: Real-time management of a multipurpose water reservoir with a heteroscedastic inflow model. Water Resour. Res. **45**(10), 10 (2009)
44. Piccardi, C.: Infinite-horizon periodic minimax control problem. J. Optim. Theory Appl. **79**, 397–404 (1993)
45. Piccardi, C., Soncini-Sessa, R.: Stochastic dynamic programming for reservoir optimal control: dense discretization and inflow correlation assumption made possible by parallel computing. Water Resour. Res. **27**(5), 729–741 (1991)
46. Powell, W.B.: Approximate Dynamic Programming. Wiley, New York (2007)
47. Read, E.G.: A dual approach to stochastic dynamic programming for reservoir release scheduling. In: Dynamic Programming for Optimal Water Resources Systems Analysis, pp. 361–372. Prentice-Hall, Englewood Cliffs (1989)
48. Saad, M., Turgeon, A., Bigras, P., Duquette, R.: Learning disaggregation technique for the operation of long-term hydroelectric power systems. Water Resour. Res. **30**(11), 3195–3203 (1994)
49. Sharma, V., Jha, R., Naresh, R.: Optimal multi-reservoir network control by two-phase neural network. Electr. Power Syst. Res. **68**, 221–228 (2004)
50. Soncini-Sessa, R., Castelletti, A., Weber, E.: Integrated and Participatory Water Resources Management. Theory. Elsevier, Amsterdam (2007)
51. Soncini-Sessa, R., Zuleta, J., Piccardi, C.: Remarks on the application of a risk-averse approach to the management of El-Carrizal reservoir. Adv. Water Resour. **13**(2), 76–84 (1991)
52. Su, Y.S., Deininger, R.A.: Generalization of White's method of successive approximations. Oper. Res. **20**(2), 318–326 (1972)

53. Trott, W.J., Yeh, W.: Optimization of multiple reservoir systems. J. Hydraul. Div. **99**, 1865–1884 (1973)
54. Tsitsiklis, J.N., Van Roy, B.: Feature-based methods for large scale dynamic programming. Mach. Learn. **22**, 59–94 (1996)
55. Turgeon, A.: A decomposition method for the long-term scheduling of reservoirs in series. Water Resour. Res. **17**(6), 1565–1570 (1981)
56. Vasiliadis, H.V., Karamouz, M.: Demand-driven operation of reservoirs using uncertainty-based optimal operating policies. J. Water Resour. Plan. Manag. **120**(1), 101–114 (1994)
57. Watkins, C.J.C.H., Dayan, P.: Q-learning. Mach. Learn. **8**(3–4), 279–292 (1992)
58. White, D.J.: Dynamic programming, Markov chains, and the method of successive approximations. J. Math. Anal. Appl. **6**, 373–376 (1963)
59. Yeh, W.: Reservoir management and operations models: a state of the art review. Water Resour. Res. **21**(12), 1797–1818 (1985)

Chapter 20
Real-Time Optimal Control of River Basin Networks

R. Evans, L. Li, I. Mareels, N. Okello, M. Pham, W. Qiu, and S.K. Saleem

20.1 Introduction

River basins are key components of water supply grids. However, unlike energy grids which are operated in closed-loop [16, 28], river basins are largely open-loop systems. One reason is the difficulty associated with the development of suitable models. Traditionally, river basin modeling efforts have focused on process-based methodologies that are potentially very accurate but not amenable to the design of feedback controllers. For control purposes river basin operators often rely on simulation-optimization and/or rule-based approaches. This method may work well for long-term planning intervals (e.g. months) but is impractical for real-term operations (e.g. hours). This limitation results in suboptimal river flows and releases

R. Evans (✉) · L. Li · N. Okello · M. Pham · W. Qiu · S.K. Saleem
National ICT Australia Ltd, Eveleigh, Australia
e-mail: rob.evans@nicta.com.au

L. Li
e-mail: li.li@nicta.com.au

N. Okello
e-mail: nickens.okello@nicta.com.au

M. Pham
e-mail: minh.pham@nicta.com.au

W. Qiu
e-mail: wanzhi.qiu@nicta.com.au

S.K. Saleem
e-mail: khusro.saleem@nicta.com.au

I. Mareels
The University of Melbourne, Melbourne, Australia
e-mail: iven.mareels@unimelb.edu.au

L. Wang, H. Garnier (eds.), *System Identification, Environmental Modelling, and Control System Design*,
DOI 10.1007/978-0-85729-974-1_20,

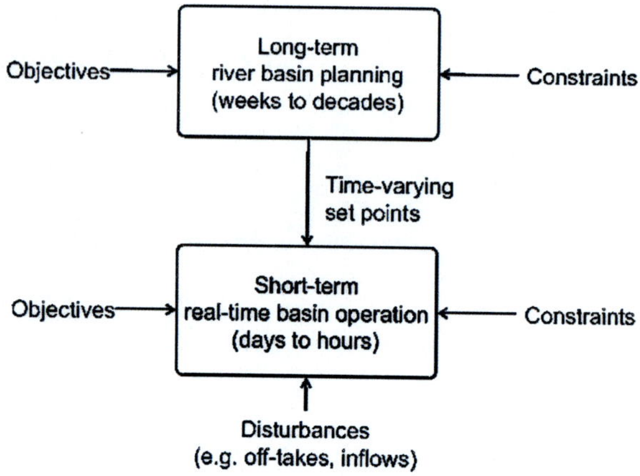

Fig. 20.1 Framework

from water storages. A systematic approach to real-time river operation is needed. Feedback control offers one solution, and is the subject of this chapter.

A common framework for describing river basin management is illustrated in Fig. 20.1 [5]. The *river planning* block generates system set-points that are usually held constant over long periods, ranging from weeks to decades. The *river operations* block tracks commanded set-points and rejects disturbances.

Real-time river basin operation is typical of large-scale control problems that have the following characteristics [23]:

1. A network structure with some kind of flow along links connecting storage units;
2. Flow is to be routed from specific sources to designated destinations;
3. Flow is subject to capacity constraints;
4. There are time-varying demand variables at the source, along the network and at the destination;
5. The links and storage units are characterized by transport lags as well as a storage component; and,
6. A communication network with limited bandwidth is used to transmit network state.

The general control problem is to specify control inputs to influence the flow in the network so as to minimize a performance criterion subject to capacity constraints and time-varying loads. River basin real-time operational objectives include in-stream flow rates and water levels in storages. This basic structure is also useful for managing water quality.

Modeling and optimization of water resources systems has a rich history [8]. The Saint-Venant equations [4] are the basis for the mathematical modeling of open water channels. These are hyperbolic partial differential equations making them difficult to use in feedback controller design. Moreover, the Saint-Venant equations however do not apply to modeling rainfall-runoff, and surface and ground-water reservoirs. Nonetheless, numerous studies have applied these equations either directly, or as starting points, to develop feedback controllers. The studies in [1] and [30] have focused on the use of decentralized PI control, [32] applied centralized LQ control to overflow regulation in sewer networks, [11] used H_∞ control, and [25] applied multivariable predictive control. The studies in [21, 26] explored the use of model predictive control. Optimal control of sewer networks has been studied extensively in [14, 15]. An alternative to the Saint-Venant equations is to exploit grey-box or data-based models derived using system identification experiments [7, 17, 33]. The key advantage of these models is that feedback controllers are easier to design [13].

With the exception of some applications in large-scale sewer network optimization, most of the studies cited above focus on modeling and control of irrigation canal networks and short river reaches. Combined simulation-optimization methods are commonly used to plan and operate river basin networks [3, 6]. This chapter builds on previous work in open canals and sewer networks to develop a framework for real-time river basin operation based on optimal control theory. The River Murray system in Australia is used as a case study.

The River Murray system [19] drains a catchment region which covers the south east corner of the Australian continent and extends over 1,060,000 km^2. The total length of the main river channel is 3,780 km and the mean discharge is 0.4 ML/sec. The system is largely fed by precipitation and snow-melt in the Australian Alps. The main consumptive demands are irrigation districts and rural populations and one major metropolitan demand site in Adelaide, South Australia. The River Murray is permanently navigable to a distance of 970 kilometers from the mouth due to a series of locks and weirs.

Water is diverted from the River Murray all year round, though demand is small in Winter. During Winter and Spring, as much water is stored as possible. Irrigation diversions normally increase progressively from August to November, but in Spring they can often be met largely from natural flows. From December to May, inflows to the river usually recede and the demand for water is largely met by controlled releases from storages.

The River Murray is operated in three modes [20]: (1) Supply mode; (2) Storing mode; and (3) Spilling mode. It is possible for different reaches of the river to be in different modes. Supply mode typically occurs during the irrigation season. The flow in the river is set to meet demands with little excess. Storing mode generally occurs when the flows in the river are in excess of that required to meet diversions, water supply, and minimum flow requirements; but which are confined within the channel. Spilling mode occurs when flow exceeds the river's channel capacity at a point as a result of runoff generated by heavy rain. This operation can be quite complex as the flow varies as tributaries join the main stream. Spilling mode occurs at a storage when there is limited airspace left in the storage and inflow rates are high. Spilling mode is also important for flood event routing. Using the River Murray system as a case study, the remainder of this chapter presents an approach for the real-time operation of entire river basin networks. In particular we derive simple low order models for storages and reaches based on real measurements. We then propose controller strategies to achieve the operational objectives while meeting various practical constraints.

20.2 Models

A schematic of the River Murray System is illustrated in Fig. 20.2. Following the methodology in [23] the river system is subdivided into sub-networks with storage capabilities. Links connecting the sub-networks are treated as pure delays. In this sense flow rates leaving a sub-network (or storage element) are control variables, whereas the volumes (or water levels) in the storage elements are state variables. In-stream flow rates further downstream from storage outlets can also be considered as state variables.

20.2.1 Storage Models

River basin storages are modeled using the continuity equation

$$\dot{V}(t) = \sum_{n \in I} q_{in,n}(t) - \sum_{m \in O} q_{out,m}(t), \tag{20.1}$$

where V is the storage volume, $q_{in,n}(t)$ and $q_{out,m}(t)$ are inflow rate and outflow rates respectively, and I and O denote the set of all inflows and outflows, respectively. The inflow is a measurement some distance upstream of the storage. The outflow depends on the specific outflow control structure. In the case of variable height overflow weirs $q_{out} = c_w h^{3/2}$ where c_w is an normalizing coefficient and h represents the depth of flow, or head, over the weir structure. In the case of an orifice opening, the outflow rate is given by $q_{out} = c_o r^2 \sqrt{d}$ where c_o is a normalizing coefficient, r is the radius of the opening and d is the depth of water, or head, over the center of the orifice. As indicated above, in this problem the storage outflows are the control variables and with the above models the resulting dynamics are clearly non-linear.

An alternative is to linearize the system through an input transformation dependent on the state of the system. This approach was originally proposed in [32] and later in [29] where the control components are defined in terms of flows. In the case of a overflow weir we define $u \triangleq c_w h^{3/2}$. The system dynamics then assume the following for

$$\dot{V}(t) = \sum_{n \in I} q_{in,n}(t) - \sum_{m \in O} u_m(t). \tag{20.2}$$

In river basin operations, storage volume is generally *inferred* from water level measured at the downstream end of a storage element, close to the outflow control point. The water level is often expressed as an elevation with respect to a common datum such as mean sea level. The function relating storage volume and water level depends on the storage element's geometry. Assuming only a single outflow structure is present, the following model for water level $y(t)$ in a storage element can be

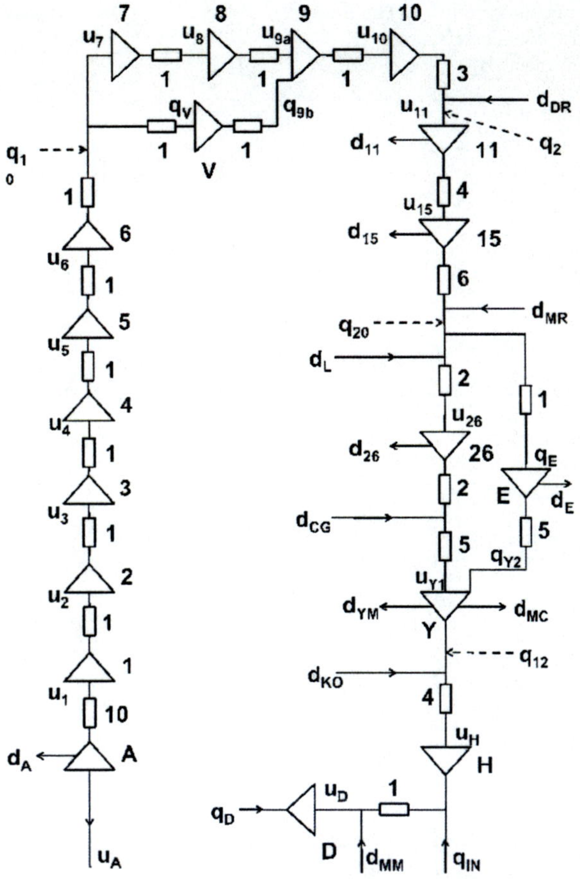

Fig. 20.2 River Murray schematic

used

$$\dot{y}(t) = \sum_{n \in I} \alpha_n(y) q_{in,n}(t) - \sum_{m \in O} \beta_m(y) u_m(t), \tag{20.3}$$

where the functions $\alpha_{(.)}(y)$ and $\beta_{(.)}(y)$ are related to the storage element's geometry. Generally these will be non-linear, for example when the storage is deep and has sloping sides. In this chapter, these functions will be assumed constant. Without this simplification, the system can still be described by a set of linearized models by selecting several operating points over the range of set-points. Gain scheduling is a popular method used for designing controllers for such systems [27]. Letting T_s denote the sample interval, and using a first order approximation for $\dot{y}$, the discrete-time model for water level is given by

$$y[k+1] = y[k] + T_s \left(\sum_{n \in I} \alpha_n q_{in,n}[k] - \sum_{m \in O} \beta_m u_m[k] \right). \tag{20.4}$$

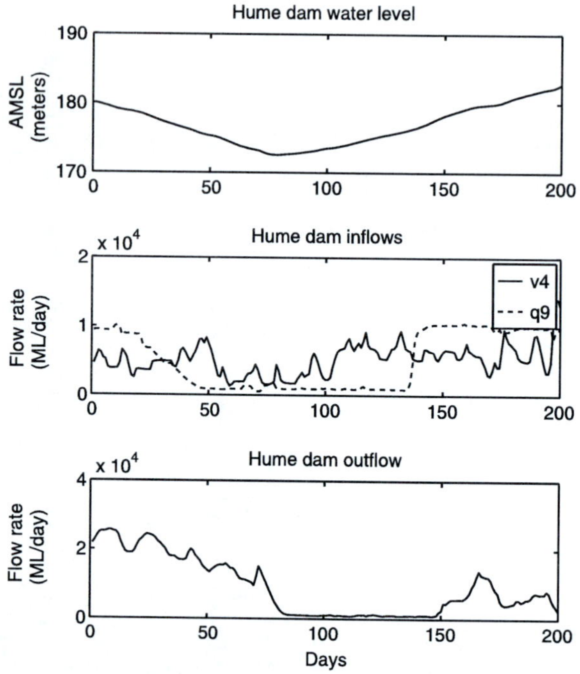

Fig. 20.3 Inflow and outflow time series for storage elements H

20.2.1.1 Parameter Estimation

This section outlines the parameter estimation for storage elements D, H, and Y in Fig. 20.2. Daily data for selected storage elements is available at [19]. Figure 20.3 illustrates the data for storage element H.

The predictor associated with the model in (20.4) for storage node D is given by

$$\hat{y}_D[k+1,\theta_D] = \hat{y}_D[k,\theta_D] + \alpha_1 q_D[k] - \beta_1 u_D[k], \tag{20.5}$$

where $T_s = 1$ day and $\theta_D = [\alpha_1\ \beta_1]^T$. For storage node H we have

$$\hat{y}_H[k+1,\theta_H] = \hat{y}_H[k,\theta_H] + \alpha_1 q_{\mathrm{MM}}[k-1] + \alpha_2 q_{\mathrm{IN}}[k] - \beta_1 u_H[k], \tag{20.6}$$

where $\theta_H = [\alpha_1\ \alpha_2\ \beta_1]^T$. For storage node Y we have

$$\begin{aligned}\hat{y}_Y[k+1,\theta_Y] = \hat{y}_Y[k,\theta_Y] + \alpha_1 d_{\mathrm{KO}}[k] + \alpha_2 d_{\mathrm{YM}}[k] + \alpha_3 d_{\mathrm{MC}} - \beta_1 u_{Y1}[k]\\ - \beta_2 u_{Y2}[k],\end{aligned} \tag{20.7}$$

where $\theta_Y = [\alpha_1\ \alpha_2\ \alpha_3\ \beta_1\ \beta_2]^T$. The coefficients are estimated using

$$\hat{\theta}_j = \arg\min_{\theta_j} \frac{1}{M} \sum_{k=k_0}^{k_0+M-1} \left(y_j[k] - \hat{y}_j[k,\theta^{(\cdot)}]\right)^2, \tag{20.8}$$

Table 20.1 Estimated parameters and mean squared water level prediction errors

Node	$\hat{\theta}_j$	Error (m^2)
D	$4.226e^{-5}, 1.922e^{-5}$	0.37
H	$1.056e^{-5}, 9.84e^{-6}, 1.01e^{-5}$	0.13
Y	$2.552e^{-5}, 2.702e^{-5}, 3.229e^{-5}, 1.996e^{-5}, 2.697e^{-5}$	$1.962e^{-5}$

Table 20.2 Storage models parameters

Storage	A_j (ha)	Capacity (GL)	AMSL (m)	Δ_y (m)	α_j
D (Dartmouth)	6,380	3,906	486	61	$3.5e^{-5}$
H (Hume)	20,019	3,038	192	15	$1.1e^{-5}$
Y (Yarrawonga)	3,933	118	125	3	$5.6e^{-5}$
E (Stevens)	4,000	120	80	3	$5.5e^{-5}$
26	1,233	37	86	3	$1.8e^{-4}$
15	1,233	37	48	3	$1.8e^{-4}$
11	1,233	37	35	3	$1.8e^{-4}$
10	1,567	47	31	3	$1.4e^{-4}$
V (Victoria)	11,283	677	27	6	$1.9e^{-5}$
9	1,067	32	27	3	$2e^{-4}$
8	800	24	25	3	$2.8e^{-4}$
7	433	13	22	3	$5.1e^{-4}$
6	1,167	35	19	3	$1.9e^{-4}$
5	1,300	39	16	3	$1.7e^{-4}$
4	1,033	31	13	3	$2.2e^{-4}$
3	1,733	52	10	3	$1.3e^{-4}$
2	1,433	43	6	3	$1.6e^{-4}$
1	2,133	64	3	3	$1e^{-4}$
A (Alexandrina)	67,167	2,015	0.75	0.75	$3.3e^{-6}$

where M is the number of data points and j is the storage index. The results are summarized in Table 20.1. Daily inflow and outflow data is only available for a limited number of the elements in Fig. 20.2. However, storage element surface areas are available, see Table 20.2. To parameterize the remaining elements we propose the following general model for all storage elements

$$y_j[k+1] = y_j[k] + \alpha_j \left(\sum_{n \in I} q_{in,n}^{(\cdot)}[k] - \sum_{m \in O} u_m^{(\cdot)}[k] \right), \tag{20.9}$$

where

$$\alpha_j = \gamma \frac{A_D}{A_j} \tag{20.10}$$

and A_j are the storage element surface areas in Table 20.2 and the scalar γ is selected based on the results in Table 20.1. Table 20.2 lists the constant scalar α_j for all storage elements, calculated according to (20.10) with $\gamma = 3.5e^{-5}$. Using (20.9) and (20.10) all storage element models are scaled relative to element D. Summarizing, the first column in Table 20.2 is the storage surface area when it is at full capacity. The second column is the volume at full capacity. The third column is the storage water level relative to sea level at full capacity, and the fourth column is the draw-down capability from the maximum water level.

20.2.2 River Reach Models

The Saint-Venant equations are a good starting point for modeling river reaches. It has been shown in [9] (see also [29]) that under relatively mild assumptions, the Saint-Venant equations can be linearized about a reference flow rate resulting in the following river reach dynamics

$$\dot{q}_{out,i} + \frac{1}{K} q_{out,i} = q_{in,i}(t - \tau_i), \tag{20.11}$$

where τ_i is the input delay, and K is the time constant. It is important to note the parameters in (20.11) vary with the reference flow rate. The above first-order system takes into account the transport delay, in-stream storage phenomena and the dispersion of the flow (or wave attenuation) as it moves downstream. In [10] the above models were used to design controllers based on the Smith Predictor. Robustness of the designs were estimated with the use of margins and also with the use of a bound on multiplicative uncertainty taking into account modeling errors. In this study the river reach model is simplified to a transport delay. As for storages described above, without this simplification, river reaches can still be described by a set of linearized models by selecting a several operating points over the range of set-points. Once again, gain scheduling can be applied. With the above simplification, in discrete-time notation we have

$$q_{out_i}[k+1] = q_{in_i}[k - \lceil \tau_i / T_s \rceil]. \tag{20.12}$$

Reference flow rates at various points along the river are indicated in Table 20.3.

20.3 Controller Objectives

Controlling a river basin network involves regulating a selected set of states around their set-points based on the operational mode of interest. For example in storage mode a river operator maintains water levels in storages at specified levels while allowing flows to take on values necessary to maintain those levels. On the other hand, in supply mode a river operator maintains constant flow rates in-stream while

Table 20.3 Reference flow rates (GL/day)

Node	Mean	Median	Mode	Max	Min
q_{IN}	4.7	5.0	4.8	8.6	1.2
q_D	0.6	0.43	0.35	3.8	0.14
d_{KO}	1.8	1.5	0.7	3.1	0.4
d_{YM}	0.76	0.68	0.62	2	0
d_{MC}	1.6	1.5	2	2.5	0
q_{12}	7.3	7.7	1.3	14.6	1.2
u_{Y1}	6.2	5.7	5.5	10.6	1.8
q_{Y2}	1.6	1.3	2	3.4	0.06
d_{CG}	0.6	0.6	0.6	2.8	0.3
d_L	0.02	0.01	0.01	0.06	0
d_E	0.2	0.2	0.03	0.3	0
d_{MR}	0.5	0.3	0.2	1.4	0.2
q_{20}	5.6	5.4	6	9.6	2.1
q_2	4.9	4.8	3.9	8.6	2.4
d_{DR}	2.8	0.8	0.06	12	0.04
q_{10}	5.4	5.4	5.4	8.3	3
u_4	5.2	5.4	3.4	8.7	2.9

allowing storage levels to take on values necessary to maintain those flows. This section summarizes the main operational objectives for the River Murray [19]:

1. Meet water demands for both consumptive use and environmental flows, expressed as a flow rate (set-point regulation);
2. Keep storage water levels close to a reference level (also set-point regulation);
3. Reject disturbances caused by urban and irrigation withdrawals and rainfall-runoff;
4. Minimize control effort by minimizing gate movement; and,
5. Maintain rate of rise and rate of fall within bounds to avoid river bank slumping.

The remainder of this chapter develops two alternative optimal control frameworks to achieve these objectives.

20.4 Centralized Controller

In this section, the River Murray control problem is formulated as a finite horizon Linear Quadratic Regulator (LQR) problem incorporating feedforward of forecast disturbances. Since this is a regulation problem, we define a set of states that are deviations from setpoints, for example $x_{i,0}[k] = y_i[k] - y_i^*[k]$, where $y_i^*[k]$ is the setpoint at node i. The same applies to flow, but with the following notation $x_{i,0}[k] = q_i[k] - q_i^*[k]$. To deal with time delays associated with flows within the network, it is necessary to define auxiliary states that *remember* past values. We therefore introduce the states $x_{j,i}[k] = u_j[k-i]$ which implies that whenever the term $u_j[k-\tau]$ appears in the model equations we introduce the following auxiliary state equations

$$x_{j,1}[k+1] = u_j[k],$$
$$x_{j,i+1}[k+1] = x_{j,i}[k], \quad i = 1, \ldots, \tau - 1$$

and the substitute $x_{j,\tau}[k]$ in all model equations where $u_j[k-\tau]$ appears [15, 31].

For the network in Fig. 20.2, the state equations associated with the first two storages and the last storage take on the form

$$\begin{aligned}
x_{D,0}[k+1] &= x_{D,0}[k] + a_D(q_D[k] - u_D[k]) + v_{sp,D},\\
x_{D,1}[k+1] &= u_D[k],\\
x_{H,0}[k+1] &= x_{H,0}[k] + a_H(x_{D,1}[k] + d_{MM}[k-1] + q_{IN}(k) - u_H[k]) + v_{sp,H},\\
x_{H,1}[k+1] &= u_H[k],\\
x_{H,i+1}[k+1] &= x_{H,i}[k], \quad i = 1, \ldots, 3,\\
&\ldots\\
x_{A,0}[k+1] &= x_{A,0}[k] + a_A(x_{1,10}[k] - u_A[k] - d_A[k]) + v_{sp,A},\\
x_{A,1}[k+1] &= u_A[k].
\end{aligned}$$

The disturbances q_D, d_{MM}, q_{IN} and d_A are inflows or offtakes while disturbances $v_{sp,A}$, $v_{sp,H}$, $v_{sp,A}$ represent water level and flow setpoints. All these disturbances are known in advance and their effects can be minimized through feedforward. Note that while the last control input $u_A[k]$ does not feed any storage within the network, it is still necessary to introduce the state $x_{A,1}[k]$ if we are to control flow and gate movement of storage A.

Using the above set of equations as an example, the state space equation for the entire network can be generated and expressed as

$$\tilde{\mathbf{x}}[k+1] = A\tilde{\mathbf{x}}[k] + B\mathbf{u}[k] + (A - I)\mathbf{x}_{SP} + B\mathbf{u}_{SP} + \mathbf{w}[k], \tag{20.13}$$

where

$$\begin{aligned}
\tilde{\mathbf{x}}[k] &= \left[x_{D,0}[k]\; x_{D,1}[k]\; x_{H,0}[k]\; x_{H,1}[k], \ldots, x_{H,4}[k], \ldots, x_{A,0}[k]\; x_{A,1}[k]\right]^T,\\
\mathbf{u}[k] &= \left[u_D[k]\; u_H[k], \ldots, u_A[k]\right]^T
\end{aligned}$$

and $\mathbf{x}_{SP}$ and $\mathbf{u}_{SP}$ are the set setpoint vectors for state $\tilde{\mathbf{x}}(k)$ and control input $\mathbf{u}(k)$, respectively. Note that all state variables are accessible and so there is no need for an observer or estimator.

For a set of selected states it is possible to achieve zero steady-state error in the presence of disturbances by including the integral of these setpoint errors in the controller criterion function. Let $\mathbf{x}_{\text{int}}[k]$ be the vector of integrated setpoint errors. If E is a matrix that maps the river network state vector $\tilde{\mathbf{x}}[k]$ to the integrated error state vector $\mathbf{x}_{\text{int}}[k]$, then the state equation for the integrated error state vector takes on the form

$$\mathbf{x}_{\text{int}}[k+1] = \mathbf{x}_{\text{int}}[k] + TE\tilde{\mathbf{x}}[k]), \tag{20.14}$$

where T is the sampling interval. A new augmented state vector $\mathbf{x}[k] = [\tilde{\mathbf{x}}^T[k]\ \mathbf{x}_{\text{int}}^T[k]]^T$ and a new state equation obtained from (20.13) and (20.14) therefore takes on the form

$$\mathbf{x}[k+1] = \mathbf{A}\mathbf{x}[k] + \mathbf{B}\mathbf{u}[k] + \mathbf{d}[k], \tag{20.15}$$

where

$$\mathbf{A} = \begin{bmatrix} A & 0 \\ E & I \end{bmatrix}, \qquad \mathbf{B} = \begin{bmatrix} B \\ 0 \end{bmatrix}, \qquad \mathbf{d}[k] = \begin{bmatrix} (A-I)\mathbf{x}_{SP} + B\mathbf{u}_{SP} + \mathbf{w}[k] \\ 0 \end{bmatrix}.$$

The quadratic cost function to be minimized is

$$\begin{aligned} J_K &= \frac{1}{2}\mathbf{x}[K]^T\mathbf{S}\mathbf{x}[K] + \frac{1}{2}\sum_{k=0}^{K-1}[\mathbf{x}^T[k]\ \mathbf{u}^T[k]]\begin{bmatrix} \mathbf{Q} & \mathbf{N} \\ \mathbf{N}^T & \mathbf{R} \end{bmatrix}\begin{bmatrix} \mathbf{x}[k] \\ \mathbf{u}[k] \end{bmatrix} \\ &= \frac{1}{2}\mathbf{x}[K]^T\mathbf{S}\mathbf{x}[K] + \frac{1}{2}\sum_{k=0}^{K-1}\mathbf{x}^T[k]\mathbf{Q}\mathbf{x}[k] + 2\mathbf{x}[k]^T\mathbf{N}\mathbf{u}[k] \\ &\quad + \mathbf{u}^T[k]\mathbf{R}\mathbf{u}[k], \end{aligned} \tag{20.16}$$

where $\mathbf{R} > 0$ and $[\begin{smallmatrix} \mathbf{Q} & \mathbf{N} \\ \mathbf{N}^T & \mathbf{R} \end{smallmatrix}]$ is a positive semi-definite matrix.

From objective 4, the physical input effort is the gate movement and so terms of the form $(u_j[k] - u_j[k-1])^2$ should be penalized, where $j = 1, 2, \ldots, N_D$ is the storage index. When expressed in terms of state variables and inputs, these terms take on the form $(u_j[k] - x_{j,1}[k])^2$. We note that these terms include both inputs and states and so we get a non-zero $\mathbf{N}$ matrix in the cost function. The weight matrices therefore take on the form

$$\begin{bmatrix} \mathbf{Q}_1 & \mathbf{N}_1 \\ \mathbf{N}_1^T & \mathbf{R}_1 \end{bmatrix} = \sum_{i=1}^{N_D} s_i s_i^T, \tag{20.17}$$

where the vectors s_i, $i = 1, 2, 3, \ldots, N_D$ have a η in the position corresponding to the $u_j(k)$ and $-\eta$ in the position corresponding to $x_{j1}(k)$ and zero elsewhere.

From control objectives 1, 2 and 4, the terms of interest are the level setpoint errors $x_{j,0}[k]$, the setpoint errors $x_{j,1}[k]$ of delayed gate outflows, the integrated setpoint errors $x_{j,\text{int}}[k]$, and the control input or gate flow errors $u_j[k]$. We denote the weights associated with the setpoint errors by γ_j, the weights associated with the delayed setpoint errors of gate outflows by ξ_j, the weights associated with the integrated setpoint errors by ζ_j, and the weights associated with gate flow errors by σ_j for $j = 1, \dots, N_D$. We therefore construct a matrix

$$\begin{bmatrix} \mathbf{Q}_2 & \mathbf{N}_2 \\ \mathbf{N}_2^T & \mathbf{R}_2 \end{bmatrix}, \tag{20.18}$$

with the weights γ_i, ξ_i, and ζ_i, and σ_j, $i = 1, \dots, N_D$ in the appropriate places.

Hence the physical control problem can be formulated in an LQ framework, with the matrices $\mathbf{Q}$, $\mathbf{R}$, and $\mathbf{N}$ given by

$$\begin{aligned} \mathbf{Q} &= \mathbf{Q}_1 + \mathbf{Q}_2, \\ \mathbf{R} &= \mathbf{R}_1 + \mathbf{R}_2, \\ \mathbf{N} &= \mathbf{N}_1 + \mathbf{N}_2. \end{aligned} \tag{20.19}$$

The control objective 3 is achieved through feedforward control while the remaining objectives can be achieved through the use of constraints. This is a basic description of the composition of the $\mathbf{Q}$, $\mathbf{R}$, and $\mathbf{N}$ matrices and assumes that each storage has a single control gate. Extension to include storages with multiple gates is straightforward.

For a given disturbance trajectory $\mathbf{w}[k]$, $k = 1, \dots, K-1$ we seek a control $\mathbf{u}(k)$ that minimizes (20.16) subject to (20.15) with initial condition $\mathbf{x}[0] = \mathbf{x}_0$. Such an LQ controller that employs feedforward for disturbance rejection can be obtained by first formulating the discrete-time Hamiltonian and then applying the *maximum principle* [24]. When the resulting difference equations are solved we obtain

$$\mathbf{u}[k] = -\mathbf{K}\mathbf{x}[k] - \mathbf{K}_d\mathbf{p}[k+1], \tag{20.20}$$

where $\mathbf{K} = (\mathbf{R} + \mathbf{B}^T\mathbf{P}\mathbf{B})^{-1}(\mathbf{B}^T\mathbf{P}\mathbf{A} + \mathbf{N}^T)$ and $\mathbf{K}_d = \mathbf{R} + \mathbf{B}^T\mathbf{P}\mathbf{B})^{-1}\mathbf{B}^T$ are the feedback and feedforward gain matrices respectively [15, 24]. The matrix $\mathbf{P}$ is the positive definite solution to the steady-state Riccati equation

$$\mathbf{P} = \mathbf{A}^T\mathbf{P}\mathbf{A} - (\mathbf{A}^T\mathbf{P}\mathbf{B} + \mathbf{N})(\mathbf{B}^T\mathbf{P}\mathbf{B} + \mathbf{R})^{-1}(\mathbf{B}^T\mathbf{S}\mathbf{A} + \mathbf{N}^T) + \mathbf{Q} \tag{20.21}$$

and the feedforward signal is

$$\mathbf{p}[k] = \mathbf{P}\mathbf{d}[k-1] + \mathbf{A}^T(\mathbf{I} - \mathbf{P}\mathbf{B}\mathbf{K}_d)\mathbf{p}[k+1]. \tag{20.22}$$

The vector $\mathbf{p}[k]$ is calculated by backward integration of (20.22) starting from terminal condition $\mathbf{p}[K+1] = 0$.

20.5 MPC Controller Design

Model predictive control (MPC) [12, 18] is one of the leading advanced control technologies in the process industries. The most attractive feature of MPC is the ability to accommodate complex performance objectives, dynamic systems and constraints in a unified framework. Similar to the process industries, the dynamics of water systems are relatively slow. Also, during control design, physical limitations and managing water level and flow within certain bounds need to be considered. MPC is a suitable controller design strategy for the current problem. Applications of MPC to water systems can be found in the literature [2, 21, 22].

While a deviation model is adopted in Sect. 20.4, in this section, for the purpose of MPC design, a slightly different non-deviation model is used, as shown below,

$$\begin{aligned}\mathbf{x}_m[k+1] &= A\mathbf{x}_m[k] + B\mathbf{u}_m[k] + w[k],\\ \mathbf{z}_m[k] &= C\mathbf{x}_m[k].\end{aligned} \tag{20.23}$$

Here $\mathbf{x}_m = \tilde{\mathbf{x}} + \mathbf{x}_{SP}$, $\mathbf{u}_m = \mathbf{u} + \mathbf{u}_{SP}$ where $\tilde{\mathbf{x}}, \mathbf{x}_{SP}, \mathbf{u}, \mathbf{u}_{SP}$ are defined in equation (20.15). The main control objectives are set-point tracking, minimizing energy consumption and disturbance rejection. As stated before, disturbance rejection can be achieved through feedforward. Denoting H_p prediction horizon and H_u control horizon, set-point tracking and minimizing energy consumption can be achieved by minimizing the following quadratic objective function at time k

$$J[k] = \sum_{i=1}^{H_p} \|\mathbf{z}_m[k+i|k] - \mathbf{r}[k+i|k]\|^2_{Q[i]} + \sum_{i=0}^{H_u-1} \|\Delta\mathbf{u}[k+i|k]\|^2_{R[i]}, \tag{20.24}$$

subject to (20.23), where $\mathbf{r}$ is a filtered version of set-point signals, $\Delta\mathbf{u}_m[i] := \mathbf{u}_m[i] - \mathbf{u}_m[i-1]$.

As introduced in Sect. 20.3, river operation is subject to many constraints. The constraints considered here comprise of hard constraints and soft constraints. Hard constraints are those which cannot be violated, for example, positive water level and flow rate, velocity of gate movement, $\mathbf{x} > 0$, $\mathbf{u} > 0$, $|\Delta\mathbf{u}_m| < \epsilon_u$. Soft constraints can be violated, but only for a very short period, to prevent the overall optimization problem infeasible, for example, the upper and lower bounds on outputs, $|\mathbf{z}_m| < \epsilon_z$. Now, a typical optimization problem at time k can be formulated as follows,

$$\textit{minimize } J[k], \textit{ subject to } (20.23) \textit{ and underlying constraints,} \tag{20.25}$$

where $J[k]$ is defined in (20.24). Once the optimal solution $\{\Delta\mathbf{u}[k|k], \Delta\mathbf{u}[k+1|k], \ldots, \Delta\mathbf{u}[k+H_u-1|k]\}$ to the optimization problem in (20.25) is obtained, only the first one $\Delta\mathbf{u}[k|k]$ is used to calculate the control action at time k, $\mathbf{u}[k] = \Delta\mathbf{u}[k|k] + \mathbf{u}[k-1]$.

20.6 Simulation Results

The following section outlines simulation results obtained using BasinCad, a computer aided design software tool for simulating river basin networks. Figures 20.4, 20.5(a), and 20.5(b) show storage water level, flow rate transients and rate of change

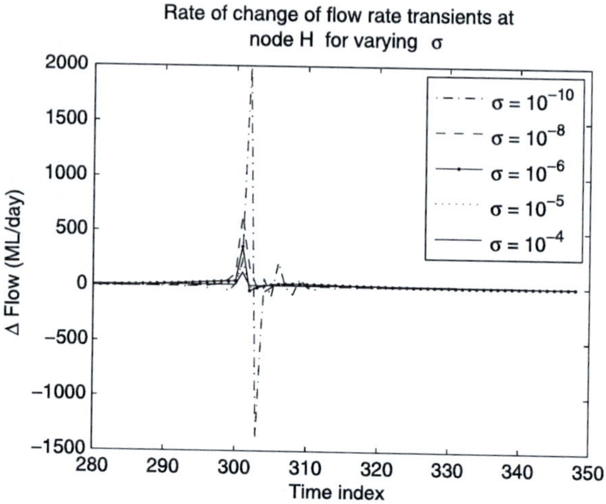

Fig. 20.4 Rate of change of flow-rate transients at storage H in response to disturbance flow q_D

of flow rate for different coefficients in weighting matrix **R**. For this example the relevant elements of **Q** were set to one. Using this figure, one can select the appropriate weighting coefficients based on actuator constraints. A unit step input disturbance with amplitude of 6 GL/day was introduced at q_D. Using Fig. 20.4 and assuming that 250 ML/day is the maximum permissible rate of change of flow rate, we select $\sigma = 10^{-6}$ for the results that follow.

Figures 20.6 and 20.7 illustrate the effect of upstream disturbances and compares the performance of the LQR and MPC controller schemes. In this example a step change at q_D of 1 GL/day is introduced at time index 300. Figures 20.6(a) and 20.6(b) illustrate the corresponding water level and flow rate transients at nodes H and A for the LQR controller. Figures 20.7(a) and 20.7(b) correspond to MPC controller. The results clearly indicate firstly the pre-release of water to accommodate the disturbance inflow. This is apparent from the rise in water levels and flow rates before time index 300. Secondly, the integral action inherent in this system smoothers out the transients as we move further downstream.

An important function for a controller is water pre-release for flood mitigation. This is demonstrated by generating a pulse disturbance of amplitude 10 GL/day over 25 days resulting a total volume of 250 GL. The immediate downstream node D has an output flow constraint set at 6 GL/day and a maximum water level of 486 m. A successful control strategy must incorporate pre-release to overcome the outflow constraint and maximum water level constraint. Figure 20.8 illustrates the water level and outflow from node D in response to the disturbance indicated by the dashed line. The key point to note is the mandatory pre-release which is evident in Fig. 20.8(a) between time indices 290 and 300.

It is well known that control of flow networks with transport lags require accurate knowledge of time delays. As indicated in the previous discussion, the transport delays in the link element change with flow rate. In this chapter we assume that these delays are constant, however this is not always the case. In the following example all link delays are overestimated in the model by one time index. Figure 20.9 shows

Fig. 20.5 Water level and flow transients at storage H in response to disturbance flow q_D

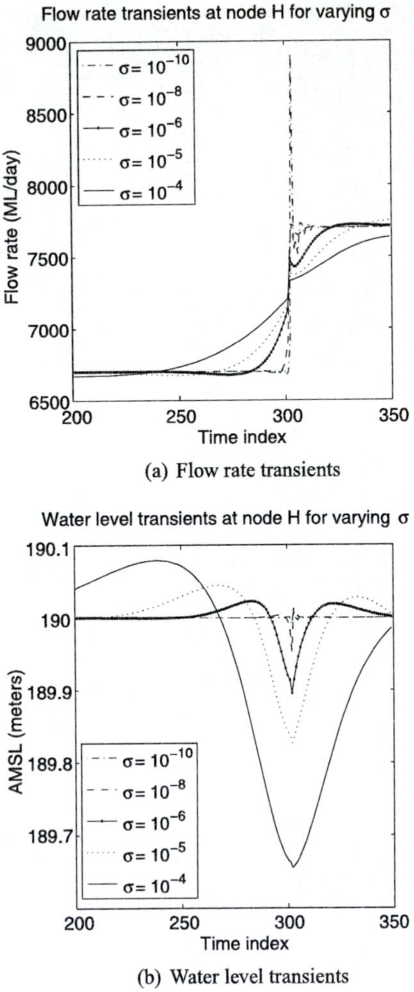

flow rate transients using LQR at storages A and H in response to delay mismatch. This can lead to potential instability at downstream nodes as illustrated in Fig. 20.9.

20.7 Conclusions and Further Work

This chapter has introduced a systematic framework of modeling and controlling river basin networks using simple linear models and optimal control principles. Two controller designed have been proposed based on LQR and MPC. This chapter has investigated the effects of disturbances, constraints, and sensitivity to transport delay and disturbance estimation. There are three important aspects that require further attention. Firstly, this chapter has assumed constant parameter linear models for

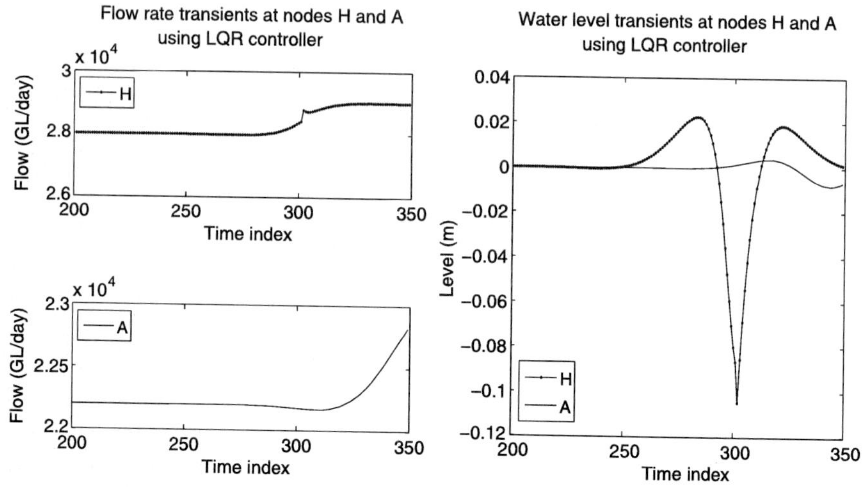

(a) Flow response (b) Level response

Fig. 20.6 LQR control: Disturbance rejection at storages A and H in response to disturbance flow q_D

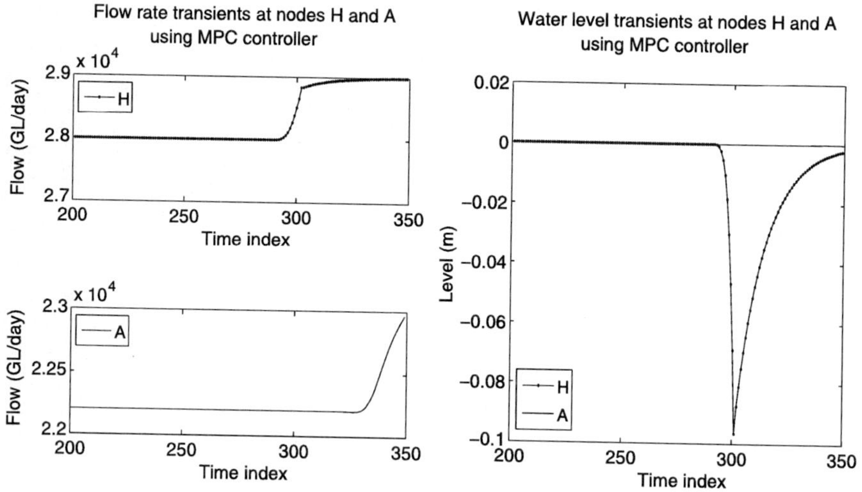

(a) Flow response (b) Level response

Fig. 20.7 MPC control: Disturbance rejection at storages A and H in response to disturbance flow q_D

both links and storage elements. In practice these elements will exhibit non-linear and time-varying characteristics. As mentioned previously, this is typical in river channels where time-delay varies with flow rate and water level changes in storages depend on geometry of the reservoir. Another example of such non-linearities is the

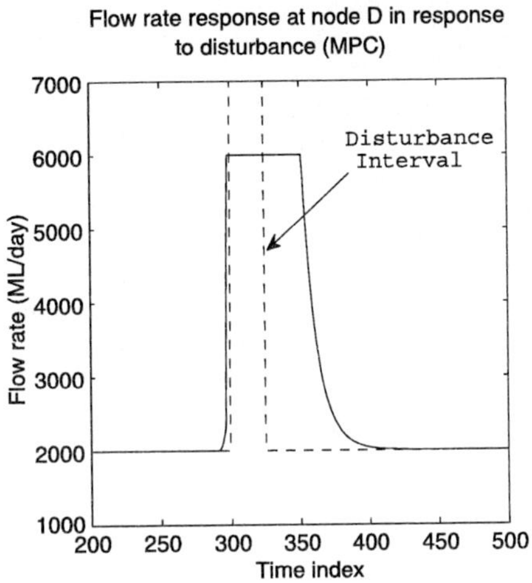

(a) Flow response

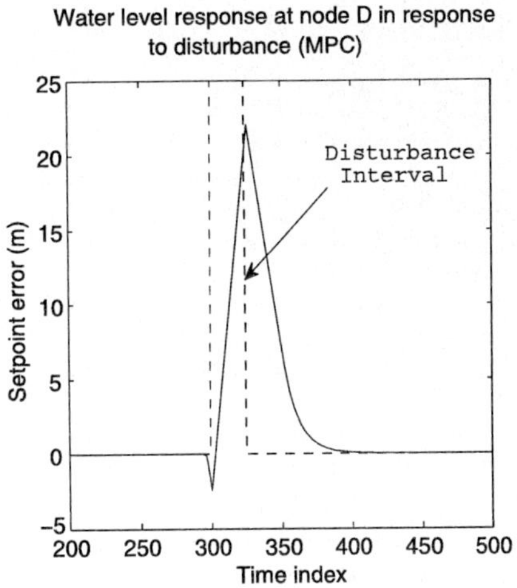

(b) Level response

Fig. 20.8 Flood mitigation using MPC

Fig. 20.9 Flow rate at storages A and H in response to delay mismatch

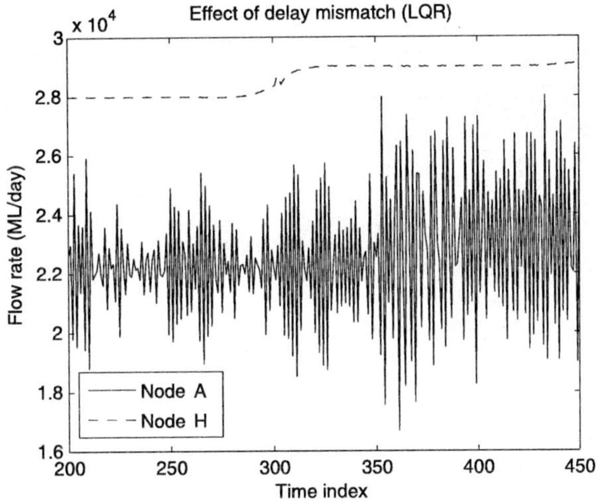

rainfall-runoff rate as a function of soil moisture. A potential practical solution to this problem is the use of gain switching mechanisms using a set of linear models that capture the relevant dynamics. This may be compatible with MPC, but perhaps not so with LQR. Secondly, as the sampling rate is increased the dimensionality of the problem may preclude the use of centralized control strategies described here. Future studies will investigate the application of distributed control to address this challenge. A third important extension of this work is to incorporate water quality and groundwater reservoirs in the problem formulation. This poses a significant modeling challenge.

References

1. Baume, J.P., Malaterre, P.O., Sau, J.: Tuning of PI controllers for an irrigation canal using optimization tools. In: Proceedings of USCID Workshop on Modernization of Irrigation Water Delivery Systems, Phoenix Arizona, USA (1999)
2. Blanco, T.B., Willems, P., De Moor, B., Berlamont, J.: Flood prevention of the demer using model predictive control. In: Proceedings of the 17th IFAC World Congress, pp. 3629–3934 (2008)
3. Bridgart, R.J., Bethune, M.: Development of RiverOperator: a tool to support operational management of river systems. In: 18th World IMACS/MODSIM Congress, pp. 3782–3788 (2009)
4. Chow, V.T.: Open-Channel Hydraulics. McGraw-Hill Book Company, New York (1988)
5. Georgakakos, A.P.: Decision support systems for integrated water resources management with an application to the Nile Basin. In: Proceedings IFAC Workshop on Modeling and Control for Participatory Planning and Managing Water Systems, Venice, Italy (2004)
6. Gilmore, R.L., Kuczera, G., Penton, D., Podger, G.: Improving the efficiency of delivering water in Australian river systems: modelling multiple paths. In: 18th World IMACS/MODSIM Congress, pp. 225–231 (2009)
7. Jakeman, A.J., Horngerger, G.M.: How much complexity is warranted in a Rainfall-Runoff model. Water Resour. Res. **29**, 2637–2649 (1993)
8. Labadie, J.W.: Optimal operation of multireservoir systems: state-of-the-art review. J. Water Resour. Plan. Manag. **130**, 93–111 (2004)
9. Litrico, X, Georges, D.: Robust continuous-time and discrete-time flow control of a dam-river system (I) modelling. Appl. Math. Model. **23**, 809–827 (1999)
10. Litrico, X., Georges, D.: Robust continuous-time and discrete-time flow control of a dam-river system (II) controller design. Appl. Math. Model. **23**, 829–846 (1999)
11. Litrico, X., Fromion, V.: H infinity control of an irrigation canal pool with a mixed control politics. IEEE Trans. Control Syst. Technol. **1**(14), 99–111 (2006)
12. Maciejowski, J.M.: Predictive Control with Constraints. Prentice Hall, London (2002)
13. Mareels, I., Weyer, E., Ooi, S.K., Cantoni, M., Yuping, Li, Nair, G.: Systems engineering for irrigation systems: Successes (2005)
14. Marinaki, M., Papageorgiou, M.: Central flow control in sewer networks. J. Water Resour. Plan. Manag. **123**(5), 274–283 (1997)
15. Marinaki, M., Papageorgiou, M.: Optimal Real-Time Control of Sewer Networks. Springer, London (2005)
16. Marwali, M.N., Keyhani, A.: Control of distributed generation systems—Part I voltages and current control. IEEE Trans. Ind. Electron. **19**(6), 1541–1550 (2004)
17. Maxwell, M., Warnick, S.: Modeling and identification of the Sevier River System. In: American Control Conference, pp. 5342–5347 (2006)
18. Mayne, D.Q., Rawlings, J.B., Rao, C.V., Scokaert, P.O.M.: Constrained model predictive control: stability and optimality. Automatica **36**(6), 789–814 (2000)
19. Murray-Darling Basin Water Resources Fact Sheet: Murray-Darling Basin Commission—July 2006. http://www2.mdbc.gov.au/. Cited July 27, 2010
20. Running the river: Murray-Darling Basin Commission. http://www2.mdbc.gov.au/. Cited July 27, 2010
21. Negenborn, R.R., Van Overloop, P.-J., Keviczky, T., De Schutter, B.: Distributed model predictive control of irrigation canals. Netw. Heterog. Media **4**, 359–380 (2009)
22. Overloop, P.J.: Model predictive control of open water systems. PhD thesis, Delft University of Technology, Delft, The Netherlands (2006)
23. Papageorgiou, M.: Optimal control of generalized flow networks. In: System Modelling and Optimization. Lecture Notes in Control and Information Sciences, vol. 59, pp. 373–382 (1984)
24. Sage, A.P., White, C.C.: Optimum Systems Control. Prentice-Hall, Upper Saddle River (1977)
25. Sawadogo, S., Malaterre, P.O., Kosuth, P.: Multivariable optimal control for on-demand operation of irrigation canals. Int. J. Syst. Sci. **1**(26), 161–178 (1995)

26. Setz, C., Heinrich, A., Rostalski, P., Papafotiou, G., Morari, M.: Application of model predictive control to a cascade of river power plants. In: Proceedings of the 17th World Congress IFAC, pp. 11978–11983 (2008)
27. Shamma, J.S., Athans, M.: Analysis of gain scheduled control for nonlinear plants. IEEE Trans. Autom. Control **35**(8), 898–907 (1990)
28. Venkat, A.N., Hiskens, I.A., Rawlings, J.B., Wright, S.J.: Distributed MPC strategies with application to power system automatic generation control. IEEE Trans. Control Syst. Technol. **16**(6), 1192–1206 (2008)
29. Weyer, E.: System identification of an open water channel. Control Eng. Pract. **9**(12), 1289–1299 (2001)
30. Weyer, E.: Decentralized PI control of an open water channel. In: IFAC, 5th Triennial World Congress, Barcelona, Spain (2002)
31. Weyer, E.: LQ control of an irrigation channel. In: Proceedings of 42nd IEEE Conference on Decision and Control, vol. 1, pp. 750–755 (2003)
32. Winn, C.B., Moore, J.B.: The application of optimal linear regulator theory to a problem in water pollution. IEEE Trans. Syst. Man Cybern. **3**, 450–455 (1973)
33. Young, P.C.: Data-based mechanistic modelling of environmental, ecological, economic and engineering systems. Environ. Model. Softw. **13**, 105–122 (1998)

Chapter 21
Modelling of Rivers for Control Design

Mathias Foo, Su Ki Ooi, and Erik Weyer

21.1 Introduction

The increase in world population and the growth of farming have created an increased demand for water. Agricultural accounts for about 70% of the world's freshwater use [42], and the operational losses in the delivery of water to farms are large. After more than a decade of drought in Southern Australia, it has become increasingly important to explore new farming practices and strategies for management of water in order to prepare for a drier future. Such a complex resource management issue calls for an interdisciplinary approach including agricultural science, engineering, ecology, hydrology, economics, social sciences, etc. The research described in this chapter is part of the project "Farms, Rivers, and Markets" (FRM), which was initiated by Uniwater, a joint research initiative by The University of Melbourne and Monash University in response to the above challenges.

As the name suggests, the project consists of three key integrated components: Farms, Rivers and Markets. The aim of the Farms project is to explore how the various sources of water can be used in flexible combinations to make farming operations more resilient. The Rivers project is concerned with the development of

M. Foo (✉)
National ICT Australia, Victoria Research Lab, Department of Electrical and Electronic Engineering, The University of Melbourne, Parkville, VIC 3010, Australia
e-mail: mfoo@ee.unimelb.edu.au

S.K. Ooi · E. Weyer
Department of Electrical and Electronic Engineering, The University of Melbourne, Parkville, VIC 3010, Australia

S.K. Ooi
e-mail: skoo@ee.unimelb.edu.au

E. Weyer
e-mail: ewey@unimelb.edu.au

L. Wang, H. Garnier (eds.), *System Identification, Environmental Modelling, and Control System Design*,
DOI 10.1007/978-0-85729-974-1_21,

systems for managing the water capable of handling the needs of irrigators and the environment in a cooperative way. The Markets project aims at developing new water products and services better suited to future demands from consumers and the environment.

Modelling and control system have important parts to play in the Rivers project since well designed control systems for river flows and levels will allow for a more efficient distribution of water leading to reduced operational water losses. In addition, it will allow for a more accurate and timely delivery of water to farmers while ensuring that the environmental and ecological water needs are satisfied.

Broadly speaking the aim of the control system is to improve water resource management and operation for the benefit of consumptive users and the environment. However, what the specific control objectives should be is not fully understood. Most likely there will be a change in farming practices due to less available water in the future which again will change the demand patterns for irrigation water. On the legislative side, higher priorities have been given to environmental water demands in the Water Act 2007 [43] in order to ensure a sustainable water supply to protect and restore environmental assets such as wetlands and streams. Part of the River project is to investigate what constitute desirable flows and water levels from an environmental and ecological point of view, and this will have an impact on the control objectives. E.g. instead of keeping the water levels at constant setpoints, the control system may in the future aim to recreate more natural flow conditions.

Under the National Water Initiative, water trading out of catchment is allowed between different entities, see [28]. Water trading allows scarce water resources to be transferred to their most productive uses, and it is anticipated that increased water trading will take place in the future. This has the implication that the demand patterns and locations will change with corresponding changes in the control objectives.

In order to design a well functioning control system, a model of the river which captures the relevant dynamics is required. Most models used to describe rivers are either too complex (partial differential equations) or operate on a too slow time scale (days and weeks) to be used for control design. An important part of the control design is therefore to find models which capture the important dynamics of the river and which are suitable for control design.

Previous work on modelling and control of irrigation channels (see e.g. [9, 20, 24, 29, 35, 44]), have demonstrated that control systems can yield significant improvements in the quality of service and water distribution efficiency. A key to this has been the use of system identification techniques to develop simple models useful for control design (see e.g. [29]) and hence particular attention will be paid to system identification of rivers. There are a number of works on modelling and control of rivers, e.g. [3, 10, 15, 18, 19, 26, 32, 36, 37, 39, 40, 46, 48]. Unlike the problem considered in this chapter, most of the modelling studies have focused on flood events and the control objectives are often connected to the operation of hydro-electric power plants.

This chapter is organised as follows. Section 21.2 describes the Broken River. Modelling and system identification for the purpose of control design are discussed in Sect. 21.3, while control design using the obtained models is considered in Sect. 21.4.

21.2 Description of the River and the Catchment

21.2.1 The Broken River

Figure 21.1 shows a map of the Broken River in Victoria, Australia. The Broken basin covers 7,724 km^2 of catchment area, and rainfall varies from 1,000 mm per year in the upper catchment to less than 500 mm per year in the lower catchment, [7]. The primary entitlements for water shares, licenses and associated commitments in the Broken system are 17,929.8 ML high-reliability water shares and 3,338.3 ML low-reliability water shares. The environment is protected with minimum flow requirements ranging from the natural flow to 25 ML/day (0.2894 m^3/s), [8].

The river originates from Lake Nillahcootie which stores 40 GL of water. Typical historical releases into the Broken River is about 15 ML/day (0.1736 m^3/s) during winter and in the range 50 to 60 ML/day (0.5787 to 0.6944 m^3/s) during summer. These releases are expected to increase due to the decommissioning of an artificial lake (not shown on the map), which also contributed to the flow in the lower parts of the river. The length of the river from Lake Nillahcootie to Gowangardie Weir (HS4) is about 75.8 km.[1] In this paper, we focus on the reach between Casey's Weir (HS3) and Gowangardie Weir (HS4) and the stretch from Lake Nillahcootie to Lake Benalla.

21.2.2 Description of the Reach Between Casey's Weir and Gowangardie Weir

Casey's Weir is a free overflow weir about 50 km downstream of Lake Nillahcootie. Upstream of the weir, there are three manually operated gates used to divert water into Broken Creek. Gowangardie Weir is also a free overflow weir located 27 km downstream of Casey's Weir. Figure 21.2 shows a side view of the reach. The water level at Casey's Weir and Gowangardie Weir are denoted by y_C and y_G respectively. They are measured with respect to the sea level, and the unit is meter Australian Height Datum (mAHD). The heights of the weirs relative to sea level are denoted by p_C and p_G. The height of the water above the weirs are called head over weir and given by $h_C = y_C - p_C$ and $h_G = y_G - p_G$ respectively. The physical parameters of the river include reach length, L, bottom slope, S_0, side slope, s, bottom width, b, top width, T, wetted perimeter, P and wetted cross sectional area, A (see Figs. 21.2 and 21.3).

[1]Obtained by approximating the rivers by straight lines between the hydraulic structures (HS) on the map.

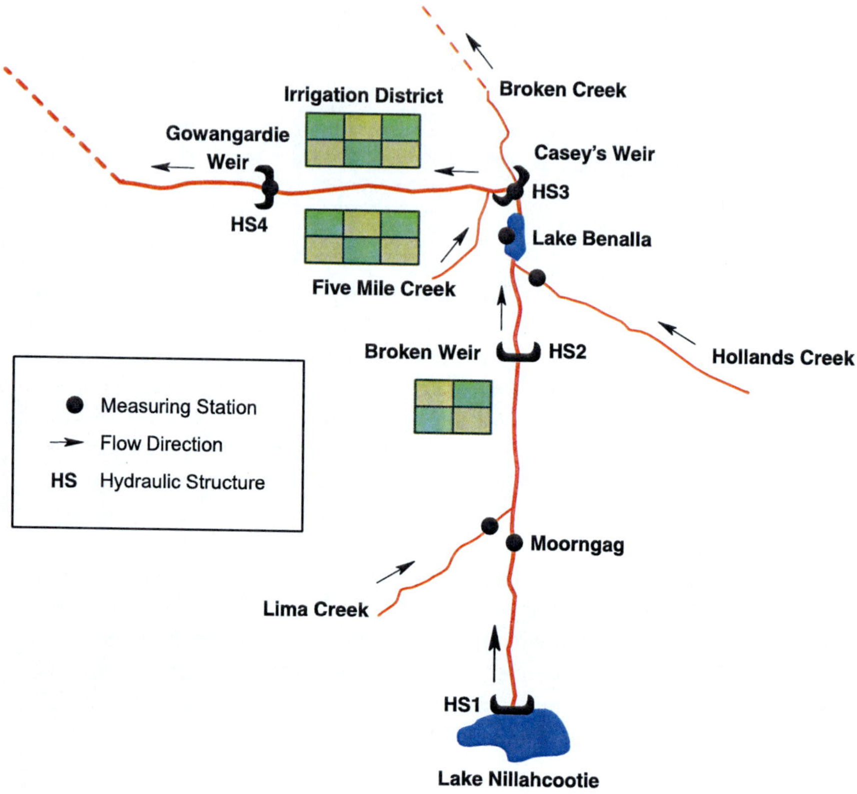

Fig. 21.1 Map of Broken River (not to scale)

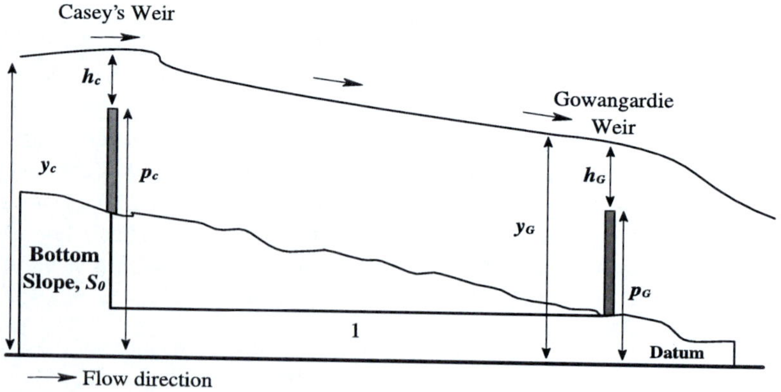

Fig. 21.2 Side view of the reach between Casey's Weir and Gowangardie Weir (not to scale)

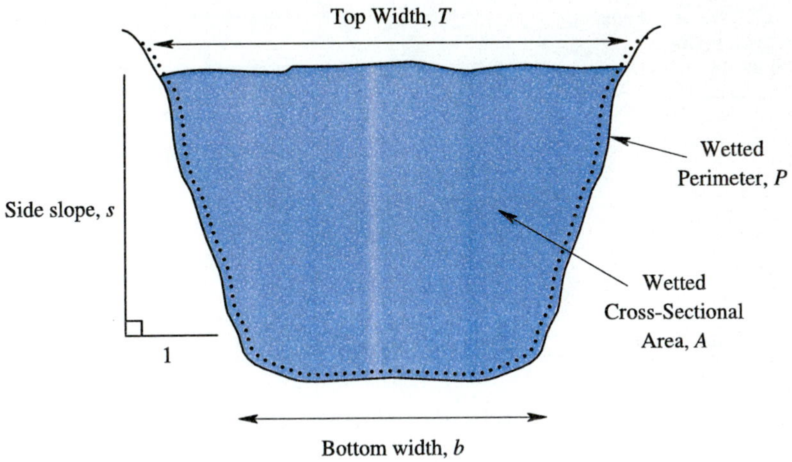

Fig. 21.3 Cross sectional view of the reach between Casey's Weir and Gowangardie Weir (not to scale)

21.3 Modelling of Rivers

In this section we consider system identification and Saint Venant equations models of river reaches, and we assess the accuracy of the models against operational data.

21.3.1 The Saint Venant Equations

Under the assumption that the flow is one-dimensional, velocity is uniform and that there are no lateral in-flows and out-flows along the river reach, the Saint Venant equations are given by (see e.g. [11])

$$\begin{aligned}
&\frac{\partial A}{\partial t}+\frac{\partial Q}{\partial x}=0,\\
&\frac{\partial Q}{\partial t}+\left(\frac{gA}{T}-\frac{Q^2}{A^2}\right)\frac{\partial A}{\partial x}+\frac{2Q}{A}\frac{\partial Q}{\partial x}+gA(S_f-S_0)=0,
\end{aligned}\tag{21.1}$$

where Q is the flow, A is the wetted cross sectional area, T is the top width, $g=9.81$ m/s^2 is the gravity constant, S_0 is the bottom slope and S_f is the friction slope. The friction slope is given by $S_f=n^2Q^2P^{4/3}A^{-10/3}$, where P is the wetted perimeter and n is the Manning friction coefficient which represents the effect of flow resistance and river roughness. Here, we assume a trapezoidal cross section, hence, $A=(b+sy)y$, $T=b+2sy$ and $P=b+2y\sqrt{1+s^2}$ where b, s and y are the bottom width, side slope and water level respectively. The boundary conditions are the flows over Casey's and Gowangardie weirs which are both sharp

Table 21.1 River parameters of the reach between Casey's Weir and Gowangardie Weir

Parameters	Values
Reach length, L	26.7 km
Bottom width, b	9.0–12.0 m
Side slope, s	2.0–3.0
Bottom slope, S_0	0.0008–0.0020
Manning friction coefficient, n	0.060–0.085

crested weirs, and thus, the flow can be approximated by [6],

$$Q_C(t) \approx c_C h_C^{3/2}(t) = c_C[y_C(t) - p_C]^{3/2}, \tag{21.2}$$

$$Q_G(t) \approx c_G h_G^{3/2}(t) = c_G[y_G(t) - p_G]^{3/2}. \tag{21.3}$$

The constants, c_C and c_G can be approximated by $0.6\sqrt{g}b_{w,C}$ and $0.6\sqrt{g}b_{w,G}$ respectively [5], where $b_{w,C}$ and $b_{w,G}$ are the width of the weirs. The equations are simulated using the Preissmann scheme, a finite difference method (see e.g. [2, 11, 14]). The initial values used for the flows and water levels are given by the steady state solution of (21.1). The input to the simulation scheme is the measured flow over Casey's Weir, which is also the upstream boundary condition, and the output is the water level at Gowangardie Weir from which the downstream boundary condition can be computed via (21.3). Based on the Hydrologic Engineering Center—River Analysis System (HEC-RAS) model in [12], the approximate river parameters for the reach are summarised in Table 21.1.

As the values in Table 21.1 are only approximate, the Saint Venant equations are calibrated against observed data. The flow and water level measurements[2] with 15 minutes sampling period for the months April to July 2001 are shown in Fig. 21.4. This period corresponds to autumn and winter where there are few withdrawals for irrigation, and the assumption that there are no lateral out-flows is more likely to be satisfied. It has been shown in [17] that for the purpose of simulating the water level at Gowangardie Weir, it is sufficient to represent the river as a straight stretch with constant geometries and to tweak the friction coefficient in order to account for the meandering of the river and the variation in the river parameters. Thus, the river parameters used are the average values of the bottom width and the side slope, which are 10.5 m and 2.5 m respectively. The average bottom slope is (161.07–137.04 m)/26700 m $\approx 9.0 \times 10^{-4}$, where 161.07 and 137.04 are the elevation in mAHD of Casey's and Gowangardie weirs respectively, while 26700 m is the length of the reach. The weir constant, c_G at Gowangardie Weir and the Manning friction coefficient, n are estimated from the data using a prediction error method

[2]Only the water levels are measured. The flows are obtained from the water levels using rating curves.

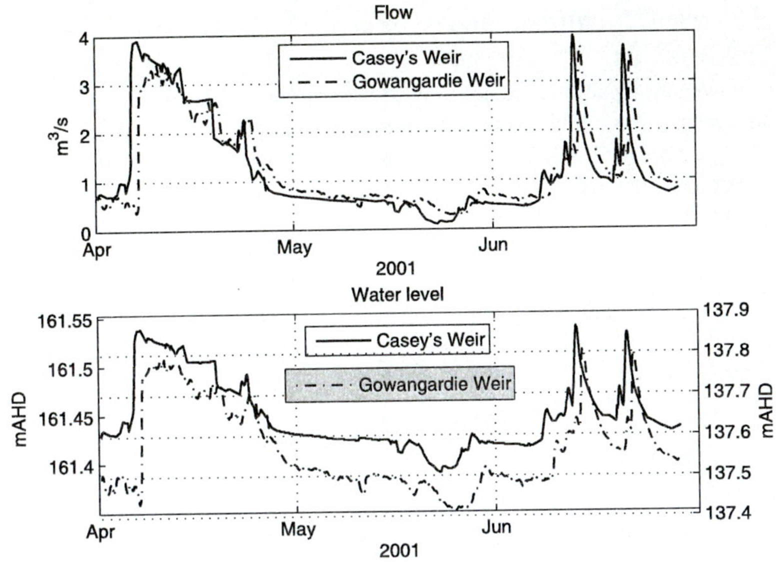

Fig. 21.4 Measurements at Casey's Weir and Gowangardie Weir from April to June 2001. *Top*: Flows. *Bottom*: Water levels at Casey's Weir (*left-axis*) and Gowangardie Weir (*right-axis*)

with quadratic criterion, i.e.

$$\hat{\theta} = \underset{\theta}{\operatorname{argmin}} \frac{1}{N} \sum_{t=1}^{N} [y_{mea}(t) - \hat{y}_{sim}(t, \theta)]^2, \tag{21.4}$$

where $N = 8640$ is the number of data points, $\theta = [c_G, n]^T$, y_{mea} is the measured water level and $\hat{y}_{sim}$ is the simulated water level using the Saint Venant equations. The estimated weir constant is $c_G = 10.10$ m$^{3/2}$/s while the estimated Manning friction coefficient is $n = 0.146$. This value for n is higher than the values given in Table 21.1. Those values of Manning friction coefficient correspond to a particular section of the river reach, and they do not include the effect of meandering. Incorporating the meandering effect results in a larger estimated value of the Manning friction coefficient. The estimation results and the accuracy of the Saint Venant equations are further discussed in Sects. 21.3.3 and 21.3.5.

The length in Table 21.1 is the length of a straight line ("as the crows fly") between Casey's and Gowangardie weirs. In [38], the length of the river itself between the two weirs is estimated to 36600 m. With this value and the corresponding bottom slope, $S_0 = 6.6 \times 10^{-4}$ the estimated Manning coefficient is $n = 0.079$ which is in agreement with the values in Table 21.1. The estimate of the weir constant is 10.11 m$^{3/2}$/s which is nearly the same as before. Using these values for L, S_0, c_G and n only leads to very minor changes to the results in Sects. 21.3.3 and 21.3.5, and hence they are not reported.

21.3.2 System Identification Approach

The Saint Venant equations are not easy to use for control design, and we therefore seek simpler models which capture the relevant dynamics for control design. From Fig. 21.4 we observe that the flow measurements show a lag between Casey's and Gowangardie weirs and this indicates that the system can be modelled as a time-delay system, i.e.,

$$Q_G(t) = Q_C(t-\tau), \tag{21.5}$$

where τ is the time delay and the subscripts 'C' and 'G' denotes the Casey's and Gowangardie weirs respectively. Using (21.2) and (21.3), equation (21.5) can be rewritten as

$$c_G h_G^{3/2}(t) = c_C h_C^{3/2}(t-\tau)$$

$$\Downarrow$$

$$c_G[y_G(t) - p_G]^{3/2} = c_C[y_C(t-\tau) - p_C]^{3/2} \tag{21.6}$$

$$\Downarrow$$

$$y_G(t) = \gamma_1 y_C(t-\tau) + \gamma_2, \tag{21.7}$$

$\gamma_1 = (c_C/c_G)^{2/3}$ and $\gamma_2 = p_G - (c_C/c_G)^{2/3} p_C$ are unknown constants, which are estimated from the observed data together with the time delay. The associated predictor for (21.7) is given by

$$\hat{y}_G(t, \theta, \tau) = \gamma_1 y_C(t-\tau) + \gamma_2. \tag{21.8}$$

Note that the parameterised mass balance model $Q_G(t) = \alpha Q_C(t-\tau)$ used in [25] leads to the same predictor, (21.8), but with different expressions for γ_1 and γ_2.

As in [25], the time delay, τ is estimated from the cross-correlation between the flow measurements at the upstream and downstream ends. The cross-correlation is shown in Fig. 21.5 and computed from

$$R_{Q'_G Q'_C}(\tau) = \frac{1}{N-\tau} \sum_{n=\tau+1}^{N} Q'_G(n) Q'_C(n-\tau), \quad \tau = 0, \pm 1, \ldots, \tag{21.9}$$

where

$$Q(n)' = Q(n) - \frac{1}{N} \sum_{j=1}^{N} Q(j), \quad N = 8640 \quad \text{and} \quad \hat{\tau} = \operatorname*{argmax}_{\tau} R_{Q'_G Q'_C}(\tau).$$

The parameter vector $\theta = [\gamma_1, \gamma_2]^T$ is estimated using least squares, i.e.

$$\hat{\theta} = \left[\sum_{t=\tau+1}^{N} \varphi(t)\varphi^T(t) \right]^{-1} \left[\sum_{t=\tau+1}^{N} \varphi(t) y_G(t) \right], \tag{21.10}$$

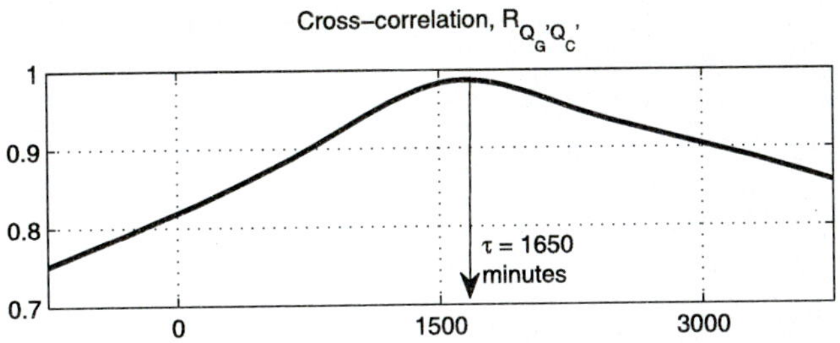

Fig. 21.5 Cross-correlation between flows at Casey's Weir and Gowangardie Weir

Table 21.2 Parameter estimates

$\hat{\gamma}_1$	$\hat{\gamma}_2$	$\hat{\tau}$
2.711	−300.05	1650 mins

where $\varphi(t) = [y_C(t-\tau), 1]^T$. The data set shown in Fig. 21.4 is used for estimation and the estimated values are given in Table 21.2.

21.3.3 Accuracy of the Saint Venant Equations and the Time-Delay Model

The accuracy of the Saint Venant equations and the time-delay model were compared on data sets not used for estimation. The data sets are from summer 2001 and winter 2002. Figure 21.6 shows the measured water levels, predicted water levels using the time-delay model and the simulated water levels from the calibrated Saint Venant equations. The mean square errors, (MSE) and the coefficient of determination, R_T^2 are given by

$$\text{MSE} = \frac{1}{N-\tau} \sum_{t=\tau+1}^{N} [y_G(t) - \hat{y}_G(t)]^2, \tag{21.11}$$

$$R_T^2 = 1 - \frac{\hat{\sigma}^2}{\sigma_Y^2}, \tag{21.12}$$

where $y_G(t)$ is the measured water level and $\hat{y}_G(t)$ is the water level predicted by the time-delay model or simulated using the calibrated Saint Venant equations. $\hat{\sigma}^2$ is variance of the model residuals (i.e. the MSE) and $\sigma_Y^2 = \frac{1}{N-\tau}\sum_{t=\tau+1}^{N}[y_G(t) - \bar{y}_G(t)]^2$ where $\bar{y}_G(t) = \frac{1}{N-\tau}\sum_{t=\tau+1}^{N} y_G(t)$. R_T^2 tells us how well the data is explained by the model. The closer the value of R_T^2 is to unity, the better the model explains the data. The MSE and R_T^2 are given in Table 21.3. From the bottom part of Fig. 21.6, it can be seen that both the time-delay model and the Saint Venant equations are accurate when compared to the measured water levels during the winter

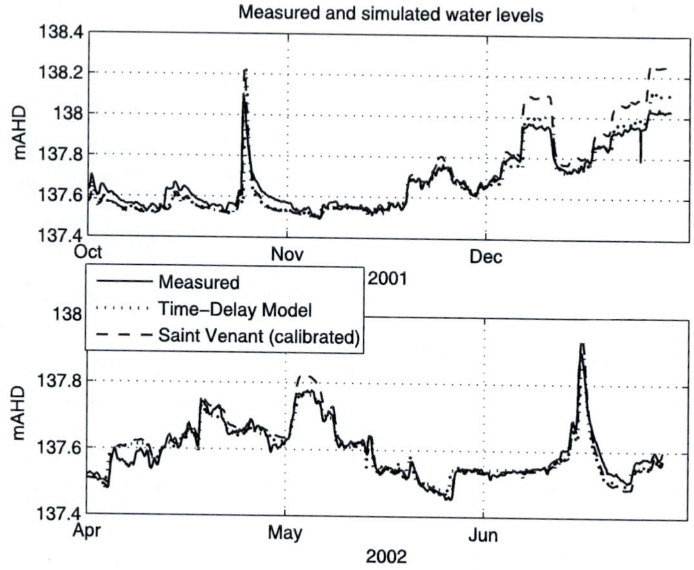

Fig. 21.6 Measured and simulated water levels at Gowangardie Weir. *Top*: Spring/summer period 2001. *Bottom*: Autumn/winter period 2002

Table 21.3 Values of MSE and R_T^2

Data period	Time-delay model		Saint Venant equations	
	MSE (10^{-3} m^2)	R_T^2	MSE (10^{-3} m^2)	R_T^2
Summer 2001	2.25	0.904	5.38	0.782
Winter 2002	0.72	0.887	0.52	0.920

period. They pick up the trends in the data very well, and the MSE values are small. The values of R_T^2 are close to unity indicating that the model explains the data well. The results from the summer period when larger volumes of water are taken from the river for irrigation are shown in the top part of Fig. 21.6. The water levels simulated using the Saint Venant equations and the time-delay model are higher than the measured water levels. This is expected as the irrigation off-takes are not taken into account in the models. The results show that both models are accurate in describing the relevant dynamics of the river systems. The time-delay model is much simpler than the Saint Venant equations, and it is preferred for control design.

21.3.4 Frequency Analysis

The Saint Venant equations and the time-delay model showed similar behaviour when compared against time domain data in Sect. 21.3.3. Many control design

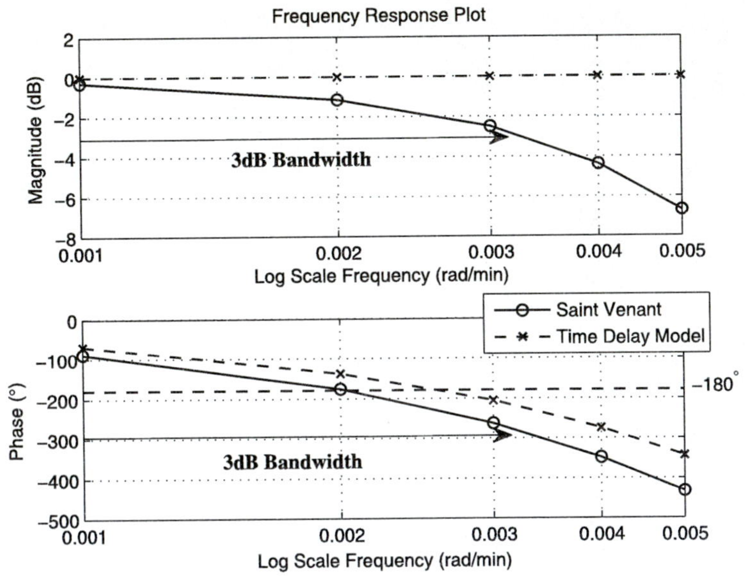

Fig. 21.7 Bode plot for the reach between Casey's Weir and Gowangardie Weir

methods are based on frequency domain considerations, and it is therefore of interest to compare the frequency domain properties of the two models. Despite being nonlinear partial differential equations, the Saint Venant equations display a nearly linear behaviour to sinusoidal flow inputs. A number of simulated sine wave tests were carried out on the Saint Venant equations where the upstream flow was given by $(0.75\sin(\omega t) + 3.00)$ m^3/s. The magnitude and the phase of the downstream flow were recorded and the Bode plot in Fig. 21.7 was obtained. In the frequency range relevant for control the frequency response of the Saint Venant equations is similar to the frequency response of the time-delay model. The 3 dB bandwidth of the system is approximately 0.0032 rad/min, but we notice that the phase shift is already more than $-180°$ at 0.0023 rad/min, indicating a dominant time delay.

21.3.5 Analysis of the Effect of Varying Flow Conditions

It is known (see e.g. [22, 41]) that the flow conditions affect the time delay in a river which again will affect the robustness margins of a control system. From the available data, sets with different flows were found. The time delays for those data sets were estimated using cross-correlation analysis as in Sect. 21.3.2. In addition, we also included the cross-correlation between the upstream flow measurements at Casey's Weir and the simulated downstream flow at Gowangardie Weir obtained from the Saint Venant equations. The estimated time delays are shown in Fig. 21.8,

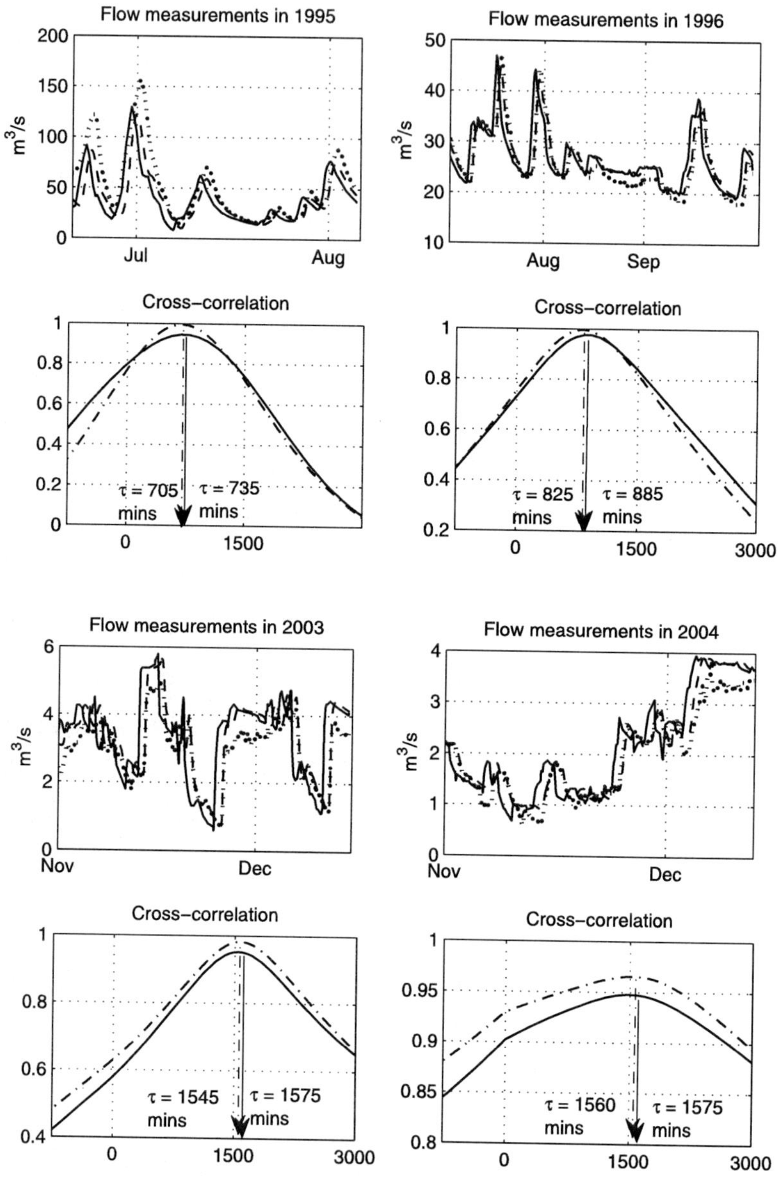

Fig. 21.8 Cross-correlations and estimated time delays. Flows plot. *Solid line*: Casey's Weir, *Dotted line*: Gowangardie Weir (measured), *Dashed*: Gowangardie Weir (simulated). Cross-correlation plots. *Solid line*: Cross-correlation from measured data. *Dash-dotted line*: Cross-correlation from measured and simulated data

and as expected the estimated time delay decreases with higher flows. In addition, there is good agreement between the cross-correlations obtained using the measured data and those obtained using the Saint Venant equations, reconfirming the accuracy of the Saint Venant equations. The varying time delay must be taken into account in the robustness specification of the controllers. This is further discussed in Sect. 21.4.

21.3.6 Discussion of the Models

21.3.6.1 Undermodelling

A number of factors such as in-flows and out-flows from creeks, rainfall, water withdrawals for irrigation, evaporation and surface-water/ground-water interactions have been ignored in the models. Here we briefly discuss the influence of these factors taking into account that the models are going to be used for control design.

Surface-water/ground-water interactions. Some stretches of the Broken River are gaining water from the ground water, while others are loosing. The reach between Casey's Weir and Gowangardie Weir is a loosing reach [1]. However, the surface-water/ground-water dynamics is slow, and it is not considered important for control. Moreover, if the surface-water/ground-water interaction is modelled as a constant in- or out-flow, $Q_{SW/GW}$, then (21.5) becomes $Q_G(t) = Q_C(t-\tau) - Q_{SW/GW}$, and we will still end up with the model structure (21.7), but with a different expression for γ_2. This is however of no importance as γ_2 is estimated from data. Furthermore a controller with integral action will reject constant unmodelled in- or out-flows.

Evaporation. The rate of evaporation is dependent on temperature, solar radiation, wind speed, atmospheric pressure, area of water surface, etc. [16]. In [46], temperature was included as an input variable in a rainfall-flow model, and the flow showed a long term dependence on temperature which could account for the effect of evaporation. As with the surface-water/ground-water interactions, the loss due to evaporation from the river is a disturbance and its effect on the levels and flows is of lesser importance for control. Evaporation from storages may be significant [13], and this may influence how the storages are operated and hence also the control objectives.

Water withdrawal from the river for irrigation. The withdrawals can be large and they can have a big impact on the predictive accuracy of the models. From a control point of view the withdrawals are load disturbances which should be rejected. The farmers order their water some days in advance, and better control can be achieved by releasing water early using feedforward action to match the amount of ordered water (see Sect. 21.4). However, as water withdrawals act as disturbances the transfer function from the in-flow we can manipulate to the water level or flow we want to control remains the same. That is, the transfer function on which a feedback control design is based remains the same, although the estimate of it may become more uncertain if there are large water withdrawals which have not been taken into account.

In- and out-flows from creeks and rainfall. If measured, the in- and out-flows from creeks can easily be included in the models (see Sect. 21.3.7), and they can also be accounted for in the controller by e.g. regarding them as part of the flow to be released by the controller. When the flows in the creeks are not measured, rainfall-runoff models (see e.g. [3, 46–49] and the references therein) are useful for estimating the additional contributions from creeks and rain. Even when the flows and water levels in creeks are measured, it may be of interest to predict flows into the future if the time delays associated with the flows in the creeks are much smaller than the time delays of the flows commanded by the control system.

21.3.6.2 Use of the Time-Delay Model for Control

Although the time-delay model gives a good representation of a river reach in the time and frequency domain as illustrated in the previous sections, some care needs to be exercised when using it for control design. One key aspect in the previous sections was that the downstream flow could not be manipulated, and the downstream flow was simply modelled as the delayed upstream flow, i.e. $Q_D(t) = Q_U(t - \tau)$. However, if the downstream flow Q_D can be set independently of the upstream flow Q_U (e.g. by regulation gates or valves), the time-delay model is obviously not going to be a good model. This point must be kept in mind if hydraulic structures are changed, e.g. if fixed weirs are replaced by regulation gates.

The time-delay model also assumes that there is little storage capabilities in the river reach in the sense that the volume of water in the reach is nearly constant. This may not be a valid assumption, particularly if the river flows through a lake as the Broken River does at Lake Benalla. In both the above cases an integrator-delay model of the type

$$\dot{V}(t) = Q_U(t - \tau) - Q_D(t)$$

seems more appropriate where V is the volume of water in (a part of) the reach. This model structure will be discussed in the next section.

21.3.7 *Integrator-Delay Models*

We consider the reach from Moorngag[3] to Lake Benalla (see Fig. 21.1). Two creeks, Lima Creek and Hollands Creek, from which we have measurements enter this reach. The data are shown in Fig. 21.9.

Due to the storage in the lake, we use the mass balance equation,

$$\dot{V}_{LB}(t) = Q_M(t - \tau_M) + Q_{LC}(t - \tau_{LC}) + Q_{HC}(t - \tau_{HC}) - Q_{LB}(t), \quad (21.13)$$

where V is the volume of Lake Benalla, Q_i and τ_i with i = M (Moorngag), LC (Lima Creek), HC (Hollands Creek) and LB (Lake Benalla) are the flows and time

[3]We have chosen Moorngag rather than Lake Nillahcootie as the upstream end of the reach simply because there are better data available from Moorngag.

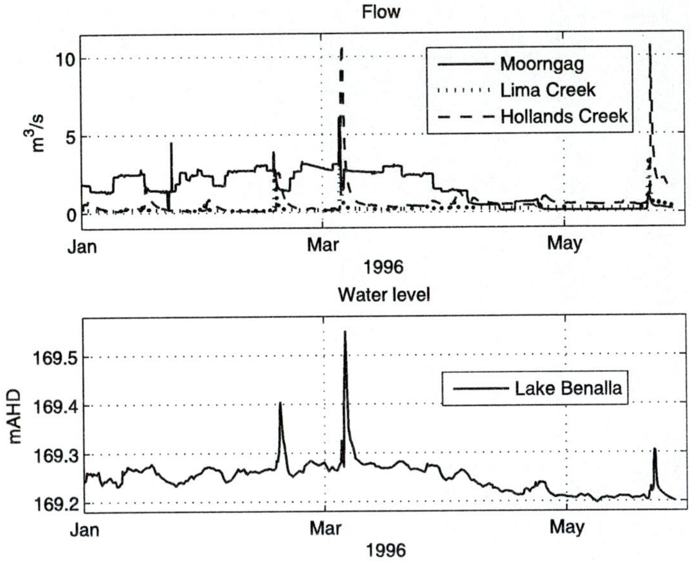

Fig. 21.9 Measurements at Moorngag, Lima Creek, Hollands Creek and Lake Benalla from January to May 1996. *Top*: Flows. *Bottom*: Water levels

delays respectively. The flow and water level relationships are obtained from rating curves. In the absence of a reliable rating curve for the out-flow of Lake Benalla for the range of levels in Fig. 21.9, a local linear relationship is used, i.e.

$$Q_{LB}(t) \approx m_{LB} y_{LB}(t) + \Delta_{LB}, \tag{21.14}$$

where y_{LB} is the water level, and m and Δ are constants. Although there are regulation gates at Broken Weir downstream of Moorngag, it usually just act as a free overfall weir and it does not need to be accounted for in the model. There is also a large channel originating just upstream of Broken Weir, but in the absence of data, out-flows through this channel are not modelled.

Substituting (21.14) into (21.13) and assuming that the water level is proportional to the volume and using an Euler approximation for the derivative, we arrive at

$$\begin{aligned} y_{LB}(t+1) = y_{LB}(t) &+ \left(\frac{T_s}{A}\right)[Q_M(t-\tau_M) + Q_{LC}(t-\tau_{LC}) + Q_{HC}(t-\tau_{HC})] \\ &- \left(\frac{T_s m_{LB}}{A}\right) y_{LB}(t) - \left(\frac{T_s \Delta_{LB}}{A}\right), \end{aligned} \tag{21.15}$$

where T_s is the sampling interval and A is the surface area of Lake Benalla. Equation (21.13) is known as an "Integrator-Delay Model" but becomes a first order model when the volume and out-flow are expressed in terms of the water level as in (21.15). The associated "Output Error" (OE) type predictor for (21.15) is

Table 21.4 Parameter estimates

$\hat{\theta}_1$	$\hat{\theta}_2$	$\hat{\theta}_3$	$\hat{\tau}_{MG}$	$\hat{\tau}_{LC}$	$\hat{\tau}_{HC}$
0.0368	−1.8670	315.90	1395 min	1380 min	525 min

$$\begin{aligned}\hat{y}_{LB}(t+1,\theta,\tau) = \hat{y}_{LB}(t,\theta,\tau) + \theta_1[&Q_M(t-\tau_M)\\ &+ Q_{LC}(t-\tau_{LC}) + Q_{HC}(t-\tau_{HC})]\\ &+ \theta_2\hat{y}_{LB}(t,\theta,\tau) + \theta_3,\end{aligned} \tag{21.16}$$

where $\theta = [\theta_1, \theta_2, \theta_3]^T = [(\frac{T_s}{A}), (-\frac{T_s m_{LB}}{A}), (-\frac{T_s \Delta_{LB}}{A})]^T$ and $\tau = [\tau_M, \tau_{LC}, \tau_{HC}]$. An OE model usually gives a good description of a system in the low frequency range (see e.g. [23, 44]) which is of most interest for control design.

As in Sect. 21.3.2, the time delays, τ_M, τ_{LC} and τ_{HC} were estimated from the cross-correlation (see (21.9)) between the measurements at Moorngag, Lima Creek, Hollands Creek and Lake Benalla. The parameter θ was estimated using a prediction error method with a quadratic criterion, i.e.,

$$\hat{\theta}_{\tau_M} = \underset{\theta_{\tau_M}}{\operatorname{argmin}} \frac{1}{N-\tau_M} \sum_{t=\tau_M+1}^{N} [y_{LB}(t) - \hat{y}_{LB}(t,\theta,\tau_M)]^2, \tag{21.17}$$

where $N = 14700$, y_{LB} is the measured water levels and $\hat{y}_{LB}$ is predicted using (21.16). The data set shown in Fig. 21.9 is used for estimation and the estimated values are given in Table 21.4. The positive value of θ_1 and the negative value of θ_2 are consistent with in-flow and out-flow. The estimated time delays are also consistent in view of the distances from Moorngag, Lima Creek and Hollands Creek to Lake Benalla (see Fig. 21.1).

The integrator-delay model and the Saint Venant equations were compared against measured data on the data sets not used for estimation. The data set are from October to December 1996 and January to March 1997. For the Saint Venant equations, two straight stretches with different geometries were used to represent this reach. The first stretch represents Moorngag to the entrance of the lake, while the second stretch represents the lake. The Manning friction coefficient for the two segments were calibrated from the data as in Sect. 21.3.1.

Figure 21.10 shows the measured water levels, the predicted water levels using the integrator-delay model and the simulated water levels using the calibrated Saint Venant equations. Both the integrator-delay model and the Saint Venant equations are accurate, and they pick up the trends in the data well. The MSE calculated using (21.11) and the R_T^2 calculated using (21.12) are shown in Table 21.5.

The MSEs for both the models are relatively small. Likewise, the values of R_T^2 indicate that both models explain the data well. Although, the integrator-delay model picks up the trends in the water levels and is accurate for flow conditions similar to those on the estimation set, it is not very accurate in predicting flow peaks, and moreover the estimate of the parameters associated with in-flows can be quite sensitive to flow peaks in the estimation data. This could be due to inaccuracies in the

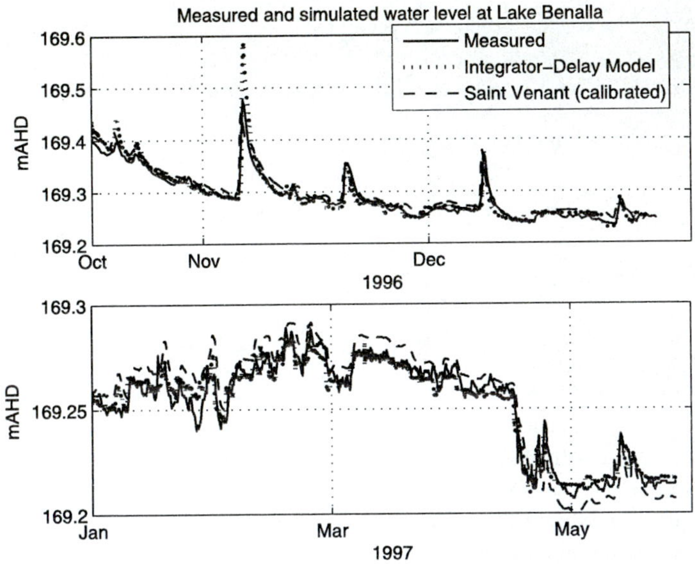

Fig. 21.10 *Top*: October to December 1996. *Bottom*: January to March 1997. The data sets are plotted separately for clarity of presentation

Table 21.5 Values of MSE and R_T^2

Data period	Integrator-delay model		Saint Venant equations		Time-delay model	
	MSE (10^{-4} m^2)	R_T^2	MSE (10^{-4} m^2)	R_T^2	MSE (10^{-4} m^2)	R_T^2
Oct–Dec 1996	2.00	0.902	2.05	0.900	1.96	0.904
Jan–May 1997	0.74	0.856	0.24	0.954	0.24	0.954

measurements and rating curves or simply due to deficiencies in the model structure. Nonetheless, this is not of a major concern for design of control system whose purpose mainly is to reduce operational losses under low flow conditions.

21.3.7.1 Time-Delay Model for the Reach Between Moorngag and Lake Benalla

The surface area of Lake Benalla is 211239.0 m^2 [38]. However, from the estimated value of θ_1 in Table 21.4, the estimated area of Lake Benalla is $A = T_s/\theta_1 = 24456.5$ m^2 ($T_s = 900$ s) which is only 12% of the area reported in [38]. One possible reason for this could be that the location of the sensor at Lake Benalla is in the middle of the lake rather than at the outlet. The smaller estimated area suggests that the "effective storage" in the lake is quite small. We therefore also consider the time-delay model for this reach. It is given by

$$Q_{LB}(t) = Q_M(t - \tau_M) + Q_{LC}(t - \tau_{LC}) + Q_{HC}(t - \tau_{HC}). \tag{21.18}$$

Substituting (21.14) into (21.18), we arrive at

$$y_{LB}(t) = \frac{1}{m_{LB}}[Q_M(t - \tau_M) + Q_{LC}(t - \tau_{LC}) + Q_{HC}(t - \tau_{HC})] - \frac{\Delta_{LB}}{m_{LB}}. \tag{21.19}$$

The associated predictor is given by

$$\hat{y}_{LB}(t, \gamma) = \gamma_1[Q_M(t - \tau_M) + Q_{LC}(t - \tau_{LC}) + Q_{HC}(t - \tau_{HC})] + \gamma_2. \tag{21.20}$$

The parameter vector $\gamma = [\gamma_1, \gamma_2]^T = [\frac{1}{m_{LB}}, -\frac{\Delta_{LB}}{m_{LB}}]^T$ is estimated using least squares. The data set shown in Fig. 21.9 is used for estimation, and the estimated parameters are $\gamma_1 = 0.0197$ and $\gamma_2 = 169.20$. Comparing (21.15) and (21.19), we see that ideally γ_1 should be equal to θ_1/θ_2 and γ_2 should be equal to $-\theta_3/\theta_2$. Using the values in Table 21.4 we found that $\theta_1/\theta_2 = 0.0197$ and $-\theta_3/\theta_2 = 169.20$, which indeed are equal to the estimated values of γ_1 and γ_2. Using the estimated parameters, we compute the MSE and R_T^2 on the data set shown in Fig. 21.10. The results are given in Table 21.5.

The value of MSE for the time-delay model is smaller and R_T^2 is larger than the corresponding values for the integrator-delay model. As the data material is limited for Lake Benalla we do not want to draw any strong conclusions, but the findings indicate that the effective storage in Lake Benalla is much smaller than what the surface area suggests.

21.3.8 Previous Work on System Identification of Rivers

There are a number of works in the literature on modelling and system identification of rivers and irrigation channels for prediction and control purposes. Most of the works which aim at finding a model relating the flow or water level at one location to flows or levels at other locations end up using linear transfer function models possibly with an input non-linearity. The use of transfer function model is not new. In [27], the model which is known as the Nash cascade flow model was introduced. This model is essentially a cascade of first order transfer functions with delay relating in-flow to out-flow.

System identification of models similar to the time-delay model were used in [25, 26]. Integrator-delay models are commonly used in control of irrigation channels, and system identification of such models for irrigation channels was considered in [44] which also considered a high order version which incorporated wave dynamics. For rivers, identification of integrator-delay models and the corresponding first order models in the water levels were considered in [3, 32, 39, 41, 50]. In [32, 50], following a Data Based Mechanistic approach, an input nonlinearity in the water level was first identified resulting in a Hammerstein type model. In [20, 21] a second order model with delay was considered based on a simplification of the Saint Venant equations while [37] considered neural network models.

21.4 Design of Controllers for the River Reach

There are a number of control design methods which have been considered for rivers for various purposes (see e.g. [4, 19, 30, 31, 33, 34, 39]). In this section the aim is to demonstrate the usefulness of the models obtained in Sect. 21.3 for control design. As mentioned in the introduction, what the control objectives should be, is still an open question, but it is reasonable to assume that they will include rejection of disturbances due to off-takes of water and keeping the water levels and flows on setpoints, possibly time-varying ones, or within a certain range. As an initial control design methodology, we therefore consider decentralised PI control [9, 45] due to its ease of design and implementation. The PI controllers will also serve as a benchmark for more advanced designs. As the aim is to illustrate the usefulness of the models, we only consider control of the reach between Casey's Weir and Gowangardie Weir. Although the controllers are designed based upon the time-delay model, the Saint Venant equations are used in the simulations.

21.4.1 Preliminary Control Design

At present, there are no regulation gates at Casey's Weir. However, one objective of the FRM project is to explore what can be achieved by employing control systems, and depending on the outcomes, making a case for improved infrastructure. Thus, we assume that the flow can be manipulated at Casey's Weir. For a number of reasons an upgrade of the infrastructure at Gowangardie Weir is unlikely, and we will therefore assume it remains a free overfall weir. Hence the flow cannot be manipulated at Gowangardie Weir, and the time-delay model (21.5) is therefore suitable for control design. The value in Table 21.2, $\tau = 1650$ minutes is used as the nominal time delay.

The water in Broken River is supplied on demand and the most suitable decentralised controller configuration for demand driven systems is the distant downstream configuration shown in Fig. 21.11. In addition, water ordered for irrigation is known four days in advance, and feedforward action from the orders are employed.

The idea behind the configuration in Fig. 21.11 is that the scheduler will release the flow corresponding to future orders from farmers 1650 minutes (the nominal time delay) before the water is required, and the PI controller will adjust for any discrepancy in the flow e.g. due to model mismatch or errors in the rating curve. From (21.3) there is a one to one relationship between the flow and water level at Gowangardie Weir, and the water level setpoint is replaced by the corresponding

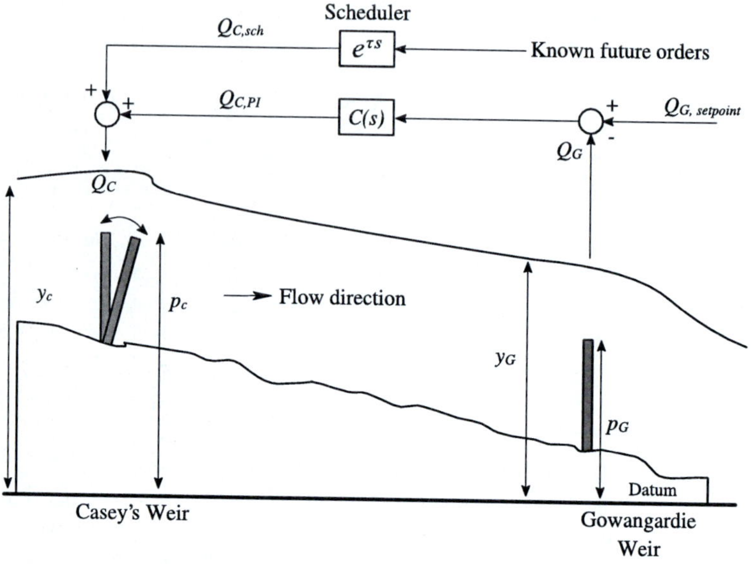

Fig. 21.11 Distant downstream control configuration with scheduler

flow setpoint. The feedback controller is a PI controller,

$$C(s) = \frac{K_p(1 + T_i s)}{T_i s}. \tag{21.21}$$

The controller takes the difference between the desired flow, $Q_{G,setpoint}$ and the actual flow, Q_G over Gowangardie Weir as input and the output, $Q_{C,PI}$ is added to the early release flow from the scheduler to give the flow over Casey's Weir, Q_C (see Fig. 21.11). The PI controller is tuned using classical frequency response methods.

Due to the variation in the time delay with the flow as shown in Sect. 21.3.5, the controller is tuned rather conservatively to ensure robustness with $K_p = 0.007$ and $T_i = 60$. The controller has a gain margin of 9.88 dB at 9.88×10^{-4} rad/min and a phase margin of 61.2° at 3.16×10^{-4} rad/min. This means that an additional 3375 minutes time delay can be tolerated before the closed loop system becomes unstable, and this is well within the range of time delays found in Sect. 21.3.5. The Bode plots of the PI controller, the time-delay model and the model with PI controller are shown in Fig. 21.12.

21.4.2 Simulation Example

Two simulation scenarios were considered. In the first scenario the flow was low and the time delay was longer than the nominal one. In the second scenario the flow was higher, and the time delay was shorter than the nominal one. The flow setpoint

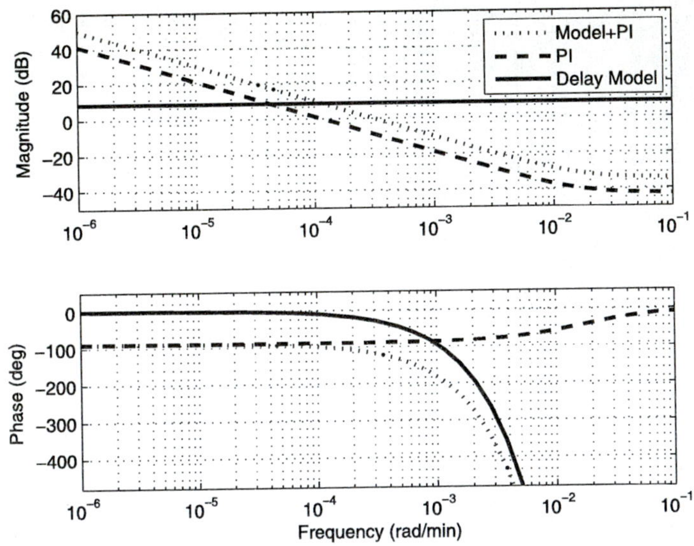

Fig. 21.12 Bode plots for the reach between Casey's Weir and Gowangardie Weir

at Gowangardie Weir was 0.2864 m^3/s (25 ML/day) and 2.3148 m^3/s (200 ML/day) under the low and high flow scenario respectively. Between time 15165 and 20925 minutes, a flow of 0.1157 m^3/s (10 ML/day) was ordered for irrigation downstream of Gowangardie Weir under the low flow scenario and a flow of 0.5787 m^3/s (50 ML/day) was ordered under the high flow scenario. Hence, the flow setpoint at Gowangardie Weir was changed accordingly in order to deliver the water requested. These flows were released by the scheduler at time 13515 minutes (1650 minutes before water was required) for 5760 minutes (see Fig. 21.13). In order to account for uncertainty in the rating curves, a 10% error in the flow over Casey's Weir was introduced in the simulations such that the actual flow was only 90% of what the controller asked for. The control configuration without the scheduler, i.e. purely feedback control was also considered, and the results are shown in Fig. 21.13.

Due to the time delay, a control solution based only on feedback is not satisfactory, and feedforward action from future known demands should be included. Under low flow (plots in the left column of Fig. 21.13), the scheduler released the required flow about 600 minutes late, and then the feedback controller increased the in-flow to compensate for the shortfall in flow. Under high flow, the scheduler released the flow about 650 minutes early due to the shorter time delay. Due to the error in the flow equation the flow released from Casey's Weir by the scheduler was short of what was required, and hence the flow over Gowangardie Weir was below setpoint. The feedback controller partly compensated for this error by increasing the flow over Casey's Weir. The controller gave acceptable performance under both flows, but obviously there is still room for improvements. For example a more advanced scheduler could take into account that the time delay varies with flow. However, the

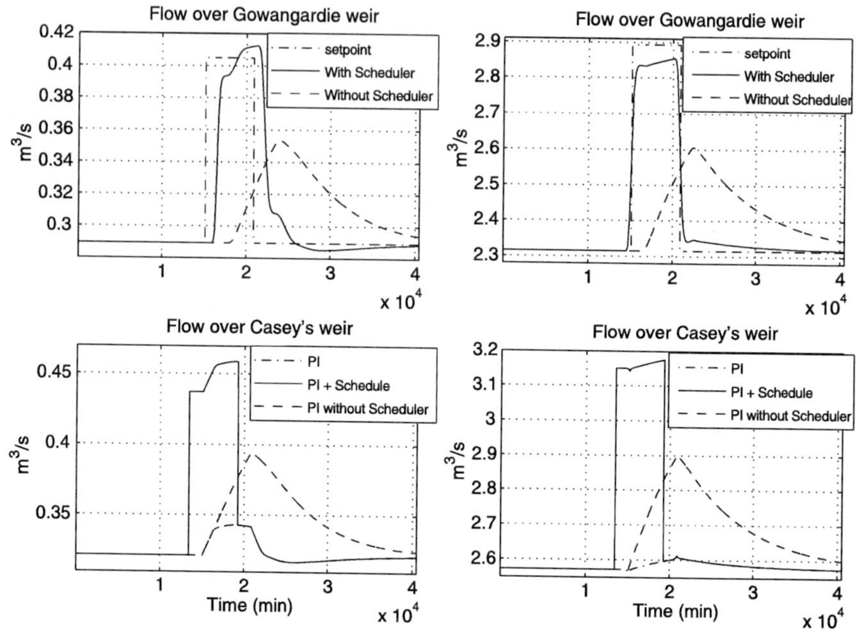

Fig. 21.13 Flows over of Casey's Weirs and Gowangardie Weir

results demonstrate that a simple time-delay model of the river reach is sufficient for control design.

21.5 Conclusions

System identification for control of the Broken River has been considered. Based on operational data and physical considerations, a time-delay model and an integrator-delay model have been proposed as suitable for control design. The models have been compared to Saint Venant equation models and experimentally verified against operational data from the Broken River. The proposed models accurately reflect the dynamics of the river important for control design.

Based on the time-delay model, control designs were carried out taking the varying time delays into account in the robustness specifications. The controller showed an acceptable performance in a simulation example based on the full Saint Venant equations further validating the usefulness of the simple time-delay model in control design.

Acknowledgements This work was supported by The Farms Rivers and Markets Project, an initiative of Uniwater and funded by the National Water Commission, the Victorian Water Trust, The Dookie Farms 2000 Trust (Tallis Trust) and the University of Melbourne and supported by the Departments of Sustainability and Environment and Primary Industry, the Goulburn Broken Catchment Management Authority and Goulburn-Murray Water. The first author also gratefully acknowledge the financial support from National ICT Australia (NICTA). NICTA is funded by the Australian Government as represented by the Department of Broadband, Communications and the Digital Economy and the Australian Research Council through the ICT Centre of Excellence program.

References

1. Adams, R., Western, A.W.: Using uncertainty analysis to improve estimates of water availability for a regulated catchment in Southern Australia. In: Proceedings of the BHS 3rd International Symposium Role of Hydrology in Managing Consequences of a Changing Global Environment, Newcastle, United Kingdom (2010)
2. Akan, A.O.: Open Channel Hydraulics. Butterworth-Heinemann Elsevier, Oxford (2006)
3. Bastin, G., Moens, L., Dierickx, P.: On line river flow forecasting with Hydromax: successes and challenges after twelve years of experience. In: Proceedings 15th IFAC Symposium on System Identification, SYSID, Saint-Malo, France, pp. 1774–1779 (2009)
4. Barjas Blanco, T., Willems, T., Chiang, P.-K., Cauwenberghs, K., De Moor, B., Berlamont, J.: Flood regulation by means of model predictive control. In: Intelligent Systems, Control and Automation: Science and Engineering, vol. 42, pp. 407–437 (2010). Part 4, Chap. 16
5. Boiten, W.: Flow measurement structures. Flow Meas. Instrum. **13**, 203–207 (2002)
6. Bos, M.G.: Discharge measurement structures. Technical report, International Institute for Land Reclamation and Improvement (IIRI), The Netherlands (1978)
7. Broken River Basin: Broken River Basin, Department of Primary Industries (2010). Available via http://www.new.dpi.vic.gov.au/fisheries/recreational-fishing/inland-angling-guide/broken-river-04. Cited 27 August 2010
8. Bulk Entitlement: Bulk entitlements and environment entitlements. Victorian Water Register, The Department of Sustainability and Environment (2010). Available via http://www.waterregister.vic.gov.au/Public/Reports/BulkEntitlements.aspx. Cited 27 August 2010
9. Cantoni, M., Weyer, E., Li, Y., Ooi, S.K., Mareels, I., Ryans, M.: Control of large-scale irrigation networks. Proc. IEEE Technol. Netw. Control Syst. **95**(1), 75–91 (2007) (special issue)
10. Castelletti, A., Soncini-Sessa, R.: Coupling real-time control and socio-economic issues in participatory river basin planning. Environ. Model. Softw. **22**, 1114–1128 (2007)
11. Chaudhry, M.: Open-Channel Flow. Prentice Hall, Englewood Cliffs (1993)
12. Cottingham, P., Stewardson, M., Roberts, J., Metzeling, L., Humphries, P., Hillman, T., Hannan, G.: Report of the Broken River scientific panel on the environmental condition and flow in the Broken River and Broken Creek. Technical report, Cooperative Research Centre for Freshwater Ecology, University of Canberra, Australia (2001)
13. Craig, I., Aravinthan, V., Baillie, C., Beswick, A., Barnes, G., Bradbury, R., Connell, L., Coop, P., Fellows, C., Fitzmaurice, L., Foley, J., Hancock, N., Lamb, D., Morrison, P., Misra, R., Mossad, R., Pittway, P., Prime, E., Rees, S., Schmidt, E., Solomon, D., Symes, T., Turnbull, D.: Evaporation, seepage and water quality management in storage dams: a review of research methods. Environ. Health **7**(3), 84–97 (2007)
14. Cunge, J.A., Holly, F.M., Verwey, A.: Practical Aspects of Computational River Hydraulics. Boston Pitman Advanced Publication Program (1980)
15. Dal Cin, C., Moens, L., Dierickx, P., Bastin, G., Zech, Y.: An integrated approach for real-time flood-map forecasting on the Belgian Meuse river. Natural Hazards **36**(1–2), 237–256 (2005)
16. Deodhar, M.J.: Elementary Engineering Hydrology. Pearson Education, Upper Saddle River (2009)
17. Foo, M., Bedjaoui, N., Weyer, E.: Segmentation of a river using the Saint Venant equations. In: Proceedings of IEEE Multiconference on Systems and Control, Yokohama, Japan (2010)
18. Glanzmann, G., von Siebenthal, M., Geyer, T., Papafotiou, G., Morari, M.: Supervisory water level control for cascaded river power plants. In: Proceedings of the 6th International Conference on Hydropower, Stavanger, Norway (2005)

19. Litrico, X.: Robust IMC flow control of SIMO dam-river open-channel systems. IEEE Trans. Control Syst. Technol. **10**(3), 432–437 (2002)
20. Litrico, X., Fromion, V., Baume, J.-P., Arranja, C., Rijo, M.: Experimental validation of a methodology to control irrigation canals based on Saint Venant equations. Control Eng. Pract. **13**(11), 1425–1437 (2005)
21. Litrico, X., Georges, D.: Nonlinear identification of an irrigation system. In: Proceedings of the 36th Conference on Decision and Control, San Diego, USA, pp. 852–857 (1997)
22. Litrico, X., Pomet, J.-B.: Nonlinear modelling and control of a long river stretch. In: Proceedings of the European Control Conference, Cambridge, UK (2003)
23. Ljung, L.: System Identification—Theory for the User, 2nd edn. Prentice Hall, Englewood Cliffs (1999)
24. Mareels, I., Weyer, E., Ooi, S.K., Cantoni, M., Li, Y., Nair, G.: System engineering for irrigation systems: successes and challenges. Annu. Rev. Control **29**, 191–204 (2005)
25. Maxwell, M., Warnick, S.: Modeling and identification of the Sevier River system. In: Proceedings of the 2006 American Control Conference, Minneapolis, USA, pp. 5342–5347 (2006)
26. Nakashima, M., Singh, K.P.: Illinois river flow system model. Contract Report 311, State Water Survey Division Surface Water Section at the University of Illinois (1983)
27. Nash, J.E.: Systematic determination of unit hydrograph parameters. J. Geophys. Res. **64**, 111–115 (1959)
28. National Water Commission: Water trading (2010). Available via http://www.nwc.gov.au/www/html/251-water-trading.asp. Cited 10 August 2010
29. Ooi, S.K., Weyer, E.: Control design for an irrigation channel from physical data. Control Eng. Pract. **16**(9), 1132–1150 (2008)
30. van Overloop, P.J., Negenborn, R.R., De Schutter, B., van de Giesen, N.C.: Predictive control for national water flow optimization in The Netherlands. In: Intelligent Systems, Control and Automation: Science and Engineering, vol. 42, pp. 439–461 (2010). Part 4, Chap. 17
31. Papageorgiou, M., Messmer, A.: Flow control of a long river stretch. Automatica **25**(2), 177–183 (1989)
32. Romanowicz, R.J., Young, P.C., Beven, K.J., Pappenberger, F.: A data based mechanistic approach to nonlinear flood routing and adaptive flood level forecasting. Adv. Water Resour., **31**(8), 1048–1056 (2008)
33. Rui, J., Chen, S.: Optimal regulation control system for cascade hydropower stations. In: Proceedings of the International Conference on Sustainable Power Generation and Supply, pp. 2425–2429 (2009)
34. Şahin, A., Morari, M.: Decentralized model predictive control for a cascade of river power plants. In: Intelligent Systems, Control and Automation: Science and Engineering, vol. 42, pp. 463–485 (2010). Part 4, Chap. 18
35. Schuurmans, J., Hof, A., Dijkstra, S., Bosgra, O.H., Brouwer, R.: Simple water level controller for irrigation and drainage canal. J. Irrig. Drain. Eng. **125**(4), 189–195 (1999)
36. Setz, C., Heinrich, A., Rostalski, P., Papafotiou, G., Morari, M.: Application of model predictive control to a cascade of river power plants. In: Proceedings of the 17th World Congress the International Federation of Automatic Control, Seoul, Korea, pp. 11978–11983 (2008)
37. Shrestha, R.R., Nestmann, F.: River water level prediction using physically based and data driven models. In: Zerger, A., Argent, R.M. (eds.) MODSIM 2005 International Congress on Modelling and Simulation: Modelling and Simulation Society of Australia and New Zealand, pp. 1894–1900 (2005)
38. SKM: Broken River and Broken Creek loss reduction concept study. Phase 1 Report, SKM, Armadale, VIC, Australia, 152 pages (2005)
39. Sohlberg, B., Sernfält, M.: Grey box modelling for river control. J. Hydroinform. **4**, 265–280 (2002)
40. Soncini-Sessa, R., Weber, E., Castelletti, A.: Integrated and Participatory Water Resources Management—Theory, vol. 1a. Elsevier, Amsterdam (2007)
41. Thomassin, M., Bastogne, T., Richard, A.: Identification of a managed river reach by a Bayesian approach. IEEE Trans. Control Syst. Technol. **17**(2), 353–365 (2009)

42. UNESCO: World water assessment program: water—a shared responsibility. Technical report, The United Nations World Water Development Report (2006)
43. Water in our environment: Water in our environment (2010). Available via http://www.environment.gov.au/water/policy-programs/environment/index.html. Cited 20 August 2010
44. Weyer, E.: System identification of an open water channel. Control Eng. Pract. **9**, 1289–1299 (2001)
45. Weyer, E.: Control of irrigation channels. IEEE Trans. Control Syst. Technol. **16**(4), 664–675 (2008)
46. Young, P.C.: Data-based mechanistic modelling of environmental, ecological, economic and engineering systems. Environ. Model. Softw. **13**, 105–122 (1998)
47. Young, P.C., Chotai, A.: Data-based mechanistic modeling, forecasting, and control. IEEE Control Syst. Mag. **21**(5), 14–27 (2001)
48. Young, P.C.: Advances in real-time flood forecasting. Philos. Trans. R. Soc., Math. Phys. Eng. Sci. **360**(1796), 1433–1450 (2002). Flood Risk in a Changing Climate, The Royal Society
49. Young, P.: Top-down and data based mechanistic modelling of rainfall-flow dynamics at the catchment scale. Hydrol. Process. **17**, 2195–2217 (2003)
50. Young, P.C., Leedal, D., Beven, K.J., Szczypta, C.: Reduced order emulation of distributed hydraulic simulation models. In: Proceedings of the 15th IFAC Symposium on System Identification, St. Malo, France (2009)

Chapter 22
Modelling Environmental Change: Quantification of Impacts of Land Use and Land Management Change on UK Flood Risk

H.S. Wheater, C. Ballard, N. Bulygina, N. McIntyre, and B.M. Jackson

22.1 Introduction

The management of water is one of society's most fundamental challenges. There is pressure on water resources to meet the needs of both society and the natural environment. Increasing population and social and economic development are increasing demand for water, while pollution and over-abstraction of water are reducing availability. At the same time, flood risk is increasing with increasing population and asset values in flood-prone areas. Superimposed on these pressures are the impacts of environmental change—changing land use and changing climate.

Given these management challenges, hydrological models have a key role to play, and the work of Peter Young has had a profound influence on the history of hydrological modelling. Increasing computing power has enabled development of models of increased complexity, from the early conceptual representation of hydrological processes in models such as the Stanford Watershed Model, to the development of so-called physics-based models, such as the Stysteme Hydrologique Europeen (SHE) model. Peter has consistently taken a systems analysis approach to the modelling problem, and argued (forcefully!) that most hydrological models are over-parameterised, and hence non-identifiable from the available data. Gradual acceptance of this view has led to a parsimonious family of models—hybrid metric-conceptual (HMC) models (as well as to Peter's well-known CAPTAIN software for Data Based Modelling and systems identification.

Some of the most challenging modelling issues arise in the prediction of the impacts of environmental change. In this chapter we review the strengths and weaknesses of alternative hydrological modelling approaches in this context and with

H.S. Wheater (✉) · C. Ballard · N. Bulygina · N. McIntyre · B.M. Jackson
Imperial College London, London SW7 2AZ, UK
e-mail: h.wheater@imperial.ac.uk

L. Wang, H. Garnier (eds.), *System Identification, Environmental Modelling, and Control System Design*,
DOI 10.1007/978-0-85729-974-1_22,

perspectives developed from Peter's work. In particular, we focus on the prediction of the effects of changing rural land use and land management, for flood risk assessment and management.

22.2 An Overview of Rainfall-Runoff Model Types

The physical processes by which rainfall is translated into river flow are complex and subject to a high degree of spatial heterogeneity. However, the data available to support modelling is typically limited to an estimate of spatial rainfall, derived from one or more gauge locations, some index of the evaporative power of the atmosphere (derived from temperature or a more complete set of meteorological variables) and an observed river flow time series. Hence the modelling of rainfall-runoff processes presents methodological challenges, and raises important issues of modelling philosophy (also addressed by Young [85]). A convenient classification of model types, after Wheater et al. [79], is presented below, with a discussion of relative strengths and weaknesses.

22.2.1 Metric Models

At the simplest level, the catchment-scale relationship between storm rainfall and stream response to climatic inputs can be represented by a volumetric loss, to account for processes such as evaporation, soil moisture storage and groundwater recharge, and a time distribution function, to represent the various dynamic modes of catchment response. In the 1930s, prior to the availability of digital computers, the unit hydrograph method was developed. In its basic form it represents the stream response to individual storm events by a non-linear loss function and linear transfer function. The simplicity of the method has provided a powerful tool for data analysis and model identification, once a set of assumptions has been adopted (identifying event response in the streamflow hydrograph and allocating rainfall losses). The method is widely used around the world; for example, its analytic capability was exploited in the UK [51] to provide methods of flood estimation for ungauged catchments. Using data from 138 UK catchments, regression relationships were defined for the model parameters as functions of storm and catchment characteristics. Similarly, Wheater et al. [77] were able to quantify potential effects of urbanisation through analysis of the differences in response of a set of catchments with varying degrees of urban development.

This data-based approach to hydrological modelling has been defined as metric modelling [79]. Such models are based primarily on observations and seek to characterise system response from those data. In principle, this limits application to the range of observed data, and effects such as catchment change cannot be directly represented. In practice, as discussed above, the analytical power of the method has enabled some effects of change to be quantified through regional analysis.

The unit hydrograph is a simple model with limited performance capability, and in general the level of model complexity that can be identified from a typical rainfall/flow data set is limited [76]. However methods of time-series analysis, which treat a body of data without isolating it into discrete events, can be used to identify more complex model structures. These are typically based on parallel linear stores, and provide a capability to represent both fast and slow-flow components of a streamflow hydrograph. These provide a powerful set of tools for a range of hydrological applications, and in particular, with updating techniques, in real-time flood forecasting [85].

22.2.2 Conceptual Models

The most common class of hydrological model in general use incorporates prior information subjectively in the form of a conceptual representation of the processes perceived to be important. The model form originated in the 1960s, when computing power first allowed integrated representation of hydrological processes, using simplified relationships, to generate continuous flow sequences. These models are characterised by parameters that usually have no direct, physically measurable identity. The Stanford Watershed Model [22], now available as the HSPF model [11], is one of the earliest examples, and, with some 16–24 parameters, one of the more complex. In application to a particular catchment, the model must be calibrated, i.e. fitted to an observed data set to obtain an appropriate set of parameter values, using either a manual or automatic procedure. However, as noted above, the information content of the available data is limited, particularly if a single performance criterion (objective function) is used, and hence the problem of non-identifiability arises, defined by Beven [7] as "equifinality". For a given model, many combinations of parameter values may give similar performance (for a given performance criterion), as indeed may different model structures. This has given rise to two major limitations. If parameters cannot be uniquely identified, then they cannot be linked to catchment characteristics, and there is a major problem in application to ungauged catchments. Similarly, it is difficult to represent catchment change if the physical significance of parameters is ambiguous.

Developments in computing power, linked to an improved understanding of modelling limitations, have led to important developments for conceptual modelling. Firstly, methods to analyse and represent parameter ambiguity have been developed. The concept of Generalized Sensitivity Analysis was introduced [63], in which the search for a unique best fit parameter set for a given data set is abandoned; parameter sets are classified as either "behavioural" (consistent with the observed data) or "non-behavioural" according to a defined performance criterion. An extension of this is the Generalised Likelihood Uncertainty Estimation (GLUE) procedure [9, 26]. Using Monte Carlo simulation, parameter values are sampled from the feasible parameter space (conditioned on prior information, as available). Based on a performance criterion, a "likelihood" measure can be evaluated for each simulation. Non-behavioural simulations can be rejected (based on a pre-selected threshold value), and the remainder assigned re-scaled likelihood values. The outputs from the runs can then be weighted and ranked to form a cumulative distribution of output time series, which can be used to represent the modelling uncertainty. This formal representation of uncertainty is an important development in hydrological modelling practice, although it should be noted that the GLUE procedure lumps together various forms of uncertainty, including data error, model structural uncertainty and parameter uncertainty. More generally, Monte Carlo analysis provides a powerful set of methods for evaluating model structure, parameter identifiability and uncertainty. For example, in a recent refinement [72], parameter identifiability is evaluated using a moving window to step through the output time-series, thus giving insight into the variability of model performance with time.

A second development is a recognition that more information is available within an observed flow time series than is indicated by a single performance criterion, and that different segments of the data contain information of relevance to different modes of model performance [78]. This has long been recognised in manual model calibration, but has only more recently been used in automatic methods. A formal methodology for multi-criterion optimisation has been developed for rainfall-runoff modelling (e.g. [29, 70, 73]). Provision of this additional information reduces the problem of equifinality and provides new insights into model performance. Modelling tool-kits for model building and Monte-Carlo analysis are currently available, which include GLUE and other associated tools for analysis of model structure, parameter identifiability, and prediction uncertainty [43, 71]. In several senses, this parsimonious conceptual modelling represents an extension of the metric concept (hence such models have been termed hybrid metric-conceptual models). There has been a progressive recognition that the first-generation conceptual models, while seeking a comprehensive and integrated representation of the component processes, are non-identifiable with typically-available data. The current generation of stochastic analysis tools allows detailed investigation of model structure and parameter uncertainty, leading to parameter-efficient models that seek to extract the maximum information from the available data. They also allow formal recognition of uncertainty in model parameters, and provide the capability to produce confidence limits on model simulations.

22.2.3 Physics-Based Modelling

An alternative modelling approach is to seek to develop “physics-based models”, i.e. models explicitly based on the best available understanding of the physical hydrological processes. Such models are based on a continuum representation of catchment processes and the equations of motion of the constituent processes are solved numerically, using a spatial mesh, normally discretized relatively crudely in catchment-scale applications due to computational limitations. They first became feasible in the 1970s when computing power became sufficient to solve the relevant coupled Partial Differential Equations [27, 28]. One of the best known models in current use is the Systeme Hydrologique Europeen (SHE) model [1, 2, 59]. These models are characterised by parameters that are in principle measurable and have a direct physical significance; an important theoretical advantage is that if the physical parameters can be determined *a priori*, such models can be applied to ungauged catchments, and the effects of catchment change can be explicitly represented. However, whether this theoretical advantage is achievable in practice is a question to which we return below.

In practice two fundamental problems arise. The underlying physics has been derived from small-scale, mainly laboratory-based, process observations. Hence, firstly, the processes may not apply under field conditions and at field scales of interest [12]. Secondly, although measurable at small scale, the parameters may not be measurable at the scales of application. An obvious example is the representation of soil water flow at hillslope scale. Field soils are characterised by heterogeneity and complexity. Macropore flow is ubiquitous, yet neglected in physics-based models, for lack of relevant theory and supporting data; the Richards' equation commonly used for unsaturated flow depends on strongly non-linear functional relationships to represent physical properties, for which there is no measurement basis at the areal scales of practical modelling interest. And field studies such as those of Pilgrim et al. [57] demonstrate that the dominant modes of process response cannot be specified *a priori*. For more detailed discussion see, for example, [6, 8].

22.3 Modelling Environmental Change: Land Use and Land Management Effects

Modelling the expected effects of catchment change represents one of the most difficult challenges for hydrological modellers; important limitations arise for each of the model types above. Here we focus on the issue of rural land use and land management change. The context is that recent floods in the UK have focused attention on the potential effects of agricultural intensification on flood risk [75]. Over recent decades agricultural intensification has been widespread across the uplands of the UK, with increases in stocking density, ploughing, reseeding and drainage of fields, use of heavy machinery, and the removal of trees from the landscape. Have the major changes that have taken place since the Second World War affected flood runoff? And if so, what is the potential of changing land management to mitigate flood risk?

Although land use and land management changes have been observed to change local runoff [45, 53, 65], quantification of catchment scale effects has proved elusive. The key methodological challenge is how to predict effects of local scale land use changes at local to catchment scales using hydrological models. In a review of the current state of knowledge about the effects of land use and management change on flood risk, O'Connell et al. [52] concluded that new modelling techniques will need to be developed in order to predict the impacts of land management on flood risk.

The application of metric models requires the availability of a set of gauged catchments spanning the range of changes of interest, and from which effects can be discriminated. This requires that (a) a signal of the effect of change is identifiable in the data, and (b) that the causes of change can be quantified as a catchment descriptor. As part of a UK review [52], a national data base of catchment flows was interrogated to see whether effects of land use change could be identified. In previous UK studies [36, 51], the effects of urbanisation could be discriminated, but when these analyses were revisited to identify more subtle effects of rural land use and land management change, these were not identifiable. One major source of difficulty is the heterogeneity of land use at catchment scale, and hence the need to identify effects of change to one part of a catchment in the presence of the mosaic of land use types at catchment scale. Another is the fact that for a given land use, effects of different land management practices may be significant, but underlying data to quantify historical changes in land management are not readily available.

Subsequently, a more targeted study [10] considered a set of catchments for which reliable data sets were available and for which significant changes in land use or land management had taken place. The conclusion was that attempts to isolate the response to these changes at catchment scale failed due to reasons including climate variability, poorly constrained spatial distribution of land management types and poor historical records of land use and land management change. Alternative ways forward were needed.

As noted above, the role of distributed physics-based models has been the subject of much debate. However, there are classes of modelling problem, such as the representation of changing catchment land use and land management, where such models can potentially offer useful insights, not readily achieved by other modelling approaches. Given that the processes modelled are highly non-linear, the parameters uncertain, particularly at the model scale, and that data on physical processes and properties are limited, important questions arise. Can effects of change be discriminated from effects of model and parameter uncertainty? If detailed process data are unavailable (as would be the case for the majority of potential applications), is there value in speculative simulation? And if surrogate data are available from donor sites, can the utility of such models be enhanced?

Conceptual or Hybrid Metric Conceptual (HMC) models also potentially have a role to play. Firstly, although the parameters of such models have no direct physical significance, there is the possibility of identifying effects through inverse modelling of detailed data sets, and also of constraining parameters using signature indices of catchment response. In the USA, for example, the US Department of Agriculture Soil Conservation Service has developed a conceptual model for event response that can be parameterised *a priori* to represent effects of crop type and soil degradation [68], with potential utility for the UK. Secondly, such models are parsimonious and computationally efficient, so the possibility arises of their use in emulating the response of computationally-demanding physics-based models for large scale applications. Where such models are constrained to a HMC form, we term this meta-modelling.

In the rest of this chapter, we report on the modelling results from a major UK research programme into the effects of rural land management on hydrological response (and in particular, flood risk). A key element of the programme has been a detailed multi-scale experimental programme at Pontbren, Wales [45], which has provided data to support the development and evaluation of new modelling methods. We explore the utility of physics-based models for data rich sites based on Pontbren, and the use of surrogate data for data poor applications, considering issues of upland peat management in the Hodder catchment, North West England. We report a new methodology in which HMC models are used as meta-models for a library of land management interventions to quantify effects of detailed local-scale interventions at catchment scale. We also report HMC application to data-poor sites, evaluating the role of regional indices of catchment response in model conditioning, and the utility of the SCS procedure for UK application. Finally we introduce results of joint work with Peter Young in the application of systems identification methods to detailed experimental data from Pontbren, and consider the strengths and weaknesses of data-based and HMC methods in that context.

22.4 Physics-Based Modeling of Land Use and Land Management Change

22.4.1 Pontbren—A Data-Rich Site

We first consider the role of physics-based models in a relatively data-rich environment, and turn to an extensive field experimental programme established at Pontbren, in the headwaters of the river Severn in Wales. The aim of the experiment was to provide multi-scale data on the effects of land management practices for a typical set of upland land management issues and interventions to support development of the new modelling approaches needed for flood risk policy and management [45, 80].

Pontbren is a farmers' cooperative, involving 10 hill farms and 1000 ha of agriculturally improved pasture (drained, ploughed, re-seeded and fertilized) and woodland (Fig. 22.1). Elevations range from 170 to 438 m AOD, and the soils are clay-rich, mainly from the Cegin and Wilcocks series, which are common in Wales. They have low permeability subsoil overlying glacial drift deposits, and are seasonally wet or waterlogged. Field drainage is ubiquitous where pasture has been improved. The predominant land use is grazing, mainly for sheep.

The Pontbren experiment arose as a result of farmers' concerns that changes to land management, and in particular changes to grazing densities and animal weights, had changed runoff response. Between the 1970s and 1990s major changes in farming intensity took place; sheep numbers increased by a factor of 6 and animal weights doubled (R. Jukes, pers.comm.). Recent farmers' initiatives have included the reduction of grazing densities and reinstatement of woodland areas and hedgerows. Research on the infiltration rates of the grazed hillslopes and woodland buffer strips (e.g. [19]) demonstrated a significant change in soil response to rainfall. Infiltration rates on the grazed pastures were extremely low, but within a few years of tree planting, soil structure and permeability in buffer strips showed significant improvement.

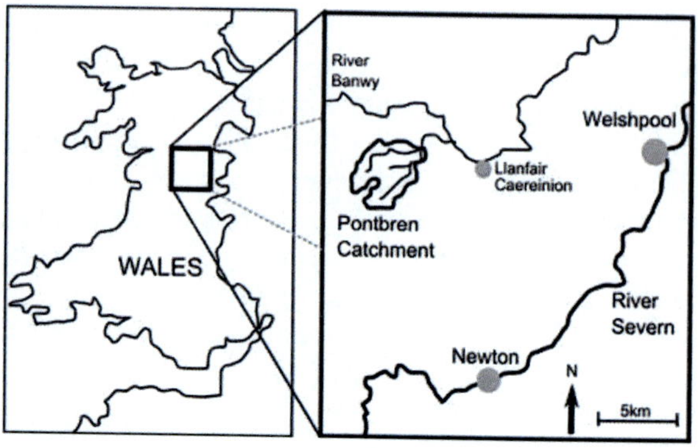

Fig. 22.1 Pontbren study site location

Details of the multi-scale experiment can be found in [45, 80]. Replicated manipulation plots have been instrumented to observe the plot-scale effects of land management change, instrumented fields and hillslopes provide data on soil water response and runoff processes (overland and drain flow) at larger scale, and multiple flow monitoring installations provide data on stream flows at scales ranging from ditches and drains to second order catchment response (12 km^2). In addition, soil physical properties have been derived from extracted soil cores and in situ infiltration tests. The modelling challenges include representing the effects of soil compaction and tree buffer strips on soil properties and runoff processes, as well as the effects of agricultural field drainage, at the scale of individual fields, and at whole catchment scale.

22.4.2 The Pontbren Physics-Based Model

A detailed, physically-based model was required, capable of representing the important hydrological processes operating at Pontbren and similar catchments, at the scale of individual fields and hillslopes. For this we developed further an Imperial College model based on Richards' equation for saturated/unsaturated soil water flow [38], to represent macropore processes and overland flow, incorporating vegetation processes (such as interception) and associated effects such as changing root depths and soil hydraulic properties, and capable of being run in 1, 2 or 3 dimensions [37]. The model has been conditioned, within a Monte Carlo-based framework of uncertainty analysis, using physically-determined soil hydraulic properties and continuous measurements of climate inputs, soil water states and runoff (as overland flow and drain flow) from the Pontbren experimental sites. Due to the highly non-linear dynamics, individual fields and hillslopes are represented at fine resolution (1 cm vertical and 1 m horizontal resolution).

The detailed model can be exercised to simulate scenarios of interest, including the planting of strips of woodland within a hillslope, and the associated changes to soil structure, evaporation processes, overland flow and drainage. Figure 22.2 illustrates the simulated response for a representative hillslope (100 m $\times$ 100 m)

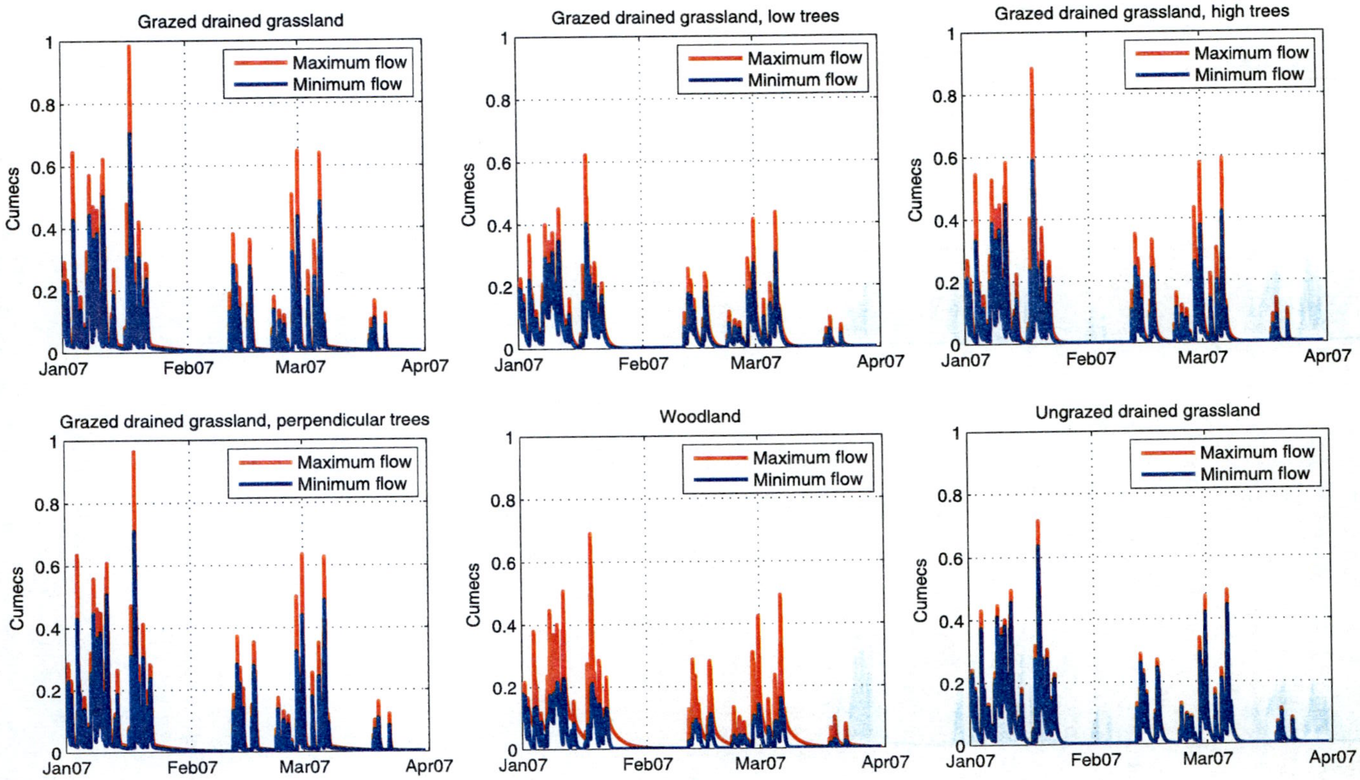

Fig. 22.2 Realisations of field-scale runoff (drain flow + overland flow) for different land use types, with uncertainty bounds

using the detailed model for a range of land management types, including grazed and ungrazed drained grassland, grassland with tree shelter belts (80 m length, 15 m width) in different locations, and full tree cover. The envelopes of response represent the range of parameter uncertainty.

While these results are instructive, some caveats remain. For example, despite the extensive field programme, there are residual uncertainties in the perceptual model, concerned with the fate of subsurface water in the tree planted areas is the fate of subsurface water. We assume here that connection to field drainage systems exists. And while the modeling allows for uncertainty in soil properties, and the model has been conditioned on field-scale data, the effects of spatial heterogeneity have not been explicitly evaluated. Another notable effect was non-stationarity in observed response associated with a hot dry summer (2006) in which soils cracked and only gradually returned to normal over the following Autumn and Winter. Nevertheless, in the absence of alternatives, the model provides a relatively sound basis for the quantification of field scale effects of these complex and spatially-localised land management options.

22.4.3 Meta-Modelling

The detailed model is computationally-intensive and not suitable for direct application at catchment scale. We have therefore developed a strategy to upscale the results in a computationally-efficient procedure. We use meta-modelling, whereby the detailed model is used to train a simpler, conceptual model that represents the response in a parsimonious and computationally-efficient manner, using basic hydrological components of loss and routing functions. This requires classification of the landscape into hydrological units, based for example on soils, land use and existing/proposed interventions. Each field in the Pontbren catchment is classified into a land use/management type, so that the corresponding set of field-scale models can be applied. The field types currently included were chosen based on dominant land use types currently within the catchment and those management changes that were perceived as likely to have an impact on flood peaks.

The detailed model is run for each member of a library of hydrological units, and hence a meta-model parameterisation is obtained for each member through the model training process. Uncertainty in parameter values is carried forward to this stage via Monte Carlo analysis. Figure 22.3 illustrates the performance of the meta-model in emulating the detailed model response for a grazed hillslope with a woodland buffer strip at the base of the slope.

22.4.4 Catchment-Scale Modelling

With a library of meta-models, the final element of the procedure is a catchment-scale semi-distributed model. We use a modular modelling structure (RRMTSD,

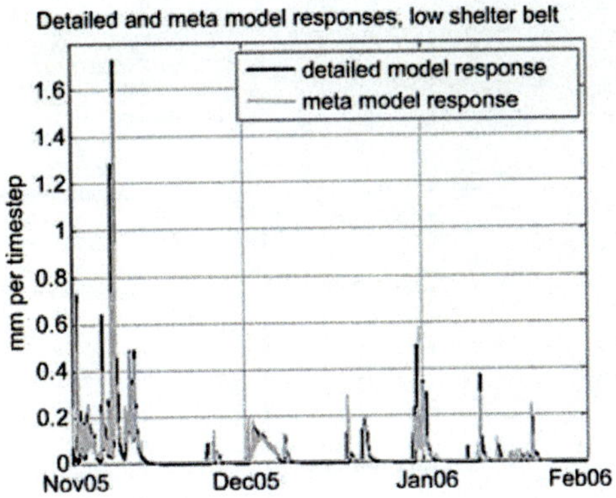

Fig. 22.3 Realisations of field-scale runoff (drain flow + overland flow) for different land use types, with uncertainty bounds

[54]), in which the meta-model elements represent individual hydrological elements, and flows are subsequently routed down the stream network. Using the semi-distributed model, the meta-model can be further conditioned on catchment-scale data to reduce parameter uncertainty.

The hydrological processes and climatological forcing data within the sub-areas are considered to be homogeneous; the degree of spatial distribution is represented mainly through the number of sub-areas. These can represent subcatchments or hydrological response units, and can incorporate the meta-model structures discussed above. Fields were chosen as the individual response units in the present application as these are an appropriate management unit when looking at the influence of land use changes. They also generally form sensible hydrological units, due to the tendency of farmers to set ditches and drainage outlets at field boundaries.

RRMTSD simulates streamflow for the uppermost sub-areas first and then adds sequentially the downstream contributions. A variety of interchangeable pre-built modules are available; others can be added, providing additional flexibility. The toolbox also allows for different optimisation methods for calibration: uniform random search, the shuffled complex evolution method [24], and local nonlinear multi-constrained methods based on simplex searching. The input data and simulated variables in every sub-area can be analysed using a variety of visualisation tools.

The overall modelling procedure provides a powerful set of modelling tools, summarised in Fig. 22.4.

We illustrate the impacts of land management change at the catchment-scale in Fig. 22.5, for a 4 km^2 Pontbren sub-catchment. The baseline is the current day land use at Pontbren, the first scenario removes the effect of the recent Pontbren tree plantings (and hence takes the catchment back to something approximating the intensive use of the early 1990s), the second adds shelter belts to all grazed grassland sites, and the third assumes the entire catchment is woodland. The median changes in flood peaks observed for the three scenarios are: removing all the trees causes up to 20% increase in flood peaks from the baseline condition, adding tree shelter belts to all grazed grassland sites causes up to 20% decrease in flood peaks from the baseline condition, and full afforestation causes up to 60% decrease in flood peaks

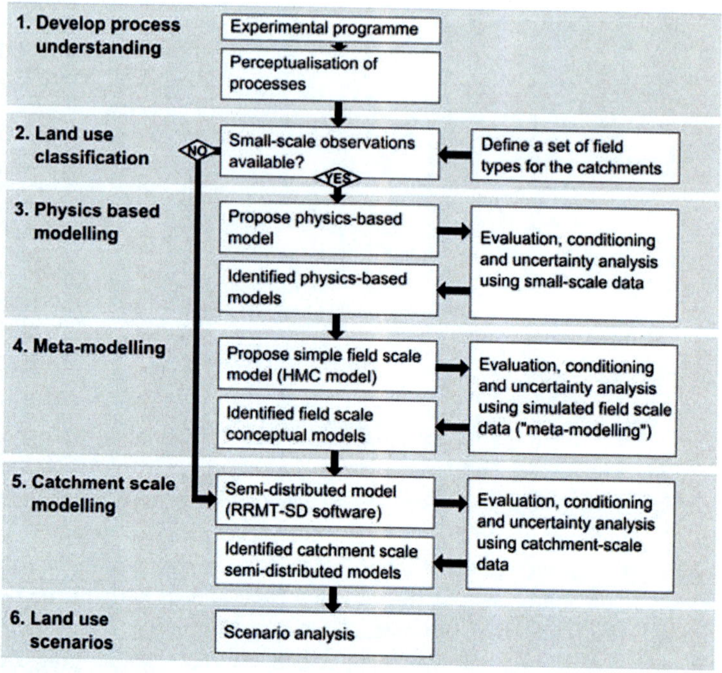

Fig. 22.4 Physics-based modelling framework for catchment-scale scenario analysis

from the baseline condition. However, these effects decrease with increasing storm return period [80].

22.4.5 Extension to Data-Poor Sites and Land Use Types

Having considered tools developed and applied to a data-rich environment, supported by an extensive field programme, we now turn to a different problem of land management, associated with the peat uplands of North-West England, and consider the role of physics-based models and the new modelling framework in a data-poor environment.

The lack of small scale data for a catchment causes problems for the methodology outlined in Fig. 22.4, given the need for data to condition the physics-based model. However, even in the absence of such data, physics-based models may still be an effective way to upscale local changes to the catchment scale, as our understanding of the impacts of land use and land management changes is largely restricted to changes in small scale processes (i.e. interception and infiltration) and physical properties (i.e. hydraulic conductivity and water retention curves). The critical question becomes—what is the role of physics-based models in data-sparse areas?

Even without hydrological measurements for a site of interest, physics-based models can be developed and tested using information about small scale hydrolog-

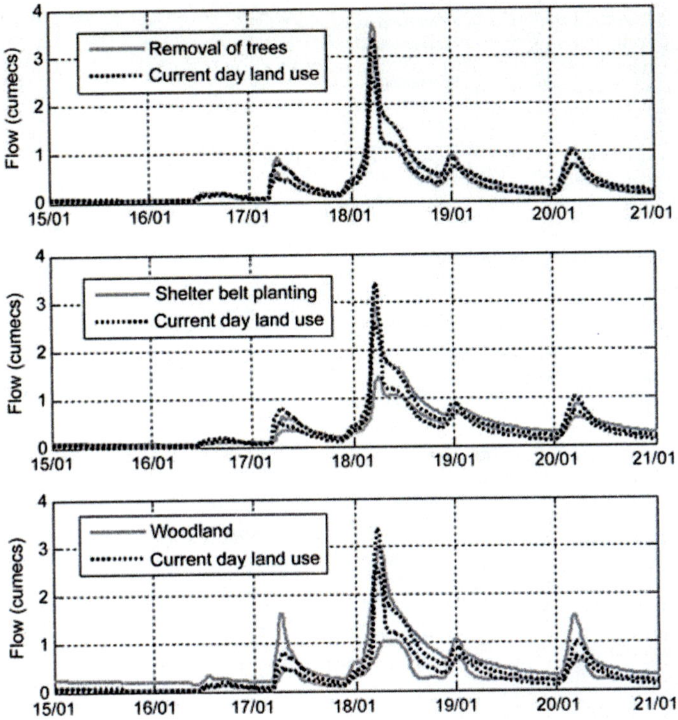

Fig. 22.5 Pontbren land management scenarios showing uncertainty

ical processes and properties from the literature, or possibly from surrogate sites, as well as qualitative information about responses through engagement with field hydrologists. By using such data to parameterise the physics-based models, uncertainty in prior parameters is likely to increase. Limited data also implies that there is a greater chance that the model structures will be poorly defined [25], thereby adding additional uncertainty to the model predictions [17]. The extent to which uncertainty can be constrained by such data is a key research question. We also note that physics-based models have the power to support the development of improved conceptual understanding of runoff processes and the dominant physical controls, and can thereby provide qualitative insights that may be of value when considering the effects of land management change and may also assist in the design of more effective monitoring programmes in order to reduce model uncertainty.

We therefore propose an alternative upscaling procedure taking into account data scarcity, shown in Fig. 22.6. Changes compared to Fig. 22.4 are shown in bold text, and dashed boxes. The key change is that data scarcity no longer automatically leads to a bypass of the physics based modelling. An alternative regionalisation approach, which by-passes the physics based modelling (stage 3), will be discussed in more detail in Sect. 22.5.

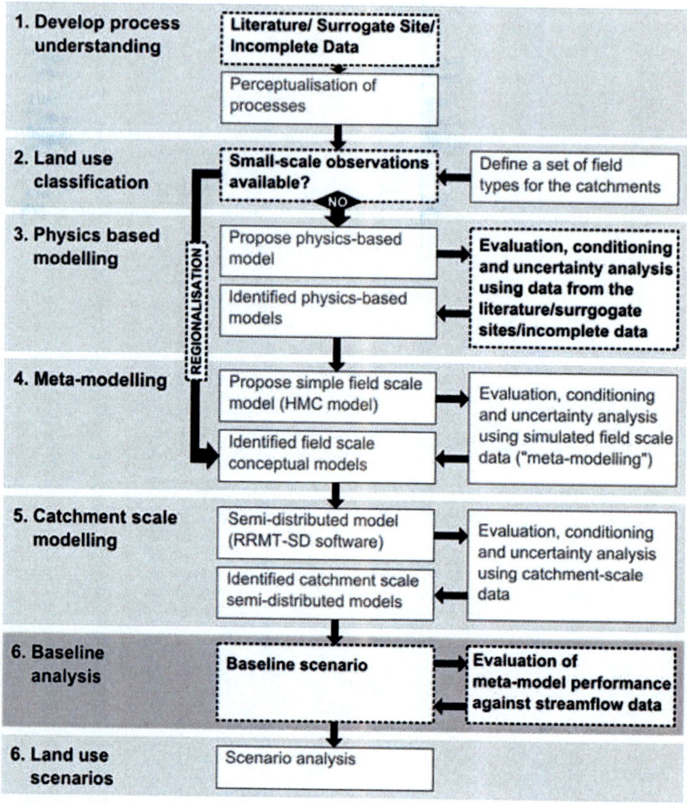

Fig. 22.6 Generalised upscaling procedure

22.4.6 Peat Management in the UK Uplands

In the UK there are approximately 2.9 Mha of upland peat, with the majority of this present as blanket peat; these areas constitute approximately 15% of the blanket peatlands globally [33]. Although historically considered to be regions of low value, the importance of peatlands in terms of carbon sequestration, ecological value and water supply is now increasingly recognised [13]. The management of peatlands has therefore become a topic of interest for a number of different stakeholders.

Almost half of the upland blanket peatlands were drained, typically using open ditch drainage, during a period of agricultural intensification across the UK in the 1960s and 1970s [64]. The intention was that water tables would be lowered to create conditions more suitable for livestock grazing [64]. The reality is that drainage generally causes only localised drawdown of the water table, while also acting as a rapid conduit for runoff. In most reported cases, the runoff response from drained blanket peatlands has reduced times to peak and increased peak flows [3, 20, 34, 60, 65]. Not only does peatland drainage cause potentially detrimental changes in the runoff response, but the practice has also been observed to lead to greater erosion in these sensitive environments [35], to changes in local ecosystems and to increases in concentrations of DOC in the runoff [81]. Due to the multiple problems perceived to be related with drainage, activities are now under way in the UK to attempt to restore these upland environments. Beginning in the 1980s, a programme of blocking peatland drains was started.

Due to the complex process interactions and relatively limited observations, there are large uncertainties about the best management practices for upland blanket peatlands; therefore suitable process-based models can potentially aid our understanding of impacts of management interventions. Although we could have used the physics-based model used in the Pontbren application (albeit modified to account for open ditch drainage), we considered that the available data could not justify such a complex model structure. We have adopted a modelling philosophy similar to that of [74], which places the main modelling effort on the representation of first-order controls on hydrological response, where these processes are identified through engagement between the modeller and the field hydrologist (or in our case, literature about field observations). In this way we developed *a priori* model structures where the key hydrological processes were included whilst working to maintain an appropriate level of complexity relative to the detail of available information concerning the system hydrological processes. To avoid over-parameterisation, minor processes were excluded or treated in a simplified manner. Full details of the peatland model development are available in [4, 5].

The drained blanket peatland model was tested against data from a surrogate site. The site had six boreholes and a weir in a peatland drain that were monitored at a high resolution over a two year period. The model was calibrated on these data, with the primary objective to examine the performance of the *a priori* model structure and to assess the identifiability of the model parameters. The model was found to perform well [5], which provided a degree of confidence that the *a priori* model structure captured the key hydrological processes for drained peatlands, particularly for peak flows. All calibrated parameters were found to be identifiable within the *a priori* parameter ranges, although some more strongly than others, and some only when including the additional borehole data. This suggests that the physical interpretation of these parameters is reasonable.

For sites that can be modelled with the same structure but different parameter values, the models were used to perform "virtual experiments" to explore aspects of hydrological response to a range of design storm events throughout the potential parameter space of UK blanket peatlands. This allows qualitative validation of the model results relative to responses reported in the literature for a range of sites, as well as providing a more general picture of the sensitivity of the flow peaks to the model parameters. Figure 22.7 demonstrates the sensitivity of the mean peak flow and time to peak to the model parameters. Parameter values are normalised by their *a priori* ranges and the x axis for the hydraulic conductivity is shown on a log scale. Details of the simulations are provided in [4].

The model behaviour was found to be consistent with observations from the literature. For example, at high hydraulic conductivities, drainage is effective in reducing peak flows; with low hydraulic conductivities (such as in peatlands), drainage

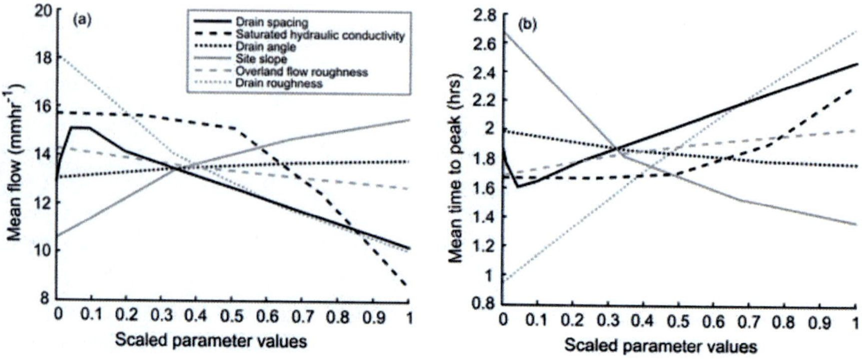

Fig. 22.7 Peatland drained model peak flow sensitivities to model parameter ranges. Mean flow rates (**a**) and mean times to peak (**b**) versus scaled parameter values

is found to increase model peak flows and decrease times to peak, with the effects generally larger in systems with closer drains and lower hydraulic conductivities [34, 60, 65]. At very close drain spacing, the peak model flows begin to reduce, suggesting that spacing contributes to both increased storage and increased conveyance. Examination of the water table profiles also shows that the spatial variation in water table depth observed in the field [21, 34] is replicated in the model. These observations provide further (albeit qualitative) validation of the model peak flow response throughout a wider parameter space.

The peatland models have then been used to perform simulations for 200 m × 200 m hillslopes, of intact, drained and blocked drain blanket peatlands for a specific upland catchment in North-West England. 100 parameter sets were selected from *a priori* parameter ranges that were restricted based on specific site knowledge (drainage maps and DEMs) and information from the literature. The flow response for the largest runoff event in the one year test period is shown in Fig. 22.8. The uncertainty bounds show the range of response for the 100 parameter sets. For the largest runoff event, the mean increase in peak flow from intact to drained peatland was 25% and the mean decrease in peak flow from drained to blocked drained peatland is 3%; the range in responses was 4–42% increase and 16% increase to 25% decrease respectively. The change in runoff response was highly dependent on local conditions and peak flow changes from drained to blocked were also dependent on the flow magnitudes, with simulations with the largest runoff in the drained simulations most likely to give larger percentage reductions in flows following drain blocking.

The *a priori* parameter ranges reflect a combination of the natural variability observed within the hydrological unit class, as well as the uncertainty about these parameter values. For example, the geometric parameters of the peatland model: the slope, drain spacing and drain angles, can be very accurately predicted for a given grid cell; therefore the *a priori* ranges used in the simulations are simply the ranges of these known parameters within the hydrological unit class and cannot be further restricted, unless the hydrological unit class is subdivided. Other parameters, however, are less readily measurable, such as the hydraulic conductivity and

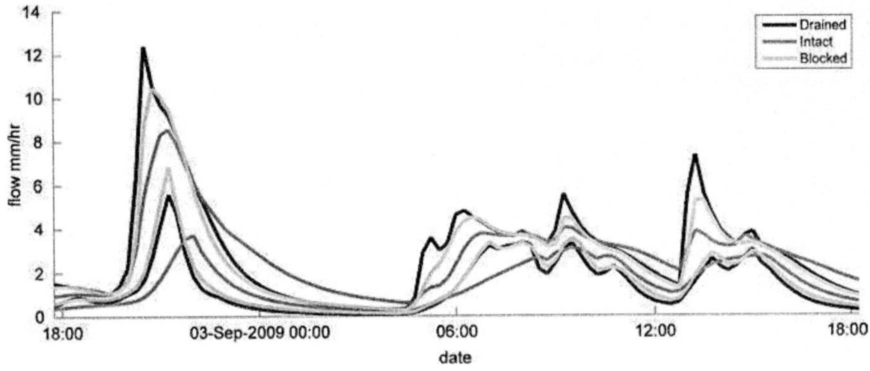

Fig. 22.8 Ensemble uncertainty bounds for a rainfall event on 2 September 2009 for (**a**) drained, (**b**) intact and (**c**) blocked drain peatlands

the overland flow and drain roughnesses; the values for individual grid cells are already uncertain, therefore their *a priori* ranges incorporate both local parameter uncertainty as well as hydrological unit class variability. The changes in peak flows for these simulations were most sensitive to the drain roughness parameter. This is related largely to the high degree of parameter uncertainty, rather than hydrological unit class variability, and therefore could potentially be constrained by detailed data. As the flow response was less sensitive to the local uncertainties of other parameters, the greatest gains in terms of reduction in ensemble uncertainty could be obtained from surface flow roughness investigations in peatlands.

Despite parameter uncertainty, the ensemble responses of the different land management types are found to be distinct; suggesting that even limited data can be used to reduce ensemble uncertainty sufficiently to allow meaningful insights into the changes in runoff response related to land management. Thus the first set of new meta-models that were added to the existing library from Pontbren, were three meta-models to describe the management of upland blanket peat, representing intact, drained and blocked drain blanket peatlands. Work is currently ongoing to examine the performance of the meta-models derived from these data-scarce physics-based models when incorporated into the catchment scale semi-distributed model.

22.5 Conceptual Modelling and Regionalisation

In this section we consider an alternative approach to the problem of estimating land use change effects in the absence of detailed data, and present a regionalisation scheme that employs indices of hydrological behavioural to constrain model parameters. First the methodology and the model used are introduced. Then parameter conditioning for the Pontbren catchment model is evaluated using the regionalised index BFI_{HOST}. BFI is a Base Flow Index, representing the proportion of the streamflow hydrograph that comprises baseflow rather than rapid stormflow, and HOST is the UK Hydrology of Soil Types classification, which has been used as a predictor of BFI for ungauged catchments. Here, speculative changes in BFI_{HOST} allow estimation of heavy grazing and afforestation effects. The conceptual model library is then expanded to include other land uses and management practices by additional parameter restriction using an "imported" behavioural index, based on the US Department of Agriculture Soil Conservation Service model, which uses a "curve number" as the key parameter, denoted here as CN_{USDA}. Finally in this section, findings and approach limitations are discussed.

22.5.1 Methods

We propose a parameter conditioning approach that uses uncertain and limited information about the catchment response in a formal Bayesian framework. This information is represented as hydrological indices that describe different aspects of the expected rainfall-runoff time series behaviour. The indices must be derived from a regionalisation procedure, thus allowing model parameter estimation for ungauged catchments both in current and future (hypothetical) conditions. In this study, we rely on two regionalised indices: Base Flow Index (BFI) and Curve Number (CN).

As noted above, BFI is the proportion of the total catchment discharge which is considered to be base flow. BFI has been regionalised in the UK as a part of the HOST classification system [14] based on the following soil characteristics: depth to gleyed/slowly permeable layer, depth to ground water, presence of a peaty surface layer, and soil substrate. CN relates rainfall volume to corresponding storm runoff volume [31, 32]. Based on data from experimental catchments, estimated values of CN_{USDA} were regionalized within the Soil Conservation Service runoff Curve Number system [67, 68] based on hydrological soil group, land use and land management.

The information is used to estimate posterior parameter distributions for a given soil class and land use management description as described in [15, 16]. In summary, the posterior likelihood of a sampled parameter set is proportional to the consistency of simulated BFI, considered alone, or BFI and CN values considered together, with the values predicted by the regionalisation method for those indices. A large sample of parameter sets and associated likelihoods defines the posterior distribution. The simulated BFI values are calculated from the continuous time simulations using the hydrograph separation procedure of [30], and the simulated CN values are calculated following [32] and [68].

The first study uses information contained in BFI_{HOST} only, and includes land use effects via BFI_{HOST} change (see below), interception and evapotranspiration changes. Two types of land use effects are evaluated: afforestation, and increased stocking density. The second study includes information from the Curve Number method (additionally to BFI_{HOST} information) to represent a much wider variety of land uses/managements [16]. To represent effects of trees and high stocking density, the first study relies on the following assumptions about BFI change. Afforestation

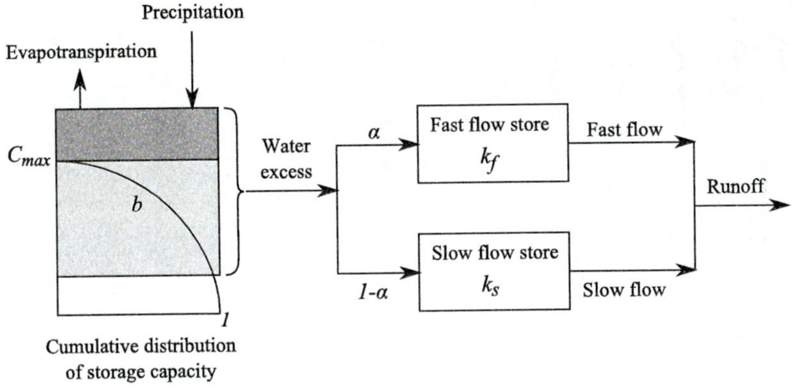

Fig. 22.9 PDM rainfall-runoff conceptual model

is assumed to lead to higher BFI, while keeping the same HOST soil type, so that the posterior includes only those parameter sets that lead to a base flow increase (with respect to the unforested BFI). Changes in interception losses associated with afforestation are estimated using a simple hard threshold bucket model, with canopy storage capacity depending on species, leaf area index, canopy cover, vegetation structure, and density [23]. Increased stocking density leads to soil structural degradation. And, following the approach of Hollis [55], degraded soil is assigned an appropriate analogue HOST class to represent the change. The rationale for the proposed changes is that soil structural degradation, in the form of topsoil and upper subsoil compaction and seasonal "capping" and sealing of soil surfaces, causes a reduction in the effective soil storage, which in turn results in increased surface runoff.

The second study adjoins CN information to BFI information to represent effects of different land uses and managements. To assign CN to each considered soil—land use combination, the British HOST soil classification (29 types) is mapped into the American USDA soil classification (4 classes) [16]. Thus, an important assumption is that the CN index can be used under conditions other than those from which it was derived.

The chosen rainfall-runoff model is the probability distributed moisture (PDM) model with two parallel linear routing stores (Fig. 22.9) [15, 49]. The choice of the PDM model has two motivations: its structural simplicity is thought appropriate given the imposed data limitations (i.e. the information used to condition the model comes from only one, or two flow indices), and it has been extensively applied to other catchments in upland Wales and other UK regions [18, 40, 41]. This model has five parameters: C_{max} is the maximum soil water storage capacity within the modelled element, b is a shape parameter defining the storage capacity distribution, k_f and k_s are fast and slow routing store residence times, and α is the proportion of the total flow going through the fast routing store.

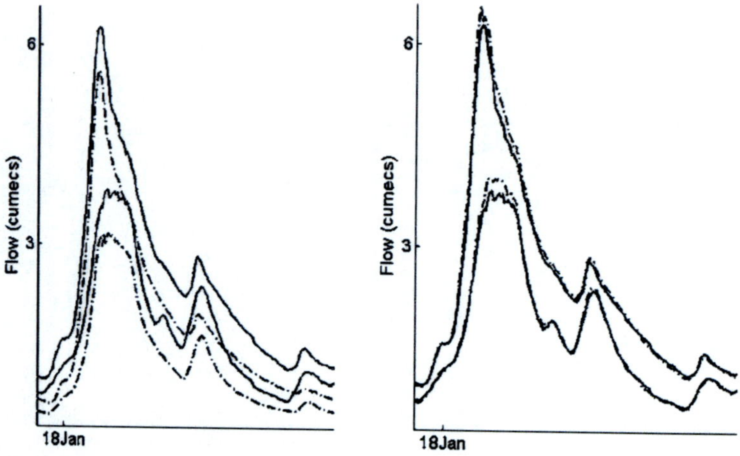

Fig. 22.10 Prediction uncertainty bounds for flows at gauge 10 due to the 18th of January, 2007 rainfall event: (**a**) afforestation, (**b**) soil degradation

22.5.2 The First Case Study—The Pontbren Catchment

Concurring with the Pontbren data time resolution we developed a 15-minute time resolution rainfall-runoff model. We discretised the Pontbren catchment into runoff generating elements (100 m × 100 m squares), so that catchment response is the integration of all the individual elemental responses (via a simple constant celerity routing). Each element response is estimated using the PDM model (see above). Our motivation behind this fine scale is to allow element-scale land management changes to be easily represented within catchment scale models. Potentially, the catchment model needs a separate set of parameters for each element. Here, it is assumed that all elements with the same BFI_{HOST} have the same set of parameter values—this number cannot exceed the number of soil types in the HOST classification (29 types).

The posterior parameter distributions restrict two (out of five) model parameters—the slow flow residence time k_s and runoff partitioning coefficient α. Low k_s values have low posterior probability, and the runoff partitioning coefficient distribution is concentrated around a value of (1-BFI_{HOST}). Model performance was estimated over a highly variable flow period of 1, January, 2007—31, March, 2007. Posterior prediction uncertainty was significantly reduced when compared to prior predictions (unrestricted parameter space). Nash-Sutcliffe statistics for the expected values of probabilistic flow predictions varied between 0.7 and 0.85 for different Pontbren subcatchments, supporting the view that BFI_{HOST} is an effective response index.

Figure 22.10 shows the predicted impacts of full afforestation and increased stocking density on runoff at the most downstream gauge in Pontbren for the 18th of January, 2007 event. Here, the solid lines represent the 90 percentiles for current conditions and the dashed lines are the corresponding results for full afforestation and soil degradation. The uncertainty in the peak flow is high compared to the expected changes, requiring more information about the model parameter values. The afforestation delayed the highest peak arrival by 15 minutes (one simulation time step), and the soil degradation scenario did not show any difference in peak flow arrival time. Full afforestation decreased peak flow by 8% (median value), and stocking intensification increased peak flow by 11% (median value).

22.5.3 The Second Case Study—The Plynlimon Catchments

The Plynlimon catchments are located in Wales, UK, and comprise the Wye and Severn River headwaters [9, 39, 44, 61]. The Wye catchment (10.55 km^2) is almost exclusively under extensively grazed grassland, while for the Severn catchment (8.7 km^2), most of the area is covered with mature coniferous forest. Both catchments are humid—the ratio of long term precipitation to potential evapotranspiration is about 5, with similar slowly permeable soil composition. Because of soil similarity, geographical proximity, and qualitatively different land uses in the catchments, the Wye and Severn catchments are ideal for Curve Number application and testing.

We used hourly data from May 1980 through June 1981, before the Severn tree-felling started and when gap-free Automatic Weather Station data are available. The simulated (by PDM) response for a catchment is the average of the responses for all relevant soil type/land use/land management combinations weighted by their relative contributing areas; this might introduce a one hour timing error at the most.

As in the previous case study, only two parameters: the slow flow residence time k_s and runoff partitioning coefficient α were restricted by the information available (BFI and CN). But, different land uses/managements (as represented by CN) introduced shifts in the parameter distributions—mainly for parameter α. Performance with respect to observed flow in all 8 subcatchments was considered generally good: the prior uncertainty was reduced by a large degree throughout the simulated periods; and probabilistic NS values [15] ranged from 0.70 to 0.81.

As an illustration of the potential applicability of the method, two simple land use change scenarios were considered: (a) the upper Severn becomes pasture in good condition; and (b) the upper Wye becomes forest in good condition. Figure 22.11 shows predictions for the event with the highest flow peak (5–6 October, 1980). Here, black lines represent 95% confidence intervals for the existing land use conditions and grey lines represent 95% confidence intervals for the scenario. The median peak flow in the Severn increases by 9% when the afforested area becomes pasture; in the Wye it reduces by 13% when the pasture land is afforested.

22.5.4 Conclusions Concerning Conceptual Model Regionalisation

In this section we presented a method to integrate regionalised information, in the form of hydrological indices, into model conditioning. We approached this task us-

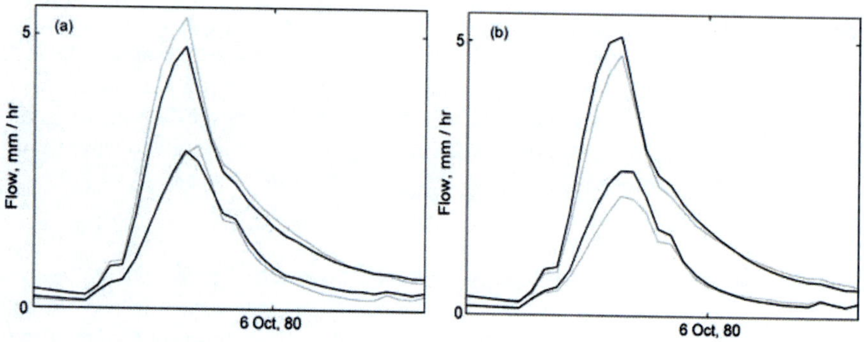

Fig. 22.11 Predictions during a large flood event. (**a**) Severn becomes pasture in good condition; (**b**) Wye becomes forest in good condition

ing either a single source of information (BFI from the HOST classification) and speculative changes due to land use/management, or combining dual sources of information (from the USDA and HOST classification systems) in a formal Bayesian framework. Applied to both the Pontbren catchment and Plynlimon paired catchment data sets, it was concluded that both CN and BFI_{HOST} are potentially valuable sources of information for hydrological modelling of ungauged catchments and the effects of land use change, if used appropriately within a stochastic modelling framework. A more extensive evaluation that introduces more sources of information and covers a range of UK conditions is recommended.

22.6 Using DBM Modelling to Identify HMC Models for Land Use Impacts Analysis

22.6.1 Previous Achievements in Using DBM as a Tool for Model Identification and Land Management Impacts Analysis

The modelling approaches so far demonstrated in this chapter—physics-based modelling, meta-modelling and regionalisation of conceptual models—may be termed "hypothetico-deductive" approaches, after [85, 86]. This is because they pre-assume a model structure and information about model parameter values based on a prior hypothesis of system functioning. For example, this is an explicit part of the meta-modelling procedure (Fig. 22.4). While the prior model structure may be refined based on testing its performance relative to observations, this refinement tends to include some speculation rather than truly letting the data speak for themselves. The need for process-based approaches, such as those we have illustrated in this chapter, is evident when considering the underlying requirement to make predictions which go beyond the range of available observations. However this comes at a price: only a limited range of possible model structures are considered; they provide little scope for detecting response modes and non-linearities which are unexpected *a priori*; and parameter and prediction uncertainty may be high due to inherent over-parameterisation.

Two goals may therefore be proposed: to improve the procedure for HMC model identification by making it more objective; to reduce parametric uncertainty through model order reduction while maintaining an adequate mechanistic basis. The DBM approach can play a major role in these aims. A variety of Peter Young's papers illustrate the power of DBM as a hydrological system identification method [85–88, 91, 93]. Peter's recent work [92] illustrates the model order reduction that can be achieved within an emulation framework, with clear applicability to formalising the model identification within the meta-modelling approach demonstrated in this chapter. His work with Renata Romanowicz and others in [10] illustrates the benefits of DBM rainfall-runoff models for seeking land use signals in time-series data, while also illustrating the role of data noise in interfering with signal detection. McIntyre and Marshall [46] applied DBM in a similar fashion to expose spatial land use signals within Pontbren. Wagener and McIntyre, in this book, apply DBM to the catchment classification problem, including land use effects. And Peter and colleagues, [58] and [54], have illustrated the role of DBM models in critically assessing conceptual hydrological model structures.

Beyond these published demonstrations of applying the DBM approach to model identification for land use and management impacts analysis, there is scope to take this contribution further. A limitation of most previous DBM hydrological applications is the *a priori* assumption that routing is a linear process. Exceptions are [62] (although this was in a real-time context without aiming for conceptualisation) and [50] (although in that case the non-linear routing was an assumption rather than inference using DBM). The assumption of linear routing is understandable given the common knowledge that the dominant non-linearity tends to be in runoff generation (which the DBM models do include), the huge saving in computer time which is usually made by assuming linearity, and the limitations in operational catchment scale rainfall-flow data which may hinder identification of routing non-linearity. Nevertheless, better understanding of non-linearity is a key element of developing HMC models suitable for land use impacts prediction. In particular, for flood studies the nature of the non-linearity, and how it is manipulated, may strongly govern peak flow rates.

Hence, in our recent joint work with Peter, we have explored how the HMC structure identification problem can be addressed using DBM methods including the disaggregation of non-linearity to runoff generation and routing components. This work is summarised in the rest of this section.

22.6.2 *Identification of Rainfall-Runoff Non-linearity Using DBM Analysis*

This work will demonstrate that the DBM approach can identify non-linear signals both in runoff generation, i.e. the dynamic nature of r/u; and in routing (i.e. the dynamic nature of $(dq/dt)/(u-q)$, where r is rainfall, u is effective rainfall and

q is runoff); and will discuss applicability to developing improved predictive HMC models.

The DBM method may be regarded as being built around a general family of transfer functions, described in the discrete time form by,

$$\hat{q}_k = \frac{b_0 + b_1 z^{-1} + b_2 z^{-2} + \cdots + b_m z^{-m}}{1 - a_1 z^{-1} - a_2 z^{-2} - \cdots - a_n z^{-n}} r_{k-\delta/\Delta t}, \tag{22.1}$$

where r is the measured model input, $\hat{q}$ is the model output, δ is the time delay between the onsets of x and $\hat{q}$, Δt is the sampling interval used in the model ($= tk - tk - 1$), k is the time-step number, z^{-r} is the backward shift operator (i.e. $z^{-r} x_k = x_{k-r}$), and a and b are parameter vectors usually estimated using instrumental variable (IV) techniques [83, 89]. The simplest, and the most common, form of (22.1) identified in hydrological applications is,

$$\hat{q}_k = \frac{b_0}{1 - a_1 z^{-1}} r_{k-\delta/\Delta t}, \tag{22.2}$$

where q is the modelled runoff and r is the input rainfall. If both a_1 and b_0 are real and positive numbers, this can be shown to be equivalent to one conceptual linear store with response time $T = -1/\ln(a_1)$ and steady state gain $G = b_0/(1 - a_1)$. Other transfer functions may also be identified, and be shown to have conceptual interpretations, for example two or three linear stores in parallel [76, 84, 89, 90, 93]). A key strength of the DBM approach is the ability to identify conceptually meaningful structures and parameter values, with the only pre-specification being the general transfer function form defined in (22.1).

A further generalisation of (22.1) can be made by introducing state dependence into the parameter vectors a and b. For the simple model in (22.2), this would lead to the form,

$$\hat{q}_k = \frac{b_0(y)}{1 - a_1(y) z^{-1}} r_{k-\delta/\Delta t}. \tag{22.3}$$

An attraction of the DBM modelling toolbox, CAPTAIN [56, 66], is the ability to identify the form of $b_0(y)$ and $a_1(y)$ using state dependent parameter estimation techniques [94]. We have shown that although these two functions are not independent, the form of $b_0(y)$ allows specification of non-linear function for estimating effective rainfall u from observed rainfall r_1 and $a_1(y)$ allows specification of the form of the non-linear routing function [47], using $y = q$. The parameters of functions $b_0(y)$ and $a_1(y)$ can then be optimised numerically.

22.6.3 A Case Study

A 3.2 km^2 subcatchment of Pontbren is used as a case study. In this subcatchment, 77% of the land area used intensively for sheep grazing, 6% is woodland, and the remainder is rough grazing, arable land, open water and paved areas. The average surface slope is 5 degrees. Soils are relatively impermeable silty clay loams. The

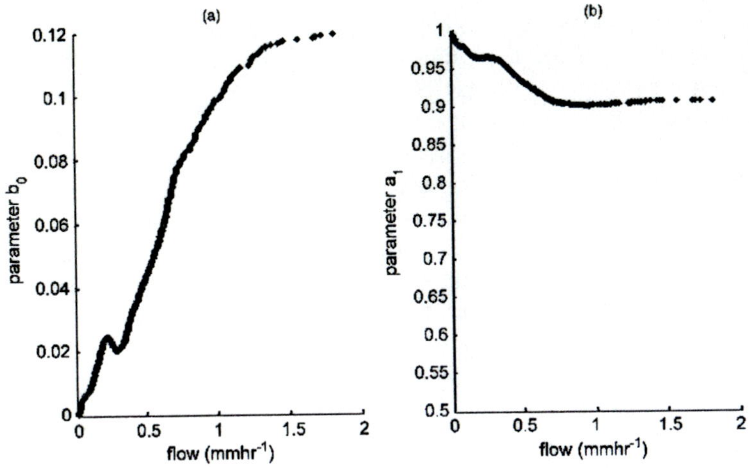

Fig. 22.12 Using the DBM analysis to inform development of HMC models

average annual rainfall is approximately 1449 mm (measured between 1st April 2007 and 31st March 2009). Rainfall is estimated from three nearby tipping bucket gauges, and flow from an in-situ Doppler gauge [45]. The time series data are aggregated to 1-hour intervals for the purpose of the modelling. The period used is 10th November 2006 to 25th January 2007, selected to be consistent with the period we previously used for plot-scale DBM analysis [47].

State dependent parameter analysis was first applied to identify the functions $b_0(q)$ and $a_1(q)$. This gave the results in Fig. 22.12. The power law form of the function $a_1(q)$ indicates non-linearity in the routing, which is consistent, except at the highest flows, with a kinematic wave model,

$$\hat{q}_k = \frac{\beta q_k^{\alpha}}{1-(1-\beta q_k^{\alpha})z^{-1}} \hat{u}_{k-\delta/\Delta t}. \tag{22.4}$$

This is derived by McIntyre et al. [47]. When coding this model, q_k is replaced by the simulated flow at the previous time step q_{k-1}, which avoids instability when zero flow is observed. The routing model in (22.4) explains the form of the function $b_0(q)$ in Fig. 22.12; however Fig. 22.12 is also consistent with the non-linearity in the runoff generation of the form,

$$\hat{u}_k = c\, q_k^{\lambda} r_k. \tag{22.5}$$

Parameters (α, β, λ and δ) are optimised using the Nash-Sutcliffe Efficiency (NSE) as the criterion (in principle the IV criterion could be applied however this would be complex without obvious benefits). c is fixed to ensure that the volume of effective rainfall is equal to the volume of observed flow. The optimised model is,

$$\hat{q}_k = \frac{0.160\hat{q}_{k-1}^{0.54}}{1-(1-0.160\hat{q}_{k-1}^{0.54})z^{-1}} \hat{u}_k, \tag{22.6}$$

$$\hat{u}_k = 0.878 q_k^{0.1} r_k. \tag{22.7}$$

The NSE value is 0.93. Interestingly, and consistent with our previous plot-scale results [47], the flow generation component is nearly linear ($\lambda = 0.1$) while the routing is relatively strongly non-linear ($\alpha = 0.54$). This result is in contrast to the linear stores generally used in catchment scale models; but is consistent with theoretically derived values of α (see [47]).

Three points are worth noting before continuing with the discussion: (1) Of course, alternative conceptualisations of this catchment may be reached, but only if a poorer performance is accepted, or if a less parsimonious model is accepted; (2) Including a mixture of drier and wetter periods in the identification period leads to a higher value of parameter λ as would be expected (e.g. [46]); (3) While (22.6) uses simulated flow $\hat{q}$ as an input, (22.7) uses observed flow q as a surrogate measure of catchment wetness. The need for an observation of q to run this model means that it cannot be used directly for scenario analysis—therefore attention is needed to the task of converting (22.7) into a predictive model.

22.6.4 *Using the DBM Analysis to Inform Development of HMC Models*

A major motivation for the DBM analysis is to assist in identification of models for predicting flow response to scenarios of land use and land management change. This includes providing insights into plot scale processes [47] which can potentially be used in development of physics-based models, and into catchment scale responses which instruct the development of HMC models to be used within the meta-modelling and regionalisation frameworks. In order to develop (22.7) into a predictive model for this purpose, rather than using q as a wetness index, the wetness needs to be simulated using an explicit soil moisture accounting model. This also provides opportunity to introduce some physical constraints into the model, such as constraining losses by potential evaporation estimates. Three HMC soil moisture accounting models were tested:

1. A version of the catchment wetness index (CWI) model [47], derived originally from the early work of Peter Young et al. [82]. The two-parameter version of this model uses an empirical factor to force runoff generation volume to equal streamflow volume, assumes a first order loss, and runoff generation is equal to wetness raised to a power. Peter has shown that this CWI model is comparable with the power law in (22.5) [86].
2. A version of the probability distributed moisture (PDM) model of [48] using a two-parameter Pareto distribution developed by Wagener et al. [69]. Losses are control by simulated evaporation rates, with the evaporation: potential evaporation ratio being proportional to the relative wetness of the catchment. This model may also be deduced from the power law model of (22.6) because both may be interpreted as representing a dynamic contributing area [42].
3. A three-parameter bucket model with storage thresholds for initiation of drain and surface flow, with an upper limit to drainage, and with evaporation equal to

potential evaporation when soil moisture is present. This was originally developed based on process understanding of experimental plots at Pontbren [37], and thus it is the only soil moisture model structure which was chosen independently of the DBM analysis.

These models are typically applied assuming linear routing, however here we apply the non-linear store identified above (22.6), and focus on assessing performance of the wetness index/loss components when optimised to NSE. The CWI and PDM models, with NSE = 0.94 and 0.93 respectively, more or less match the performance of the DBM runoff generation model. The third model, while performing less well with NSE = 0.90, has the conceptual attraction of distinguishing between surface flow and drain flow generation. This distinction is potentially important for predicting the effects of changes to soil and drainage associated with land management, and hence this model, with linear routing, has been used so far for the Pontbren meta-modelling. The DBM analysis has indicated that non-linear routing should be implemented in this model, and that the lumped treatment of soil wetness may be improved by adding a dynamic contributing area concept such as is implicit to the CWI and PDM models. Clearly, the DBM analysis would first need to be repeated for different subcatchments and periods including validation tests.

22.6.5 Concluding Upon the Value of DBM Modeling for Land Use Impacts Analysis

The DBM modelling framework has been applied previously to identifying spatial and temporal signals in hydrological response, and linking these to land use and land management change [10, 46], and there is scope for broadening such analyses to the catchment classification problem (Wagener and McIntyre, this book). DBM analysis has recently been used for emulation of complex physics-based models [92], with clear applicability to further formalising our meta-modelling strategy. Our recent work (McIntyre et al. in review) has illustrated the insights into plot scale responses provided by a DBM analysis, which can inform development of small-scale physics-based models. And finally, the DBM analysis in this chapter has provided initial indications of how the HMCs used in our meta-modelling may be improved to better represent observed non-linearity.

22.7 Conclusions

In this chapter, we have reflected critically on the role of alternative modelling approaches and philosophies, guided by Peter Young's work. We consider the practical problem of prediction of the effects of land use and land management change on hydrological response at scales from individual fields to catchments, which challenges current methodologies. While the role of physics-based models has received much critical discussion in the literature, we show that with data support from a detailed field experimental programme, at Pontbren, Wales, useful predictions of the effects of complex interventions (for example planting spatially-localised tree shelter belts) can be made at field scale, albeit with relatively large confidence intervals. These models are too demanding computationally to be applied at catchment scale, hence we use simpler Hybrid Metric Conceptual (HMC) meta-models to emulate their behaviour and hence produce catchment-scale estimates of the impacts of different land management strategies.

We explore the role of physics-based models in a data-scarce situation through consideration of the management of upland peat in the UK. A simplified physics-based model is conditioned on surrogate data, and provides a useful tool to explore scenarios of land management change. Current work is extending this through meta-modelling to catchment scale.

We also consider an alternative approach, using a conceptual modelling framework, constrained using regionalised indices of hydrological response. A regionalised Base Flow Index, dependent on soil type, has considerable power in constraining ungauged catchment simulations; by speculating on the effects of land management interventions on soils we produce estimates of the catchment-scale effects. In addition we consider the use of US Curve Number methods. These have been developed on the basis of small scale experiments and contain tabulated guidance on the effects of soil structural change. By mapping US to UK soils we provide an additional source of information with which to constrain catchment-scale simulations. Despite the limited data support for these methods, results to date are convincing, and generally consistent with the physics-based upscaling approach. Clearly further work is desirable to establish the validity of these methods for more general application, but in the absence of alternatives, the method shows great promise for national applicability.

Finally, we apply Peter's algorithms to Data-Based Modelling, using our Welsh experimental data. It could of course be argued that this should have been the first, not the last, model analysis; the results show the power of DB modelling in providing insight into experimental data and appropriate model structures. The specific results have significant implications for the representation of flow routing in hydrological models; more generally the results illustrate the major potential for Peter's DBM analysis to be applied to high quality experimental data to improve understanding of appropriate model structures and their identifiability.

Acknowledgements FRMRC, FREE, Pontbren cooperative. This research was sponsored by NERC Grant NE/F001061/1 and EPSRC Grants EP/FP202511/1 and GR/S76304/01.

References

1. Abbott, M.B., Bathurst, J.C., Cunge, J.A., O'Connell, P.E., Rasmussen, J.: An introduction to the European Hydrological System—Systeme Hydrologique Europeen, SHE. 1. History and philosophy of a physically-based, distributed modelling system. J. Hydrol. **87**, 45–59 (1986)

2. Abbott, M.B., Bathurst, J.C., Cunge, J.A., O'Connell, P.E., Rasmussen, J.: An introduction to the European Hydrological System—Systeme Hydrologique Europeen, SHE. 2. Structure of a physically-based, distributed modelling system. J. Hydrol. **87**, 61–77 (1986)
3. Ahti, E.: Ditch spacing experiments in estimating the effects of peatland drainage on summer runoff. In: Proceedings of the International Symposium on Influence of Man on Hydrological Regime, Helsinki. IAHS-AISH Publication, vol. 130, pp. 49–53 (1980)
4. Ballard, C., McIntyre, N., Wheater, H.S.: Peatland drain blocking—can it reduce peak flood flows. In: Proceedings of BHS 2010 International Conference, Newcastle, UK (2010)
5. Ballard, C.E., McIntyre, N., Wheater, H.S., Holden, J., Wallage, Z.E.: Hydrological modelling of drained blanket peatland. J. Hydrol. **407**, 81–93 (2011)
6. Beven, K.J.: Changing ideas in hydrology: the case of physically-based models. J. Hydrol. **105**, 157–172 (1989)
7. Beven, K.J.: Prophecy, reality and uncertainty in distributed hydrological modelling. Adv. Water Resour. **16**, 41–51 (1993)
8. Beven, K.J.: How far can we go in distributed hydrological modelling? Hydrol. Earth Syst. Sci. **5**(1), 1–12 (2001)
9. Beven, K.J., Binley, A.: The future of distributed models—model calibration and uncertainty prediction. Hydrol. Process. **6**(3), 279–298 (1992)
10. Beven, K.J., Young, P., Romanowicz, R., O'Connell, P.E., Ewen, J., O'Donnell, G.M.O., Homan, I., Posthumus, H., Morris, J., Hollis, J., Rose, S., Lamb, R., Archer, D.: Analysis of historical data sets to look for impacts of land use and management change on flood generation. Defra R&D Final Report FD2120. Defra, London (2008)
11. Bicknel, B.R., Imhoff, J.C., Kittle, J.L., Jobes, T.H., Donigian, A.S.: Hydrological Simulation Program—FORTRAN HSPF Version 12 User's Manual. AQUA TERRA Consultants Mountain View, California 94043 (2001)
12. Binley, A.M., Beven, K.J.: A physically based model of heterogeneous hillslopes 2. Effective hydraulic conductivities. Water Resour. Res. **25**(6), 1227–1233 (1989)
13. Bonn, A., Allott, T.E.H., Hubacek, K., Stewart, J.: Introduction: drivers of change in upland environments: concepts, threats and opportunities. In: Bonn, A., Allott, T.E.H., Hubacek, K., Stewart, J. (eds.) Drivers of Change in Upland Environments, Routledge, Oxon, pp. 1–10 (2009)
14. Boorman, D., Hollis, J., Lilly, A.: Hydrology of Soil Types: A Hydrologically-Based Classification of the Soils of the United Kingdom. Institute of Hydrology, Wallingford (1995)
15. Bulygina, N., McIntyre, N., Wheater, H.S.: Conditioning rainfall-runoff model parameters for ungauged catchments and land management impacts analysis. Hydrol. Earth Syst. Sci. **13**(6), 893–904 (2009)
16. Bulygina, N., McIntyre, N., Wheater, H.S.: Bayesian conditioning of a rainfall-runoff model for predicting flows in ungauged catchments and under land use changes. Water Resour. Res. (2010). doi:10.1029/2010wr009240
17. Butts, M.B., Payne, J.T., Kristensen, M., Madsen, H.: An evaluation of the impact of model structure on hydrological modelling uncertainty for streamflow simulation. J. Hydrol. **298**(1–4), 242–266 (2004)
18. Calver, A., Crooks, S., Jones, D., Kay, A., Kjeldsen, T., Reynard, N.: National river catchment flood frequency method using continuous simulation. DEFRA (2005)
19. Carroll, Z.L., Bird, S.B., Emmett, B.A., Reynolds, B., Sinclair, F.L.: Can tree shelterbelts on agricultural land reduce flood risk? Soil Use Manag. **20**, 357–359 (2004)
20. Conway, V.M., Millar, A.: The hydrology of some small peat-covered catchments in the Northern Pennines. J. Inst. Water Eng. **14**, 415–424 (1960)
21. Coulson, J.C., Butterfield, J.E.L., Henderson, E.: The effect of open drainage ditches on the plant and invertebrate communities of moorland and on the decomposition of peat. J. Appl. Ecol. **27**(2), 549–561 (1990)
22. Crawford, N.H., Linsley, R.K.: Digital simulation in hydrology: Stanford Watershed Model IV. Tech. Rpt 39, Stanford University, California (1996)
23. David, J.S., Valente, F., Gash, J.: Evaporation of intercepted rainfall. In: Anderson, M.G. (ed.) Encyclopedia of Hydrological Sciences. Wiley, New York (2005)

24. Duan, Q., Sorooshian, S., Gupta, V.K.: Shuffled complex evolution approach for effective and efficient global minimization. J. Optim. Theory Appl. **76**(3), 501–521 (1993)
25. Ebel, B.A., Loague, K.: Physics-based hydrologic-response simulation: seeing through the fog of equifinality. Hydrol. Process. **20**(13), 2887–2900 (2006)
26. Freer, J., Beven, K.J., Abroise, B.: Bayesian uncertainty in runoff prediction and the value of data: an application of the GLUE approach. Water Resour. Res. **32**, 2163–2173 (1996)
27. Freeze, RA: Role of subsurface flow in generating surface runoff. 2: Upstream source areas. Water Resour. Res. **8**, 1272–1283 (1972)
28. Freeze, R.A., Harlan, R.L.: Blueprint for a physically-based, digitally simulated hydrologic response model. J. Hydrol. **9**, 237–258 (1969)
29. Gupta, H.V., Sorooshian, S., Yapo, P.O.: Towards improved calibration of hydrological models: multiple and non-commensurable measures of information. Water Resour. Res. **34**(4), 751–763 (1998)
30. Gustard, A., Bullock, A., Dickson, J.: Low flow estimation in the United Kingdom. Report no 101, Institute of Hydrology, Wallingford (1992)
31. Hawkins, R.H.: The importance of accurate curve numbers in the estimation of storm runoff. Water Resour. Bull. **11**(5), 887–891 (1975)
32. Hawkins, R.H.: Asymptotic determination of runoff curve numbers from data. J. Irrig. Drain E **119**(2), 334–345 (1993)
33. Holden, J., Chapman, P.J., Labadz, J.C.: Artificial drainage of peatlands: hydrological and hydrochemical process and wetland restoration. Prog. Phys. Geogr. **28**(1), 95–123 (2004)
34. Holden, J., Evans, M.G., Burt, T.P., Horton, M.M.: Impact of land drainage on peatland hydrology. J. Environ. Qual. **35**(5), 1764–1778 (2006)
35. Holden, J., Gascoign, M., Bosanko, N.R.: Erosion and natural revegetation associated with surface land drains in upland peatlands. Earth Surf. Processes Landf. **32**(10), 1547–1557 (2007)
36. Institute of Hydrology: Flood Estimation Handbook. CEH Wallingford, Wallingford (1999)
37. Jackson, B.M., Chell, J., Francis, O., Frogbrook, Z., Marshall, M., McIntyre, N., Reynolds, B., Solloway, I., Wheater, H.S.: The impact of upland land management on flooding: insights from a multi-scale experimental and modelling programme. J. Flood. Risk Man. **1**(2), 71–80 (2008)
38. Karavokyris, I., Butler, A.P., Wheater, H.S.: The development and validation of a coupled soil-plant-water model (SPW1). Nirex Safety Series Report, NSS/R225 (1990)
39. Kirby, C., Newson, M.D., Gilman, K.: Plynlimon Research: The First Two Decades. Institute of Hydrology, Wallingford (1991)
40. Lamb, R., Kay, A.L.: Confidence intervals for a spatially generalized, continuous simulation flood frequency model for Great Britain. Water Resour. Res. (2004). doi:10.1029/2003WR002428
41. Lee, H., McIntyre, N., Wheater, H.S., Young, A.: Selection of conceptual models for regionalization of the rainfall-runoff relationship. J. Hydrol. **312**(1–4), 125–147 (2005)
42. Lees, M.J.: Data-based mechanistic modelling and forecasting of hydrological systems. J. Hydroinform. **2**, 15–34 (2000)
43. Lees, M.J., Wagener, T.: A Monte-Carlo Analysis Toolbox (MCAT) for Matlab—User Manual. Imperial College, UK (1999)
44. Marc, V., Robinson, M.: The long-term water balance (1972–2004) of upland forestry and grassland at Plynlimon, mid-Wales. Paper presented at McCulloch Symposium on a View from the Watershed Revisited held at the General Assembly of the European-Geosciences-Union, European Geosciences Union, Vienna, Austria (2006)
45. Marshall, M.R., Francis, O.J., Frogbrook, Z.L., Jackson, B.M., McIntyre, N., Reynolds, B., Solloway, I., Wheater, H.S., Chell, J.: The impact of upland land management on flooding: results from an improved pasture hillslope. Hydrol. Process. **23**(3), 464–475 (2009)
46. McIntyre, N., Marshall, M.R.: Identification of rural land management signals in runoff response. Hydrol. Process. (2010). doi:10.1002/hyp.7774

47. McIntyre, N., Young, P.C., Orellana, B., Marshall, M.R., Reynolds, B., Wheater, H.S.: Identification of nonlinearity in rainfall-flow response using data-based mechanistic modelling. Water Resour. Res. **47**, W03515 (2011)
48. Moore, R.J.: The probability-distributed principle and runoff production at point and basin scales. Hydrol. Sci. J. **30**, 273–297 (1985)
49. Moore, R.J.: The PDM rainfall-runoff model. Hydrol. Earth Syst. Sci. **11**(1), 483–499 (2007)
50. Mwakalila, S., Campling, P., Feyen, J., Wyseure, G., Beven, K.J.: Application of a data-based mechanistic modelling (DBM) approach for predicting runoff generation in semi-arid regions. Hydrol. Process. **15**, 2281–2295 (2001)
51. NERC: Flood Studies Report, vols. I–V, Natural Environment Research Council, London, UK (1975)
52. O'Connell, P.E., Beven, K.J., Carney, J.N., Clements, R.O., Ewen, J., Fowler, H., Harris, G., Hollis, J., Morris, J., O'Donnell, G.M.O., Packman, J.C., Parkin, A., Quinn, P.F., Rose, S.C., Shepher, M., Tellier, S.: Review of Impacts of Rural Land Use and Management on Flood Generation. Report A: Impact Study Report. R&D Technical Report FD2114/TR, DEFRA, London, UK. 152 pages (2004)
53. O'Connell, P.E., Ewen, J., O'Donnel, G.M.O., Quinn, P.: Is there a link between agricultural land-use and flooding?. Hydrol. Earth Syst. Sci. **11**, 96–107 (2007)
54. Orellana, B., McIntyre, N., Wheater, H.S., Sarkar, A., Young, P.C.: Comparison of lumped rainfall-runoff modelling approaches for a semiarid basin. In: Proceedings of "Water Environment Energy and Society", New Delhi, 12–16 January 2009, pp. 713–721 (2009)
55. Packman, J., Quinn, P., Hollis, J., O'Connell, P.E.: Review of impacts of rural land use and management on flood generation. Short term improvement to the FEH rainfall-runoff model: technical background, pp. 1–66, DEFRA (2004)
56. Pedregal, D.J., Taylor, C.J., Young, P.C.: System Identification, Time Series Analysis and Forecasting. The Captain Toolbox. Handbook v2.0, Centre for Research on Environmental Systems and Statistics, Lancaster University, UK (2007)
57. Pilgrim, D.H., Huff, D.D., Steele, T.D.: A field evaluation of subsurface and surface runoff: II. Runoff processes. J. Hydrol. **38**(3–4), 319–341 (1978)
58. Ratto, M., Young, P.C., Romanowicz, R., Pappenberger, F., Saltelli, A., Pagano, A.: Uncertainty, sensitivity analysis and the role of data based mechanistic modeling in hydrology. Hydrol. Earth Syst. Sci. **11**, 1249–1266 (2007)
59. Refsgaard, J.C., Storm, B., Clausen, T.: Systeme Hydrologique Europeen (SHE): review and perspectives after 30 years development in physically-base hydrological modelling. Hydrol. Res. **41**(5), 355–377 (2010)
60. Robinson, M.: Changes in catchment runoff following drainage and afforestation. J. Hydrol. **86**(1–2), 71–84 (1986)
61. Robinson, M., Dupeyrat, A.: Effects of commercial timber harvesting on streamflow regimes in the Plynlimon catchments, mid-Wales. Hydrol. Process. **19**(6), 1213–1226 (2005)
62. Romanowicz, R.J., Young, P.C., Beven, K.J.: Data assimilation and adaptive forecasting of water levels in the river Severn catchment, United Kingdom. Water Resour. Res. (2006). doi:10.1029/2005WR004373
63. Spear, R.C., Hornberger, G.M.: Eutrophication in Peel inlet, II, Identification of critical uncertainties via generalised sensitivity analysis. Water Resour. Res. **14**, 43–49 (1980)
64. Stewart, A.J., Lance, A.N.: Moor-draining: a review of impacts on land use. J. Environ. Manag. **17**(1), 81–99 (1983)
65. Stewart, A.J., Lance, A.N.: Effects of moor-draining on the hydrology and vegetation of northern Pennine blanket bog. J. Appl. Ecol. **28**(3), 1105–1117 (1991)
66. Taylor, C.J., Pedregal, D.J., Young, P.C., Tych, W.: Environmental time series analysis and forecasting with the Captain toolbox. Environ. Model. Softw. **22**, 797–814 (2007)
67. USDA: Urban hydrology for small watersheds. Technical Release 55, 2nd edn., NTIS PB87-101580. US Department of Agriculture Soil Conservation Service, Springfield, Virginia (1986)
68. U.S. Soil Conservation Service: Hydrology. In: Nat. Eng. Handbook, Sec. 4:547 p. U.S. Govt. Print Off. Washington, DC (1972)

69. Wagener, T., Boyle, D.P., Lees, M.J., Wheater, H.S., Gupta, H.V., Sorooshian, S.: A framework for the development and application of hydrological models. Hydrol. Earth Syst. Sci. **5**(1), 13–26 (2001)
70. Wagener, T., Boyle, D.P., Lees, M.J., Wheater, H.S., Gupta, H.V., Sorooshian, S.: A framework for the development and application of hydrological models. In: Proc. BHS 7th National Symp., Newcastle upon Tyne, pp. 3.75–3.81 (2000)
71. Wagener, T., Lees, M.J., Wheater, H.S.: A Rainfall-Runoff Modelling Toolbox (RRMT) for Matlab—User Manual. Imperial College, UK (1999)
72. Wagener, T., McIntyre, N., Lees, M.J., Wheater, H.S., Gupta, H.V.: Towards reduced uncertainty in conceptual rainfall-runoff modelling: dynamic identifiability analysis. Hydrol. Process. **17**, 455–476 (2003)
73. Wagener, T., Lees, M.J., Wheater, H.S.: Reducing conceptual rainfall-runoff modelling uncertainty. In: Proc. of Workshop on "Runoff Generation and Implications for River Basin Modelling", Freiburg, Germany (2000)
74. Weiler, M., McDonnell, J.: Virtual experiments: a new approach for improving process conceptualization in hillslope hydrology. J. Hydrol. **285**(1–4), 3–18 (2004)
75. Wheater, H.S.: Flood hazard and management: a UK perspective. Philos. Trans. R. Soc. A **364**, 2135–2145 (2006)
76. Wheater, H.S., Beck, M.B., Kleissen, F.M.: Identifiability of conceptual hydrochemical models. Water Resour. Res. **26**(12), 2979–2992 (1990)
77. Wheater, H.S., Shaw, T.L., Rutherford, J.C.: Storm runoff from small lowland catchments in South West England. J. Hydrol. **55**, 321–337 (1982)
78. Wheater, H.S., Bishop, K.H., Beck, M.B.: The identification of conceptual hydrological models for surface water acidification. J. Hydrol. Process. **1**, 89–109 (1986)
79. Wheater, H.S., Jakeman, A.J., Beven, K.J.: Progress and directions in rainfall-runoff modelling. In: Jakeman, A.J., Beck, M.B., McAleer, M.J. (eds.) Modelling Change in Environmental Systems, pp. 101–132. Wiley, New York (1993)
80. Wheater, H.S., Reynolds, B., McIntyre, N., Marshall, M.R., Jackson, B.J., Frogbrook, Z.L., Solloway, I., Francis, O.J., Chell, J.: Impacts of land management on flood risk: FRMRC RPA2 at Pontbren. FRMRC Final Report and UFMO (2008)
81. Worrall, F., Armstrong, A., Holden, J.: Short-term impact of peat drain-blocking on water colour, dissolved organic carbon concentration, and water table depth. J. Hydrol. **337**(3–4), 315–325 (2007)
82. Whitehead, P.G., Young, P.C., Hornberger, G.H.: A systems model of stream flow and water quality in the Bedford-Ouse river—Part l: Stream flow modelling. Water Res. **13**, 1155–1169 (1976)
83. Young, P.C.: Recursive Estimation and Time-Series Analysis. Springer, New York (1984)
84. Young, P.C.: Data-based mechanistic modelling of environmental, ecological, economic and engineering systems. Environ. Model. Softw. **13**, 105–122 (1998)
85. Young, P.C.: Advances in real-time flood forecasting. Philos. Trans. R. Soc. A **360**, 1433–1450 (2002)
86. Young, P.C.: Top-down and data-based mechanistic modelling of rainfall-flow dynamics at the catchment scale. Hydrol. Process. **17**, 2195–2217 (2003)
87. Young, P.C.: Rainfall-runoff modeling: transfer function models. In: Anderson, M.G. (ed.) Encyclopedia of Hydrological Sciences, vol. 3, part II, pp. 1985–2000. Wiley, Hoboken (2005)
88. Young, P.C.: Real time flow forecasting. In: Wheater, H.S., Sorooshian, S., Sharma, K.D. (eds.) Hydrological Modelling in Arid and Semi-Arid Areas, pp. 113–137. Cambridge University Press, Cambridge (2008)
89. Young, P.C.: The refined instrumental variable method: unified estimation of discrete and continuous-time transfer function models. J. Eur. Syst. Autom. **42**, 149–179 (2008)
90. Young, P.C.: Gauss, Kalman and advances in recursive parameter estimation. J Forecasting **30**, 104–146 (2011) (special issue celebrating 50 years of the Kalman Filter)
91. Young, P.C., Beven, K.J.: Data-based mechanistic modelling and the rainfall-flow nonlinearity. Environmetrics **5**, 335–363 (1994)

92. Young, P.C., Ratto, M.: A unified approach to environmental systems modelling. Stoch. Environ. Res. Risk Assess. **23**, 1037–1057 (2009)
93. Young, P.C., Jakeman, A.J., Post, D.A.: Recent advances in data-based modelling and analysis of hydrological systems. Water Sci. Technol. **36**, 99–116 (1997)
94. Young, P.C., McKenna, P., Bruun, J.: Identification of non-linear stochastic systems by state dependent parameter estimation. Int. J. Control **74**, 1837–1857 (2001)

Chapter 23
Hydrological Catchment Classification Using a Data-Based Mechanistic Strategy

Thorsten Wagener and Neil McIntyre

'In plain words, Chaos was the law of nature; Order was the dream of man.' Henry Adams (1918)

23.1 Introduction

An important task in any field of science is to perpetually organize the body of knowledge gained by scientific inquiry. Classification or taxonomy is an essential component of such organization, whereby we attempt to organize the entity of interest into homogeneous or similar classes. If the object of classification is a natural entity (rather than, for example, human-made objects), then classification becomes the search for a theory about the basis of natural order and to make sense of the heterogeneous world around us [9]. In this sense classification is not merely the creation of a filing system, but a rigorous scientific inquiry into the causes of similarities and dissimilarities of a particular entity of interest (e.g. organisms in biology). Some sciences (such as biology and fluid mechanics) have made great strides in establishing classification systems, which have led to rapid advances in their theoretical foundations [5], while other sciences (such as hydrology) are younger and therefore less advanced in understanding how their knowledge set could be similarly organized or generalized.

T. Wagener (✉)
Department of Civil and Environmental Engineering, The Pennsylvania State University, University Park, PA 16802, USA
e-mail: thorsten@engr.psu.edu

N. McIntyre
Department of Civil and Environmental Engineering, Imperial College London, London SW72AZ, UK
e-mail: n.mcintyre@imperial.ac.uk

L. Wang, H. Garnier (eds.), *System Identification, Environmental Modelling, and Control System Design*,
DOI 10.1007/978-0-85729-974-1_23,

In the field of hydrology, the catchment forms an entity of major interest, which provides a sensible (though of course not the only possible) unit for a hydrological classification system. A catchment is typically defined as the drainage area that contributes water to a particular point (usually) along a channel network, based on its surface topography. Catchments form landscape elements across spatial scales that integrate all aspects of the hydrologic cycle within a defined area that can be studied, quantified, and acted upon [36]. Catchments vary widely in terms of their landscape characteristics (e.g., vegetation, topography, soils, geology) and in terms of the climatic characteristics of the region they are located in (e.g., precipitation, temperature, radiant energy). Despite the degree of uniqueness and complexity that each catchment exhibits [2], we generally assume that some level of organization and therefore a degree of predictability of the functional behavior of a catchment exists [5, 30], which may be a result of natural self-organization or co-evolution of the climate, soils, vegetation and topography. While we are relying heavily on the assumption of hydrological similarity, and therefore on the ability to transfer information from one region to a similar one, the science of hydrology has thus far not established a common catchment classification system that would provide order and structure to the global assemblage of these heterogeneous spatial units (see detailed discussions in [18, 36]).

Identifying and categorizing catchments based on their dominant functional characteristics, i.e. based on their hydrological behavior, is one strategy to quantify the degree of similarity that may exist between catchments. Understanding how and why certain functional behavior occurs in a given catchment would ultimately shed new light on the reasons for the similarity or dissimilarity that is exhibited between catchments [8]. Grigg [10, 11] lists three main reasons for the classification of geographical data: (1) to give names to things, i.e. the main classification step; (2) to permit transfer of information, i.e. regionalization of information; (3) to permit development of generalizations, i.e. to develop new theory. In the light of increasing concerns about non-stationarity of the responses of hydrologic systems [21, 37], we add a fourth reason for the need for a robust catchment classification system, namely: (4) to provide a first order environmental change impact assessment, i.e., the hydrologic implications of climate, land use and land cover change.

All four of the above aspects have to be objectives of any catchment classification system to ultimately achieve order, new understanding and predictive power. As mentioned above, classification implies that entities (e.g. catchments) with similar characteristics belong to the same group, while dissimilar entities form separate groups. The first question to be addressed is therefore how one should define hydrologic similarity or dissimilarity in a catchment classification system. Strategies for classification in the past have largely focused on physical similarity (e.g., similarity in physical characteristics, or how the catchments look) or on similarity of some (narrow) characteristic of the streamflow record (how the catchments behave within a given, somewhat narrow context). Below we argue that both approaches fall short in achieving all of our objectives, and that the general idea of catchment function [30, 36] can bridge the gap between them and in this way help fulfill the needs of a more general classification system.

A wide range of previous studies has grouped catchments based on the similarity in physical characteristics, such as soils, land cover, or topography. Winter [38] developed a catchment classification system that is based on the idea of hydrologic landscapes, which are defined on the basis of similarity of climate, topography and geology, assuming that catchments that are similar with respect to these three criteria will behave similarly in a hydrological sense. In a similar manner, Buttle [4] suggests that, within a hydro-climatic region, three factors should provide first-order controls on the streamflow response of catchments: (1) typology—hydrologic partitioning between vertical and lateral pathways, (2) topology—drainage network connectivity, and (3) topography—hydraulic gradients as defined by basin topography. These studies then make the (often) implicit assumption that the physical (climate and landscape) characteristics considered are the dominant controls on the 'hydrologic behavior' of a catchment and are therefore sufficient to group catchments that are hydrologically similar. In general, these strategies are based on the expected mapping or assumed relationships between physical, climatic and hydrological response characteristics, including an eventual test of the predictive power of these relationships. The general availability of physical and climatic characteristics enables the widespread implementation of such schemes. However, such assumed relationships might not always be a complete or even sufficient explanation of the inter-catchment variability. To fully permit (hydrologic) information transfer and to achieve a generalization of the relationships between catchment attributes, climate and hydrologic responses (i.e. steps (2) and (3) in our objectives listed above), an explicit quantitative assessment of such relationships is required, whereas implicit assumptions will prove to be ultimately insufficient (although they can be good starting point for the analyses).

An alternative strategy to the above physically based classification scheme relies on the analysis of some aspect of the streamflow response. Assessing similarity in terms of certain streamflow characteristics, such as river regime, has been particularly useful in aquatic ecology, due to their importance for maintenance of aquatic habitats [22]. For example, Haines et al. [12] classified river regimes in terms of seasonality of flow, Olden and Poff [22] with respect to ecologically relevant flow characteristics, Krasovskaia [15] based on entropy and Krasovskaia et al. [16] with respect to inter-annual variability of streamflow. This is by no means an exhaustive list of studies of this type, but merely a representative sample. These studies provide a grouping with respect to similarity of a specific hydrologic response, and thus represent similarity only in a narrow sense. In particular, they suffer from the fact that they do not attempt to connect the hydrologic responses back to both climate and landscape characteristics that caused them, and hence do not achieve all of our objectives, as stated above.

Black [3] introduced the idea of hydrologic function, defined as the actions of the catchment exerted on the precipitation it collects. Wagener et al. [35, 36] expanded on this idea by viewing catchments as non-linear space-time filters, which perform a set of common hydrologic functions, broadly consisting of the partitioning, storage, and release of water. Using the definitions introduced in those previous papers, partitioning is defined as the process whereby incoming precipitation is partitioned at the land surface into several components (e.g. infiltration, interception and surface runoff). Storage refers to the mechanism by which any of the components of the incoming precipitation is held in temporary storage before its eventual release from the catchment, which arises due to the fact that the rate of arrival of precipitation is greater than the rate of release of water from the catchment. Types of storage within a catchment include soil moisture (unsaturated zone), groundwater (saturated zone), surface storage (lakes, wetlands), above surface storage in vegetation or leaf litter (interception zone), as well as snow and ice (though one could see the latter also as external to the catchment since it is climate driven). Release of stored water is defined as the pathway (and state) through which water ultimately leaves the catchment. Examples include evaporation, transpiration, surface runoff, and groundwater flow. Aspects of the release function are of course crucial with regard to several water resource questions, such as the amount of blue water (mainly in rivers and aquifers) and green water (soil moisture and subsequent evapotranspiration) [7]. These dominant functions of a catchment are of course captured in hydrological models that reflect this dominant behavior as well as possible.

In this paper we attempt to utilize data-based mechanistic modeling [44, 45] to classify and group a large number of catchments across the Eastern US. This approach relates to what has been described as a downward or top-down modeling framework for hydrologic analysis [14, 31, 32]. The general idea is to identify the appropriate level of model complexity from input-state-output data of environmental systems with the objective to understand dominant controls on the system's response. The DBM approach put forward by Young [45] is one particularly elegant and statistically consistent framework in which such a downward strategy can be implemented. In the DBM approach to modeling the hydrology of catchments, the (acknowledged) non-linear response of the catchment is usually broken into a non-linear loss function and a linear routing component. The appropriate level of non-linearity of the loss function and the required complexity of the routing function can then be estimated as discussed in detail below. The DBM approach has previously been applied to distinguishing dominant modes and response parameters between catchments [1, 20], although only for small groups of catchments in the UK for the specific goal of land use impacts analysis. Despite its clear potential for doing so, the DBM method has not yet been used to contribute to a more general catchment classification scheme.

In the study presented here, we analyze 278 catchments distributed across the Eastern USA using a DBM strategy. We attempt to understand the catchment similarity that can be found with respect to both model parameters (if the same model structure is applied) and with respect to model structures identified as most suitable. Finally, we relate the identified structures and parameters to available physical and climatic catchment-scale characteristics to see whether a further generalization of our result is possible.

23.2 Data-Based Mechanistic (DBM) Modelling

The DBM framework is based on a set of methods of statistical identification of transfer functions, and the subsequent decomposition of the transfer functions into structures and parameter values which have a conceptual interpretation. Commonly, in the rainfall-runoff context, the identified structure is a system of linear hydrological storages in parallel and series, although bypasses, feedbacks and non-linear responses may also be identified. Non-linear components of the system can be identified through time and/or state dependent parameter estimation. The modeler may then make hypotheses about the physical processes, which could lead to the identified structure and set of parameter values. The ethos behind the method is that any hypothesis-making should come after the information is extracted from the data, not beforehand as is common in conceptual or physics-based modeling.

The attractions of DBM modeling over more conventional conceptual modeling include: (1) using DBM modeling, the reliance on prior assumptions about hydrological behavior is minimal, and the insights delivered by the method may go beyond prior expectations [19, 25]; (2) the DBM transfer function framework and associated parameter estimation methods are computationally efficient and thus allow rapid assessment of large amounts of data (e.g. [20]), and can provide efficient emulation of higher order models [47]; (3) DBM models are identified in a statistically defendable manner with consistent, transparent and testable underlying assumptions; (4) the requirement for models to be statistically identifiable means that the DBM approach guarantees parsimonious models.

The DBM approach is described extensively elsewhere (e.g. [43]) and the description here is limited to the immediately relevant components. The general form of a single input discrete time linear transfer function is:

$$\hat{y}_k = \frac{b_0 + b_1 z^{-1} + b_2 z^{-2} + \cdots + b_m z^{-m}}{1 - a_1 z^{-1} - a_2 z^{-2} - \cdots - a_n z^{-n}} u_{k-\delta/\Delta t}, \tag{23.1}$$

where u is the model input and y is the model output, a and b are parameter vectors, δ is a time-lag parameter, k is the time-step number, and z is the backward shift operator (so that for example (i.e. $z^{-1}x_k = x_{k-1}$)). The linear relationship in (23.1) means that, in the context of rainfall-runoff modeling, where y is the estimated streamflow at the catchment outlet, it is usual for u to be the estimated catchment average effective rainfall with the assumption that linear routing applies.

Equation (23.1) defines a family of model structures depending on how many terms are included (i.e. the values of m and n), and for any one catchment it is usual to seek just one structure which is considered most applicable. Naturally, we may seek the model structure which gives the best fit between modeled and observed streamflow, measured for example using the coefficient of determination applied to the residuals (R_T^2) ((6.54) in [24]) which in this case is the same as using the NSE. This may favor model structures with large numbers of parameters and poor parameter identifiability, therefore it is common to also assess parsimony, for example using Young's Information Criterion (YIC) [41], with the intention that both a good fit and good identifiability should be achieved. The final criterion, central to the DBM ethos, is that the identified model structure should be decomposable into a conceptually plausible combination of hydrological storages, feedbacks and bypasses [39].

Several forms of (23.1) have been identified for different catchments. The structure

$$\hat{y}_k = \frac{b_0 + b_1 z^{-1}}{1 - a_1 z^{-1} - a_2 z^{-2}} u_{k-\delta/\Delta t} \tag{23.2}$$

(with the same definitions as for (23.1)) has been identified in numerous hydrological studies [45]. With two denominator and two numerator parameters, this is referred to as a [2 2 δ] structure. This transfer function can often be decomposed into two parallel linear stores with real and positive response time parameters, consistent with the routing presumed in many conceptual rainfall-runoff models. In many cases, however, the response times associated with this structure are not physically plausible (i.e. they are complex or negative), in which case a simpler [1 1 δ] structure, equivalent to only one linear store, is likely to be preferred:

$$\hat{y}_k = \frac{b_0}{1 - a_1 z^{-1}} u_{k-\delta/\Delta t}. \tag{23.3}$$

This single store model tends to ensure physically plausible parameters, and usually improved YIC values over more complex models. The identification of more than two storages is uncommon although possible (e.g. [19, 39, 40]). The addition of a parallel bypass pathway to either (23.2) or (23.3) allows an instantaneous response to be superimposed [45], for example applied to (23.3) it would result in the [1 2 δ] structure:

$$\hat{y}_k = \frac{b_0 + b_1 z^{-1}}{1 - a_1 z^{-1}} u_{k-\delta/\Delta t}. \tag{23.4}$$

The response time, steady state gain and bypass flow of the catchment may be estimated from the transfer function parameters. For example, for the single store model of (23.4), the response time of the store (T) would be,

$$T = \frac{-1}{\ln(a_1)} \tag{23.5}$$

and the steady-state gain of the model ($G = y_k/u_k$ at steady state) would be,

$$G = \frac{b_0 + b_1}{1 - a_1} \tag{23.6}$$

and the proportion of flow bypassing the store (P) would be,

$$P = \frac{b_1}{b_1 - \frac{b_0 a_1 + b_1}{1 - a_1}}. \tag{23.7}$$

A corresponding derivation can be done from (23.3), which gives T for each of the two stores (one relatively small value T_q which may be interpreted as the stormflow response time and a larger value T_s representing the baseflow response time), the gain G for each store and the split between the two stores q equal to the ratio of the gain for the stormflow response divided by the sum of the two gains [43, p. 2211].

As effective rainfall u cannot itself be measured, it is usually estimated using a non-linear model. Various conceptual soil moisture models are available and may be applied prior to the transfer function identification (e.g. [13, 43]). However, it has been found that a more empirical method, which uses observed flow as a wetness index, often performs better (also see [17, 26, 39, 42]),

$$\hat{u}_k = c\, r_k\, q_k^{\lambda}. \tag{23.8}$$

For example, $\lambda = 0$ represents a linear loss model while $\lambda = 1$ represents large variability in the runoff generation between wet and dry periods. Parameter λ can be estimated by trial and error or optimization [17, 43]. Parameter c is usually fixed so that the cumulative volume of effective rainfall is equal to the observed volume of streamflow, so that we may expect the result $G = 1$. A limitation in the use of (23.6) is that observed flow is required, therefore while this model can be used for system identification, it cannot be used for prediction, although short-term forecasting applications are possible (e.g. [34]).

In summary, the runoff is estimated using (23.6) or another suitable non-linear model and a suitable linear transfer function belonging to the family of models defined by (23.1). The DBM approach may be significantly extended by including data-based identification of non-linear components [19]; by estimation of an error model which would explicitly allow for the presence of colored noise [43]; and the introduction of continuous-time transfer functions to remove parameter bias associated with the discrete time approximation [46]. For the purpose of the case study below, where 278 catchments are screened as part of a classification procedure, the simple implementation of DBM using (23.1) and (23.8) is adopted.

23.3 Case Study: Data

A total of 278 catchments were used in this study, spanning the eastern half of the United States (Fig. 23.1). The studied catchments range in size from 67 km^2 to 10,096 km^2, and show aridity indices between 0.41 and 3.3, demonstrating the heterogeneity of the dataset (Table 23.1). They cover type 1 eco-regions 5, 8 and 9, which are defined as Northern Forests, Eastern Temperate Forests, and Great Plains, respectively [23].

Time-series of hydrologic variables at a daily time-step used in this study were provided by the MOPEX project [6]. Streamflow (from USGS) and precipitation are daily observed variables. Results shown are based on analysis of 10 years of daily data between 1970 and 1979. The National Climate Data Center (NCDC) provided the precipitation data used in the MOPEX database, and hence in this study. Minimum acceptable precipitation gauge density within each catchment was defined following the equation,

$$N_k = 0.6\, A^{0.3}, \tag{23.9}$$

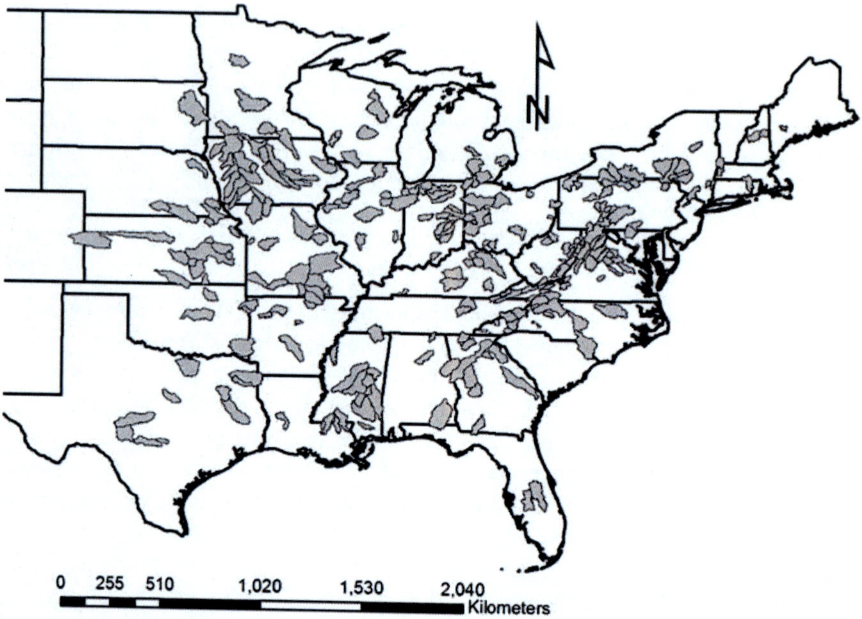

Fig. 23.1 Map showing the 278 study catchments, distributed across the eastern half of the United States

Table 23.1 Selected physical and climatic catchment attributes of the study basins

Name	Units	Mean value	Min value	Max value
Mean annual precipitation (P)	[mm]	1011	490	1900
Mean annual potential evaporation (PE)	[mm]	915	660	1620
Aridity index (PE/P)	[–]	0.97	0.41	3.3
Mean annual streamflow	[mm]	373	12	1280
Drainage area	[km^2]	2924	67	10096

where N is the number of precipitation gauges and A is the area of the catchment (km^2) [29]. The number of precipitation gauges calculated by (23.9) is the required minimum to capture the heterogeneity of storm events to estimate reliable spatially averaged precipitation values for a catchment. The use of this guideline provides mean areal precipitation estimates at each time step and should result in less than 20% error 80% of the time [28].

23.4 Case Study: Methods

A relatively simple implementation of the DBM approach, using the family of linear models defined by (23.1) and an assumed non-linear model, (23.8), are adopted.

Due to the significant snowfall in most of the catchments, the non-linear model is extended to include a one-parameter degree-day snowmelt model. Precipitation falling when the air temperature is less than zero is taken to be snow. The rate of snow melt m (mm day^{-1}) is given by $m_k = \theta K_k$ when snow cover is present and air temperature K_k (°C) is above freezing, otherwise $m_k = 0$. Parameter θ is a degree-day factor (mm °C^{-1}) estimated as explained below.

Firstly, the most basic DBM models thought to be useful for classification purposes are implemented, with the view that the models with the fewest degrees of freedom may best expose spatial signals in the system response. Then the degrees of freedom in the model optimization are increased in order to look for new and/or more physically interpretable differences between catchments. Thus, the following three levels of modeling were implemented:

Level 1. The transfer function structure is fixed to one linear store with zero time delay (denoted by [1 1 0], equivalent to (3) with $\delta = 0$). The non-linearity parameter λ is fixed at a value of 0.4 (after McIntyre and Marshall [20]) and θ is fixed at a value of 8 (after some initial modeling which showed this to be amongst the best of the uniformly applied values). Parameter c in (23.8) is fixed so that the volume of effective rainfall is equal to the volume of observed streamflow over the 10-year period. The transfer function parameters are optimized using the simplified refined instrumental variable (SRIV) technique within the CAPTAIN toolkit ([24, 33]; available from http://www.es.lancs.ac.uk/cres/captain/).

Level 2. The transfer function structure remains fixed at [1 1 0]; but both the transfer function parameters and the non-linear model parameters are optimized. While the transfer function parameters are optimized using SRIV, λ and θ are optimized using a non-linear least squares optimization. At each iteration in the non-linear procedure the SRIV-optimal linear transfer function is identified—this guarantees that the optimal combination of linear and non-linear model is found. c is again defined by volume balance.

Level 3. The transfer function parameters and structure are optimized. The transfer function structure is optimized by repeating the parameter optimization for all possible combinations of m, n and δ from [1 1 0] to [3 3 1], resulting in 18 tested model structures, including various combinations of routing stores in parallel and in series. The optimization of the model structure is conditional on the optimum values of θ and λ which were found in Level 2—this makes the procedure more computationally tractable than attempting to simultaneously optimize the non-linearity parameters with the structure. The best model structure is taken to be the one with the least squares solution, equivalent to the highest R_T^2 value, subject to the constraint that the optimized parameter values must be physically plausible. Using the R_T^2 criterion without requiring physical plausibility would be unsuitable because the highest R_T^2 values are (almost) always achieved by the most complex model structure tested. The requirement for physical plausibility limits complexity because model structures beyond a certain complexity, depending on the nature of the catchment (see results below), produce G and/or T and/or P values which are physically implausible (negative or complex). A possible further constraint would be the YIC (or a comparable criterion), which penalizes high covariance in the parameter estimates and hence promotes parsimonious models; however this would require some subjective decision about the suitable compromise between R_T^2 and YIC, which proved difficult to implement successfully over the large sample of catchments.

For each of these three levels of analysis, the transfer function parameters were transformed to the physically interpretable forms (e.g. (23.5)–(23.7)). The variability of the parameter values and the routing structure across the 278 catchments was then examined. This consisted of assessing the nature and strength of the relationships between parameter values/structures and six selected physical catchment descriptors (PCDs). The selected PCDs are those identified by [27] to be potentially important in describing differences in catchment function: Aridity (annual average potential evaporation divided by annual average rainfall), Number of cold days (defined by $K < 0°\text{C}$), Percentage tree coverage, Percentage sand in the soil, Catchment area, and Average elevation.

23.5 Case Study: Results

From the Level 1 analysis (using the [1 1 0] model with default values of λ and θ), the variability of the identified response time over space is shown in Fig. 23.2, and over the six other PCDs is shown in Fig. 23.3. There is only a slight trend towards worse performance in the more snowy (more northern) catchments, indicating that the simple spatially uniform degree-day model reasonably (but not entirely) limited influence of snow on performance. The clearest linear univariate relationships identified were between $\log(T)$ and percentage sand, and between $\log(T)$ and area; significant relationships between $\log(T)$ and the other variables were also present although not clearly visible. Multiple linear regression showed that these six variables together explained 46% of the variability of $\log(T)$.

The Level 2 analysis (using the [1 1 0] model with optimized values of λ and θ) provided the opportunity to explore influences on the degree of non-linearity. Figures 23.4 and 23.5 show that the degree of non-linearity (λ) is related to location and to all six PCDs. Due to the correlations between PCDs, this is not straightforward to interpret. However it is proposed that the primary physical reason for the variability in λ is aridity (climate), with more arid catchments tending to be more non-linear, since this has the strongest univariate effect ($R^2 = 0.18$). Another potential influence indicated by Figs. 23.4 and 23.5 is elevation (with high, steep catchments tending to be more linear, potentially due to less non-linear storage effects). The influences on response time T are essentially the same as for the Level 1 analysis.

The Level 3 analysis allowed us to explore the influence of catchment type on model structure. A single linear store was preferred for 113 of the 278 catchments, two parallel stores for 163 catchments, three parallel stores for one catchment, and no suitable structures were found for one catchment. In all cases the time lag δ was zero. For all but one of the single store models, and two of the parallel flow models, a bypass was identified. This is because the addition of a bypass almost always improves the fit due to the improved performance at high flows without compromising

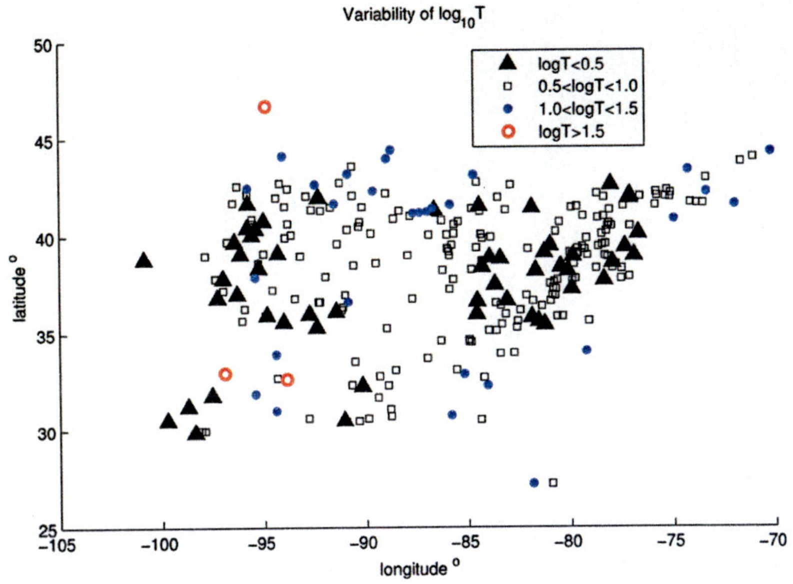

Fig. 23.2 [1 1 0] model with default λ and θ: spatial variability of T

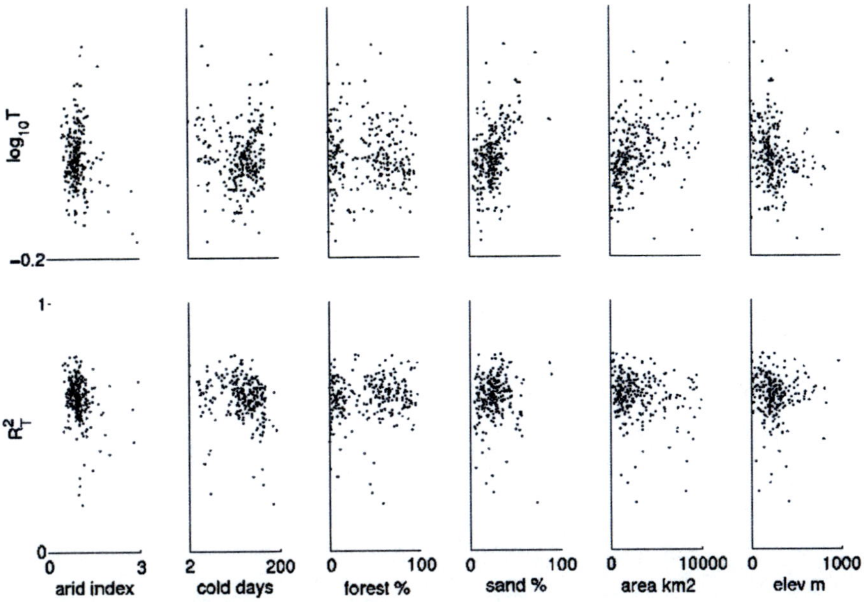

Fig. 23.3 [1 1 0] model with default λ and θ: variability of T and R_T^2 with PCDs

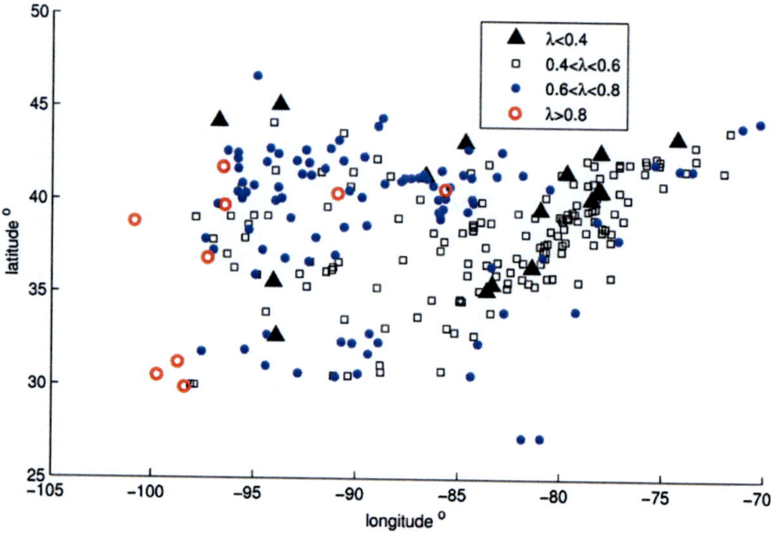

Fig. 23.4 [1 1 0] model with optimized λ and θ: spatial variability of λ

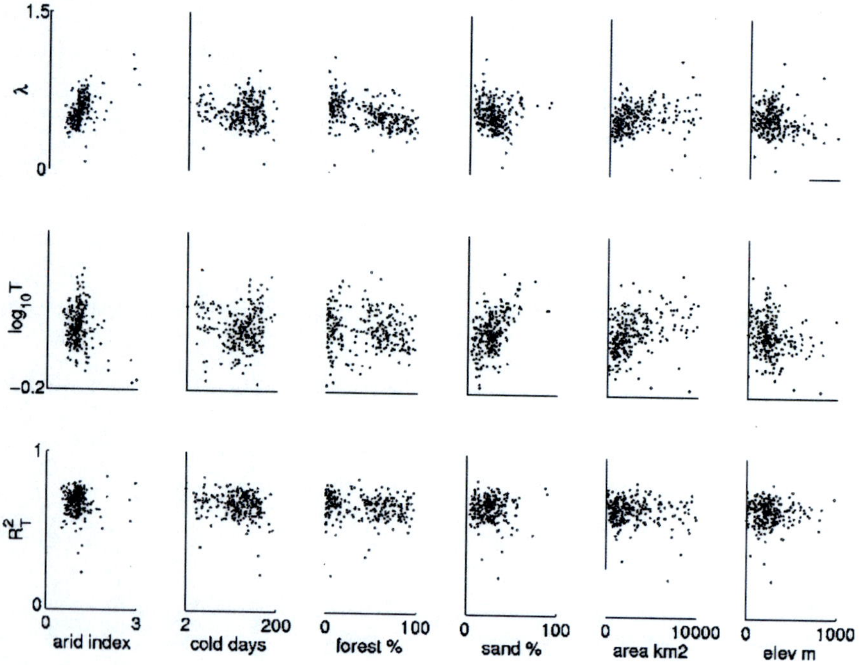

Fig. 23.5 [1 1 0] model with optimized λ and θ: variability of λ, T and R^2_T with PCDs

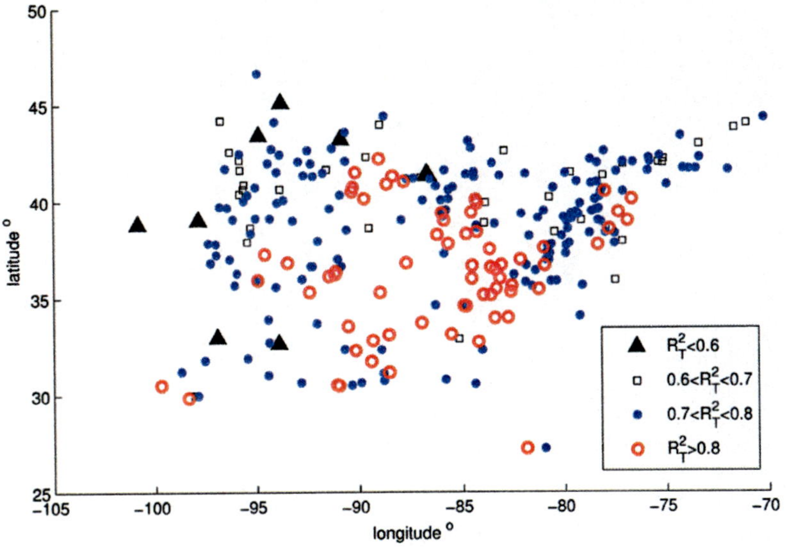

Fig. 23.6 [n m δ] model with pre-optimized λ and θ: spatial variability of R_T^2

the physical plausibility of the model (recalling that the fitting process allowed the fit to be improved by adding more parameters so long as the model remained physically plausible). Figure 23.7 shows the spatial variability of the optimized model structure, while Fig. 23.6 maps their performance distribution. This indicates that the Appalachian chain and the northeastern US in general are best modeled by two parallel stores, the flatter humid catchments in the south favor a single store while the picture in the rest of the country is more mixed. The 277 catchments for which models were identified were separated into the two groups: those that favored a single store, and those that favored parallel stores. For both groups, Fig. 23.8 (top two rows) shows the frequency of catchments across each of the six PCDs (frequency is plotted as number of catchments within a bin divided by the total number of catchments in that group to make the subplots comparable). The aim is to indicate which of the six PCDs influences the identified model structure: if a PCD has no influence then the two frequency distributions should be the same within the bounds of sampling error. The strongest inferences from this plot are: more arid catchments tend to prefer a single store; warmer catchments prefer a single store; lower elevation (and less steep) catchments prefer a single store. Figure 23.8 also includes plots of the model parameter values against the PCDs. The proportion of flow going to the quick store (q) is only plotted for the parallel store models. It is most related to the elevation and the area of the catchment ($R^2 = 0.12$ for each) and also weakly related to percentage sand ($R^2 = 0.04$), percentage forest ($R^2 = 0.09$) and number of cold days ($R^2 = 0.06$), although the significance of these relationships cannot be seen in the scatter plots. Catchments at higher elevations and with high proportions of forest do not show small values of q, and hence require a mix of quick and slow contributions for a good fit.

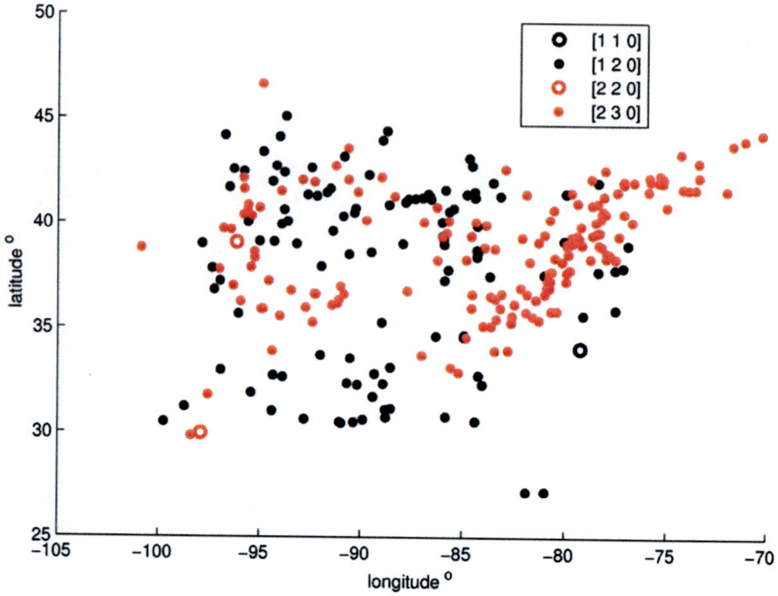

Fig. 23.7 [n m δ] model with pre-optimized λ and θ: spatial variability of structure. The single [3 3 0] model is omitted for clarity

23.6 Case Study: Discussion and Conclusions

Catchment classification remains a significant challenge for hydrologists, with available schemes not providing a sufficient basis for consistently distinguishing between different types of hydrological behavior. The inconsistencies and overlaps within available general classification schemes are intellectually unsatisfying and do not provide the sought level of support for simulation of hydrology of ungauged basins or for environmental change assessment. Thus, despite this being a well-established general area of research, there remains considerable interest and activity in seeking better classifications. One opportunity for improvement lies in distinguishing between catchments according to their functional behavior as embedded in selected response signatures (rather than their physical properties, just streamflow characteristics alone, or the values of parameters which represent these properties within a conceptual model).

The idea of classification via response signatures requires a practical and consistent method of describing a range of signatures from large samples of catchments. The data-based mechanistic (DBM) approach to time-series modeling, developed by Peter Young and colleagues throughout the last 40 years, seems an eminently suitable approach. The DBM approach is designed to extract the dominant modes (signatures) of a system's response; its value for this in the hydrological context has previously been established (albeit not for classification purposes). The DBM approach, with its formal statistical basis, can be applied consistently; and the estimation algorithms within the DBM toolbox, CAPTAIN, allow efficient dominant

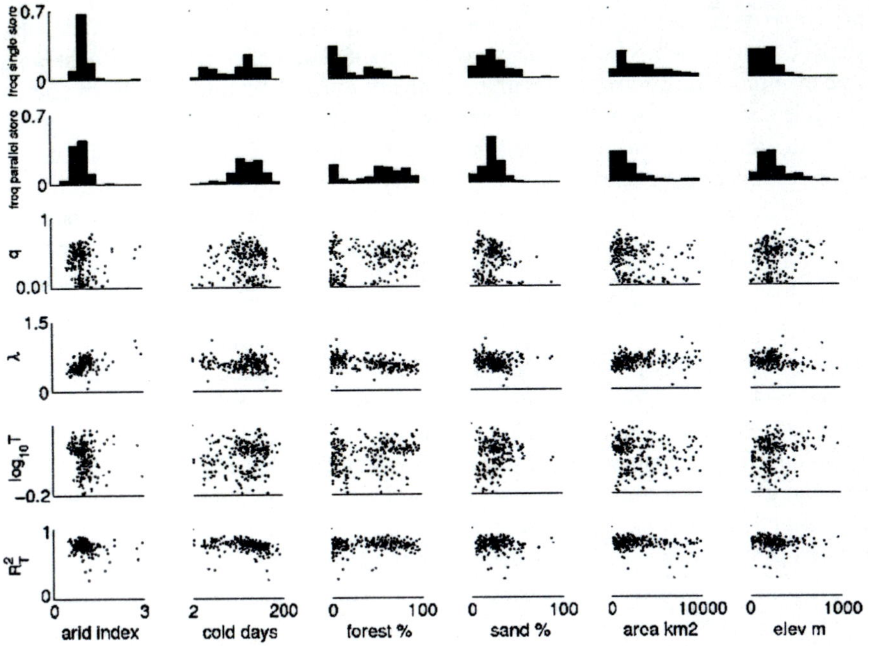

Fig. 23.8 [n m δ] model with pre-optimized λ and θ: relative frequency of single store and parallel store models (top two rows of plots) over each PCD; and variability of q, λ, T and R_T^2 with PCDs

mode and parameter estimation over very large numbers of catchments. Furthermore, the method of DBM modeling—to eradicate or at least minimize the reliance on prior perceptions of how a catchment functions—provides an opportunity to uncover responses and classifications which were unexpected a priori.

Hence the DBM approach was applied in this chapter to 278 catchments distributed across the Eastern USA, with the aim of exploring whether the catchments may be classified according to their dominant mode responses. This includes identifying both the type of response (the transfer function structure) and the scale of the response (the associated parameter values). In terms of the type of response, the DBM dominant mode identification distinguished between catchments best described by one routing store, and those best described by two routing stores in parallel. A significant regional pattern emerged (Fig. 23.7), reflecting the influences of aridity, elevation (steepness) and temperature (Fig. 23.8). In terms of parameter estimates, the most interesting variability between catchments is that in response non-linearity. Significant regional patterns in the non-linearity parameters emerged, and reasonable physical explanations were proposed.

The results give the impression that DBM could be fruitfully applied towards the catchment classification goal. The classification of catchments by their dominant modes of response, over regional to national scales, could provide new insights into how catchments differ functionally, beyond those achievable by merely applying parameter estimation to a pre-conceived hydrological model. In principle, the parameter estimates might then be seen as secondary classification measures. Our study, however, merely states the hypothesis that DBM modeling is a useful step in catchment classification, along with preliminary steps towards testing the hypotheses: it is not conclusive and further investigation is recommended.

Using the same data set as used here, further investigation could include using state dependent parameter analysis to identify the nature of the non-linearities in the model, as opposed to presuming a power law. How can routing non-linearity be used to aid classification; and how do controls on effective rainfall generation differ over catchments? Such questions should be examined using a more advanced level of DBM analysis. Whatsmore, there are a number of limitations in the case study which need to be addressed. The runoff coefficient is implicit in the value of parameter c, but this parameter is difficult to interpret because it has the dual role of also making (23.8) dimensionally consistent. An explicit runoff coefficient could be added. There is potential for interaction between parameter values and model structures, for example the single store models tended to have higher non-linearity, potentially compensating for the fact that they are missing flow components [19]: this complication requires further exploration. Another particular limitation of the method is that model structure identification was conditional on pre-optimized non-linearity parameters, however these parameters were found to be not very sensitive to model structure. Finally, the interpretation of the results could benefit from a deeper multi-variate analysis to inspect how combinations of physical properties explain functional similarity.

In conclusion, the DBM modeling approach developed since the 1970s by Peter Young and colleagues provides a consistent and efficient framework for classifying catchments with respect to their dominant modes of hydrological response and associated parameter estimates. The potential of the DBM method in this role has been demonstrated in this chapter by identification of time series models for 278 catchments across the eastern USA.

Acknowledgements We would like to thank Prof. Peter Young for supporting both of us since we first met while we were PhD students. His inquisitive nature, his openness and his friendly nature provided a great example to us. We'd like to thank the Symposium organizers and the editors of this book. Thanks to Keith Sawicz for producing Fig. 23.1. Partial funding to TW was provided by an EPA STAR Early Career Award. The CAPTAIN toolbox (www.es.lancs.ac.uk/cres/captain) was used under license from Lancaster University.

References

1. Beven, K., Young, P., Romanowicz, R., O'Connell, E., Ewen, J., O'Donnell, G., Homan, I., Posthumus, H., Morris, J., Hollis, J., Rose, S., Lamb, R., Archer, D.: Analysis of historical data sets to look for impacts of land use and management change on flood generation. Defra R and D Final Report FD2120. Defra, London (2008)
2. Beven, K.J.: On the uniqueness of place and process representations in hydrological modeling. Hydrol. Earth Syst. Sci. **4**(2), 203–212 (2000)
3. Black, P.E.: Watershed functions. J. Am. Water Resour. Assoc. **33**, 1–11 (1997)
4. Buttle, J.: Mapping first-order controls on streamflow from drainage basins: the T3 template. Hydrol. Process. **20**, 3415–3422 (2006)

5. Dooge, J.C.I.: Looking for hydrologic laws. Water Resour. Res. **22**(9), 46–58 (1986)
6. Duan, Q., Schaake, J., Andreassian, V., Franks, S., Gupta, H.V., Gusev, Y.M., Habets, F., Hall, A., Hay, L., Hogue, T.S., Huang, M., Leavesley, G., Liang, X., Nasonova, O.N., Noilhan, J., Oudin, L., Sorooshian, S., Wagener, T., Wood, E.F.: Model Parameter Estimation Experiment (MOPEX): overview and summary of the second and third workshop results. J. Hydrol. **320**, 3–17 (2006)
7. Falkenmark, M., Rockstrom, J., Savenije, H.: Balancing Water for Humans and Nature: The New Approach in Ecohydrology. Earthscan, London (2004)
8. Gottschalk, L.: Hydrological regionalization of Sweden. Hydrol. Sci. J. **30**(1), 65–83 (1985)
9. Gould, S.J.: Wonderful Life: The Burgess Shale and the Nature of History. Norton and Company, New York (1989)
10. Grigg, D.B.: The logic of regional systems. Ann. Assoc. Am. Geogr. **55**, 465–491 (1965)
11. Grigg, D.B.: Regions, models and classes. In: Models in Geography, pp. 467–509. Methuen, London (1967)
12. Haines, A.T., Finlayson, B.L., McMahon, T.A.: A global classification of river regimes. Appl. Geogr. **8**(4), 255–272 (1988)
13. Jakeman, A.J., Littlewood, I.G., Whitehead, P.G.: Computation of the instantaneous unit hydrograph and identifiable component flows with application to two small upland catchments. J. Hydrol. **117**, 275–300 (1990)
14. Klemes, V.: Conceptualization and scale in hydrology. J. Hydrol. **65**, 1–23 (1983)
15. Krasovskaia, I.: Entropy-based grouping of river flow regimes. J. Hydrol. **202**(1–4), 173–191 (1998)
16. Krasovskaia, I., Gottschalk, L., Kundzweicz, Z.W.: Dimensionality of Scandinavian river flow regimes. Hydrol. Sci. J. **44**(5), 705–723 (1999)
17. Lees, M.: Data-based mechanistic modelling and forecasting of hydrological systems. J. Hydroinform. **2**, 15–34 (2000)
18. McDonnell, J.J., Woods, R.A.: On the need for catchment classification. J. Hydrol. **299**, 2–3 (2004)
19. McIntyre, N., Young, P.C., Orellana, B., Marshall, M., Reynolds, B., Wheater, H.S.: Identitication of nonlinearity in rainfall-flow response using data-based mechanistic modelling. Water Resour. Res. (2011). doi:10.1029/2010WR009851
20. McIntyre, N., Marshall, M.: Identification of rural land management signals in runoff response. Hydrol. Process. (2010). doi:10.1002/hyp.7774
21. Milly, P.C.D., Betancourt, J., Falkenmark, M., Hirsch, R.M., Kundzewicz, Z.W., Lettenmaier, D.P., Stouffer, R.J.: Stationarity is dead: Whither water management? Science **319**, 573–574 (2008)
22. Olden, J.D., Poff, N.L.: Redundancy and the choice of hydrologic indices for characterizing streamflow regimes. River Res. Appl. **19**, 101–121 (2003)
23. Omernik, J.M.: Ecoregions of the conterminous United States. Ann. Assoc. Am. Geogr. **77**, 118–125 (1987)
24. Pedregal, D.J., Taylor, C.J., Young, P.C.: System Identification, Time Series Analysis and Forecasting. The Captain Toolbox. Handbook v2.0, Centre for Research on Environmental Systems and Statistics, Lancaster University, UK (2007)
25. Ratto, M., Young, P.C., Romanowicz, R., Pappenberger, F., Saltelli, A., Pagano, A.: Uncertainty, sensitivity analysis and the role of data based mechanistic modeling in hydrology. Hydrol. Earth Syst. Sci. **11**, 1249–1266 (2007)
26. Romanowicz, R.J., Young, P.C., Beven, K.J.: Data assimilation and adaptive forecasting of water levels in the river Severn catchment, United Kingdom. Water Resour. Res. **42**, W06407 (2006). doi:10.1029/2005WR004373
27. Sawicz, K.: Catchment classification. Unpublished Masters Thesis, Department of Civil and Environmental Engineering, The Pennsylvania State University, USA (2009)
28. Schaake, J., Cong, C., Duan, Q.: The US MOPEX data set, large sample basin experiments for hydrological model parameterization. In: IAHS Publ., vol. 307, pp. 9–26 (2006)

29. Schaake, J.C., Duan, Q., Smith, M., Koren, V.: Criteria to select basins for hydrologic model development and testing. Preprints, 15th Conf. on Hydrology, Long Beach, CA, Amer. Meteor. Soc., 10–14 January 2000, Paper P1.8 (2000)
30. Sivapalan, M.: Pattern, process and function: Elements of a unified theory of hydrology at the catchment scale. In: Anderson, M. (ed.) Encyclopedia of Hydrological Sciences, pp. 193–219. Wiley, London (2005)
31. Sivapalan, M., Young, P.C.: Downward Approach to Hydrological Model Development. Encyclopedia of Hydrological Sciences (2005)
32. Sivapalan, M., Bloschl, G., Zhang, L., Vertessy, R.: Downward approach to hydrological prediction. Hydrol. Process. **17**, 2101–2111 (2003). doi:10.1002/hyp.1425
33. Taylor, C.J., Pedregal, D.J., Young, P.C., Tych, W.: Environmental time series analysis and forecasting with the Captain toolbox. Environ. Model. Softw. **22**, 797–814 (2007)
34. Vaughan, M., McIntyre, N.: The utility of data-based mechanistic flood forecasting models. Water Manag. (2011, in press)
35. Wagener, T., Sivapalan, M., McGlynn, B.L.: Catchment classification and services—toward a new paradigm for catchment hydrology driven by societal needs. In: Anderson, M.G. (ed.) Encyclopedia of Hydrological Sciences. Wiley, New York (2008)
36. Wagener, T., Sivapalan, M., Troch, P.A., Woods, R.A.: Catchment classification and hydrologic similarity. Geogr. Comput. **1/4**, 901–931 (2007)
37. Wagener, T., Sivapalan, M., Troch, P.A., McGlynn, B.L., Harman, C.J., Gupta, H.V., Kumar, P., Rao, P.S.C., Basu, N.B., Wilson, J.S.: The future of hydrology: an evolving science for a changing world. Water Resour. Res. **46**, W05301 (2010). doi:10.1029/2009WR008906
38. Winter, T.C.: The concept of hydrologic landscapes. J. Am. Water Resour. Assoc. **37**, 335–349 (2001)
39. Young, P.C., Jakeman, A.J., Post, D.A.: Recent advances in data-based modelling and analysis of hydrological systems. Water Sci. Technol. **36**, 99–116 (1997)
40. Young, P.C.: Real time flow forecasting. In: Wheater, H.S., Sorooshian, S., Sharma, K.D. (eds.) Hydrological Modelling in Arid and Semi-Arid Areas, pp. 113–137. Cambridge University Press, Cambridge (2008)
41. Young, P.C.: Recursive estimation, forecasting and adaptive control. In: Leondes, C.T. (ed.) Control and Dynamic Systems. Advances in Theory and Applications, vol. 30, pp. 119–166. Academic Press, San Diego (1990)
42. Young, P.C.: Advances in real-time flood forecasting. Philos. Trans. R. Soc., Math. Phys. Eng. Sci. **360**, 1433–1450 (2002)
43. Young, P.C.: Top-down and data-based mechanistic modelling of rainfall-flow dynamics at the catchment scale. Hydrol. Process. **17**, 2195–2217 (2003)
44. Young, P.C., Lees, M.J.: The active mixing volume: a new concept in modelling environmental systems. In: Barnett, V., Turkman, K. (eds.) Statistics for the Environment, pp. 3–43. Wiley, Chichester (1993)
45. Young, P.C.: Data-based mechanistic modelling of environmental, ecological, economic and engineering systems. Environ. Model. Softw. **13**, 105–122 (1998)
46. Young, P.C., Garnier, H.: Identification and estimation of continuous-time data-based mechanistic (DBM) models for environmental systems. Environ. Model. Softw. **21**, 1055–1072 (2006)
47. Young, P.C., Ratto, M.: A unified approach to environmental systems modeling. Stoch. Environ. Res. Risk Assess. **23**, 1037–1057 (2009)

Chapter 24
Application of Optimal Nonstationary Time Series Analysis to Water Quality Data and Pollutant Transport Modelling

Renata J. Romanowicz

24.1 Introduction

This chapter addresses the application of a data-based systems approach to two closely related issues, pollutant transport and water quality modelling. The aim is the presentation of on-going research on data-based models and a comparison of the results with physically based approaches using worked case examples.

It is well known that environmental systems are poorly defined and difficult to model, because of a lack, or limited number, of observations of the inputs to the system, spatial variability and heterogeneity of the processes involved, scale issues (the process variability requires a small spatial scale with a relatively large time scale required), uncertainty of climatic variables influencing the system, and difficulties in reproducing the experiments. Among many approaches to modelling, the most popular are, the so called physically-based and black-box models that describe an input-output relationship.

Black-box models can be simple or complex in terms of the number of parameters involved. The Artificial Neural Network (ANN) modelling approach [15] involves a great number of parameters and is not statistically efficient. On the contrary, a black-box model in the form of a Stochastic Transfer Function (STF), being the basic tool of the Data Based Mechanistic (DBM) approach introduced by Peter Young [32], involves the minimum number of parameters that are necessary from the point of view of the goal of modelling and the available data.

The Multiple-Input-Single-Output (MISO) STF model, including the Box-Jenkins time series model [4], is equivalent to a linear stochastic difference equation. This approach introduces simplifications into the description of usually nonlinear

R.J. Romanowicz (✉)
Institute of Geophysics, Polish Academy of Sciences, ul. Ksiecia Janusza 64, 01-452 Warsaw, Poland
e-mail: Romanowicz@igf.edu.pl

L. Wang, H. Garnier (eds.), *System Identification, Environmental Modelling, and Control System Design*,
DOI 10.1007/978-0-85729-974-1_24,

physical processes that allow for the application of statistically efficient methods of analysis. The derived models are fit for purpose and provide the best use of the available data. The application of physically-based models, on the other hand, requires an extensive pool of observations that are usually not available. Therefore, the complexity of the process description is not supported by the experimental data and details are lost due to a lack of knowledge of the physical parameters involved [9]. In addition, due to the complexity of the physical models, an estimation of their predictive uncertainty is difficult. In contrast, much simpler, but well defined, black-box stochastic time series models give robust predictions over a wide range of input variability together with estimates of the predictive uncertainty.

The inability to reach a physical interpretation of statistical models is one of the main criticisms of that approach. The introduction of the concept of inductive, Data Based Mechanistic (DBM) modelling [33] is an answer to that criticism. This concept advocates an inductive, top-down approach to environmental modelling, where the best identifiable model structure is first inferred from the available data in an inductive way. This black-box, but statistically efficient, approach is followed by a physical interpretation of the model structure and its parameters. The approach was introduced to water quality and pollutant transport in the 1970s [31], although its underlying philosophy was formulated much later [33].

The main challenge of the DBM approach lies in the search for a physical explanation of the model structure. This is the area where the knowledge and expertise of the scientists involved in the detailed studies of environmental processes are of the utmost value. We shall show here that a similar linear dynamic model can be used to describe the solute transport processes in the river and the biological processes governing the oxygen concentrations. There may be different lines of explanation used depending on the aim of the research and the background of the researcher. One of the possible approaches is an idea of the dominant mode of behaviour of the dynamic environmental systems [33]. The "dominant mode of behaviour" can be described as a typical mode of working of the system for most commonly occurring input patterns. This approach is well known and widely studied in control science. In the environmental sciences, due to the wide ranges of variability of environmental processes, this approach is not popular and a reductionist approach is preferred [32]. The argument often used is that environmental processes are so complex that simple linear models cannot give appropriate solutions. However, in the search for general models of environmental systems, such as rainfall-runoff, solute transport or water quality processes, the scientist applies the general goodness of fit criteria that, by definition, average the model's behaviour towards typical input patterns. As a result, a linear approximation of the process dynamics works well, despite the process complexity. The proof can be found in many examples given by Peter Young and his co-workers in articles published over the last 30 years. In certain cases, in order to secure the linear model assumptions (i.e. a linear relationship between model variables), nonlinear transformation of the process variables is required. Following the DBM philosophy, these approximations should have a physical justification.

This chapter addresses Peter Young's contributions to water quality and pollutant transport modelling against a wider background of recent developments in the area.

Professor Young's main interest lies in data-based modelling and in looking for a simple representation of complex environmental processes that would fulfil the conditions of identifiability and the best use of available observations. His engineering approach ensures that the aims of his research are reached in the most efficient way whilst the scientific rigour of his work assures the validity of the approaches undertaken. Professor Young's main contribution lies in the introduction of a unified, statistical systems approach to environmental modelling ([34], and references therein). Its success is due largely to the application of recursive estimation methods that allows the model parameters to be updated sequentially while working through the data [35]. This in turn enables an estimation to be made of time-variable model parameters of non-stationary and nonlinear processes [32]. The work presented in this chapter was carried over many years and followed the development of methods of pollutant transport modelling. Its contributions include many aspects of water pollution modelling, from the simplest advection dispersion modelling through water quality models to the parameterisation of the dependence of models parameters on flow.

In the second subsection we present an application of DBM methods to solute transport, based on tracer experiments and compare the results with a physically-based model. It is a typical input-output process, in the sense that both input and output variables have the same physical meaning of pollutant concentration. The calibration of both physically-based and DBM pollutant transport models for a river tracer study is usually performed under steady and known flow values. The model parameters depend on two main processes, longitudinal advection by turbulent flow and molecular cross-sectional diffusion with diffusivity depending on flow. It will be shown that the parameterisation of the dependence of the DBM model parameters on discharge allows it to be applied to discharges different to the calibration runs, as well as under gradually varying (unsteady) flow conditions and in water quality assessment. The application of STF methods to the analysis of tracer experiments and comparison with physical model results will be illustrated using the River Narew (North Poland) case study, whilst the procedure of transformation of model parameters that take into account their dependence on flow and an application of the obtained model to transport of pollutant in varying flow conditions is illustrated using the Murray Burn (UK) experiments.

The third subsection deals with water quality modelling in rivers, which usually employs both rainfall-runoff and flow routing models and transport modelling in order to derive the time-variable concentration profiles [29, 30]. In what follows we shall present an application of a multi-rate STF procedure to model concentrations of oxygen at one river cross-section, without the involvement of transport models. The evolution of oxygen concentration is modelled using temperature, radiation and pH measurements from the same site as input variables. These inputs are related to biological processes that control the oxygen production and depletion. The procedure will be illustrated using the River Elbe case study. The summary and conclusions of the applications of STF modelling tools will be given in the last Sect. 24.4.

24.2 Solute Pollutant Transport

In nature, the transport of passive and conservative pollutant in a river is governed by the combined action of the advection and diffusion processes. The assumption of passiveness enables the separation of the flow process from the pollutant transport. A full description of the transport process involves 3-D equations describing the changes of pollutant concentrations due to changes in the velocity of the flow and volume [19]. Analytical solutions to the idealized cases can be found [22], but in real-life problems of transport and mixing, simplified numerical solutions are commonly applied, due to the problem's complexity and the lack of observations regarding boundary conditions. The common approach is to average the velocity field. The introduction of simplifications into the process description is an important task for the modeller. It should be preceded by a detailed analysis of the goals of the modelling, the available information about the processes involved and the observational data base, as well as the computer resources.

24.2.1 Physically-Based Models: Advection Dispersion Models (ADE) and One-Dimensional Transport with Inflow and Storage Model (OTIS)

The transport of a conservative soluble pollutant along a uniform channel with a steady flow is usually described by the well-known Advection-Dispersion Equation (ADE) [7]. This model was successfully applied to many engineering applications (e.g. [26]). A review of the ranges of its applicability in small lowland rivers is given by [24], who show that the natural process studied follows Fickian behaviour at distances greater than 80 to 100 times the river width. The main problems discussed in the literature regarding the applicability of ADE to describe dispersion in natural rivers point to a quicker decrease of the concentration maximum than predicted by ADE, the nonlinear growth of the variance of concentration distribution with time and the existence of longer tails of concentration distribution at sufficiently long distances, than those that follow from the balance between advection and dispersion. The main reasons for the violation of dispersion process laws as described by ADE are thought to be the influence of a laminar sublayer, the irregular geometry of the channel producing so-called dead-zones, and heterogeneous velocity profiles [24].

The One-dimensional Transport with Inflow and Storage model (OTIS) was introduced by [2] to describe the long tails of the concentration profiles observed in real cases in a physically meaningful way. The OTIS model is formed by writing mass balance equations for the stream channel and the so-called storage zone. Water in the storage zone is considered immobile relative to water in the stream channel. The exchange of solute mass between the stream channel and the storage zone is modelled as a first-order mass transfer process. Under steady flow conditions, conservation of mass for the stream channel and storage zone yields [2, 18].

$$\frac{\partial C}{\partial t} = -U\frac{\partial C}{\partial x} + \frac{\partial D \partial C}{\partial x} + \alpha\left(C_s - C\right) + \frac{q_L}{A}\left(C_L - C\right), \qquad (24.1)$$

$$\frac{dC_s}{dt} = \alpha \frac{Q}{UA_s} (C - C_s), \tag{24.2}$$

where: C—solute concentration in the stream [g/m^3], t—time [s], Q—flow [m^3/s], A—the main channel cross-sectional area [m^2], x—distance downstream [m], D—the coefficient of longitudinal dispersion [m^2/s], C_S—the concentration in the storage zone [g/m^3], α—the exchange coefficient [1/s] and A_S—the storage zone cross-sectional area [m^2], q_L—lateral volumetric inflow rate [m^3/s], C_L—solute concentration in lateral inflow [g/m^3], U—mean cross-sectional flow velocity in x direction [m/s].

The variables in (24.1)–(24.2) are defined as averaged over the channel cross-section and the equation is valid when the solute is well mixed. The physically-based models are formulated in a deterministic set-up and their prediction uncertainty is evaluated using Monte Carlo (MC) based techniques. These involve simple MC sampling e.g. with the application of Generalised Likelihood Uncertainty Estimation (GLUE) methods [3] or Markov Chain Monte Carlo (MCMC) approaches, e.g. [21].

24.2.2 DBM Modelling Approach: Active Mixing Volume AMV Model

As an alternative to the transient storage model, the Active Mixing Volume (AMV) model was introduced by [36]. This concept is a further extension of the earlier Advection-Dispersion with dead zones ADZ approach [1]. The AMV model structure is identified and the parameters are estimated from the observed time series data (i.e. temporal concentration profiles) using system identification techniques [35]. The method is stochastic and the model parameters, including the residence time of the tracer transport process, have the form of random variables, thus allowing for derivation of their dependence on flow in a stochastic form. Obviously, the AMV model parameters also depend on discharge, as shown by [28, 36, 38].

In the AMV model the change of solute concentration in a river reach is described as a Stochastic Transfer Function model [13, 35]:

$$Cout_k = \frac{B(z^{-1})}{A(z^{-1})} Cin_{k-\delta}; \tag{24.3}$$

$$Cobs_k = Cout_k + \xi_k \tag{24.4}$$

where Cin_k is the concentration at the upstream end of the river reach at a sample time $k\Delta t$, $Cout_k$ is the estimated concentration at the downstream end of the river reach, $Cobs_k$ is the measured concentration at the downstream end of the river reach, z^{-1} is the backshift operator, and δ is a pure, advective time delay of $\delta\Delta t$ time units, A and B are polynomials of the backshift operator of the order 'm' and 'n'

respectively, and ξ_k represents the combined effect of all stochastic inputs to the system, including measurement noise. A and B are given by:

$$B(z^{-1}) = b_0 + b_1 z^{-1} + \cdots + b_m z^{-m}, \tag{24.5}$$

$$A(z^{-1}) = 1 + a_1 z^{-1} + \cdots + a_n z^{-n}. \tag{24.6}$$

The time series transfer function model described by (24.3)–(24.6) is equivalent to a stochastic difference equation:

$$Cout_k + a_1 Cout_{k-1} + \cdots + a_n Cout_{k-n} = b_0 Cin_{k-\delta} + \cdots + b_m Cin_{k-m-\delta} + \zeta_k. \tag{24.7}$$

The order of the STF model describing the transport of solutes in a river reach is described by the triad $[n, m, \delta]$ and is determined in a statistical time series analysis technique using the recursive-iterative simplified, refined, instrumental variable (SRIV) method [35] which is available in the CAPTAIN Toolbox developed at Lancaster University (www.es.lancs.ac.uk/cres/captain/). The toolbox functions and their use in a wide range of environmental and engineering applications is described in the manual [13]. The SRIV method gives the estimates of parameter uncertainty and the uncertainty of its predictions and these two factors are used during the choice of best model structures. However, the choice of model structure and its parameters depends not only on statistical criteria but also on the physical interpretation of the model structure and its parameters. The approach can be summarized as follows:

- Prepare (normalize) input and output data in the form of equal length columns; input data should have no gaps, but gaps filled with NaNs may be present in output.
- Apply rivid CAPTAIN function to identify the model structure, applying two goodness of fit criteria simultaneously—*YIC* and $R_T{}^2$.
- Apply riv CAPTAIN function to estimate the AMV model parameters.
- Check the model roots for the physical interpretation of the model (only values lying between 0 and 1 are feasible).
- Check the autocorrelation of the model residuals using the acf MATLAB function.

In the case where the model residuals are highly auto-correlated, a full recursive instrumental variable (RIV) approach including the noise model should be used. The model may have first order, and then it can be directly related to the ADE model. The parameters (a_1, b_0) of that model can be used to calculate the residence time T_{res} and a mean travel time of a pollutant $\bar{t}$:

$$T_{res} = \frac{\Delta t}{\ln(-a_1)} \tag{24.8}$$

and mean travel time defined as:

$$\bar{t} = \Delta t \delta + T_{res}. \tag{24.9}$$

The b_0 parameter can be related to mass conservation (or steady state gain), $SSG = b_0/(1 + a_1)$. The model can also be of second order and can be decomposed into first order models using fractional analysis, thus representing different pathways with different residence times [36]. For each of the pathways of the AMV model, the mean travel time of the pollutant and the dispersive coefficient [10] can be compared with the parameters derived from the physically-based models ADE and OTIS.

24.2.3 Case Studies

The application of the data Based Mechanistic modelling approach to solute transport will be illustrated using two separate case studies. The Upper River Narew case study [17] will be used to compare the AMV model results with the numerical solution of the OTIS model and solute transport in unsteady flow conditions will be illustrated using the Murray Burn tracer experiments [27].

24.2.3.1 Upper Narew Case Study

The case study illustrating an application of AMV model to pollutant transport and comparison with OTIS model results is based on a tracer experiment performed in a multi-channel system of the Upper River Narew in northeast Poland. The experiment is described in detail by [17]. It consisted of the injection of a solution of Rhodamine WT upstream the 17 km long river reach. The concentration profiles were measured at 5 cross-sections downstream, using the field Turner Design fluorometer. Water samples were also collected at sampling points. The transient storage model OTIS was calibrated using the tracer experiment data. As the OTIS model is deterministic, the Generalised Likelihood Uncertainty Estimation (GLUE) method [3] was applied to derive the uncertainty of model predictions [11].

The same tracer data were used to derive AMV models for each of the cross-sections. The best identified models were first order, as shown in Table 24.1. The identification of the model's structure and estimation of their parameters for each of the cross-sections were performed using the SRIV algorithm from the CAPTAIN toolbox, following the procedure outlined in Sect. 24.2.2. The measurements from the 4th cross-section had to be omitted due to their bad quality (no mass conservation of the tracer).

Apart from the model structure, the table gives the values of goodness of fit criteria $(R_T)^2$, the *YIC* criterion, steady state gain *SSG*, residence time constants T_{res} (24.8) and mean travel times $\bar{t}$ (24.9). The scheme of the whole analysed river reach, equivalent to the distributed OTIS model, is shown in Fig. 24.1. The advective delays for each of the models are measured in the relation to each model input. Each of the boxes represents a separate river reach, e.g. "$2N - 3N$" represents the reach between the 2nd and 3rd cross-section where the measurements were taken.

The OTIS and AMV model predictions together with 0.95 confidence limits derived by the AMV model, for the calibration stage, are shown in Fig. 24.2.

A comparison of the model results shows that both give very similar predictions. However, the AMV model required much less computation effort, as its predictions included both the estimated concentration values and their uncertainty.

Table 24.1 The AMV models structure for the river sections described in Fig. 24.1

Section	$2N-3N$	$3N-5N$	$5N-6N$	$6N-7N$
$[n, m, \delta]$	[1, 1, 29]	[1, 1, 18]	[1, 1, 29]	[1, 1, 46]
$(R_T)^2$	0.9996	0.9994	0.9970	0.9934
YIC	−19	−18	−15	−15
SSG	0.95	0.93	0.80	0.91
$T_{res}[h]$	0.24	0.40	0.39	1.43
$\bar{t}[h]$	2.65	1.90	2.80	5.25

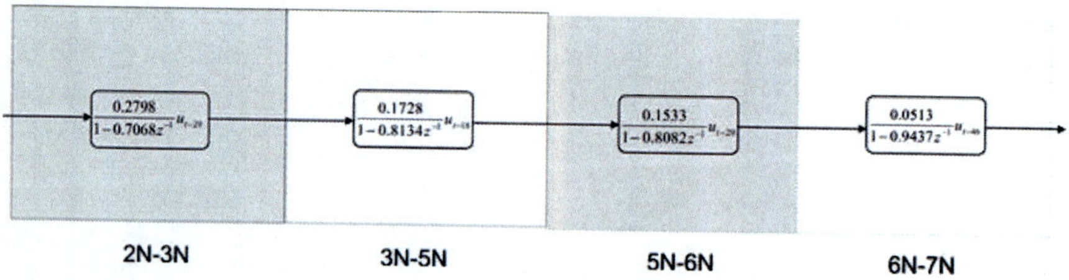

Fig. 24.1 Schematic representation of the AMV system for four modelled reaches of the River Narew located within the National Narew Park

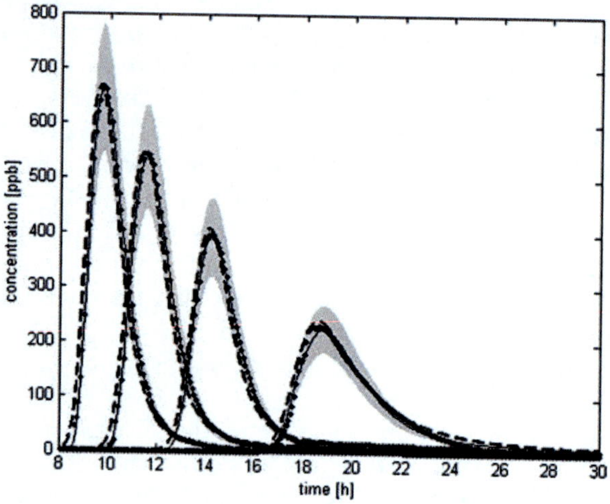

Fig. 24.2 Comparison of AMV (*continuous line*) and OTIS (*dashed line*) model predictions, observations of concentrations of Rhodamine WT at 4 cross-sections in the National Narew Park are marked with *dots*, *shaded area* denotes 0.95 confidence limits

24.2.3.2 Solute Transport in Unsteady Flow Conditions: Murray Burn Case Study

An application of the Data Based Mechanistic approach to solute transport in unsteady flow conditions will be illustrated using the Murray Burn tracer experiments [27]. The dispersion and travel times of a pollutant along the river depend on the flow conditions and in particular on the discharge. The nature of that dependence is complex and usually is derived from a number of tracer experiments [6, 8].

Studies of pollutant transport under varying flow conditions require either combined numerical modeling of flow and transport equations, or these equations are solved in sequence. The main difficulty in modeling this sort of problem lies in the lack of observations necessary for model calibration and validation. In this subsection we present the application of an AMV model, calibrated using a series of tracer experiments, for predicting solute transport in unsteady flow conditions with the assumption that mixing of the pollutant is fast enough and the concentrations at the cross-sections downstream are the sum of impulse responses related to the concentrations upstream at varying flow rates. The idea of this approach originates from the basic integration characteristics of linear transfer function models. The simplifying assumptions, required by the proposed approach, are consistent with those made in the previous studies, e.g. [20].

The Murray Burn, where the experiments were performed, is a small stream flowing through the Heriot-Watt University Campus at Riccarton in Edinburgh, UK. The experiments are described in [27]. Each experiment consisted of the gulp injection of a known mass of Rhodamine WT tracer and the measurement of tracer concentrations at up to four cross-sections below the point of injection, which were performed using a calibrated Turner Designs fluorometer. Each experiment was carried out under steady flow conditions. Only first two cross-sections were used in this research. The results presented form part of the research described in detail in [12] and [16].

An outline of the procedure for the development of an "unsteady flow" AMV model and evaluating its predictive uncertainty is as follows:

- Stochastic calibration of AMV models for a range of flow values.
- Parameterisation of a mean and variance of posterior distribution of the AMV model parameters on flow using the power law.
- Uncertainty analysis of parameterised model using GLUE with variance adjusted for heteroscedastic errors.
- Running the parameterised AMV model with input concentration values in the form of a series of impulses of a given amplitude, related to each measured flow value.
- Numerical integration (summation) of the model impulse responses into the total response concentration profile.

Following the above procedure, the AMV model structure and its parameters are estimated from the input observed tracer concentrations from the first cross-section and the output tracer concentrations from the second cross-section. 14 out of 18 experiments were used for the calibration and 4 experiments were used for the validation. The estimated stochastic transfer function models were of 1st order (i.e. $n = m = 1$) for all the experiments, given by (24.3)–(24.6).

In order to account for the observation errors varying with the value of flow, the MC analysis was performed with flow values following a heteroscedastic distribution and the AMV model parameters derived for each generated value of flow. Additionally, in the inner loop, the AMV model parameters were also sampled randomly following the distribution estimated by the SRIV routine.

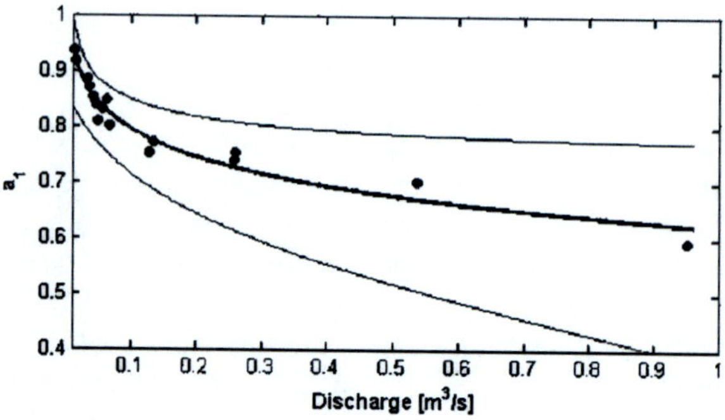

Fig. 24.3 AMV model parameter a_1 as a function of flow; *dotted lines* denote 0.95 confidence limits

The dependence of AMV model denominator (parameter a_1) on flow is shown in Fig. 24.3.

We chose for the analysis only the experiments with no mass losses and therefore the AMV models could be further simplified by substituting $b_0 = 1 + a_1$. Apart from the parameter a_1, also advective time delay of the model depends on flow and has been parameterised using the power law [16]. After the parameterization, the new AMV model has the form:

$$Cout_k = \frac{1 + x_a Q^{p_a}}{(1 - x_a Q^{p_a})z^{-1}} Cin_{k - x_\delta Q^{p_\delta}}; \qquad Cobs_k = Cout_k + \xi_k. \qquad (24.10)$$

It is now not linear in its new parameters, $x_a, p_a, x_\delta, p_\delta$, so the uncertainty of its predictions should be estimated using the MC based uncertainty analysis. The GLUE approach was used here. The parameter values were varied following a normal distribution with mean value equal to the median values and with variances estimated from the parameterization. In order to account for the heteroscedasticity we applied the weighting of the likelihood function following the methodology outlined by [23] and [14].

Following the "unsteady flow" scheme outlined above, for each discrete flow measurement at a given time instant, the pollutant at the input of the power-law parameterised AMV model has a form of an impulse with the amplitude ascribed to that time instant. The resulting ensemble concentration profiles (Fig. 24.4, dashed lines) are summed up to obtain the total concentration profile at the cross-section downstream. The approach was tested using flow measurements from the Murray Burn during autumn 2003, varying in the range 0.47–0.54 m^3/s, with input concentrations taken from the tracer experiment run at the steady flow of 0.128 m^3/s. The MC sampling of the power law parameters and flow values was applied with the same a priori distributions as those used for the estimation of uncertainty of model parameterisation. The uncertainty bounds of the model predictions in the unsteady flow conditions were obtained using the weights obtained from the model calibration stage (i.e. only the uncertainty of parameterisation was taken into account). The total predictive uncertainty could be assessed only if observations of the pollutant transport under varying flow conditions were available.

In Fig. 24.4 the total estimated concentration profiles for unsteady flow are shown by continuous lines with 0.95 confidence limits marked by shaded area. The concen-

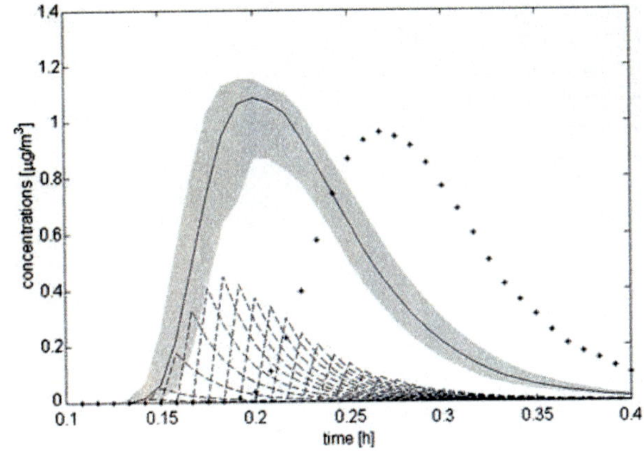

Fig. 24.4 Modelling transport in unsteady flow conditions using AMV model; observed concentrations at steady flow 0.128 m^3/s are shown by *dots*, *continuous line* denotes AMV model predictions in unsteady flow with 0.95 confidence limits denoted by *shaded area*; impulse responses are shown by *dashed lines*

tration profiles obtained for the steady flow of 0.128 m^3/s are shown by the black dots. Due to the higher flow values, the concentration profile starts earlier and is more asymmetrical, with higher maximum concentration peak than the steady-flow profile. Unfortunately, we do not have observations that allow for the validation of the proposed procedure and on-going research will show how useful this approach may be to model pollutant transport in rivers.

24.3 Water Quality Modelling

In a natural river system, the transport of a pollutant is a complex process and some adequate approximations must be made to predict the concentration distribution. There are many numerical modelling tools available for water quality prediction. They differ in the type of simplifications introduced, from plug-flow reactors that assume the dominance of advection, to continuously stirred reactors, assuming the dominance of diffusion. In between these two extremes there are models that use various numerical approximations to the governing transport equations. A review of the methods is given, for example, by [22].

In this section we shall present an application of Data Based Mechanistic approach to water quality modelling using the River Elbe case study as an example. This work was done during the author's involvement in the IMPACT EU project (2000–2003), together with Peter Young and colleagues from GKSS [5]. The first subsection presents the modelling of total oxygen using Multi Input Single Output (MISO) STF model with radiation, temperature and pH as input variables, the second presents the MISO STF model in a multi-rate set-up.

The available observations consist of 4 year-long (1997–2000) hourly measurements of total oxygen DO, acidity measure, pH, water temperature T and radiation R at the Geesthacht Weir station, Germany. The measurements have some missing samples. However, these are very good data in comparison with many water quality data sets found in the literature. The data were initially analysed using the Dynamic Harmonic Regression (DHR) techniques [37] in the CAPTAIN toolbox.

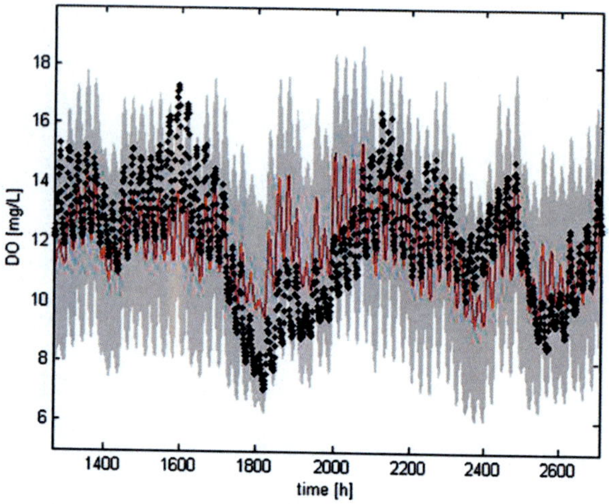

Fig. 24.5 Validation of MISO STF model for *DO* with radiation and temperature as input variables, *dotted lines* denote observed values, *continuous red line* denotes the simulation and *shaded areas* show the 0.95 confidence limits, year 1997

24.3.1 STF Model of Total Oxygen Using Temperature and Radiation as Input Variables

In order to find a black box model equivalent to the biological processes governing the water quality in the river, we should use temperature and radiation as input variables. This choice is not the only one possible, of course, and the further analysis discussed below shows that some other input variables provide a superior explanation of the data. This first MISO STF model is calibrated on the 1999 data. The same as before SRIV routine from CAPTAIN toolbox is applied (Sect. 24.2.2), with a vector B in (24.3) replaced by a matrix with dimensions corresponding to the number of input variables. We applied all the "growing season" period (from April until October). The best identified model has the form [1 1 1 4 1], which reads: $n = 1$; $m = [1\ 1]$; $d = [4\ 1]$ in $[n\ m\ d]$ triad:

$$DO_k + 0.9265DO_{k-1} = -0.0157T_{k-4} + 0.6850R_{k-1}, \tag{24.11}$$

where DO_k denotes total oxygen concentrations [mg/L] at sample time k, T_k denotes temperature [°C], and R_k denotes radiation [mJ].

This model has a coefficient of determination $R_T{}^2 = 0.39$, i.e. 39% of the variance of the *DO* observations are explained by the model on the basis of the temperature and radiation observations). Validation was performed on both 1997 and 2000 year data sets: the results for the year 1997 are shown in Fig. 24.5, with $R_T{}^2 = 0.41$, where a continuous line denotes the simulations and a dotted line denotes the observed values. Also shown are 0.95 confidence limits for the predictions, denoted by shaded areas. The analysis of autocorrelation function of model residuals was done using a Matlab function acf. The analysis indicates that the residuals are highly correlated. That means that a great deal of data variability is not explained by a linear model based only on temperature and radiation. Higher order models were also examined but their structure was not physically feasible (i.e. the denominator roots were complex variables). Therefore these models were not fulfilling the assumptions of Data Based Mechanistic approach [33]. The model reproduces mean variations of oxygen concentrations but fails to reproduce the extreme values (both low and high oxygen levels).

The STF model (24.11) used in this first example is linear and based solely on the external forcing variables, not taking into account variations of nutrient concentrations in the river. According to water quality specialists, extreme values of oxygen often result from the sudden changes in nutrient concentration, in particular, the changes of nutrient resources in the upper reaches of the river. Therefore the extreme oxygen concentration values are related to algal growth in the river, which introduces a feedback mechanism.This kind of information is not present in the external forcing variables, temperature and radiation. Consequently, it was decided to introduce pH as an additional input variable. This choice was made following the results of previous analysis based on the relations between DO concentrations and the other water quality indicators. Moreover, pH is easily measured and has conservative properties, so it constitutes a reliable indicator of water quality.

24.3.2 DO Model Using Temperature, Radiation and pH as Input Variables

The acidity measure pH is a logarithmic indicator of the amount of carbon ions in water since algae use carbon during their growth, thus changing water acidity. The pH values during pre-unification times were much lower than after German unification due, in part, to the low algal population in the water. Hence, the introduction of an exponent of pH measurements should provide information about the biological processes taking place in water.

As before, the whole growing period from 1999 was used for the calibration, using the SRIV algorithm from CAPTAIN toolbox. The resulting STF model has the following form [1 1 1 1 1 3 4]:

$$\begin{aligned} DO_k + 0.5199 DO_{k-1} &= -0.331 T_{k-1} + 3.89 R_{k-3} \\ &\quad + 0.0015 \exp(pH_{k-4}). \end{aligned} \tag{24.12}$$

This model gave better goodness of fit than the previous model, both for the calibration and validation stages (year 1997) (${R_T}^2 = 0.68$ and ${R_T}^2 = 0.77$, respectively). The validation results for this model are shown in Fig. 24.6.

Due to its additive structure, it is possible to distinguish the components of the model output due to each of the input variables (Fig. 24.7). These results show that temperature governs mainly the average variations of the oxygen concentrations; radiation is responsible for the diurnal pattern of the changes in DO; and pH models the remaining variations of DO. The analysis of the model residuals shows a high degree of autocorrelation, as in the case of two input variables. This DO model could be extended by the noise model, using RIV, but it seems to be not very practical for the off-line predictions.

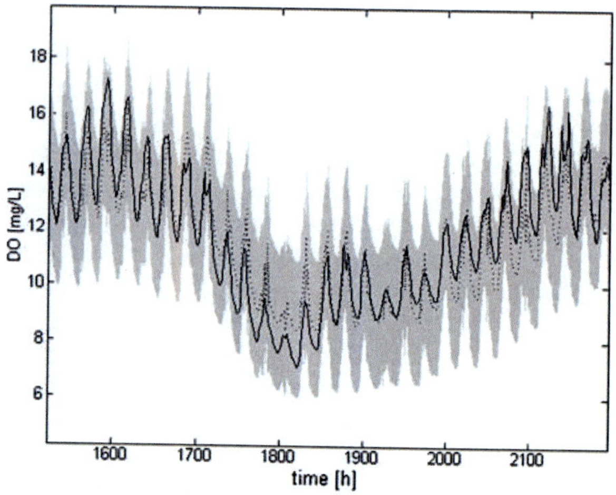

Fig. 24.6 MISO STF model for *DO* with exp(*pH*), temperature and radiation as an input variables, validation 1997; *dotted lines* denote observed values, *continuous line* denotes the simulation and *shaded areas* show the 0.95 confidence limits

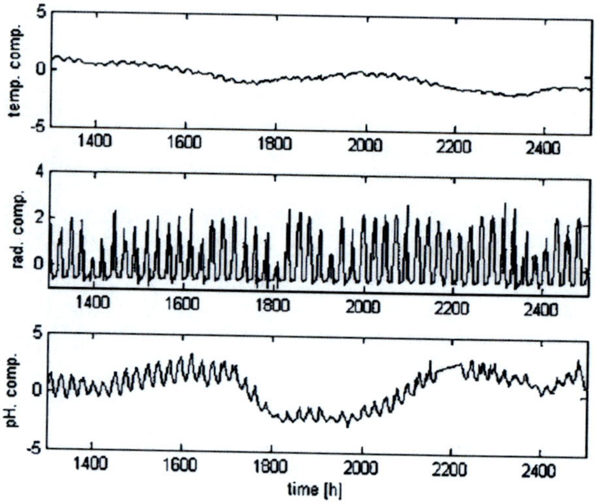

Fig. 24.7 Temperature (*upper panel*), radiation (*middle panel*) and exp(*pH*) (*lower panel*) components of MISO STF model for *DO*, validation 1997

24.3.3 A Multi-rate STF Model

The results of modelling using discrete transfer function methods can depend on the interval of sampling [25]. The analysis of the time constant obtained from model (24.12) indicates that different input variables may require different time constants in the model. Further analysis of the three input STF model, allowing the three STFs to have different time constants, yielded time constant estimates of about 10 hours for temperature, 20 hours for radiation and 30 days for *pH*. This strongly suggests that a larger sampling interval is required to model the *pH* transfer function relationship. Consequently, a new model was estimated in which temperature and radiation were based on hourly sampling interval and *pH* was sampled every 2 hours. This 'multi-rate' model consists of three STF sub-models: the first two with a one hour sampling interval and temperature and radiation, respectively, as inputs; and the second with 2 hour sampling interval and non-linearly transformed pH as input. The choice of this sampling interval gave best validation results and best conditioning of model parameter uncertainty.

The following MISO model was estimated using the iterative ‘relaxation’ procedure based on the SRIV estimation algorithm in CAPTAIN toolbox:

$$DO_k = -\frac{0.0084}{1-0.98134z^{-1}}T_{k-4} + \frac{0.1888}{1-0.9813z^{-1}}R_{k-1} + \frac{0.9126-0.9175Z^{-1}}{1-0.9972Z^{-1}}\exp(pH)_{k-1}. \tag{24.13}$$

This equation is presented in a rather unusual nomenclature to denote the multi-rate nature of the relationship: the lower case z^{-1} denotes the backward shift operator for a sampling interval of one hour; while the upper case Z^{-1} denotes the operator for a sampling interval of 2 hours. The resulting equation form of the first two STF functions (upper line of (24.13)) is as follows:

$$DO_k + 0.9724DO_{k-1} = -0.0084T_{k-4} + 0.1888R_{k-1} + \eta_k, \tag{24.14}$$

where the sampling interval is one hour. A second sub-model is then defined that considers the error η_k as the output and the exponentially transformed pH as the input, i.e.,

$$\eta_k - 0.9972\eta_{k-1} = 0.9126\exp(pH)_{k-1} - 0.9175\exp(pH)_{k-2}, \tag{24.15}$$

with the sampling interval in this case being 2 hours. This multi-rate model (24.13) has a structure similar to the previous model (24.12) but the introduction of different sampling intervals has provided a slight improvement in the goodness of fit function ($R_T{}^2$= 0.84). The procedure consists of multiple iterations with estimation of one of the model parameters using SRIV with the other model parameters fixed. Six iterations were required to obtain model convergence.

The validation of the model (24.13) was performed using the data from 2000 and the results are shown in Fig. 24.8. The coefficient of determination for the validation stage is now ($R_T{}^2$= 0.68) and this is a bit worse than previous MISO model results. However, the model structure distinguishes different patterns of process behaviour in response to “external” (i.e. temperature and radiation) and “internal” (pH) inputs. The residence time for radiation and temperature is equal to 1.8 days, while the pH has residence time of 27 days.

The estimation and validation results for this final, multi-rate STF model indicate that it is more general and flexible than the models identified previously and describes the water quality processes equally well. The differences in the time constants that necessitated the use of the multi-rate model are an interesting feature of the process and require further study. However, the model could serve as a useful basis for subsequent DBM modelling studies.

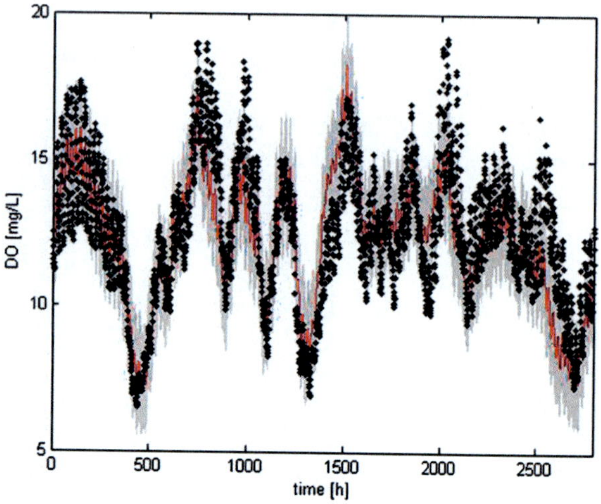

Fig. 24.8 Validation of multi-rate MISO STF model for *DO* with radiation and temperature as input variables; *dotted lines* denote observed values, *continuous red line* denotes the simulation and *shaded areas* show the 0.95 confidence limits; year 2000

24.4 Conclusions and Discussion

The aim of this section was to present the application of DBM methods developed by Peter Young over the last 30 years to solute transport and water quality modelling using case studies from Poland, the UK and Germany. Solute transport, representing a typical input-output process is illustrated using the Narew case study, where an AMV model was compared with a physically-based OTIS model. A methodology for coping with (slowly varying) unsteady flow was presented using the Murray Burn case study. The approach is based on a numerical integration of impulse responses of an AMV model, which incorporates the parameterisation of the model's parameters on flow. The AMV approach is an attractive alternative to the more complex, simultaneous distributed modeling of flow and transport in unsteady flow conditions. It can easily be extended to incorporate the flow model, run in parallel with the pollutant transport model. However, further work is necessary to verify the approach, ideally using observations of solute transport under varying flow conditions. The water quality application of the DBM tools was illustrated using the River Elbe case study. The MISO STF models were applied to simulate oxygen concentrations at one cross-section of the river, using external variables, temperature and radiation and water acidity index as a measure of biological feedback. We compared the influence of the internal (biological) input on model performance and the influence of external factors. The model was extended by a multi-rate application of SRIV methods that give a better physical explanation but worse goodness of fit of the model simulations for the validation case.

Acknowledgements I would like to thank my colleagues from GKSS, Germany, for further use of the data from the River Elbe (among others, Ulrich Callies and Wilhelm Petersen). My collaborators Marzena Osuch, Jaroslaw Napiorkowski and Pawel Rowinski (Institute of Geophysics, PAS, Poland) and Steve Wallis (Heriot Watt University, Edinburgh, UK) are thanked for their help in the River Narew and Murray Burn case studies.

References

1. Beer, T., Young, P.C.: Longitudinal dispersion in natural streams. J. Environ. Eng. **109**(5), 1049–1067 (1983)
2. Bencala, K.E., Walters, R.A.: Simulation of solute transport in a mountain pool-and riffle stream: a transient storage model. Water Resour. Res. **19**, 718–724 (1983)
3. Beven, K.J., Binley, A.: The future of distributed models: model calibration and uncertainty prediction. Hydrol. Process. **6**, 279–298 (1992)
4. Box, G.E.P., Jenkins, G.M.: Time Series Analysis: Forecasting and Control. Holden-Day, San Francisco (1970)
5. Callies, U., Scharfe, M., Blöcker, G., Romanowicz, R., Young, P.C.: Normalisation involving mechanistic models: benefits compared to purely statistical approaches. Technical report, SCA Project IST 1999-11313, Deliverable 7 (2002)
6. Deng, Z.-Q., de Lima, J., Singh, V.P.: Transport rate-based model for overland flow and solute transport: Parameter estimation and process simulation. J. Hydrol. **315**, 220–235 (2005)
7. Fischer, H.B.: The mechanisms of dispersion in natural systems. J. Hydraul. Div. **93**, 187–216 (1967)
8. Harvey, J.W., Wagner, B.J.: Quantifying hydrologic interactions between streams and their sub-surface hyporheic zones. In: Streams and Ground Waters, pp. 200–221. Academic Press, San Diego (2000)
9. Jakeman, A.J., Hornberger, G.M.: How much complexity is needed in a rainfall-runoff model? Water Resour. Res. **29**, 2637–2649 (1993)
10. Lees, M.J., Camacho, L.A., Chapra, S.C.: On the relationship of transient-storage and aggregated dead zone models of longitudinal solute transport in streams. Water Resour. Res. **36**, 213–224 (2000)
11. Osuch, M.: Modelowanie przepływu i migracji wybranych zanieczyszczeń na odcinku Narwiańskiego Parku Narodowego. PhD thesis, Instytut Geofizyki Polskiej Akademii Nauk (2008)
12. Osuch, M., Romanowicz, R.J., Wallis, S.: Uncertainty in the relationship between flow and parameters in models of pollutant transport. Publ. Inst. Geophys., Pol. Acad. Sci. **E-10**(406), 127–138 (2008)
13. Pedregal, D.J., Taylor, C.J., Young, P.C.: System Identification, Time Series Analysis and Forecasting. The Captain Toolbox. Handbook v 1.1 July 2004. CRES, Lancaster University (2004)
14. Petersen-Overleir, A.: Accounting for heteroscedasticity in rating curve estimates. J. Hydrol. **292**, 173–181 (2004)
15. Piotrowski, A., Wallis, S.G., Napiórkowski, J.J., Rowinski, P.M.: Evaluation of 1-d tracer concentration profile in a small river by means of multi-layer perceptron neural networks. Hydrol. Earth Syst. Sci. **11**, 1883–1896 (2007)
16. Romanowicz, R.J., Osuch, M., Wallis, S.: Modelling of pollutant transport in rivers under unsteady flow. In: IAHR2010 Proceedings (2010)
17. Rowiński, P.M., Napiórkowski, J.J., Szkutnicki, J.: Transport of passive admixture in a multichannel river system—the upper narew case study. Part 1. Hydrological survey. Ecol. Hydrobiol. **3**, 371–379 (2003)
18. Runkel, R.L., Broshears, R.E.: One-dimensional transport with inflow and storage (otis): a solute transport model for small streams. Technical report, Boulder, Co., University of Colorado, CADSWES Technical Report 91-01, 85 p. (1991)
19. Rutherford, J.C.: River Mixing. Wiley, New York (1994)
20. Sincock, A.M., Wheater, H.S., Whitehead, P.G.: Calibration and sensitivity of a river water quality model un-der unsteady flow conditions. J. Hydrol. **277**, 214–229 (2003)
21. Smith, P., Beven, K., Tawn, J., Blazkova, S., Merta, L.: Discharge-dependent pollutant dispersion in rivers: estimation of aggregated dead zone parameters with surrogate data. Water Resour. Res. **42**, W04412 (2006). doi:10.1029/2005WR004008
22. Socolofski, R.P., Jirka, A.: Environmental fluid mechanics, engineering lectures. Technical report, Institut fur Hydromechanik, Universitat Karlsruhe, 76128-Karlsruhe, Germany (2002)

23. Sorooshian, S., Dracup, J.A.: Stochastic parameter estimation procedures for hydrologic rainfall-runoff models: correlated and heteroscedastic cases. Water Resour. Res. **16**(2), 430–442 (1980)
24. Sukhodolov, A.N., Nikora, V.I., Rowinski, P.M., Czernuszenko, W.: A case study of longitudinal dispersion in small lowland rivers. Water Environ. Res. **69**, 1246–1253 (1997)
25. Tych, W., Pedregal, D.J., Young, P.C., Davies, J.: An unobserved component model for multirate forecasting of telephone call demand: the design of a forecasting support system. Int. J. Forecast. **18**, 673–695 (2002)
26. Vuksanovic, V., De Smedt, F., Van Meerbeeck, S.: Transport of polychlorinated biphenyls (pcb) in the scheldt estuary simulated with the water quality model wasp. J. Hydrol. **174**, 1–10 (1996)
27. Wallis, S., Manson, R.: Modelling solute transport in a small stream using discus. Acta Geophys. Pol. **53**(4), 501–515 (2005)
28. Wallis, S.G., Guymer, I., Bilgi, A.: A practical engineering approach to modelling longitudinal dispersion. In: Proceedings of the International Conference on Hydraulics and Environmental Modelling of Coastal, Estuarine and River water. Gower Technical, Bradford (1989)
29. Whitehead, P., Young, P.C.: Water quality in river systems: Monte-Carlo analysis. Water Resour. Res. **15**, 451–459 (1979)
30. Whitehead, P., Young, P.C., Hornberger, G.: A system model of stream flow and water quality in the Bedford-Ouse river—1. stream flow modelling. Water Res. **13**, 1115–1169 (1979)
31. Young, P.C.: A general theory of modelling for badly defined dynamic systems. In: Modeling, Identification and Control in Environmental Systems, Amsterdam, pp. 103–135 (1978)
32. Young, P.C.: Data-based mechanistic modelling of environmental, ecological, economic and engineering systems. Environ. Model. Softw. **13**, 105–122 (1998)
33. Young, P.C.: Data-based mechanistic modelling, generalised sensitivity and dominant mode analysis. Comput. Phys. Commun. **117**, 113–129 (1999)
34. Young, P.C., Minchin, P.E.H.: Environmental time series analysis. Int. J. Biol. Macromol. **13**, 190–201 (1991)
35. Young, P.C.: Recursive Estimation and Time-Series Analysis. Springer, Berlin (1984)
36. Young, P.C., Lees, M.: The active mixing volume: a new concept on modelling environmental systems. In: Statistics for the Environment, pp. 3–44. Wiley, New York (1993)
37. Young, P.C., Pedregal, D.J., Tych, W.: Dynamic harmonic regression. J. Forecast. **18**, 369–394 (1999)
38. Young, P.C., Wallis, S.G.: Solute transport and dispersion in channels. In: Channel Network Hydrology, pp. 129–173. Willey, New York (1993)

Chapter 25
Input-Output Analysis of Phloem Partitioning Within Higher Plants

Peter E.H. Minchin

25.1 Introduction

All higher plants have two vasculature systems, xylem and the phloem. Xylem is involved in transport of water, mineral ions, and metabolites especially root-derived hormones from the roots to the shoot, which includes leaves, flowers, and fruit. The phloem pathway between the mature photosynthesising leaves (or "source" leaves, as they are the main source of all carbohydrate) to all parts of a plant (sinks) the store of utilise this carbohydrate is the topic of this chapter. Phloem transport consists of bulk flow of solution, driven by the large hydrostatic pressure gradient generated at the source leaves through active transport of the carbohydrate into the flow pathway and the concomitant osmotic flow of water [33], and carbohydrate unloading as the sink. Water moves passively (i.e. no energy is involved) in and out of the phloem pathway under to influence of osmosis. Within the phloem pathway the hydrostatic pressure is in the order of 1 MPa (*circa* 10 atmospheres) [8], or greater: damaging this tissue during preparation for microscopy, sap sampling, or any invasive measurements leads to a surge of flow, resulting in immediate blockage. This results not only in visual artefacts but also in problems in carrying out dynamic studies of its function.

Carbohydrate is the major substrate for all plant growth, and its partitioning between competing sinks (fruit, vegetative growth, storage) determines harvest yield [7]. Increased harvest yield of modern crops is a direct consequence of increased partitioning of available carbohydrate to the organs of agronomic importance, e.g. grain, and fruits. In a mature apple tree on a dwarfing rootstock, about 70 percent of a season's growth is invested in the fruit, the harvested fraction. In many other crops, this fraction is only 30 percent. A lot of effort involved in crop

P.E.H. Minchin
The New Zealand Institute for Plant and Food Research Limited, Te Puke, 412 No. 1 Rd, RD2, Te Puke 3182, New Zealand

L. Wang, H. Garnier (eds.), *System Identification, Environmental Modelling, and Control System Design*,
DOI 10.1007/978-0-85729-974-1_25,

management, and hence production cost, goes into control of unwanted vegetative vigour which is competing with the fruit for carbohydrate.

All plants can be considered as a network of sources (e.g. mature leaves) and sinks (e.g. roots, new shoots, flowers/fruit) [9, 35] and one aspect of crop management is to maximise growth, i.e. carbohydrate partitioning, to the part of economic value, which may not be the natural growth pattern of the plant in the wild.

Qualitative rules have been worked out to describe partitioning within plants (e.g. [34, 35]), but we do not understand the processes that control partitioning, and hence have little quantitative description. Qualitatively we know that a sink is usually supplied by a nearby source, such that roots are supplied by the lower leaves, the shoot apex by the apical leaves, and a fruit by nearby leaves. This is consistent with the concept of supply via the route of least resistance, although phloem flow resistance has never been measured. There are many claims that this resistance is not a major factor limiting growth (e.g. [10, 34]), while Thornley [32] has argued that all partitioning models must start with the irreducible framework of transport and chemical conversion.

Experiments involving altering the size and number of sinks, and sources, have demonstrated that there is a hierarchy of supply: under limited supply, seeds have the highest priority and storage has the lowest priority. In general, the priority order is:

$$\textit{seeds} > \textit{fleshy fruit parts} = \textit{shoot apices} = \textit{leaves} > \textit{cambium} > \textit{roots} > \textit{storage}.$$

A probable reason for lack of progress in the understanding of partitioning is that it is difficult to measure and many techniques are destructive. Hence it has not been possible to obtain detailed time sequences of partitioning before and after any experimental treatment. Changes in dry weight are the simplest measure of partitioning, but require destructive measurements using multiple plants: in addition, the statistical methods used to overcome plant-to-plant variation can hide small or short term effects. Another confounding effect is that after a treatment, the modified plant will adapt to its new source-sink configuration and so partitioning measurements immediately (minutes) after a treatment reflect flow changes induced in the pre-treatment system, while later measurements reflect the adapting or adapted system.

Much work has been done using the radioactive tracer $^{14}CO_2$ applied to individual leaves. Typically, destructive measurement after tracer application has been the norm except in the special case when the monitored sink is an immature leaf. An immature leaf is thin enough for the radiation from imported ^{14}C to be measured in vivo (e.g. [6]), but the other sinks (roots) and the transport pathway cannot be simultaneously followed. Geiger's laboratory produced the first detailed experimental results on phloem function using this approach, much of it on sugar beet.

Non-destructive measurements enable both the initial and the later responses to be observed on the same plant. The isotope ^{11}C, applied as $^{11}CO_2$, has been used as a tracer that can be measured in vivo because its decay radiation is able to penetrate many centimetres of plant tissue. The difficulty with this isotope of carbon is that it has a very short half-life, just 20.4 minutes, so must be prepared on site using a cyclotron or other nuclear particle accelerator [27, 28]. This isotope is now used extensively in positron emission tomography (PET) medical imaging but this is still not readily available to plant scientists.

25.2 In Vivo Measurement of Phloem Flow

The short-lived property of ^{11}C, does provide a number of advantages. With short-lived tracers, the same plant can be used for many tracer loadings, because there will be no tracer carry over between labellings, provided these are far enough apart. This allows a plant to be used as its own control, overcoming problems of plant-to-plant variability, that are introduced when measurements are destructive and several plants need to be used to complete experiments. In addition, in vivo monitoring of the tracer enables flow to be followed with a very fine time resolution. Typically 1-minute count times have been used. In vivo measurement enables dynamic studies: several tracer profiles can be collected from different points within a source-sink network, allowing changes in the flow patterns to be followed. A typical setup to monitor the movement of tracer pulses along a stem is shown in Fig. 25.1A, with the observed temporal profiles resulting from a single pulse label of ^{11}C shown in Fig. 25.1B and C. Useful amounts of tracer last for 150 min (7 half-lives), before the tracer amounts become low and noisy because of the Poisson statistics of nuclear decay. The half-life corrected tracer curves, Fig. 25.1C, continually increase to the end of the measurement because the detectors are observing a terminal sink, that is, a sink that imports and cannot export tracer. Waiting until a pulse has completely decayed before re-labelling allows decay correction for each pulse, but results in gaps in the tracer data. More frequent labelling results in overlap of consecutive pulses (Fig. 25.1D), resulting in a continuous flow of tracer through the system; however, half-life correction is not possible as the tail of one pulse is overlaid with the early part of the next. However, input-output analysis of a continuous string of pulses is no more difficult than for a single pulse (see below) and allows one to follow the time variation of the model parameters over as long a period as one continues the tracer excitation. Experiments have been done involving up to 16 pulses in a single day, applied at hourly intervals [3]. This can be continued for several days to follow any adaptive responses.

Transport of labelled photosynthate from the site of fixation into various sinks has provided some of the first insights into the short-term distribution of photosynthate. The effects of experimental treatments designed to perturb this distribution provide insight on possible mechanisms driving this flow. For example, during collection of the data shown in Fig. 25.1D, the temperature of the segment of stem for which detector 2 is an input and detector 3 an output was warmed from 25 to 35°C, while the segments on either side (between detectors 1 and 2, and between 3 and 4) were held constant at 25°C. Care was taken to ensure that each of the detectors was set up with uniform sensitivity to tracer within their field of view. This ensured that once tracer was within view, subsequent movement did not result in a change in observed count rate. Then, change in tracer count rate within any field of view was a result of tracer movement into the field of view or tracer decay.

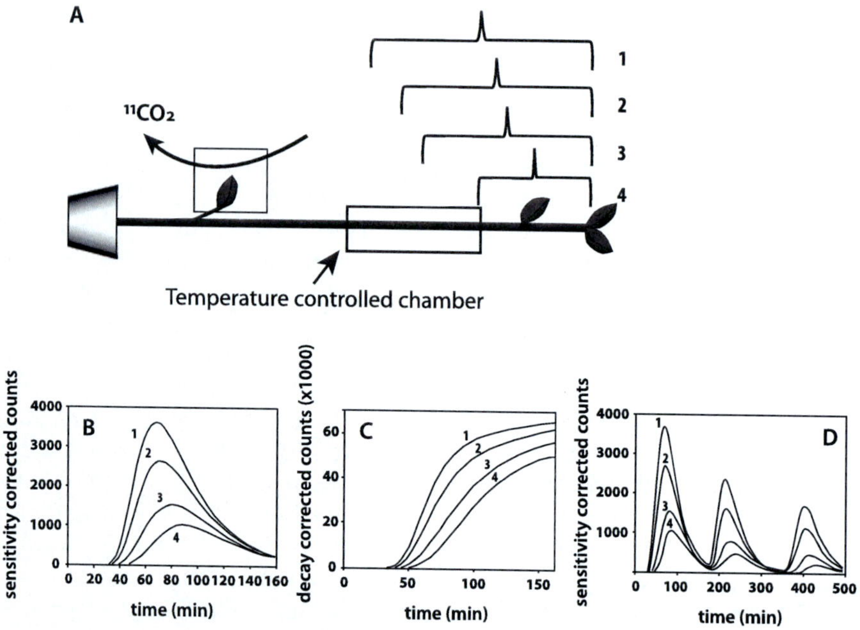

Fig. 25.1 (**A**) Schematic showing how movement of ^{11}C-labelled photosynthate produced in a mature leaf to the developing shoot apex of a bean is made. A mature source leaf is enclosed in a clear chamber for tracer labelling. A pulse of $^{11}CO_2$ is applied to this leaf and after a few minutes this is flushed with air continuously pumped through the chamber to maintain leaf photosynthesis. In this example, a segment of the stem was enclosed in a temperature-controlled chamber to allow the effect of stem temperature to be measured. The fields of view of 4 radiation detectors are shown by the brackets and labelled 1...4, achieved by careful placement of lead radiation shielding. (**B**) Typical temporal tracer profiles observed after a single pulse of $^{11}CO_2$ was applied to the mature leaf at time zero, corrected to give equal sensitivity to all the detectors. (**C**) Data from B that have been corrected for isotope decay. (**D**) Three overlapping tracer pulses, where B and C are just the first pulse, used to follow tracer movement over longer periods than a single pulse allows. The vertical dotted line shows when the temperature of the 20-cm length of stem, with input observed by detector 2 and output by detector 3, was raised from 25 to 35°C while the 1-cm segments 1–2 and 3–4 were held at 25°C

Figure 25.2 is a schematic diagram showing how measurements of tracer distribution from a single labelled leaf of a barley seedling to the two competing sinks of a split root system were made. One radiation detector, labelled A in Fig. 25.2, observed the tracer within the entire plant minus the labelled leaf, and so monitored the amount of tracer mobilised out of the labelled leaf and available for distribution between competing sinks. Detectors B and C monitored tracer accumulating within each root-half. As in the example above, care was taken to ensure that each of these detectors was set up such that they had uniform sensitivity to tracer within their field of view. A possible route for tracer loss from a terminal sink was via respiration, which would result in $^{11}CO_2$ loss from the plant tissue. During the short life of the ^{11}C tracer, respiration of labelled photosynthate did occur, but only consumed a small fraction of the tracer within a region [4]. With a longer lived tracer, respira-

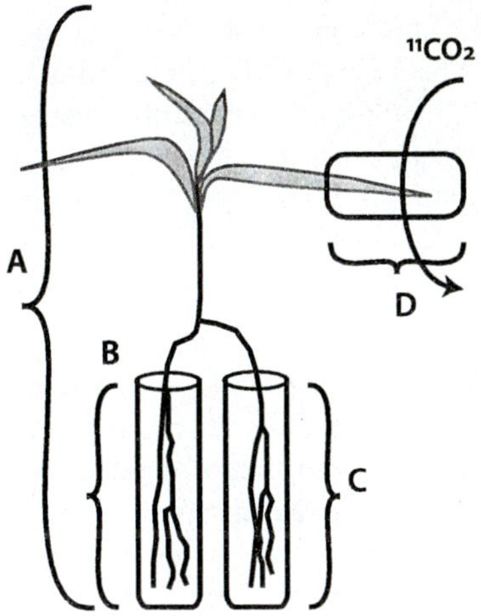

Fig. 25.2 Layout of radiation detectors to monitor partitioning between each half of a split root barley seedling. Detector D monitors the labelled leaf, A the entire plant except for the labelled leaf, B and C one half of a split root. Tracer labelling is done to the tip of one leaf, which is enclosed in a clear plastic chamber, which before and after labelling has air continuously pumped through it to maintain photosynthesis

tion would have become a significant route of loss, and so would xylem transport. This is another advantage of working with a short-lived tracer, but clearly is a major disadvantage if one is interested in pathways with time constants equal to several half-lives of the tracer.

25.3 Quantitative Analysis of in Vivo Tracer Profiles

In the early ^{11}C work (e.g. [11, 13, 16, 20, 21]), detectors observing a short segment of plant stem through a slit collimator were used to monitor input into a region further downstream. This type of detection is typically referred to as a slit detector. Given there is leakage from the phloem transport pathway, a slit detector is measuring both the tracer within the phloem transport pathway as well as the locally leaked and accumulated tracer, without distinction. Once it was realised that a slit detector was not giving a good measure of input into the region below, we changed to using integrating detectors to observe tracer movement into a terminal sink, as described above [20, 21]. This mode of monitoring tracer is typically described as a sink detector, as once tracer arrives into its field of view it cannot move out again. With large segments of plant, it is not possible to achieve uniform sensitivity with a single detector, so multiple detectors are often used, and the sensitivity-corrected count rates added to produce the effect of a sink detection. With a sink detector, the only way tracer can move into the field of view is via phloem transport across the defining surface: any movement within the field of view, either through phloem transport or unloading from the phloem transport pathway, will not result in any change in observed count rate. Thus, a sink detector observes the total import into the region of sensitivity, which is the integral of the input. This requires that any attenuation of the monitored radiation by the sink tissue is small, or corrections can be made for it.

In the example illustrated in Fig. 25.2, detector A observes tracer input to the whole plant, and detectors B and C each measure the inputs into a segment of the split root, or equivalently the output from the more apical part of the plant. So, representing the count rate seen by detector A (the input) as u_k, and detector B (the output) as y_k at times k, the general input-output equation is:

$$y_k = -a_1 y_{k-1} - a_2 y_{k-2} - \cdots - a_n y_{k-n} + b_0 u_k + b_1 u_{k-1} + \cdots + b_m u_{k-m}, \tag{25.1}$$

where $a_1 \ldots a_n, b_0 \ldots b_m$ are constants that need to be estimated from the data.

This equation describes the movement of tracer from the field of view of detector A into that of detector B. To relate this to the movement of labelled photosynthate, the observed tracer levels need to be decay-corrected. However, use of multiple overlapping pulses (e.g. Fig. 25.1D) to increase the observation time beyond that possible with just one tracer pulse, means half-life correction of the raw data is no longer possible. Dynamic modelling by using (25.1) offers a simple solution. Decay-correction of the data involves correcting the observed count rates back to what they would have been at some standard time, usually the start of the measurements. This can be achieved by multiplying the count rate observed at time t by $\exp(\lambda t)$, where λ is the isotope's decay constant and related to the isotope half life $t_{1/2}$ by $\lambda = \log_e(2)/t_{1/2}$. The isotope ^{11}C has a half-life of 20.4 minutes, so for ^{11}C $\lambda = 0.033978/\text{min}$. When (25.1) is fitted to decay-corrected data, with a sampling time of T, then the common term $\exp(\lambda k T)$ can be cancelled, giving

$$y_k = -a_1^* y_{k-1} - a_2^* y_{k-2} - \cdots - a_n^* y_{k-n} + b_0^* u_k + b_1^* u_{k-1} + \cdots + b_m^* u_{k-m}, \tag{25.2}$$

with

$$a_i^* = a_i \exp(\lambda i T); \qquad b_i^* = b_i \exp(\lambda i T). \tag{25.3}$$

That is, for each time k the raw tracer data are decay-corrected to this time and not to the start of the experiment. Then (25.2) fits the observed data to the general input-output equation (25.1) and half-life corrections are made to the estimated parameter values according to (25.3).

Minchin and Thorpe [19] showed that the same input-output equation holds for summed data, using the same model parameters. Hence the summed input and summed output count rates, that is, with both detectors in 'sink' mode, enables the dynamics of the unobservable tracer input and output to be obtained. From the time of this work we work with summed data.

Radioactive decay follows Poisson statistics, so the variance in the mean value of the observed decay rate at some specified time is the square root of the mean count rate at that time. Hence observed count rates, and half-life corrected rates, suffer from heteroscedasticity, i.e. a variable variance. The least squares method of parameter estimation requires that the data have constant variance, and that the dependent variable (y_k in (25.1), (25.2)) is error free. With tracer data, neither of these requirements is satisfied. Error associated with the dependent variable results in least squares estimations of the model parameter values being biased [14, 36]. A very neat and simple way to prevent the bias problem is to use the instrumental variable (IV) modification to the least squares algorithm. The requirements of an IV are that it be highly correlated with the observed output y_k and uncorrelated with the noise associated with y_k. Young [36, 37] introduced the iterative method where the least squares parameter estimates were initially used to calculate an IV for a subsequent estimation and then the updated parameters were used to calculate an improved IV with this loop continuing until the model parameters converged. In practice, this usually requires 2–3 iterations. Young [36, 37] also introduced recursive parameter estimation, where the parameters and their covariance matrix at time k are calculated from the estimate at time $k-1$ and the data at time k.

Fitting of (25.1) or (25.2) to a data set involves both model identification and parameter estimation, processes which in practice cannot be separated. Model identification involves testing various values of n and m, of (25.1) or (25.2), to find which specific values best describe the data. Recursive IV fitting of data to the input-output (25.1), or (25.2), has been found to be very robust [37, 38] in that convergence to a set of parameter values is often possible when conventional block data fitting of the same equation would result in problems in matrix inversion due to collinearity. The second advantage of recursive estimation is that it provides a simple method to test if the model parameters are time invariant. If the system under investigation is time invariant, then it must be describable by a set of constant parameters. This can be achieved by giving the recursive algorithm a bad memory for earlier data, so that at a specific time the current parameter estimates are based only upon the more recent data. This results in the early data having an exponentially reducing effect on the current parameter estimates and can therefore drift, if the recent data require this. If the model is badly chosen or the system is not time invariant, the parameter values tend to wander. Young and his many collaborators have written extensively in this area and have developed sophisticated algorithms, which will be briefly summarised below.

Recursive estimation is used, with its extension to IV, as a means of eliminating estimation bias. The resultant estimates are used to construct data filters for further iterative estimation. As these filters are based upon the current estimates of the input-output model parameters they result in improved estimates and reduced estimated parameter variances of observed data. The simplified form of this approach, the simplified refined IV algorithm (SRIV), is not quite as robust as the IV method but, because it has lower parameter variances, is often preferred. Extensive simulation work has found that the statistical nature of the noise is unimportant to IV and SRIV estimation provided that the system input is independent of the noise, even when the noise is highly structured [37]. Both the IV and SRIV approaches overcome potential issues of heteroscedasticity: in our experience the model fits are very good, with little sign of heteroscedasticity in the model errors.

The input-output model is a parsimonious description of changes in shape of the tracer profile moving through the plant, independent of the actual shape of the tracer profiles used in determining it. It embodies all the information about the system that is available from the stimulating tracer profile, i.e. the input [25]. From this model, several physiologically meaningful parameters can be calculated, free from any mechanistic assumptions, being based solely upon the data. For example, the transfer function, derived from the input-output model, gives the distribution of tracer transit times. The bulk flow transit time is given by the time delay in the transfer function, while the average transit time can be calculated from the distribution of transit times. The system gain is interpreted as the fraction of the decay-corrected tracer that enters a field of view and eventually leaves. One minus the gain is then the fraction of the decay-corrected tracer that enters a field of view and never leaves. That is, the fraction lost between the input and output, which for transport along a stem is the phloem leakage within this length. In Fig. 25.1, the system gain is the fraction of labelled photosynthate that enters the field of view of the input detector and eventually leaves the field of view of the output detector. With a bifurcating transport system (e.g. Fig. 25.2), the system gain is the fraction of labelled photosynthate partitioned to the observed sink.

Input-output analysis of in vivo ^{11}C-photosynthate transport within a plant has provided the first, and only, direct measurement of phloem leakage [19], and the most detailed measurements of phloem partitioning between competing sinks (e.g. [5, 23, 24, 26, 30]), as well as statistically sound estimates of both the bulk flow and average transit times, as well as the distribution of transit times.

25.4 Examples of ^{11}C-Tracer Results

Input-output analysis has provided the first detailed information available about phloem leakage in the long-distance transport pathway, sink function which has lead into a mechanistic understanding of the control of partitioning. Examples of the input-output analysis will now be presented: firstly, an example involving flow along the stem of a climber bean plant, which resulted the only detailed information available on phloem leakage within the long-distance transport pathway. Then, an example of sink function based upon the measurement of import into a terminal sink is described, and finally an example of in vivo measurement of phloem partitioning between two competing sinks that led to the first mechanistic understanding of the control of partitioning.

Schematic diagrams of the plant and radiation detector setups are given in the figures, but do not include the lead or tungsten needed for radiation shielding of each detector. This radiation shielding is needed to create a well defined field of view for each detector. In the experimental setup, the plant is often buried in this shielding and not only difficult to see but also to get adequate light to. In some experiments mirrors have been needed to get light to the labelled leaf. Modern positron emission tomography cameras (e.g. [12, 29]) do not require such shielding, as they provide the spatial coordinates of each observed radiation event, enabling the selection of the regions of interest after the measurements are made. This gives a great deal of flexibility, but the volume within which sensitivity is uniform is limited, restricting their use to small plants.

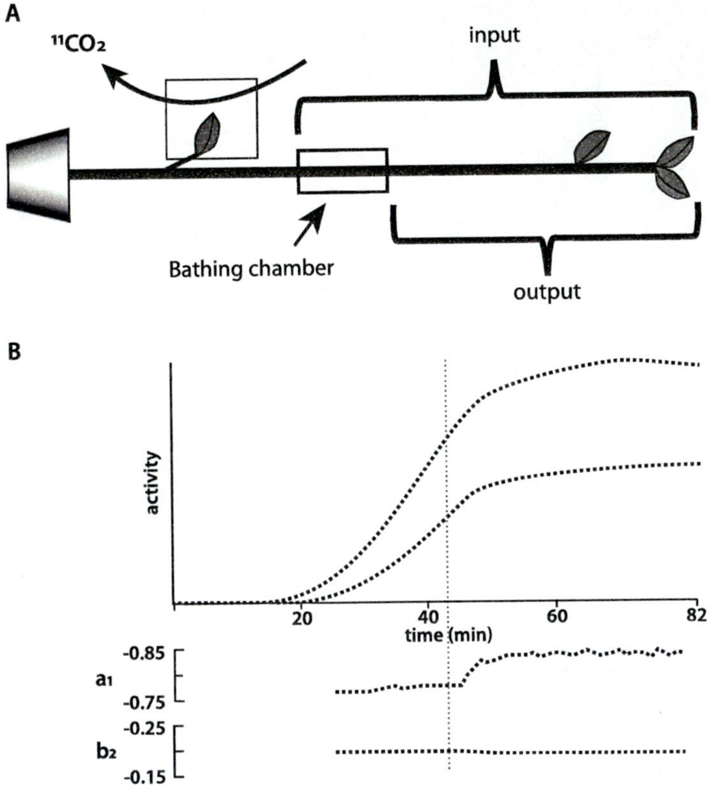

Fig. 25.3 Measurement of leakage/reloading within the stem of bean. (**A**) Schematic diagram of measurement configuration showing the leaf labelled with $^{11}CO_2$, a chamber on the lower segment of the stem, and the location of the input and output radiation detectors. (**B**) The observed input and output tracer profiles are shown with the time the inhibitor was applied to the chamber as a vertical dotted line, and below are the estimated input-output model parameters. Redrawn from [19] with permission

25.4.1 Stem Leakage

Input-output analysis of ^{11}C-labelled photosynthate transport through the stem of a climber bean provided quantitative data on leakage and reloading along the transport pathway [19]. Earlier mechanistic modelling of ^{11}C-labeled photosynthate movement through the stem of soybean [2] demonstrated net leakage along the transport pathway and estimated a rate constant as well as the speed of bulk flow.

Using a single tracer pulse, and input-output analysis, we measured the net loss of tracer within segment of a bean stem (Fig. 25.3). Application of a reagent known to inhibit reloading, enabled calculation of the unloading rate: this being the difference between the inhibited and uninhibited tracer loss. In this example [19] the unloading flux was 6.0%/cm of the incoming flux, and reloading was 3.2%/cm of the incoming flux, giving a net unloading of 2.8%/cm before the inhibitor was applied. This leaked

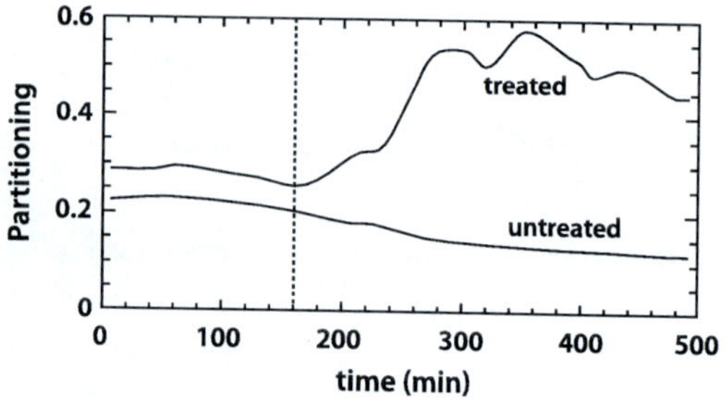

Fig. 25.4 Estimated partitioning of ^{11}C-photosynthate between the two halves of a split root barley seedling (see Fig. 25.2 for measurement setup). The vertical line shows the time when galactose (final concentration of 50 mM) was added to the root solution of the treated root segment. From [31], with permission

photosynthate would have been used in local growth and possibly laid down in short-term storage. Short-term storage of carbohydrate has an important role in buffering changes in source supply and sink utilisation [17, 18].

25.4.2 Sink Function

The flux of photosynthate into a sink depends upon the sink's ability to utilise this photosynthate. Treatments altering a sinks importing capacity gives information on the importing mechanisms within the sink, but this requires in vivo measurement of import so as to observe the 'before' and 'after' treatment flows. This is an application to which ^{11}C is well suited. Plant roots are a very convenient sink to study because roots can be grown hydroponically, making it very easy to apply chemical and physical treatments. In addition, a hydroponic root can be separated into two fractions, allowing one fraction to be used as a control while treatments are applied to the other (Fig. 25.2).

An example of experimentally altering the importing capacity of a sink is the application of galactose. This is a naturally occurring plant sugar, which when applied to the roots of barley seedlings immediately induced a huge, but temporary, increase in import of labelled photosynthate into the roots (Fig. 25.4). Root elongation showed a similar response. No other sugars tested were able to similarly enhance import [4]. Only the Poaceae (grasses) reacted this way: all other species tested showed a decline in root import upon application of galactose. The specificity of the large increase in phloem transport was explained by the unique biochemistry of cell wall formation in Poaceae, where the applied galactose results in transient starvation of glucose. This induced a large demand for phloem import [31].

25.4.3 Sink Competition

A plant has several sinks in competition for available photosynthate from a common source. For example in a barley seedling, the roots and the shoot compete for

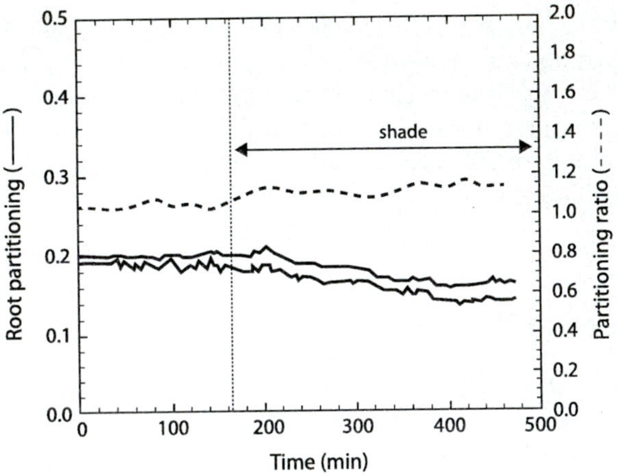

Fig. 25.5 Estimated partitioning of ^{11}C photosynthate to each of the root halves of a split root barley seedling. The shoot was shaded from the time indicated by the dotted vertical line. From [24], with permission

available carbohydrate. Barley plants usually have 7 seminal roots, each originating at the base of the shoot and competing with one another. When the root is split into 2 parts, say 4 and 3 seminal roots respectively, each part receives an amount of photosynthate in proportion to the amount of the total root in each fraction. When the shoot is shaded, so that there is a reduced supply of photosynthate, the fraction of the available photosynthate supplied to the roots is reduced (Fig. 25.5) with the ratio of partitioning fractions to each root part being unchanged. That is, each root fraction experiences equal fractional change in photosynthate supply. This has been described by the root fractions being equivalent sinks [22], in that they behave similarly to a change in available photosynthate. This is not unexpected as each root fraction have the same physiology.

The fraction partitioned to the root plus the fraction partitioned to the shoot must sum to unity, so if the partitioning to the root is reduced by shading the shoot, we deduce that the fraction of recently exported photosynthate supplied to the shoot must increase. That is, partitioning of available photosynthate increases to the shoot at the expense of the root. The root and shoot sinks are thus in-equivalent sinks, in that they are not treated equally when the supply of photosynthate changes. This probably reflects physiologically differences between the two sink types because of different mechanisms being involved, or possibly the same mechanisms but different isoenzymes involved. This difference in root and shoot response to changes in supply of available photosynthate results from the plant attempting to maintain its physical shoot to root ratio [1]. When photosynthate supply is reduced by shading, the shoot activity has reduced relative to the root activity, so the plant compensates by increasing photosynthate supply to the shoot to support increased shoot growth, to offset reduced shoot activity. Once the shading is removed (i.e. balance is regained), partitioning to the roots returns to its original level. The corresponding phenomenon is seen when the roots are pruned: partitioning to the shoot is temporally reduced in favour of the roots until increased root growth returns the balance. This, and related work, led to the development of a simple mechanistic model of sink competition [22], where the observed priority behaviour was an emergent property. This work demonstrated that photosynthate partitioning is not solely a function of the sink, but a holistic property of the source, the long-distance pathway, and the sink. Extension of this mechanistic model to incorporate further, known, mechanistic complexity [15] did not significantly alter the overall behaviour of the initial minimalist version, or the holistic nature of partitioning.

25.5 Summary

Investigation into the function of phloem vasculature requires experiments to be carried out on whole plants, as invasive techniques result in immediate blockage of the transport system. Radioactive tracers provide a method of in vivo measurement of vascular transport and changes in its function in response to experimental perturbations, without inducing blockage. The most convenient radioactive tracer to label photosynthate is ^{14}C, as this has a long half-life, and so can be bought and used when convenient, but it has very limited use for in vivo measurements, as the emitted radiation can only penetrate about 50 microns of plant tissue. Another isotope of carbon, ^{11}C, is readily observed in vivo but its short-half life makes interpretation of its tracer measurements more difficult, as it never reaches isotopic equilibrium within the experimental plant. Mechanistic modelling has been very successful in interpreting ^{11}C data [2], and able to extract a transport speed and net pathway leakage. Input-output analysis of ^{11}C tracer profiles has enabled non-mechanistic interpretation of changes in shape of a tracer profile as it moves through the phloem system. This has enabled the separation of phloem loading, long distance transport and phloem unloading processes, allowing each to be studied separately, which was not possible using mechanistic approaches or with data collected by destructive tracer measurements (for more detail see [16]). Input-output analysis has enabled mechanistic-free estimates of the distribution of speeds involved (i.e. dispersion), pathway leakage, and partitioning between competing sinks. The variable parameter algorithm has been shown to be a very powerful tool in following changes in these physiological parameters, either natural diurnal changes or those induced by experimental treatments. Analyses that use an instrumental variable and data-filtering (e.g. SRIV) algorithms have proven to be very robust and able to take care of potential issues of parameter bias and heteroscedasticity. Applications of these methods have resulted in the first mechanistic model describing sink interactions within plants, and have shown that sink priority is an emergent property of this sink interaction model.

Acknowledgements I am indebted to Drs J.H. Troughton and W.F. Pickard who encouraged me right from the beginning to develop my understanding of input-output analysis, and to Professor P.C. Young for his encouragement and collaboration for over 30 years. Several of his students have had a big input into both data analysis and development of control systems used in the $^{11}CO_2$ tracer facility. Dr M.R. Thorpe has been a long time collaborator and the sabbatical time of Prof J.F. Farrar spent in our laboratory introduced us to a lot of new plant applications.

References

1. Brouwer, R.: Distribution of dry matter in the plant. Neth. J. Agric. Sci. **10**, 361–376 (1962)
2. Evans, N.T.S., Ebert, M., Moorby, J.: A model for the translocation of photosynthate in the soybean. J. Exp. Bot. **14**, 221–231 (1963)
3. Farrar, J.F., Minchin, P.E.H.: Carbon partitioning in split root systems of barley: relation to metabolism. J. Exp. Bot. **42**, 1261–1269 (1991)
4. Farrar, J.F., Minchin, P.E.H., Thorpe, M.R.: Carbon import into barley roots: stimulation by galactose. J. Exp. Bot. **45**, 17–22 (1994)
5. Farrar, J.F., Minchin, P.E.H., Thorpe, M.R.: Carbon import into barley roots: effects of sugars and relation to cell expansion. J. Exp. Bot. **46**, 1859–1865 (1995)
6. Fondy, B.R., Geiger, D.R.: Effect of rapid changes in sink-source ratio on export and distribution of products of photosynthesis in leaves of *Beta vulgaris L.* and *Phaseolus vulgaris L.* Plant Physiol. **66**, 945–949 (1980)
7. Gifford, R.M., Evans, L.T.: Photosynthesis, carbon partitioning, and yield. Annu. Rev. Plant Physiol. **32**, 485–509 (1981)
8. Gould, N., Minchin, P.E.H., Thorpe, M.R.: Direct measurements of sieve element hydrostatic pressure reveal strong regulation after pathway blockage. Funct. Plant Biol. **31**, 987–993 (2004)
9. Grossman, L., DeJong, T.M.: Peach: A simulation model of reproductive and vegetative growth in peach trees. Tree Physiol. **14**, 329–345 (1994)
10. Heuvelink, E.: Re-interpretation of an experiment on the role of assimilate transport resistance in partitioning in tomato. Ann. Bot. **78**, 467–470 (1996)
11. Jahnke, S., Stocklin, G., Willenbrink, J.: Translocation profiles of 11C-assimilates in the petiole of *Marsilea quadrifolia L.* Planta **153**, 56–63 (1981)
12. Jahnke, S., Menzel, M.I., van Dusschoten, D., Roeb, G.W., Buhler, J., Minwuyelet, S., Blumler, P., Temperton, V.M., Hombach, T., Streun, M., Beer, C., Khodaverdi, M., Ziemons, K., Coenen, H.H., Schurr, U.: Combined MRI-PET dissects dynamic changes in plant structures and functions. Plant J. **59**, 634–644 (2009)
13. Kays, S.J., Goeschl, J.D., Magnuson, C.E., Fares, Y.: Diurnal changes in fixation, transport, and allocation of carbon in the sweet potato using ^{11}C tracer. J. Am. Soc. Horticult. Sci. **112**, 545–554 (1987)
14. Kendall, M.G., Stuart, A.: The Advanced Theory of Statistics, vol. II. Griffen, London (1961)
15. Lacointe, A., Minchin, P.E.H.: Modelling phloem and xylem transport within a complex architecture. Funct. Plant Biol. **35**, 772–780 (2008)
16. Minchin, P.E.H., Troughton, J.H.: Quantitative interpretation of phloem translocation data. Annu. Rev. Plant Physiol. **31**, 191–215 (1980)
17. Minchin, P.E.H., Thorpe, M.R.: Apoplastic phloem unloading in the stem of bean. J. Exp. Bot. **35**, 538–550 (1984)
18. Minchin, P.E.H., Ryan, K.G., Thorpe, M.R.: Further evidence of apoplastic unloading into the stem of bean. Identification of the phloem buffering pool. J. Exp. Bot. **35**, 1744–1753 (1984)
19. Minchin, P.E.H., Thorpe, M.R.: Measurement of unloading and reloading of photo-assimilate within the stem of bean. J. Exp. Bot. **38**, 211–220 (1987)
20. Minchin, P.E.H., Grusak, M.A.: Continuous in vivo measurement of carbon partitioning within whole plants. J. Exp. Bot. **39**, 561–571 (1988)
21. Minchin, P.E.H.: System estimation in plant physiology. In: Young, P.C. (ed.) Concise Encyclopedia of Environmental Systems, pp. 570–579. Pergamon Press, Oxford (1993)
22. Minchin, P.E.H., Thorpe, M.R., Farrar, J.F.: A simple mechanistic model of phloem transport which explains sink priority. J. Exp. Bot. **44**, 947–955 (1993)
23. Minchin, P.E.H., Farrar, J.F., Thorpe, M.R.: Partitioning of carbon in split root systems of barley: effect of temperature of the root. J. Exp. Bot. **45**, 1103–1109 (1994)
24. Minchin, P.E.H., Thorpe, M.R., Farrar, J.F.: Short-term control of root:shoot partitioning. J. Exp. Bot. **45**, 615–622 (1994)

25. Minchin, P.E.H., Lees, M.J., Thorpe, M.R., Young, P.C.: What can tracer profiles tell us about the mechanisms giving rise to them? J. Exp. Bot. **47**, 275–284 (1996)
26. Minchin, P.E.H., Thorpe, M.R., Wunsche, J., Palmer, J.W., Picton, R.F.: Carbon partitioning between apple fruits: short- and long-term responses to available photosynthate. J. Exp. Bot. **48**, 1401–1406 (1997)
27. More, R.D.: Production of short-lived isotopes. In: Minchin, P.E.H. (ed.) Short-Lived Isotope in Biology. Proceedings of an International Workshop on Biological Research with Short-Lived Isotopes, Wellington. DSIR Bulletin, vol. 238 (1985). (Available from P.E.H. Minchin)
28. More, R.H., Troughton, J.H.: Production of ^{11}C with a 3-MeV Van de Graaff accelerator. Int. J. Appl. Radiat. Isot. **23**, 344–345 (1972)
29. Shinpei, M., Shu, F., Hiroshi, U., Noriko, I.S., Tamikazu, K.: A new visualization technique for the study of the accumulation of photoassimilates in wheat grains using $[^{11}C]CO_2$. Appl. Radiat. Isot. **64**, 435–440 (2006)
30. Thorpe, M.R., Minchin, P.E.H.: Continuous monitoring of fluxes of photoassimilate in leaves and whole plants. J. Exp. Bot. **42**, 461–468 (1991)
31. Thorpe, M.R., MacRae, E.A., Minchin, P.E.H., Edwards, C.M.: Galactose stimulation of carbon import into roots is confined to the Poaceae. J. Exp. Bot. **50**, 1613–1618 (1999)
32. Thornley, J.H.M.: Modelling allocation with transport/conversion processes. Silva Fenn. **31**, 341–355 (1997)
33. van Bel, A.J.E.: The phloem, a miracle of ingenuity. Plant Cell Environ. **26**, 125–149 (2003)
34. Wardlaw, I.F.: The control of carbon partitioning. New Phytol. **116**, 341–381 (1990)
35. Wright, C.J.: Interactions between vegetative and reproductive growth. In: Wright, C.J. (ed.) Manipulation of Fruiting. Butterworths, London (1989)
36. Young, P.: Recursive approaches to time series analysis. Bull. Inst. Math. Appl. **10**, 209–224 (1974)
37. Young, P.: Recursive Estimation and Time-Series Analysis. An Introduction. Springer, Berlin (1984)
38. Young, P.C., Minchin, P.E.H.: Environmental time-series analysis: modelling natural systems from experimental time-series data. Int. J. Biol. Macromol. **13**, 190–201 (1991)

Chapter 26
Chaos Theory for Modeling Environmental Systems: Philosophy and Pragmatism

Bellie Sivakumar

26.1 Introduction

There have been two dominant approaches for environmental systems modeling: deterministic and stochastic. According to the deterministic approach, systems can be described fairly accurately by deterministic mathematical equations based on well-known scientific laws, provided sufficient detail can be included to explain the underlying physical processes. According to the stochastic approach, systems do not adhere to any deterministic principles and, thus, can be described only by probability distributions based on probability concepts. Either approach has its own merits for environmental modeling, having solid foundations in scientific principles/philosophies, verifiable assumptions for specific situations, and the ability to provide reliable results. For instance, the deterministic approach has merits considering the 'permanent' nature of the Earth, ocean, and the atmosphere and the 'cyclical' nature of the associated processes, whereas the merits of the stochastic approach lie in the facts that environmental phenomena exhibit 'complex and irregular' structures and that we have only 'limited ability to observe' the detailed variations.

In view of these, the question of whether the deterministic or the stochastic approach is better for environmental modeling is meaningless. Such a question is really a philosophical one that has no general answer, but it is better viewed as a pragmatic one, which has an answer only in terms of specific situations [9]. These

B. Sivakumar (✉)
The University of New South Wales, Sydney, NSW 2052, Australia
e-mail: s.bellie@unsw.edu.au

B. Sivakumar
University of California, Davis, CA 95616, USA
e-mail: sbellie@ucdavis.edu

L. Wang, H. Garnier (eds.), *System Identification, Environmental Modelling, and Control System Design*,
DOI 10.1007/978-0-85729-974-1_26, © Springer-Verlag London Limited 2012

specific situations must be viewed in terms of the system, process, scale, and purpose of interest. For some situations, both approaches may be equally appropriate; for some other situations, the deterministic approach may be more appropriate; and for still others, the stochastic approach. It is also reasonable to contend that the two approaches are complementary to each other, since oftentimes both deterministic and stochastic properties are intrinsic to environmental processes, though scale plays a defining role. For example, there is significant determinism in river flow in the form of seasonality and annual cycle, whereas stochasticity is brought by the interactions of various mechanisms involved and by their different degrees of nonlinearity.

These observations seem to suggest that a coupled deterministic-stochastic approach, incorporating both the deterministic and the stochastic components, would most likely yield better outcomes compared to when either approach adopted independently. Although the need for this combinatorial approach was recognized more than 40 years ago [67], there is not much evidence in the literature that points out to any serious effort to this end. Recently, I have argued (e.g. [42]) that 'chaos theory,' with its three underpinning concepts of (1) nonlinear interdependence, (2) hidden determinism and order, and (3) sensitivity to initial conditions, can bridge the gap between our extreme views of determinism and stochasticity and also offer a balanced and more realistic middle-ground perspective for modeling environmental systems. The appropriateness of these concepts to environmental systems and the potential role chaos theory can play in their modeling may be realized from the following situations: (1) nonlinear interactions are dominant among the components and mechanisms in the hydrologic cycle; (2) determinism and order are prevalent in daily temperature and annual river flow; and (3) contaminant transport in surface and sub-surface waters is highly sensitive to the time at which the contaminants were released. The first represents the 'general' nature of environmental processes, whereas the second and third represent their 'deterministic' and 'stochastic' natures, respectively.

The finding that 'complex and random-looking' behaviors are not necessarily the outcomes of complex systems but can also be from simple nonlinear deterministic systems with sensitivity to initial conditions (i.e. chaos) has far reaching implications in environmental modeling, since most outputs from such systems (e.g. time series of rainfall, river flow, water quality) are typically 'complex and random-looking.' One crucial implication of this finding is the need, first of all, to identify the dynamic nature of the given system towards selecting an appropriate modeling approach, as opposed to the current practice of simply resorting to a particular approach based on certain preconceived notion (determinism or stochasticity) that may or may not be valid. This (i.e. identification of system's dynamic nature), in fact, has been an important objective of most chaos theory studies in environmental systems. Although there is no question that chaos studies have offered important insights about the dynamic nature of environmental systems and their modeling, there continue to be skepticisms and criticisms on such studies on the basis of some potential limitations in chaos concepts and methods (time series based) to real environmental systems (i.e. spatio-temporal) and the associated time series (e.g. small data size, presence of noise, large number of zeros). Some of these criticisms indeed have merits, but others are simply a result of flawed lines of thinking (e.g. [42, 43]).

It should be obvious, nevertheless, that a fundamental understanding of chaos theory is a pre-requisite for an honest assessment on the usefulness of the theory to serve as a bridge between our deterministic and stochastic views for environmental modeling. At the same time, it must also be noted that our knowledge of chaos theory is very limited, which is not surprising considering that it is relatively new to environmental science and engineering, especially when weighed against our long-standing and far more established deterministic and stochastic theories. In view of these, the objectives of this chapter are: (1) to detail the development of chaos theory and identification methods; (2) to review the applications of chaos theory to environmental systems; and (3) to highlight the need for a down-to-earth pragmatic view of the philosophy of chaos theory for a more balanced and middle-ground approach for environmental modeling. The rest of the chapter is organized as follows. First, a brief history of the development of chaos theory is presented. Next, some basic chaos identification methods are described, with examples of their applications to synthetic time series. Then, applications of chaos theory to environmental systems is reviewed, and progress and pitfalls are underlined. Finally, the need for pragmatism in chaos philosophy in environmental systems is highlighted, through analysis of real environmental time series.

26.2 Chaos Theory: A Brief History

[The name] 'chaos theory' may be both enticing and confusing. It is enticing because it brings a fascinating and counterintuitive perspective on 'complex' systems, i.e. revealing simplicity in complexity. It is confusing because of the unusual meaning the word 'chaos' takes in modern scientific literature against its meaning in traditional and common usage.

In common parlance, the word 'chaos' typically means a state lacking order or predictability; in other words, chaos is synonymous to 'randomness.' In modern dynamic systems science literature, however, chaos refers to 'random-looking' determinism with sensitivity to initial conditions; therefore, chaos and randomness are quite different. This latter definition has important implications for system modeling and prediction: randomness is irreproducible and unpredictable, while chaos is reproducible and predictable in the short term (due to determinism) but irreproducible and unpredictable only in the long term (due to sensitivity to initial conditions).

The roots of chaos theory date back to about 1900, in the studies of Henri Poincaré on the problem of the motion of three objects in mutual gravitational attraction, the so-called 'three-body problem.' Poincaré found that there can be orbits which are non-periodic, and yet not forever increasing nor approaching a fixed point. Despite this interesting finding, chaos theory remained in the background during the entire first half of the twentieth century, perhaps due to lack of computational power. However, the invention of high-speed computers in the 1950s changed this situation for the better, as computers allowed experimentation with equations in a way that was impossible before, especially the process of repeated iteration of mathematical formulas to study nonlinear dynamic systems.

Such experiments led to Edward Lorenz's discovery, in 1963, of chaotic motion on a 'strange attractor' [24]. Lorenz studied a simplified model of convection rolls in the atmosphere to gain insight into the notorious unpredictability of the weather. He found that the solutions to his equations never settled down to equilibrium or to a periodic state; instead, they continued to oscillate in an irregular, aperiodic fashion. Moreover, when the simulations were started from two slightly different initial conditions, the resulting behaviors became totally different. The implication was that the system was inherently unpredictable—tiny errors in measuring the current state of the atmosphere would be amplified rapidly. But Lorenz also showed that there was structure (in the chaos)—when plotted in three dimensions, the solutions to his equations fell onto a butterfly-shaped set of points.

The 1970s witnessed the main developments in chaos theory. Ruelle and Takens [36] proposed a new theory for the onset of turbulence in fluids, based on abstract consideration about 'strange attractors.' May [25] found examples of chaos in iterated mappings arising in population biology, and stressed on the pedagogical importance of studying simple nonlinear systems, to counterbalance the often misleading linear intuition fostered by traditional education. Study of other simple nonlinear mathematical models, such as the Henon map [14] and the Rössler system [35], also revealed the hidden beauty of chaos. Beautiful 'strange attractors' that described the final states of these systems were produced and studied, and routes that lead a dynamic system to chaos were discovered. Feigenbaum [8] discovered that there are certain universal laws governing the transition from regular to chaotic behavior; completely different systems can go chaotic in the same way. His work established a link between chaos and phase transitions. The study of chaos then moved to the laboratory. Ingenious experiments were set up and chaotic behavior was studied in fluids, mechanical oscillators, and semiconductors (e.g. [22, 58, 60]). The experiments elevated chaos theory from being just a mathematical curiosity and established it as a physical reality.

The positive outcomes from these laboratory experiments encouraged search for chaos outside the 'controlled' space—in Nature. However, the search also presented an enormous challenge, as one had to deal with an 'uncontrolled' system whose mathematical formulation was not always known accurately. Nevertheless, advances in computational power and measurement technology facilitated development, in the 1980s, of a new set of mathematical techniques for chaos identification and prediction. Understandably, most of these techniques were based on (or designed for) time series, with concepts of data reconstruction, dimensionality, entropy, and predictability (e.g. [7, 12, 13, 27, 59, 66]).

Since their developments, the techniques have been employed for identification and prediction of chaos in many real systems, including atmospheric, biological, ecological, economic, engineering, environmental, financial, political, and social. The studies are already too numerous to list, and also growing by the day. Examples of notable books on chaos theory and its applications are those by Tsonis [62], Strogatz [57], Abarbanel [1], and Kantz and Schreiber [18]. For a more general and non-mathematical description of chaos theory, the reader is referred to [10] and, to some extent, [11].

26.3 Identification of Chaos

26.3.1 Limitations of Linear Tools

In the analysis of time series for identification of system properties, it is customary to use autocorrelation function (ACF) and power spectrum. The autocorrelation function characterizes the dynamic properties of a time series through determination of the degree of dependence present in the values. For a purely stochastic process, the ACF fluctuates randomly about zero, indicating that the process at any certain instance has no 'memory' of the past at all. For a periodic process, the ACF is also periodic, indicating the strong relation between values that repeat over and over again. For signals from a chaotic process, the ACF is expected to decay exponentially with increasing lag, because the states are neither completely dependent nor completely independent of each other. The power spectrum is particularly useful for identifying the regularities/irregularities in a time series. For a purely stochastic process, the power spectrum oscillates randomly about a constant value, indicating that no frequency explains any more of the variance of the sequence than any other frequency. For a periodic or quasi-periodic sequence, only peaks at certain frequencies exist, measurement noise adds a continuous floor to the spectrum and, thus, in the spectrum, signal and noise are readily distinguished. Chaotic signals may also have sharp spectral lines but even in the absence of noise there will be continuous part (broadband) of the spectrum, which is an immediate consequence of the exponentially decaying autocorrelation function.

Although autocorrelation function and power spectrum provide compelling distinctions between stochastic and periodic (or quasi-periodic) signals, they are not reliable for distinguishing between stochastic and chaotic signals. This is demonstrated herein through their application to two artificially generated time series (Fig. 26.1(a) and (b)) that look very much alike (both 'complex' and 'random') but nevertheless are the outcomes of systems (equations) possessing significantly different dynamic characteristics. The first series (Fig. 26.1(a)) is the outcome of a pseudo random number generation function:

$$X_i = \mathrm{rand}() \tag{26.1}$$

which yields independent and identically distributed numbers (between 0 and 1). The second (Fig. 26.1(b)) is the outcome of a fully deterministic two-dimensional map [14]:

$$X_{i+1} = a - X_i^2 + bY_i; \qquad Y_{i+1} = X_i, \tag{26.2}$$

which yields irregular solutions for many choices of a and b, but for $a = 1.4$ and $b = 0.3$, a typical sequence of X_i is chaotic.

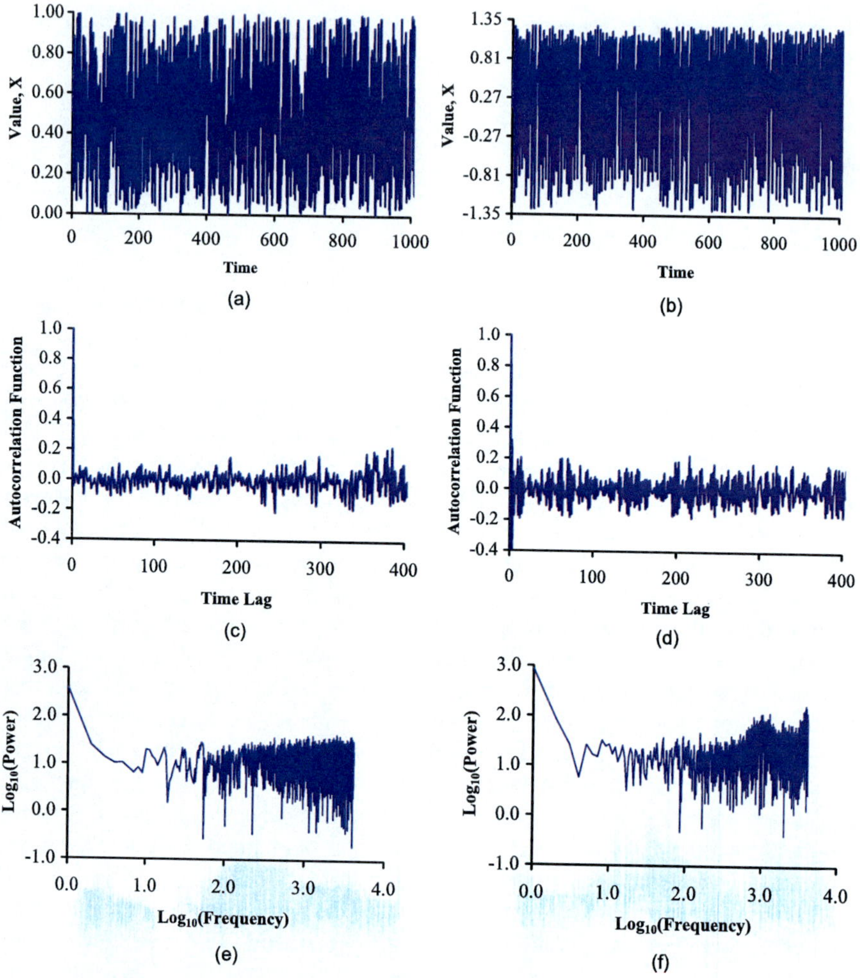

Fig. 26.1 Random vs. chaotic data: (**a**) and (**b**) time series; (**c**) and (**d**) autocorrelation function; and (**e**) and (**f**) power spectrum

Figure 26.1(c) and (d) shows the ACFs for these two series, while the power spectra are presented in Fig. 26.1(e) and (f). It is clear that both tools fail to distinguish between the two series. The failure is not just qualitative, but also quantitative: for both series, the time lag at which ACF first crosses zero is equal to 1 (no exponential decay for the chaotic series) and the spectral exponent is equal to 0 (pure randomness in the dynamics of both). Therefore, it is fair to say that linear tools may not be sufficient for characterization of real systems, as such systems are inherently nonlinear and are often sensitively dependent on initial conditions. In what follows, two simple nonlinear tools, namely phase space reconstruction and correlation dimension, are explained and their superiority over the above linear tools demonstrated.

26.3.2 Phase Space Reconstruction

Phase space is essentially a graph, whose coordinates represent the variables necessary to completely describe the state of the system at any moment (e.g. [27]). The trajectories of the phase space diagram describe the evolution of the system from some initial state and, hence, represent the history of the system. The 'attractor' of these trajectories in the phase space provides at least important qualitative information on the system properties.

Given a single-variable time series, X_i, where $i = 1, 2, \ldots, N$, a multi-dimensional phase space can be reconstructed as:

$$\mathbf{Y}_j = (X_j, X_{j+\tau}, X_{j+2\tau}, \ldots, X_{j+(m-1)\tau}), \tag{26.3}$$

where $j = 1, 2, \ldots, N - (m-1)\tau$; m is the dimension of the vector $\mathbf{Y}_j$, called embedding dimension; and τ is the delay time. A correct phase space reconstruction in a dimension m allows interpretation of the system dynamics in the form of an m-dimensional map, f_T, as:

$$\mathbf{Y}_{j+T} = f_T(\mathbf{Y}_j), \tag{26.4}$$

where $\mathbf{Y}_j$ and $\mathbf{Y}_{j+T}$ are vectors of dimension m, describing the state of the system at times j (current state) and $j + T$ (future state), respectively. With (26.4), the task is basically to find an appropriate expression for f_T (e.g. F_T) to predict the future.

To demonstrate its utility for system identification, Fig. 26.2(a) and (b) presents the phase space plots for the above two series. These diagrams correspond to reconstruction in two dimensions ($m = 2$) with delay time $\tau = 1$, i.e. the projection of the attractor on the plane $\{X_i, X_{i+1}\}$. For the first set, the points are scattered all over the phase space (i.e. absence of an attractor), a clear indication of a 'complex' and 'random' nature of the underlying dynamics and potentially of a high-dimensional system. On the other hand, the projection for the second set yields a very clear attractor, indicating a 'simple' and 'deterministic' (yet non-repeating) nature of the underlying dynamics and potentially of a low-dimensional system.

26.3.3 Correlation Dimension

The dimension of a time series is, in a way, a representation of the number of dominant variables present in the evolution of the corresponding dynamic system. Correlation dimension is a measure of the extent to which the presence of a data point affects the position of the other points lying on the attractor in phase space. The correlation dimension method uses the correlation integral (or function) for determining the dimension of the attractor and, hence, for distinguishing between low-dimensional and high-dimensional systems.

Many algorithms have been formulated for the estimation of the correlation dimension of a time series, but the Grassberger-Procaccia algorithm [12] has been the

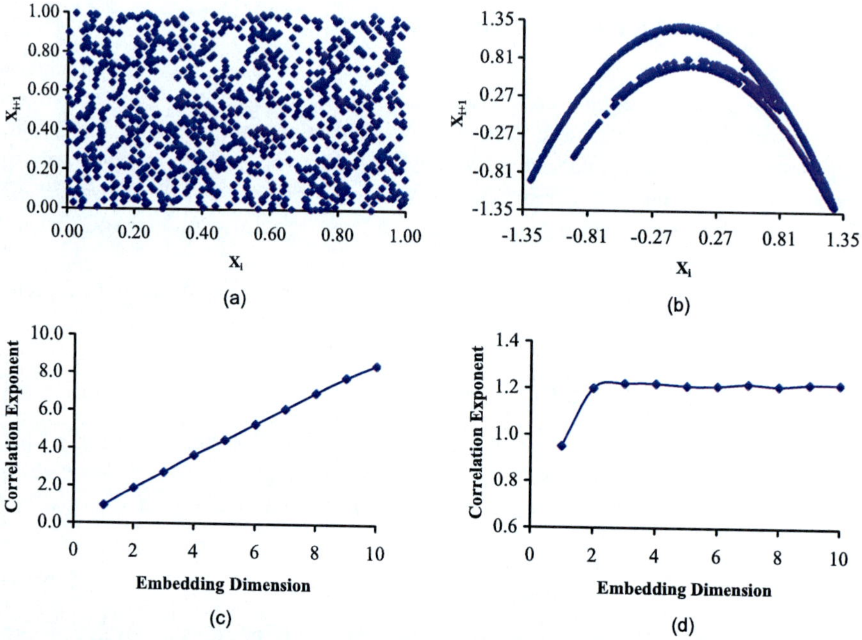

Fig. 26.2 Random vs. chaotic data: (**a**) and (**b**) phase space; (**c**) and (**d**) correlation dimension

most popular. The algorithm uses the concept of phase space reconstruction for representing the dynamics of the system from an available single-variable time series (26.3). For an m-dimensional phase space, the correlation function, $C(r)$, is given by

$$C(r) = \lim_{N\to\infty} \frac{2}{N(N-1)} \sum_{i,j} H(r - |\mathbf{Y}_i - \mathbf{Y}_j|), \quad 1 \le i < j \le N, \tag{26.5}$$

where H is the Heaviside step function, with $H(u) = 1$ for $u > 0$, and $H(u) = 0$ for $u \le 0$, where $u = r - \|\mathbf{Y}_i - \mathbf{Y}_j\|$, r is the vector norm (radius of sphere) centered on $\mathbf{Y}_i$ or $\mathbf{Y}_j$. If the time series is characterized by an attractor, then $C(r)$ and r are related according to:

$$C(r) \approx \alpha r^{\upsilon}, \quad r \to \infty, \quad N \to \infty \tag{26.6}$$

where α is a constant and υ is the correlation exponent or the slope of the $\mathrm{Log}\,C(r)$ versus $\mathrm{Log}\,r$ plot. The slope is generally estimated by a least square fit of a straight line over a certain range of r (scaling regime) or through estimation of local slopes between r values.

The distinction between low-dimensional (and perhaps deterministic) and high-dimensional (and perhaps stochastic) systems can be made using the υ versus m plot. If υ saturates after a certain m and the saturation value is low, then the system is generally considered to exhibit low-dimensional deterministic dynamics. The saturation value of v is defined as the correlation dimension (d) of the attractor, and the nearest integer above this value is generally an indication of the number of variables dominantly governing the dynamics. On the other hand, if v increases without bound with increase in m, the system under investigation is generally considered to exhibit high-dimensional stochastic behavior.

To demonstrate the utility of the dimension concept, Fig. 26.2(c) presents the correlation dimension results for the first set, whereas those for the second set are shown in Fig. 26.2(d). In each case, embedding dimensions from 1 to 10 are used for phase space reconstruction. It is clear that the first set is the outcome of an infinite-dimensional system, i.e. absence of saturation in dimension, whereas the second set is the outcome of a low-dimensional system [with a correlation dimension value of 1.22].

26.3.4 Other Methods

Other chaos identification methods that have found widespread applications include the nonlinear prediction method (e.g. [7]), the false nearest neighbor algorithm [19], the Lyapunov exponent method [66], the Kolmogorov entropy method [13], and the surrogate data method for detection of nonlinearity [61], among others.

The nonlinear prediction method is primarily used for prediction. However, identification of chaos can be made by assessing the prediction accuracy against the parameters involved in the prediction method (embedding dimension, lead time, and neighbors). This is termed as the 'inverse approach' [3, 4]. In the prediction method, the f_T domain in (26.4) is sub-divided into many sub-sets (neighborhoods), each of which identifies some approximations, F_T, valid only in that sub-set. In this way, the underlying system dynamics are represented step by step locally in the phase space. The false nearest neighbor algorithm provides information on the minimum embedding dimension of the phase space required for representing the system dynamics. Lyapunov exponents are the average exponential rates of divergence or convergence of nearby orbits in the phase space. Kolmogorov entropy is the mean rate of information created by the system. The surrogate data method involves generation of substitute data in accordance to the probabilistic structure underlying the original data and rejection of the null hypothesis that the original data have come from a linear stochastic process.

26.4 Environmental Applications

'Environmental systems' generally refer to all living and non-living systems that naturally occur on Earth. Consequently, they may encompass a wide range (depending upon the context, nature, purpose, and scale of interest), and include atmospheric, biologic, ecologic, geographic, geologic, hydrologic, and any of their combinations (e.g. biogeosciences; ecohydrology; hydrogeology). Chaos theory has found numerous applications in each and every one of these systems/sub-systems, and any attempt to list all such studies is next to impossible. The focus herein is limited only to applications to hydrologic systems, including some river-related studies that may also fall within the area of geomorphology. However, such studies are equally applicable to, or at least can be interpreted for, other systems as well.

There already available in the literature are extensive review studies on chaos theory applications to hydrologic systems (and geophysical systems at large), including details of complexity, nonlinearity, and chaos, and their sources and roles (e.g. [28, 29, 39, 42, 44]). A journal special issue exclusively focusing on the status and future challenges in the study of nonlinear deterministic dynamics in hydrologic systems has also been published [55]. Therefore, for reasons of overlaps/repetitions and space constraints, details of chaos applications in hydrology are not reported herein. Only a brief account of the general developments of chaos theory in hydrology is presented.

Studies on chaos theory applications to hydrologic systems started in the late 1980s, and have been growing ever since. Very early applications focused mainly on the identification and prediction of chaos in rainfall, river flow, and lake volume time series (e.g. [2, 32, 34, 65]). Subsequently, chaos theory was applied for other purposes, such as scaling and data disaggregation, missing data estimation, and reconstruction of system equations (e.g. [6, 40, 49, 77]), and other processes, such as rainfall-runoff and sediment transport (e.g. [41, 45, 48]). They also addressed the important issues perceived to influence the outcomes of chaos methods when applied to real hydrologic data, including data size, data noise, and zero values (e.g. [30, 31, 40, 46, 47, 50, 52, 64]). Further, they compared chaos theory with others (e.g. stochastic methods, neural networks) for prediction (e.g. [17, 23, 51, 52]).

During the last few years, studies have applied chaos theory to either advance the earlier ones or address yet other hydrologic processes and problems, including groundwater contamination, parameter estimation, and catchment classification (e.g. [5, 15, 16, 33, 53, 54, 56]), while at the same time also continuing investigations into the potential problems with chaos identification methods (e.g. [20, 43]). Recent and current applications include the assessment of rainfall dynamic behavior under impacts of climate change (e.g. [21]).

26.5 Progress and Pitfalls

The above review makes it abundantly clear that there has been a noticeable progress in the applications of chaos theory to environmental systems, despite the fact that the theory is still in a fairly exploratory stage when compared to the far more established deterministic and stochastic theories. The inroads we have made in recent years in the areas of scaling, groundwater contamination, parameter estimation, catchment classification, and climate change, are particularly significant, since these are arguably among the most important topics in environmental studies at the current time. Therefore, there is every reason to believe that chaos theory applications to environmental systems will continue to grow.

The review further brings to light some important merits of chaos theory in the study of environmental systems. First, in the absence of knowledge of system equations (deterministic theories require system equations), chaos theory offers a more simplified view of environmental phenomena when compared to the view offered by stochastic theories. Second, chaos theory has been found to provide better results than some other theories (stochastic theories, neural networks) in environmental predictions, especially in the short-term, although this cannot be generalized. Third, with its fundamental concepts of nonlinear interdependence, hidden order and determinism, and sensitivity to initial conditions, chaos theory can connect the deterministic and stochastic theories and serve as a more reasonable middle-ground between these two dogmatic and extreme views of nature.

There, however, also remain critical challenges. Among them, two general ones are noteworthy: (1) improving our understanding of the largely less-understood chaos concepts and methods for environmental applications; and (2) finding ways to integrate the chaos concepts with one or more other scientific concepts towards better environmental modeling and forecasting. The former is important to avoid 'blind' applications of chaos methods to real environmental systems, which are often constrained in terms of data quantity and quality; it will also help to more accurately interpret the outcomes of such methods and eliminate 'false' claims. The latter is important for taking advantage of the merits of different approaches for their 'collective utility' to solve environmental problems rather than their 'individual brilliance' as perceived.

Notwithstanding this progress and promise, one cannot ignore the potential limitations in the applications of chaos theory to environmental systems. The limitations are system-dependent and data-dependent, as is normally the case with any other theory. Therefore, it is hard to point out here each and every instance of when, where, how, and why they occur. However, some common system and data properties that are likely to give rise to problems in the applications of these methods and interpretations of the outcomes (and vice-versa) may be more easily identified. A few of such problems/inadequacies are highlighted next.

A significant majority of chaos studies in environmental systems are essentially applications of methods based on reconstruction of single-variable time series, such as the correlation dimension method. Although these methods provide useful information, they are still, at best, crude 'one-dimensional' approximations to the complex three- and four-dimensional spatio-temporal environmental problems. What is also required, therefore, are methods that have more fundamental conceptualization of environmental systems and processes. This would certainly offer avenues to establish important links that may exist between the data (observed at specific spatial and temporal scales) and the system physics (occurring across all scales). A similar concern is also on the lack of explanation on the connection between the specific parameters used in the chaos methods and the components of the environmental system under study. For example, is the delay time in the embedding procedure related to any system component, and how?

A fundamental assumption in the formulation of chaos identification methods is that the time series is infinite and noise-free. This assumption has, in fact, been the basis for much of the criticisms on chaos studies in environmental systems, since data from such systems are not only finite but also often short and are always contaminated with noise. A more specific argument is this. When the data size is smaller than the minimum required (the minimum data size is often linked to embedding dimension or correlation dimension) or when the noise is higher than a certain level (even as low as 2 percent), the methods may yield inaccurate estimation of the invariants. For example, the correlation dimension may be underestimated when data size is small (e.g. [26]), while it may be overestimated when the noise level is high (e.g. [38]). This means that the outcomes may indicate the presence of chaotic behavior when actually it is absent, and vice-versa.

Chaos identification methods are generally designed for data that are regularly sampled, i.e. equal sampling interval. In certain situations, however, observations of environmental systems/processes are (or can be) made only at irregular sampling intervals. These situations may be necessitated by the measurement technology available, measurement cost, human resources, and other relevant factors. Therefore, applications of chaos methods to such data sets are fraught with difficulties, especially since the essential first step of chaos analysis involves the delay embedding phase space reconstruction procedure (e.g. [59]), wherein the delay time is often taken as a suitable multiple of the sampling time.

Studies reveal that the presence of a large number of zeros in a time series could significantly influence the outcomes of chaos identification methods, such as underestimation of the correlation dimension (e.g. [20, 40, 63]). This problem can turn out to be very serious in environmental applications, since zero values are a common occurrence in environmental time series (e.g. rainfall, flow), especially at finer resolutions. The fact that zero values are intrinsic to the system dynamics and, thus, must not be removed in data analysis (possible exceptions may exist, such as in data disaggregation; see [49]) makes the problem only more complicated. It is also important to recognize that this problem is not just limited to zeros but can be a much wider one, since it is simply a question of 'repetition' of one or more values and that such repetitions may occur in many different ways depending on the system (e.g. minimum streamflow, average temperature, minimum/maximum water level in a reservoir, daily suspended sediment load).

Many of the ideas and methods of chaos theory attempt to represent the 'dimensionality' of the system (more specifically, time series) under consideration. This, in turn, is used to identify the nature of system dynamics and to select an appropriate type of model (chaotic or stochastic), among other purposes. The dimensionality of the time series is also often linked to the 'complexity' of the system. However, the definitions of 'dimensionality' and 'complexity' as well as the relationships between them are often hazy, to put it mildly. This situation only complicates the interpretation of the outcomes of chaos identification methods.

26.6 Philosophy and Pragmatism

It is clear, from the discussion so far, that chaos theory is appropriate, and can even be better than other theories, for modeling environmental systems, but at the same time also possesses some potential limitations. Consequently, while there have been appreciations and more applications, there have been skepticisms and criticisms as well (e.g. [20, 37]). Some of these skepticisms and criticisms indeed have merits, but others are simply a result of flawed lines of thinking (see [39, 42, 43, 50, 52] for details). In view of these, a careful discussion balancing the philosophy of chaos theory and the pragmatism of its applications to environmental systems is necessary. Comparisons of the limitations of chaos theory with those of the other theories for environmental systems could also help put this discussion on an even stronger footing.

Almost every environmental system/process exhibits unique characteristics. For example, the characteristics of a mountainous terrain are entirely different from those of a flat terrain, and so are the process generating mechanisms and the magnitudes of events. Further, in a general sense, environmental data possess some unique properties when compared to data in some other fields. For example, rainfall and streamflow in arid and semi-arid areas typically contain a significantly large proportion of zeros, a property that is not commonly observed in most other natural and physical systems. Consequently, some modeling approaches and techniques that perform well for other systems may not be appropriate for environmental systems. In view of this, the wise thing to do is to exercise utmost caution in applying chaos theory (and other theories) to environmental systems and in interpreting the results.

Although there are some potential limitations in the applications of chaos methods to real systems, that alone should not preclude any and all chaos studies in environmental systems. It is also important to note that almost all of such limitations are equally applicable to most other modeling approaches/techniques for environmental systems as well. Some examples are as follows: (1) many of the methods, including stochastic time series ones, are still crude 'one-dimensional' approximations to the complex three- and four-dimensional spatio-temporal environmental problems; (2) stochastic methods generally require long environmental time series to yield reliable results; (3) a log-normal analysis is not appropriate, or at least very challenging, for environmental time series consisting of zero values; and (4) despite the sophistication, stochastic methods are still not able to establish links between data and system physics, and many of the parameters in such methods have no relevance to the system characteristics and physics at all.

In view of these observations, any strict adherence to the assumptions of chaos identification methods (e.g. infinite and noise-free time series) may not be helpful for studying environmental systems, just as the situation with respect to other approaches. What is needed instead is a more down-to-earth pragmatic view of chaos studies, emphasizing the following: (1) understanding the potential limitations of chaos identification methods; (2) careful consideration of the environmental system, data, and problem for study; and (3) utmost caution in applications of chaos methods and interpretation of the results. It must be noted that such a pragmatic approach has indeed been advocated/followed in many, if not all, of the chaos studies in environmental systems, with clear indications of their positives and negatives (e.g. [39, 42–44, 49, 52, 56]).

As mentioned earlier, chaos theory, with its underpinning concepts, can bridge the gap between our extreme views of determinism and stochasticity and offer a balanced middle-ground perspective for modeling environmental systems. In what follows, this point is studied from a different angle. The basic idea in this is that environmental systems can exhibit a whole range of behaviors, anywhere from purely deterministic to purely stochastic, but oftentimes between these two extremes, i.e. chaotic. This is supported through analysis of four river flow time series, representing different geographic regions, climatic conditions, catchment characteristics, and scales. These four time series are: daily flows from the Mississippi River and from the Kentucky River in the USA, and monthly flows from the Salmon River in the USA and from the Göta River in Sweden. A brief account of these four series is presented next, followed by their analysis and results.

The Mississippi River is one of the largest rivers in the world, and flow data are measured at numerous locations. The present study considers a sub-basin station at St. Louis, Missouri (USGS station No. 07010000), which is situated at 38°37′03″ latitude and 90°10′58″ longitude. The drainage area of this sub-basin that falls within the Mississippi River basin is 251,230 km^2. For the present analysis, daily flow spanning a period of 20 years (January 1, 1961–December 31, 1980) are considered. The Kentucky River is a tributary of the Ohio River, and has a drainage area of about 18,000 km^2. For the present study, daily flow observed at the gaging station near Winchester, Kentucky (USGS station No. 03284000) are considered. This station is situated at 37°53′41″ latitude and 84°15′44″ longitude. The sub-basin has a drainage area of 10,244 km^2. Flow data observed over a period of 30 years (January 1, 1960–December 31, 1989) are analyzed. The Salmon River basin is situated in the state of Idaho, at 44°59′13″ latitude and 115°43′30″ longitude. The drainage area of this basin is 35,094 km^2. For the present study, monthly flow over a period of 62 years (1932–1993) are considered. Consistent with the 'water years', the records start in October 1931 and end in September 1993 and are average monthly values. The Göta River basin is located in the south of Sweden between 55° and 60°N and 12.9° and 16°E. The drainage area is 50,132 km^2. For this study, monthly flow over a period of 131 years (January 1807–December 1937) are considered.

Figure 26.3(a) to (d) presents the time series plots of the four data sets. All the four series exhibit highly 'irregular and complex' structures, and neither are they helpful to make distinctions nor do they offer clues on the nature of the dynamics. Therefore, it is not possible to construe whether the flow series are the outcomes of deterministic dynamic systems or stochastic ones. However, these time series plots reveal important information about some other characteristics, such as extreme events ('peaks' and 'dips') and/or annual cycles.

Figure 26.4(a) to (d) shows the autocorrelation function plots for the four flow series. The ACFs generally indicate slow decays with increasing lag (i.e. temporal persistence), suggesting that the dynamic properties of the underlying systems are certainly not stochastic. However, it is also difficult to construe that the dynamics are deterministic. Although the ACFs reveal certain periodicity and/or annual cycle (especially for the Salmon River), such do not provide any clues as to whether the underlying dynamic properties are deterministic or stochastic in nature. The lag times at which the ACF first crosses the zero line are found to be 198 and 95 for the daily flow series from the Mississippi River and the Kentucky River, and 3 and 20

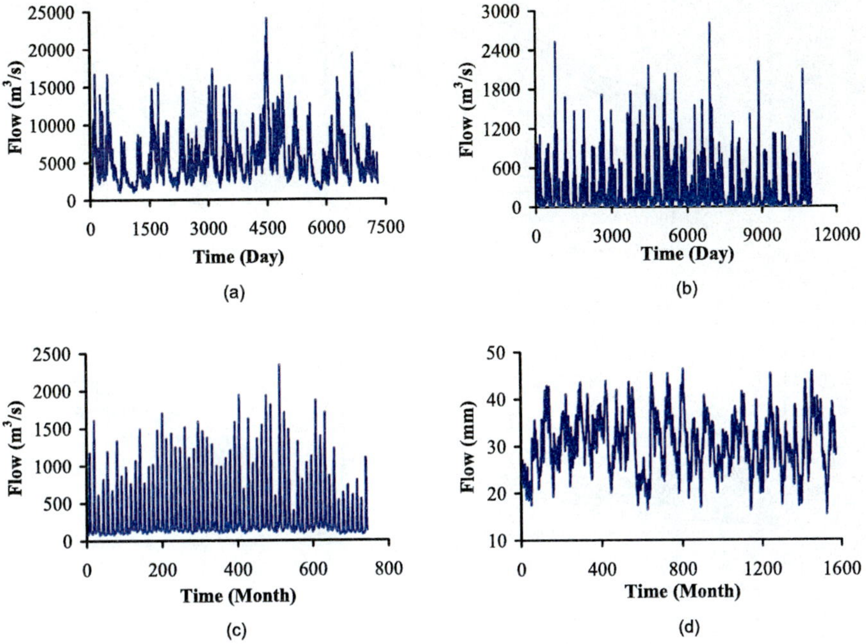

Fig. 26.3 Time series: (**a**) daily flow from the Mississippi River at St. Louis, Missouri, USA; (**b**) daily flow from the Kentucky River near Winchester, Kentucky, USA; (**c**) monthly flow from the Salmon River in Idaho, USA; and (**d**) monthly flow from the Göta River in Sweden

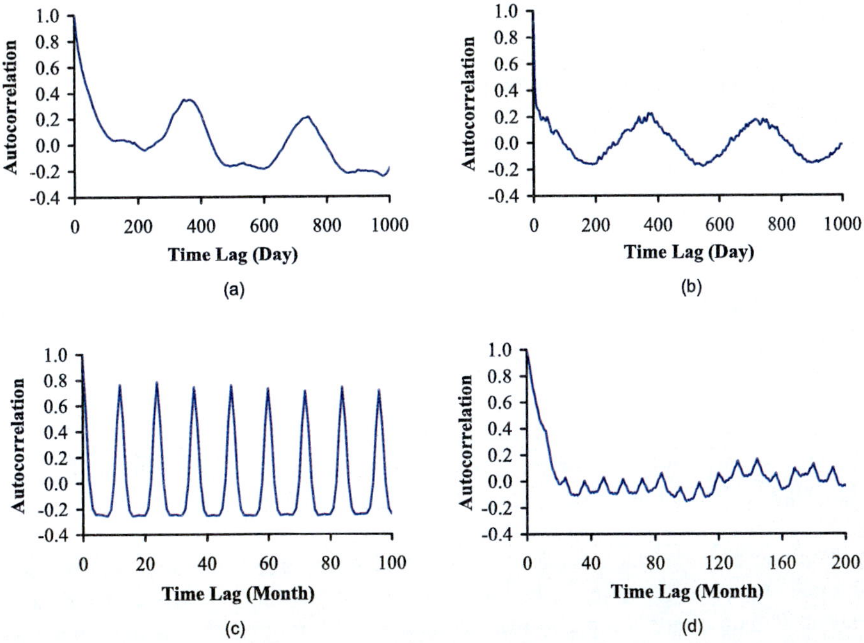

Fig. 26.4 Autocorrelation functions for the river flow series in Fig. 26.3

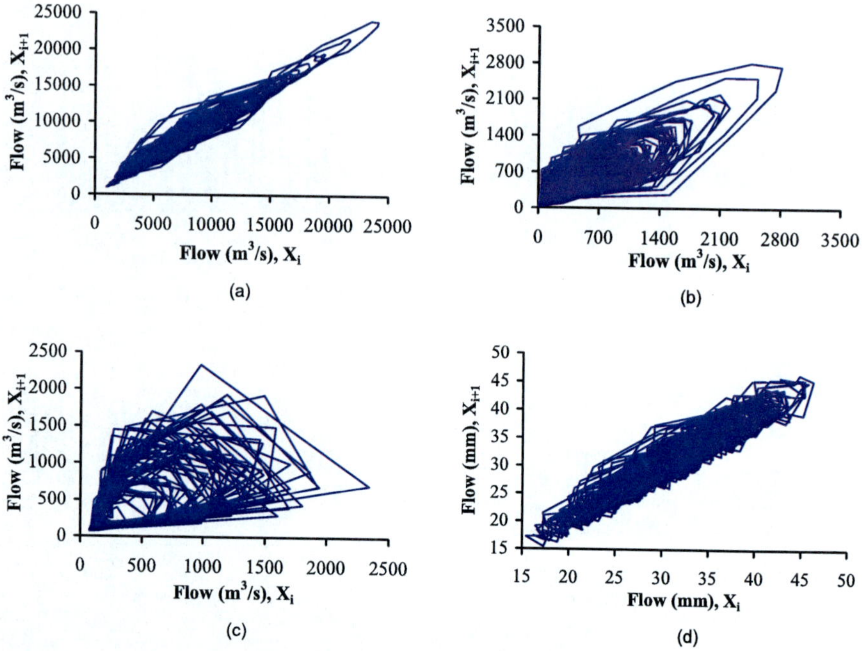

Fig. 26.5 Phase space diagrams for the river flow series in Fig. 26.3

for the monthly flow series from the Salmon River and the Göta River, respectively. These values seem to suggest some seasonal patterns in the flow dynamics, but even such an interpretation may be valid only for the first three rivers (six or three months, as the case may be), since the time period for the Göta River is almost as long as two years.

Figure 26.5(a) to (d) presents the two-dimensional phase space plots for the four flow series. For the Mississippi River flow series, the phase space diagram exhibits a clear attractor in a well-defined region, suggesting that the system dynamic properties are simple and certainly not stochastic. The phase space diagram for the Kentucky River flow series also shows a reasonably clear attractor, though not as clear as that for the Mississippi River flow series; this seems to suggest that the underlying dynamics may be simple and are certainly not stochastic. The flow series from the Salmon River also seems to exhibit a reasonably clear attractor, although the region of attraction is much larger for the dynamics to be simple; however, it is also difficult to say that the dynamics are stochastic. The phase space diagram for the Göta River flow series shows a clear attractor in a well-defined region, suggesting that the underlying dynamics are simple and certainly not stochastic.

The phase space diagrams for the four river flow series generally indicate the absence of stochastic nature in the underlying system dynamics. They also indicate that the dynamics are 'simple' (or less complex) in all four cases, although the 'extent of simplicity' certainly varies, with the flow dynamics in the Mississippi River

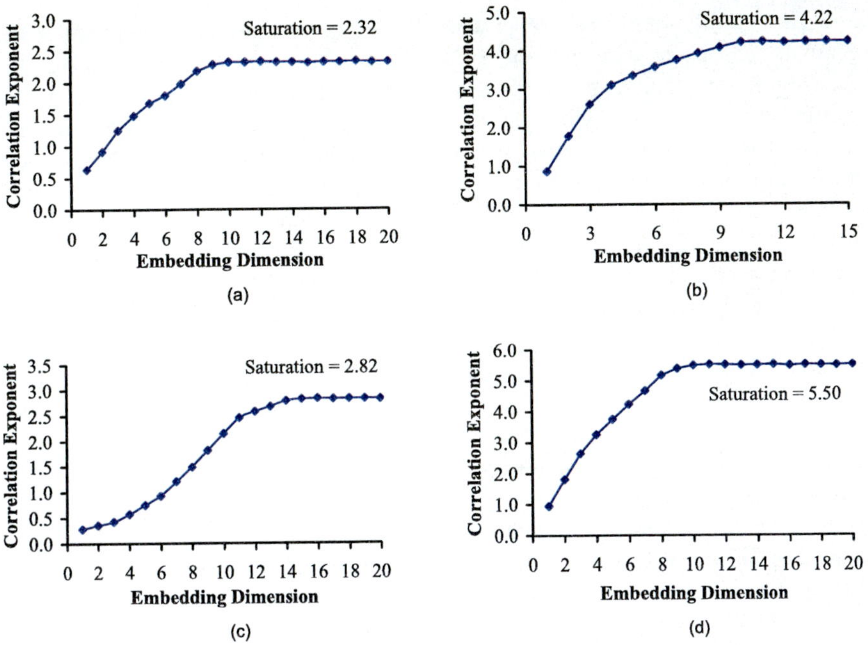

Fig. 26.6 Correlation dimension results for the river flow series in Fig. 26.3

and in the Göta River being simpler than the other two, and with the Salmon River flow dynamics being the most complex of all. These results are indeed encouraging, since they seem to suggest that simpler models may be sufficient, and may be preferred to complex models. At the same time, however, they also do not offer any convincing information to construe that the flow dynamics are deterministic; after all, 'simple' does not (necessarily) mean 'deterministic.'

Figure 26.6(a) to (d) shows the correlation dimension results for the four flow series. Saturation in correlation exponent is observed for all the four series, suggesting that the dynamics are certainly not stochastic. Further, the correlation dimension value is found to be small (less than 6) for all the series, suggesting the low-dimensional, and simple to medium-complexity, nature of the underlying dynamics. Looking closely at the correlation dimension values, the Mississippi River flow dynamics ($d = 2.32$) and the Salmon River flow dynamics ($d = 2.82$) seem simpler than the others, each dominantly governed by just three variables. The Göta River flow dynamics, on the other hand, seem to have the highest level of complexity ($d = 5.50$), with the number of dominant governing variables being six. The flow dynamics in the Kentucky River shows a level of complexity somewhere between these two cases (with $d = 4.22$), with an indication that there are five dominantly governing variables. For any of these four series, while the observation that only a small number of variables dominantly govern the flow dynamics is encouraging (from the viewpoint of model complexity), there is no definitive evidence to say that the dynamics are indeed deterministic.

Although the phase space reconstruction and correlation dimension results are generally consistent with each other (i.e. absence of stochastic dynamics, simple- to medium-complexity, no definitive evidence for determinism), there are some inconsistencies as well. For example, the phase space reconstruction suggests that the Salmon River flow series is the most complex among the four, but the correlation dimension method indicates that the Göta River flow series has the highest level of complexity. The observations of very low correlation dimension for the Salmon River flow and the very clear attractor for the Göta River flow only raise further questions on the ability of the phase space reconstruction and/or the correlation dimension method to provide definitive conclusions. All these point out that caution needs to be exercised in employing these methods and interpreting the outcomes. When such is done, these nonlinear tools can indeed provide important information on the nature and complexity of the system dynamics, as presented earlier for the synthetic series and also as indicated by the consistent results for the Mississippi River flow series (the clearest attractor as well as the lowest correlation dimension, among the four series).

Nevertheless, the results from the correlation dimension method also help answer a major criticism on chaos studies in real environmental systems, i.e. data size. For example, a correlation dimension value of 5.5 is observed for the Göta River flow series with a data length of as few as 1572 values, which is significantly higher than the correlation dimension of 2.32 for the Mississippi River flow series with a data length of as many as 7305 values. These results clearly reveal that the crucial point in regards to data size is whether the data series is long enough to capture the essential features of the underlying system dynamics, rather than any ill-conceived notion of a relationship between data size and embedding (or correlation) dimension. The present results thus further strengthen the conclusions made, through different means, by earlier studies on the effect of data size (e.g. [39, 43, 50, 52]). It is also relevant to note that none of the four series has zero values, thus also eliminating the possibility of the correlation dimension to be underestimated; for instance, the minimum flow value for Mississippi River is 980 (m^3/s).

26.7 Closing Remarks

During the past two decades or so, chaos theory has been finding increasing applications in environmental systems. However, there have also been continuing skepticisms and criticisms on such studies, on the basis of some potential limitations in chaos identification and prediction methods. This chapter has attempted to offer a balanced view between the philosophy of chaos theory on the one hand and the need for pragmatism in its application to environmental systems on the other. A systematic procedure has been followed, involving: (1) investigation of the reliability of chaos identification methods through their applications to two synthetic time series whose characteristics are known a priori; (2) review of chaos studies in environmental systems, and the progress and pitfalls; and (3) application of chaos methods to four real environmental (river flow) time series and interpretation of the results.

The results for the four river flow time series, using both linear and nonlinear tools, with necessary 'foundations' through analysis of synthetic deterministic and stochastic time series to facilitate interpretations, reveal that the dynamic properties of the systems underlying the flow series are neither deterministic nor stochastic. Rather, the nature of the systems' dynamic properties falls somewhere in between these two extremes, and dominantly governed by three to six variables, depending upon the system. With our knowledge that nonlinearity and sensitive dependence are inherent characteristics of environmental processes and that river flow processes possess, among others, correlation, seasonality, and annual cycle (depending upon the river basin characteristics together with the space and time scales under consideration), it is indeed reasonable to construe that the dynamic properties of the systems underlying the river flow series studied herein (and many others) are clearly a combination of deterministic and stochastic.

In view of these observations, the general question of whether the deterministic approach or the stochastic approach is better for environmental modeling is meaningless. Consequently, any theory that is based purely either on determinism or on stochasticity is probably a misconception of the workings of environmental processes, and only their combination would be appropriate. This is where chaos theory could play a vital role, with its fundamental principles (nonlinear interdependence, hidden determinism and order, and sensitivity to initial conditions) being clearly relevant for environmental systems and processes and also providing a balancing middle-ground approach between the deterministic and the stochastic views. Another important thing to note, especially from the viewpoint of model complexity, is that the question is not whether environmental systems exhibit deterministic dynamics or stochastic dynamics (since they are a combination) but whether a low-dimensional model is sufficient or a high-dimensional model is required. As for the four river flow time series studied herein, the correlation dimension analysis reveals that significant portions of the dynamic complexities of the river systems arise as a result of nonlinear interactions among three to six variables (that are most likely interdependent), depending upon the system. This type of information could play a crucial role in the formulation of a coupled deterministic-stochastic approach.

It is also relevant to note that there are other theories and concepts that may be coupled or integrated with chaos theory to formulate a more general framework for environmental modeling. One such concept is the data-based mechanistic (DBM) modeling concept, introduced by Peter Young. Although the term 'data-based mechanistic modeling' was first used only in the 1990s [73], the basic concepts had been developed over many years before that. For example, the concepts were first introduced in the early 1970s [68] and applied within a hydrologic context with application to river water quality modeling [71]. Since then, the DBM concepts have been strengthened further and also applied to many different systems in diverse areas (e.g. [69, 70]). The DBM concepts take into account many of the salient characteristics of environmental systems and processes, including nonlinearity (e.g. [72]) and simplicity in complexity (e.g. [74, 76]). The DBM approach may also offer, in its own way, a unified view of environmental systems, through combining the deductive and inductive philosophies of environmental modeling [75].

In view of the strengths of chaos theory and the DBM concepts (commonalities as well as differences), coupling the two seems to be a promising way to formulate a much-needed general framework for environmental modeling. This, however, is easier said than done. The fact that researchers on either side of the aisle are not that familiar with the ideas of the other concept certainly complicates this thought. There is also an understandable reluctance among the researchers to couple the two concepts, or any two concepts for that matter. Part of this reluctance may have scientific merits, but other non-scientific factors often play important roles too. It is my hope that the future will witness serious efforts towards formulation of a general framework for environmental modeling. I also hope that such a framework will have, among others, DBM concepts as an important component, which would be a fitting tribute to Peter Young, for all his contributions to the study of environmental systems.

Acknowledgements This work was financially supported in part by the Korean Research Foundation funded by the Korean Government (MEST) (KRF-2009-D0010).

References

1. Abarbanel, H.D.I.: Analysis of Observed Chaotic Data. Springer, New York (1996)
2. Abarbanel, H.D.I., Lall, U.: Nonlinear dynamics of the Great Salt Lake: system identification and prediction. Clim. Dyn. **12**, 287–297 (1996)
3. Casdagli, M.: Nonlinear prediction of chaotic time series. Physica D **35**, 335–356 (1989)
4. Casdagli, M.: Chaos and deterministic versus stochastic nonlinear modeling. J. R. Stat. Soc. B **54**(2), 303–328 (1992)
5. Dodov, B., Foufoula-Georgiou, E.: Incorporating the spatiotemporal distribution of rainfall and basin geomorphology into nonlinear analysis of streamflow dynamics. Adv. Water Resour. **28**(7), 711–728 (2005)
6. Elshorbagy, A., Simonovic, S.P., Panu, U.S.: Estimation of missing streamflow data using principles of chaos theory. J. Hydrol. **255**, 123–133 (2002)
7. Farmer, D.J., Sidorowich, J.J.: Predicting chaotic time series. Phys. Rev. Lett. **59**, 845–848 (1987)
8. Feigenbaum, M.J.: Quantitative universality for a class of nonlinear transformations. J. Stat. Phys. **19**, 25–52 (1987)
9. Gelhar, L.W.: Stochastic Subsurface Hydrology. Prentice-Hall, Englewood Cliffs (1993)
10. Gleick, J.: Chaos: Making of a New Science. Penguin Books, New York (1987)
11. Goerner, S.J.: Chaos and the Evolving Ecological Universe. Gordon and Breach, Langhorne (1994)
12. Grassberger, P., Procaccia, I.: Measuring the strangeness of strange attractors. Physica D **9**, 189–208 (1983)
13. Grassberger, P., Procaccia, I.: Estimation of the Kolmogorov entropy from a chaotic signal. Phys. Rev. A **28**, 2591–2593 (1983)
14. Henon, M.: A two-dimensional mapping with a strange attractor. Commun. Math. Phys. **50**, 69–77 (1976)
15. Hossain, F., Sivakumar, B.: Spatial pattern of arsenic contamination in shallow wells of Bangladesh: regional geology and nonlinear dynamics. Stoch. Environ. Res. Risk Assess. **20**(1–2), 66–76 (2006)
16. Hossain, F., Anagnostou, E.N., Lee, K.H.: A non-linear and stochastic response surface method for Bayesian estimation of uncertainty in soil moisture simulation from a land surface model. Nonlinear Process. Geophys. **11**, 427–440 (2004)

17. Jayawardena, A.W., Gurung, A.B.: Noise reduction and prediction of hydrometeorological time series: dynamical systems approach vs. stochastic approach. J. Hydrol. **228**, 242–264 (2000)
18. Kantz, H., Schreiber, T.: Nonlinear Time Series Analysis. Cambridge University Press, Cambridge (1997)
19. Kennel, M.B., Brown, R., Abarbanel, H.D.I.: Determining embedding dimension for phase space reconstruction using a geometric method. Phys. Rev. A **45**, 3403–3411 (1992)
20. Koutsoyiannis, D.: On the quest for chaotic attractors in hydrological processes. Hydrol. Sci. J. **51**(6), 1065–1091 (2006)
21. Kyoung, M.S., Kim, H.S., Sivakumar, B., Singh, V.P., Ahn, K.S.: Dynamic characteristics of monthly rainfall in the Korean Peninsula under climate change. Stoch. Environ. Res. Risk Assess. **25**(4), 613–625 (2011)
22. Linsay, P.S.: Period doubling and chaotic behaviour in a driven anharmonic oscillator. Phys. Rev. Lett. **47**, 1349–1352 (1981)
23. Lisi, F., Villi, V.: Chaotic forecasting of discharge time series: a case study. J. Am. Water Resour. Assoc. **37**(2), 271–279 (2001)
24. Lorenz, E.N.: Deterministic nonperiodic flow. J. Atmos. Sci. **20**, 130–141 (1963)
25. May, R.M.: Simple mathematical models with very complicated dynamics. Nature **261**, 459–467 (1976)
26. Nerenberg, M.A.H., Essex, C.: Correlation dimension and systematic geometric effects. Phys. Rev. A **42**(12), 7065–7074 (1990)
27. Packard, N.H., Crutchfield, J.P., Farmer, J.D., Shaw, R.S.: Geometry from a time series. Phys. Rev. Lett. **45**(9), 712–716 (1980)
28. Phillips, J.D.: Sources of nonlinearity and complexity in geomorphic systems. Prog. Phys. Geogr. **26**, 339–361 (2003)
29. Phillips, J.D.: Deterministic chaos and historical geomorphology: a review and look forward. Geomorphology **76**, 109–121 (2006)
30. Phoon, K.K., Islam, M.N., Liaw, C.Y., Liong, S.Y.: A practical inverse approach for forecasting of nonlinear time series analysis. J. Hydrol. Eng. **7**(2), 116–128 (2002)
31. Porporato, A., Ridolfi, R.: Nonlinear analysis of river flow time sequences. Water Resour. Res. **33**(6), 1353–1367 (1997)
32. Puente, C.E., Obregon, N.: A deterministic geometric representation of temporal rainfall. Results for a storm in Boston. Water Resour. Res. **32**(9), 2825–2839 (1996)
33. Regonda, S., Sivakumar, B., Jain, A.: Temporal scaling in river flow: can it be chaotic? Hydrol. Sci. J. **49**(3), 373–385 (2004)
34. Rodriguez-Iturbe, I., De Power, F.B., Sharifi, M.B., Georgakakos, K.P.: Chaos in rainfall. Water Resour. Res. **25**(7), 1667–1675 (1989)
35. Rossler, O.E.: An equation for continuous chaos. Phys. Lett. A **57**, 397–398 (1976)
36. Ruelle, D., Takens, F.: On the nature of turbulence. Commun. Math. Phys. **20**, 167–192 (1971)
37. Schertzer, D., Tchiguirinskaia, I., Lovejoy, S., Hubert, P., Bendjoudi, H.: Which chaos in the rainfall-runoff process? A discussion on 'Evidence of chaos in the rainfall-runoff process' by Sivakumar et al. Hydrol. Sci. J. **47**(1), 139–147 (2002)
38. Schreiber, T., Kantz, H.: Observing and predicting chaotic signals: is 2 percent noise too much. In: Kravtsov, Yu.A., Kadtke, J.B. (eds.) Predictability of Complex Dynamical Systems. Springer Series in Synergetics, pp. 43–65. Springer, Berlin (1996)
39. Sivakumar, B.: Chaos theory in hydrology: important issues and interpretations. J. Hydrol. **227**(1–4), 1–20 (2000)
40. Sivakumar, B.: Rainfall dynamics at different temporal scales: a chaotic perspective. Hydrol. Earth Syst. Sci. **5**(4), 645–651 (2001)
41. Sivakumar, B.: A phase-space reconstruction approach to prediction of suspended sediment concentration in rivers. J. Hydrol. **258**, 149–162 (2002)
42. Sivakumar, B.: Chaos theory in geophysics: past, present and future. Chaos Solitons Fractals **19**(2), 441–462 (2004)
43. Sivakumar, B.: Correlation dimension estimation of hydrologic series and data size requirement: myth and reality. Hydrol. Sci. J. **50**(4), 591–604 (2005)

44. Sivakumar, B.: Nonlinear dynamics and chaos in hydrologic systems: latest developments and a look forward. Stoch. Environ. Res. Risk Assess. **23**, 1027–1036 (2009)
45. Sivakumar, B., Jayawardena, A.W.: An investigation of the presence of low-dimensional chaotic behavior in the sediment transport phenomenon. Hydrol. Sci. J. **47**(3), 405–416 (2002)
46. Sivakumar, B., Liong, S.Y., Liaw, C.Y., Phoon, K.K.: Singapore rainfall behavior: chaotic? J. Hydrol. Eng. **4**(1), 38–48 (1999)
47. Sivakumar, B., Phoon, K.K., Liong, S.Y., Liaw, C.Y.: A systematic approach to noise reduction in chaotic hydrological time series. J. Hydrol. **219**(3/4), 103–135 (1999)
48. Sivakumar, B., Berndttson, R., Olsson, J., Jinno, K.: Evidence of chaos in the rainfall-runoff process. Hydrol. Sci. J. **46**(1), 131–145 (2001)
49. Sivakumar, B., Sorooshian, S., Gupta, H.V., Gao, X.: A chaotic approach to rainfall disaggregation. Water Resour. Res. **37**(1), 61–72 (2001)
50. Sivakumar, B., Berndtsson, R., Olsson, J., Jinno, K.: Reply to 'Which chaos in the rainfall-runoff process?' by Schertzer et al. Hydrol. Sci. J. **47**(1), 149–158 (2002)
51. Sivakumar, B., Jayawardena, A.W., Fernando, T.M.G.H.: River flow forecasting: use of phase-space reconstruction and artificial neural networks approaches. J. Hydrol. **265**(1–4), 225–245 (2000)
52. Sivakumar, B., Persson, M., Berndtsson, R., Uvo, C.B.: Is correlation dimension a reliable indicator of low-dimensional chaos in short hydrological time series? Water Resour. Res. (2002). doi:10.1029/2001WR000333
53. Sivakumar, B., Wallender, W.W., Puente, C.E., Islam, M.N.: Streamflow disaggregation: a nonlinear deterministic approach. Nonlinear Process. Geophys. **11**, 383–392 (2004)
54. Sivakumar, B., Harter, T., Zhang, H.: Solute transport in a heterogeneous aquifer: a search for nonlinear deterministic dynamics. Nonlinear Process. Geophys. **12**, 211–218 (2005)
55. Sivakumar, B., Berndtsson, R., Lall, U.: Nonlinear deterministic dynamics in hydrologic systems: present activities and future challenges. Nonlinear Processes. Geophys. (2006)
56. Sivakumar, B., Jayawardena, A.W., Li, W.K.: Hydrologic complexity and classification: a simple data reconstruction approach. Hydrol. Process. **21**(20), 2713–2728 (2007)
57. Strogatz, S.H.: Nonlinear Dynamics and Chaos: With Applications to Physics, Biology, Chemistry, and Engineering. Perseus Books, Cambridge (1994)
58. Swinney, H.L., Gollub, J.P.: Hydrodynamic instabilities and the transition to turbulence. Phys. Today **31**(8), 41–49 (1998)
59. Takens, F.: Detecting strange attractors in turbulence. In: Rand, D.A., Young, L.S. (eds.) Dynamical Systems and Turbulence. Lecture Notes in Mathematics, vol. 898, pp. 366–381. Springer, Berlin (1981)
60. Teitsworth, S.W., Westervelt, R.M.: Chaos and broad-band noise in extrinsic photoconductors. Phys. Rev. Lett. **53**(27), 2587–2590 (1984)
61. Theiler, J., Eubank, S., Longtin, A., Galdrikian, B., Farmer, J.D.: Testing for nonlinearity in time series: the method of surrogate data. Physica D **58**, 77–94 (1992)
62. Tsonis, A.A.: Chaos: From Theory to Applications. Plenum Press, New York (1992)
63. Tsonis, A.A., Triantafyllou, G.N., Elsner, J.B., Holdzkom, J.J. II, Kirwan, A.D. Jr.: An investigation on the ability of nonlinear methods to infer dynamics from observables. Bull. Am. Meteorol. Soc. **75**, 1623–1633 (1994)
64. Wang, Q., Gan, T.Y.: Biases of correlation dimension estimates of streamflow data in the Canadian prairies. Water Resour. Res. **34**(9), 2329–2339 (1998)
65. Wilcox, B.P., Seyfried, M.S., Matison, T.M.: Searching for chaotic dynamics in snowmelt runoff. Water Resour. Res. **27**(6), 1005–1010 (1991)
66. Wolf, A., Swift, J.B., Swinney, H.L., Vastano, A.: Determining Lyapunov exponents from a time series. Physica D **16**, 285–317 (1985)
67. Yevjevich, V.M.: Misconceptions in hydrology and their consequences. Water Resour. Res. **4**(2), 225–232 (1968)
68. Young, P.C.: Recursive approaches to time-series analysis. Bull. Inst. Math. Appl. **10**, 209–224 (1974)
69. Young, P.C.: Data-based mechanistic modeling of environmental, ecological, economic and engineering systems. Environ. Model. Softw. **13**, 105–122 (1998)

70. Young, P.C.: The data-based mechanistic approach to the modelling, forecasting and control of environmental systems. Annu. Rev. Control **30**, 169–182 (2006)
71. Young, P.C., Beck, M.B.: The modelling and control of water quality in a river system. Automatica **10**, 455–468 (1974)
72. Young, P.C., Beven, K.J.: Data-based mechanistic modeling and rainfall-flow non-linearity. EnvironMetrics **5**(3), 335–363 (1994)
73. Young, P.C., Lees, M.J.: The active mixing volume: a new concept in modelling environmental systems. In: Barnett, V., Turkman, K. (eds.) Statistics for the Environment, pp. 3–43. Wiley, Chichester (1993)
74. Young, P.C., Parkinson, S.: Simplicity out of complexity. In: Beck, M.B. (ed.) Environmental Foresight and Models: A Manifesto, pp. 251–294. Elsevier, Oxford (2002)
75. Young, P.C., Ratto, M.: A unified approach to environmental systems modeling. Stoch. Environ. Res. Risk Assess. **23**, 1037–1057 (2009)
76. Young, P.C., Parkinson, S.D., Lees, M.: Simplicity out of complexity: Occam's razor revisited. J. Appl. Stat. **23**, 165–210 (1996)
77. Zhou, Y., Ma, Z., Wang, L.: Chaotic dynamics of the flood series in the Huaihe River Basin for the last 500 years. J. Hydrol. **258**, 100–110 (2002)

Part IV
Control System Design

Chapter 27
Linear and Nonlinear Non-minimal State Space Control System Design

C. James Taylor, Arun Chotai, and Wlodek Tych

27.1 Introduction

Control systems are classically analysed by means of continuous-time or discrete-time Transfer Function (TF) models, represented in terms of either the Laplace transform (s-operator) or the backward shift operator (z^{-1}) respectively. Here, closed-loop stability and satisfactory transient responses are obtained through graphical techniques such as the Evans Root Locus method, which allow for the specification of closed-loop pole-zero patterns [1]. Alternatively, a frequency domain approach is utilised involving Nyquist and Bode diagrams [2].

By contrast, modern control systems are usually derived from precise algorithmic computations, often involving numerical optimisation [3, 4]. Here, one very important concept is the idea of state space, which originates from the state-variable method of describing differential equations. While the TF model approach is concerned only with input-output characteristics, the state space approach also provides a description of the internal behaviour of the system. For mechanical systems, the states are often defined to represent physical characteristics, such as the positions and velocities of a moving body. Alternatively, the state space formulation is derived from a TF model. In this regard, the practical examples mentioned below have

C.J. Taylor (✉)
Engineering Department, Lancaster University, Lancaster, UK
e-mail: c.taylor@lancaster.ac.uk

A. Chotai · W. Tych
Lancaster Environment Centre, Lancaster University, Lancaster, UK

A. Chotai
e-mail: a.chotai@lancaster.ac.uk

W. Tych
e-mail: w.tych@lancaster.ac.uk

L. Wang, H. Garnier (eds.), *System Identification, Environmental Modelling, and Control System Design*,
DOI 10.1007/978-0-85729-974-1_27,

utilised a Matlab compatible toolbox, CAPTAIN, for the identification and estimation of suitable control models from experimental data [5]. This toolbox has evolved from Professor Peter Young's research on time series analysis, forecasting and control; see e.g. [6–8] among other citations below. Hence, in a book prepared in his honour, it is a pleasure to acknowledge his tremendous contributions as a researcher and teacher, and as a good friend to the present authors.

The state space formulation of control design is, perhaps, the most natural and convenient approach for use with computers. It has the advantage over TF methods of allowing unified treatment of both univariate and multivariable systems. The approach allows for the implementation of powerful state variable feedback control designs, including pole assignment and optimal control. Traditionally, however, the state space approach has one major difficulty: the state vector is not normally available for direct measurement. Therefore, much research effort has been applied to the development of state observer techniques, notably including the Kalman Filter [9] and Luenberger observer [10]. The observer is employed to generate a surrogate state vector which converges asymptotically to the true state vector and so can be used in the implementation of state variable feedback control [3, 4].

In this chapter, we re-examine an alternative approach, based on the definition of a Non-Minimal State Space (NMSS) form, in which the dimension of the state vector is dictated by the complete structure of the TF model. This contrasts with minimal state space descriptions, which only account for the order of the denominator and whose state variables, therefore, usually represent combinations of the input and output signals. The non-minimal state vector is composed only of those variables which can be directly measured and stored in the digital computer. In the discrete-time case, these are the present and past sampled values of the output variables and the past sampled values of the input variables.

Various authors have considered NMSS models, including Young & Willems [11], Young et al. [7], Hesketh [12], Taylor et al. [13], Gonzalez et al. [14] and Gawthrop et al. [15]. However, of particular interest is the Proportional-Integral-Plus (PIP) control algorithm pioneered by Professor Young and his colleagues, including the present authors [7, 16–18]. Successful practical applications include environmental systems [19–21], heavy construction machinery [22, 23] and nuclear decommissioning robots [24]. The present tutorial chapter focuses on PIP design for single-input, single-output models, represented in both linear and nonlinear form. However, the basic approach is readily extended into multivariable and model-predictive systems, hence the chapter also gives brief pointers to these areas.

27.2 System Identification

We aim to regulate the behaviour of a controlled output variable $y(k)$, typically a velocity, torque or some other measured variable, where the argument k indicates that $y(k)$ is sampled in time, with a constant sampling interval of Δt time units. At each sampling instant, the control algorithm updates the control input variable $u(k)$, representing the actuator. For example, permanent magnet DC motors are commonly used to provide motion for a wide variety of electromechanical devices, including robotic manipulators, disk drives and machine tools. They convert direct current (DC) electrical energy into rotational mechanical energy [4].

If the electrical time constant is at least an order of magnitude faster than the mechanical time constant, then the behaviour of a DC motor can be approximated by a first order, scalar difference equation, in which a_1 and b_1 are constant coefficients: $y(k) = -a_1 y(k-1) + b_1 u(k-1)$. In this case, $y(k)$ represents the sampled shaft velocity and $u(k)$ the applied voltage. The upper left subplot of Fig. 27.1 shows the step response of this model using $a_1 = -0.9393$ and $b_1 = 3.2102$, compared with experimental data for an illustrative motor, with $\Delta t = 0.01$ seconds. Substituting for the backward shift operator z^{-1}, yields the equivalent TF representation,

$$y(k) = \frac{b_1 z^{-1}}{1 + a_1 z^{-1}} u(k). \tag{27.1}$$

Here, z^{-i} is the basic discrete-time operator used in this chapter, defined as follows, $z^{-i} y(k) = y(k-i)$. It is clear that z^{-i} denotes a delay of i samples.

The construction industry generally uses *hydraulic* actuators for heavy lifting. Here, the control problem is made difficult by a range of factors that include highly varying loads, speeds and geometries. A commercial mini-tracked excavator and a laboratory model have been used to investigate such issues at Lancaster University. In this regard, the laboratory excavator bucket dynamics are well approximated by the model (27.1), in which $y(k)$ represents the joint angle and $u(k)$ the control input voltage, with a typical $\Delta t = 0.1$ seconds [25]. The upper right subplot of Fig. 27.1 compares experimental data for bucket position with the response of the TF model (27.1) using $a_1 = -1$ and $b_1 = 0.0498$. By contrast, the slew (horizontal) joint has a pure time delay of 0.2 seconds hence, with $\Delta t = 0.1$ seconds, the model becomes,

$$y(k) = \frac{b_2 z^{-2}}{1 + a_1 z^{-1}} u(k). \tag{27.2}$$

Generalising yields the following nth order TF model, where a_i and b_i $(i = 1, 2, \ldots)$ are constant coefficients,

$$y(k) = \frac{b_1 z^{-1} + b_2 z^{-2} + \cdots + b_m z^{-m}}{1 + a_1 z^{-1} + a_2 z^{-2} + \cdots + a_n z^{-n}} u(k). \tag{27.3}$$

Any pure time delay of $\delta > 1$ samples is accounted for by setting $b_1, \ldots, b_{\delta-1} = 0$. For example, $n = 1$, $m = 2$ and $b_1 = 0$ yields equation (27.2), in which $\delta = 2$ samples.

Finally, consider a second order model based on equation (27.3) with $n = m = 2$, $\delta = 1$ and illustrative numerical values for the coefficients,

$$y(k) = \frac{b_1 z^{-1} + b_2 z^{-2}}{1 + a_1 z^{-1} + a_2 z^{-2}} u(k) = \frac{0.5 z^{-1} - 0.4 z^{-2}}{1 - 0.8 z^{-1} + 0.15 z^{-2}} u(k). \tag{27.4}$$

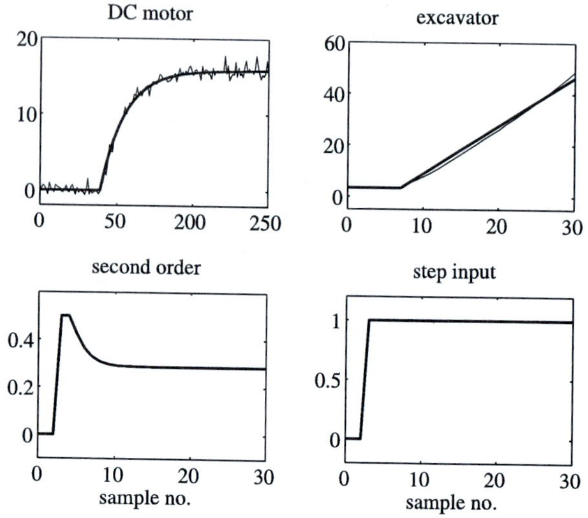

Fig. 27.1 Open-loop step response. *Top subplots*: first order TF model (*thick trace*) fitted to laboratory data for DC motor speed (*left*) and excavator bucket angle (*right*). *Lower subplots*: response of the second order TF model (27.4) to a unit step input. All variables are plotted against sample number

The unit step response is shown by the lower left subplot of Fig. 27.1. Note that the stability of a discrete-time TF model is usually analysed in the z-domain [3, 4]. Multiplying the denominator polynomial by z^2 and setting this equal to zero yields the characteristic equation $z^2 - 0.8z + 0.15 = (z - 0.3)(z - 0.5) = 0$. The system is defined by two poles, i.e. $p_1 = 0.3$ and $p_2 = 0.5$, both lying inside the unit circle on the real axis of the complex z-plane, hence the system (27.4) is stable. Such pole positions are important when we consider closed-loop control later in the chapter.

For practical control system design, (27.3) is usually identified from measured input-output data, collected either from planned experiments or during the normal operation of the plant. In the present chapter, such analysis utilises the instrumental variable methods pioneered by Professor Young [6], since they have proven robust for the examples considered here. An appropriate model structure also needs to be identified, i.e. the order of the polynomials (n, m) and the pure time delay δ. In this regard, the two main statistical measures utilised in CAPTAIN are the coefficient of determination R_T^2 and the Young Identification Criterion (YIC). Since the present chapter focuses on control system design, details of these system identification methods are omitted: refer to the citations above.

27.3 Minimal and Non-minimal State Space Models

This section illustrates the difference between minimal and non-minimal state variable feedback. For brevity, generic forms are omitted, with the discussion focusing instead on an illustrative worked example. In this regard, consider the following state space representation of the second order model (27.4),

$$\begin{bmatrix} y(k) \\ y(k-1) \\ u(k-1) \end{bmatrix} = \begin{bmatrix} -a_1 & -a_2 & b_2 \\ 1 & 0 & 0 \\ 0 & 0 & 0 \end{bmatrix} \begin{bmatrix} y(k-1) \\ y(k-2) \\ u(k-2) \end{bmatrix} + \begin{bmatrix} b_1 \\ 0 \\ 1 \end{bmatrix} u(k-1), \quad (27.5)$$

$$y(k) = \begin{bmatrix} 1\,0\,0 \end{bmatrix} \mathbf{x}(k). \quad (27.6)$$

Here, (27.5) and (27.6) are termed the state and observation equation respectively. Examination of the difference equation obtained from the TF model (27.4),

$$y(k) = -a_1 y(k-1) - a_2 y(k-2) + b_1 u(k-1) + b_2 u(k-2) \quad (27.7)$$

verifies that the equation pair {(27.5), (27.6)}, holds as one particular representation of the system. In fact, this is an example of a *non-minimal* state space (NMSS) model, since there are three state variables, i.e. the order of the state vector is greater than the order of the TF model. In fact, the (regulator) NMSS representation of the general TF model (27.3) has $n + m - 1$ states, as follows,

$$\mathbf{x}(k)^T = \begin{bmatrix} y(k)\ y(k-1)\ \ldots\ y(k-n+1)\ u(k-1)\ u(k-2)\ \ldots\ u(k-m+1) \end{bmatrix}. \quad (27.8)$$

Section 27.4.2 develops the (servomechanism) NMSS model in full. Such non-minimal models contrast with the following *minimal* representations of the examplar TF model (27.4), which are based on the definition of n state variables, here denoted $x_1(k)$ and $x_2(k)$. The controllable canonical form is,

$$\begin{bmatrix} x_1(k) \\ x_2(k) \end{bmatrix} = \begin{bmatrix} -a_1 & -a_2 \\ 1 & 0 \end{bmatrix} \begin{bmatrix} x_1(k-1) \\ x_2(k-1) \end{bmatrix} + \begin{bmatrix} 1 \\ 0 \end{bmatrix} u(k-1), \quad (27.9)$$

$$y(k) = \begin{bmatrix} b_1\ b_2 \end{bmatrix} \mathbf{x}(k) \quad (27.10)$$

and the observable canonical form,

$$\begin{bmatrix} x_1(k) \\ x_2(k) \end{bmatrix} = \begin{bmatrix} -a_1 & 1 \\ -a_2 & 0 \end{bmatrix} \begin{bmatrix} x_1(k-1), \\ x_2(k-1) \end{bmatrix} + \begin{bmatrix} b_1 \\ b_2 \end{bmatrix} u(k-1), \quad (27.11)$$

$$y(k) = \begin{bmatrix} 1\,0 \end{bmatrix} \mathbf{x}(k). \quad (27.12)$$

The state space representations developed above, namely the equation pairs {(27.5), (27.6)}, {(27.9), (27.10)} and {(27.11), (27.12)}, all describe the same essential input-output relationship specified by the TF model (27.4) and equivalent difference equation (27.7).

The advantage of the controllable canonical form is that it is straightforward to construct and yields a convenient structure for the computation of a pole assignment control algorithm (Sect. 27.3.1). Here, it can be verified that the state vector is implicitly formed from the delayed and filtered output variable, where the filter is defined by the numerator polynomial of the TF model. For (27.9),

$$x_1(k) = y(k-1)/\left(b_1z^{-1} + b_2z^{-2}\right); \qquad x_2(k) = y(k-2)/\left(b_1z^{-1} + b_2z^{-2}\right). \tag{27.13}$$

The observable canonical form is so called because it is a particularly convenient state space model to employ when formulating either a deterministic observer or a stochastic Kalman Filter [4]. Straightforward algebra shows that, for (27.11),

$$x_1(k) = y(k); \qquad x_2(k) = -a_2y(k-1) + b_2u(k-1). \tag{27.14}$$

By contrast, the non-minimal state vector in (27.5) is composed explicitly from the sampled values of the input and output variables, and does not include the parameters. For the present example, the states are $y(k)$, $y(k-1)$ and $u(k-1)$. The characteristic equation for this NMSS model is determined from $\det|\lambda\mathbf{I} - \mathbf{A}| = 0$, where $\mathbf{A}$ is the 3×3 state transition matrix in (27.5). Substituting for the numerical values (27.4) and evaluating the determinant yields $\lambda(\lambda^2 - 0.8\lambda + 0.15)$, hence the eigenvalues are $\lambda = 0.5$, 0.3 and 0. Two of these are equivalent to those of the minimal state space model and are also the poles of the TF model (27.4). The eigenvalue at the origin owes its existence to the additional state $u(k-1)$.

27.3.1 State Variable Feedback and Pole Assignment

The minimal state variable feedback control law associated with either the controllable {(27.9), (27.10)} or observable {(27.11), (27.12)} forms is,

$$u(k) = -l_1x_1(k) - l_2x_2(k) + k_dy_d(k), \tag{27.15}$$

where $y_d(k)$ is the command input, while the control gains l_1, l_2 and k_d are chosen by the designer to ensure a satisfactory response. For the purposes of this example, the command is introduced as a straightforward open-loop element with gain k_d, as shown in Fig. 27.2. For the controllable canonical form, using (27.13) and (27.15),

$$u(k) = -l_1y^*(k-1) - l_2y^*(k-2) + k_dy_d(k), \tag{27.16}$$

where $y^*(k) = y(k)/(\hat{b}_1z^{-1} + \hat{b}_2z^{-2})$ in which $\hat{b}_1$ and $\hat{b}_2$ are the estimated parameters associated with the numerator polynomial of the control model. We have introduced this notation to emphasize the inevitable mismatch between the true system (or plant) and the estimated TF model used for control system design. Hence, putting together the control algorithm (27.16) and system representation (27.4), the block diagram of the closed-loop control system with potential model mismatch is illustrated by Fig. 27.2. However, at the control *design* stage, we assume zero plant-model mismatch, i.e. $\hat{b}_1 = b_1$ and $\hat{b}_2 = b_2$. In this case, block diagram or algebraic reduction of Fig. 27.2 yields the following closed-loop TF,

$$y(k) = \frac{k_d(b_1z^{-1} + b_2z^{-2})}{1 + (a_1 + l_1)z^{-1} + (a_2 + l_2)z^{-2}}y_d(k). \tag{27.17}$$

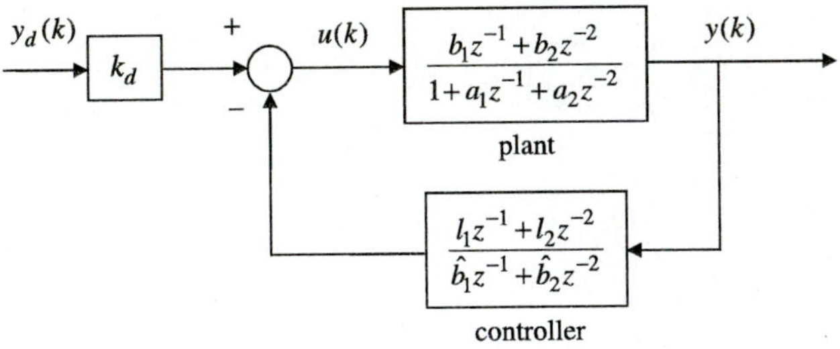

Fig. 27.2 Unity gain (minimal) regulator based on the controllable canonical form

The denominator of a TF defines its stability and transient dynamic behaviour. Therefore, we consider the characteristic polynomial as follows,

$$1 + (a_1 + l_1)\, z^{-1} + (a_2 + l_2)\, z^{-2}. \tag{27.18}$$

When using the controllable canonical form, it is a trivial matter to obtain a desired characteristic polynomial $D(z^{-1})$. For example, if we require,

$$D(z^{-1}) = 1 + d_1 z^{-1} + d_2 z^{-2}, \tag{27.19}$$

where d_1 and d_2 are chosen by the designer, (27.18) and (27.19) yields $l_1 = d_1 - a_1$ and $l_2 = d_2 - a_2$. The characteristic equation $D(z^{-1}) = 0$ is equivalent to $D(z) = z^2 + d_1 z + d_2 = 0$. Since the roots of $D(z) = 0$ are known as the closed-loop poles, the approach is called state variable feedback pole assignment. Finally, k_d is selected independently to obtain unity steady state gain (Sect. 27.3.2).

Considering now the NMSS case, based on the three state variables in (27.5), the state variable feedback control law is,

$$u(k) = -f_0 y(k) - f_1 y(k-1) - g_1 u(k-1) + k_d y_d(k), \tag{27.20}$$

where f_0, f_1 and g_1 are the state feedback control gains and k_d is the command input gain. Hence, the unity gain NMSS regulator in block diagram form is illustrated by Fig. 27.3. Block diagram or algebraic reduction of Fig. 27.3 yields,

$$y(k)/y_d(k) = \frac{k_d(b_1 z^{-1} + b_2 z^{-2})}{1 + (a_1 + g_1 + b_1 f_0) z^{-1} + (a_2 + a_1 g_1 + b_2 f_0 + b_1 f_1) z^{-2} + (a_2 g_1 + b_2 f_1) z^{-3}}. \tag{27.21}$$

The closed-loop characteristic polynomial is third order, one more than for minimal design. Initially, this may appear to be a weakness of the NMSS solution, since we have to design for third order closed-loop dynamic behaviour. However, this argument is fundamentally flawed, because the problem of model mismatch has not yet been considered. In fact, evaluation of the closed-loop TF for the minimal design

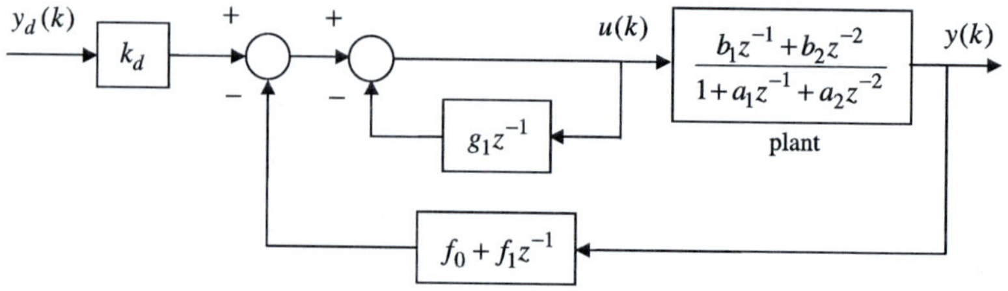

Fig. 27.3 Unity gain NMSS regulator

in Fig. 27.2, this time with $\hat{b}_1 \neq b_1$ and $\hat{b}_2 \neq b_2$, shows that

$$y(k)/y_d(k)$$

$$= \frac{k_d(b_1 z^{-1} + (\frac{b_1\hat{b}_2}{\hat{b}_1} + b_2)z^{-2} + \frac{b_2\hat{b}_2}{\hat{b}_1}z^{-3})}{1 + (\frac{\hat{b}_2}{\hat{b}_1} + a_1 + \frac{l_1 b_1}{\hat{b}_1})z^{-1} + (\frac{a_1\hat{b}_2}{\hat{b}_1} + a_2 + \frac{l_1 b_2}{\hat{b}_1} + \frac{l_2 b_1}{\hat{b}_1})z^{-2} + (\frac{a_2\hat{b}_2}{\hat{b}_1} + \frac{l_2 b_2}{\hat{b}_1})z^{-3}}. \tag{27.22}$$

The minimal closed-loop system is also third order! Note that the NMSS solution is the rather simpler TF already quoted: the parameter estimates do not appear in (27.21), since the model is not utilised in the NMSS controller of Fig. 27.3. This result is clearly significant in practical terms because there will always be mismatch between the estimated model and the real world.

27.3.2 Numerical Example with Model Mismatch

Assigning the closed-loop poles to real values of, say 0.7 and 0.8, the desired characteristic polynomial is,

$$D(z^{-1}) = (1 - 0.7z^{-1})(1 - 0.8z^{-1}) = 1 - 1.5z^{-1} + 0.56z^{-2}. \tag{27.23}$$

Using the numerical values (27.4), equating like coefficients from (27.18) and (27.23), yields $l_1 = -0.7$ and $l_2 = 0.41$. Assuming no model mismatch, (27.22) reduces to,

$$\frac{y(k)}{y_d(k)} = \frac{k_d(0.5z^{-1} - 0.8z^{-2} + 0.32z^{-3})}{1 - 2.3z^{-1} + 1.76z^{-2} - 0.448z^{-3}} = \frac{k_d(0.5z^{-1} - 0.4z^{-2})}{1 - 1.5z^{-1} + 0.56z^{-2}}. \tag{27.24}$$

The poles of the left hand side TF are 0.8, 0.8 and 0.7. Cancelling one of the zeros (i.e. roots of the numerator polynomial) with a pole, yields the second order system shown on the right hand side. The steady state gain (obtained by setting $z^{-1} = 1$) is $k_d/0.6$, hence $k_d = 0.6$ yields unity gain when there is no model mismatch.

Table 27.1 Comparison of NMSS and minimal state variable feedback control with model mismatch. Design and Poles refer to the user selected and model mismatch closed-loop pole positions respectively. Zeros refers to the closed-loop zeros with model mismatch

Design		Poles		Zeros	
Minimal	NMSS	Minimal	NMSS	Minimal	NMSS
0.8	0.8	1.1573	0.8573	0.8000	0.7238
0.7	0.7	0.6632	0.6777	0.7238	–
–	0	0.5545	0	–	–

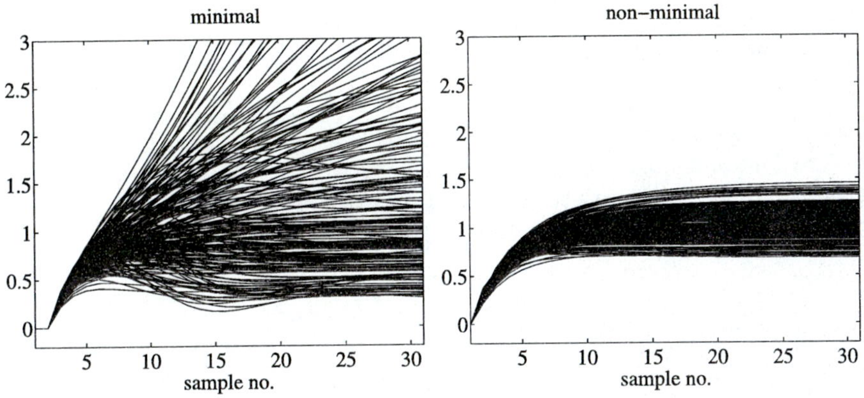

Fig. 27.4 Monte Carlo Simulation using the unity gain controllable canonical form (*left*) and NMSS (*right*) regulators. The closed-loop unit step response is plotted against sample number

We will constrain the NMSS solution to the same closed-loop as (27.24). This is achieved by assigning two of its poles to 0.7 and 0.8 (as in the minimal case) and the additional pole to zero, i.e. the origin of the complex z-plane. Equating the denominator coefficients of (27.21) and (27.23) yields $g_1 = -0.8$, $f_0 = 0.2$ and $f_1 = -0.3$. In this case, the NMSS and the minimal controllers produce identical closed-loop responses, with the output asymptotically approaching the command level at steady state. However, consider the effect of the following (arbitrary) mismatched values for the system parameters,

$$a_1 = -0.84; \qquad a_2 = 0.1425; \qquad b_1 = 0.525; \qquad b_2 = -0.38. \tag{27.25}$$

The new closed-loop poles and zeros shown in Table 27.1, are obtained by substituting the gains and mismatched system parameters into (27.21) and (27.22). The control model coefficients $\hat{b}_1 = 0.5$ and $\hat{b}_2 = -0.4$ are unchanged. In this case, the pole-zero cancellation of the minimal solution does not occur and the system is third order. Furthermore, one of the poles is outside the unit circle, so that the minimal solution is unstable! By contrast, the NMSS poles are all within the unit circle.

Finally, Fig. 27.4 shows an evaluation of these algorithms using Monte Carlo simulation. The minimal solution is much less robust than the NMSS equivalent, with numerous unstable realisations. For more information about evaluating robustness using Monte Carlo simulation, with the model uncertainty estimated from experimental data, see e.g. references [20, 21].

27.3.3 Transformations and Constrained NMSS Control

Let us now consider the observable canonical representation {(27.11), (27.12)}. Utilising the control algorithm (27.15) and substituting for the state vector (27.14) yields,

$$u(k) = -l_1 y(k) + 1.5 l_2 y(k-1) - 0.4 l_2 u(k-1) + k_d y_d(k). \tag{27.26}$$

The non-minimal (27.20) and minimal (27.26) algorithms take the same structural form, although in the NMSS case there is an additional independent control gain.

The following transformation constrains the NMSS solution into exactly the same control law as the (observable) minimal design [17],

$$\begin{bmatrix} f_0 \; f_1 \; g_1 \end{bmatrix} = [l_1 \; l_2] \begin{bmatrix} 1 & 0 & 0 \\ 0 & -a_2 & b_2 \end{bmatrix}. \tag{27.27}$$

Equation (27.27) is equivalent to assigning the $(m-1)$ extra poles in the NMSS case to the origin. Reference [17] develops a similar transformation for the generalised model (27.3), and shows how to constrain the NMSS solution using either pole assignment or linear quadratic optimal control. However, assigning the extra poles to the origin is not necessarily the best solution: one advantage of the NMSS approach is that these extra poles can be assigned to desirable locations anywhere on the complex z-plane. This is a particularly useful feature in the optimal control case [26].

Furthermore, minimal models cannot automatically handle the case when $m > n$, whereas the NMSS form has no problems in this regard. For this tutorial example, we have deliberately selected a TF model where $n = m = 2$ and $\delta = 1$. However, examination of equations {(27.9), (27.10)} and {(27.11), (27.12)} shows that the minimal formulation of the problem can only deal with cases when $m > n$ by changing the state vector to dimension m and setting the trailing denominator parameters $a_{n+1}, a_{n+2}, \ldots, a_m$ to zero. Of course, such an approach is implicitly non-minimal anyway, hence it makes rather more sense to utilise the general NMSS model from the start.

In conclusion, NMSS design provides a flexible and logical approach to state variable feedback because the state space model is built in the simplest possible way and yet provides the greatest degrees of freedom in the final design. Unfortunately, the regulator control systems considered so far do not necessarily track the command input when there is model uncertainty. This problem is illustrated in Fig. 27.4, in which the command input for each realisation is unity. In the following section, therefore, the NMSS model is augmented by an integral-of-error state variable, in order to provide the required Type 1 servomechanism performance.

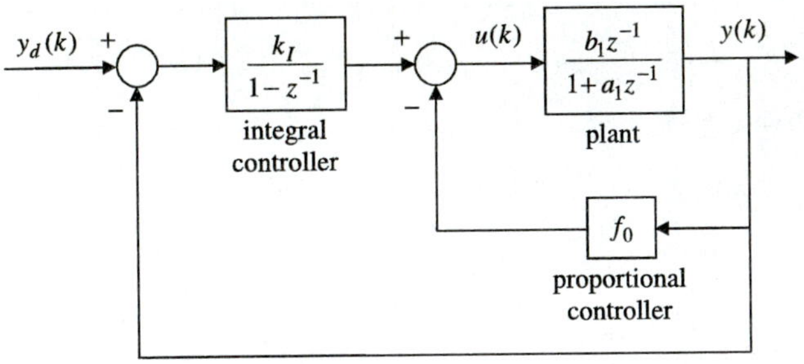

Fig. 27.5 Proportional-Integral control of a first order TF model

27.4 Linear Proportional-Integral-Plus Control

An important feedback algorithm widely used in industrial control systems is the Proportional-Integral-Derivative (PID) controller. An early citation is Challender et al. [27] but the algorithm remains the focus of much practical and theoretical development into the current decade. Let us consider, however, a special case of the PID controller: the 'two-term' Proportional-Integral (PI) control algorithm. One particular PI control structure, applied to the TF model (27.1), is illustrated in Fig. 27.5. Examination of Fig. 27.5 shows that the PI control algorithm is,

$$u(k) = \frac{k_I}{1-z^{-1}}\left(y_d(k) - y(k)\right) - f_0 y(k). \tag{27.28}$$

Substituting for z^{-1} and rearranging yields the difference equation form of the algorithm, suitable for implementation on a digital computer,

$$u(k) = u(k-1) + k_I\left(y_d(k) - y(k)\right) - f_0\left(y(k) - y(k-1)\right). \tag{27.29}$$

The closed-loop TF determined from Fig. 27.5 is,

$$y(k) = \frac{k_I b_1}{1 + (f_0 b_1 + a_1 - 1 + k_I b_1)z^{-1} + (-a_1 - f_0 b_1)z^{-2}} y_d(k). \tag{27.30}$$

The steady state gain of the closed-loop system is unity. Hence, if the closed-loop system is stable, the output will converge asymptotically to a constant command input. If f_0 and k_I are selected so that the closed-loop poles lie at the origin of the complex z-plane, i.e. $p_1 = p_2 = 0$, then we obtain the deadbeat response illustrated by Fig. 27.6 (left). Here, the output signal $y(k)$ follows a step change in the command input after just one sampling instant, the fastest theoretical response of a discrete-time control system. By contrast, the response shown in Fig. 27.6 (right) is obtained by selecting f_0 and k_I so that the closed-loop poles form a complex conjugate pair with $p_1 = 0.6 + 0.3j$ and $p_2 = 0.6 - 0.3j$.

The latter of these two controllers yields a slower speed of response and deliberately incorporates a small temporary overshoot of the command level, which

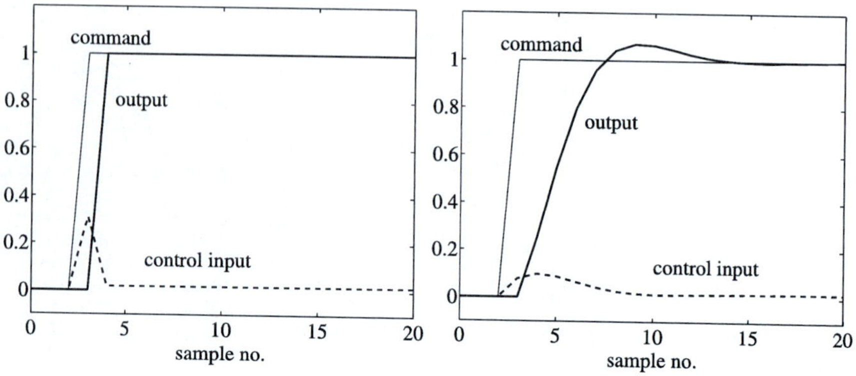

Fig. 27.6 Proportional-Integral control of a first order TF model showing a deadbeat response (*left*) and a response using complex poles (*right*)

is sometimes desirable in practice e.g. it is more likely that the desired steady state level is quickly achieved despite practical limitations in the system, such as mechanical friction effects. In practical applications, it is also more robust to uncertainty in the model parameters and generates a less severe control input signal than deadbeat design. The associated characteristic polynomial of the closed-loop system is,

$$D(z) = (z - 0.6 + 0.3j)\,(z - 0.6 - 0.3j) = z^2 - 1.2z + 0.45. \tag{27.31}$$

Following a similar approach to Sect. 27.3.1, the control gains are determined by equating (27.31) with the denominator of (27.30), in either the z or z^{-1} domain. To illustrate, consider again the TF model (27.1) with $a_1 = -0.9393$ and $b_1 = 3.2102$, representing the speed of a DC motor. The simultaneous equations are straightforwardly solved to obtain $f_0 = 0.1524$ and $k_I = 0.0779$. The characteristic polynomial for a deadbeat response is $D(z) = (z-0)(z-0) = z^2$ which, for the DC motor example, yields $f_0 = 0.2926$ and $k_I = 0.3115$. Using these two sets of control gains, the PI controller yields the responses illustrated by Fig. 27.6.

Now consider the 'servomechanism' NMSS form associated with the model for a DC motor (27.1). Here, $\mathbf{x}(k) = [y(k)\ z(k)]^T$ where $z(k)$ represents the discrete-time integral (summation) of the error between the output $y(k)$ and the command input $y_d(k)$, an integral-of-error state variable: $z(k) = z(k-1) + y_d(k) - y(k)$ or,

$$z(k) = \frac{1}{1 - z^{-1}}(y_d(k) - y(k)), \tag{27.32}$$

where the TF will be recognised as a discrete-time integrator. Using the model (27.1), $z(k) = z(k-1) + y_d(k) + a_1 y(k-1) - b_1 u(k-1)$, hence the NMSS model is,

$$\begin{bmatrix} y(k) \\ z(k) \end{bmatrix} = \begin{bmatrix} -a_1 & 0 \\ a_1 & 1 \end{bmatrix} \begin{bmatrix} y(k-1) \\ z(k-1) \end{bmatrix} + \begin{bmatrix} b_1 \\ -b_1 \end{bmatrix} u(k-1) + \begin{bmatrix} 0 \\ 1 \end{bmatrix} y_d(k), \tag{27.33}$$

$$y(k) = [1\ 0] \begin{bmatrix} y(k) \\ z(k) \end{bmatrix}. \tag{27.34}$$

The state variable feedback control algorithm associated with this model is called the Proportional-Integral-Plus or PIP controller. Here,

$$u(k) = -\mathbf{k}\mathbf{x}(k) = -\begin{bmatrix} f_0 & -k_I \end{bmatrix} \begin{bmatrix} y(k) \\ z(k) \end{bmatrix} = -f_0 y(k) + k_I z(k), \qquad (27.35)$$

where $\mathbf{k} = [f_0 \ -k_I]$ is the control gain vector. Using (27.32) and (27.35), the control law for this system is given by (27.28) and Fig. 27.5. In other words, for the simplest TF model (27.1), the PIP formulation is equivalent to a digital PI control system. Nevertheless, one advantage of the PIP approach is already clear: although the control algorithm retains the straightforward implementation structure of classical PI control, it has been derived within the framework of state variable feedback design. Hence, the vagaries of manual tuning can now be replaced by state variable feedback techniques such as optimal or robust design, as discussed later.

27.4.1 Linear PIP Control of Laboratory Excavator

Returning to the laboratory excavator mentioned in Sect. 27.2, the model (27.1) with $a_1 = -1$ and $b_1 = 0.0498$ is identified from experimental data. Here, $y(k)$ represents the bucket joint angle, while $u(k)$ is a scaled voltage. Solving the pole assignment problem for $p_1 = 0.8139 + 0.1539j$ and $p_2 = 0.8139 - 0.1539j$, yields $f_0 = 6.303$ and $k_I = 1.171$. These closed-loop poles have been obtained using linear quadratic optimal design, as described later. The response to a time varying command input is illustrated by Fig. 27.7. Here, $y_d(k)$ has been determined as part of the high-level control objectives for this device, namely digging a trench in the sandpit [25].

In fact, the excavator arm consists of four hydraulically actuated joints, also including the 'boom', 'dipper' and 'slew' angles. In this case, statistical identification

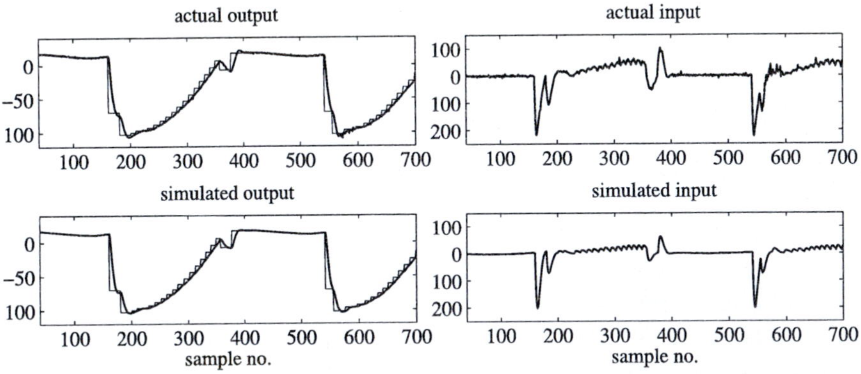

Fig. 27.7 PIP control of laboratory excavator bucket angle [28]. *Left subplots*: joint angle in degrees (*thick traces*) and command input (*thin traces*). *Right subplots*: control input signals (*scaled voltage*). The *upper subplots* show experimental data and the *lower subplots* show the equivalent simulated response based on the TF model (27.1). Sampling interval $\Delta t = 0.1$ seconds

from experimental data, utilising a uniform sampling rate of $\Delta t = 0.1$ seconds, suggests that a first order model provides an approximate representation of all four joints, with $m = \delta = 1$ for the dipper and bucket, while $m = \delta = 2$ for the boom and slew. In other words, the boom and slew joint dynamics are well represented by (27.2), whilst the dipper and bucket are modelled using (27.1).

For the case that $\delta = 2$, but still utilising the PI controller shown in Fig. 27.5, it can be shown that the closed-loop system is third order. With only two control gains, the pole assignment problem cannot be solved, i.e. the designer cannot arbitrarily assign the closed-loop poles. Similar results emerge for PI control of second or higher order systems. Therefore, in order to design for the general TF model (27.3), the following subsection utilises the generalised NMSS model introduced earlier.

27.4.2 Linear PIP Control of the Generalised TF Model

The NMSS servomechanism representation of the generalised TF model (27.3) is,

$$\mathbf{x}(k) = \mathbf{F}x(k-1) + \mathbf{g}u(k-1) + \mathbf{d}y_d(k); \qquad y(k) = \mathbf{h}\mathbf{x}(k) \tag{27.36}$$

in which,

$$\mathbf{F} = \begin{bmatrix} -a_1 & -a_2 & \ldots & -a_{n-1} & -a_n & b_2 & b_3 & \ldots & b_{m-1} & b_m & 0 \\ 1 & 0 & \ldots & 0 & 0 & 0 & 0 & \ldots & 0 & 0 & 0 \\ \vdots & \vdots & \ddots & \vdots & \vdots & \vdots & \vdots & \ddots & \vdots & \vdots & \vdots \\ 0 & 0 & \ldots & 1 & 0 & 0 & 0 & \ldots & 0 & 0 & 0 \\ 0 & 0 & \ldots & 0 & 0 & 0 & 0 & \ldots & 0 & 0 & 0 \\ 0 & 0 & \ldots & 0 & 0 & 1 & 0 & \ldots & 0 & 0 & 0 \\ \vdots & \vdots & \ddots & \vdots & \vdots & \vdots & \vdots & \ddots & \vdots & \vdots & \vdots \\ 0 & 0 & \ldots & 0 & 0 & 0 & 0 & \ldots & 1 & 0 & 0 \\ a_1 & a_2 & \ldots & a_{n-1} & a_n & -b_2 & -b_3 & \ldots & -b_{m-1} & -b_m & 1 \end{bmatrix}$$

while $\mathbf{g} = [b_1\, 0 \ldots 0\, 1\, 0 \ldots 0\, {-b_1}]^T$, $\mathbf{d} = [0 \ldots 0\, 1]^T$ and $\mathbf{h} = [1\, 0 \ldots 0]$. Finally, the $n+m$ dimensional non-minimal state vector $\mathbf{x}(k)$ is,

$$\mathbf{x}(k) = \left[\, y(k)\; y(k-1) \ldots y(k-n+1)\; u(k-1) \ldots u(k-m+1)\; z(k) \,\right]^T. \tag{27.37}$$

The PIP control law takes a similar form to (27.35), here with the $n+m$ dimensional gain vector, $\mathbf{k} = [f_0\ f_1\ \cdots\ f_{n-1}\ g_1\ \cdots\ g_{m-1}\ {-k_I}]$. In block-diagram terms, the controller can be implemented as shown in Fig. 27.8, where it is clear that it can be considered as one particular extension of the PI controller, in which the PI action is, in general, enhanced by the higher order forward path and feedback compensators,

$$\begin{aligned} G_1(z^{-1}) &= g_1 z^{-1} + \cdots + g_{m-1} z^{-(m-1)}; \\ F_1(z^{-1}) &= f_1 z^{-1} + \cdots + f_{n-1} z^{-(n-1)}. \end{aligned} \tag{27.38}$$

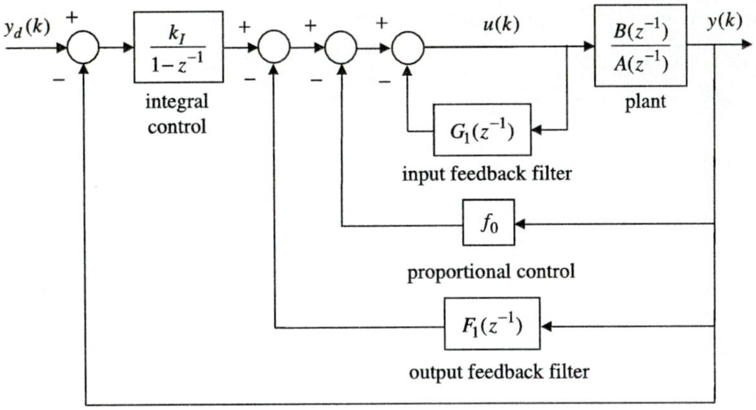

Fig. 27.8 Proportional-Integral-Plus (PIP) control

However, because it exploits a state space framework, PIP control allows for well-known state variable feedback design strategies, such as optimisation in terms of a Linear-Quadratic (LQ) cost function,

$$J = \frac{1}{2}\sum_{i=0}^{\infty}\{\mathbf{x}(i)\mathbf{Q}\mathbf{x}(i) + ru^2(i)\} \tag{27.39}$$

where $\mathbf{Q}$ is a $n+m$ symmetric matrix and r is a scalar weight. Here, the control gains are obtained from the steady state solution of the discrete-time matrix Riccati equation [3, 4]. Risk sensitive and stochastic solutions can be similarly developed [17, 29]. Due to the special structure of the non-minimal state vector, the elements of the LQ weighting matrices have particularly simple interpretation. In fact, good performance is often obtained by straightforward manual tuning of the diagonal terms associated with the input and output signals, as is the case for the excavator. Alternatively, the PIP control system is ideal for incorporation within a multi-objective optimisation framework, where satisfactory compromise can be obtained between conflicting objectives such as robustness, overshoot, rise times and multivariable decoupling [16, 30].

To illustrate this control framework, consider the servomechanism NMSS state equation for the first order model with two samples time delay (27.2),

$$\begin{bmatrix} y(k) \\ u(k-1) \\ z(k) \end{bmatrix} = \begin{bmatrix} -a_1 & b_2 & 0 \\ 0 & 0 & 0 \\ a_1 & -b_2 & 1 \end{bmatrix} \begin{bmatrix} y(k-1) \\ u(k-2) \\ z(k-1) \end{bmatrix} + \begin{bmatrix} 0 \\ 1 \\ 0 \end{bmatrix} u(k-1) + \begin{bmatrix} 0 \\ 0 \\ 1 \end{bmatrix} y_d(k). \tag{27.40}$$

In this case, $b_1 = 0$, $G_1(z^{-1}) = g_1 z^{-1}$ and $F_1(z^{-1})$ is not required. The closed-loop system is third order and there are three control gains (f_0, g_1 and k_I), hence the pole assignment problem can be solved.

The regulator NMSS state equation for the second order model (27.4) is given by (27.5), whilst the servomechanism NMSS version is,

$$\begin{bmatrix} y(k) \\ y(k-1) \\ u(k-1) \\ z(k) \end{bmatrix} = \begin{bmatrix} -a_1 & -a_2 & b_2 & 0 \\ 1 & 0 & 0 & 0 \\ 0 & 0 & 0 & 0 \\ a_1 & a_2 & -b_2 & 1 \end{bmatrix} \begin{bmatrix} y(k-1) \\ y(k-2) \\ u(k-2) \\ z(k-1) \end{bmatrix} + \begin{bmatrix} b_1 \\ 0 \\ 1 \\ -b_1 \end{bmatrix} u(k-1). \tag{27.41}$$

Here, the control polynomials in Fig. 27.8 are $F_1(z^{-1}) = f_1 z^{-1}$ and $G_1(z^{-1}) = g_1 z^{-1}$. The closed-loop system is fourth order and there are four control gains. It becomes rather inefficient to calculate the pole assignment solution 'by-hand' for such higher order systems, but there are a number of algorithmic approaches available [7], another advantage of using a state space formulation.

27.5 Nonlinear Proportional-Integral-Plus Control

Any inherent nonlinearities encountered in PIP control systems are usually addressed by simulation-based adjustment of the pole positions or LQ weighting matrices. However, recent research has instead utilized a quasi-linear State Dependent Parameter (SDP) model structure, in which the parameters are functionally dependent on other variables in the system. Numerous articles have described an approach for the identification of such models: see e.g. [5, 8] and the references therein.

Initial research utilised the SDP model and conventional linear methods to update a PIP control law at each sampling instant, as suggested by Young [31]. To illustrate, analysis of data for the laboratory excavator dipper joint reveals limitations in the linear model (27.1). In particular, SDP identification suggests that a_1 and b_1 can be expressed as functions of the applied voltage [25],

$$a_1(k) = 0.52 \times 10^{-6} u_{k-2}^2 - 1; \qquad b_1(k) = -0.048 \times 10^{-3} u_{k-1} + 0.0293, \tag{27.42}$$

where the $a_1(k)$ and $b_1(k)$ notation is introduced to represent the time varying nature of the parameters. Typical implementation results are illustrated in Fig. 27.9, where the SDP-PIP controller is obtained by solving the LQ problem (27.39) at each sample. Here, the SDP-PIP response closely follows the theoretical design response, whilst the fixed gain linear PIP controller is slower (because of model mismatch).

Such an approach relates closely to State Dependent Riccati Equation (SDRE) methods; see e.g. [32] and the references therein. Unfortunately, while some theoretical advances have been made regarding the asymptotic stability of the approach, the conditions obtained can be difficult to check and/or fulfill. By contrast, the following subsection describes a new pole assignment algorithm, based on the NMSS model, that avoids these difficulties and ensures closed-loop stability.

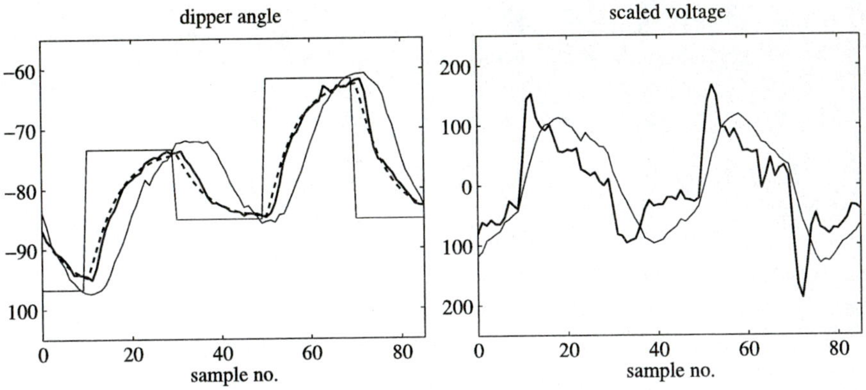

Fig. 27.9 PIP control of laboratory excavator dipper angle [28]. *Left*: command input in degrees (*sequence* of step changes), theoretical response (*dashed*), nonlinear SDP-PIP (*thick*) and linear PIP (*thin*). *Right*: scaled input voltages. Sampling interval $\Delta t = 0.1$ seconds

27.5.1 Nonlinear Pole Assignment: Background

Consider the following nth order SDP model,

$$y(k) = -a_1(k)y(k-1) - \cdots - a_n(k)y(k-n) + b(k)u(k-\delta) \qquad (27.43)$$

where $\delta \geq 1$ is the time delay, while $a_i(k)$ and $b(k)$ are state dependent parameters. Although this model represents a subset of the entire class of SDP models, it has proven particularly useful in practical applications [25]. The states are typically derived from the input (such as (27.42) for the excavator) and output signals, but could also be a function of other measured variables. The NMSS model is,

$$\mathbf{x}(k+1) = \mathbf{F}(k)\mathbf{x}(k) + \mathbf{g}(k)u(k) + \mathbf{d}y_d(k+1); \qquad y(k) = \mathbf{h}\mathbf{x}(k) \quad (27.44)$$

where the $n+\delta$ non-minimal state vector is,

$$\mathbf{x}(k) = \left[y(k) \;\ldots\; y(k-n+1)\; u(k-1) \;\ldots\; u(k-\delta+1)\; z(k)\right]^T. \qquad (27.45)$$

Following a similar approach to Sect. 27.4.2, the state transition matrix and other vectors are straightforward to define [18, 25]. The control law is,

$$u(k) = -\mathbf{c}(k)\mathbf{x}(k), \qquad (27.46)$$

where $\mathbf{c}(k) = [f_0(k) \;\ldots\; f_{n-1}(k)\; g_1(k) \;\ldots\; g_{\delta-1}(k)\; -k_I(k)]$ is the state dependent control gain vector. Applying the control algorithm (27.46) to the open-loop NMSS model (27.44), yields the closed-loop control system,

$$\mathbf{x}(k+1) = \mathbf{A}(k)\mathbf{x}(k) + \mathbf{d}y_d(k+1); \qquad y(k) = \mathbf{h}\mathbf{x}(k), \qquad (27.47)$$

where $\mathbf{A}(k) = \mathbf{F}(k) - \mathbf{g}(k)\mathbf{c}(k)$. To develop a nonlinear pole assignment algorithm, define a $n+\delta$ square matrix $\mathbf{D}$ with user specified (arbitrary) eigenvalues p_i

($i = 1, \ldots, n+\delta$). An example of $\mathbf{D}$ is given in Sect. 27.5.2 below. The eigenvalues of $\mathbf{D}$ are equivalent to the roots of the closed-loop characteristic polynomial,

$$D(z^{-1}) = 1 + d_1 z^{-1} + \cdots + d_{n+\delta} z^{-(n+\delta)}, \tag{27.48}$$

where d_i are design coefficients. Taylor et al. [24, 33] develop an algorithm for determining $\mathbf{c}(k)$ such that, for an externally specified command $y_d(k)$, equations (27.47) yield a closed-loop output response defined by equation (27.48). With $\delta > 1$, as for the boom and slew joints of the laboratory excavator, $\mathbf{g}(k) = \mathbf{g}$ is time invariant. In this case, a transformation of the state vector (27.45) is required, i.e. $\bar{\mathbf{x}}(k) = \mathbf{T}(k)\mathbf{x}(k)$ where $\mathbf{T}(k)$ is a $n+\delta$ square matrix. The general form of $\mathbf{T}(k)$ is omitted for brevity but an example is given in Sect. 27.5.2 below. The transformed open-loop model is,

$$\mathbf{T}(k+1)\mathbf{x}(k+1) = \mathbf{F}(k)\mathbf{T}(k)\mathbf{x}(k) + \mathbf{g}u(k) + \mathbf{d}y_d(k+1);$$
$$y(k) = \mathbf{h}\mathbf{T}(k)\mathbf{x}(k).$$

Substituting from (27.46) and rearranging yields,

$$\mathbf{x}(k+1) = \mathbf{T}^{-1}(k+1)\left(\mathbf{F}(k) - \mathbf{g}\mathbf{c}(k)\right)\mathbf{T}(k)\mathbf{x}(k) + \mathbf{T}^{-1}(k+1)y_d(k+1). \tag{27.49}$$

Equating the closed-loop state transition matrix above with $\mathbf{D}$ yields,

$$\mathbf{T}^{-1}(k+1)\mathbf{F}(k)\mathbf{T}(k) - \mathbf{D} = \mathbf{T}^{-1}(k+1)\mathbf{g}\mathbf{c}(k)\mathbf{T}(k). \tag{27.50}$$

With a suitable transformation, the first n and last $\delta - 1$ rows of (27.50) consist of zeros [24, 33]. By equating the $(n+1)$th row of (27.50), and solving the resultant set of $n+\delta$ simultaneous equations off-line, $\mathbf{c}(k)$ is obtained for implementation on-line. This approach yields the same control gains as those developed algebraically in another recent article [18]. However, the present discussion has obtained the controller directly from the NMSS model, facilitating more straightforward stability analysis, as illustrated below.

27.5.2 *Nonlinear Pole Assignment: Worked Example*

Consider a first order SDP model based on (27.43) with $\delta = 3$ samples time delay, i.e. $y(k) = -a_1(k)y(k-1) + b(k)u(k-3)$. The fourth order NMSS form (27.44) is defined by $\mathbf{g} = [0\ 1\ 0\ 0]^T$, $\mathbf{d} = [0\ 0\ 0\ 1]^T$, $\mathbf{h} = [1\ 0\ 0\ 0]^T$ and,

$$\mathbf{x}(k) = \begin{bmatrix} y(k) \\ u(k-1) \\ u(k-2) \\ z(k) \end{bmatrix}; \qquad \mathbf{F}(k) = \begin{bmatrix} -a_1(k+1) & 0 & b(k+1) & 0 \\ 0 & 0 & 0 & 0 \\ 0 & 1 & 0 & 0 \\ a_1(k+1) & 0 & -b(k+1) & 1 \end{bmatrix}. \tag{27.51}$$

The transformation and design transition matrices are,

$$\mathbf{T}(k) = \begin{bmatrix} 1 & 0 & 0 & 0 \\ 0 & \frac{1}{b(k+2)} & \frac{a_1(k+2)}{b(k+2)} & 0 \\ \frac{a_1(k+1)}{b(k+1)} & 0 & \frac{1}{b(k+1)} & 0 \\ 0 & 0 & 0 & 1 \end{bmatrix}; \qquad \mathbf{D} = \begin{bmatrix} 0 & 0 & 1 & 0 \\ d_4 & -1-d_1 & (-1-d_1-d_2) & \tilde{d} \\ 0 & 1 & 0 & 0 \\ 0 & 0 & -1 & 1 \end{bmatrix} \tag{27.52}$$

where $\tilde{d} = 1 + d_1 + d_2 + d_3 + d_4$. Substituting these into (27.50) and equating yields,

$$f_0(k) = \frac{-a_1(k+1)(1+d_1+d_2-a_1(k+2)[1+d_1-a_1(k+3)])-d_4}{b(k+3)}, \tag{27.53}$$

$$g_1(k) = \frac{b(k+2)(1+d_1-a_1(k+3))}{b(k+3)}, \tag{27.54}$$

$$g_2(k) = \frac{b(k+1)(1+d_1+d_2-a_1(k+2)[1+d_1-a_1(k+3)])}{b(k+3)}, \tag{27.55}$$

$$k_I(k) = \frac{1+d_1+d_2+d_3+d_4}{b(k+3)}. \tag{27.56}$$

Examination of the time indices above shows that the algorithm requires a forward shift of the parameters. In this regard, it is important to recall their state dependent form. For many engineering devices, such as the laboratory excavator, these parameters are functions of the *delayed* input and output signals, hence a forward shift does not usually cause problems, i.e. a prediction of the SDP is not required.

To analyse the closed-loop response, the state variable feedback controller based on these gains is substituted into the open-loop model. Here, it can be verified that the transition matrix in (27.47) is $\mathbf{A}(k+1) = \mathbf{T}(k+1)\mathbf{D}\mathbf{T}^{-1}(k)$, where $\mathbf{T}(k)$ and $\mathbf{D}$ are defined by (27.52). Hence, pre-multiplying the state equations in (27.47) by $\mathbf{T}^{-1}(k+1)$ yields $\tilde{\mathbf{x}}(k+1) = \mathbf{D}\tilde{\mathbf{x}}(k) + \mathbf{T}^{-1}(k+1)\mathbf{d}y_d(k+1)$, where $\tilde{\mathbf{x}}(k) = \mathbf{T}^{-1}(k)\mathbf{x}(k)$. Noting that $\mathbf{T}^{-1}(k+1)\mathbf{d} = \mathbf{d}$, successive substitutions yields,

$$\tilde{\mathbf{x}}(k) = \mathbf{D}^4\tilde{\mathbf{x}}(k-4) + \mathbf{D}^3\mathbf{d}y_d(k-3) + \mathbf{D}^2\mathbf{d}y_d(k-2) + \mathbf{D}\mathbf{d}y_d(k-1) + \mathbf{D}^0\mathbf{d}y_d(k).$$

Note from the characteristic polynomial (27.48) and the Cayley-Hamilton theorem that $\mathbf{D}^4 + d_1\mathbf{D}^3 + d_2\mathbf{D}^2 + d_3\mathbf{D} + d_4\mathbf{I} = \mathbf{0}$, i.e. a matrix of zeros. Hence, taking $\tilde{\mathbf{x}}(k) + d_1\tilde{\mathbf{x}}(k-1) + d_2\tilde{\mathbf{x}}(k-2) + d_3\tilde{\mathbf{x}}(k-3) + d_4\tilde{\mathbf{x}}(k-4)$ and re-arranging yields,

$$\begin{aligned}\tilde{\mathbf{x}}(k) &= -d_1\tilde{\mathbf{x}}(k-1) - d_2\tilde{\mathbf{x}}(k-2) - d_3\tilde{\mathbf{x}}(k-3) - d_4\tilde{\mathbf{x}}(k-4)\\ &\quad + \left(\mathbf{D}^3 + d_1\mathbf{D}^2 + d_2\mathbf{D} + d_3\mathbf{I}\right)\mathbf{d}y_d(k-3) + \left(\mathbf{D}^2 + d_1\mathbf{D} + d_2\mathbf{I}\right)\mathbf{d}y_d(k-2)\\ &\quad + (\mathbf{D} + d_1\mathbf{I})\,\mathbf{d}y_d(k-1) + \mathbf{d}y_d(k)\end{aligned} \tag{27.57}$$

from which the time response of each state can be determined for given $y_d(k)$. Furthermore, since the observation equation is $y(k) = \mathbf{h}\tilde{\mathbf{x}}(k) = \mathbf{h}\mathbf{T}^{-1}(k)\mathbf{x}(k) = \mathbf{h}\mathbf{x}(k)$, the transformation does not affect the first element of the state vector and it is a trivial matter to obtain the output response from (27.57). In this regard, note that $\mathbf{hId} = 0$, $\mathbf{hDd} = 0$, $\mathbf{hD}^2\mathbf{d} = 0$ and $\mathbf{hD}^3\mathbf{d} = 1 + d_1 + d_2 + d_3 + d_4$, hence,

$$y(k) = -d_1y(k-1) - d_2y(k-2) - d_3y(k-3) - d_4y(k-4) + \tilde{d}y_d(k-3). \tag{27.58}$$

Expressed as a discrete-time TF, this has the desired characteristic polynomial (27.48), time invariant scalar numerator and a time delay of $\delta = 3$ samples, as required. In other words, the nonlinear terms are eliminated in the closed-loop and so the nature of the state dependency does not influence the theoretical response. Naturally this result assumes zero model mismatch, the same assumption as for linear pole assignment design. The robustness to model mismatch and disturbances is the subject of current research by the authors. However, simulation and experimental results support the practical utility of the approach [18, 24].

Finally, using the open-loop state equation (27.44) and following a similar substitution approach to Kuo [3], yields the controllability matrix for this example,

$$\mathbf{S}(k) = \left[\mathbf{g},\ \mathbf{F}(k+2)\mathbf{g},\ \mathbf{F}(k+2)\mathbf{F}(k+1)\mathbf{g},\ \mathbf{F}(k+2)\mathbf{F}(k+1)\mathbf{F}(k)\mathbf{g}\right], \tag{27.59}$$

which is non-singular if and only if $b(k) \neq 0, \forall k$. Although discussion of controllability is beyond the scope of the present chapter, it is sufficient to note that the nonlinear pole assignment problem can always be solved if $b(k) \neq 0$.

27.6 Extensions and Interpretations

This section briefly reviews a number of other extensions to basic NMSS design.

1. *Continuous-time design.* This chapter is limited to the discrete-time domain. However, models of physical systems are often derived as differential equations on the basis of natural laws; and they are characterized by parameters that can have a prescribed physical meaning. In order to move from such a differential equation model to its discrete-time equivalent, it is necessary to utilise some method of transformation. The resultant model and its associated parameter values are functions of the sampling interval. By contrast, continuous-time models are defined by a unique set of parameters and may be adapted for irregular sampling periods. Hence, it is logical to also consider delta-operator [16] and continuous-time [15, 34] versions of the NMSS approach.

2. *Multivariable control.* The general state space formulation of PIP control facilitates straightforward extension to the multivariable case [16, 19, 26]. Here, the system is characterized by multiple control inputs that affect the state and output variables in a potentially complicated and cross-coupled manner. Multivariable PIP design must take account of this natural cross-coupling and generate control inputs which ameliorate any unsatisfactory aspects of the system behaviour that arise from it. In this regard, the multi-objective optimisation framework mentioned in Sect. 27.4.2 has proven extremely valuable in practice.
3. *Model Predictive Control.* PIP control can be constrained to yield exactly the same algorithm as both Generalised Predictive Control (GPC) and minimal state, Linear Quadratic Gaussian (LQG) design methods [17]. In the more general Model Predictive Control (MPC) case, a number of recent articles have also utilised NMSS models; e.g. [14, 30, 35]. In contrast to basic PIP control, this approach has the advantage of explicitly handling system constraints at the design stage. Reference [30], for example, describes a framework for performance tuning of MPC using goal-attainment methods. Here, simulation experiments again suggest that NMSS models offer better design flexibility in some cases and hence can yield improved performance in comparison to minimal MPC.

27.7 Conclusion

This chapter has used worked examples to re-visit some of the most fundamental results in the theory of non-minimum state space (NMSS) control system design. NMSS design does not need any form of state reconstruction, since it always involves full state feedback based solely on measured input and output signals. In fact, the NMSS form appears to be the most obvious and straightforward way to represent a transfer function in state space terms. Minimal models represent the same system in a less intuitive manner, requiring each state to be formed from various, often rather abstract, combinations of the input and output signals. The NMSS approach has the further advantage in that it can inherently handle high order numerator polynomials and long time delays.

Nonetheless, the chapter has shown how the NMSS model can be transformed into a minimal form, while the associated control law can be constrained to yield the same closed-loop system as minimal design when there is no model mismatch. Even in this constrained form, however, non-minimal design can be more robust than state variable feedback based on some minimal models. By contrast, the observable canonical form yields a control system with a similar *structure* to the NMSS based controller. When we assign the additional poles in the NMSS case to the origin, these control algorithms are exactly the same. However, it is apparent from both simulation and practical case studies, that we would not normally wish to constrain NMSS design in this way. Rather, it is advantageous to utilise the additional poles in the NMSS case to provide extra flexibility and design freedom.

The discussion above has focused on single-input, single-output models, where the NMSS representation is used to develop Proportional-Integral-Plus (PIP) control systems. The approach is readily extended to stochastic, risk sensitive, model-predictive, multivariable and nonlinear systems. For example, recent research has utilised NMSS models with state dependent parameters. Practical examples of this nonlinear approach have included environmental control in buildings, construction robotics and nuclear decommissioning [18, 24, 25]. As discussed in Sect. 27.5, the control gains are updated at each sampling instant on the basis of the latest parameter values. Alternatively, in the pole assignment case, algebraic solutions can be derived off-line to yield a nonlinear PIP control algorithm that is relatively straightforward to implement on a digital computer, using a standard hardware-software arrangement. The references listed below provide much more detail about all these methods, including a wide range of practical and theoretical results.

Acknowledgements The authors welcome this opportunity to contribute to a book in honour of Professor Peter Young. We have particular reason to celebrate Peter's friendship, ideas and advice over many years. We are also grateful to Essam Shaban [28] and Philip Cross.

References

1. Evans, W.R.: Analysis of control system. AIEE Trans. **67**, 547–551 (1948)
2. Bode, H.W.: Feedback amplifier design. Bell Syst. Tech. J. **19**, 42 (1940)
3. Kuo, B.C.: Digital Control Systems. CBS Publishing Japan Ltd, Tokyo (1980)
4. Dorf, R.C., Bishop, R.H.: Modern Control Systems. Pearson Prentice Hall, Upper Saddle River (2008)
5. Taylor, C.J., Pedregal, D.J., Young, P.C., Tych, W.: Environmental time series analysis and forecasting with the Captain Toolbox. Environ. Model. Softw. **22**, 797–814 (2007)
6. Young, P.C.: Recursive Estimation and Time Series Analysis. Springer, Berlin, (1984)
7. Young, P.C., Behzadi, M.A., Wang, C.L., Chotai, A.: Direct digital and adaptive control by input-output, state variable feedback pole assignment. Int. J. Control **46**, 1867–1881 (1987)
8. Young, P.C.: Stochastic, dynamic modelling and signal processing: time variable and state dependent parameter estimation. In: Fitzgerald, W.J. (ed.) Nonlinear and Nonstationary Signal Processing. Cambridge University Press, Cambridge (2000)
9. Kalman, R.E.: A new approach to linear filtering and prediction problems. ASME Trans. J. Basic Eng. **83**, 95–108 (1960)
10. Luenberger, D.G.: Observing the state of a linear system. IEEE Trans. Mil. Electron. **8**, 74–80 (1964)
11. Young, P.C., Willems, J.C.: An approach to the linear multivariable servomechanism problem. Int. J. Control **15**, 961–979 (1972)
12. Hesketh, T.: Linear quadratic methods for adaptive control—a tutorial. Control Systems Centre Report 765, UMIST, Manchester, UK (1992)
13. Taylor, C.J., Young, P.C., Chotai, A., Whittaker, J.: Non-minimal state space approach to multivariable ramp metering control of motorway bottlenecks. IEE Proc., Control Theory Appl. **145**(6), 568–574 (1998)
14. Gonzalez, A.H., Perez, J.M., Odloak, D.: Infinite horizon MPC with non-minimal state space feedback. J. Process Control **19**, 473–481 (2009)
15. Gawthrop, P.J., Wang, L., Young, P.C.: Continuous-time non-minimal state-space design. Int. J. Control **80**, 1690–1697 (2007)
16. Chotai, A., Young, P.C., McKenna, P., Tych, W.: PIP design for delta-operator systems (parts I and II). Int. J. Control **70**, 123–168 (1998)

17. Taylor, C.J., Chotai, A., Young, P.C.: State space control system design based on non-minimal state-variable feedback: further generalisation and unification results. Int. J. Control **73**, 1329–1345 (2000)
18. Taylor, C.J., Chotai, A., Young, P.C.: Nonlinear control by input-output state variable feedback pole assignment. Int. J. Control **82**, 1029–1044 (2009)
19. Young, P.C., Lees, M., Chotai, A., Tych, W., Chalabi, Z.S.: Modelling and PIP control of a glasshouse micro-climate. Control Eng. Pract. **2**(4), 591–604 (1994)
20. Taylor, C.J., Young, P.C., Chotai, A., Mcleod, A.R., Glasock, A.R.: Modelling and PIP control design for free air carbon dioxide enrichment systems. J. Agric. Eng. Res. **75**, 365–374 (2000)
21. Taylor, C.J., Leigh, P.A., Chotai, A., Young, P.C., Vranken, E., Berckmans, D.: Cost effective combined axial fan and throttling valve control of ventilation rate. IEE Proc., Control Theory Appl. **151**(5), 577–584 (2004)
22. Gu, J., Taylor, C.J., Seward, D.W.: The automation of bucket position for the intelligent excavator LUCIE using the Proportional-Integral-Plus (PIP) control strategy. J. Comput.-Aided Civil Infrastruct. Eng. **12**, 16–27 (2004)
23. Shaban, E.M., Ako, S., Taylor, C.J., Seward, D.W.: Development of an automated verticality alignment system for a vibro-lance. Autom. Constr. **17**, 645–655 (2008)
24. Taylor, C.J., Chotai, A., Robertson, D.: State dependent control of a robotic manipulator used for nuclear decommissioning activities. In: IEEE International Conference on Intelligent Robots and Systems (IROS), Taipei, Taiwan (2010)
25. Taylor, C.J., Shaban, E.M., Stables, M.A., Ako, S.: Proportional-Integral-Plus (PIP) control applications of state dependent parameter models. IMECHE Proc., J. Syst. Control Eng. **221**, 1019–1032 (2007)
26. Taylor, C.J., Shaban, E.M.: Multivariable proportional-integral-plus (PIP) control of the ALSTOM nonlinear gasifier simulation. IEE Proc., Control Theory Appl. **153**(3), 277–285 (2006)
27. Challender, A., Hartree, D.R., Porter, A.: Time lag in a control system. Philos. Trans. R. Soc. Lond., Ser. A, Math. Phys. Sci. **235**, 415–444 (1936)
28. Shaban, E.M.: Nonlinear control for construction robots using state dependent parameter models. PhD thesis, Lancaster University, Engineering Department (2006)
29. Taylor, C.J., Young, P.C., Chotai, A.: PIP optimal control with a risk sensitive criterion. In: UKACC International Control Conference, Exeter, UK (1996)
30. Exadaktylos, V., Taylor, C.J.: Multi-objective performance optimisation for model predictive control by goal attainment. Int. J. Control **83**, 1374–1386 (2010)
31. Young, P.C.: A general approach to identification, estimation and control for a class of nonlinear dynamic systems. In: Friswell, M.I., Mottershead, J.E. (eds.) Identification in Engineering Systems, Swansea, UK, pp. 436–445 (1996)
32. Banks, H.T., Lewis, B.M., Tran, H.T.: Nonlinear feedback controllers and compensators: a state-dependent Riccati equation approach. Comput. Optim. Appl. **37**, 177–218 (2007)
33. Taylor, C.J., Chotai, A., Burnham, K.J.: Controllable forms for stabilising pole assignment design of generalised bilinear systems. Electron. Lett. **47**(7), 437–439 (2011)
34. Wang, L., Young, P.C., Gawthrop, P.J., Taylor, C.J.: Non-minimal state-space model-based continuous-time model predictive control with constraints. Int. J. Control **82**, 1122–1137 (2009)
35. Wang, L., Young, P.C.: An improved structure for model predictive control using non-minimal state space realisation. J. Process Control **16**(4), 355–371 (2006)

Chapter 28
Simulation Model Emulation in Control System Design

C.X. Lu, N.W. Rees, and Peter C. Young

28.1 Introduction

In *Dominant Mode Analysis* (DMA) [13], the data-based modeling tools employed for the analysis of real data are used to identify and estimate a reduced order 'emulation' of the high order simulation model, based on data obtained from planned experiments carried out on the simulation model with a fixed set of 'nominal' parameters. Although this *nominal emulation model* reproduces the dynamic behaviour of its high order progenitor with very high accuracy, it does not allow for the emulation of the simulation model if the parameters of the latter are changed from these nominal values. In order to address this limitation, the present chapter considers how the DMA can be extended to develop a more complete *Dynamic Emulation Model* (DEM; also called a 'meta-model') that maps the relationship between the two models in a more complete manner.

Emulation modeling of this type is discussed fully in [23, 24] (see also Chap. 16 in this book) and it is illustrated diagrammatically in Fig. 28.1. The complete dynamic emulation model behaves like the high order simulation model and so it can

C.X. Lu (✉) · N.W. Rees
School of Electrical Engineering and Telecommunications, University of New South Wales, Sydney, Australia
e-mail: c.lu@unsw.edu.au

N.W. Rees
e-mail: n.rees@unsw.edu.au

P.C. Young
Systems and Control Group, Lancaster Environment Centre, Lancaster University, Lancaster, UK
e-mail: p.young@lancaster.ac.uk

P.C. Young
Fenner School of Environment and Society, Australian National University, Canberra, Australia

L. Wang, H. Garnier (eds.), *System Identification, Environmental Modelling, and Control System Design*,
DOI 10.1007/978-0-85729-974-1_28,

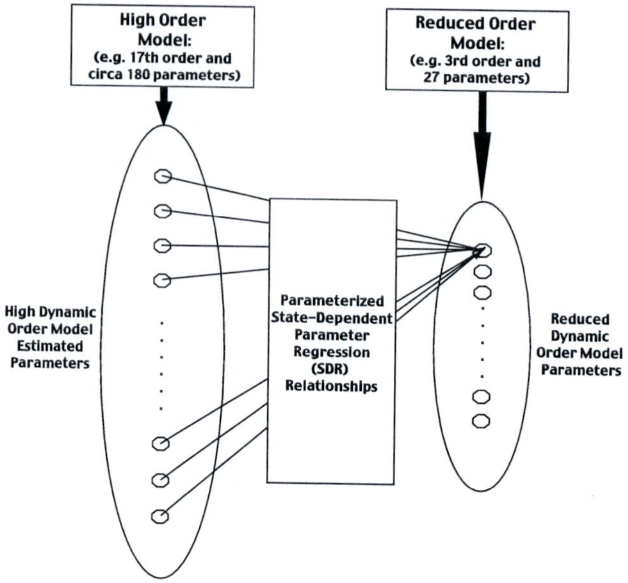

Fig. 28.1 The process of dynamic emulation model synthesis (note that the example referred to in this diagram is the third order emulation of a 17th order general equilibrium econometric model estimated for the Euro Area)

usefully replace it in certain applications. For example, it can assist in (or even replace) conventional sensitivity analysis, which is normally required to investigate which of the many parameters that characterize the high order simulation model are most important in defining the model's dynamic behaviour. And it can take the place of the simulation model in applications such as real-time flood forecasting [2] or control systems analysis and design. It is the use of emulation modeling in this latter context that we consider in the present chapter. However, while complete dynamic emulation modeling has been applied recently in economics [24] and hydrology [19, 23], the illustrative automatic control application described in the present chapter has been limited so far to the use of a nominal emulation model. In this sense, the example represents the first stage of a larger project, with the prospect of extending the nominal emulation and control system design described here to a complete DEM and control system design in the future. In fact, the results obtained in this example are useful in their own right, demonstrating significant improvements introduced by the present nominal DEM and multivariable control system design and suggesting that the implementation using full DEM and associated scheduled adaptive control should be straightforward.

28.2 Dominant Mode Analysis

It has been known for many years that the response of a high order, linear dynamic model is almost always dominated by a relatively few modes of dynamic behaviour, with many of the eigenvalues having little effect on the output response. This is illustrated well by the 'dispersion analysis' of [5]. Although quite different theoretically, this can be compared with principal component analysis, since it reveals the percentage of the output response that is explained by each dynamic mode of the model, as defined by its associated eigenvalue.

As Liaw shows, the part of the model defined by these dominant modes can be extracted and used as a reduced order model. However, a reduced order model of the same order can also be estimated using dominant mode analysis, but the dominant modes are not constrained to be exactly the same as those obtained analytically using Liaw's approach. In effect, the DMA leads to a reduced order model that captures not only the analytical dominant mode behaviour, but also some of the less dominant modal characteristics. In this way, for any given reduced order model order, the unconstrained DMA model will tend to explain the output response much better than the constrained Liaw model.

In the case of reduced order linearized models, the DMA methodology involves experiments in which the complex, physically-based simulation model is perturbed about some defined equilibrium or operating point, using an input signal (or signals) that will reveal all the dominant behavioural modes. A low order, normally multi-input, TF model is then identified and estimated from the resulting set of simulated input-output data using the RIV/RIVC model identification algorithms [12, 16]. As might be expected from dynamic systems theory, a low order linear model obtained in this manner reproduces the quasi-linear behaviour of the original nonlinear model about the operating point almost exactly for small perturbations. Perhaps more surprisingly, however, the dominant mode model can often also mimic the larger perturbation response (see e.g. [21, 22] and the later practical application). It is important to note that, in the case of general nonlinear systems, it is not possible to ensure that the experiments and associated DMA will reveal all of the complex simulation model response characteristics. In the absence of any theory in this connection (which would be very difficult in the case of general nonlinear models), all that can be done is to ensure that the experimentation on the simulation model is as comprehensive as possible over the user-defined parameter ranges. This implies the need for the planning of the dynamic simulation experiments that are required for the DMA. One advantage in the emulation context, however, is that the complex model is known exactly, so that designing such planned experiments to include optimal input perturbations (see e.g. [4]) may be possible. For the practical example presented later, however, no attempt has been made yet to optimize the inputs.

Finally, it should be noted that, if a reduced order, linearized model representation is not sufficiently effective in explaining the high order, nonlinear model behaviour, then a nonlinear description is essential. In this situation, the linear TF identification considered in the present chapter can be replaced by nonlinear *State-Dependent Parameter* (SDP) transfer function modeling [14, 15, 20]. These references provide a number of simulated examples (e.g. the chaotic logistic growth equation and the cosine map) and real examples (e.g. signals from the giant axon of a squid and the limit cycle behaviour of blowfly populations) that include systems exhibiting complex limit cycling and chaotic behaviour. In addition, SDP relations can exhibit sharp discontinuities, thresholds, harsh limits and even hysteresis: for example, [17] identify a hydrological DBM model where hysteresis associated with the effects of snow-melt is identified by SDP analysis. It is clear, therefore, that if the high order model response had these various kinds of nonlinear dynamic characteristics, then it is likely that such SDP models would be required for emulation.

The major difference between other approaches and the procedure described in the present chapter is the method of approximating the large dynamic simulation model. In our DMA approach, this approximation is based on the same model form used by most large simulation models: namely differential equations or their discrete-time equivalents. As a result, the reduced order model produced by DMA is likely to be more parametrically efficient than these other approximations since it provides an explicit dynamical representation in terms of the time constants, natural frequencies and steady state gains that characterize many large dynamic simulation models.

28.3 Emulation Modeling and Control of a Multivariable Power Plant System

A large Simulink model of the multivariable power plant system including its controllers has been developed by the two UNSW authors and is shown in Fig. 28.2, using a core nonlinear drum model known as the Åström-Bell model [1]. The Simulink modules represent important sections of the power plant, including the fuel system, feedwater system, boiler, superheater, throttle valve, turbine, reheater and generator. The open loop boiler system is unstable so it is necessary to have three major individual PID control loops for power output (MW controller), boiler drum level (Level Controller) and throttle pressure (Pressure Controller) introduced to stabilize and operate the system. For such a highly coupled, 3×3 multivariable, nonlinear system, as shown in Fig. 28.3, the control performance in many power plants is not adequate using the standard SISO PID arrangement [9]. In particular, a large water level variation known as 'shrink and swell' in the boiler drum during normal load change operation may trip the power plant [7]. Multivariable control strategy is a natural solution [10]. And since a typical linear modeling synthesis of a power plant boiler would generate a model of 14 to 20 or higher orders [8], it is not realistic to use such a model for control design and a good lower order emulation model of the system is an obvious alternative.

28.3.1 Dynamic Emulation Modeling

It is the system shown in Fig. 28.3 that is the object of the emulation analysis, where the overall power plant system block diagram is represented by three setpoints u_1 u_2 u_3 and three outputs y_1 y_2 y_3; and where the three PID controllers are part of

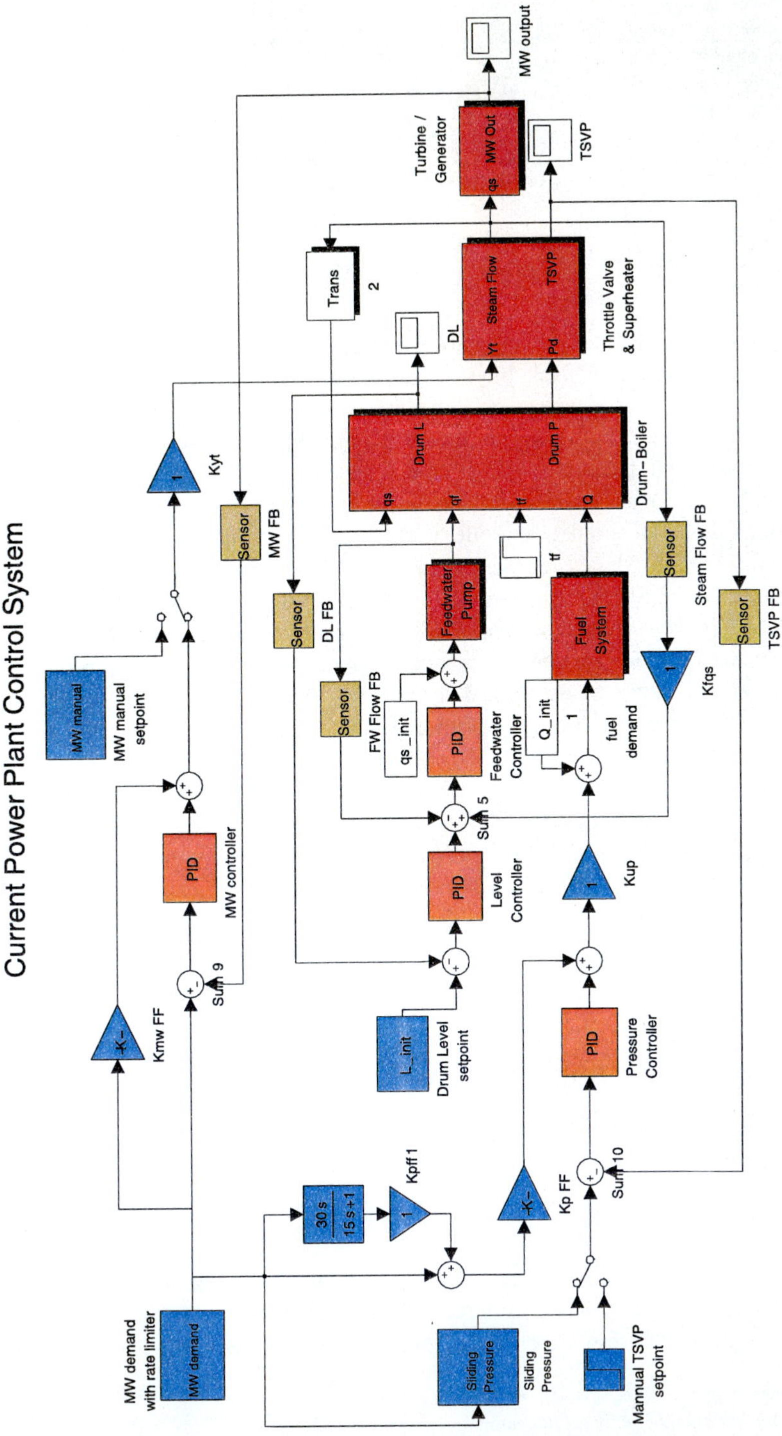

Fig. 28.2 Rees-Lu Simulink Model of the power plant system under standard, single channel PID control

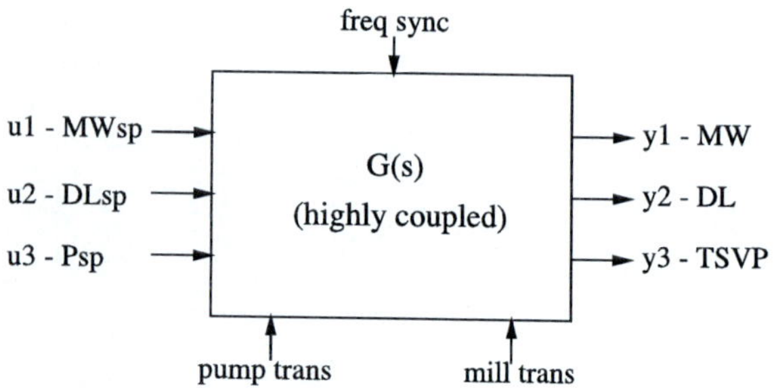

Fig. 28.3 Power plant system major inputs/outputs: the system has 3 setpoints u_1 u_2 u_3 and 3 outputs y_1 y_2 y_3, the system dynamics including 3 PID controllers all in $G(s)$

the process dynamics $G(s)$. This system is perturbed by the normalized unity step command inputs shown as black, blue and red dashed lines in Figs. 28.4, 28.5, 28.6. The resulting measured responses from the large simulation model are shown dotted in each of these figures but these are obscured by the low order, nominal emulation model responses, plotted as a full black line, which almost perfectly match the high order simulation model responses, as required.

Although the large simulation model is nonlinear, model structure identification shows that a linear model is able to explain the perturbational responses well, as we see in Figs. 28.4–28.6. This linear emulation model is a third order, continuous-time, multiple MISO form, with a common denominator polynomial for each MISO sub-model. The third order structure and associated parameters of these sub-models are identified and estimated by the hybrid continuous-time RIVC algorithm, using the simpler SRIVC option of the rivcbjid (structure identification) and rivcbj (parameter estimation) routines in the CAPTAIN Toolbox[1] for Matlab™. This simpler option applies because there is no noise on the large model simulation data except that induced by the extremely small residual that normally results from this kind of DMA exercise. The first sub-model for the MW power output, given below, illustrates the nature of the emulation model: the other two sub-models are of a similar form.

$$y_1(t) = \frac{B_{11}(s)}{A_1(s)}u_1(t) + \frac{B_{12}(s)}{A_1(s)}u_2(t) + \frac{B_{13}(s)}{A_1(s)}u_3(t), \tag{28.1}$$

where

$$A_1(s) = s^3 + 0.059921s^2 + 0.00042915s + 2.712 \times 10^{-6},$$
$$B_{11}(s) = 0.048289s^2 + 0.00036434s + 2.7192 \times 10^{-6},$$
$$B_{12}(s) = 0.00018577s^2 + 4.2822 \times 10^{-7}s - 2.0187 \times 10^{-10},$$
$$B_{13}(s) = 0.0013011s^2 + 5.8307 \times 10^{-5}s - 2.854 \times 10^{-9}.$$

The validation of the full multivariable model, as formed by the three MISO sub-models, is carried out by applying three smoothed random input commands concurrently. Figure 28.7 shows the results for y_1. As can be seen, the emulation for

[1]See http://www.es.lancs.ac.uk/cres/captain/.

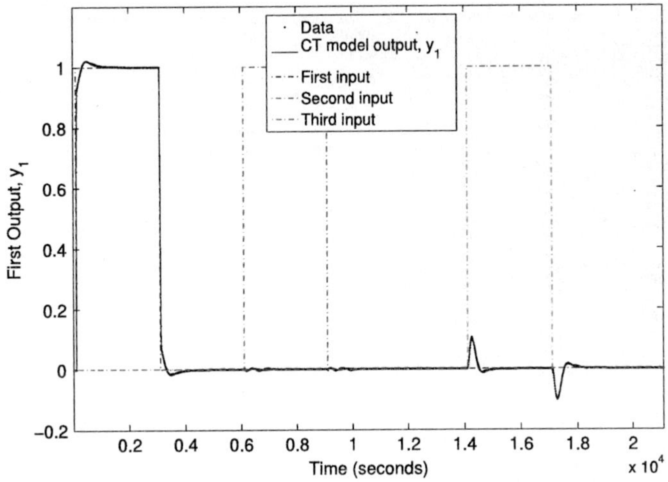

Fig. 28.4 Perturbation inputs, system response and emulation results: channel 1

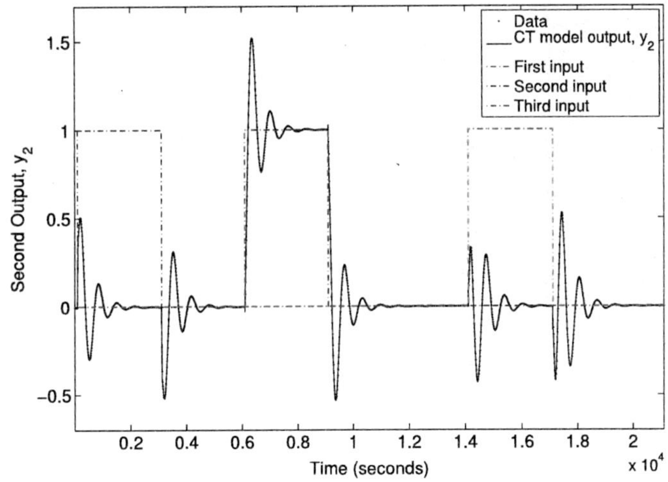

Fig. 28.5 Perturbation inputs, system response and emulation results: channel 2

this variable is almost perfect and the other two channels that are not shown have equally good fits, so providing confidence that model is a sensible basis for control system design.

28.3.2 Multivariable LQ-PIP Control System Design

The combined three MISO sub-models constitute the full, continuous-time, multi-variable emulation model of the boiler system. One advantage of this model is that

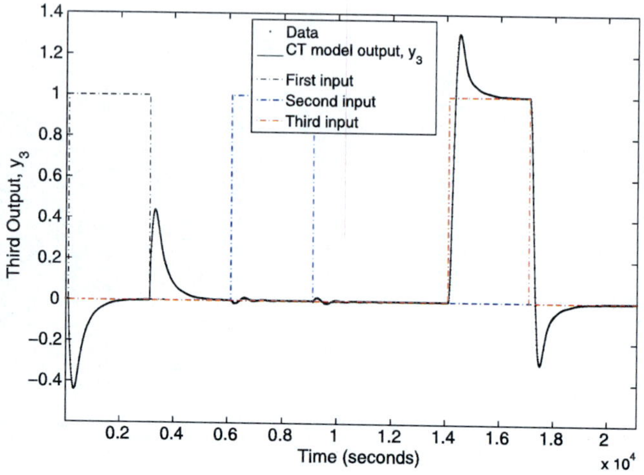

Fig. 28.6 Perturbation inputs, system response and emulation results: channel 3

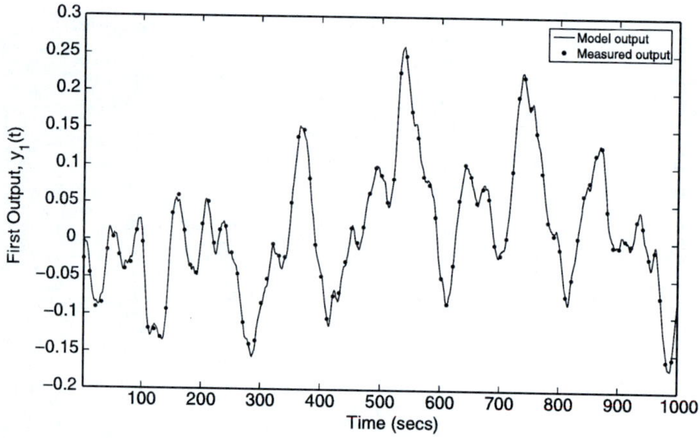

Fig. 28.7 Emulation validation: y_1 with concurrent inputs on all 3 channels

it can be converted into discrete-time models at different sampling intervals Δt, so that the effect of the sampling interval on digital control can be evaluated quite easily. These discrete-time MISO models are in a form that can be used immediately for multivariable *Proportional Integral Plus* (PIP) control system design: see, for example, [11] and the prior references therein. In this regard, the PIP routines in the CAPTAIN Toolbox are first used to convert the model into the multivariable *Non-Minimal State Space* (NMSS) model form required for the subsequent three channel, multivariable PIP control system design. The initial control system design studies have been based on the 'forward-path' (FP-PIP) implementation of the PIP multivariable controller using the *Linear-Quadratic* (LQ-PIP) design option, as shown

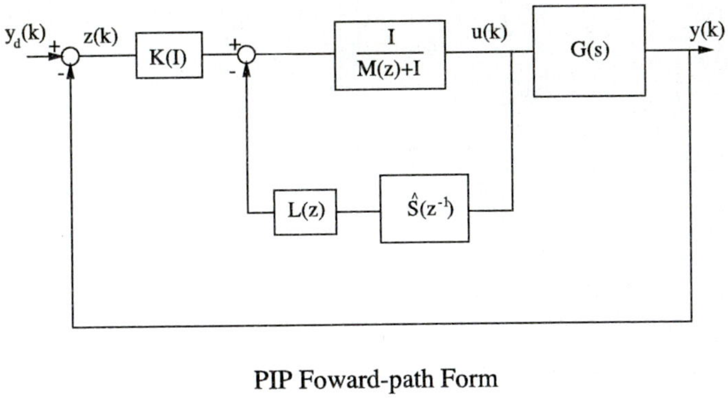

Fig. 28.8 PIP control forward-path form (FP-PIP) (see e.g. [11] and the prior references therein)

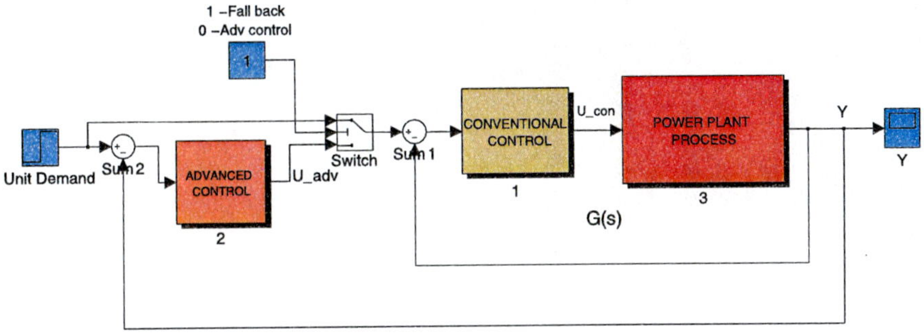

Fig. 28.9 FP-PIP control implementation schematic

in Fig. 28.8. This controller is then implemented as shown in Fig. 28.9 to trim the setpoints.

28.3.3 Multivariable LQ-PIP Control Results

The power plant control simulation setting is shown in Fig. 28.10, where the FP-PIP controller is being integrated into the original nonlinear simulator. This simulator is then subjected to the following operating scenario from its nominal 300 MW (60% load) steady state conditions:

1. The MW setpoint is ramped up by 100 MW at 3000 s to 400 MW and again at 4500 s to full load 500 MW, all at the rate of 10 MW/min.
2. The setpoint is then ramped down by 150 MW at 6500 s and again at 8000 s; and finally ramped down another 100 MW at 9500 s to reduce the MW output to 100 MW, all at the rate of 10 MW/min.

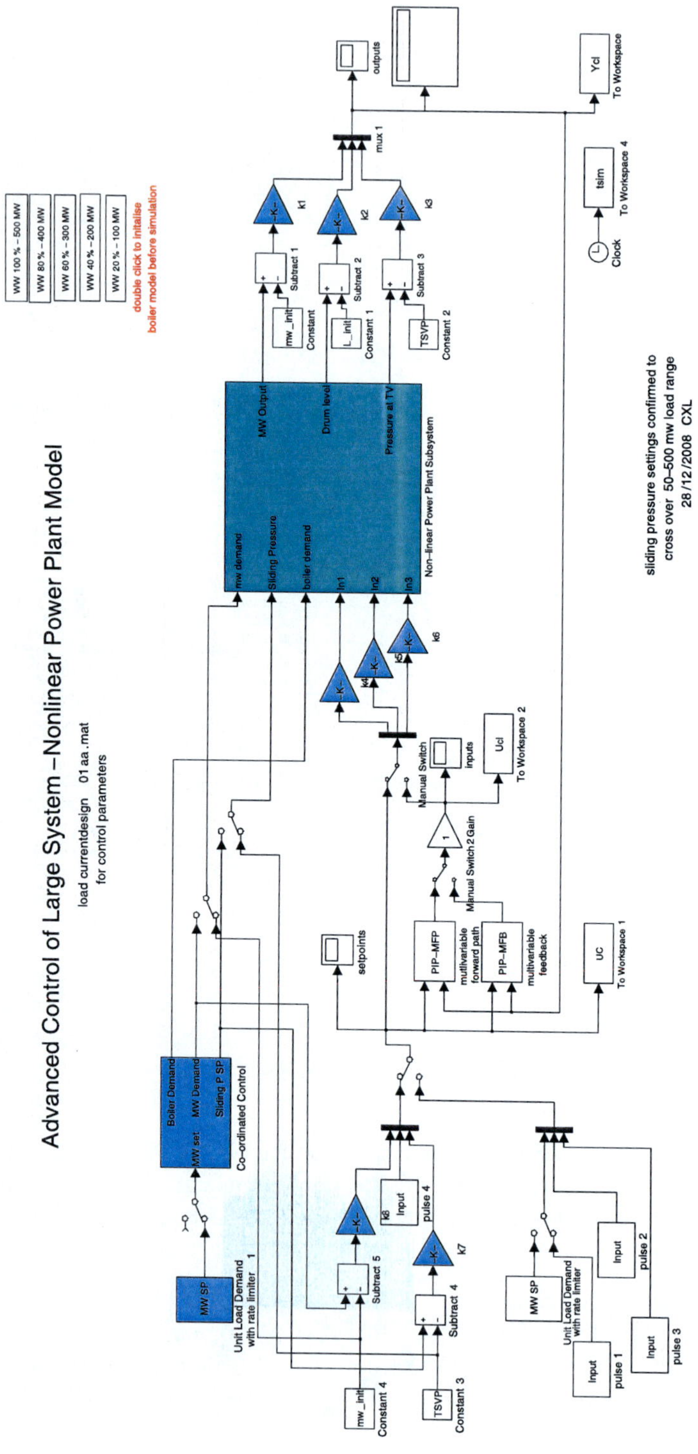

Fig. 28.10 PIP controlled boiler simulation

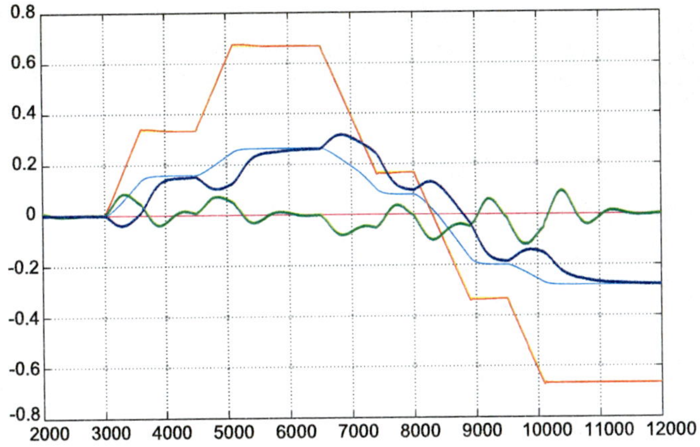

Fig. 28.11 Rees-Lu boiler simulation model: Best PID simulation results. MW output—*red*, setpoint—*yellow*; drum level output—*green*, setpoint—*magenta*; boiler pressure output—*blue*, setpoint—*light blue*

3. A sliding pressure setpoint change, which is a function of the MW setpoint values in order to optimise the efficiency of the boiler/turbine from the nonlinear steam thermodynamic property.
4. The drum level setpoint is always set at a constant level.

Figures 28.11 and 28.12 compare the setpoint tracking performance, in the above scenario tests, of the standard three channel PID controlled system, shown in Fig. 28.11, and the multivariable PIP closed loop control, shown in Fig. 28.12, where the y-axis is the % of changes based on 300 MW and the x-axis is the time in seconds. In particular, Fig. 28.11 shows the best performance that could be obtained by tuning the three channel PID controlled system responses: we see that the MW response (red) is very good, as required by the network management (otherwise the performance would be unacceptable). On the other hand, the other two outputs, boiler pressure (blue) and the drum level (green), have large variations caused by the load changes. At each point where the MW is required to rise, the pressure (blue) has a large, delayed drop behind the sliding pressure setpoint change (light blue); and it is similarly delayed in the opposite direction when MW is going down. Finally, the drum level (green) develops 'shrink and swell' at the load changing times (see earlier discussion), as the controller is unable to maintain the setpoints.

The PID control results can be compared with the PIP closed loop control results shown in Fig. 28.12, under the same scenario, where the initial tuning on the diagonal elements of the weighting matrix for the LQ-PIP (see e.g. [11]) are:

$$w_y = \begin{bmatrix} 2900 \\ 1 \\ 500 \end{bmatrix}; \qquad w_u = \begin{bmatrix} 125 \\ 1 \\ 1 \end{bmatrix}; \qquad w_z = \begin{bmatrix} 16 \\ 1 \\ 1 \end{bmatrix}.$$

These weights have been tuned from the default diagonal weighting values using a systematic, on-line tuning method developed by the first author [6] to minimise the

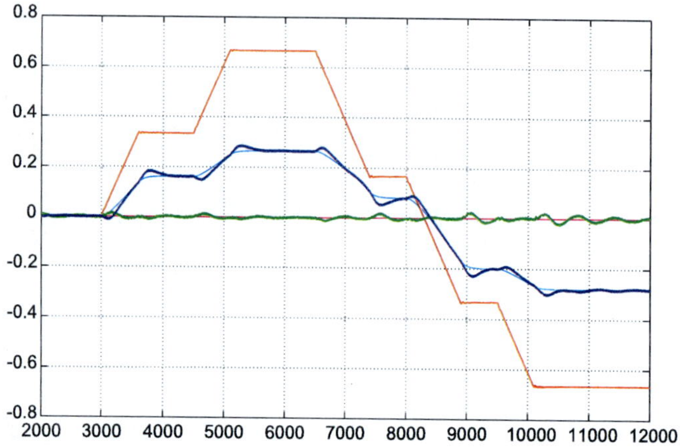

Fig. 28.12 Rees-Lu boiler simulation model: Initial multivariable FP-PIP simulation results. MW output—*red*, setpoint—*yellow*; drum level output—*green*, setpoint—*magenta*; boiler pressure output—*blue*, setpoint—*light blue*

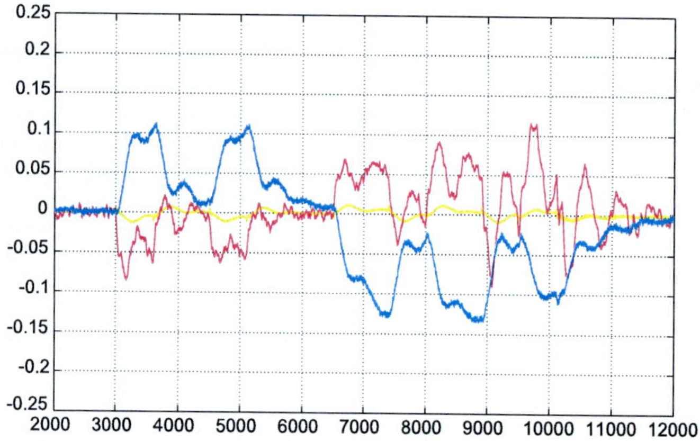

Fig. 28.13 Multivariable PIP control trim signals to PID controllers: u1 (for MW)—*light blue*; u2 (for drum level)—*yellow*; and u3 (for pressure)—*magenta*

error cost indices. The resulting improvement in performance is large when compared with the PID results: the boiler pressure (blue) is much tighter and, more importantly, the 'shrink and swell' of the drum water level (green) is well under control; while MW (red) follows the load change profile almost perfectly.

The multivariable PIP controller's outputs, which are the 'trim adjustments' added to the power plant coordinated setpoints of the existing PID controllers, are shown in Fig. 28.13.

The PIP controller's improvement in performance is obvious.

1. At the points where MW is ramped up, the pressure loop (light blue) is pre-fired to prevent the delayed drag, which compresses the drum level swell through cou-

pling; at the same time, MW (magenta) has been turned down to avoid overshoot caused by pressure rise; and, finally, the drum level has been trimmed down to control the swell.

2. At the points where MW is ramped down, the PIP control adjustments do the opposite in order to maintain the tight control.

28.3.4 Note

These are the first stage results obtained with only manipulation of the purely diagonal weightings associated with the quadratic cost function used in LQ-PIP control system design analysis for this study. No attempt has been made, so far, to utilize the decoupling power of the off-diagonal elements associated with the integral-of-error state variables in the NMSS model [3], nor investigate whether Delta-operator PIP control system design [3, 18] might offer some advantages. Such possibilities are being investigated in the continuing research and development programme.

28.4 Conclusions

This chapter has presented the concept of large simulation model emulation and outlined the methodological tools developed by the third author and his colleagues at Lancaster for identifying and estimating both 'nominal' and 'complete' emulation models and how these can be applied to a real industrial control problem investigated by the first and the second authors at the University of New South Wales, Sydney, Australia. It has shown how the large, nonlinear, power boiler system simulation model developed by the first two authors can be emulated by a linear, multivariable, third order, nominal emulation model. This emulation model reproduces the multivariable response of the simulation model very well indeed and can be used for multivariable control system design. This is illustrated by the initial design of a multivariable LQ-PIP control system and its successful implementation as an 'outer-loop' control system that considerably improves the multivariable performance of the 'inner-loop' PID controlled boiler system.

Surprisingly, although this LQ-PIP control system design is based on the nominal linear, third order emulation model, it is able to maintain good control over a wide range of load changes. Although it is noticeable that the PIP control performance of the drum water level is not as good at low load range (below 200 MW to 100 MW) as the range above it, this performance would be perfectly adequate in practice. However, it is well known in power plant control that there is a significant characteristic change at the lowest loads, so that the conventional PID controllers need to be switched into a different controller structure: see [7].

The development of a complete dynamic emulation model for the power plant model is under way. This involves varying the internal parameters of the large simulation model in order to obtain a series of plant input-output data on which to base the full emulation model analysis. It is hoped that a LQ-PIP controller based on such a complete dynamic emulation model can be operative over a whole range of the nonlinear power plant simulator, thus eliminating the need to switch the inner-loop PID control structure at low loads.

Acknowledgements Part of this project was carried out during Peter Young's visits to the UNSW as Visiting Professor. He is grateful for the financial assistance received on these visits.

References

1. Åström, K.J., Bell, R.D.: Drum-boiler dynamics. Automatica **36**, 363–378 (2000)
2. Beven, K.J., Young, P.C., Leedal, D.: Computationally efficient flood water level prediction (with uncertainty). In: Proceedings European Conference on Flood Risk Management, Oxford (2008)
3. Chotai, A., Young, P.C., McKenna, P.G., Tych, W.: Proportional-Integral-Plus (PIP) design for delta operator systems: Part 2, MIMO systems. Int. J. Control **70**, 149–168 (1998)
4. Goodwin, G.C., Payne, R.L.: Dynamic System Identification: Experiment Design and Data Analysis. Academic Press, New York (1977)
5. Liaw, C.M.: Model reduction of discrete systems using the power decomposition method. Proc. Inst. Electr. Eng. D **133**, 30–34 (1986)
6. Lu, C.: Advanced control of power plant—a practical approach. PhD thesis, University of New South Wales, Australia (2009, submitted)
7. Lu, C., Rees, N., Donaldson, S.: The use of the Åström-Bell model for the design on drum level controllers in power plant boilers. In: Proceedings of the 16th IFAC World Congress (2005)
8. McDonald, J., Kwatny, H.: Design and analysis of boiler turbine generator controls using optimal linear regulator theory. IEEE Trans. Autom. Control **18**(3), 202–209 (1973)
9. Rees, N.: Advanced power plant control for large load changes and disturbances. In: IFAC/CIGRE Symposium on Control of Power Systems and Power Plants, pp. 641–649 (1997)
10. Rees, N., Lu, C.: Some thoughts on the advanced control of electrical power plants. Trans. Inst. Meas. Control **24**(2), 87–106 (2002)
11. Taylor, C.J., Chotai, A., Young, P.C.: State space control system design based on non-minimal state variable feedback: further generalization and unification results. Int. J. Control **73**, 1329–1345 (2000)
12. Young, P.C.: Recursive Estimation and Time-Series Analysis. Springer, Berlin (1984). New revised edition in preparation for publication in 2011
13. Young, P.C.: Data-based mechanistic modelling, generalised sensitivity and dominant mode analysis. Comput. Phys. Commun. **117**, 113–129 (1999)
14. Young, P.C.: Stochastic, dynamic modelling and signal processing: time variable and state dependent parameter estimation. In: Fitzgerald, W.J., Walden, A., Smith, R., Young, P.C. (eds.) Nonlinear and Nonstationary Signal Processing, pp. 74–114. Cambridge University Press, Cambridge (2000)
15. Young, P.C.: The identification and estimation of nonlinear stochastic systems. In: Mees, A.I. (ed.) Nonlinear Dynamics and Statistics, pp. 127–166. Birkhäuser, Boston (2001)
16. Young, P.C.: The refined instrumental variable method: unified estimation of discrete and continuous-time transfer function models. J. Eur. Syst. Autom. **42**, 149–179 (2008)
17. Young, P.C., Castelletti, A., Pianosi, F.: The data-based mechanistic approach in hydrological modelling. In: Castelletti, A., Sessa, R.S. (eds.) Topics on System Analysis and Integrated Water Resource Management, pp. 27–48. Elsevier, Amsterdam (2007)

18. Young, P.C., Chotai, A., McKenna, P.G., Tych, W.: Proportional-Integral-Plus (PIP) design for delta operator systems: Part 1, SISO systems. Int. J. Control **70**, 123–147 (1998)
19. Young, P.C., Leedal, D., Beven, K.J.: Reduced order emulation of distributed hydraulic models. In: Proceedings 15th IFAC Symposium on System Identification SYSID09, St Malo, France (2009)
20. Young, P.C., McKenna, P., Bruun, J.: Identification of nonlinear stochastic systems by state dependent parameter estimation. Int. J. Control **74**, 1837–1857 (2001)
21. Young, P.C., Parkinson, S.: Simplicity out of complexity. In: Beck, M.B. (ed.) Environmental Foresight and Models: A Manifesto, pp. 251–294. Elsevier, Oxford (2002)
22. Young, P.C., Parkinson, S., Lees, M.J.: Simplicity out of complexity: Occam's razor revisited. J. Appl. Stat. **23**, 165–210 (1996)
23. Young, P.C., Ratto, M.: A unified approach to environmental systems modeling. Stoch. Environ. Res. Risk Assess. **23**, 1037–1057 (2009)
24. Young, P.C., Ratto, M.: Statistical emulation of large linear dynamic models. Technometrics **53**, 29–43 (2011)

Chapter 29
Predictive Control of a Three-Phase Regenerative PWM Converter

Dae Keun Yoo, Liuping Wang, and Peter Gawthrop

29.1 Introduction

One of the key components in a renewable energy system is a three phase regenerative PWM (Pulse-Width-Modulation) converter that will perform both AC to DC and DC to AC conversions. In AC to DC conversion, the converter draws currents from the main electrical grid, to supply the power to a load. When the system is operating in a regenerative mode (i.e. DC to AC conversion), the converter injects the current into the main electrical grid, from a DC power source, such as the renewable energy generator.

For the proper and safe operation of the converter, a control system must satisfy the following two main control objectives. First, DC-link bus voltage has to be kept at the pre-defined voltage level under various loads. Depending on the applications, a load may range from a simple resistive type load (i.e. constant load current), to a more generic load, such as multiple motor drives with a common DC bus. The second control objective is to operate the converter in unity power factor. By maintaining the operation at the unity power factor, power losses (i.e. reactive power) are minimized while drawing and regenerating the power from the grid. In the literature, the most widely adopted control structure is the cascaded synchronous PI control [2, 5, 8] which typically employs outer and inner control loops for voltage and currents respectively.

D.K. Yoo (✉) · L. Wang
RMIT University, Victoria 3000, Australia
e-mail: dkyo@hotmail.com

L. Wang
e-mail: liuping.wang@rmit.edu.au

P. Gawthrop
School of Engineering, University of Glasgow, Glasgow, UK
e-mail: Peter.Gawthrop@glasgow.ac.uk

L. Wang, H. Garnier (eds.), *System Identification, Environmental Modelling, and Control System Design*,
DOI 10.1007/978-0-85729-974-1_29,

Departing from the cascaded PI control and single-loop system configuration, in this paper, the interactions between the plant dynamics are taken into consideration by using the model predictive control (MPC) technology [9, 12]. Although the predictive based controller is not an entirely new concept for this type of converters [7], MPC is becoming an attractive solution to control the power converter of this type. A recent work by [10], showed a concept, refer to as a finite state predictive control, where the optimal control solution is chosen through model prediction and cost function minimization by considering the finite switching combination. In this work, setting in the environment of continuous-time control system, a linearized model of the converter (known as state averaged model) is used to predict the response of the model, and based on the model prediction, the design objective is to find the derivative of the control input $\dot{u}(t)$ that minimizes the quadratic cost function. The structural and design differences between the current work and the work by [10] are that the predictive controller is designed using a continuous-time model and the discretization occurs at the implementation stage, and the model used for prediction is embedded with two integrators. The continuous-time design framework permits a fast sampling rate without the complication of model becoming ill-conditioned, hence produces a better dynamic closed-loop performance. With the integrators embedded in the design model, the derivatives of the control signals are optimized, leading to the simplified implementation procedures. Moreover, the proposed model predictive controller, DC bus voltage and i_q current are controlled directly by computing the optimum switching inputs that minimizes the error function between the predicted and measured DC bus voltage and i_q current. Furthermore, it is illustrated in this paper that the original cost function is modified by including a prescribed degree of stability factor in the design to improve the transient response.

29.2 Process Description and Plant Model

A typical three phase regenerative PWM converter is shown in Fig. 29.1. A three phase source (E_a, E_b, E_c) (i.e. main electrical grid) is connected to line inductors L_s with a equivalent series resistance, shown as R. The line reactor is the integral part of the rectifier, which provides a boosting feature of DC-link bus voltage, as well as suppressing the harmonics in the input AC currents. Followed by the line inductors are the six bi-directional switches, which have the ability to conduct current from the main three phase grid to DC-link bus when the device is used as a rectifier, and convert the DC voltage to AC currents when it is used as an inverter. These switches are made of semi-conductor devices, which can only operate in either ON or OFF states. Thus, a PWM module is often employed to produce the desired switching outputs. Finally, a DC-link capacitor (C_{dc}) is connected between the positive and negative of the DC bus, which acts as a voltage source to a load.

In this work, several assumptions are made about the operation of the converter. First, it is assumed that all switches are ideal and operate in a continuous conduction mode (CCM), and the grid voltage is symmetric and balanced with the following

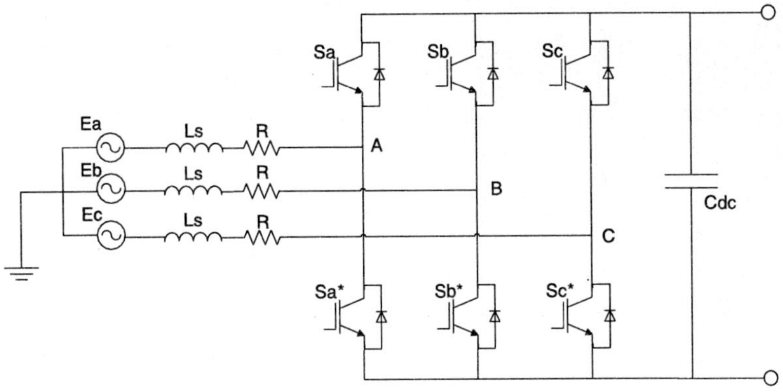

Fig. 29.1 A three phase PWM rectifier

mathematical description,

$$E_a = E_m \cos(\omega t), \tag{29.1}$$

$$E_b = E_m \cos\left(\omega t - \frac{2\pi}{3}\right), \tag{29.2}$$

$$E_c = E_m \cos\left(\omega t + \frac{2\pi}{3}\right), \tag{29.3}$$

where $\omega = 2\pi f$, f is 50 Hz. Furthermore, the system is assumed to be a three wire system, thus the sum of the three phase currents is equal to zero according to KCL,

$$i_a + i_b + i_c = 0. \tag{29.4}$$

Based on the assumptions, the dynamic behaviour of the converter is described by the non-linear state-space model, [11],

$$\begin{bmatrix} \frac{d}{dt} i_a \\ \frac{d}{dt} i_b \\ \frac{d}{dt} i_c \\ \frac{d}{dt} V_{dc} \end{bmatrix} = \begin{bmatrix} -\frac{R}{L_s} & 0 & 0 & \frac{S_a'}{L_s} \\ 0 & -\frac{R}{L_s} & 0 & \frac{S_b'}{L_s} \\ 0 & 0 & -\frac{R}{L_s} & \frac{S_c'}{L_s} \\ \frac{S_a}{C_{dc}} & \frac{S_b}{C_{dc}} & \frac{S_c}{C_{dc}} & 0 \end{bmatrix} \begin{bmatrix} i_a \\ i_b \\ i_c \\ V_{dc} \end{bmatrix} + \begin{bmatrix} \frac{E_a}{L_s} \\ \frac{E_b}{L_s} \\ \frac{E_c}{L_s} \\ 0 \end{bmatrix}, \tag{29.5}$$

where L_s is the inductance (mH), R is the resistance (Ω), C_{dc} is the capacitance (μF). Sa, Sb and Sc are the sinusoidal functions satisfying the following equations,

$$S_a = \frac{m}{2} \cos(\omega t - \psi) + \frac{1}{2}, \tag{29.6}$$

$$S_b = \frac{m}{2} \cos\left(\omega t - \psi - \frac{2\pi}{3}\right) + \frac{1}{2}, \tag{29.7}$$

$$S_c = \frac{m}{2}\cos\left(\omega t - \psi + \frac{2\pi}{3}\right) + \frac{1}{2}, \tag{29.8}$$

where m is the magnitude of the modulation and ω is the same as before. It is clearly seen from (29.5) that the dynamic system is nonlinear and time varying.

In order to create a control scheme that is simple and robust, it is necessary to transform the above (nonlinear and time-variant) model into a synchronous frame axis so to take advantage of linear time-invariant system. This transformation, namely synchronous frame transformation, is obtained by applying the transformation matrix given below,

$$T = \begin{bmatrix} \cos(\omega t) & \cos(\omega t - \frac{2\pi}{3}) & \cos(\omega t - \frac{4\pi}{3}) \\ -\sin(\omega t) & -\sin(\omega t - \frac{2\pi}{3}) & -\sin(\omega t - \frac{4\pi}{3}) \end{bmatrix}. \tag{29.9}$$

As a result, the dynamic equations of the converter in synchronous frame axis are expressed as,

$$\begin{aligned} L_s \frac{di_d}{dt} &= -Ri_d + \omega L_s i_q + e_d - v_d, \\ L_s \frac{di_q}{dt} &= -v_q - Ri_q - \omega L_s i_d, \\ C_{dc}\frac{dv_d}{dt} &= \frac{3}{4}(S_d i_d + S_q i_q) - i_L, \end{aligned} \tag{29.10}$$

where e_d is a grid source voltage, i_d, i_q are the input currents and v_d, v_q denotes control inputs, which are defined as below,

$$v_d = S_d * (v_{dc}/2), \tag{29.11}$$

$$v_q = S_q * (v_{dc}/2), \tag{29.12}$$

S_d and S_q are switching functions. Note that with the switching functions S_d and S_q as control variables, (29.10) become a set of bilinear equations. Further simplification of the model is obtained via linearization of the bilinear model at operating conditions, and the resulted linear time-invariant model is shown as

$$\begin{bmatrix} \dot{i}_d \\ \dot{i}_q \\ \dot{v}_{dc} \end{bmatrix} = \begin{bmatrix} -\frac{R}{L_s} & \omega & -\frac{S_{do}}{2L_s} \\ -\omega & -\frac{R}{L_s} & -\frac{S_{qo}}{2L_s} \\ \frac{3S_{do}}{4C_{dc}} & \frac{3S_{qo}}{4C_{dc}} & 0 \end{bmatrix} \begin{bmatrix} i_d \\ i_q \\ v_{dc} \end{bmatrix} + \begin{bmatrix} \frac{-v_{dco}}{2L_s} & 0 \\ 0 & \frac{-v_{dco}}{2L_s} \\ \frac{3i_{do}}{4C_{dc}} & \frac{3i_{qo}}{4C_{dc}} \end{bmatrix} \begin{bmatrix} S_d \\ S_q \end{bmatrix}, \tag{29.13}$$

where $S_{do}, S_{qo}, V_{dco}, i_{do}$ and i_{qo} represents steady state equivalent solutions. Specifically, supposing that at steady state operating condition, the converter maintains a target DC bus voltage with unity power factor, in other words both the magnitude of i_{qo} and v_q are assumed to be zero, then the steady state values of the parameters in the linear model are selected as $i_{qo} = 0$, $v_q = 0$, $V_{dco} = V_{ref}$. For switching functions S_{do}, S_{qo} are computed as [6],

$$S_{do} = \frac{2(e_d - Ri_d)}{V_{dco}}, \tag{29.14}$$

$$S_{qo} = \frac{-2\omega L_s i_d}{V_{dco}}. \tag{29.15}$$

Furthermore, the sum of switching functions must satisfy the following limit to avoid saturation.

$$S_d^2 + S_q^2 = \left[\frac{3}{4}\cos(30^\circ)\right]^2 = \frac{4}{3}. \tag{29.16}$$

As it was also pointed out in, [6], there is also an upper limit in a load current, which is,

$$I_L = \frac{3e_d^2}{8RV_{dc}}. \tag{29.17}$$

By choosing the load current to satisfy the inequality (29.17) constraints in I_L, a steady state solution of i_d can be obtained as follows.

$$i_d = \frac{1}{2}\left[\frac{e_d}{R} \pm \sqrt{\left(\frac{e_d}{R}\right)^2 - \frac{8V_{dco}I_L}{3R}}\right]. \tag{29.18}$$

29.3 Model Predictive Control Design

The objective of the model predictive control system is to regulate the DC bus voltage at a desired value specified by the applications while maintaining unity power factor. With this objective, the outputs of the regenerative power supply are chosen to be the voltage of the DC bus V_{dc} and the current i_q. When the system has a unity power factor, $i_q = 0$, which is the set-point signal for this output.

The model used in the design of the model predictive controller is

$$\dot{X}_m(t) = A_m X_m(t) + B_m u(t), \tag{29.19}$$

$$y(t) = C_m X_m(t), \tag{29.20}$$

where A_m, B_m and X_m are defined as

$$A_m = \begin{bmatrix} -\frac{R}{L_s} & \omega & -\frac{S_{do}}{2L_s} \\ -\omega & -\frac{R}{L_s} & -\frac{S_{qo}}{2L_s} \\ \frac{3S_{do}}{4C_{dc}} & \frac{3S_{qo}}{4C_{dc}} & 0 \end{bmatrix}, \qquad B_m = \begin{bmatrix} \frac{-v_{dco}}{2L_s} & 0 \\ 0 & \frac{-v_{dco}}{2L_s} \\ \frac{3i_{do}}{4C_{dc}} & \frac{3i_{qo}}{4C_{dc}} \end{bmatrix},$$

$$C_m = \begin{bmatrix} 0 & 1 & 0 \\ 0 & 0 & 1 \end{bmatrix}, \qquad X_m = \begin{bmatrix} i_d \\ i_q \\ v_{dc} \end{bmatrix}, \qquad u = \begin{bmatrix} S_d \\ S_q \end{bmatrix}.$$

In the operation of the converter, there are low frequency disturbances and harmonic distortion, thus integrators are needed in the controller. To embed the integrator, two auxiliary variables are chosen as

$$z(t) = \dot{x}_m(t),$$
$$y(t) = C_m x_m(t),$$

and based on them, a new state variable vector is defined as $x(t) = [z(t)^T \ \ y(t)^T]^T$. With these auxiliary variables, in conjunction with the original plant model, the augmented state space model is defined as:

$$\begin{bmatrix} \dot{z}(t) \\ \dot{y}(t) \end{bmatrix} = \overbrace{\begin{bmatrix} A_m & o_m^T \\ C_m & o_{q\times q} \end{bmatrix}}^{A} \begin{bmatrix} z(t) \\ y(t) \end{bmatrix} + \overbrace{\begin{bmatrix} B_m \\ o_{q\times m} \end{bmatrix}}^{B} \dot{u}(t), \tag{29.21}$$

$$y(t) = \underbrace{\begin{bmatrix} o_m & I_{q\times q} \end{bmatrix}}_{C} \begin{bmatrix} z(t) \\ y(t) \end{bmatrix}, \tag{29.22}$$

where $I_{q\times q}$ is the identity matrix with dimensions 2×2; $o_{q\times q}$ is a 2×2 zero matrix, $o_{q\times m}$ is a 2×2 zero matrix, and o_m is a 2×3 zero matrix.

29.3.1 Prediction and Optimization

In the core of model predictive control algorithm, the prediction of future states is constructed within a moving horizon window, followed by selection of a cost function and optimization of the cost function to obtain the future control trajectory. Following the same framework of continuous-time predictive control [12], for $0 \leq \tau \leq T_p$ (T_p is the prediction horizon), the derivative of control signal $\dot{u}(\tau)$ with two inputs is expressed as

$$\dot{u}(\tau) = \begin{bmatrix} \dot{u}_1(\tau) \ \dot{u}_2(\tau) \end{bmatrix}^T$$

and the input matrix B is partitioned as

$$B = \begin{bmatrix} B_1 & B_2 \end{bmatrix},$$

where B_1 and B_2 are the first and second columns of the B matrix. With this formulation, each input signal is described with a Laguerre function expansion. Namely, by choosing two continuous-time Laguerre function vectors $L_1(\tau)$ and $L_2(\tau)$ with dimensions N_1 and N_2, the derivative of the control signal $\dot{u}(\tau)$ is represented by

$$\dot{u}(\tau) = \begin{bmatrix} L_1^T(\tau) & o_{L2}^T \\ o_{L1}^T & L_2^T(\tau) \end{bmatrix} \begin{bmatrix} \eta_1 \\ \eta_2 \end{bmatrix},$$

where o_{L1} and o_{L2} are the zero column vectors with the same dimensions as $L_1(\tau)$ and $L_2(\tau)$. In addition, both $L_1(\tau)$ and $L_2(\tau)$ satisfy the differential equation as below, with their own scaling factors and number of terms (p and N parameters)

$$\dot{L}(\tau) = A_p L(\tau) \tag{29.23}$$

where

$$A_p = \begin{bmatrix} -p & 0 & \dots & 0 \\ -2p & -p & \dots & 0 \\ \vdots & \vdots & \ddots & 0 \\ -2p & \dots & -2p & -p \end{bmatrix}$$

and $L(0)$ is the $N \times 1$ vector with each element equal to $\sqrt{2p}$.

Assume that at the current time, say t_i, the state variable vector $x(t_i)$ is measured. Then at the future time τ, $\tau > 0$, the predicted state vector, denoted by $x(t_i + \tau \mid t_i)$ is described by the following equation

$$x(t_i + \tau \mid t_i) = e^{A\tau} x(t_i) + \int_0^\tau e^{A(\tau-\gamma)} \begin{bmatrix} B_1 L_1^T(\gamma) & B_2 L_2^T(\gamma) \end{bmatrix} d\gamma \begin{bmatrix} \eta_1 \\ \eta_2 \end{bmatrix}.$$

To simplify the notation, let the convolution integral be denoted as

$$\phi(\tau)^T = \int_0^\tau e^{A(\tau-\gamma)} \begin{bmatrix} B_1 L_1^T(\gamma) & B_2 L_2^T(\gamma) \end{bmatrix} d\gamma,$$

where $\phi(\tau)^T$ can be easily computed by solving a set of linear algebraic equations (see [12]). With $\eta^T = [\eta_1^T \; \eta_2^T]$, the prediction of future states is expressed as

$$x(t_i + \tau \mid t_i) = e^{A\tau} x(t_i) + \phi(\tau)^T \eta. \tag{29.24}$$

In general terms, the cost function used in predictive control has the form

$$J = \int_0^{T_p} x(t_i + \tau \mid t_i)^T Q x(t_i + \tau \mid t_i) \, d\tau + \eta^T R_L \eta, \tag{29.25}$$

where Q and R_L are symmetric positive definite and positive semi-definite matrices, written as $Q > 0$ and $R_L \geq 0$ respectively. By substituting the predicted state variables into the cost function, this cost function becomes

$$J = \eta^T \Omega \eta + 2\eta^T \Psi x(t_i) + \text{constant}, \tag{29.26}$$

where the quantities of Ω and Ψ are

$$\Omega = \left\{ \int_0^{T_p} \phi(\tau) Q \phi(\tau)^T \, d\tau + R_L \right\}; \qquad \Psi = \int_0^{T_p} \phi(\tau) Q e^{A\tau} \, d\tau.$$

Consider now the unconstrained minimization with respect to the parameter vector η of the general cost function (29.26) in the absence of hard constraints. Then the minimizing η is the least squares solution

$$\eta = -\Omega^{-1} \Psi x(t_i). \tag{29.27}$$

By the principle of receding horizon control, the optimal control $\dot{u}(t)$ for the unconstrained problem at time t_i is

$$\dot{u}(t_i) = \begin{bmatrix} L_1^T(0) & o_{L2}^T \\ o_{L1}^T & L_2^T(0) \end{bmatrix} \begin{bmatrix} \eta_1 \\ \eta_2 \end{bmatrix}. \tag{29.28}$$

With the derivative of the control signal computed, the actual control signal is written as

$$u(t_i) = u(t_i - \Delta t) + \dot{u}(t_i) \Delta t, \tag{29.29}$$

where Δt is the sampling interval.

29.4 Predictive Control with a Prescribed Degree of Stability

This section investigates issues associated with tuning the performance of the proposed predictive controller. The outcome is the modification of the predictive control system with a prescribed degree of stability to achieve desired closed-loop performance.

The class of commonly used cost functions in the design of model predictive control is associated with the errors between the output of the converter $y(t)$ and the reference signal $r(t)$ and it has the form [3, 4],

$$J = \int_0^{T_p} \{(r(t_i) - y(t_i + \tau \mid t_i))^T (r(t_i) - y(t_i + \tau \mid t_i)) + \dot{u}(\tau)^T R \dot{u}(\tau)\} \, d\tau.$$

This selection of cost function provides a simple solution to the choice of design parameters in the model predictive controller, where the Q matrix in the general cost function (29.25) is selected as $Q = C^T C$. In the absence of constraints, it is known [12] that if exponential data weighting is employed then the predictive controller converges to the corresponding linear quadratic regulator with sufficiently large prediction horizon T_p and large N_1 and N_2 (where the last two parameters are the numbers of terms included in the Laguerre function expansion). Therefore, the closed-loop poles of the predictive control system will follow the stable branches of the dual root-locus, dictated by the choice of the weight coefficient r_w ($R = r_w I$).

$$\det\left(I + \frac{1}{r_w}\frac{G(s)G(-s)}{s(-s)}\right) = 0, \tag{29.30}$$

where $G(s) = C_m(sI_{n_1} - A_m)^{-1}B_m$ is the Laplace transfer function of the converter. The stable branches of the dual root-locus provide limited options for the desired closed-loop eigenvalues in the design. In order to overcome the performance limitation, consider the general cost function

$$J = \int_0^{T_p} \left[x(t_i + \tau \mid t_i)^T Qx(t_i + \tau \mid t_i) + \dot{u}(\tau)^T R\dot{(\tau)}\right] d\tau, \tag{29.31}$$

where $Q \geq 0$, $R > 0$. With this general form cost function, the resulting closed-loop poles of the predictive control system will not necessarily obey the root-locus rule given above. In the application of regenerative PWM converter the matrix Q is 5×5, containing 25 elements. It is very difficult and time consuming to select the individual element in Q to achieve desired closed-loop performance, because not only the individual elements themselves affect the closed-loop performance, but also their combinations are important to the effect. Furthermore, the formulation of the predictive control problem has led to an augmented system state matrix (A) that has 2 poles on the origin of the complex plane. As a result, the predictive control system is numerically ill-conditioned. Therefore, there is a need to improve the numerical conditioning and develop a systematic way to tune its closed-loop performance.

One approach to predictive control with a prescribed degree of stability has been developed in [12] where the resulting design also overcomes the numerical ill-conditioning problem. It is essential to use a stable model in the predictive computation and the strategy here is to select an exponential weighting α and $A - \alpha I$ in this computation, where $\alpha > 0$ for the regenerative power supply. In the case where the plant is unstable with all its eigenvalues lying to the left of the line $s = -\varepsilon$ line in the complex plane, where $\varepsilon > 0$, $\alpha > \varepsilon$ is required.

Once the exponential weighting factor α is selected, the eigenvalues of the matrix $A - \alpha I$ are fixed. Since this matrix is stable for an appropriate choice of α, the prediction of the state variables is numerically well conditioned and prediction horizon T_p is selected sufficiently large to capture the transformed state variable response. In general, if the eigenvalues of $A - \alpha I$ were further away from the imaginary axis in the complex plane, then a smaller T_p can be used.

The use of exponential data weighting alters the original closed-loop performance as specified by the cost function weighting matrices Q and R, and in order to compensate for this variation, the Q matrix is replaced by

$$Q_\alpha = Q + 2\alpha P, \tag{29.32}$$

where P is the solution of the Riccati equation

$$PA + A^T P - PBR^{-1}B^T P + Q = 0. \tag{29.33}$$

To achieve a prescribed degree of stability β, the P matrix is found as the solution of the Riccati equation, [1]

$$P(A + \beta I) + (A + \beta I)^T P - PBR^{-1}B^T P + Q = 0 \tag{29.34}$$

and Q_α from

$$Q_\alpha = Q + 2(\alpha + \beta)P. \tag{29.35}$$

29.5 Experimental Results

29.5.1 Experimental Set-Up

The test-bed shown in Fig. 29.2 has been developed for this work to validate the proposed control design in the real hardware environment. As shown in this figure, the three phase regenerative PWM converter laboratory set-up consists of a step-down transformer (see mark (2)) that is used to reduce the line voltage from the main grid voltage (see mark (1)) of 415 V to 30 V. From the transformer, there are three line reactors (see mark (3)) connected between the converter and the transformer. The converter (see mark (4)) is made up of largely three components: a soft-start circuit, a number of sensors and a switching module. The soft-start circuit mainly provides a starting mechanism to limit the in-rush current when the DC-link capacitor is fully discharged at the start. The sensors include AC current sensors and DC bus voltage sensors. The switching module consists of six IGBT devices including freewheeling diode. The real-time model predictive controller is developed using xPC target (see mark 5), and finally a DC-link load is connected to the system (see mark 6).

The overall control software and PWM switching output is executed in xPC target (shown as (5) in figure). xPC target is Simulink's real-time toolbox which allows easy and seamless transition from a Simulink model to a real-time executable code. The system parameters used in the experimental set-up are $V_{ac} = 30$ V, $L_s = 8.9$ mH, $R_{dc} = 20 - 60\ \Omega$, $C_{dc} = 296\ \mu$F, and the reference DC link bus voltage is set to 65 V. The other parameters such as R, p and N are chosen according to the guidelines given in [12].

29.5.2 Comparison Study with and Without Prescribed Degree of Stability

To illustrate the significance of the prescribed degree of stability used in the design of model predictive control, a comparison study is done between the case where the

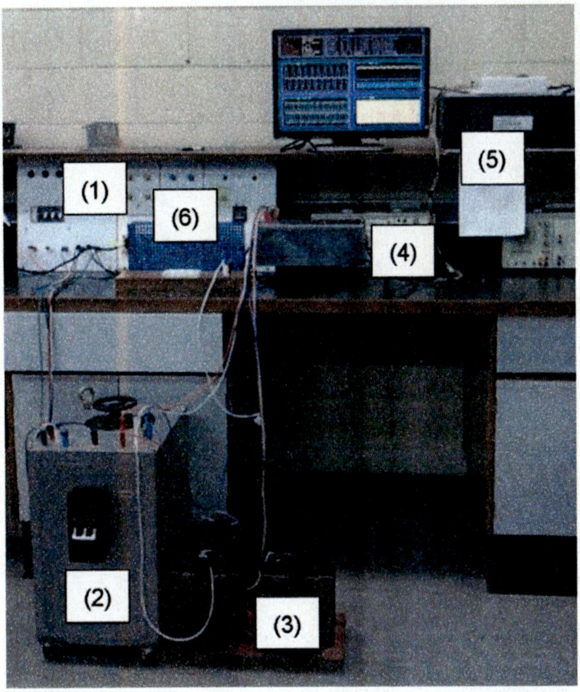

Fig. 29.2 Experimental set-up

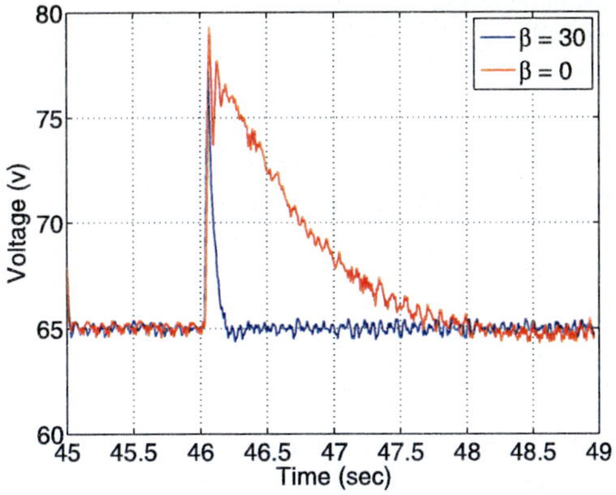

Fig. 29.3 Comparison of DC bus voltage response to a step load change with $\beta = 0$ and $\beta = 30$

prescribed degree of stability $\beta = 0$ and the case where $\beta = 30$. With the value of $\beta = 30$, all closed-loop eigenvalues of the predictive control system lie on the left of -1 line on the complex plane. In the experimental results shown in Fig. 29.3, a step load change of DC link resistance from 20 Ω to 40 Ω occurs at around 46 second, and the transient responses of the DC bus voltage are compared. It is seen from this figure that it took about 0.1 second for the DC bus voltage to return to the reference signal when $\beta = 30$, whilst when $\beta = 0$ it took at least more than 3 seconds for DC bus voltage to get to the vicinity of the reference signal. The results clearly show that the transient response of DC bus voltage to a step load change is

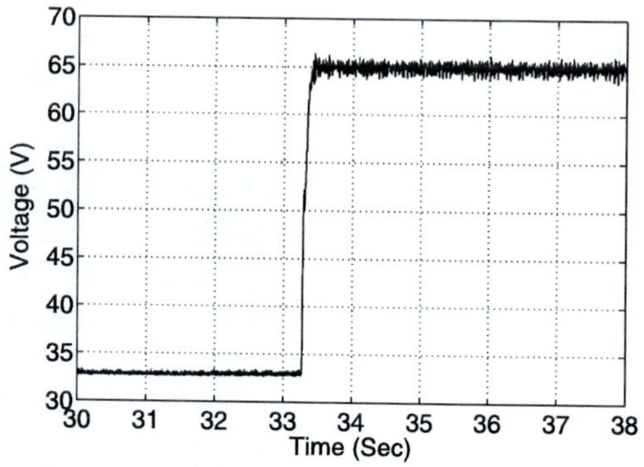

Fig. 29.4 DC bus voltage response to a step input in rectification mode

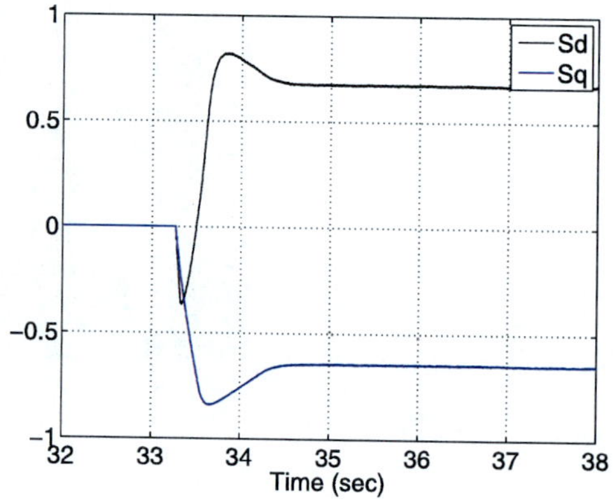

Fig. 29.5 Response of S_d and S_q in rectification mode

greatly improved for the case of $\beta = 30$. From hereafter, the experimental results are obtained with the prescribed degree of stability ($\beta = 30$).

29.5.3 Experimental Results for Rectification Mode

The control objective is keep the DC bus voltage at 65 V, and i_q current at zero for unity power factor. In the experimental testing, prior to the rectification mode, the converter is operating as a diode rectifier where the switching functions of IGBT are disabled and the current is only conducting through the freewheeling diode of IGBT. At the time (around $t = 33.3$ second) when the rectification mode is switched on, the predictive controller is activated to boost the DC bus voltage from 35 V to 65 V. Figures 29.4, 29.5, 29.6, 29.7 show that the closed-loop responses of the outputs $V_{dc}(t)$ and $i_q(t)$, as well as the state variable $i_d(t)$. It is seen from these figures that it took about 0.1 second for both output signals ($V_{dc}(t)$ and $i_q(t)$) and the state i_d to complete the closed-loop transient responses. In the rectification mode, Fig. 29.7 also shows that while drawing an extra current from the grid, i_q is well maintained around zero which results in zero phase shift between phase voltage and

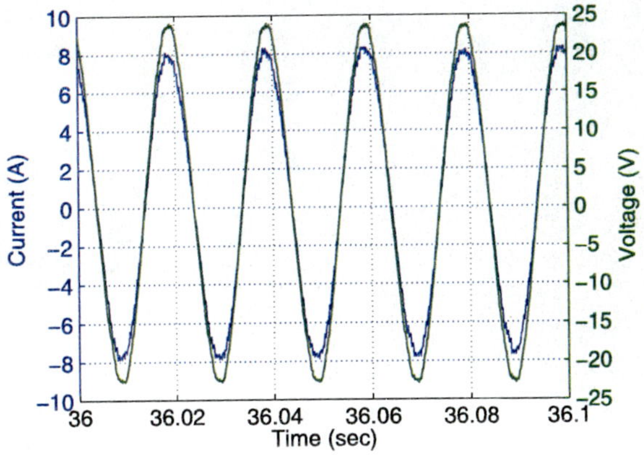

Fig. 29.6 Phase voltage and current in rectification mode

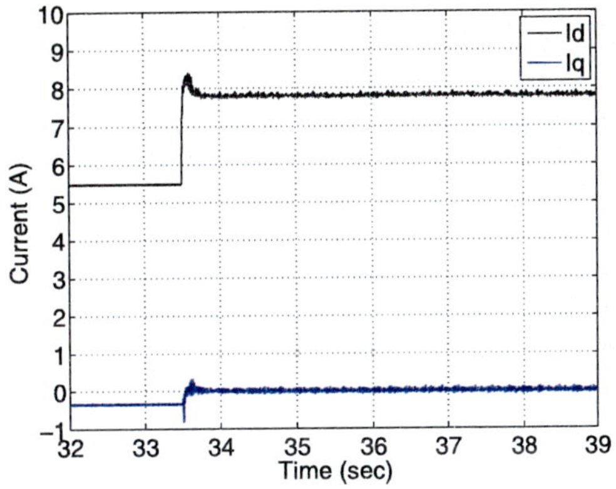

Fig. 29.7 I_d and I_q current in rectification mode

phase current (i.e. unity power factor) and i_d is increased to a new steady-state value according to the required DC bus voltage level. To confirm the reality of unity power factor, Fig. 29.6 shows phase A voltage and current in rectification mode, which indeed indicates the zero phase shifting between the phase voltage and current.

29.5.4 Experimental Results for Regeneration Mode

The control objective is to keep the DC bus voltage at 65 V, and i_q current at zero for unity power factor. Prior to the regeneration mode, the converter is operating in rectification mode. At around 70 second, as shown in Fig. 29.8, an extra current is injected in the DC bus, which resulted in initial overshoot of DC bus voltage and the predictive controller regulates the DC bus voltage around 65 V. Figure 29.9 shows the closed-loop responses of i_d and i_q in regeneration mode. It is seen that the steady-state value of i_d is negative that indicates that the current flow is reversed compared to the rectification mode. In this case the extra current injected into DC bus is converted to AC currents which feeds back into main grid. For unity power

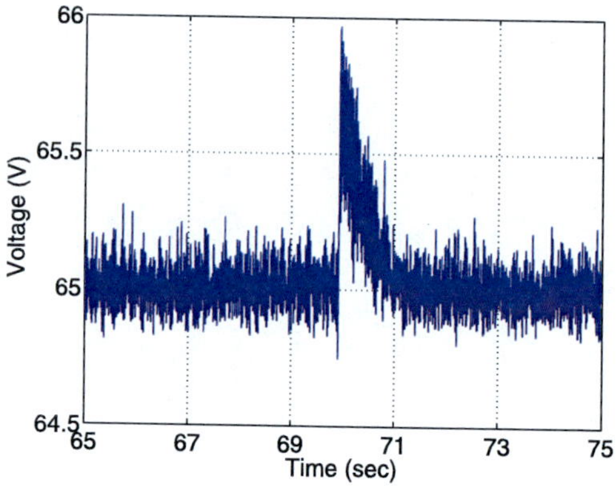

Fig. 29.8 DC bus voltage in regeneration mode

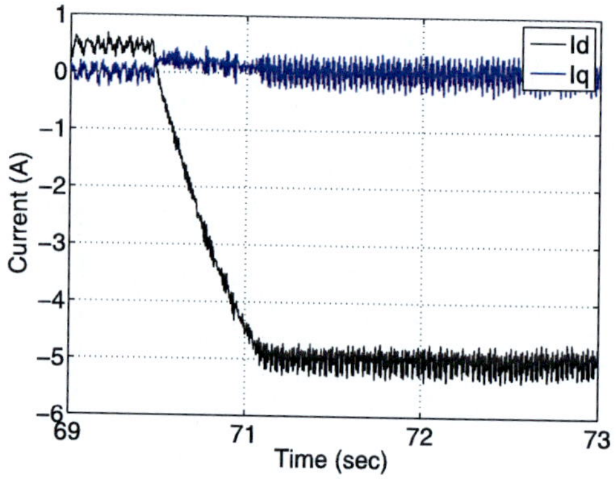

Fig. 29.9 Responses of i_d and i_q in regeneration mode

factor operation, i_q is still maintained around zero. This is evident via the plots of phase A voltage and current in regeneration mode as shown in Fig. 29.10, where it is seen that the phase A voltage and its current have 180° of phase shift.

29.5.5 Experimental Results for Disturbance Rejection

The control objective is keep the DC bus voltage at 65 V, and i_q current at zero for unity power factor, while load disturbance occurs. A series of step changes in the load are simulated by inserting or removing an extra DC load resistance in the circuit. At around 43.7 sec in Fig. 29.11, the DC link resistance is changed from 20 Ω to 40 Ω, corresponding to the case where the load current is decreased from 3.2 A to 1.6 A, where as in Fig. 29.12, around 50.2 sec, the resistance is changed from 40 Ω to 20 Ω. In both figures, it shows the closed-loop response of the DC bus voltage where it is seen that the predictive controller rejects the disturbance in about 0.1 second. Furthermore, the closed-loop responses of the i_d and i_q currents are also shown, where it is seen that the i_q current is kept well around zero while the steady-state value of i_d changes according to the load requirements.

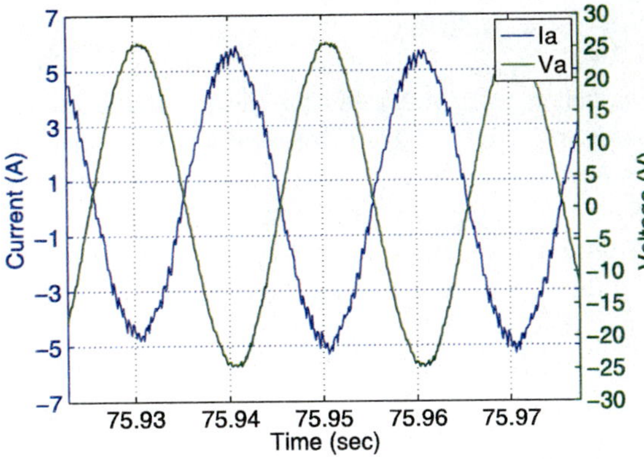

Fig. 29.10 Phase A voltage and current in regeneration mode

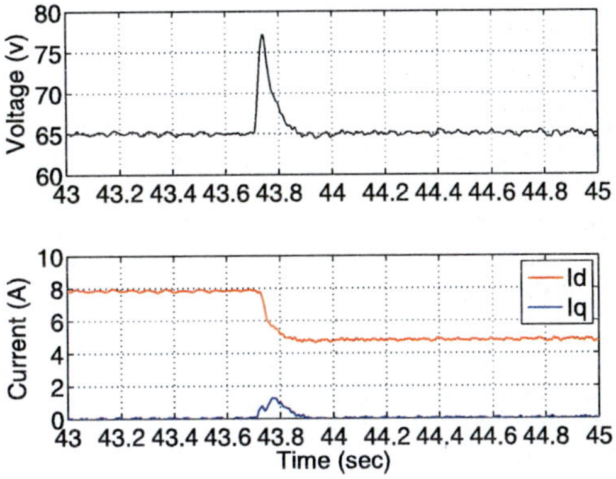

Fig. 29.11 DC bus voltage, i_d and i_q response to a step load change of R_{dc} from 20 Ω to 40 Ω

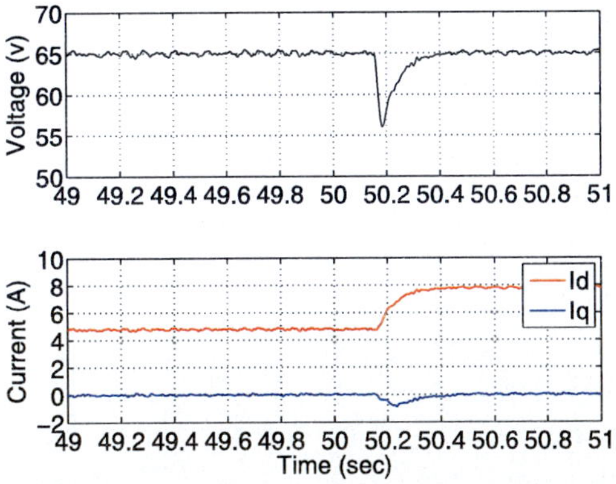

Fig. 29.12 DC bus voltage, i_d and i_q response to a step load change of R_{dc} from 40 Ω to 20 Ω

29.6 Conclusions

This paper has investigated the design and implementation of a continuous-time model predictive control system for a regenerative power supply. In particular, the proposed approach included a prescribed degree of stability in the algorithm that overcomes the performance limitation caused by the existing right-half-plant zero in the system, also provided an effective tuning parameter for the desired closed-loop performance. Experimental results show that the design algorithm and implementation are successful.

References

1. Anderson, B.D.O., Moore, J.M.: Linear Optimal Control. Prentice-Hall, Hemel Hempstead (1971)
2. Blasko, V., Kaura, V.: A new mathematical model and control of a three-phase ac-dc voltage source converter. IEEE Trans. Power Electron. **12**(1) (1997)
3. Clarke, D.W., Mohtadi, C., Tuffs, P.S.: Generalized predictive control. Part 1: The basic algorithm. Part 2: Extensions and interpretations. Automatica **23**, 137–160 (1987)
4. Cutler, C.R., Ramaker, B.L.: Dynamic matrix control-a computer control algorithm. Presented at the Meeting of the American Institute of Chemical Engineers, Houston, Texas (1979)
5. Kazmierkowski, M.P., Malesani, L.: Current control techniques for three-phase voltage-source pwm converters: survey. IEEE Trans. Indust. Electron. **45**(5) (1998)
6. Komurcugil, H., Kukrer, O.: Lyapunov-based control for three-phase pwm ac/dc voltage-source converters. IEEE Trans. Power Electron. **13**(5) (1998)
7. Kouro, S.: Model predictive control-a simple and powerful method to control power converters. IEEE Trans. Indust. Electron. **56**(6) (2009)
8. Liserre, M., Aquilla, A., Blaabjerg, F.: An overview of three-phase voltage source active rectifiers interfacing the utility. In: IEEE Bologna Power Tech. Conference (2003)
9. Maciejowski, J.: Predictive Control with Constraints. Prentice Hall, New York (2000)
10. Kennel, R., Quevedo, D., Cortes, P., Kazmierkowski, M.P., Rodriguez, J.: Predictive control in power electronics and drives. IEEE Trans. Indust. Electron. **55**(12) (2008)
11. Dewan, S.B., Wu, R., Slemon, G.: Analysis of an ac-to-dc voltage source converter using PWM with phase and amplitude control. IEEE Trans. Indust. Appl. **27**(2) (1991)
12. Wang, L.: Model Predictive Control System Design and Implementation Using MATLAB. Springer, Berlin (2009)

Chapter 30
SSpace: A Flexible and General State Space Toolbox for MATLAB

Diego J. Pedregal and C. James Taylor

30.1 Introduction

Numerous publications by the present authors, often in collaboration with Professor Peter C. Young, have made an extensive use of state space models in a wide range of contexts; e.g. [7, 13, 14, 16–20, 22–24, 27]. These articles demonstrate the versatility and generality of the state space framework. Although now considered as a 'classical' tool (especially in engineering and economics), the approach is in remarkably good health as an area of active research, judging by the enormous number of journal articles and books. With regard to software tools, there are also numerous options, with some packages for state space modelling freely downloadable over the Internet and many others available commercially.

The initial idea of building the `SSpace` toolbox was born many years ago, during a postdoctoral visit to Lancaster University by the first author, under the direction of Professor Young. At this time, the CAPTAIN toolbox for MATLAB was being developed by Professor Young and colleagues, including the present authors [23]. It became clear that a very general and flexible tool for state space systems could support the research activities of the authors, and time has proven this to be true. In this regard, the genesis of the present `SSpace` toolbox was for personal research use. However, it has subsequently been developed into a more formal and systematic tool, hence it is now being offered to the wider modelling community.

D.J. Pedregal (✉)
E.T.S. de Ingenieros Industriales and Institute of Applied Mathematics to Science and Engineering (IMACI), University of Castilla, La Mancha, Ciudad Real, Spain
e-mail: Diego.Pedregal@uclm.es

C.J. Taylor
Engineering Department, Lancaster University, Lancaster, UK
e-mail: c.taylor@lancaster.ac.uk

L. Wang, H. Garnier (eds.), *System Identification, Environmental Modelling, and Control System Design*,
DOI 10.1007/978-0-85729-974-1_30,

`SSpace` is implemented within the MATLAB software environment, hence interaction with other toolboxes is straightforward. It consists of a relatively small number of functions and so provides a user-friendly but exhaustive analysis of time series data. In particular, the state space model is provided in a very general form, with specific structures addressed in a flexible and intuitive manner. Almost any state space model may be implemented by appropriate user coding, but a growing number of templates are offered, in order to make routine tasks easier.

Section 30.2 reviews the general state space framework on which `SSpace` is based. This is followed in Sects. 30.3 and 30.4 by an overview of the toolbox and an explanation of how specific models are implemented. Section 30.5 provides a number of worked examples, of varying degrees of complexity, with the conclusions in Sect. 30.6.

30.2 General State Space Framework

In order to reduce *a priori* constraints, the following very general state and observation equations form the core of `SSpace`:

$$\begin{aligned} &\text{State equations:} \quad \mathbf{x}_{t+1} = \mathbf{\Phi}_t \mathbf{x}_t + \mathbf{\Gamma}_t \mathbf{u}_t + \mathbf{E}_t \mathbf{w}_t \\ &\text{Observation equations:} \quad \mathbf{z}_t = \mathbf{H}_t \mathbf{x}_t + \mathbf{D}_t \mathbf{u}_t + \mathbf{C}_t \mathbf{v}_t \end{aligned} \tag{30.1}$$

where $\mathbf{z}_t$ and $\mathbf{u}_t$ are the $m \times 1$ and $k \times 1$ vectors of output and input data, respectively (we assume that T observations are available of each of these variables); $\mathbf{w}_t$ and $\mathbf{v}_t$ are vectors of zero mean Gaussian noise, with dimension $v \times 1$ and $h \times 1$, and covariance matrices $\mathbf{Q}_t$ and $\mathbf{R}_t$, respectively; and $\mathbf{x}_t$ is the $N \times 1$ dimensional state vector, in which the initial state is independent of the noise. The remaining elements in (30.1) are system matrices with appropriate dimensions, i.e.,

$$\begin{aligned} &\mathbf{\Phi}_t : N \times N; \qquad &&\mathbf{\Gamma}_t : N \times k; \\ &\mathbf{E}_t : N \times v; &&\mathbf{H}_t : m \times N; \\ &\mathbf{D}_t : m \times k; &&\mathbf{C}_t : m \times h. \end{aligned}$$

Salient properties of the model (30.1) include: (i) all the elements are matrices or vectors, hence it is capable of representing Single-Input, Single-Output (SISO), Multiple-Input, Single-Output (MISO) and MIMO dynamic systems; (ii) all the elements are potentially time-varying, even the noise covariance matrices; and (iii) state and observation noise is correlated through the matrix $\mathbf{S}_t = E(\mathbf{v}_t \mathbf{w}_t^T)$.

With the assumption of Gaussian stochastic disturbances, the estimation problem consists of finding the first and second order moments (i.e. mean and covariance) of the state vector, conditional on all the data in a sample. The tools that allow us to perform this operation within a stochastic state space framework, are the Kalman Filter (KF) and Fixed Interval Smoothing (FIS) algorithms [2, 11].

For a data set of T samples, the KF runs forwards in time to yield a 'filtered' estimate of the state vector at every sample t. FIS is applied after this filtering pass and runs backwards in time, to generate a 'smoothed' estimate of the state vector which, at every sample t, is based on all T samples of the data. In `SSpace`, the formulation of these algorithms derives from: [4, 6, 8–10, 19, 26]. In particular, the forward recursions (KF) are:

$$\begin{aligned}
\mathbf{F}_t &= \mathbf{H}_t\hat{\mathbf{P}}_{t|t-1}\hat{\mathbf{H}}_t^T + \mathbf{C}_t\mathbf{R}_t\mathbf{C}_t^T, \\
\mathbf{G}_t &= \mathbf{\Phi}_t\hat{\mathbf{P}}_{t|t-1}\hat{\mathbf{H}}_t^T + \mathbf{E}_t\mathbf{S}_t\mathbf{C}_t^T, \\
\mathbf{K}_t &= \mathbf{G}_t\mathbf{F}_t^{-1}, \\
\hat{\mathbf{x}}_{t+1|t} &= [\mathbf{\Phi}_t - \mathbf{K}_t\mathbf{H}_t]\hat{\mathbf{x}}_{t|t-1} + \mathbf{K}_t\mathbf{z}_t + [\mathbf{\Gamma}_t\mathbf{u}_t - \mathbf{K}_t\mathbf{D}_t\mathbf{u}_t], \\
\hat{\mathbf{P}}_{t+1|t} &= \mathbf{\Phi}_t\hat{\mathbf{P}}_{t|t-1}\mathbf{\Phi}_t^T - \mathbf{K}_t\mathbf{G}_t^T + \mathbf{E}_t\mathbf{Q}_t\mathbf{E}_t^T.
\end{aligned} \tag{30.2}$$

The backward recursions (FIS) are:

$$\begin{aligned}
\hat{\mathbf{x}}_{t|N}^T &= \hat{\mathbf{x}}_{t|t-1} + \hat{\mathbf{P}}_{t|t-1}^T\mathbf{r}_{t-1}, \\
\hat{\mathbf{P}}_{t|N}^T &= \hat{\mathbf{P}}_{t|t-1} - \hat{\mathbf{P}}_{t|t-1}^T\mathbf{R}_{t-1}\hat{\mathbf{P}}_{t|t-1}^T, \\
\mathbf{r}_{t-1} &= \mathbf{H}_t^T\mathbf{F}_t^{-1}\mathbf{v}_t + \bar{\mathbf{\Phi}}_t^T\mathbf{r}_t \quad \text{with } \mathbf{r}_N = \mathbf{0}, \\
\mathbf{R}_{t-1} &= \mathbf{H}_t^T\mathbf{F}_t^{-1}\mathbf{H}_t + \bar{\mathbf{\Phi}}_t^T\mathbf{R}_t\bar{\mathbf{\Phi}}_t \quad \text{with } \mathbf{R}_N = \mathbf{0}, \\
\bar{\mathbf{\Phi}}_t &= \mathbf{\Phi}_t - \mathbf{K}_t\mathbf{H}_t.
\end{aligned} \tag{30.3}$$

The application of the recursive KF/FIS algorithms requires knowledge of all the system matrices, together with the noise covariance matrices. Depending on the particular structure of the model, there will be a number of elements that are known *a priori* (usually zeros and ones). Normally, however, there will also be some unknown elements. This hyper-parameter estimation problem is approached using time domain Maximum Likelihood (ML) optimisation. Assuming that all the disturbances are normally distributed, the log-likelihood function is computed using the KF via 'prediction error decomposition' [8, 21]. In this case:

$$\log L = -\frac{mT}{2}\log 2\pi - \frac{1}{2}\sum_{t=1}^{T}\log|\mathbf{F}_t| - \frac{1}{2}\sum_{t=1}^{T}\mathbf{v}_t^T\mathbf{F}_t^{-1}\mathbf{v}_t. \tag{30.4}$$

When maximising (30.4), the problem of defining initial conditions for the state vector and its covariance matrix needs to be resolved. The most popular solution in the literature is to define *diffuse priors*, e.g. zero values for the initial state vector and large values for the diagonal elements of its covariance matrix. This is the default initialization in `SSpace`, although other initial conditions may be chosen.

30.3 SSpace Overview

We believe that the key strength of `SSpace` is the flexibility with which models are specified. In this regard, unique features include:

- models are formulated by user-coded MATLAB functions, allowing full flexibility to both specify the model and to introduce bespoke requirements, such as imposing constraints on certain parameters;
- it is possible to define system matrices as strings, opening up the possibility of using some types of nonlinear model with standard state space tools;
- variance intervention [12] may be introduced for any model type.

Furthermore:

- analysis is performed with relatively few functions, hence the user needs to recall just a few function names;
- missing observations are treated automatically by signaling them with the usual MATLAB `NaN` variable (Not-a-Number);
- because of the common MATLAB environment, `SSpace` can be utilised to support other toolboxes, such as CAPTAIN and ECOTOOL, in which powerful graphical and statistical interfaces are implemented for a range of predefined model types [15, 23].

Table 30.1 shows a simplified list of functions, similar to that obtained by the command `help SSpace`. In general, help information is displayed by entering the function name without any input and output arguments. The main functions are in the first block of Table 30.1. Here, `fis` provides the Fixed Interval Smoother; `SSestim` is used for Maximum Likelihood model estimation; `SSpaceini` is an editable file that controls the global operation of the toolbox (optimisation convergence criteria, the appearance of tabular results, warning messages, etc.); `SSpacedemos` activates the tutorials and demos. The second set of functions listed in Table 30.1 allow for the conversion of predefined model types into state space form, whilst the third block lists various support functions for building new models (see Sect. 30.4 for details).

The general state space form in (30.1) can be directly exploited with `SampleSS`. In addition, combinations of models are possible with the help of `SampleCAT`, opening up a wide range of possibilities, in which the imagination of the user is the only limit. Clearly, this flexibility yields dangers for ill-advised users. `SSpace` allows the user to specify a model structure that is not identifiable, or that has other complications. The open-ended `SSpace` approach is, therefore, both a strength and (for inexperienced users) a possible weakness.

Furthermore, coding a new model in state space form can be rather cumbersome, hence a number of predefined models are already included in `SSpace`. Most of these have been developed under the influence of Professor Young and are available in other packages, such as CAPTAIN [23]. The authors gratefully acknowledge Professor Young's numerous research contributions and publications in this area. To illustrate, some of these models are briefly described in the following paragraphs.

Table 30.1 Simplified table of contents for the SSpace Toolbox

Main functions	
fis	Fixed Interval Smoother for SS systems
SSestim	Estimation of general SS models
SSpaceini	File that controls general options of SSpace toolbox
SSpacedemos	Run SSpace demos
Predefined models	
SampleSS	State Space general system
SampleCAT	Concatenation of State Space systems
SampleDAR	Dynamic AutoRegression
SampleLR	Linear Regression (static)
SampleDLR	Dynamic Linear Regression
SampleDHR	Dynamic Harmonic Regression
SampleBSM	Basic Structural Model
SampleTF	Transfer Function
SampleVARMAX	Vector AutoRegressive Moving Average with eXogenous variables
Building new models	
catsystem	Concatenation into a single model
components	Estimate components for BSM and DHR models
constrain	Transform any vector in a vector of values between given limits
varmatrix	Builds a semidefinite positive matrix from a vector of values
confband	Confidence bands of forecasts
vdif	Differentiation of a vector of variables
corrmatrix	Builds correlation matrix from covariance matrix
evalfun	Evaluates objective function at a given point

`SampleDAR` provides a template to deal with scalar dynamic autoregressions of order p, in which the parameters may be modelled as simple or integrated random walks, i.e. the state and observation equations in (30.1) become:

$$z_t = a_{1,t} z_{t-1} + \cdots + a_{p,t} z_{t-p} + v_t;$$
$$a_{i,t} = \frac{w_{i,t}}{(1-B)} \quad \text{or} \quad a_{i,t} = \frac{w_{i,t}}{(1-B)^2}. \tag{30.5}$$

`SampleLR` is a template to deal with scalar multivariate linear regressions. It is used either independently or in multivariate models by block concatenation. Since there are no dynamics in this model, the state equation is redundant and the formulation utilises only the observation equation in (30.1), i.e. $\mathbf{z}_t = \mathbf{D}_t \mathbf{u}_t + \mathbf{v}_t$. By contrast, `SampleDLR` is a combination of the two previous models: time varying parameters are included, but in a scalar static linear regression, i.e. dynamic linear regression,

$$z_t = a_{1,t} u_t + \cdots + a_{k,t} u_{k,t} + v_t;$$

$$a_{i,t} = \frac{w_{i,t}}{(1-B)} \quad \text{or} \quad a_{i,t} = \frac{w_{i,t}}{(1-B)^2}. \tag{30.6}$$

SampleDHR is a template for multivariate dynamic harmonic regression [27],

$$\begin{aligned} \mathbf{z}_t &= \mathbf{T}_t + \mathbf{SE}_t + \mathbf{v}_t \\ &= \mathbf{T}_t + \sum_{t=1}^{P/2} \left[\rho_i \mathbf{A}_{i,t} \cos(2\pi it/P) + \rho_i \mathbf{B}_{i,t} \sin(2\pi it/P)\right] + \mathbf{v}_t, \end{aligned} \tag{30.7}$$

$$\begin{aligned} T_{k,t} &= \frac{w_{k,t}}{(1-B)} \quad \text{or} \quad T_{k,t} = \frac{w_{k,t}}{(1-B)^2} \quad \text{or} \quad T_{k,t} = \frac{w_{k,t}}{(1-B)} + \frac{w_{k,t}}{(1-B)^2}, \\ a_{j,i,t} &= \frac{w_{j,i,t}}{(1-B)} \quad \text{or} \quad a_{j,i,t} = \frac{w_{j,i,t}}{(1-B)^2}, \\ b_{j,i,t} &= \frac{w^*_{j,i,t}}{(1-B)} \quad \text{or} \quad b_{j,i,t} = \frac{w^*_{j,i,t}}{(1-B)^2}, \end{aligned} \tag{30.8}$$

where P is the number of observations per year associated with the seasonal component; $\mathbf{T}_t$ is a vector of trend components, whose generic element $T_{k,t}$ is modelled as a random walk, integrated random walk or local linear trend (listed from left to right above) and is independent of the rest of components; ρ_i are damping factors; $\mathbf{A}_{i,t}$ and $\mathbf{B}_{i,t}$ are vectors of time parameters associated to the i-th harmonic, whose generic elements $a_{j,i,t}$ and $b_{j,i,t}$ behave as either simple or integrated random walks.

Another option for seasonal models is SampleBSM, which is a template for the multivariate basic structural model of Harvey [8]. SampleTF is a template for multivariate transfer function models, with polynomials represented by the backward shift operator (arbitrary orders). Finally, SampleVARMAX is a template for VARMA models with eXogenous inputs, for which special cases include: the scalar ARIMA model, linear regression, ARX, VARX and VARMA, among others.

30.4 Model Implementations in SSpace

Assuming that a separate identification stage has been undertaken and posterior validation is carried out on the innovations, model implementation using SSpace comprises of four steps:

1. Specify the model in state space form.
2. Create a function that translates this model into MATLAB code. The function SampleSS may be used as a template for entering new models from scratch. Alternatively, one of the predefined models Sample* is used (Sect. 30.3).
3. Estimate unknown parameters (SSestim).
4. Determine optimal estimates of the states and innovations (fis).

We will illustrate these steps with the following random walk plus noise model, which is useful when the signal is considered as a moving mean with noise:

$$z_t = \frac{w_t}{(1-B)} + v_t; \qquad \text{var}(w_t) = Q; \qquad \text{var}(v_t) = R, \tag{30.9}$$

where Q and R are unknown parameters representing the (positive) noise variances.

30.4.1 Specify Model (Step 1)

One state space representation for (30.9) is:

$$\begin{aligned} &\text{State equation:} \quad x_{t+1} = x_t + w_t, \\ &\text{Observation equation:} \quad z_t = x_t + v_t. \end{aligned} \tag{30.10}$$

Comparison with the general system (30.1) provides the required values of the system matrices, i.e. $\mathbf{\Phi}_t = \mathbf{E}_t = \mathbf{H}_t = \mathbf{C}_t = 1$, while $\mathbf{\Gamma}_t$ and $\mathbf{D}_t$ do not exist because the model has no inputs and so are represented as empty `[]` matrices in `SSpace`.

30.4.2 Translate Model into MATLAB Code (Step 2)

The function `SampleSS` is edited and saved with a new filename, `example1`. All the system matrices are defined with obvious reserved words (`Phi` to represent Φ), whilst `Inter` and `PInter` are additional variables to define variance interventions (Sect. 30.4.4). For system matrices varying in time, three dimensional matrices should be used, in which the third dimension is time. Additional MATLAB code may be included as required (and there are typically two or more input arguments), but nothing should be removed from the standard template.

Listing 30.1 shows the `SampleSS` template and Listing 30.2 the modified version. Here, the input argument `p` is a vector of parameters, i.e. Q and R. By default, both state and observation noises are considered independent. Furthermore, the first element in the vector `p` has been assigned to the matrix `Q`, while the second is assigned to `R`. Since both must be positive values, the system matrices are defined as powers of 10. An alternative and equivalent definition is: `Q=varmatrix(p(1))`. For brevity in this chapter, the formating has been minimised in the various listings: the original `SSpace` templates are based on more conveniently spaced code.

30.4.3 Estimate Unknown Parameters (Step 3)

For application to an arbitrary data set, the unknown parameters are estimated using `SSestim`, following the syntax in Listing 30.3. Here, the variables `p` and `covpar`

Listing 30.1: Standard template `SampleSS` associated with (30.1)

```
function [Phi, Gam, E, H, D, C, Q, R, S, ...
  Inter, PInter]= SampleSS(p)
Phi=[]; Gam=[]; E=[]; H=[]; D=[]; C=[];
Q=[]; R=[];
S=zeros(size(E, 2), size(R, 2)); Inter=[]; PInter=[];
```

Listing 30.2: `SampleSS` adapted for (30.10)

```
function [Phi, Gam, E, H, D, C, Q, R, S, ...
  Inter, PInter]= example1(p)
Phi=1; Gam=[]; E=1; H=1; D=[]; C=1;
Q=10.^p(1); R=10.^p(2);
S=zeros(size(E, 2), size(R, 2)); Inter=[]; PInter=[];
```

Listing 30.3: Calling syntax for `SSestim` function

```
[p, covpar]= SSestim(z, u, model, p0, ...)

INPUTS:
  z:       Output data (m x T  or  T x m)
  u:       Input data  (Nu x T  or  T x Nu)
  model:   Cell or string {mfun, x0, P0}
           mfun is a string with the Matlab function
                name defining the model
           x0 is the initial state vector
           P0 is the initial covariance of states
  p0:      Vector of starting values for parameters in
                the search (np x 1)
  ...:     Additional inputs to model

 OUTPUTS:
  p:       Optimal values of parameters (np x 1)
  covpar: Covariance of parameters (np x np)
```

contain the optimal parameter estimates and their covariance matrix, respectively. Input models may also be defined as a cell array of the form `{mfun, x0, P0}`, where `mfun` is the function name (`example1`). In this case, the initial state vector and its covariance matrix conditions for filtering are given by `x0` and `P0`. If no initial conditions are specified, a *diffuse prior* is assumed.

Listing 30.4: Calling syntax for `fis` function

```
[inn,zhat,xhat,Pz,Px]=fis(z,u,model,p,SMOOTH,...)

INPUTS:
  z:       Output data (m x T  or  T x m)
  u:       Input data  (Nu x T  or  T x Nu)
  model:   Cell or string {mfun, x0, P0}
           mfun is a string with the Matlab function
                 name defining the model
           x0 is the initial state vector
           P0 is the initial covariance of states
  p:       Vector of model parameters (np x 1)
  SMOOTH:  Smoothing (1) versus Filtering (0)
  ...:     Additional inputs to model

 OUTPUTS:
  inn:     Matrix of innovations (m x T)
  zhat:    Matrix of filtered or smoothed fit values
  xhat:    Matrix of filtered or smoothed states
  Pz:      Matrix of covariance matrices of
                 innovations (m x m x T)
  Px:      Matrix of covariance matrices of states
                 (Ns x Ns x T)
```

30.4.4 Estimate the States, Innovations, etc. (Step 4)

Once the parameters have been estimated, `fis` is used to obtain the optimal estimates of the state vector and its covariance matrix, see Listing 30.4. Additional operations such as interpolation, forecasting and signal extraction are also performed using this function. Forecasts are generated by simply adding `NaN` values at the end of the output series. The function calling syntax is very similar to `SSestim`.

Application of the model (30.9) to annual flow readings for the Nile river at Aswan from 1871 to 1970 initially yields the results illustrated by the top panel of Fig. 30.1. An interesting challenge for this time series, is to evaluate whether the construction of the Aswan in 1899 (observation 29) leads to a significant decline in the river flow [5, 23]. When a variance intervention is considered, the model clearly detects a mean break in 1899, as shown by the lower panel. This is achieved by revising the last line of Listing 30.2 with `Inter=29; PInter= 1e4` and saving the file as `example2`. The associated script to generate the graphs are shown in Listing 30.5, which assumes that the data are pre-loaded into variable `z`.

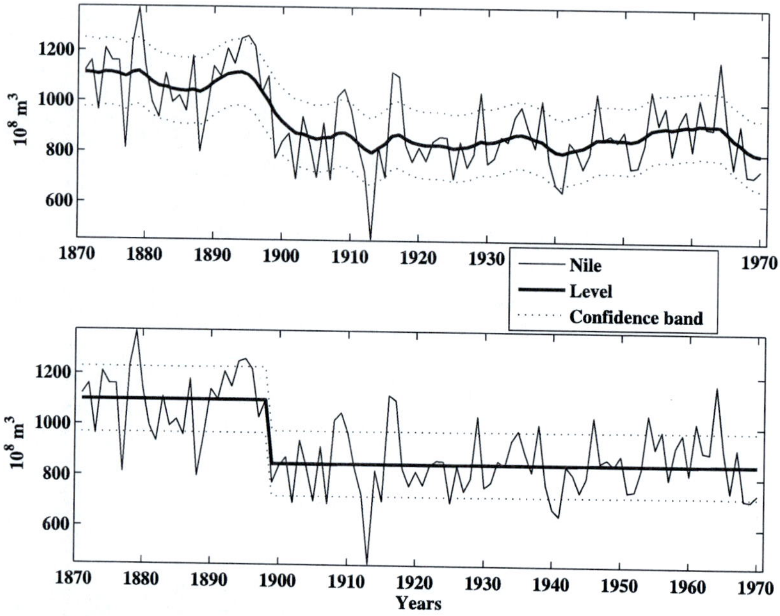

Fig. 30.1 Analysis of Nile river data based on Listing 30.5

Listing 30.5: Script to model Nile river data and generate Fig. 30.1

```
t= (1871 : 1970)';
[p, covp]= SSestim(z, [], 'example1');
[inn, zhat, xhat, Pz]= fis(z, [], 'example1', p);
plot(t, [z zhat' confband(zhat, Pz, 1)])

[p, covp]= SSestim(z, [], 'example2');
[inn, zhat, xhat, Pz]= fis(z, [], 'example2', p);
plot(t, [z zhat' confband(zhat, Pz, 1)])
```

30.5 Examples

This section highlights the versatility of `SSpace` and is intended as a tutorial introduction. The results may be replicated by copying the code into MATLAB.

30.5.1 Univariate Unobserved Components Models

Earlier research by the first author and Professor Young has included the development of a class of Unobserved Components Models that has now been successfully

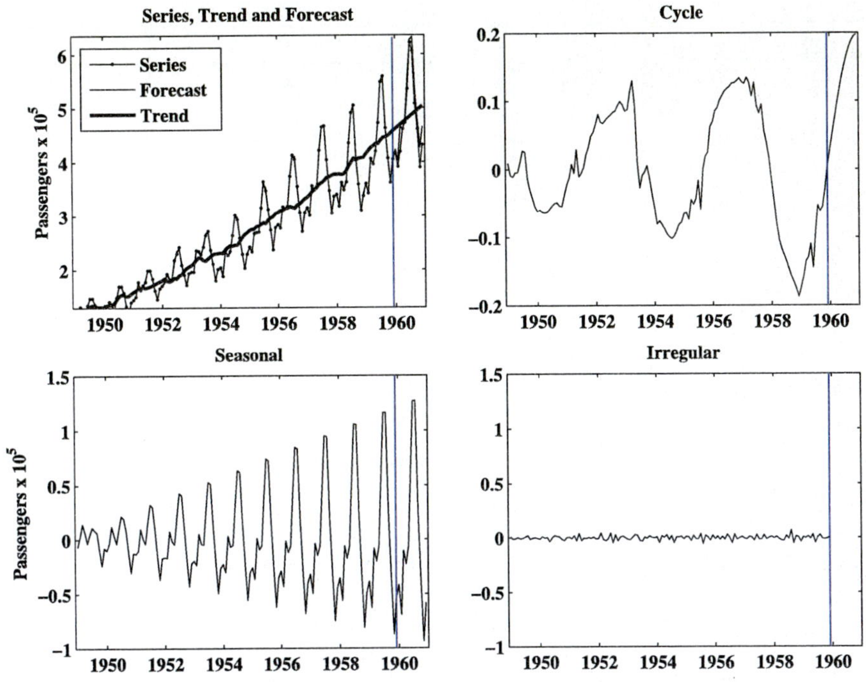

Fig. 30.2 Analysis of air passengers series based on Listings 30.8 and 30.9

applied in a range of scientific disciplines, across the environmental sciences, engineering and economics. The univariate Dynamic Harmonic Regression (DHR) quoted in (30.7) is an important element of these developments. Hence, the first example shows how to implement the DHR model in `SSpace` using the template `SampleDHR`, applied to the well known air passengers series [1]. These data are illustrated by the thin trace on the top left panel of Fig. 30.2.

The template `SampleDHR` in Listing 30.6 defines basic parameters for the trend, seasonal and irregular components. Listing 30.7 provides a similar template for a monthly time series with a local linear trend model. Other components, like linear regression, transfer functions, or ARMA terms may be added by system combinations (Sect. 30.5.5). The least intuitive part of this code is probably the trend model, hence the expanded help for this in `SampleDHR`. In particular, the only parameters that have to be defined are the number of outputs `m` in the model and the covariance matrix `Qt` of the state trend noise. In this regard, the default model is a local linear trend but an integrated random walk may be specified instead using `Et= [0; I]` and `Qt= varmatrix()`.

Listing 30.8 is a nonstandard DHR model structure that has been developed for the purposes of this example. It consists of a model estimated on the original time series (no log transformation is taken) with parameters that behave as integrated random walks, in order to model the increasing amplitude. We assume the existence of a business cycle of about four years, but the exact period is estimated jointly with

Listing 30.6: `SampleDHR` function

```
function [Phi, Gam, E, H, D, C, Q, R, S, ...
   Inter, PInter]= SampleDHR(p, T)

% SS trend model
%      x(t+1)= Phit * x(t) + Et * w(t)
%       z(t)  =   Ht * x(t)
%       Qt= COV(w(t));
% m is the number of outputs
m= []; I= eye(m); O= zeros(m);
Phit = [I I;O I]; Ht = [I O]; Et = [I O; O I];
Qt = blkdiag(varmatrix(), varmatrix());

% Seasonal/cyclical DHR components
Periods = []; Rho = []; Qs = repmat(varmatrix(), , );

% RW (h=1) or IRW (h=2) parameters
h = [];

% Covariance matrix of irregular component
R = varmatrix();
```

the rest of the parameters. The reader may need to spend some time considering Listing 30.8 in order to discover all the elements of the model—a full explanation is omitted here for brevity. Finally, Listing 30.9 is a script to use the `SSpace` toolbox with this function, in which the air passengers data is pre-loaded as the variable name `airpas`. The results are illustrated in Fig. 30.2.

30.5.2 Multivariate Unobserved Components Models

The DHR models in the previous example are readily extended to the multivariate case. Here, components across time series may be correlated by means of the noise covariance matrices, but are independent of the remaining components, e.g. the trends are correlated among themselves, but are independent of seasonal components, and so on. In this regard, the template `SampleDHR` shown in Listing 30.6 above is straightforwardly updated, as shown by Listing 30.10 for a vector of three quarterly time series. In this case, `m=3` and any variance elements in the model are now covariance matrices of dimension 3. These are assumed to be symmetrical and positive semidefinite, hence only 6 different coefficients have to be estimated (`varmatrix` provides a parameterisation of an unconstrained vector of parameters into a covariance matrix); `Periods` and `Rho` are the block matrices used in (30.7).

Listing 30.7: `SampleDHR` adapted for a monthly time series with local linear trend

```
function [Phi, Gam, E, H, D, C, Q, R, S, ...
  Inter, PInter]= airpasdhr1(p, N)

% SS trend model
%     x(t+1)= Phit * x(t) + Et * w(t)
%      z(t)  =  Ht * x(t)
%      Qt= COV(w(t));
% m is the number of outputs
m= 1; I= eye(m); O= zeros(m);
Phit = [I I;O I]; Ht = [I O]; Et = [I O; O I];
Qt = blkdiag(varmatrix(p(1)), varmatrix(p(2)));

% Seasonal/cyclical DHR components
Periods=[12 6 4 3 2.4 2]; Qs =repmat(10.^p(3),1,6);
Rho = [1 1 1 1 1 1];

% RW (h=1) or IRW (h=2) parameters
h = 1;

% Covariance matrix of irregular component
R = 10.^p(4);
```

The model in Listing 30.10 is estimated on the log transform of the quarterly UK energy consumption data (coal, coal + other and gas) from the first quarter of 1960 to the last quarter of 1986 [8]. Listing 30.11 generates the component estimates shown in Fig. 30.3. In addition to the graphical output, correlations among components are shown in Table 30.2. Such a decomposition allows the user to analyze the correlation among time series in component terms: here, high correlations appear in the trends and trend slopes; one high negative correlation appears in the seasonal components; and small correlations are detected on the irregulars.

30.5.3 Time Aggregation

There are numerous situations in which data are collected at different sampling intervals, sometimes because of the nature of the data or sometimes because of problems in taking the measurements. It can also be useful to investigate the relationship between two or more variables that are measured at different sampling intervals. In such cases, one solution is to use 'time aggregation', an approach that fits naturally into a state space framework. If the missing data can be replaced by missing observations, the problem is easy to solve using `NaN` variables. The more interesting case,

Listing 30.8: `SampleDHR` adapted for air passengers data

```
function [Phi, Gam, E, H, D, C, Q, R, S, ...
  Inter, PInter]= airpasdhr2(p, N)

% SS trend model
%     x(t+1)= Phit * x(t) + Et * w(t)
%      z(t) =  Ht * x(t)
%      Qt= COV(w(t));
% m is the number of outputs
m= 1; I= eye(m); O= zeros(m);
Phit = [I I;O I]; Ht = [I O]; Et = [I O; O I];
Qt = blkdiag(varmatrix(p(1)), varmatrix(p(2)));

% Seasonal/cyclical DHR components
Periods = [constrain(p(5), [34 50]) 12 6 4 3 2.4 2];
Rho = [constrain(p(6), [0 1]) 1 1 1 1 1 1];
Qs = [10.^p(7) repmat(10.^p(3), 1, 6)];

% RW (h=1) or IRW (h=2) parameters
h = 2;

% Covariance matrix of irregular component
R = 10.\^{p}(4);
```

Listing 30.9: Script to model air passengers data and generate Fig. 30.2

```
z= airpas/100;
p0= [-1;-1;-1;-1;2;5;-1];
model= 'airpasdhr2';
p= SSestim(z(1 : 132), [], model, p0, 132);
C= components([z(1 : 132); nan(12, 1)], [], model, p,
           1, [], [], [], 144);
plot([z C(:, 1:2)]); plot(t, C(:, 3))
plot(t, C(:, 4)); plot(t, C(:, 5))
plot([z(133:end) C(133:end, 1)])
```

is when missing observations have to add up (or have to be combined in some way) to match the values available at the end of a period. For example, accidents might be recorded at different time intervals. The total for one year could be the addition of either twelve months or four quarters, depending on data availability.

Listing 30.10: `SampleDHR` adapted for a vector of three quarterly time series

```
function [Phi, Gam, E, H, D, C, Q, R, S, ...
  Inter, PInter]= energydhr(p, N)

% SS trend model
m= 3; I= eye(m); O= zeros(m);
Phit = [I I;O I]; Ht = [I O]; Et = [I O; O I];
Qt = blkdiag(varmatrix(p(1:6)), varmatrix(p(7:12)));

% Seasonal/cyclical DHR components
Periods = repmat([4 2], 3, 1);
Rho=ones(3,2); Qs=repmat(varmatrix(p(13:18)),1,2);

% RW (h=1) or IRW (h=2) parameters
h = 1;

% Covariance matrix of irregular component
R = varmatrix(p(19:24));
```

Listing 30.11: Script for estimation of a multivariate DHR model

```
t= (1960 : 0.25 : 1986.75)';
p= SSestim(z, [], 'energydhr', [], 108);
[inn, zhat, xhat, Pz]=...
  fis(z, [], 'energydhr', p, 1, 108);
C= components(z, [], 'energydhr', p, ...
  [], [], [], [], 108);
plot(t, [z permute(C(:, 2,1:3), [1 3 2])])
plot(t, permute(C(:, 3, 1:3), [1 3 2])
plot(t, permute(C(:, 4, 1:3), [1 3 2])
```

Table 30.2 Correlation among components of UK energy data

Variables	Trends	Slopes	Seasonal	Irregular
coal vs. coal+other	−0.687	0.994	−0.168	0.344
coal vs. gas	−0.930	−0.929	−0.044	0.340
coal+other vs. gas	0.901	−0.963	−0.977	0.059

In order to simplify the exposition, we will consider (30.1) with constant system matrices and without inputs for the shortest time interval:

$$\mathbf{x}_{t+1} = \mathbf{\Phi x}_t + \mathbf{E w}_t; \qquad \mathbf{z}_t = \mathbf{H x}_t + \mathbf{C v}_t. \tag{30.11}$$

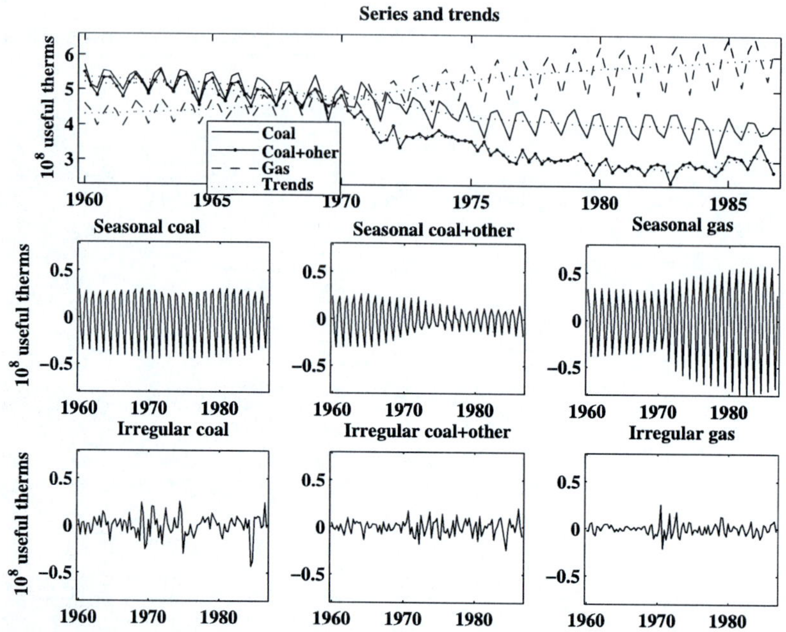

Fig. 30.3 Analysis of UK energy data based on Listings 30.10 and 30.11

The observation equations are now incorporated into the transition equations:

$$\begin{bmatrix} \mathbf{z}_{t+1} \\ \mathbf{x}_{t+1} \end{bmatrix} = \begin{bmatrix} \mathbf{0} & \mathbf{H\Phi} \\ \mathbf{0} & \mathbf{\Phi} \end{bmatrix} \begin{bmatrix} \mathbf{z}_t \\ \mathbf{x}_t \end{bmatrix} + \begin{bmatrix} \mathbf{C} & \mathbf{HE} \\ \mathbf{0} & \mathbf{E} \end{bmatrix} \begin{bmatrix} \mathbf{v}_t^* \\ \mathbf{w}_t \end{bmatrix}; \qquad \mathbf{z}_t = \begin{bmatrix} \mathbf{I}\ \mathbf{0} \end{bmatrix} \begin{bmatrix} \mathbf{z}_t \\ \mathbf{x}_t \end{bmatrix}, \tag{30.12}$$

with $\mathbf{v}_t^* = \mathbf{v}_{t+1}$. This constraint is not a problem when the transition and observation noise terms are uncorrelated. By defining a cumulator variable [8] we can tell the system when the data points are available. The cumulator variable takes into account the position of the data and the fact that all the time series are flow variables, namely: $C_t = 0$ for t equal to the next observation following any available data point; and $C_t = 1$ otherwise. The final system, including the time aggregation, is (30.12) with the transition matrix replaced by:

$$\begin{bmatrix} C_t \otimes \mathbf{I} & \mathbf{H\Phi} \\ \mathbf{0} & \mathbf{\Phi} \end{bmatrix}. \tag{30.13}$$

Extensions to handle models with inputs are straightforward. The procedure is quite general, in the sense that the initial state space can be any type of model.

Consider fatal occupational accidents in Spain between December 1998 and March 2009, illustrated by the left panel of Fig. 30.4 [3]. The data have been recorded at highly irregular time intervals: annually at the beginning, but irregularly afterwards, including quarterly, bi-monthly and increasing to monthly by the end of the series. Listing 30.12 shows the time aggregated model, in which `maccidents`

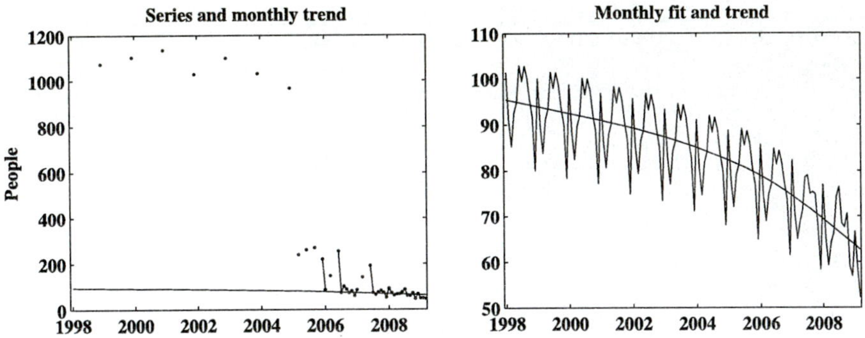

Fig. 30.4 Analysis of occupational accidents using Listings 30.12 and 30.13

Listing 30.12: Time aggregation model for occupational accidents

```
function [Phi, Gam, E, H, D, C, Q, R, S, ...
  Inter, PInter]= accidents(p, z)
[Phi1, Gam1, E1, H1, D1, C1, Q1, R1, S1]= ...
  maccidents(p);
Ns= size(Phi1, 1); N= length(z);
Phi= repmat([1 H1*Phi1; zeros(Ns, 1) Phi1], [1 1 N]);
Phi(1, 1, find(~isnan(z)))= 0;
E= [C1 H1*E1; zeros(Ns, 1) E1];
H= [1 zeros(1, Ns)]; C= 0; R= 0;
Q= blkdiag(R1, Q1); S= zeros(size(E, 2), size(R, 1));
Gam= []; D= []; Inter= []; PInter= [];
```

is another state space system specified for monthly observations. It is not shown here because of space constraints, but takes the form of a basic structural model developed from `SampleBSM`. The transition matrix `Phi` is now time varying, hence it is a three dimensional matrix. Listing 30.13 shows how to use this model, whilst the right hand panel of Fig. 30.4 illustrates the fitted values at the monthly time intervals.

30.5.4 Nonlinear Systems

This example illustrates how string system matrices may be used in `SSpace`, in order to develop unusual models. The case analyzed here is the quarterly rate of change of the US GNP from the first quarter of 1947 until the last quarter of 1990, as illustrated by Fig. 30.5 [25]. There are more values above zero than below, with zero being a value what marks the difference between an expansion or a recession of the economy. This observation suggests that different models may be applicable

Listing 30.13: Script to model occupational accidents in Spain and to generate Fig. 30.4

```
t= (1998 : 1/12 : 2009.2)';
p= SSestim(z, [], 'accidents', [], z);
[inn, zhat, xhat, Pz]= ...
  fis(z, [], 'accidents', p, 1, z);
Trend= xhat(2, :)';
Seasonal= sum(xhat(4:2:end, :))';
plot(t, [z Trend])
plot(t, [z Trend Trend+Seasonal])
```

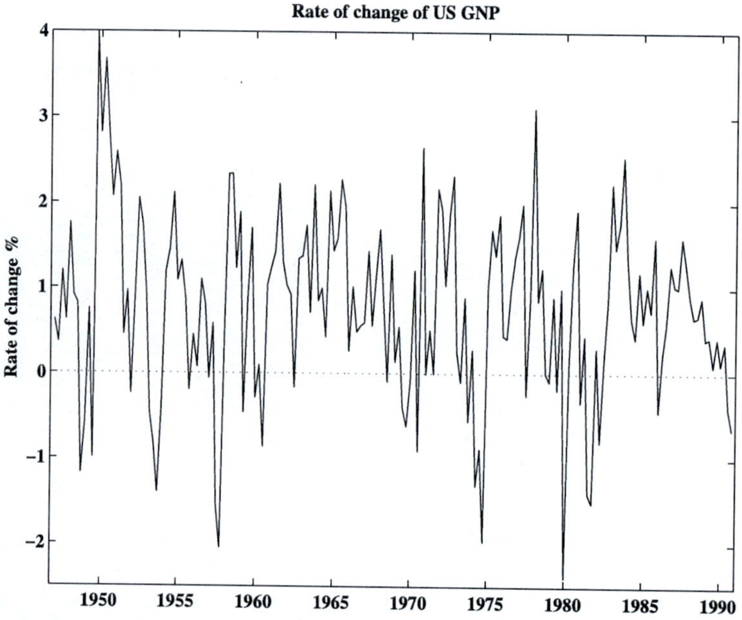

Fig. 30.5 Quarterly rate of change of US GNP

to the time series when the data are above or below zero. For the purposes of this example, a Threshold AR (TAR) nonlinear model of order two is selected:

$$z_t = \begin{bmatrix} c_1 \; \phi_1^1 \; \phi_2^1 \end{bmatrix} \begin{pmatrix} 1 \\ z_{t-1} \\ z_{t-2} \end{pmatrix} + v_{1,t}, \quad z_{t-2} > 0, \tag{30.14}$$

$$z_t = \begin{bmatrix} c_2 \; \phi_1^2 \; \phi_2^2 \end{bmatrix} \begin{pmatrix} 1 \\ z_{t-1} \\ z_{t-2} \end{pmatrix} + v_{2,t}, \quad z_{t-2} \leq 0. \tag{30.15}$$

Listing 30.14: `SampleSS` template adapted for estimation of a TAR(2) model

```
function [Phi, Gam, E, H, D, C, Q, R, S, ...
  Inter, PInter]= gnptar(p)
Phi= 0; Gam= [0 0 0]; E= 0; H= [0]; C= 1; Q= 0;
D= 'D= p(1:3)''; if t>2, if z(t-2)>0, ...
  D= p(5:7)''; end, end';
R= 'R=10.\^{p}(4); if t>2, if z(t-2)<=0, ...
  R=10.\^{p}(8); end, end';
S= zeros(size(E, 2), size(R, 1));
Phider= []; Hder= []; Inter= []; PInter= [];
```

Listing 30.15: Script to model US GNP data

```
data= lag(z, [0 1 2]);
u= [ones(174, 1) data(:, 2 : 3)];
[p, covp, inn]= SSestim(data(:, 1), u, 'gnptar');
```

Table 30.3 AR(2) constant parameter estimates and TAR(2) estimates in two regimes

Variables	AR(2)	TAR(2) Regime 1	TAR(2) Regime 2
Constant	0.0077	−0.0030	0.0037
First delay	−0.2858	0.4412	0.3134
Second delay	−0.1893	−0.6972	0.2024
Residual variance $\times 10^4$	0.9766	0.7623	1.4538

The `SampleSS` template is modified as shown in Listing 30.14. The important point here, is the definition of matrices `D` and `R` as strings, in which the particular values at each sample depend on the output variable. The code necessary to estimate this model on the GNP data is given by Listing 30.15. Note that there are three inputs to the model: a column of ones and two columns representing the delayed output variable. Finally, Table 30.3 highlights the differences in parameter estimates between the two regimes, and also with respect to the standard linear AR(2) model. These results support the choice of the TAR(2) model.

30.5.5 System Combinations

Using the state space framework, several different models can be conveniently concatenated into a single model. Typical examples include: Transfer Function with colored noise; Unobserved Components with added linear terms, affecting the inputs by means of Transfer Functions or a linear regression; and Unobserved Com-

Listing 30.16: `SampleCAT` template for state space system concatenation

```
function [Phi, Gam, E, H, D, C, Q, R, S, ...
  Inter, PInter]= SampleCAT(p)
[Phi1, Gam1, E1, H1, D1, C1, Q1, R1, S1]= Afnc(p());
[Phi2, Gam2, E2, H2, D2, C2, Q2, R2, S2]= Bfnc(p());
[Phik, Gamk, Ek, Hk, Dk, Ck, Qk, Rk, Sk]= kfnc(p());
[Phi, Gam, E, H, D, C, Q, R, S]= ...
catsystem({Phi1,Gam1,E1,H1,D1,C1,Q1,R1,S1},...
          {Phi2,Gam2,E2,H2,D2,C2,Q2,R2,S2},...
          {Phik,Gamk,Ek,Hk,Dk,Ck,Qk,Rk,Sk});
Inter= []; PInter= [];
```

ponents with coloured noise. Concatenation of the transition equations requires a block concatenation approach. If the output variables are different for each model, the observation equations also have to be concatenated block-wise. By contrast, if the output variables are common (the usual case), the overall matrices $\mathbf{H}_t$ and $\mathbf{D}_t$ are obtained by horizontal concatenation, leaving the overall $\mathbf{C}_t$ associated with only one of the models.

`SSpace` allows concatenation with the help of the `catsystem` function or by using the `SampleCAT` template, shown in Listing 30.16. The power of `catsystem` resides in the fact that it works with time varying systems. Note that `Afnc`, `Bfnc` etc. represent any state space system defined by other MATLAB functions. To avoid the proliferation of m-files, it is convenient to define these functions immediately below `SampleCAT` and saved in the same file.

30.6 Conclusions

This chapter has presented a new MATLAB toolbox for generic analysis of State Space models, called `SSpace`. The toolbox is available from the authors on request. It is intended for a wide audience, including professional practitioners, researchers and students, indeed anyone involved in the analysis of time series, forecasting or signal processing. `SSpace` is composed of a relatively small number of powerful functions. These provide the tools for general state space estimation, including: specification of models, maximum likelihood estimation, filtering and smoothing. Particular models may be explicitly implemented using MATLAB code written by the user. However, many templates for common model types are already included in the toolbox. In this chapter, the toolbox has been demonstrated in action for several case studies, including some classical examples and other less familiar ones.

The toolbox has a number of salient features. Firstly, it is user-oriented, in that just a few MATLAB functions are sufficient for an exhaustive analysis of time series. Secondly, the state space system used is rather general. In particular, all the system matrices are time varying, correlation between observed and transition noises are allowed, variance interventions and string system matrices are possible, and block concatenation of state space models is straightforward. Finally, the manner in which particular models are specified is exceptionally flexible, since each model is an independent MATLAB function. This approach allows for a number of interesting possibilities. For example, alternative parameterizations of a model are possible, as are the imposition of linear or nonlinear constraints on parameters. These properties should make the toolbox particularly interesting for those in need of non-standard models, for which commercial software is not necessarily available.

Acknowledgements This work was supported by the Junta de Comunidades de Castilla-La Mancha grant PII1I09-0209-6050.

References

1. Box, G.E.P., Jenkins, G.M., Reinsel, G.C.: Time Series Analysis Forecasting and Control, 3rd edn. Prentice Hall, Englewood Cliffs (1994)
2. Bryson, A.E., Ho, Y.C.: Applied Optimal Control, Optimization, Estimation and Control. Blaisdell Publishing, Waltham (1969)
3. Carnero, C., Pedregal, D.J.: Modelling and forecasting occupational accidents of different severity levels in Spain. Reliab. Eng. Syst. Saf. **95**, 1134–1141 (2010)
4. Casals, J., Jerez, M., Sotoca, S.: Exact smoothing for stationary and non-stationary time series. Int. J. Forecast. **16**, 59–69 (2000)
5. Cobb, G.W.: The problem of the Nile: conditional solution to a change point problem. Biometrika **65**, 243–251 (1978)
6. de Jong, P.: Smoothing and interpolation with the state space model. J. Am. Stat. Assoc. **84**, 1085–1088 (1989)
7. Exadaktylos, V., Silva, M., Aerts, J.-M., Taylor, C.J., Berckmans, D.: Real-time recognition of sick pig cough sounds. Comput. Electron. Agric. **63**, 207–214 (2008)
8. Harvey, A.C.: Forecasting Structural Time Series Models and the Kalman Filter. Cambridge University Press, Cambridge (1989)
9. Durbin, J., Koopman, S.J.: Time Series Analysis by State Space Methods. Oxford University Press, London (2001)
10. Jazwinski, A.H.: Stochastic Processes and Filtering Theory. Springer, Berlin (1970)
11. Kalman, R.E.: A new approach to linear filtering and prediction problems. ASME Trans., J. Basic Eng. D **83**, 95–108 (1960)
12. Ng, C.N., Young, P.C.: Recursive estimation and forecasting of non-stationary time series. J. Forecast. **9**, 173–204 (1990)
13. Pedregal, D.J., Carnero, C.: State space models for condition monitoring. A case study. Reliab. Eng. Syst. Saf. **91**, 171–180 (2006)
14. Pedregal, D.J., Carnero, C.: Vibration analysis diagnostics by continuous-time models: a case study. Reliab. Eng. Syst. Saf. **94**, 244–253 (2009)
15. Pedregal, D.J., Contreras, J., Sánchez, A.: ECOTOOL: a general MATLAB forecasting toolbox with applications to electricity markets. In: Pardalos, P.M., Pereira, M.V.F., Iliadis, N.A., Rebennack, S., Sorokin, A. (eds.) Handbook of Networks in Power Systems, pp. 69–104. Springer, Berlin (2010)
16. Pedregal, D.J., Dejuán, O., Gómez, N., Tobarra, M.A.: Modelling demand for crude oil products in Spain. Energy Policy **37**, 4417–4427 (2009)
17. Pedregal, D.J., Pérez, J.J.: Should quarterly government finance statistics be used for fiscal surveillance in Europe? Int. J. Forecast. **26**, 794–807 (2010)

18. Pedregal, D.J., Trapero, J.R.: Mid-term hourly electricity forecasting based on a multi-rate approach. Energy Convers. Manag. **51**, 105–111 (2010)
19. Pedregal, D.J., Young, P.C.: Statistical approaches to modelling and forecasting time series. In: Clements, M., Hendry, D. (eds.) Companion to Economic Forecasting. Blackwell, Oxford (2002)
20. Pedregal, D.J., Young, P.C.: Development of improved adaptive approaches to electricity demand forecasting. J. Oper. Res. Soc. **59**, 1066–1076 (2008)
21. Schweppe, F.: Evaluation of likelihood function for Gaussian signals. IEEE Trans. Inf. Theory **11**, 61–70 (1965)
22. Taylor, C.J., Chotai, A., Young, P.C.: Nonlinear control by input-output state variable feedback pole assignment. Int. J. Control **82**, 1029–1044 (2009)
23. Taylor, C.J., Pedregal, D.J., Young, P.C., Tych, W.: Time series analysis and forecasting with the Captain Toolbox. Environ. Model. Softw. **22**, 797–814 (2007)
24. Taylor, C.J., Shaban, E.M., Stables, M.A., Ako, S.: Proportional-Integral-Plus (PIP) control applications of state dependent parameter models. IMECHE Proc., Part I, J. Syst. Control Eng. **221**(17), 1019–1032 (2007)
25. Tsay, R.S.: Testing and modeling threshold autoregressive processes. J. Am. Stat. Assoc. **84**, 231–240 (1989)
26. Young, P.C.: Recursive Estimation and Time Series Analysis. Springer, Berlin (1984)
27. Young, P.C., Pedregal, D.J., Tych, W.: Dyanmic harmonic regression. J. Forecast. **18**, 369–394 (1999)

Index

L. Wang, H. Garnier (eds.), *System Identification, Environmental Modelling, and Control System Design*,
DOI 10.1007/978-0-85729-974-1,

O

P